PRODUCT DESIGN AND
PROCESS ENGINEERING

**McGRAW-HILL
BOOK COMPANY**
New York
St. Louis
San Francisco
Düsseldorf
Johannesburg
Kuala Lumpur
London
Mexico
Montreal
New Delhi
Panama
Paris
São Paulo
Singapore
Sydney
Tokyo
Toronto

BENJAMIN W. NIEBEL

Head of Department and Professor of Industrial Engineering
The Pennsylvania State University

ALAN B. DRAPER

Associate Professor of Industrial Engineering
The Pennsylvania State University

Product Design and Process Engineering

This book was set in Modern 8A by Maryland Composition Incorporated.
The editors were B. J. Clark and J. W. Maisel;
the cover was designed by Joseph Gillians;
the production supervisor was Leroy A. Young.
New drawings were done by Felix Cooper.
The Maple Press Company was printer and binder.

Library of Congress Cataloging in Publication Data

Niebel, Benjamin W
    Product design and process engineering.

    1. Design, Industrial. 2. Production engineering.
3. Manufacturing processes. I. Draper, Alan B., joint
author. II. Title.
TS171.N53   620'.004'2   74-4294
ISBN 0-07-046535-5

**PRODUCT DESIGN AND
PROCESS ENGINEERING**

34567890MAMM7987

CONTENTS

Preface   xi

Chapter 1 **Creative Thinking and Organizing for Product Innovation**   1

The product-design function. The process-design function. Locating ideas for new products. Selecting the right product. Qualifications of the production-design engineer. Creative thinking. Curiosity and imagination. Ideas generate ideas. Taking time to think. Using a systematic procedure for product innovation. Opportunities for the production-design engineer.

Chapter 2 **Criteria for Product Success**   18

Areas to be studied preparatory to design. Functional design. The value of appearance. Principles and laws of appearance. Incorporating quality and reliability into the design. Man-machine considerations. Designing for ease of maintenance.

Chapter 3 **Cost and Product Development**   36

Sources of funds for development cost. Product costs. Estimating product costs. Kinds of cost procedures. Cost reduction.

Chapter 4 **Properties and Behavior of Materials** 60

Physical properties of materials. Relation of physical properties to electron population and distribution. Heat capacity. Thermal conductivity. Electrical conductivity. Magnetic properties. Mechanical properties of materials. Tensile properties of materials. Hardness. Impact strength. Fatigue behavior and endurance limit. Creep and stress rupture. Comparison of materials. Relative cost of engineering materials.

Chapter 5 **Enhancement of the Properties of Meterials** 98

Hardening and strengthening mechanisms: strain hardening, control of grain size, solid state reactions, solid state transformations in ferrous alloys. Heat treatment of ferrous alloys: softening operations—stress relief, normalizing, annealing, spheroidizing; fundamentals of hardening—hardness and hardenability, microstructure of heat-treated steels, TTT diagrams; hardening operations—through hardening, surface hardening. Heat treatment of aluminum alloys: precipitation hardening, designation of heat-treated alloys, heat-treating operations, heat-treating equipment.

Chapter 6 **Ferrous Alloys** 133

Iron-carbon equilibrium diagram. Cast irons. Properties of cast irons: gray iron, malleable iron, ductile iron, white iron, alloyed irons. Carbon steel. Alloy steel. Tool steel. Stainless steel. Selection of steel. Endurance limit of steel. Metallurgical factors in selecting steels. Selection of alloying elements. Specification of purchased steel. Considerations for fabrication. How design and processing affect the selection of steel.

Chapter 7 **Nonferrous Metals** 185

Importance of nonferrous alloys. Aluminum and its alloys. Designation of wrought and casting alloys. Classification of commercial shapes. Magnesium. Copper and its alloys. Properties and uses. Brasses, properties and uses. Bronzes, properties and use. Nickel and its alloys, properties and use. Molybdenum alloys. Cobalt alloys. Titanium. Low-melting alloys.

Chapter 8 **Plastics** 223

The chemistry of plastics. General properties of plastics. Effect of temperature. Effects of humidity. Effects of light. Weight. Electrical resistivity. Fabrication. Effects of oxygen. Effects of loading. Thermoplastics. Acrylic resins. Aniline-formaldehyde resins. Polyamide resins. Polystyrenes. ABS materials. Fluorocarbons.

Acetal. Thermosetting plastics. Phenol formaldehyde. Urea formaldehyde. Melamine formaldehyde. Polyesters. Epoxies. Design considerations of plastics in general. Rubber. Synthetic rubbers.

Chapter 9   Ceramics and Powdered Metals   271

Ceramic groups. Whiteware. Glass. Refractories. Designing with ceramics. Carbon. Designing with carbon. Powder metallurgy. Designing for powdered metallurgy.

Chapter 10   Basic Manufacturing Processes: Liquid State   296

Fundamentals of casting: materials, weight and dimensional limitations, the casting process, behavior of alloys during solidification. Major casting processes: sand-casting processes, precision-casting processes, reusable-mold processes, full-mold process. Design for the casting process: pattern design, cores and core making, inserts, other considerations in molding and casting. Design for casting quality: general design, gating-system design, riser design. Design example: design details for a cast connecting rod.

Chapter 11   Basic Manufacturing Processes: Solid State   415

Fundamentals of hot working: fiber flow lines, structural integrity, strength, forgeability. Hot-working processes: forging, upsetting, extruding, rolling. Hammer and press forging: closed-die forging, hydraulic-press forging, high-velocity forming, upsetting, extrusion, hot rolling and tube forming from solid bars. Design for hot-working processes: design factors, standard specifications for forgings, design for upset forging, die design considerations.

Chapter 12   Basic Manufacturing Processes: Plastics   445

Compression molding. Transfer molding. Injection molding. Extrusion casting. Cold molding. Thermoforming. Blow molding. Machining from solid stock. Mold design guides.

Chapter 13   Secondary Manufacturing Processes: Material Removal   460

Fundamentals of material removal: background, machine tools, machining parameters—feed, speed, depth of cut, cutting-tool geometry, cutting-tool materials. Analysis of metal cutting: tool wear and tool life, machinability, analysis of metal-cutting forces, force relationships, stress analysis, shear strain, velocity relations, energy considerations, power requirements, tool wear and tool life, cutting fluids. Basic metal-cutting operations: basic machining operations—turning, planing, drilling, milling, grinding. Production machining operations: turning, milling, hole-making operations. Machining of special contours: tapers, threads, gears. Grinding operations: cutting action of grinding wheels, grinding wheels, grinding equipment, centerless grinding, thread grinding, internal grinding, surface grinding, cutoff. Fine abrasive machining: polishing and buffing, honing, lapping, superfinishing. Machining of hard materials: electric-discharge machining, chemical milling, electrochemical machining, ultrasonic machining.

Chapter 14 | **Secondary Manufacturing Processes: Forming**   **566**

Cold-working fundamentals: common characteristics of cold-working, properties of materials used in press-working, lubrication. Shearing: punching, blanking, and stamping operations. Forming processes: fundamentals of forming and bending, bending operations, special forming operations. Compression processes: rolling, extrusion, other compression processes. Drawing: fundamentals of drawing, drawing operations, stretch forming, spinning. Die design and limitations. Comparison of presses.

Chapter 15   **Decorative and Protective Coatings**   **643**

Cleaning. Acid pickling. Solvent cleaning. Alkaline cleaning. Electrolytic cleaning. Emulsion cleaning. Ultrasonic cleaning. Organic finishes. Oil paint. Enamels. Varnishes. Shellac. Stain. Luminescent paint. Metal-flake paints. Pearl essence. Crystal finish. Wrinkle finish. Crackle finish. Hammered finish. Flock finishing. Application methods. Inorganic coatings. Porcelain enamels. Ceramic coatings. Metallic coatings. Electroplating. Metalizing. Solder sealing. Hot-dip coating. Immersion coating. Vapor-deposited coating. Conversion coatings. Phosphate coatings. Chromate coatings. Anodic coatings. Design criteria for selection of decorative and protective coatings.

Chapter 16   **Joining Processes**   **672**

Joining processes: production joining processes, basic welding process. Fundamentals of welding: effect of heat—residual stress, distortion, relieving stress; weldability—steels, stainless steels, aluminum alloys; welding versus other processes. Welding processes: arc welding (consumable electrodes); shielded-metal arc (stick-electrode) welding; semiautomatic welding, flux cored electrodes, Automatic welding, submerged-arc welding, electroslag welding; arc welding (nonconsumable electrodes); gas tungsten-arc welding (TIG), plasma-arc welding, electron-beam welding; resistance welding—spot welding, projection welding, seam welding, flash-butt welding, percussion welding, stud welding; friction welding; brazing, soldering, and cold welding—brazing, induction brazing, braze welding, soldering, cold welding. Design of Weldments: welding symbols, joint design, weld stress calculations, design of a fillet weld size treating the weld as a line. Fasteners: nuts, bolts and screws, rivets, metal stitching. Adhesives: adhesive bonding, rubber adhesives. Joining plastics.

Chapter 17   Reliability and Quality Control   750

Quality-control procedures. Inspection and test equipment. Automatic gaging and control. Nondestructive inspection. Surface-finish measurement and evaluation. Statistical control. Manufacturing reliability. Proability the tool of reliability. Reliability operations. Developing a quality-control and reliability program.

Chapter 18   Planning the Optimum Operation Sequence   779

Classifying the secondary operations. Critical operations. Placement operations. Tie-in operations. Protection operations. Operation-sequence principles. Developing the operation sequence. Locating the work in the tooling. Supporting the work. Clamping the work.

Chapter 19   Patents   791

Classes of exclusive rights. Patents. Combination versus aggregation. Novelty and utility. Design patents. Patent disclosure. Patent application steps. Patent office prosecution. Sale of patent rights. Trademarks. Copyrights.

Appendix   Comparative Prices of Selected Engineering Materials   810

Index   819

PREFACE

This book has been written to provide a modern text for engineering and technical students associated with either the functional design of the product or the design of the manufacturing process. Thus, the text is design-oriented. It provides a background of how to design so as to help ensure product success. In so doing, the book provides fundamental information on engineering materials and processes so that the product designer can capitalize on the most favorable materials and the methods to transform them into the specifications of his design. The principal constraints associated with product design are developed so as to determine their influence on the design itself. These include: quantity to be produced, delivery requirements, product reliability, the price the potential customer will pay, the value of appearance, human engineering, product function, and preventive maintenance.

This volume will be a suitable text (at the sophomore–junior level) for students in the various engineering and technology curricula. It should be of particular value to mechanical, industrial, electrical, and aeronautical students concerned with design. This book will also be a valuable reference text for the product-design and manufacturing departments in industry. Design, manufacturing, and tool engineers will find this book helpful in the designing and processing of a product to meet competition from both United States and foreign industries. The practicing production-design engineer need not concentrate on Chap. 1, which is primarily written for the student of production design.

The authors wish to express their deep appreciation to the many design engineers and engineering educators who gave helpful suggestions and provided a large amount of the specific material as well as constructive criticism that led to the completion of this text. Notable among these are Mr. E. N. Baldwin, former George Westinghouse Professor; and Professor C. A. Ellsworth of the Department of Industrial Engineering at The Pennsylvania State University.

BENJAMIN W. NIEBEL

ALAN B. DRAPER

PRODUCT DESIGN AND
PROCESS ENGINEERING

1

CREATIVE THINKING AND ORGANIZING FOR PRODUCT INNOVATION

Industrial technology and the products designed and produced are the bases for any advanced economy. The life blood of any individual product-producing enterprise is the continual introduction of new products or product improvements, without which that enterprise will stagnate and ultimately fail. These new products and product improvements must satisfy potential customer needs or demands, and must recognize the pressure from competition, domestic and foreign.

The object of this book is to present fundamentals underlying the design and profitable production of a marketable commodity. A design engineer may create on paper a device of excellent functional utility or sales appeal; but if that product is to become a reality, it must be produced at a practical cost in sufficient time to satisfy the customer. Thus it must be produced from available and advantageous materials, methods, processes, and equipment. Also, it must be competitive in quality, performance, appearance, and service life. In order to accomplish these objectives, the successful design engineer must be acquainted with these related factors or he must collaborate closely with those who specialize in these aspects of the overall problem.

Designing for production includes the work of two distinct functions:

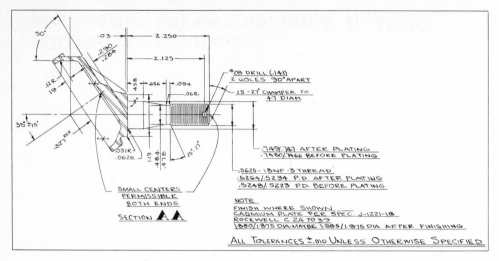

FIGURE 1-1
Representative product design is the development of the housing shown above.

product design and process design. The product-design function involves the development of specifications of a product that will be functionally sound, have eye appeal, and will give satisfactory performance for an adequate life. The process-design function includes developing the method of manufacture of the product so that it can be produced at a competitive price. This work will include planning the sequence of operations and inspections to be performed; the design of the jigs, fixtures, gages, and special equipment needed to produce the work; and the establishment of allowed elemental times for performing the work (see Figs. 1-1 and 1-2).

Thus, designing for production not only includes the designing of a product for economical manufacture, but also the design, specification, or creation of tools, equipment, methods, and manufacturing information for its production. Designing for production provides the product specifications, methods specifications, and physical facilities whereby the production organization can proceed with manufacture. In an enterprise of moderate or large magnitude, many individuals concerned with designing for production may be specialists in comparatively limited fields—methods engineers, tool engineers, cost engineers, and writers of specifications. But each specialist must perform his responsibilities in harmony with the overall objective of developing and manufacturing the product. Even the product design involves functional specialization. We are primarily concerned here with the correlation of product design with the manufacturing problem. In this book we shall adopt the title "production-design engineer" for the individual who is thus associated with the function of designing for production.

FORM 364 REV. DIRECT LABOR CREDIT CARD MASTER SET (FIRST SIDE)

CLASS	DESCRIPTION		DWG. NO.	PART NO.									
132	Housing		J-1470-38	MR-36-2A									
ROUTING						DATE			ISSUED BY				
36-09-48-Heat Treat-48-21-48-11-08-09-11-16-48-Cad.Plt.						Rev.3-16-56			Ind.Eng.				
MATERIAL													
(1) J-1470-11R (1) J-1470-51					48-14-48- 35								
CREDIT CARD DISTRIBUTION	A 1	B 1	C 1	D 1	E 1	F	G 1	H	I 1	J	K	L	M

OP. NO.	OPERATION	MACHINE	DEPT.	SET-UP	TOTAL SET-UP	UNIT TIME	TOTAL TIME
1	Grind Flash around Ears Ga. F-1961	Stand Grinder	09			4.17	4.17
2	Inspect		48				
3	Heat Treat-Spec. J-1222#8		Out	Purchase			
4	Inspect (Rockwell)		48				
5	Shotblast		21				
6	Inspect (Rockwell)		48				
7B	Face Air Chuck FC-5061A-1	W & S #3	11C*	1.00		1.85	
8B	Turn Arbor F-619	W & S #3	11C*	1.00		5.08	
9A	Turn Inside Fix. F-674.FD-6623A	W & S #3	11C*	3.00		13.30	
10A	Turn Stem & Lugs L.Fix. F-628-2R	W & S #3	11C*	1.75	6.75	15.83	36.06
11	Rockwell		48				
12 B	Drill 2 Lug Holes, Stamp Heat Code (3) 295 @ .004 Jig F-6001A Ga. FA-679, FD-6883A-5 & 10,	C.B. D.P. Ga.FA-804 (2)	08C* 13/32 H	.46 S.T.S. Drill		5.49	

FIGURE 1-2
Representative process design is the development of the operation card shown.

Thus the work of the production-design engineer as presented in this text will be principally product design or process design. It is the intent of this book to bring out the necessity of close collaboration between these two production-design areas so that more goods of better quality may be produced at less cost.

LOCATING IDEAS FOR NEW PRODUCTS

Ideas that lead to the development of new products may come from several sources. Usually ideas come from company executives, company sales force, customer suggestions, government agencies, and research laboratories. Perhaps the first step that should be taken by a concern that is actively seeking new products is to prepare a list of sources that may yield worthwhile suggestions. Certainly included in this list would be United States Patent Office, Department of Commerce; advertising agencies; commercial laboratories; patent attorneys and brokers; firms going out of business; trade associations; technical journals and periodicals.

It should be apparent that not every idea necessarily leads to a new product that is placed on the market. However, every new product represents the crystallization of someone's idea. It has been estimated that on the average it takes about 500 possibilities at the idea stage for every new product that is placed in production. The majority of ideas are usually eliminated at new-product conferences or after a preliminary laboratory investigation or economy study.

SELECTING THE RIGHT PRODUCT

After a firm has developed a list of ideas that appear to have product potential, it will need to select those that will most likely lead to success. This selection procedure takes place at new-product conferences. Those attending the new-product conference should be representatives of sales, product engineering, manufacturing engineering, and marketing.

Among the persons called upon to consider new ideas, some are always resistant to change. Their immediate reactions are negative. They dread adjustment to the new and different. Those participating in new product conferences should be practical men of vision who are up to date in their respective fields and who have the imagination and courage to work on an innovative idea that has promise. Twenty-five years ago little future was envisioned for nuclear-reactor power plants. As late as 1964 Westinghouse had sold no power plants of this type, and only 7 in 1966 and 14 in 1967. The future for nuclear-reactor plants today is bright, yet men without vision said no when this product was introduced.

At the new product conference, it is a good idea to record a profile of the various influencing factors for each new product being considered. Some of the more important factors include:

1 The utility value of the product
2 The need for the product
3 The product's sales appeal
4 The advantages and improvements over similar products on the market
5 The size of the potential market
6 The patentability of the product
7 R&D costs
8 Setup and tool costs
9 The profit potential
10 The suitability of the company's engineering talent and production facilities
11 The suitability of the company's sales force and means of distribution
12 The strength of the company compared to the competition
13 The expected life of the product
14 The compatibility of the product with other company products

For example, a profile of a given idea may appear as shown in Fig. 1-3.

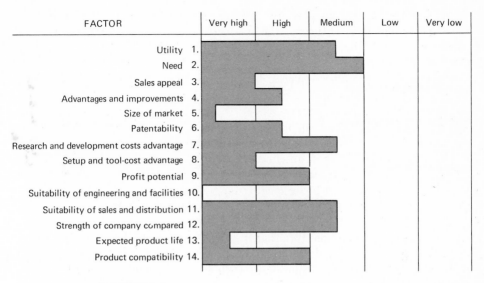

FIGURE 1-3
New-product profile record.

QUALIFICATIONS OF THE PRODUCTION-DESIGN ENGINEER

The engineer associated with production-design activities must be a versatile, creative, and well-informed person. He not only must be well grounded in the essential physical sciences that underlie engineering, but he also must have a comprehensive knowledge of a considerable variety of materials and of industrial processes. He must be familiar with the organization and functioning of industrial concerns and the human factors involved in a manufacturing enterprise. He must realize that progress is made through working with others, and that honesty, acceptance of responsibility, and the meeting of commitments cheerfully form a basis for good relationships with his fellow workers. He must know that high production rates are achieved when manufacturing operations can be carried on in safety and relative comfort by satisfied shop personnel. He must also be conscious that appearance is a factor in selling a product and either design his product with that in mind or get help in the problem of achieving good appearance without sacrificing utility. Most important of all, he must be cost conscious, as almost all actions are based eventually on cost.

Creative Thinking

Designing for production is a creative occupation. Creative thinking is not, of course, confined to a particular field or to a few individuals, but is possessed in varying degrees by people in many occupations: the artist sketches, the newspaper writer promotes an idea, the teacher encourages student development, the scientist develops a theory, the production design engineer develops improved manufacturing processes or applies improved materials to the creation of a better product.

Creativeness implies newness, but it is as often concerned with the improvement of old products as it is with the creation of new ones. In engineering, the newly created thing must be useful; it should be of benefit to people, yet should not be so much of an innovation that others will not purchase it. A "how-to-make-some-thing-better" attitude, tempered with good judgment, is an essential characteristic of an effective, creative engineer. This characteristic differentiates the soundly creative engineer from the crackpot.

Billions of dollars are being spent today by industry and the federal government on creative thinking in research and development programs, as contrasted to the few million dollars spent at the beginning of the twentieth century. Directors of these programs state that it is impossible to determine the ultimate value of a creative project. Many projects that seem hopeless often become profitable and sometimes revolutionize a whole industry. The main problem is to confine creative effort to the interests of the organization supporting the development work. Creative work is limited by the cost of placing the ideas into production. It has been stated that one creative engineer can keep 2,000 men working. Without creative thinking an industry gradually dies. Creative thinking brings new and improved products, better processes, higher quality, increased sales, steady work, and higher profits. For example, the Dow Chemical Company is the result of creative thinking on the part of Herbert Dow. He founded and developed the organization that helped establish the chemical industry in America in the face of competition from the European cartels. Mr. Dow's aim in business was to keep creating new products or better ways to make old ones. He was one of the first American industrial chemists to see that creation is industry's big job and to accept the fact that research and development require time, money, and—most important of all—patience and courage in the face of frustration.

Is creative ability born in an individual or is a person able to develop this ability? Both parts of this question can be answered in the affirmative. Certainly some people are born with more creativeness than others, just as certain people are born with a higher IQ than others. Also, it is possible to develop creative ability much as it is possible to develop mental and physical skills.

In order to develop creative ability related to production design activities, it is necessary to have an understanding of what creative engineering really involves. Creative engineering is, in part, a combination of the following:

1   A curious and imaginative mind
2   A broad background of fundamental knowledge
3   An enthusiastic desire to do a complete and thorough job of solution once a problem has been defined

Knowledge of fundamental principles of physics, chemistry, mathematics, and engineering subjects is a good foundation for creative thinking. Factual knowledge in the field of endeavor is also necessary. Young engineers cannot be expected to create designs of manufactured parts unless they have had contact with many kinds of parts and have studied their uses and methods of manufacture. However, knowledge is only a basis for creative thinking and does not necessarily stimulate it. The inherent personal characteristics of curiosity, intuitive perception, ingenuity, initiative, and persistence produce an effective creative thinker. Curiosity seems to stimulate more ideas than does any other personal characteristic.

Curiosity and Imagination

One aid to development or restoration of curiosity is to train oneself to be observant. An engineer, especially, should be observant of objects about him that have been created by man. He must ask how the object is made, of what materials it is constructed, why it was designed of a particular size and shape, why and how it was finished as it was, and how much it cost. These observations lead the creative thinker to see ways in which it can be improved, or to devise a better object to take its place. In the world of competitive industry, he may also be led to see a way of reducing its cost. Observation often leads to a revolutionary idea that may satisfy a public need. Witness the experience of DeForrest, who was led to develop the thermionic electron tube by observing the effects of an electrical discharge on a nearby gas jet. This observation, seemingly irrelevant to what he was doing, caused him to wonder why the jet behaved in an unexplained manner. This was an incentive to develop the relationship of heat to electron flow and, having arrived at the explanation of that phenomenon, he progressed to the invention of the three-element electron tube, which is the core of radio broadcasting.

Closely associated with curiosity is inventive imagination. This trait, which is frequently indistinguishable from curiosity, can also be developed and strengthened through proper use and exercise. By taking materials from our imagination and putting them together in new combinations we are developing our creative imagination. For example, try to visualize:

1 An automobile with two back wheels and one front wheel
2 A power-driven sled
3 A toothbrush with cleaning fluid contained in the handle
4 A screwdriver with a flashlight tip
5 An adjustable box wrench
6 An edible beer can
7 A cigarette box that also serves as an ash tray
8 A fishing lure that is remotely controlled

By bringing together, in the mind, various combinations of known materials, we are using inventive imagination to develop new designs. It is not possible to visualize absolutely new material, to go beyond the bounds of one's own experience. Inventive imagination takes place only by putting together known materials in a new way. Thus, when a painter creates a new picture, or a sculptor develops a new figure, or an engineer conceives a new design, his inventive imagination is combining facts already in his mind.

IDEAS GENERATE IDEAS

One significant creative idea usually opens up fields of activities that lead to many new ideas. As a student, Dr. C. R. Hanna, former associate director of the Westinghouse Research Laboratories, was interested in acoustics. This led him to research work on loudspeakers when radios were first produced, and resulted in the development of both large and small speakers with various types of power units. The use of mechanical equivalents for electric currents and electric circuit equivalents for mechanical devices led to many variations and combinations of electrical and mechanical devices. The microphones, pickup devices, and speakers that were developed used mechanical or electrical actuators. When the sound movie came into existence, it was troubled by excessive noise in the system used for recording, pickup, and broadcasting. When fundamental principles of electrical and mechanical filtering and damping were incorporated in the camera and projector, the present-day high-fidelity sound movies were obtained. These achievements in the field of sound attracted many similar problems to Hanna's department, such as speed regulation, damping, and elimination of noise in other types of apparatus. As a result, improved speed regulators, quiet rotating equipment, and an automobile shock absorber were developed. The control of large amounts of power by very sensitive electric circuits, mechanical devices, and a hydraulic valve system was a natural step in the solution of damping problems and power control. The control of speed and stability of equipment led to the use of the gyroscope, which is used to keep guns on the target when mounted in tanks, on ships, and in airplanes. The master stabilizer equipment on battleships and the automatic pilot for airplanes and guided missiles are some of the later applications of the foregoing systems. A less spectacular but equally significant application is the use of a gyroscope to regulate the speed of a relatively slow-moving part, such as that of a gearless elevator motor as it decelerates to a stop. Damping systems have been applied to railroad passenger cars in order to reduce the sway and road shock. The railroad cars also can be tilted to compensate for centrifugal forces in going around a curve. Thus it can be seen that possibilities of an idea are endless and most profitable to the person who recognizes its significance. A single idea may become the basis of one's life work.

TIME TO THINK

Many creative thinkers have discovered that it is necessary to take time out from their everyday activities to concentrate on problems of interest. They find that in a state of mental relaxation ideas come to them more readily. Notable among these was David Ross, inventor of the Ross steering gear. Ross was inactivated by a critical illness shortly after his graduation as an engineer from Purdue University. During this period he conceived the idea of the cam type of automobile steering gear, upon which was founded an important and profitable industry. He was very much interested in the development of student engineers, and one of his cardinal beliefs was that the young engineer should be given time to think. An accident early in life impaired the hearing of Thomas Edison. As a result, he was not distracted by ordinary things about him and was able to think through his problems.

Unless one takes time to think, one will be unable to focus his mind along a definite line. By taking time to think, one is able to concentrate upon a problem. Concentration necessitates keeping the mind logically thinking about one problem or idea for a period of time. Concentration can be developed in much the same way as imagination and curiosity. Thus, the mind can be trained to remain upon a subject for an indefinite length of time. It has been said that genius is nothing but infinite power of concentration.

ENERGY AND PERSISTENCE

Men like Ross and Edison were not dreamers, but men of energy and persistence who used ingenuity and initiative in developing their ideas. As soon as they had an idea, they put it down on paper, sketched it in detail, and improved it while the idea was fresh in their minds. They worked long, continuous hours, often until exhausted. This physical effort and recording of ideas clarified their thinking, stabilized their perspective, and eliminated foolish and undesirable ideas. In time each developed a systematic procedure for solving his problems.

SYSTEMATIC PROCEDURE

The road taken by the creative thinker in developing a new product is long, expensive, and difficult, beset with many detours and byways. A systematic procedure for the development of a product will reduce the time and effort of the engineer, who promotes the development from its early stages to completion. He sees it through design; selection of materials; setting up of operations and processes; selection of equipment; design of tools; formulation of specifications, operation sheets, and sales data; use of standards; adoption of ideas; obtaining of patents; and many other functions. All of these items are governed by economic or cost requirements and are developed by a systematic procedure.

The systematic procedure that will give results through creative thinking begins with the establishment of the problem area. Being able to accurately and completely define the problem area is a long step toward problem solution. Certainly solution cannot begin until the problem is presented. In order to be able to state the problem, the engineer must have acute powers of observation and be able to associate ideas and facts in new relationships. He should utilize the questioning approach and have no misgivings as to change. He should not feel at all reluctant to "not conform" or be considered "illogical." He should have no fears of being ridiculed, but should have personal confidence and acceptance of his own inherent creative ability.

After the problem area has been established, the next step is the collecting of pertinent facts and ideas. This usually starts with a survey of the trade literature to determine what already has been done. The local library usually can provide much helpful information through the use of current periodicals and reference text materials. An important source of facts and ideas is the United States Patent Office. Here, issued patents may be purchased at a nominal cost in the general field of the established problem area. The issued patents will present vividly the thinking and developments that have taken place to date in the problem area.

Once the pertinent facts have been established, then ideas need to be crystallized. Existing relationships in the problem area will need to be considered as well as the development of new relationships. An idea is of value only when used and accepted by associates, management, and the public. The person whose idea is accepted—not necessarily the person who had the idea first—deserves the credit. Merit alone does not promote an idea.

Although an incubation period is necessary, a creative idea should be crystallized as soon as possible. This is accomplished by recording the idea in sketches, drawings, and written descriptions. After the thought is fairly well crystallized, discussion with friends and associates in order to obtain their reactions helps to assure that the idea is practical. Models and tests soon prove whether the idea is valuable, and also provide patent protection. By guiding the young engineer in crystallizing his idea, industry will save time and money, and may profit greatly from the sales of the new product.

The schedule of events that should take place between the development of ideas to the transformation of suitable resources into useful products is common to all projects. An understanding of this chain of events will give the student an appreciation of the many difficulties that must be overcome in new-product development. The flow chart given in Fig. 1-4 will help clarify the procedure that should be followed.

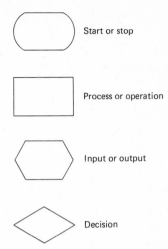

Start or stop

Process or operation

Input or output

Decision

FIGURE 1-4
Flow chart of events that take place between the development of new ideas
and the manufacture of useful products.

OPPORTUNITIES FOR THE PRODUCTION-DESIGN ENGINEER

Most of the great inventions were made by men in their early twenties. Marconi, McCormick, the Wright brothers, Morse, Edison, Westinghouse, and the men mentioned previously in this chapter made their principal contributions at an early age. Inventive capacity is a young person's trait. New ideas take root best in young people's minds before other interests crowd them out. They do not know that "it cannot be done," but go ahead and do the impossible. The high percentage of young engineers being utilized in research, development, and design departments of progressive industries is significant.

Production-design work in both the areas of product and process design will provide a rewarding lifetime job. The engineer knows that his work will make possible better products for more people at less cost. He will have rewarding personal contacts with the shop, sales, and service forces, and with his fellow engineers, research workers, designers, shop supervisors, and managers. The field of production design has come to offer some of the best opportunities afforded in industry. It may involve just a routine operation wherein apparatus is designed or modified to suit customer requirements, or it may be a comprehensive program including—in addition to design—sales analysis, research, invention, experimentation, and field tests which result in the development of a product. It may include the development of special equipment for the manufacturer of a product; the training of operators, construction and service men, and salesmen; and co-operation in the promotion of sales through advertising and demonstration.

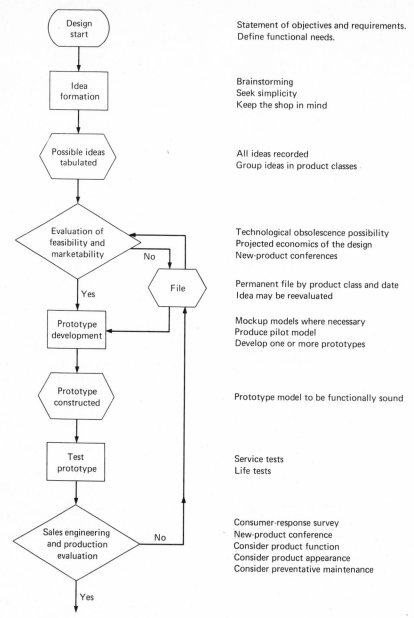

FIGURE 1-4 (Continued)

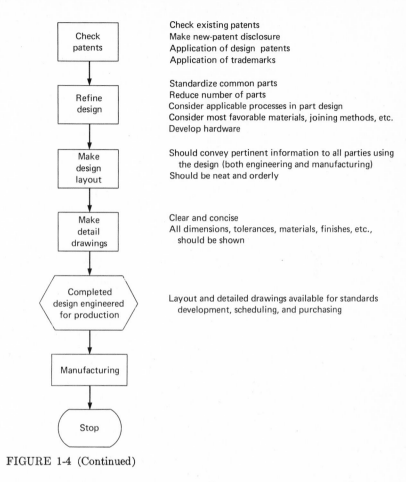

FIGURE 1-4 (Continued)

Production-design positions may involve the development or improvement of a line of products or a part, such as the engine, chassis, or body of an automobile. These improvements are accumulated and at periodic intervals new designs are introduced. Tools, equipment, and buildings must then be arranged for the production of the new design or new product.

In some instances, product design may involve the creation of large integrated units, such as an entire plant producing chemicals, pharmaceuticals, gas, or energy. For example, a steel rolling mill may be controlled numerically from the heating of the billet through the many stages of rolling until the final finished sheet is made. The adjustments of speed, size, and heat are all made automatically. The refinement of petroleum products is completely automated in many locations. Power plants are started up automatically, steam is generated,

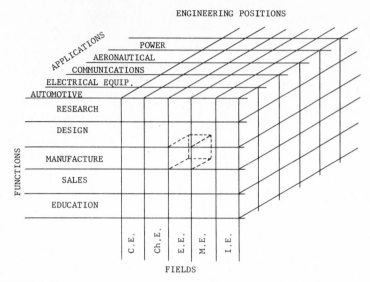

FIGURE 1-5
Areas of employment utilizing the services of engineering.

the turbine is brought up to speed, and the generator develops the power without the aid of an operator. Today, many components are made by processing the raw material into the final product without a hand touching the controls of the processing equipment. As American industry becomes more and more automated, the production-design engineer will find greater demand for his highest skills.

In the very small plant, all the work related to designing a product and planning its production may be the responsibility of one individual. As a plant grows, more product lines may be taken on and these immediately introduce the use of new and different materials and processes. This usually results in two separate activities: the design of the product and the design or planning of the production of the product. As an organization continues to grow, the size of these two activities increases and they become more decentralized. Within the product-design group, functional specialists, such as the vibration engineer and the electrical designer, have come into being. These may be further specialized by application, such as automotive, aircraft, power, or appliance. This same specialization will prevail in the process-design section. Thus there may be a machine shop planner, inspection planner, finishing planner, joining planner, or assembly planner. All these activities must be coordinated.

Engineering positions in industry are classified according to the following:

1   Fields (such as industrial, mechanical, electrical, civil, and chemical)

2 Functions (such as research, manufacturing, sales, education and training; and their subdivisions, for example, development, design, application, time study, methods, tool, maintenance, and service)

3 Applications (such as electrical manufacturing, aeronautical, communications, power, and railroad)

Figure 1-5 illustrates the various areas in which engineers are utilized. For example, the dotted cube represents electrical engineering positions in the manufacturing function of the automotive industry. There may be hundreds of positions within the dotted cube and all other cubes indicated on the diagram.

Production design is responsible for the cooperative efforts of the functional design department and the manufacturing engineering department. It serves as the means whereby barriers of misunderstanding and lack of knowledge as to processes, materials, and the functions of the product are overcome. There has been a definite trend recently toward the realization of the need of departmental cooperation, and also for organization and exchange of pertinent information. The work of production design is being recognized.

SUMMARY

When is a product "designed for production"?

To satisfy the requirements of designing for production, a product must be functionally sound, it must have sales appeal, and it must be competitive in price. In order for a product to be made economically, it must be designed so that the most appropriate materials and processes will be utilized.

An engineer cannot do an effective job of product design unless he knows or is supplied with adequate information as to how his designs will be produced. Therefore, the principal problem of engineering for production is sound functional design plus the selection of the materials and the processes to be used so that the new product can be produced on a competitive basis.

In choosing these materials and processes, the functional designer must make many modifications and changes in his original conception. The shape, color, size, tolerance requirements, texture, weight, and the functional design itself may be affected before the ideal design is developed that is functionally sound, has sales appeal, and is economical to produce within the required time.

QUESTIONS

1 What is meant by "designing for production"?

2 What two distinct functions are included in designing for production?

3 Why does the process-design function not include the development of standard tools?

4   In the drawing shown in Fig. 1-1, why is the tolerance on the stud diminished after the plating operation?

5   Define pitch diameter.

6   Why is it part of the responsibility of the process-design function to include standard elemental times on the operation card?

7   Give some of the requirements for engineers entering the production-design field.

8   How are engineering positions in industry classified?

9   When is a product designed for production?

10 Why is the work of the production-design engineer a creative occupation?

11 How is creative work limited?

12 For what is Dow known?

13 What are the principal requirements of creative thinking?

14 Is an engineering background necessary in order to do real creative thinking? Why?

15 What inherent personal characteristics produce an effective creative thinker?

16 Why is inventive capacity a young person's trait? Give several examples where discovery was the result of a young person's effort.

17 What does creative engineering involve?

18 What step follows the establishment of the problem area in the procedure for development of creative designs?

19 What particular areas of employment are of interest to you? Locate these areas on Fig. 1-5.

20 What products are made in a particular automatic plant?

21 When would it be desirable to make mock-up models?

22 What is the difference between a pilot model and a prototype model?

PROBLEMS

1   Most toasters today have approximately the same external appearance. Novel or unusual features should result in marketing advantages. Explain how you could design a toaster that would provide toast with the owner's initials or name legibly shown on each piece of toast. This new feature should not increase the cost of the toaster materially. Sketch your design and estimate the increase in the unit cost as a result of the new feature.

2   Redesign an existing rotary lawn mower to do double duty as a combined leaf mulcher and fertilizer spreader in the fall season. If possible, design the device in such a way that it could be sold as an attachment for use on equipment already in the field. The device should sell for $10 to $25 to achieve the maximum market potential. This study should be in the nature of a feasibility study with a number of possible sketches rather than detailed designs. It would be expected that several sessions be spent in gathering ideas before any of them are carried on to the more-detailed design stage.

SELECTED REFERENCES

ALLEN, MYRON S.: "Morphological Creativity," Prentice-Hall, Englewood Cliffs, N.J., 1962.

BUHL, HAROLD R.: "Creative Engineering Design," Iowa State University Press, Iowa City, Iowa, 1960.

GORDON, WILLIAM J.: "Synectics, The Development of Creative Capacity," 1st ed., Harper, New York, 1968.

HARRISBERGER, LEE: "Engineersmanship, A Philosophy of Design," Brooks/Cole, Belmont, Calif., 1966.

OSBORNE, ALEX F.: "Applied Imagination," Scribner, New York, 1957.

VON FANGE, EUGENE: "Professional Creativity," Prentice-Hall, Englewood Cliffs, N.J., 1959.

2

CRITERIA FOR PRODUCT SUCCESS

With so much dependence on the success of new products, it is surprising that only one out of five new products put on the market in the United States proves to be successful. This is especially disconcerting in view of the cost of new-product failure. A recent survey by the LaSalle Extension University indicates that the minimum cost of bringing a single new product to the point of market introduction, on the average, is $150,000. This four-to-one failure-to-success ratio was confirmed recently by the Bjorksten Research Laboratories in a study showing that of 27,000 new products introduced in a given year by American firms in manufacturing categories only about 5,400 proved successful.

Dr. Johan Bjorksten brought out that the principal reasons for the failure of new products is due to lack of technical expertise related to the new product. This lack of expertise may be inadequate experience with applicable new materials, inadequate experience with applicable processes, or inadequate experience with field testing or market analysis.

In many cases, however, failure can result from unanticipated activities or reactions of competitors (their new or improved products or price reductions, for example). This is an application for "game theory," which is the name given to a particular model of competitive systems. There are three elements considered in game theory: (1) the number of players (companies in the suggested example); (2) the strategies available to each player (the product mix of each company in our example); and (3) the payoff function. The payoff represents the gain or improvement to each "player," which is usually monetary although not necessarily so, and this gain is a function of the set of decisions made by each player.

Before a product is designed for production, it should undergo a thorough analysis to determine if the probabilities for its success are of sufficient magnitude to warrant the outlay of capital that would be necessary to get into production. It has been stated that the engineer today can design almost any desired product. His designs can make products more efficient, less costly, and have a longer working life. However, these factors alone do not assure that the potential customers will buy the product. The design of a successful product must start with the customer. The design must include the features that the customer wants. In order to determine these features, several areas of influence should be studied. This preliminary analysis is usually the prerogative of the sales and marketing departments, but will be mentioned here so that the production-design engineer will have a better understanding of all the intangibles that affect the success of a product.

The areas that should be examined preparatory to design include the following:

1 An analysis of the market in order to determine its size, the nature of the customers, and possible trends

2 Evaluation of the competition to determine its extent and strength and to get a clear picture of the pricing situation

3 Appraisal of distribution to determine if the product can be sold through the company's regular channels of distribution and with its existing sales organization

4 Determination of how much advertising and promotion will be needed to introduce the product

5 Appraisal of the effect of the product on the existing business to determine if it will increase or decrease the sale of existing products

6 Determination of financial requirements in order to learn how much investment may be required to handle development, manufacture, and marketing of the product and what the probable returns will be

All the above areas should be examined to appraise the idea from standpoints of customer, market, and the overall business. Once this is done, and if it appears sound to embark upon the idea, objectives should be established and responsibilities assigned.

Too frequently the product designer is cognizant only of function when he is designing a product. It is the intent of this chapter to present all the important factors that must be considered in the product-design function.

If any product is to maintain a lasting satisfactory sales appeal it must:

1 Have a sound functional design
2 Have eye appeal
3 Have quality characteristics, both in material and workmanship
4 Provide for convenient maintenance
5 Be competitive in price
6 Be delivered to the customer in time to meet his needs

Of the above requirements, 1, 2, 3, and 4 are principally the responsibility of the product-design function and 5 and 6 are, to a large extent, the responsibility of the process-design function. To assure that all six of the criteria are met, it is necessary that product design and process design work closely together and in harmony.

The product designer can obtain much helpful information as to the specification of optimum geometry, tolerances, and finish from the process designer. Likewise the process designer can receive a great deal of assistance from the product designer as to locating points, material characteristics, and product use. Without the combined knowledge of these two sources, it is difficult, if not impossible, to engineer a new design for production.

FUNCTIONAL DESIGN

As a new product is conceived or an old product is improved, the major objective is to provide a commodity that will meet some need or render some service in a manner superior to that of a former product. Sound functional design assures that the product will satisfactorily operate for a reasonable period of time in the manner intended. Thus the materials going into the design will have been thoroughly checked for physical characteristics, such as strength, stiffness, and weight, as well as for such service characteristics as corrosion resistance and conductivity. Sound functional design takes into consideration all those details that affect the operation of a product. In other words, functional design assures a design that will work and accomplish the purpose for which it was intended.

Functional design in itself does not take into consideration ease of manufacture or maintenance, and it frequently neglects eye appeal and human engineering. It must always be remembered that a design which is impractical to manufacture is just as unsatisfactory as a design which will function improperly. It is also important that the design appeal to the esthetic sense and be made so as to be effectively used with due consideration to human physical and mental limitations.

It is not the purpose of this text to delve into the "how" of good functional design, as this subject alone would fill a bookshelf. However, since the "pure" functional designer so frequently gives insufficient consideration to appearance, quality, and human engineering, these criteria, which so greatly affect sales, will be discussed. Shape, color, texture, quality, and consideration of human dimensions are more intimately related to the work of the product designer than to that of the process designer.

VALUE OF APPEARANCE

Engineers must appreciate the value of appearance and its relation to the function, cost, and sales of the product. This modern trend of thinking was apparent at a recent machine-tool show. Visitors gathered around the "attractive" machines, which had good appearance built into the design, and tended to pass by less-attractive machines designed for function alone. An operator takes pride in a machine with a good appearance. His interest results in lower maintenance costs and better-quality products. Appearance has become a major factor in all types of household appliances, machinery, process equipment (such as baking and canning machinery), locomotives, and power equipment. Better balance is now being attained between functional design, materials, finishes, colors, processes, methods of manufacture, maintenance, and appearance. Fundamental principles and laws govern the shapes and colors for esthetic appeal. Designing "appearance" into a product is within the province of the engineer because "Appearance must be built into a product, not applied to a product."

The appearance of a product makes deep impressions on a buyer: it may suggest power and speed in a locomotive, durability in a machine tool, cleanliness in hospital equipment, precision in an instrument. When two products have approximately the same functions, the same cost, and the same time required for delivery, the product with the better appearance will have the most sales appeal.

The importance of appearance to the auto industry is illustrated by the fact that the original Chrysler Air-Flow and the Buick bulging-body designs failed to attract the public. This awakened the auto industry to the importance of appearance. If the appearance of a car produced by Chevrolet in quantities exceeding a million a year should fail to appeal to the customer, losses would be disastrous. Consequently, the auto industry spends large sums on market research to find the style and features that customers prefer. Styling is also affected by changes in public taste and by the passage of time (as for clothing). A style, such as automobile tail fins, can become too common or outdated. And style identifies the individual make of product. The importance of appearance is probably one of the reasons that the recently hired chief engineer of one automobile manufacturing company is a man with an artistic background.

The engineers of the Hamilton Watch Company have recognized the problem of appearance as well as quality and cost. Watchmaking is an old art in which sound customs and principles have been established for many

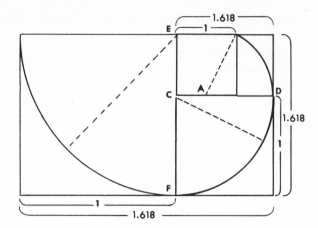

FIGURE 2-1

The golden-mean rectangle was used by the ancient Greeks as a tool in proportioning their buildings. Starting with a square, a radius is drawn about point *A* which bisects the side of the square and extends to one of the far corners of the square. The sides of the rectangle thus formed have the proportions of 1.618 to 1. The long side of this rectangle (*CD*) and each long side of succeeding rectangles (*EF*) become a radius, which, when extended, forms a new rectangle of like proportions. The result is a series of shapes which combine rhythm and variety.

years. The Hamilton Watch Company has realized the value of these customs and principles and has acted on them. Their interest in the history of watchmaking led their engineers to study the history of the form and appearance of watches. They also began to study appearance from the viewpoint of the artist and found that in ancient days the proportions of the "golden rectangle" had been established as the form most pleasing to the eye (Fig. 2-1). Since the Hamilton watch had been produced in many shapes and forms, it was decided to check the sales of various styles and their artistic characteristics. The survey proved that the wristwatches built closest to the proportion of the golden rectangle had the greatest sale. The proportions used apparently were the result of good judgment because the golden rectangle had not been considered when the watch designs were created. The new Hamilton wristwatches have the proportions of the golden rectangle. When an engineer enters a well-established field, he should be careful about discarding well-tried principles of design and should analyze the sales record of the products to determine the functions, form, and color that have appealed to the customer.

Another example of the use of principles and laws evolved over hundreds of years is in the creation of printing type. The magazines of today, both technical and popular, use hundreds of different styles of type in their advertising material and usually one distinctive type for their reading material. Each style has been created by an outstanding artist. The perfection of the art is not always realized. Letters are made in accordance with certain laws of proportion; the spacing, the size of lines, and the proportion of top to bottom are all worked out with mathematical precision. Originals are made on a large scale and reduced to the size of print. The large originals, although made according to the rules of proper proportion, always require slight modifications by the artist to obtain balance, spacing, legibility, and proper characteristics. Molds and dies are made to extreme accuracy to secure the proper effect.

PRINCIPLES AND LAWS OF APPEARANCE

A design that has appealing appearance has marked sales advantages over one that does not. The designer must keep in mind that the ultimate consumer or user of his creation is probably an ordinary person, and it is he to whom the designer must appeal and not an esthete. All men have a sense of beauty to some degree. We learn through experience to appreciate pleasing shapes and contours and colors that blend harmoniously.

The basic form of a drill press, a milling machine, or an orange squeezer is determined by its function, utility, and many other factors. This basic form can be modified and given a pleasing appearance by considering:

1 Unity—simplification of form; proportional relationship; repetition
2 Interest—emphasis; contrast; rhythm
3 Balance—symmetry
4 Surface treatment—color; texture

It must be remembered that men spend a lifetime perfecting the use of these laws and principles which have been established through the centuries. The successful industrial designer does not use them mechanically; he interprets the feelings and desires of the customers by applying many variations in accordance with sound principles. In complicated, elaborate, and doubtful cases, the help of an industrial designer can be obtained. The industrial designer is trained to create forms that attract attention and appeal to the potential market.

Unity

Unity in product design means that the form of the product is such that people will like it instinctively. If the design has unity, then all components are blended together to make a complete and self-contained design. The three qualifications

FIGURE 2-2
Reflection and absorption of light by white, gray, and black surfaces.

leading to "unity" are (1) simplicity of form, (2) proportional relationship, and (3) repetition.

When the eye sees a new thing, it immediately records its general geometry, such as a cube, a cylinder, a cone, a pyramid, a sphere, or an ellipsoid. These simple forms lead to the development of all designs.

Proportion has to do with the size relation of one part of a design to other components within the design or to the entire design as a whole. Unity in the design is achieved principally by proportion.

Repetition, as the name implies, has to do with the regular recurrence of identical or similar components. Thus, the regular occurrence of such parts as wheels, buttons, and trim will lead to the blending together of the various components in the design.

Interest

Through "interest" the designer uses contrasting elements placed and controlled so as to attract and hold attention. To accomplish this, he calls for emphasis, contrast, and rhythm in his designs.

Emphasis is accomplished through prominence in size or color. Thus, if through "repetition" there is a series of three button groupings, emphasis can be acquired by appreciably increasing the size of one of each of the sets of buttons and/or by brightening its color.

Contrast is used to show differences among two or more components or to stimulate emphasis on a given component of a design. Contrast is obtained through size, color, and location. To give a feeling of strength or safety, a member or section may be designed so that its location with a similarly shaped section gives a feeling of durability in view of "contrast" with the smaller, more insignificant component.

Rhythm in design is signified by a regular occurrence of elements. Thus, due to arrangements of components, the eye will move along a design and see a regular pattern of similarly shaped components. Simple rhythm will include alternating small and large shapes, while more complicated rhythms may involve periodic combinations of shape, size, and color.

Balance

Balance of design gives the viewer an immediate impression of stability. Balance is inherent in our makeup—we tend to look for this property in all objects with which we come in contact.

Symmetry is the most used method of giving balance to a design. When two halves of an object are identical, the halves are classified as being symmetrical. A design can also be symmetrical when developed around a central point. Thus, concentric discs within a wheel are examples of radial symmetry.

Surface Treatment

When an individual "sees," patterns of light are reflected from the object he is viewing to his eye. The surface of the object being viewed will have certain qualities which will either aid or detract in the perceptive process. For example, the reflection of light by white surfaces is considerably greater than gray surfaces and gray surfaces will reflect more than black surfaces (Fig. 2-2). The reflective ability for light striking a surface is known as value. White surfaces are at the top of the value scale and black, with theoretically no light-reflecting ability, is at the bottom, with all colors falling in between these limits.

The form that the designer achieves can be appreciated to a large extent by sight and touch: the surface treatment is the principal medium through which these senses can work. The two factors that affect sight from the standpoint of surface treatment are color and surface texture.

Color   Color and texture have selling power. The psychological effects of color on people are very real. Yellow is the accepted color for butter; therefore, margarine must be made yellow in order to appeal to the appetite. Meat cooked in 45 seconds (s) on an electronic grill did not appeal to the customer because it lacked a seared, brown, "appetizing" surface. A special attachment had to be designed to sear the beef steak. Girls in an air-conditioned midwestern plant complained of feeling cold although the temperature was maintained at 72°F. When the white walls of the plant were painted a warm coral color, all complaints ceased. Workers in a factory complained that boxes were too heavy until the plant engineer had the old boxes repainted a light green. Several men said to the foreman the next day, "Say, those new lightweight boxes sure make a difference."

Sales are conditioned by colors. People recognize a company's products instantly by the pattern of colors used on packages, trademarks, letterheads, trucks, and buildings. Research has shown that the preference in color is influenced by nationality, location, and climate. Sales of a product formerly made in one color were increased when several colors suitable to the different customer demands were supplied.

Table 2-1 EMOTIONAL AND PSYCHOLOGICAL SIGNIFICANCE OF THE PRINCIPAL COLORS

Color	Characteristics
Yellow	Has the highest visibility of any color under practically all lighting conditions. It tends to instill a feeling of freshness and dryness. It also can give the sensation of wealth and glory; yet can suggest cowardice and sickness.
Orange	Tends to combine the high visibility of yellow and the vitality and intensity characteristic of red. It attracts attention more than any other color in the spectrum. It gives a feeling of warmth, and frequently has a stimulating or cheering effect.
Red	A high-visibility color, having intensity and vitality. It is the physical color associated with blood. It suggests heat, stimulation, and action.
Blue	A low-visibility color. It tends to lead the mind to thoughtfulness and deliberation. It tends to be a soothing color, although it can promote a depressing mood.
Green	A low-visibility color. It imparts a feeling of restfulness, coolness, and stability.
Purple and Violet	Low-visibility colors. They are associated with pain, passion, suffering, heroism, etc. They tend to bring the feeling of fragility, limpness, and dullness.

Standard colors carried by the manufacturer should be used. These colors should be practical and appropriate, and the lacquer or paint should be durable and easily applied. The range of these standard colors is great enough to meet the needs of the discriminating color designer.

The six combinations of color that provide the sharpest contrast when used together, as shown by tests of legibility, are black on yellow, green on white, red on white, blue on white, white on blue, and black on white. Other combinations that provide sharp contrast are yellow on black, white on red, white on green, and white on black.

Table 2-1 illustrates the typical emotional effects and psychological significance of the principal colors.

Figure 2-3 illustrates pairs of colors that will give harmonious hues; also, those colors that are complementary in nature may be discerned.

Texture Texture is the pattern of contrasts in light reflections that identify the surface. The influence of surface texture on customers is as significant as

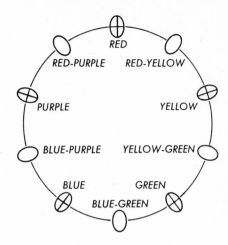

FIGURE 2-3
The color wheel is based on the Munsell color system. Five primary and five
secondary colors or hues are indicated. Pairs of adjacent colors on this wheel
are known technically as "harmonious" hues. Colors opposite to each other
are complementary.

color. There has been a definite trend toward mirror finishes, for example, glass-
fronted stores, Koroseal fabrics, and chrome finish on automobiles and appliances.

There are other finishes that create interest and stimulate sales appeal,
such as hammered-finished types, which simulate the mottled texture of hammered metals, and Rigid-Tex, the rigidized metals rolled in any of a number of
textured patterns and finished in a variety of colors. Rigid-Tex, concealing
finger marks through its pattern designs, is particularly well adapted to dic-
tating machines, which are subject to much handling. It is available in blue,
neutral, gray, dusty rose, beige, green, chartreuse, turquoise, and golden yellow—
colors well suited to office use.

INCORPORATING QUALITY INTO THE DESIGN

Quality cannot be inspected into a product but must be designed and built into
it. Tolerances, machine finish, material specifications, and performance requirements must be specified so as to assure a product that is of sufficient quality.
Tolerances and specifications that are more rigid than necessary should not be
assigned. There is a tendency for product designers to establish tolerances and

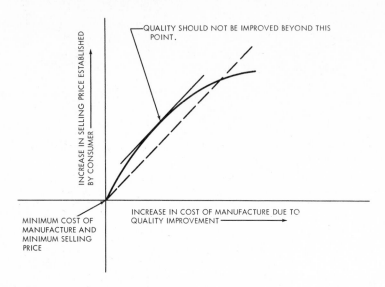

FIGURE 2-4
Relationship between quality improvement and selling price.

specifications that make the cost of manufacture prohibitive. There must be a balance between satisfactory quality and competitive cost of production.

The production-design engineer should realize that almost no market exists for the best possible product. The customer usually is desirous of a "less perfect" product at a much lower price. Each product has its ratio of increments of cost and selling price. Starting with a given product, having minimum acceptable quality, each increase in quality will have a measurable increase in the cost of manufacture. This increase in quality will result in larger permissible selling price up to a certain point. As long as the permissible price increase is greater than the cost increase of quality improvement, there will be a profitable investment in quality. This relationship is shown in Fig. 2-4.

If substandard parts begin to show up in a commodity, the engineer should find the source of trouble and endeavor to correct it by eliminating the substandard material, tools, or equipment, and by assisting and training the operator. After one large corporation made a survey of the causes of its products' poor quality, it found that quality had been designed into the product but management was guilty by not having provided better tooling, adequate instruction of operators, and other items under their control. Operators have a responsibility for errors caused by operator negligence, indifference, carelessness, and generally poor employee attitude. In the past, the feeling has been that all rejections were due to poor employee attitudes, loss of job pride, and a desire on the part of the worker to make as much money as possible without regard for product quality. Most of the items on the following checklist are management's responsibility:

A   Personnel
    1   Did the workman have the required information and instruction regarding:
        a   Specifications?
        b   Machine operation?
        c   Setups?
        d   Tools, jigs, and fixtures?
        e   Materials?
        f   Measuring instruments?

    2   Has the workman been properly trained?
    3   Does the workman have the proper attitude?
    4   Does the workman have any physical handicaps?
    5   Has the workman done the job before?

B   Design
    6   Was the design satisfactory?
    7   Were design problems referred to the engineer?
    8   Did he take any action?
    9   If so, did it correct the difficulty?

C   Specifications
    10   Were the specifications given on the drawing?
    11   Was the drawing the latest submitted?
    12   Was the process and finish specification book available to everyone?
    13   Was this book up to date?
    14   Were the specifications correct?
    15   Was the process inspector consulted?
    16   Was the manufacturing information correct?

D   Materials
    17   Were the materials those which were specified?
    18   Were the materials inspected beforehand?
    19   Was the material a substitute?
    20   If substitute, was it approved by engineers?
    21   Was the material defective?

E   Machine
    22   Was the proper machine used?
    23   Was it the type of machine called for on the routing sheet?

24   Was the machine in proper condition?
25   Was the machine properly used?

F   Fixture (or jig)

26   Was the proper fixture used?
27   Was it the fixture specified?
28   Was the fixture in good condition?
29   Was the fixture used properly?

G   Tools

30   Were the proper tools used?
31   Were the tools in good condition?
32   Were they checked before being used?
33   Were the tools used properly?

H   Material handling

34   Were the proper handling facilities used?
35   Were the handling facilities properly used?
36   Were the handling facilities in good condition?

I   Working conditions

37   Were the atmospheric conditions satisfactory?
38   Were the lighting conditions satisfactory?
39   Was the area around the work place clean (good housekeeping)?
40   Were the working conditions conducive to a good attitude toward quality?

J   Job history

41   Is there a record of this job having been done before?
42   Was there any trouble with it before?
43   If so, is the defect the same?
44   If so, is it any worse now?
45   If so, have any changes been made in the job?
46   Did the change correct the trouble?
47   If there is no previous history of trouble, have any changes been made in the job since it was last done?

MAN-MACHINE CONSIDERATIONS

Wherever a human being and a product work together, the functional design engineer should consider those factors influencing the operator's ability to do his job. Such data as human response time under various conditions, the amount of information that can be processed per unit of time, the size and arrangement

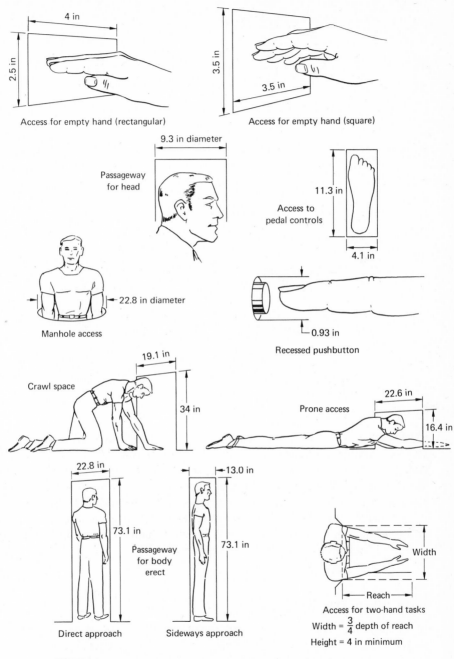

FIGURE 2-5
Minimum dimensions for easy access.

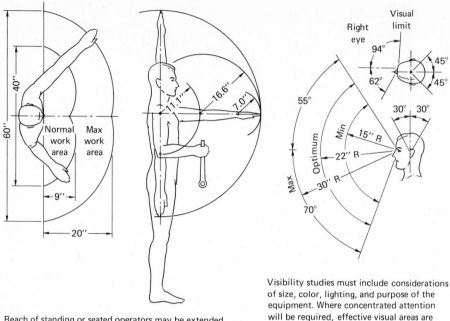

Reach of standing or seated operators may be extended beyond these limits by lateral motion or pivoting at the waist. Drawings are based on the small man (2½ percentile), body rigid and stationary.

Visibility studies must include considerations of size, color, lighting, and purpose of the equipment. Where concentrated attention will be required, effective visual areas are considerably reduced.

FIGURE 2-6

of controls, etc., have an impact on his performance. As humans, we have both physical and mental limitations that need to be considered. Notable limitations include reaction time; muscular response to both static and dynamic forces; dimensional limitation due to size of hand, reach of arm, step length of leg, angle of rotation of neck; and, the fatigue factor, mental and physical.

Some of the more important minimum dimensions for easy access are given in Fig. 2-5, and Figs. 2-6 and 2-7 provide many of the human dimensions that are of importance to the functional designer.

EASE OF MAINTENANCE

Practically every product produced requires periodic maintenance in order to assure satisfactory performance. Wherever moving parts are involved, lubrication must be provided or else wear will be excessive. In any event, moving parts do wear and will require replacement. Of course, stationary parts need maintenance and replacement also, especially when subjected to corrosion, load, heat,

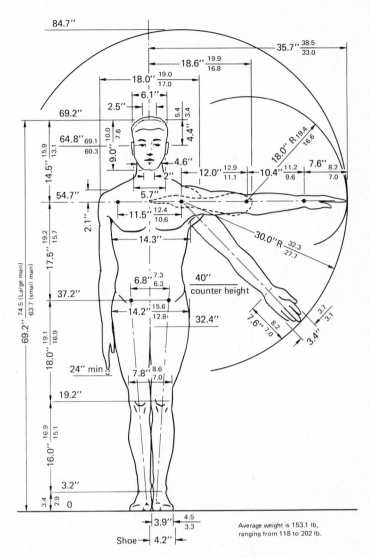

Human dimensions of the average adult male based on
the 2½ to 97½ percent range of measured subjects.

FIGURE 2-7
Human dimensions of the average adult male.

and other factors that shorten service life. In order for a design to be completely
acceptable, it must not only be functionally sound, economically priced, and
have sales appeal, but it also must readily allow normal maintenance and repair.
Some of the lack of success of recent automobile styles has been due to the
neglect to provide ease of maintenance.

One of the recent models of washing machines was designed so that the bottom cover had to be removed, the transmission dropped, and several other components removed in order to install a new drive belt. Needless to say, the life of this design was short. Customers who find it necessary to pay exorbitant repair bills simply because ease of maintenance was not provided for in the design can hardly be expected to purchase a given style again.

Oil wells, grease points, and fan and drive belts should be readily accessible. In the design of any assembly, the ease of replacement of the various components should be directly proportional to the service life. Thus, a bearing that has a normal service life of 1 year on an assembly that should provide 15 years of operating performance should be located so that it can be periodically replaced.

SUMMARY

A good functional design may have a disappointing sales volume if basic sales considerations have been neglected. The product must not only work satisfactorily but it must stimulate sufficient interest in the customer to cause him to buy. It is essential that the production design engineer have an appreciation of the value of appearance and understand the influence of proportion, repetition, emphasis, contrast, rhythm, symmetry, color, and texture. The basic laws followed by the artist should be adapted by the production design engineer.

A successful product is a quality product. High quality must be designed and built into a product and then maintained so that the customer receives the product as designed. It is important that the production design engineer understand the relationship between cost of manufacture and quality. He must not specify dimensions, finishes, and other specifications that are impractical to maintain. He must maintain a balance in the quality of all the components comprising the product. Thus it would be foolishness to specify quality requirements that would allow a service life of 100 years on one part of an assembly and let other components have a service life of but 10 years.

Product designers should always be conscious of the ultimate user of the product. When the point of contact between product and the people becomes a point of friction, then the production-design engineer has failed. However, if the users of the product are made safer, more comfortable, or more efficient through the use of the product, then the production-design engineer has succeeded.

Sound designs must permit normal maintenance and general overhaul. A design that necessitates costly maintenance and repairs will lose out to the one that allows easy replacement of worn parts and readily permits normal maintenance, providing the other criteria of the two designs are comparable.

Of course, even if all the above characteristics are incorporated in a design, it will have a limited market unless it is competitive in price and can be delivered when the customer needs it. These latter two points, which are largely the responsibility of the process-design function, will be discussed in subsequent chapters.

QUESTIONS

1   What six characteristics are fundamental in order to assure successful sales volume?
2   What is meant by functional design?
3   In what type of products is appearance of particular value?
4   What are the proportions of the "golden rectangle"?
5   What would be the length of an appealing rectangle that has a width of 1 in?
6   In styling a product, what is meant by "unity"?
7   How can interest be instilled in the design of a product?
8   What does the characteristic of balance impart to a product?
9   What is meant by "value" when referred to a surface?
10   What are the psychological effects of color on people?
11   What combinations of color provide sharp contrast?
12   Name some typical mirror finishes.
13   What finishes tend to create interest?
14   Why must quality be designed and built into a product rather than inspected into a product?
15   In what ways can management control quality?
16   What principles should the designer keep in mind relative to product maintenance?
17   What are the chances of a new product succeeding on the open market? List 10 designs that you are familiar with that did not succeed.
18   What would be the recommended radius of the arc generating an area of vision requiring concentrated attention?
19   What notable human limitations should the production-design engineer be cognizant of?

SELECTED REFERENCES

McCORMICK, ERNEST J.: "Human Factors Engineering," McGraw-Hill, New York, 1964.
MORGAN, CLIFFORD T.: "Human Engineering Guide to Equipment Design," McGraw-Hill, New York, 1963.

3

COST AND PRODUCT DEVELOPMENT

All final decisions are usually based on cost. No matter how perfect the final design may be, no matter how much eye appeal it has, it will not sell unless the price is in line with competitive products. The relation between product design and price is obvious, yet this connection is too often overlooked by the functional designer. The better a product's features and design, the higher the price it will command. Conversely, a product that suffers in comparison with its competitors must be sold at a lower price, if at all.

No one price can be considered as being ideal for a given product class. If its features are changed within limits, it may be possible to raise or lower the selling price. All features are not of equal value when it comes to attracting customers. Likewise, all products do not respond similarly with changes in like features. For example, Fig. 3-1 illustrates the quality sensitivity of two different products as related to sales. The two products have a wide disparity with relation to potential sales upon a given percent improvement in quality.

Selecting the manufacturing processes and recommending specific materials to be incorporated in the design are the responsibilities of the process-design section. This work must be done in conjunction with the work of the

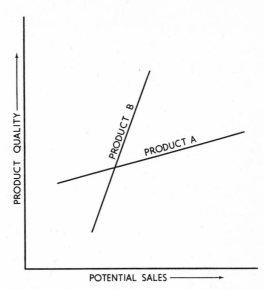

FIGURE 3-1
Quality sensitivity of two different products.

functional designer so that the objective of designing for production can be realized.

Since cost influences many decisions of all those associated with designing for production, it is quite important that the engineer have not only an appreciation for, but an understanding of, cost.

Two kinds of costs are of vital interest to the engineer: development costs and the cost of the product. Development costs include cost of sales' analysis, research and engineering (such as engineering and draftsmen time), models, and tests. They include cost of tools, equipment, plant rearrangement, and obsolete stock; cost of training sales and service personnel; and promotional expenses such as advertising and instruction and service manuals. The cost of the product is composed of the cost of each component of the product. Knowing these costs enables the engineer to choose between materials and processes so as to obtain the lowest product cost.

Costs are notably related to quantity produced or expected sales and prices. An appraisal of all essential cost considerations may be termed *economic analysis*. A preliminary economic analysis is a basis for the commencement of any development project. As development proceeds and results become more evident or conditions change, additional economic appraisals may be called for; and a final cost analysis should precede authorization for actual production in quantity.

SOURCE OF FUNDS FOR DEVELOPMENT COST

Appropriations for development expense are based upon a percentage of the sales dollar. If sales are high, more money is available; if sales are low, the development budget may suffer and every effort will be made to spend the money on the development that gives the best promise of increasing sales and profits. Frequently special customer orders require engineering development expense which is charged directly to the customer's order and estimated as part of the cost of the job. This development work on customers' orders sometimes gives employment to a large percentage of the engineering department.

Expenditures on development are reported to management periodically. The engineering department usually has the responsibility of holding expenses within appropriations. Expenditures concern each department. The shop may desire to spend $10,000 on new equipment and tools to reduce the cost of a part. This expenditure should have the approval of the engineering department because the engineer may be contemplating changing the part from sheet metal to plastic. Also, the sales department should be consulted; it may realize that this part will have greatly reduced sales in the next months. The shop should approve expenditures to guard against obsolescence of existing tools, equipment, skills, and parts.

The formality of approvals, issuance of orders and reports, and use of proper charge numbers is irksome to most engineers, especially to young people who are impatient and want to get the development completed. However, the quickest way to get a project accomplished is to work through proper channels and let each man do his own job. By not using proper channels and spending

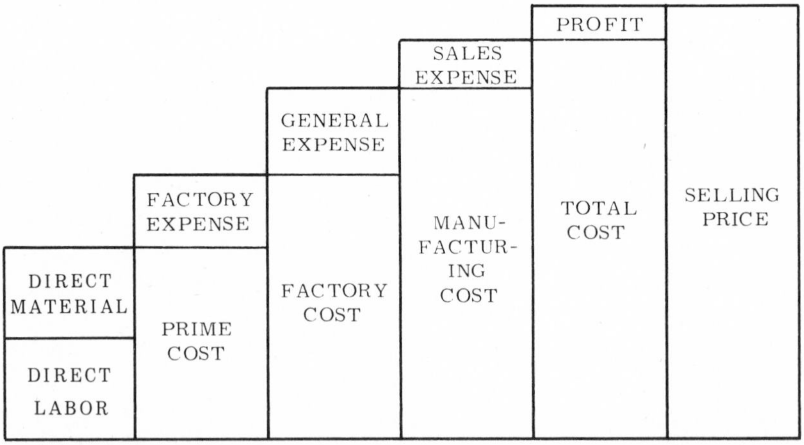

FIGURE 3-2
Elements of cost and profit entering into the development of selling price.

time on the duties of others, the engineer is failing to do his own job of developing the product.

PRODUCT COSTS

"What does it cost?" is the eternal question asked the production design engineer by a supervisor or manager. There are many other important factors to be considered, such as capacity, appearance, and ease of repair, but cost enters the picture sooner or later in choosing materials, processes, and functions of design, and in negotiating a sales contract with the customer. Only through a knowledge of costs can profits be made, and since profits are the reason for the existence of business, a knowledge of costs is most essential. The structure of product cost is given in Fig. 3-2.

ITEM OF COST ENTERING INTO THE DEVELOPMENT OF SELLING PRICE

First cost or prime cost is usually thought of as including the cost of direct material and direct labor. When factory expense is added to prime cost, the factory cost is obtained. Factory expense includes such items as rent, light, heat, power, factory supplies, and factory indirect labor. The manufacturing cost is made up of factory cost plus general expense. General expense includes such items as engineering, purchasing, production, methods, office costs, and depreciation. Thus, general expense will include both supplies and salaries in the administrative areas of the business. Sales expense involves all those items of cost necessary to sell and deliver the product. Here is included advertising expense, bad debt expense, and salesmen's salaries and commissions. The total cost of the product is then determined by adding a selling expense to the manufacturing cost. After the total cost is determined, a reasonable profit is added in order to determine the selling price.

However, price is not always determined by cost; it is frequently determined by what the buyer is able and willing to pay. It is influenced considerably by the lowest competitive price, and is determined at the point of sale, not in the factory. If price is less than cost, the company must take steps to reduce the cost or cease manufacture. Often a company does not know its costs and is unknowingly losing money for itself and competitors. Recently trade associations have educated small companies in cost procedures and thus maintained the market price above actual costs.

An understanding of the basis of costs will enable the production-design engineer to use good judgment in the selection of materials, processes, and functions to create the best product. An increase in perfection from 90 to 99

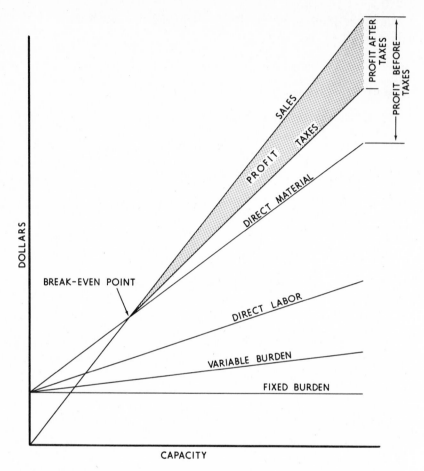

FIGURE 3-3
Break-even chart indicating relationship between cost, sales, profit or loss, and volume.

percent may result in a 50 percent increase in development and product cost and destroy the sales value of the product. Cost, quality, and degree of perfection should be carefully considered in order to obtain the greatest profit over a period of time. Cost is still the deciding factor. The relationship between cost, sales, profit or loss, and volume is best revealed on what is known as a *break-even chart*. Figure 3-3 illustrates a typical break-even chart.

Cost data are obtained from the accounting department and from suppliers through the purchasing department. Cost data are subject to error, for costs are usually compiled by clerks who frequently have limited knowledge of actual values and cannot detect mistakes. Therefore it is necessary to know from past experiences anchor points on costs which an engineer can use to verify the cost figures he receives. The costs per pound of castings, essential raw materials,[1] and completed apparatus, and the cost of various processes and operations are good anchor points. The cost of operations such as drilling, milling, and boring are best acquired from the time-study engineers who have devised quick estimating methods from long experience. If costs seem out of line, the engineer can investigate to make certain the data are correct. The use of costs is his responsibility and poor decisions cannot be blamed on cost department records.

For example, in a line of machines, it has been known that the cost of material is approximately 40 percent and that of labor is 60 percent of the total cost. A new machine with some improvements has a cost of 30 percent material and 70 percent labor. The engineer realizes that this is different from the usual cost ratio. He investigates and discovers that a cost clerk used $0.20/lb instead of $0.40/lb on the major casting. After the correction, the cost of material became 40 percent and the labor cost 60 percent.

ESTIMATING PRODUCT COSTS

Both the product- and process-engineering functions involve the estimating of time and material required for the manufacture of a product. The buildup of these estimates into a selling price is usually the prerogative of the sales department after cost figures relative to the application of labor, material, and expense rates are received from the accounting department.

After the product-design section has developed drawings and bills of material for the product to be produced, the process-planning section estimates the time required for each operation of each component entering into the product. Formulas, standard data, and fundamental motion times as developed by the standards department allow the process-design group to evaluate comparative methods rapidly. As each process is assigned or designed, the estimated setup time and each piece production time is recorded. After all direct labor operations are estimated and amounts of direct material entering into each product are shown, the estimate in time and material is converted into money by multiplying by the correct conversion factors.

KINDS OF COST PROCEDURES

The two most common forms of cost procedures are actual or recorded costs and standard or predetermined costs. Neither actual nor standard costs are true

[1] See Appendix I.

costs, as both are based on assumptions as to yearly volume of business, taxes, heat, light, rent, and many other expenses that are prorated to each unit.

Actual Cost

Under the actual cost system, the actual costs of material and direct labor are charged directly to the order. The factory overhead may be prorated entirely on the basis of labor costs or may be applied on both material and labor. The common method is to apply overhead only on labor costs. In some cases the actual labor rate of the person working on the job is charged directly to the order; however, the average rate is usually used. The material cost plus the direct labor cost plus the factory overhead cost equals the factory cost. To this is added an arbitrary cost of administrative functions, which covers engineering, sales, management, taxes, and other expenses. The factory cost plus administrative and sales expense equals the total cost. The difference between the price and total cost is the profit or loss.

Therefore, there is no such thing as an actual cost. It is true that the actual material and possibly the actual labor is charged to the job. It could be possible to have a manufacturing unit making one product for a period of one year. Thus, a true cost could be determined. Still the expense, depreciation, and maintenance of the equipment and buildings are arbitrary. Actual costs usually are compiled long after the job is delivered. It is difficult to use actual cost data to negotiate future sales contracts and set a price; therefore, accountants have developed a procedure that establishes standard costs.

Standard Costs

Standard costs are estimated using sound assumptions as to material, labor, and overhead expenses. Each one of these items is under budget control, and variances from the standards are reported to management periodically. Thus management knows the current status of costs and can take action before jobs are completed or before damage is done. Costs can be predicted for sales negotiation by using standard costs of material, overhead, and labor. The labor values are based on time values established or estimated by the time-study engineers. Through the use of standard costs, guesswork is reduced and a standard that measures performance is established. Each part and assembly has a standard cost card on which is recorded the standard material, labor, and overhead costs, and the total cost.

Standard Material Costs

Standard material costs are based on actual costs, which include selling price plus cost of transportation. The cost analyst looks over the history of the part

Table 3-1   COST OF CASTING #3275A-1

Date	No. on order	Cost each	Transpor-tation cost each	Total cost each	Accumu-lated direct cost
Feb. 1	10	$2.50	$1.00	$3.50	$3.50
Feb. 15	35	2.25	.50	2.75	2.92
Mar. 1	50	2.00	.40	2.40	2.64
Mar. 15	10	2.50	1.00	3.50	2.73
Apr. 11	100	2.00	.25	2.25	2.50
Apr. 15	Standard cost $2.50				

and, with the aid of production and purchasing departments, sets a standard price for the material. An example of this is given in Table 3-1 which gives the cost of casting #3275A-1.

Every casting #3275A-1 received is placed in inventory at its actual cost. When the casting is used on an order, the order is charged with the standard cost ($2.50) and the inventory is credited with the actual cost; and a material variance account is debited or credited with the difference between actual cost and standard cost. Thus the order for March 15 would debit the material variance account by $7.70 and the order on April 11 would credit the account by $25. Standard costs are reviewed and changed periodically. A major price change requires considerable clerical work to review all cost records of parts. The engineer, in studying material costs, should use the actual cost of each material before making a final decision to change the material or process, such as a change from sand-casting to die-casting. The standard costs may be misleading because the comparison may not be on the same basis.

Standard Labor Costs

Standard labor costs are based upon the average labor rate of the operators for a machine, work station, group, or department. By means of job descriptions and classifications the average rate for each operator in a group or department can be determined. If the costing rates are based on a group including several grades of jobs ranging from $3.00 to $4.00 an hour, the normal composition of the group is recorded and the average rate is calculated. The same procedure applies to a department (see Table 3-2).

Again, a labor variance account may be established to record deviations from standards. The actual rate of the operator times the number of hours he

Table 3-2   LABOR COSTS

	Class of operator	Rate	Number required	Accumulated average
	A	$4.00	1	$4.00
Group B-12	B	3.60	3	3.70
Screw machines	C	3.40	5	3.54
Standard labor rate	D	3.20	3	3.45
$3.42	E	3.00	1	3.42

works in the group are charged to the group, and the actual hours times the standard labor rate ($3.42) are credited to the group. The difference is reflected in the labor variance account (see Table 3-3). A figure of $20.06 would be credited to the labor variance account and shown as a "below-standard" variance.

Therefore, if an operation in the group requires a class E man who is paid $3.00 an hour, the operation will be actually charged at $3.42 an hour. If this same operation is performed in another group having a costing rate of $3.00 an hour, there will be no difference in cost to the company, as the same class of operator at $3.00 an hour will be used. An hour's work with the same base rate of pay costs the same whether it is in one group or another, regardless of the group costing rate. Therefore, averages may be misleading and cost based on actual operations and rates of pay should decide competitive methods of manufacture.

Table 3-3   LABOR VARIANCE

Man no.	Rate	Hours in group	Accumulated total
1,043	$3.80	40	$152.00
937	3.60	35	278.00
874	3.30	44	423.20
939	3.20	38	544.80
720	3.10	40	668.80
1,136	3.00	36	776.80
Total		233 × $3.42 =	$796.86
Labor variance			−$20.06

JOB COST SYSTEM

It is important that the production-design engineer have an understanding of the flow of cost data in a typical job cost system. With this understanding he is in a much better position to talk with the accountants in relation to cost analysis of a given product design. Also, he will be better able to plan processes so as to minimize costs by routing to areas that do not carry heavy burden charges in the form of fixed charges and indirect labor. He will be better able to take steps to reduce costs in areas where they appear disproportionate to the factory.

In any job cost system, debit and credit are names given to left-side and right-side entries in the general ledger. For every left-side entry, there must be an equal and offsetting right-side entry.

Assets reflect debit balances and include such items as cash, receivables, and inventories. Liabilities are shown by credit balances and include payables as well as accrued items. Net worth is shown as a credit balance on common stock, preferred stock, and surplus. The relationship between assets, liabilities, and net worth is:

$$\text{Assets} = \text{liabilities} + \text{net worth}$$

The principal accounts maintained in a job cost system include: raw

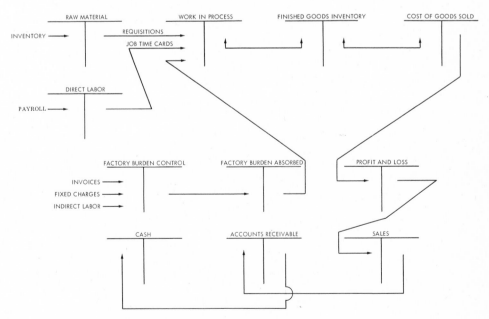

FIGURE 3-4
Flow of entries in typical job cost system.

material, direct labor, factory burden control, factory burden absorbed, work in process, finished goods inventory, cost of goods sold, cash, accounts receivable, sales, and profit and loss.

The flow of entries in these accounts is illustrated in Fig. 3-4. The production-design engineer should note that a sale is always a credit to the sales account and a debit to the accounts receivable account. The engineer should study carefully the entries made in the various T accounts and understand the relationship between debit and credit balances and their significance to the status of the enterprise.

COST-REDUCTION ACTIVITIES

Since costs are so vital to decisions made in designing for production, the people in engineering and manufacturing should understand the accounting system. Moreover, a definite plan for recording any savings, both actual and estimated, should be made to serve as an incentive to further efforts in designing for lower cost.

For example (Table 3-4) when budgets and costing rates are based on

Table 3-4   REDUCTION IN DIRECT LABOR ON A PART MAY CAUSE HOURLY OVERHEAD TO EXCEED HOURLY BUDGETED OVERHEAD

	Old method	New-method carbide tools (overhead based on established hourly rate)	New-method carbide tools (overhead based on true cost)
Allowed time for turning sheave in vertical boring mill	5.00 h	2.50 h	2.50 h
Direct labor rate per hour	$2.00	$2.00	$2.00
Direct labor cost	$10.00	$5.00	$5.00
Overhead cost for inspection, repairs, material handling, etc., per hour	$4.00	$4.00	$8.00*
Cost of overhead	$20.00	$10.00	$20.00
Cost of direct labor and overhead	$30.00	$15.00	$25.00

Apparent savings per piece = $30 minus $15 = $15
Actual savings per piece = $30 minus $25 = $5

* Since overhead cost per piece will tend to remain constant, the actual overhead cost for the new method will be the same as for the old method.

direct labor hours, the foreman who is progressive and introduces new and improved methods may be penalized. He may have a part which he is turning on a vertical turret lathe in 5 h. By the use of tungsten-carbide tools, he can reduce the time to $2\frac{1}{2}$ h. This is 50 percent less direct labor over which he must spread the additional cost of maintenance (since the machine is worked harder) and the additional material handling, inspection, production clerks, and supplies, which vary according to the actual number of parts handled rather than the direct labor. When his budgets go into the red, he can only point out that he is turning parts out 50 percent faster than before and therefore his expenses are higher in proportion to the direct labor. Suppose by introducing carbide tooling, M-2 H.S.S. drills, and other improvements throughout his shop, he reduced all of his direct labor cost 50 percent and still produced the same number of parts. His budgets are based on direct labor. Therefore, adjustments in budgets must be made as improvements are made. When forced to increase budgets and costing rates, management should help by getting better handling facilities and rearranging equipment in order to reduce overhead expense. The efficient shop usually has a low direct labor and a high overhead rate. A low costing rate may not indicate an efficient shop. When making cost reductions, the management should study the efficiency of the overhead departments and obtain overall reductions in cost. The engineer should be aware of these conditions and assist in obtaining a true picture of the product cost.

DISTINCTION BETWEEN COST REDUCTION AND COST CONTROL

From the foregoing description, it is evident that *cost reduction* is accomplished by changes in tools, processes, and design that eliminate operations, reduce the cost of the material, and use less expensive machines and fewer skilled operators. The costs of changes are justified by the decrease or difference in cost per unit. The savings per unit must pay for the change in a reasonable time. A separate record of such savings and expenditures for making the improvements must be kept for management's information. *Cost control* has no connection with cost reduction. Cost-control procedures inform management that expenses are or are not according to plan. Rising or lowering costs of labor and material are reflected in variance accounts. Although these factors are often out of the control of management and supervision, it is their obligation to combat rising costs by training more efficient operators and obtaining less expensive materials. Lower costs permit reduction of selling prices and eventual broadening of the market.

Many expenses under the control of supervision can be reduced. The engineer, through his knowledge of these expenses, can save indirect costs. For example, he may know that one part is used on several active assemblies and could be stocked. By suggesting that the part be stocked, he saves the cost of setup. If he understands the operations used to make the part, he can usually suggest to the methods engineering department a better process of manufacture which is more economical in view of the increased number of parts on the order.

CONTRACT NEGOTIATION FOR EQUIPMENT

If there are standard costs on all parts, the cost of any item previously manufactured can be quickly determined. Most contract negotiations are for products similar to previously sold equipment. If the cost of similar equipment is known, the variations can be closely estimated and added or subtracted from the cost of the similar product. If necessary, the variation can be sketched by the engineering department and cost of material, labor, tools, patterns, and equipment can be more closely estimated.

The sales price includes: (1) the cost of material, labor, and shop overhead; (2) the cost of engineering development, tools, and patterns for the particular job; (3) administrative and selling costs; (4) allowances to cover variances; and (5) profits.

Under the standard cost plan, the cost of the contract can be checked with the estimate as soon as drawings, manufacturing routings, and time values are available. Thus management can know whether the estimated costs need to be adjusted before the parts are manufactured. At this stage it is often possible to make improvements that will increase profits or prevent losses. This would be impossible under the actual or recorded cost system.

INFLUENCE OF COST ESTIMATES ON PROFITS

The contract price of the equipment is based largely on the estimated cost of the product, tools, patterns, and engineering development. These estimated costs are usually the responsibility of the engineer. If the estimates are too high, the customer may refuse to buy and there will be less work in the shop and fewer profits. If too low, actual costs will exceed estimates and estimated profits will decrease or turn into a loss. In either case, management looks to the engineer for the answer to the problem of obtaining more profits. Estimated costs must be under control so expenditures do not exceed estimates and so that future estimates will be close to actual costs.

The factory cost of a product, and the cost of engineering development, tools, and patterns, may be satisfactory, and yet other costs of administration, sales, service, and construction may be responsible for loss of business or decrease in profits. Therefore it is imperative that the production-design engineer thoroughly understand the cost picture, as well as exercise ingenuity in design and selection of materials and processes to obtain minimum costs.

INFLUENCE OF ACTIVITY ON COSTS

As the number of pieces per year increases, the activity may be great enough to install a continuous production line in which parts are made with a minimum amount of setup time. The engineer can make cost reductions by obtaining a straight-line production on parts which have increased activity.

If 2 million cars are sold per year and there are 6 cylinders per engine and 4 piston rings per cylinder, it is necessary to produce 48 million piston rings per year. A saving of $0.01 on each ring equals $480,000 per year. Therefore, in order to obtain the minimum cost, considerable engineering effort can be profitably used beginning with raw material to the installation of the finished process. It is not wise "to put all eggs in one basket," so it is necessary to divide high activity parts among different units, divisions, or outside suppliers to assure a constant supply. Also competition tends to improve operations and reduces costs beyond anything forecast as possible.

There is constant competition between materials and processes based on costs which are influenced by the number of pieces made during a period of time.

The activity of parts affects the amount of time the equipment is operated as compared with the hours available. The ratio of hours operated to hours available has considerable effect upon costs. Here is an extreme example:

A large hydraulic press, including the hydraulic pumps and building to house it, costs over $1 million. It is known that depreciation, maintenance, and interest on investment amount to about 10 percent ($100,000) per year. There are normally 2,000 working hours a year in one shift (8 h/day × 5 = 40 h/week × 50 = 2,000 h/year). Three shifts would be 6,000 h/year available; $100,000 ÷ 6,000 = $16.66/h, the minimum cost of the machine in a 24-h day. Actually, the sales department can sell only enough to keep the machine busy 8 h a day. Therefore, the machine costs $100,000 ÷ 2,000 = $50/h. If sales decrease, the cost of the machine per hour will increase and make it difficult to obtain business at a profit.

The responsibility of the sales department to keep equipment loaded is great when increased or decreased costs due to activity are considered. A big automobile assembly plant is equivalent to one big machine and must be operated for at least two shifts in order to obtain reasonable selling prices. The entire auto industry is based on a two-shift operation to obtain the prices on cars today.

When activity of products such as airplanes, cars, locomotives, and tractors increases during their lives, or when the estimated number of units of a newly developed product ranges from low to high activity, the emphasis on the various

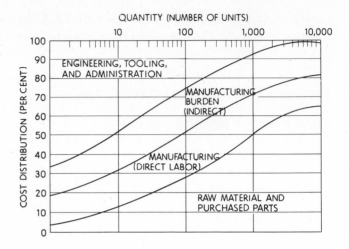

FIGURE 3-5

Cost-distribution factors charted in relation to quantity to indicate changing emphasis as production increases.

factors of costs may change as shown in Fig. 3-5. When quantities are low, the proportion of development cost is high as compared to the cost of manufacturing overhead, direct labor, raw material, and purchased parts. The development cost includes design, preparation of drawings, writing manufacturing information, designing and building tools, testing, inspection, and many other items incident to placing the first parts into production. As the number of units increases, the emphasis centers in reduction of overhead, labor, and material costs. When quantities are low, expenditures on tools and engineering refinements will net less return in cost reduction. When quantities are high, an expenditure of engineering effort will result in reduced labor and material costs which will give large returns for even a small savings per unit.

PURCHASED PARTS VERSUS MANUFACTURED PARTS

One of the most difficult decisions for a production-design engineer is whether to make a part in the shop or purchase it. Often the supplier's cost is less than the factory cost of the part; in this case it is generally economical to purchase from the supplier. However, the department may possibly have an overhead charge which has no relation to the manufactured part. If this expense is disregarded, the factory cost may be favorable. Given the same equipment and materials and an overhead charge applicable to the product, the shop should be able to meet competition. Progressive companies encourage competition between divisions, and encourage suppliers to bid in order to improve the products and reduce costs. Good accounting practice should avoid unrealistic accounting procedures that distort costs and cause unfair competition.

INCLUSION OF COST REDUCTIONS IN THE DEVELOPMENT PROGRAM

The greatest opportunity for cost reduction and increased profits is an effective development program which takes ideas of the members of an organization and translates them into a salable product within a minimum length of time.

The profit history of most products will reveal a single peak and two valleys. The first valley is due to setup costs, development costs, and costs of

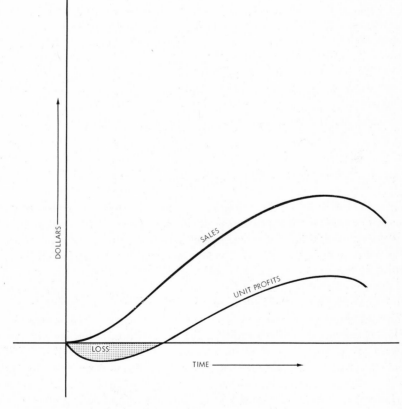

FIGURE 3-6
Typical unit profit history of two new products.

getting started. The second valley is caused by price competition or development of similar but new competitive products. This relationship is shown in Fig. 3-6.

Usually the shop men become vitally interested in the project when they see it coming down the production line, and then it may be too late to incorporate desirable features for operations, assembly, and tooling. Any changes at this time are expensive, although they still may be profitable. That expense could have been eliminated if the information were available to the engineer sooner.

The risks of price changes, competitive products, or reduced markets make it imperative that the great expenditures of money, time, and effort in developing a product be recovered quickly. This means that management must make decisions *without delay*, sales must furnish accurate data *without delay*, and the engineer must furnish information and requisition material in time for model and production manufacture. The manufacturing department must build the experimental models, tools, and production models *without delay*, so tests and improvements can be made and incorporated in the design.

For example, a company had a development project that meant $500 a day to them. A tight schedule was set up for completion, and through the cooperation of management, engineering, manufacturing, and sales, the project was completed 1 month ahead of schedule, saving approximately $10,000.

There are many ways to organize for effective development, each suitable for any organization and group of personalities, but the greatest savings are made by expediting the program. Let us take a hypothetical example that can be duplicated in many plants.

DEVELOPMENT PROJECT OF COMPANY X

Competition has forced a redesign of a line of products. Since many improvements can be incorporated in the old design, preliminary specifications, cost of engineering, tooling, and promotional activities for a new design are estimated for management's approval. Since management usually is not close to the project, considerable time is taken by department heads in answering many questions. Eventually supervision decides to take the project to top management, which has to wait for approval at the next board of director's meeting. The project is approved three months later. The time taken is represented by the bar in step 1 at the bottom of Fig. 3-7. The time for each succeeding step is shown in like manner. Overlapping or operations performed in parallel are shown. The dollars indicated at each step comprise the total expenditure for the particular item. Table 3-5 gives the same information on expenditure of time and money. The development program proceeds as indicated in Fig. 3-7 and Table 3-5.

The development program results in a savings of $10 per unit. The sales

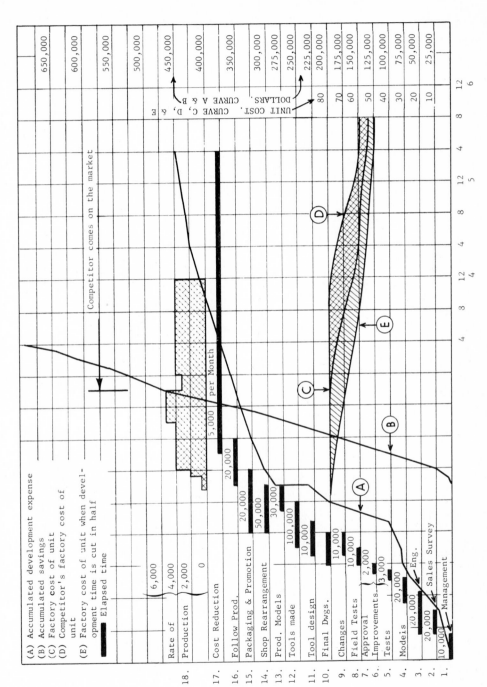

FIGURE 3-7
Company X development project.

Table 3-5   COMPANY X DEVELOPMENT PROJECT

Expenditure of time and money	Total time, months	Overlap with previous item	Actual time (total time minus overlap)	Additional time, months	Cost
1 Management	3	0	3	3	$ 10,000
2 Sales surveys and determination specifications for new products	3	0	6	3	20,000
3 Initial development in engineering department	5	3	8	2	20,000
4 Experimental models	3	1	10	2	20,000
5 Tests	1	0	11	1	3,000
6 Improvements and check tests	1	0	12	1	2,000
7 Approval of design for manufacturer—12 months later					
8 Field test and reaction of trade	2	0	14	2	10,000
9 Improvements incorporated (some patterns and tools changed)	3	1	16	2	10,000
10 Final drawings released for production with improvements—16 months later	4	4	16	0	——
11 Tool designs made	4	3	17	1	10,000
12 Tools made	6	3	20	3	100,000
13 Production material ordered. Production models made. Production starts 22 months later	3	1	22	2	30,000
14 Shop rearrangements	6	6	22	0	50,000
15 Packaging and promotion material	8	6	24	2	20,000
16 Follow-up production	6	2	28	4	20,000
Totals				28	$325,000

17 Cost-reduction program. Life of design. $50,000 per year.

price has not changed, and with the improved product, the share of market is the same or better. Curve B in Fig. 3-7 represents the accumulated savings made as units are sold.

To recover $325,000 takes the sales of 32,500 units; these were produced and sold 10 months after start of production. This rate of production or sales gives a saving of $140,000 on the first year's production (Table 3-6). From the chart it can be seen that it took 32 months to recover the development expenditure, which is quite a risk. If the development time could be reduced by one half (from 22 months to 11 months), the time for recovery of the development money would be spread over 11 months plus approximately 10 months for production of 32,500 units. Thus the time for recovery of development would be 21 months instead of 32 months.

Note that the rate of sales reached 5,000 per month due to more salable features. Therefore, there would be 11 more months to produce at the rate of 1,200 per week or a saving of $550,000 (5,000 × 11 × $10 = $550,000), which is much more than the original development cost of $325,000.

However, under normal conditions, competition receives word that a new development is under way and in turn develops and introduces its new design one year after company X starts production. Competition meets most of the new features and has some additional improvements to sell at the same time,

Table 3-6 RATE OF SALES AFTER START OF
PRODUCTION

Month after start of production	Production	Savings at $10 each
1	500	$ 5,000
2	2,000	20,000
3	4,000	40,000
4	4,000	40,000
5	4,000	40,000
6	4,000	40,000
7	4,000	40,000
8	4,000	40,000
9	5,000	50,000
10	5,000	50,000
10-month total	36,500	$365,000
11	5,000	$50,000
12	5,000	50,000
Total	46,500	$465,000
Cost of development		$325,000
Net savings first year after production		$140,000

thus cutting the sales volume of company X. Sales of company X drop to 3,000 per month, but in the meantime company X has made savings by installing new automatic machines and obtaining skilled workmen and better sources of supply. It is able to reduce costs another $2.00 and still a savings of $10 per unit can be made. It has incorporated additional sales features and improved the quality of the product due to field experience, better know-how, and a better-informed sales force. Therefore, its sales pick up.

The cost of development has probably been about the same for the competitor as for company X. Its progress in making improvements follows the same pattern as that of company X, as shown in curves C and D of Fig. 3-7. Again, if development time is cut in half, a great advantage over competition can be gained for 4 or 5 years, as shown in curve E.

Suppose the situation were reversed and company X had to catch up with competition. Company X would probably cut its development time in half. Why not do it in either case?

An effective development program is most profitable when reasonably expedited. In this case 1 month's time equals $50,000. There are other advantages besides those mentioned previously. When new designs and improvements are rapidly introduced:

1   The latest advancement in materials and processes can be incorporated.
2   The newly purchased equipment will be utilized to a greater extent.
3   Production loads on present equipment can be distributed better.
4   Time values that are out of line can be improved.
5   Engineers can make further advances in less time.
6   Field troubles with resulting loss of sales can be cleared up with savings in service charges and field trips.

COOPERATION IN THE DEVELOPMENT PROGRAM

The foregoing description illustrates the vital part managers and shop supervisors have in the development program. Often the development program is considered as the engineer's responsibility while the shop assists in making tools and models, the attitude being that production is the main purpose of the shop and these other items are secondary. However, if there is real interest, teamwork, and proper organization, decisions will be made; shop ideas will be suggested; designs will be adapted to the best materials, equipment, and processes; and models will be made quickly so that everyone can see the advantages and disadvantages, and improvements can be accepted readily.

These questions might be asked:

1   Is management being informed constantly (daily) of field troubles, competitor advantages, and possible improvements so that approval for the development project will be expedited?

2   Is there an artificial barrier between engineering and manufacturing?

3   Are all the "brains of the organization" in the product before the design shapes into final form? Has the engineer had an opportunity to absorb the ideas, especially those from the shop?

4   Are capable men loaded down with supervisory duties and so prevented from contributing to development projects and making prompt decisions?

5   Can ideas go from the operator and shop supervisor to the engineer rather than ideas always coming down from the engineer?

6   Can confidence between the experimental department and engineer be increased so that models can be made quickly, with less paper work and fewer alibi records?

7   Can specifications for materials and purchased parts be released sooner to enable the purchasing department to obtain suppliers and negotiate for best prices?

8   Can preliminary drawings be released to pattern, tool, and equipment suppliers for preliminary estimates and recommendations?

9   Can tool designs be made before the final release of manufacturing information? It is easier to make changes on paper than in actual tools.

10   Can more activities be paralleled rather than worked on in series?

11   Can engineers obtain readily, through the purchasing staff, information about new materials, cost potentials, and purchasable services prior to design decisions?

There are many more questions that could be asked, and many more phases of manufacturing, sales, and engineering that could be brought into this subject of engineering for production, such as:

1   The use of motion study for improving the design of machines to obtain better operation.

2   The use of quick-process tools for making die parts for experimental parts (for example, laminated and cast plastic tools).

3   The proper organization of an experimental department:

a   Should development work, including tools and methods, be under engineering or divided between engineering and manufacturing?

b   Should engineers select suppliers and arrange for purchasing of materials and parts?

There are many methods of developing a product that vary according to the ingenuity of the engineer. For example, one engineer runs his equipment at high speed and overloads until something breaks. He then makes this part stronger or revises the design and applies greater speed and loads. Other parts show weakness and are strengthened. By this experimental procedure he is able to improve a design.

SUMMARY

A product that is successful from the standpoint of distribution is one that is functionally sound, has eye appeal, readily permits normal maintenance and periodic overhaul, and is competitive in price. It is essential that the process-design engineer work closely with the product-design engineer so that these objectives can be realized.

Costs are the basis of action within an organization. They are the ultimate criterion of the value of a product. Direct costs are a good basis for judgment; but indirect costs, overhead or burden or tool expense, may have more influence than material and direct labor costs. Errors can creep into calculations made by clerks and persons unfamiliar with the product; therefore, the engineer must have sufficient data to check costs in order to detect errors. The many variables encountered in cost considerations make it necessary to consider each case on its individual merit.

QUESTIONS

1   What two kinds of cost are of vital interest to the engineer?
2   How are funds obtained for development costs?
3   Illustrate graphically how selling price is derived.
4   How is manufacturing cost distinguished from factory cost?
5   What items of burden are included in factory expense? In general expense? In sales expense?
6   Explain why price is not always determined by cost.
7   How does the process designer accurately estimate direct labor times?
8   What are the two most common forms of cost procedures?
9   How is factory overhead prorated?
10   What is the function of the "material variance" account?
11   What are the disadvantages of allocating factory overhead on a percentage of direct labor?
12   How can the production design engineer save indirect costs?
13   What is the "return on investment" in percent of the money invested in the development of the product X? How does it vary from year to year?
14   What length of time should be planned to recover the complete development cost of a product? Why?
15   How large an original investment could you justify on a new design having a life of 3 years and an estimated profit of $10,000 at the end of the first year, $15,000 at the end of the second year, and $10,000 at the end of the third year? Management expects a return of at least 20 percent on designs lasting from 2 to 5 years.
16   What is involved in the preliminary economic analysis?

17 Illustrate the relationship between selling price and quality sensitivity of several products with which you are familiar.

18 Would medical instruments be more sensitive to quality or to quantity from a sales volume standpoint? Why?

PROBLEMS

1 The development cost of a new product is estimated to be $10,000 with a tooling cost of $6,500. Sales of the product have been estimated to be between 2,500 and 3,000 units the first year. Industrial engineering has estimated the manufacturing cost to be $16.75 per unit. If the selling price is established at $27.50 per unit, what is the break-even point? How much profit will the plant make if the selling cost is taken as 50 percent of the manufacturing cost? If the manufacturing cost rose to $21.35 per unit because of a change in the union contract and an inflation of 10 percent in raw materials, what must the new selling price be to make the same profit margin as originally planned?

2 A manufacturer of small electric hand tools such as drills, sanders, hedge clippers, and buffers had sales of $2,000,000 last year but made a gross profit of only $200,000. The main cause of the problem appeared to be failure to increase the sales price, although labor cost had increased to $750,000 and raw material had risen to $800,000 per year. At a staff meeting, the production manager suggested that an increase of 25 percent in the selling price of their product line would place the company in a good profit position again. But the vice president of sales disagreed strongly, claiming that such a large increase in price would drive the company out of the market. After discussion both men finally agreed that the net result of such a price increase would be a 10 percent decrease in sales and production. The chief engineer proposed another alternative—plant modernization and cost reduction. By making some long overdue improvements in the operating equipment and the purchase of some more efficient machines at an estimated cost of $200,000 for both items, he estimated that costs could be reduced by 25 percent. Analyze the data presented and make a report to the president showing him which proposal is the better, and explain why.

SELECTED REFERENCES

BIERMAN, HAROLD, AND ALLAN R. DREBIN: "Managerial Accounting," Macmillan, New York, 1968.
HEIMER, ROGER C.: "Management for Engineers," McGraw-Hill, New York, 1958.
JELEN, F. C.: "Cost and Optimization Engineering," McGraw-Hill, New York, 1970.
MOORE, FRANKLIN G.: "Manufacturing Management," Irwin, Homewood, Ill., 1973.

4

PROPERTIES AND BEHAVIOR OF MATERIALS

INTRODUCTION

The designer or engineer is usually more interested in the behavior of materials under load or when in an environment such as a magnetic field than he is in why they behave as they do. Yet the better he understands the nature of materials and the reasons for their physical and mechanical properties the more quickly and wisely will he be able to choose the proper material for a given design. Generally, a material property is the measured magnitude of its response to a standard test performed according to a standard procedure in a given environment. In engineering materials the loads are mechanical or physical in nature and the properties are recorded in handbooks or, for new materials, are made available by the supplier. Frequently such data is tabulated for room-temperature conditions only, so when the actual service conditions are at sub-freezing or elevated temperatures, he will need more information.

The properties of material are sometimes referred to as structure-sensitive as compared to structure-insensitive properties. In this case structure-insensitive properties include the traditional physical properties: electrical and thermal conductivity, specific heat and density, and magnetic and optical properties. The structure-sensitive properties include the tensile and yield strength, hardness, and impact, creep, and fatigue resistance. It is recognized that some sources maintain that hardness is not a true mechanical property because it varies somewhat with the characteristics of the indentor and therefore it is a technological test. It is well known that other mechanical properties vary significantly with rate of loading, temperature, geometry of notch in impact testing, and the size and geometry of the test specimen. In that sense all mechanical tests of material properties are technological tests. Furthermore, since reported test values of materials properties are statistical averages, a commercial material frequently would have a tolerance band of ± 5 percent or more of any published value.

In the solid state, materials can be classified as metals, polymers, ceramics, and composites. Any particular material can be described by its behavior when subjected to external conditions. Thus, when it is loaded under known conditions of direction, magnitude, rate, and environment, the resulting responses are called mechanical properties. There are many possible complex interrelationships among the internal structure of a material and its service performance (Fig. 4-1). Mechanical properties such as yield strength, impact strength, hardness, creep and fatigue resistance are strongly structure-sensitive, i.e., they depend upon the arrangement of the atoms in the crystal lattice and on any imperfections in that arrangement, whereas the physical properties are less structure-sensitive. These include electrical, thermal, magnetic, and optical properties. The latter do depend in part upon structure; for example, the resistivity of a metal increases with the amount of cold work. Physical properties depend primarily upon the relative excess or deficiency of the electrons that establish structural bonds and upon their availability and mobility. Between the conductors with high electron mobility and insulators with no free electrons, precise control of the atomic architecture has created semiconductors which can have a planned modification of their electron mobility. Similarly, advances in solid state optics have led to the development of the stimulated emission of radiant energy in the microwave spectrum (masers) and in the visible spectrum (lasers).

In studying the general structure of materials, one may consider three groupings: first, atomic structure, electronic configuration, bonding forces, and the arrangement of the aggregations of atoms; second, the physical aspect of materials, including properties such as electrical and thermal conductivity, specific heat, and magnetism; and third, their macroscopic properties such as their mechanical behavior under load, which can be explained in terms of impurities and imperfections in the lattice structure and the procedures used to modify that behavior.

PHYSICAL PROPERTIES OF MATERIALS

In the selection of materials for industrial applications, many engineers normally refer to their average macroscopic properties, as determined by engineering

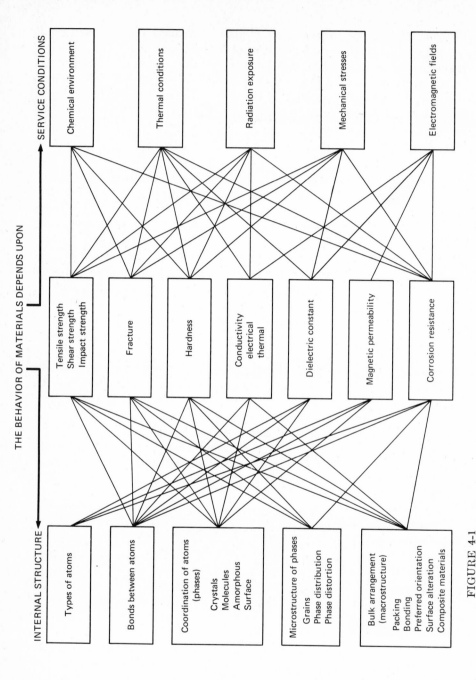

FIGURE 4-1
Interrelationships between material behavior, internal structure, and environment.

tests, and seldom are concerned with microscopic considerations. Others, because of their specialty or the nature of their positions, have to deal with microscopic properties most of the time.

The *average* properties of materials are those involving matter in bulk with its flaws, variations in composition, and variations in density caused by manufacturing fluctuations. *Microscopic* properties pertain to atoms, molecules, and their interactions. These aspects of materials are studied for their direct applicability to industrial problems and also so that possible properties in the development of new materials can be estimated.

In order not to become confused by apparently contradictory concepts when dealing with the relationships between the microscopic aspects of matter and the average properties of materials, it is wise to consider the principles that account for the nature of matter at the different levels of our awareness. These levels are: the commonplace, the extremely small, and the extremely large. The commonplace level deals with the average properties mentioned, and the principles involved are those set forth by classical physics. The realm of the extremely small is largely explained by means of quantum mechanics, whereas that of the extremely large is dealt with by relativity.

Relativity is concerned with very large masses, such as planets or stars, and large velocities that may approach the velocity of light. It is also applicable to smaller masses, ranging down to subatomic particles, when they move at high velocities. Relativity has a definite place in the tool boxes of nuclear engineers and electrical engineers who deal with particle accelerators. For production engineers relativity is only of academic interest and is mentioned here for the sake of completeness.

Application of Concepts from Quantum Mechanics

Quantum mechanics, once restricted to the academic halls, has now become a bread-and-butter topic. It generally deals with particles of atomic or subatomic sizes. But from the understanding of the behavior of these particles comes a better understanding of such phenomena as thermal conductivity, heat capacity, electric conductivity . . . and the very existence of transistors and thermistors.

Quantum mechanics is a complex subject, largely outside the scope of this text, yet a brief mention of two of its most important concepts may aid the production-design engineer in the study of the basic characteristics of materials. One of these is Planck's quantum hypothesis, later extended by Einstein and others. The other is Pauli's exclusion principle.

The quantum hypothesis postulates that the bound energy state of particles of very small size cannot be represented by a continuous function but is discrete in nature. Thus between any two permissible energy states of a bound particle there exists a region of states that are forbidden.

Each of the electrons of any atom has a particular energy level, E_1, E_2,

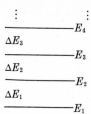

FIGURE 4-2
Concept of energy levels of planetary electrons.

E_3, etc. For an electron to go from one energy state to a higher one, it must receive sufficient energy to jump through the forbidden energy regions. Thus if an electron at energy level E_2 is to go to the higher-energy level E_3, it must receive an energy ΔE_2 (Fig. 4-2).

For it to go to energy level E_4 it must receive an energy of $\Delta E_2 + \Delta E_3$. If an electron receives insufficient energy to jump a forbidden energy region, it will get rid of its extra energy in the form of electromagnetic radiation and remain in its original energy state. By the same token, if an electron at, say, energy level E_1 receives energy greater than ΔE_1 but smaller than $\Delta E_1 + \Delta E_2$, it will be able to go to E_2 and the excess energy over ΔE_1 will be emitted as electromagnetic radiation.

All matter has a natural spontaneous tendency to be at the lowest possible energy state. Consequently, the electrons of any atom fill the lowest permissible energy levels and only are excited to upper empty levels when they are given energy by means of interaction with electromagnetic radiation, particle bombardment (electrons, protons, etc.), an electric field, or by thermal excitation (collision with neighboring atoms brought about by an increased amplitude of atomic vibration caused by an increase in temperature).

Thus, if an electron is dislodged from its normal energy level it can only go to a higher unoccupied level or entirely out of the atom. On doing this, it leaves an empty level that will be immediately occupied either by the original dweller or by any of the electrons at a higher energy; in any case, the electron that drops down will release its excess energy in the form of electromagnetic radiation. This in turn will leave its own original level vacant and there will be more transitions until all the normal levels are filled again.

It may be recalled that the number of electrons in any atom is equal to the number of protons in the nucleus. If for any reason an atom is stripped of some or all of its electrons, free electrons in the vicinity of the ion will immediately fall into the empty energy levels, emitting their excess energy in the form of radiation.

The preceding discussion states that each electron in an atom is associated with a particular energy state—but there can be only one electron at each

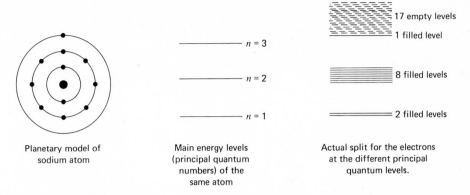

Planetary model of
sodium atom

Main energy levels
(principal quantum
numbers) of the
same atom

Actual split for the electrons
at the different principal
quantum levels.

FIGURE 4-3
Relationship between the planetary electrons of the sodium atom and their
energy levels.

particular energy level. This is explained by Pauli's exclusion principle, which
states that no two bound particles in an interacting system can exist at exactly
the same energy state. Thus when there are two electrons in one of the electron
"shells" of an atom (principal quantum numbers), that normally could be
considered to be at the same energy, their actual individual energy levels are
different although rather close to one another. By the same token, when there
are 8 electrons in a "shell," they have eight separate energy levels. When two
or more atoms are brought together, there is a further shift and splitup of energy
levels.

The maximum permissible number of electrons in the first "shell" (principal
quantum number) is 2, in the second 8, in the third 18, in the fourth 32, etc.
Then for an isolated atom there will be 2 individual levels at the first principal
level, 8 at the second, and so on (Fig. 4-3). If two atoms are brought together
(Fig. 4-4), then the first principal level will have 4 energy states, the second 16,
and so forth. When many atoms are brought together, there will be many
electrons at any principal quantum level and the spacing between individual
energy levels will be so small that there will be in effect energy bands rather than
individual levels. Note that there are 6.023×10^{23} atoms in one gram-atom,
which is Avogadro's number.

It must be understood that each kind of atom has its own particular
energy-level arrangement and that energy splits and band formations only take
place where there would be a coincidence of energy levels, such as is illustrated
in Fig. 4-4. But if there are two different atoms, A and B, and their energy levels
do not coincide, the energy-level arrangement of their combination will be a
superposition rather than a level shifting (Fig. 4-5).

In the same way, if there is a large group of atoms of the same kind and

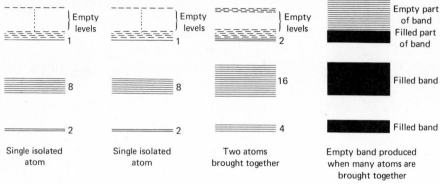

FIGURE 4-4
Formation of energy bands in a solid.

one different atom (as an impurity atom) is brought into the group, there will be a shift in the levels that coincide and a superposition where there is no coincidence (Fig. 4-6). It is to be noted that when there is this type of superposition it is the same as introducing permissible steps in the forbidden energy regions. By means of these steps the electrons may jump to higher-energy levels without requiring as much energy at any one time.

If the bound arrangement of a given solid is such that all the energy levels in a certain band are filled, whereas the energy levels in the next band higher up are entirely empty, the material is an insulator. If this material were to conduct, the valence electrons would require acceleration to gain some energy above the ground state—but in such a case there are no immediately adjacent empty higher-energy levels, only a wide forbidden energy gap. To jump that energy gap would require considerably more energy than is provided in the normal mode of excitation, and therefore the material is an insulator—but it could conduct if a sufficient input of energy were supplied.

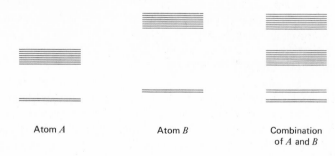

FIGURE 4-5
Superposition of energy levels.

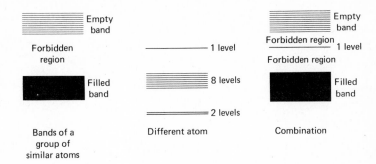

FIGURE 4-6
Superposition of an impurity atom on the band structure of similar atoms.

If the valence electrons of a solid were to occupy a partially empty band, these electrons could be accelerated with very little energy and the material would be classed as a conductor.

When the valence electrons of a solid occupy a filled band but the forbidden energy gap is small, of the order of kT (where k is Boltzmann's constant, 1.38×10^{-16} ergs/°K and T is absolute temperature, °K) at ordinary temperatures, then it is not difficult for the valence electrons to jump to the empty band. Such a material is known as an intrinsic semiconductor.

In the case of insulators and semiconductors, the top filled band is known as the valence band and the higher adjacent empty band is known as the conduction band.

In conductors, the higher-energy electrons are already in the conduction band and can move more freely at the least excitation. On the other hand, insulators require considerable energy for an electron in the valence band to go to the conduction band. Any insulator may be made to conduct electricity if its electrons are sufficiently energized. The electrons that can move about freely in the conduction band are viewed as "free electrons" that are not associated with any atom in particular. These electrons are considered principally responsible for the electrical and thermal conductivity of matter.

Before entering into a detailed discussion of the physical properties of materials it is convenient to review another mechanism that also intervenes in the evaluation of these properties. This is the oscillatory motion of atoms or molecules that are bound in a solid. In gases and liquids, the atoms are free to move about, to a greater or lesser extent, respectively, but in a solid, the atoms or molecules are restricted to oscillation about fixed points.

The atoms or molecules of a solid above absolute zero always vibrate about fixed centers. The higher the temperature, the greater the amplitude of the oscillation, and this amplitude of oscillation has a direct effect on the conductivity of solids. The oscillation stores energy and this energy is directly related to the heat capacity of solid materials.

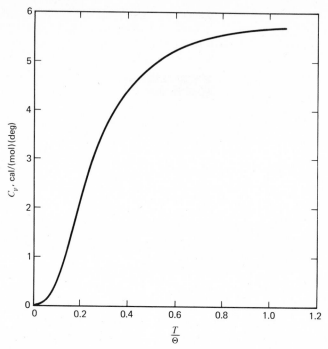

FIGURE 4-7
Heat capacity of a solid (in three dimensions), according to the Debye approximation. (*Redrawn from Charles Kittel, "Introduction to Solid State Physics,"* *2d ed., Wiley, New York, 1956.*)

De Broglie's relationship between moving particles and wave motion is not only applicable to the theory of relativity and quantum physics but also to modern physics as a whole. De Broglie conceived that a moving particle can be associated with a wave motion and that a wave propagation can be considered as a particle. This viewpoint can account for the diffraction of particles such as electrons and neutrons—a phenomenon that was formerly considered an exclusive property of wave motion. It also accounts for the particlelike collision of photons.

Heat Capacity

The heat capacity of a solid depends on the amplitude of oscillation of the particles about their centers of equilibrium. The greater the temperature the greater the amplitude and consequently the greater will be the heat capacity. It is measured in terms of energy per unit mass. (Recall that the specific heat

Table 4-1   DEBYE TEMPERATURE FOR
SOME ELEMENTS

Substance	θ, °K
Beryllium	1160
Magnesium	406
Molybdenum	425
Tungsten	379
Iron	467
Nickel	456
Copper	338
Zinc	308
Aluminum	418
Lead	94.5

SOURCE: From Charles Kittel, "Introduction to Solid State Physics," 2d ed., Wiley, New York, 1960.

of a substance is the ratio of its heat capacity to the heat capacity of water; it is a dimensionless quantity.) This variation is represented as a continuous function in different ranges; however, it is actually a discrete function, since the particles cannot vibrate in a continuous mode but in quantized ones.

Starting from a very low temperature the heat capacity increases proportionally to the third power of the temperature and tapers off to a constant value of $3R$ [about 6 cal/(mol)(°K)] for any substance after passing a critical temperature called the Debye temperature, θ (Fig. 4-7 and Table 4-1).

As the gram-molar values for different substances vary, their heat capacity per pound will vary even though their heat capacities per gram-mole are the same.

Thermal Conductivity

The thermal conductivity of a solid may be viewed from the standpoint of the free electron theory. As these electrons drift due to a difference of thermal potential they may travel along until they interact with an atom. This travel is denoted as the mean free path. When the atoms are lined up with low amplitude of oscillation (at low temperatures), the electrons may travel a fairly long distance without interference (a large mean free path); but as the temperature of the solid increases so does the amplitude of oscillation and the electrons will interact with the atoms more often (a small mean free path). The larger the mean free

CONDUCTORS:
THE LOWER THE TEMP
THE HIGHER THE
THERMAL CONDUCTIVITY

INSULATORS:
THE HIGHER THE TEMP
THE HIGHER THE
THERMAL CONDUCTIVITY

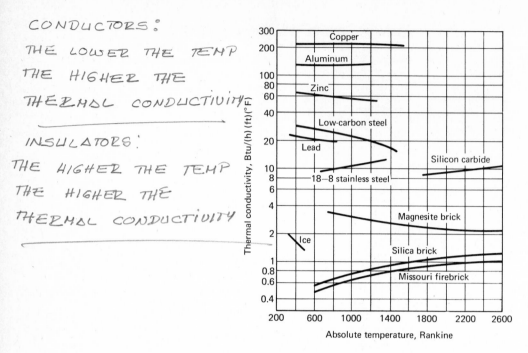

FIGURE 4-8
Variation of the thermal conductivity of solids with absolute temperature.

path, the larger the thermal conductivity of the material. Thus at low temperatures the thermal conductivity of conductors is usually larger than at higher temperatures (Fig. 4-8).

In the case of a thermal insulator, the electrons have to be given considerable energy before they can jump into the conduction band and act as conductors. Therefore, insulators gain in conductivity as the temperature increases in spite of the interference due to the increased oscillation of the atoms.

Another way to look at thermal conductivity is on the basis of wave propagation. It has been determined that the transit of heat through a body can be considered as the propagation of waves. From De Broglie's relationship these waves may be treated as particles and the whole phenomenon studied as the scattering of these particles, which are called phonons. Using this model of phonon scattering, it has been determined that the isotope distribution in a material has a bearing on the scattering of the phonons, for the minority isotopes act as scattering centers thus reducing the thermal conductivity of the material.

Electrical Resistivity The phenomenon of electrical conductivity is similar to that of thermal conductivity. They are both related to the drift of free elec-

Table 4-2   RESISTIVITY OF VARIOUS ELEMENTS

Element	Resistivity ρ, $\mu\Omega\cdot$cm	Density, g/cm³	Element	Resistivity ρ, $\mu\Omega\cdot$cm	Density, g/cm³
Silver	1.59	10.60	Titanium	11.5	4.50
Copper	1.673	8.89	Selenium	12.0	4.30
Aluminum	2.655	2.70	Mercury	94.1	13.55
Magnesium	4.46	1.74	Carbon	1.375×10^3	2.25–3.52
Tungsten	5.5	18.60	Silicon	1×10^5	2.35–2.92
Zinc	5.92	7.04	Germanium	6.0×10^7	5.46
Nickel	6.89	8.60	Boron	1.8×10^{12}	2.45–2.54
Iron	9.71	7.85			

SOURCE: Theodore Baumeister and Lionel S. Marks "Standard Handbook for Mechanical Engineers," 7th ed., McGraw-Hill, New York, 1967.

trons. But in many cases the electrical conductivity is replaced by its inverse, resistivity, for practical engineering calculations. Thus the resistance R in ohms of a conductor at a given temperature is calculated by

$$R = \rho \frac{l}{A}$$

where ρ = resistivity of conductor material, $\Omega\cdot$cm
   l = length of conductor, cm
   A = cross-sectional area, cm²

Table 4-2 gives the resistivity of various metallic elements and semiconductors at room temperature arranged in order of increasing resistivity. Note that the semiconductors have high resistivity in the unaltered state.

The resistance of pure metals increases with temperature. It may be calculated at any temperature below the melting point by means of the relationship

$$R = R_1[1 + \alpha_1(t - t_1)]$$

where α = temperature coefficient of resistance
   t = temperature

The subscript (1) refers to the reference temperature values. Thus if the resistance R_1 and the temperature coefficient α_1 have been evaluated at temperature t_1, the resistance R at temperature t can be calculated (Table 4-3).

Semiconductors   In the case of conductors, the electrons are already in the conduction band and can drift at the least excitation. On the other hand, the

Table 4-3   TEMPERATURE COEFFICIENTS OF RESISTANCE

Initial temperature, °C	Increase in resistance per °C		Initial temperature, °C	Increase in resistance per °C	
	Copper	Aluminum		Copper	Aluminum
0	0.00427	0.00439	25	0.00385	0.00396
5	0.00418	0.00429	30	0.00378	0.00388
10	0.00409	0.00420	40	0.00364	0.00373
15	0.00401	0.00411	50	0.00352	0.00360
20	0.00393	0.00403			

electrons of insulators have to receive considerable energy before their electrons can jump from the valence band to the conduction band. When this happens it is referred to as a breakdown of the insulator.

When some impurity atoms are included in some materials, this introduces intermediate permissible steps in the forbidden band and then the valence electrons can jump out of the valence band with much less excitation than in the pure insulator. Materials arranged in this manner are known as extrinsic semiconductors.

The motion of the positive hole that remains in the valence band con-tributes to the conductivity of the material as well as the motion of the electrons that have reached the conduction band.

When the impurity atoms introduce vacant levels close to the valence band, electrons can jump from the valence band to these levels, leaving behind their positive holes. The free levels are called acceptor levels. Materials arranged

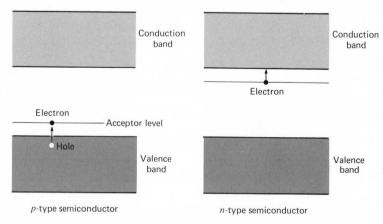

FIGURE 4-9
Band representation of p- and n-type semiconductors.

in this way are called p-type semiconductors for the $positive$ hole responsible for the electrical conductivity (Fig. 4-9).

The n-type semiconductors are those in which the impurity atoms introduce filled energy levels close to the lower edge of the conduction band. In this case, the electrons from these levels can be easily executed into the conduction band. The superposed levels are called donor levels. The n designation is due to the $negative$ charge introduced into the conduction band.

Impurities with loosely bonded electrons introduce donor levels in semiconductors; that is, they can easily contribute negative charges to the empty conduction band of the parent material, and thus they produce n-type semiconductors. Common "impurities" used to produce n-type semiconductors are arsenic, phosphorus, and antimony. The typical parent material for n-type semiconductors is germanium. The p-type impurities are those that introduce acceptor levels in a semiconductor—and permit the production of positive "holes" in the valence band. Examples of these elements are gallium, boron, and indium. Typical parent materials are silicon and germanium.

The calculation of resistances or conductivities of semiconductors are outside the scope of this book. It is suggested that interested students consult standard works concerning the physics of electrical engineering.

Magnetic Properties of Materials

A moving electric charge produces a magnetic field which is concentric with its direction of motion. If the charge moves in a circle, the magnetic field concentrates in the center of the circle and produces a magnetic dipole (Fig. 4-10). A moving charge within an atom contributes to the magnetic behavior of a material. The magnetic flux B induced in a material depends on how well the tiny dipoles line up when an external magnetic field H is applied. The ratio of

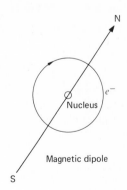

FIGURE 4-10
Schematic representation of a magnetic dipole.

B to H is known as the magnetic permeability M of the material:

$$M = \frac{B}{H}$$

MAGNETIC FLUX INDUCED
——————————————
EXTERNAL MAGN. FIELD

Solids exhibit three major types of magnetic behavior: diamagnetism, paramagnetism, and ferromagnetism. Diamagnetism is observed in materials which have complete outer electron rings, such as the elements in the eighth column of the periodic table. Such elements normally have zero *net* magnetic moment. The presence of an external field induces an opposite weak magnetism. The magnetic permeability of diamagnetic materials is less than unity.

Paramagnetic materials have permanent microscopic magnetic moments caused by the spin and orbital motions of the electrons. Paramagnetism is found in all atoms with partially filled inner shells. It is also produced from the spin of the free electrons which are present in metals. The microscopic magnets readily line up when the material is placed in a magnetic field. The permeability of paramagnetic substances is greater than unity.

Para- and diamagnetic behavior are briefly mentioned only for the sake of completeness, since the magnitude of the effects is so small when compared with the magnitude for ferromagnetic materials. Ferromagnetism is the important magnetic property of an engineering material. It stems from the spontaneous alignment of the small atomic magnets into magnetic domains. The permeability of ferromagnetic materials is much greater than unity. This magnetic behavior is not obvious in soft magnetic materials, such as pure iron, because all their magnetic domains are oriented at random. In permanent magnets there is a greater number of domains with a particular orientation than there is without a domain. The magnetic domains can be aligned in the presence of a magnetic field H, thus producing a magnetic flux B. The rotation of the magnetic domains must overcome internal friction within the magnetic field, therefore band H is not linearly proportional. If H is cycled and B and H are plotted, a hysteresis curve whose internal area is proportional to the energy consumed per cycle is obtained. Engineers considering materials for use in electromagnetic circuits should select ferromagnetic materials having a narrow hysteresis loop.

Most materials are nonmagnetic. Only iron, nickel, cobalt, ferric oxide, and alloys containing chromium, nickel, manganese, or copper are magnetic to any extent. High-silicon iron is the most permeable material and is widely used in magnetic circuits such as those in transformers.

MECHANICAL PROPERTIES OF MATERIALS

Once the important physical properties of a material have been established, mechanical properties such as yield strength and hardness must be considered.

Mechanical properties are structure-sensitive in the sense that they depend upon the type of crystal structure and its bonding forces, and especially upon the nature and behavior of the imperfections which exist within the crystal itself or at the grain boundaries.

An important characteristic which distinguishes metals from other materials is their ductility and ability to be deformed plastically without loss in strength. In design, 5 to 15 percent elongation provides the capacity to withstand sudden dynamic overloads. In order to accommodate such loads without failure, materials need dynamic toughness, high moduli of elasticity, and the ability to dissipate energy by substantial plastic deformation prior to fracture.

In order to predict the behavior of a material under load, engineers require reliable data on the mechanical properties of materials. Handbook data is available for the average properties of common alloys at 68°F. In design, the most frequently needed data are tensile yield strength, hardness, modulus of elasticity, and yield strengths at temperatures other than 68°F. Designers less frequently use resistance to creep, notch sensitivity, impact strength, and fatigue strength. Suppliers' catalogs frequently give more recent or complete data.

Production-engineering data which are seldom found in handbooks include strength-to-weight ratios, cost per unit volume, and resistance to specific service environments.

A brief review of the major mechanical properties and their significance to design is included to ensure that the reader is familiar with the important aspects of each test.

Tensile Properties of Materials

When a material is subjected to a tensile or compressive load of sufficient magnitude, it will deform at first elastically and then plastically. The deformation is elastic if after the load is removed the material returns to its original shape and dimensions; if not, the material has undergone plastic in addition to elastic deformation. The *engineering* stress-strain diagram presents the elastic data for a material, whereas the *natural* stress-strain diagram can present both elastic and plastic properties in a form which is useful to the designer and the engineer.

a. Engineering Stress-Strain Relationships

In the elastic range, deformation is proportional to the load which caused it. The proportionality is stated by Hooke's law in terms of the stress S and strain e:

$$S = Ee \tag{1}$$

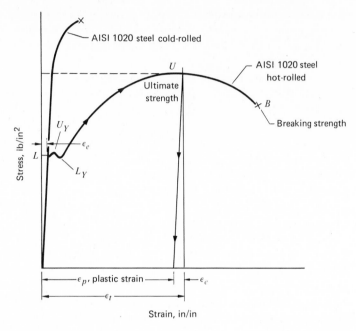

FIGURE 4-11
Engineering stress-strain diagram for AISI 1020 hot-rolled and cold-rolled steel.

where E = Young's modulus, the constant of proportionality between stress and strain, lb/in^2

S = stress = P/A = load/initial cross-sectional area, lb/in^2

e = engineering strain = $\Delta l/l_0$ = change in length/original length, in/in

If in the process of pulling a tensile specimen to failure, the stress is plotted as a function of the strain at selected strains, the engineering stress-strain curve results (Fig. 4-11). Note that the stress is plotted as a function of the strain contrary to the usual procedure of plotting the dependent variable as a function of the independent variable.

The straight part of the stress-strain curve is the proportional range described by Hooke's law. The upper limit L of this range is known as the proportional limit. In the case of annealed low-carbon steel there follows a region, the yield point, where the strain increases without any significant increase in stress. As indicated in the diagram, there can be an upper yield point, U_Y, and a lower yield point, L_Y. The maximum stress to which a material may be subjected is labeled U and the breaking point is denoted by B.

When a material is stressed beyond the elastic limit a permanent deformation results. However, this permanent deformation is associated with an elastic one. Notice that if the material of Fig. 4-11 is stressed to point P, upon the removal of the load so that the stress returns to zero, the plastic (permanent) strain is obtained following downward from point P along a line parallel to the proportional line. Here we have at point P the total strain is ε_t, composed of an elastic component ε_e and a plastic one ε_p.

The area under the curve in Fig. 4-11, up to the proportional limit, represents the potential energy per unit volume (resilience) stored in an elastically deformed body; on the other hand, the total area under the curve up to the point B represents the total energy per unit volume (modulus of toughness) required to break the specimen.

* b. Natural Stress-Strain Relationships

In many engineering designs, only the elastic behavior of a material is significant, so the engineering stress-strain curve was commonly used to depict the behavior of a material up to yield point. The yield strength of a material is calculated by dividing the load at the yield point by the *original* cross-sectional area of the test specimen. However when dealing with a material which is stressed beyond the elastic range, as in plasticity or metal forming, it is apparent that a way should be found to define the relationships in terms of the changing cross-sectional area. Thus the natural or true stress, σ, was defined as the load at any instant divided by the cross-sectional area *at that instant*. Thus,

$$\sigma = \frac{P_i}{A_i} \tag{2}$$

Similarly we may define the true strain as the change in linear dimension divided by the instantaneous value of the length. Thus:

$$\varepsilon = \int_{l_0}^{l_i} \frac{dl}{l} = \ln \frac{l_i}{l_o} \tag{3}$$

Typical natural stress-strain curves for steel and brass are given in Fig. 4-12.

The plastic region of the true stress-strain diagram may be expressed by

$$\sigma = \sigma_0 \varepsilon_t{}^m = \sigma_0 (\varepsilon_e + \varepsilon_p)^m \tag{4}$$

where σ = true stress
  σ_0 = work-hardening constant (a constant of the material)
  ε_t = total strain = elastic strain ε_e + plastic strain ε_p
  m = a constant of the material

Equation (4) is known as the strain-hardening equation. If both Eqs. (1) and

* For a more detailed treatment of True Stress-True Strain Relationships, see J. Datsko, *MATERIAL PROPERTIES AND MANUFACTURING PROCESSES*, J. Wiley (1966).

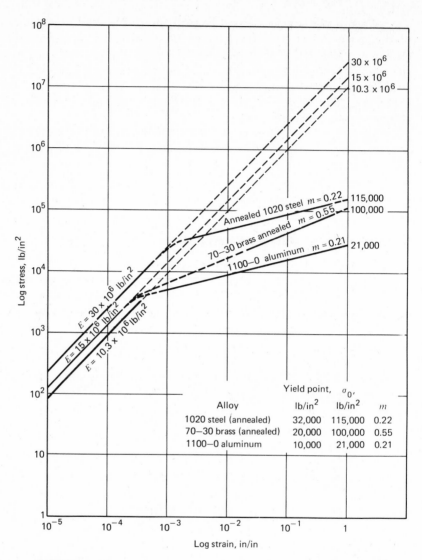

FIGURE 4-12
True-stress–true-strain curves for AlSl 1020 steel, 70–30 brass, and 1100–0 aluminum.

(4) are expressed in logarithmic form, the result is

$$\ln S = \ln E + \ln \varepsilon_e \tag{5}$$

$$\ln \sigma = m \ln \varepsilon_t + \ln \sigma_0 = m \ln (\varepsilon_e + \varepsilon_p) + \ln \sigma_0 \tag{6}$$

These two equations plotted on log-log paper yield two straight lines with a discontinuity about the region of the yield point (Fig. 4-12). It will be noted that Eq. (5), for the elastic zone, is the equation of a straight line with a slope of unity (45°) and an intercept of $\ln E$. Equation (6) is also a straight line but with intercept $\ln \sigma_0$ and slope m.

c. Correlation of Engineering and Natural Stresses and Strains

Hooke's law also applies for the elastic range of the true stress-strain diagram. The relationship is expressed in this case by

$$S = Ee \tag{7}$$

where E = Young's modulus, as for Eq. (1)

$\quad S$ = engineering stress = P/A_0 = loading force/original
$\qquad\qquad\qquad\qquad\qquad\qquad\qquad$ cross-sectional area $\tag{8}$

$\quad e$ = engineering strain = $(l_i - l_0)/l_0$ = change of length/original
$\qquad\qquad\qquad\qquad\qquad\qquad\qquad\qquad$ length $\tag{9}$

It is convenient to derive a relationship between true stress-strain and engineering stress-strain. For true stress, there is the relationship

$$\sigma = \frac{P}{A_i} \quad \text{[see Eq. (4)]}$$

As the volume of the material remains constant during the loading, $A_i l_i = A_0 l_0$; thus:

$$A_i = A_0 \frac{l_0}{l_i} \tag{10}$$

Consequently,

$$\sigma = \frac{P}{A_0} \frac{l_i}{l_0} \tag{11}$$

Substituting Eq. (6) into Eq. (9) results in:

$$\sigma = S \frac{l_i}{l_0} \tag{12}$$

which is the relationship between true stress and engineering stress.

On the other hand, true strain is defined as

$$\varepsilon = \int_{l_0}^{l_i} \frac{dl}{l} = [\ln l]_{l_0}^{l_i} = \ln \frac{l_i}{l_0} \tag{13}$$

Recalling from Eq. (7) that the engineering strain is

$$e = \frac{l_i - l_0}{l_0} = \frac{l_i}{l_0} - 1 \tag{14}$$

there results:

$$\frac{l_i}{l_0} = e + 1 \tag{15}$$

Substituting Eq. (15) into Eq. (13) gives

$$\varepsilon = \ln (e + 1) \tag{16}$$

the relationship between the true strain and the engineering strain.

It is to be noted that the strain corresponding to the elastic limit of most engineering materials is very small, usually smaller than 0.005; consequently,

$$\frac{l_i}{l_0} \approx 1 \qquad \text{and} \qquad \ln (e + 1) \approx e \tag{17}$$

Therefore,

$$E = \frac{\sigma}{\varepsilon} = \frac{S(l_i/l_0)}{\ln (e + 1)} \approx \frac{S}{e} \tag{18}$$

It is worthwhile for the practical engineer to be conscious of the differences between the basic stress-strain diagrams for different materials, and also of how the stress-strain diagram for one material varies according to its mechanical or thermal treatment. Also, it must be emphasized again that the strain at the elastic limit for most engineering materials is very small. Most textbook presentations represent the elastic part of the stress-strain curve tilted completely out of proportion for ease of illustration.

Hardness

Hardness is an engineering property that is related to the wear resistance of a material, its ability to abrade or indent another material, or its resistance to permanent or plastic deformation. The selection of the appropriate hardness test is dependent upon the relative hardness of the material being tested and the amount of damage that can be tolerated on the surface of the test specimen. Table 4-4 provides information relative to 10 hardness standards, each having application for certain types of materials. Hardness can be measured by three

Table 4-4

BRINELL   } FERROUS MAT'LS
ROCKWELL }
KNOOP
VICKERS
SCLEROSCOPE
DUROMETER
MOH

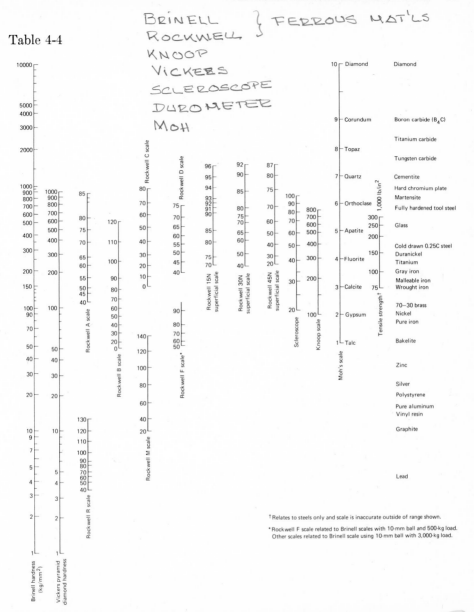

† Relates to steels only and scale is inaccurate outside of range shown.

* Rockwell F scale related to Brinell scales with 10-mm ball and 500-kg load. Other scales related to Brinell scale using 10-mm ball with 3,000-kg load.

types of tests: (1) indentation—Brinell, Rockwell, Knoop, Durometer; (2) rebound or dynamic—Scleroscope; and (3) scratch—Moh. These 10 standards together with the materials to which they apply are as follows:

1 Brinell—ferrous and nonferrous metals, carbon and graphite
2 Rockwell B—nonferrous metals or sheet metals
3 Rockwell C—ferrous metals

Test	Indenter	Shape of indentation		Load	Formula for hardness number
		Side view	Top view		
Brinell	10-mm sphere of steel or tungsten carbide	(side view, diameter D, d)	(top view, d)	P	$BHN = \dfrac{2P}{\pi D[D - \sqrt{D^2 - d^2}\,]}$
Vickers	Diamond pyramid	$136°$	(top view, d_1)	P	$VHN = 1.72P/d_1^{\,2}$
Knoop microhardness	Diamond pyramid	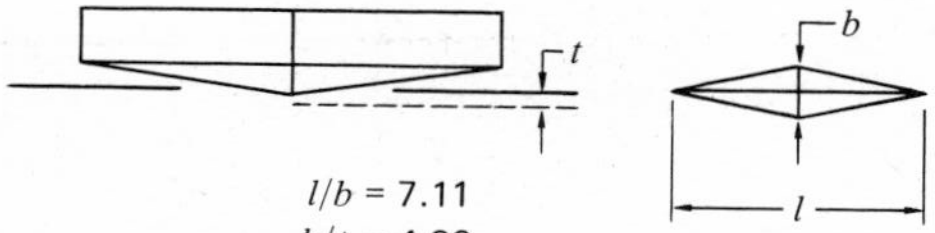 $l/b = 7.11$ $b/t = 4.00$	(top view, b, l)	P	$KHN = 14.2P/l^2$
Rockwell					
A	Diamond cone	$120°$ (t)		60 kg	$R_A =$
C				150 kg	$R_C =$ $\}$ 100–500t
D				100 kg	$R_D =$
B	$\frac{1}{16}$-in-diameter steel sphere	(t)		100 kg	$R_B =$
F				60 kg	$R_F =$ $\}$ 130–500t
G				150 kg	$R_G =$
E	$\frac{1}{8}$-in-diameter steel sphere			100 kg	$R_E =$

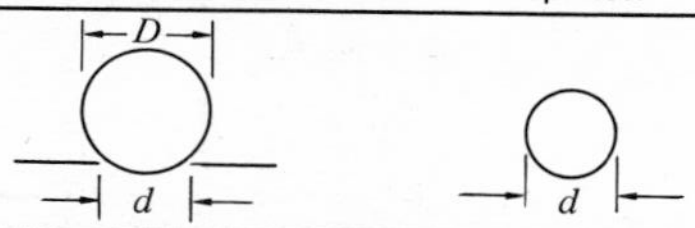

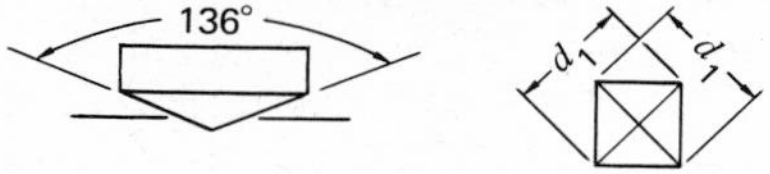

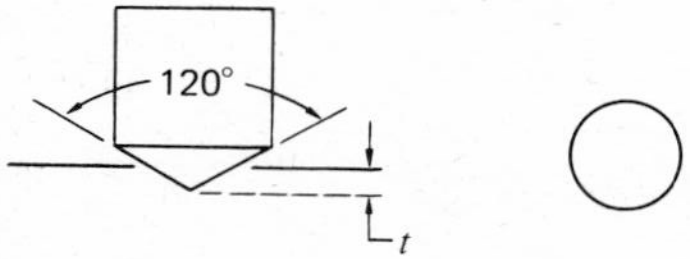

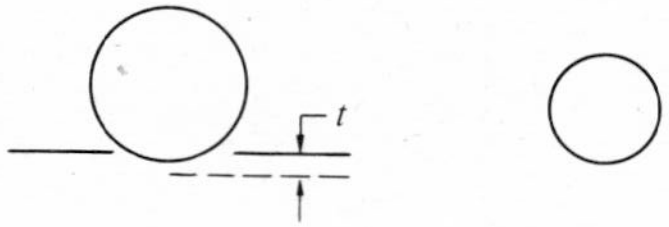

FIGURE 4-13

Comparison of hardness tests. (*Redrawn from H. W. Hayden, W. G. Moffatt, and V. Wulff. "The Structure and Properties of Materials," Wiley, New York, 1965.*)

4 Rockwell M—thermoplastic and thermosetting polymers
5 Rockwell R—thermoplastics and thermosetting polymers
6 Knoop—hard materials produced in thin sections or small parts
7 Vicker's diamond pyramid hardness—all metals
8 Durometer—rubber and rubberlike metals
9 Scleroscope—primarily used for ferrous alloys
10 Moh—minerals

Characteristics of the various hardness tests are compared in Fig. 4-13.

The Brinell hardness test is applicable to most metals and their alloys. This test provides a number related to the area of the permanent impression made by a ball indentor, usually 10 mm in diameter, pressed into the surface of the material under a specified load. The larger the Brinell number, the harder the material. The Brinell number (Bhn) is computed as follows:

$$\text{Bhn} = \frac{P}{(\pi D/2)(D - \sqrt{D^2 - d^2})}$$

where P = applied load, kg
 D = diameter of ball, mm
 d = diameter of impression, mm
 Bhn = Brinell hardness, kg/mm²

Rockwell hardness is also an indentation test used on both metals and plastics. The Rockwell number is derived from the net increase in depth of an impression of a standard indentor as the load is increased from a fixed minor load to a major load and then returned to the minor load. The Rockwell C test is used primarily for ferrous alloys, whereas the Rockwell B procedure is followed for softer or thinner alloys where a minimum indentation is preferred. Rockwell M numbers are applied to harder plastics.

The Knoop and Vickers hardness tests measure the microhardness of small areas of a specimen. They are especially suited for measuring the hardness of very small parts, thin sections, and individual grains of a material using a microscope equipped with a measuring eyepiece. Both tests impose a known load on a small region of the surface of a material for a specified time. In the case of the Knoop test, the indentor is a diamond with a length-to-width ratio of about 3:1, whereas the Vickers indentor is a square diamond pyramid. Both values of hardness are calculated by dividing the applied load by the projected area of the indentation as measured by the microscope in terms of the diagonals of the indentation.

The Shore Scleroscope is a rebound device which drops a ball of standard mass and dimensions through a given distance. The height of the rebound is measured. As the hardness of the test surface increases, the rebound height increases because less energy is lost in plastic deformation of the test surface.

The Durometer hardness test is used in conjunction with elastomers

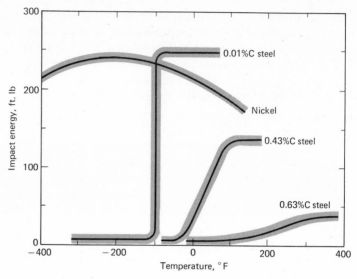

FIGURE 4-14
Impact-test results for several alloys over a range of testing temperatures.
(*Redrawn from H. W. Hayden, W. G. Moffatt, and V. Wulff, "The Structure
and Properties of Materials," Wiley, New York, 1965.*)

(rubber and rubberlike materials). Unlike other hardness tests, which measure
plastic deformation, this test measures elastic deformation. Here an indentor of
hardened steel is extended into the material being tested. A hardness value of
100 represents zero extension of the indentor; and at zero reading, the indentor
extends 0.100 in plus 0.000 in minus 0.003 in beyond the presser foot which
surrounds the area being measured.

Moh hardness values are used principally in the designation of hardness
of minerals. The Moh scale is an arbitrary scale of hardness based upon 10
selected minerals. The scale, along with values of these 10 minerals, are: talc, 1;
rock salt or gypsum, 2; calcite, 3; fluorite, 4; apatite, 5; feldspar, 6; quartz, 7;
topaz, 8; corundum, 9; diamond, 10. According to the Moh hardness scale a
fingernail has a hardness of 2, annealed copper 3, and martensite 7. However,
the Moh scale has too few values to be of use in the metalworking field because
four values cover the range of the softest to the hardest metals.

In view of the wide variation in the characteristics of engineering materials,
one hardness test for all materials is not practical but conversion from one
hardness scale to another can be done in some cases. The Brinell and Rockwell B
and S scales can be converted from one to the other. To approximate equivalent
hardness numbers, the alignment chart shown in Table 4-4 can be used. This
chart has not included a Durometer scale, since correlation between Durometer

and Brinell values is inconsistent. The Durometer test measures the elastic properties of a material, whereas all other indentation hardness tests measure the plastic properties of a material.

Impact Properties

In some designs dynamic forces are likely to cause failure. For example an alloy may be hard and have high compressive strength and yet be unable to withstand a sharp blow. In particular, low-carbon steels are susceptible to brittle failure at certain temperatures (Fig. 4-14). Experience has demonstrated that the impact test is sensitive to the brittle behavior of such alloys. Most impact tests use a calibrated hammer to strike a notched or unnotched test specimen. In the former the test result is strongly dependent on the base of the notch where there is a large concentration of triaxial stresses which produce a fracture with little plastic flow. The impact test is particularly sensitive to internal stress producers such as inclusions, flake graphite, second phases, and internal cracks.

  The results from an impact test are not easily expressed in terms of design requirements because it is not possible to determine the triaxial stress conditions at the notch. There also seems to be no general agreement on the interpretation or significance of the result. Yet the impact test has proved especially useful in

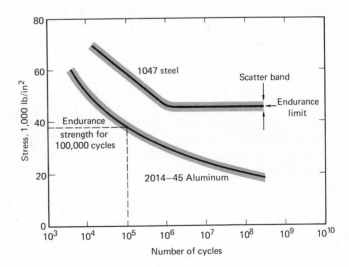

FIGURE 4-15
Typical fatigue-failure curves for Al and 1047 steel. (*Redrawn from H. W. Hayden, W. G. Moffatt, and V. Wulff, "The Structure and Properties of Materials," Wiley, New York, 1965.*)

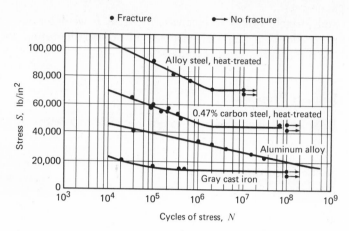

FIGURE 4-16
Fatigue (*S-N*) curves and endurance limits for various alloys. (*Redrawn from A. G. Guy, "Elements of Physical Metallurgy," Addison-Wesley, Reading, Mass., 1960.*)

defining the temperature at which steel changes from brittle to ductile behavior. Low-carbon steels are particularly susceptible to brittle failure in a cold environment such as the North Atlantic. There were cases of Liberty ships of World War II vintage splitting in two as a result of brittle behavior.

Fatigue Properties and Endurance Limit

Although yield strength is a suitable criterion for designing components which are to be subjected to static loads, for cyclic loading the behavior of a material must be evaluated under dynamic conditions. The fatigue strength or endurance limit of a material should be used for the design of parts subjected to repeated alternating stresses over an extended period of time. As would be expected, the strength of a material under cyclic loading is considerably less than it would be under a static load (Fig. 4-15). The plot of stress as a function of the number of cycles to failure is commonly called an *S-N* curve. It is interesting to note that for specimens of SAE 1047 steel there is a stress called the endurance limit below which the material has an infinite life ($> 10^8$ cycles), i.e., the steel would not fail regardless of exposure time (Fig. 4-16). In contrast, the 2014-T6 aluminum alloy has no limiting stress value.

Fatigue data is inherently more variable than tensile test data. In part the scatter is caused by variation in surface finish and environment (Fig. 4-17). Polished specimens of the same material give significantly better life than machined or scaly surfaces. Since most fatigue failures initiate at surface notches,

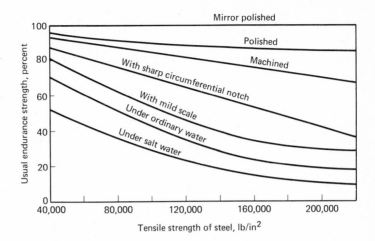

FIGURE 4-17
Effect of surface condition and environment on the endurance strength of
steel (Karpoo and Ballens). (*Redrawn from A. G. Guy, "Elements of Physical
Metallurgy," Addison-Wesley, Reading, Mass., 1960.*)

fatigue behavior and notch sensitivity are closely related. Thus mechanical or
other treatments which improve the integrity of the surface, or add residual
compressive stresses to it, improve the endurance limit of the specimen.

Several standard types of fatigue-testing machines are commercially available. The results of extensive fatigue-life testing programs are now available
in the form of *S-N* curves which are invaluable for comparing the performance
of a material which is expected to be subjected to dynamic loads.

Creep and Stress Rupture

As metals are exposed to temperatures within 40 percent of their absolute
melting point, they begin to elongate continuously at low load. A typical creep
curve is a plot of the elongation of a wire subjected to a tensile load at a given
temperature against time (Fig. 4-18). Creep is explained in terms of the interplay
between work hardening and softening from recovery processes such as dislocation climb, thermally activated cross slip, and vacancy diffusion. During
the primary creep the rate of work hardening decreases because the recovery
processes are slow, but during secondary creep both rates are equal. In the
third stage of creep, grain-boundary cracks or necking may occur which reduce
the net cross-sectional area of the test specimen. Creep is sensitive to both the
test temperature and the applied load. Increases in temperature speed up
recovery processes and an increased load raises the whole curve.

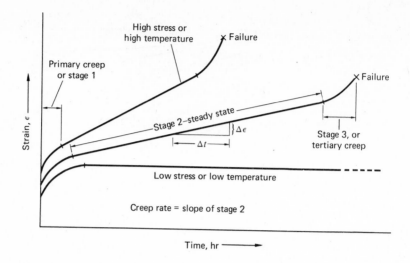

FIGURE 4-18
Typical creep curve showing the stages of creep at constant temperature.
Stage 2 determines the useful life of the material.

In the stress rupture test, the specimen is held under an applied load at a given temperature until it fails. Then a series of curves are plotted which can be helpful to the designer when he must consider high-temperature applications.

COMPARISON OF THE PROPERTIES OF MATERIALS

The materials commonly used in manufacturing can be divided into four groups in order to contrast their properties: ferrous, nonferrous, thermosetting, and thermoplastic. In some cases those four groups are not particularly appropriate, since, for example, dielectric strength is not applicable to metals. On the other hand, polymers have such low yield strengths that they fare poorly when compared with even the weakest metal on the basis of strength.

The bar charts in Fig. 4-19 compare selected physical properties of the four classes of materials including dielectric strength, resistivity, coefficient of thermal expansion, thermal conductivity, and specific weight. It can be readily seen that mica and glass have up to four times as much breakdown strength as polymers or rubber. But some thermoplastics exceed all other materials in electrical insulation, with most other materials fairly close to second place. The coefficient of thermal expansion of polymers is an order of magnitude greater than that of metals, whereas the thermal conductivity of metals far exceeds that of all plastics. The specific weight of ferrous alloys is four to five times that of polymers and three times that of aluminum alloys.

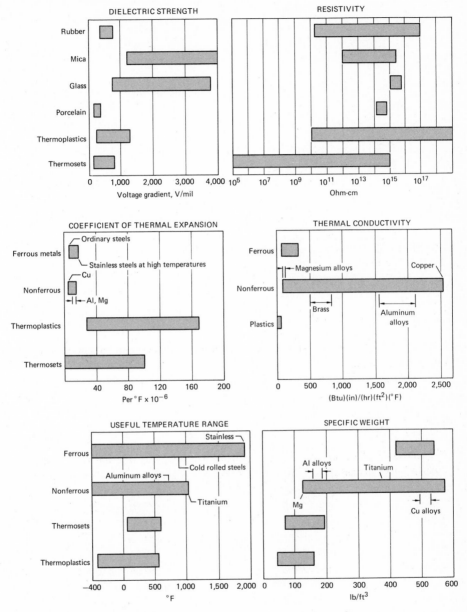

FIGURE 4-19
Comparison of the physical properties of materials.

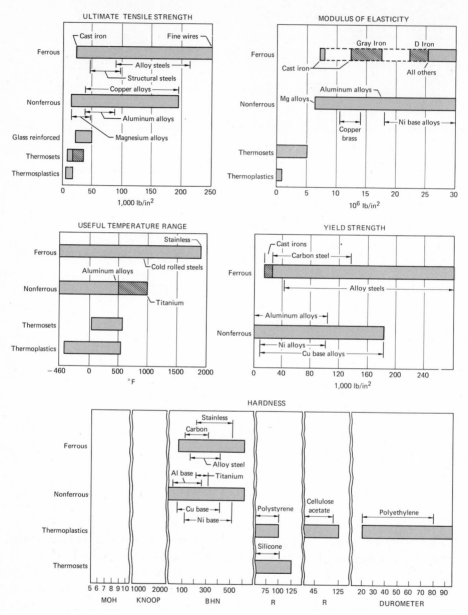

FIGURE 4-20
Comparison of the mechanical properties of materials.

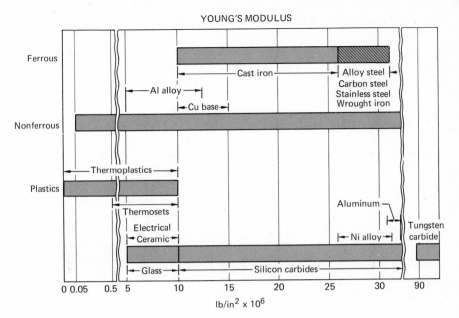

FIGURE 4-21
Comparison of the elastic moduli of materials.

The same groups of materials were compared with respect to their mechanical properties including ultimate strength, yield strength, modulus of elasticity, hardness, and useful temperature range (Fig. 4-20). The ultimate tensile strength of metals far exceeds that of polymers—by a factor of 25 for unreinforced materials and a factor of 8 for the woven glass reinforced resins. In the case of the modulus of elasticity, the best thermosets are below the lowest nonferrous materials with a maximum of 5×10^6 lb/in² (Fig. 4-21). The ferrous materials range from 12×10^6 for class 25 gray iron to 30×10^6 for most steels. Tungsten carbides are off the graph with values up to 90 to 95×10^6 lb/in². Therefore tungsten carbide, despite its cost and weight, is useful for long, cantilever-beam, boring-tool holders because of its low deflection under load. In the comparison of yield strengths metals are far superior to polymers. Nonferrous materials range from a few hundred lb/in² to 80,000 lb/in² for aluminum and 175,000 lb/in² for certain copper- and nickel-based alloys. Ferrous alloys range from 20,000 lb/in² for certain gray irons to 280,000 lb/in² or more for particular alloy steels.

The usable range of temperature for steel ranges from $-460°F$ to almost 2000°F for specific stainless steels. Aluminum alloys can withstand temperatures from 300 to 500°F, and some titanium-reinforced polymers are useful up to 400 to 900°F, but the vast majority are good only to 200°F. Hardness is the most difficult property to use for making valid comparisons because the deformation of plastics and elastomers under an indentor is different from metals. As a group polymers are far softer than metals. Ferrous and nickel-base alloys range from 100 to 600 BHN, which is a tremendous range of values.

RELATIVE COST OF ENGINEERING MATERIALS

The design engineer is responsible for the initial specification of a material, yet in many cases he spends little time checking alternatives. Many designers feel that their obligation ends with the functional design, so they tend to minimize the importance of material selection. The reasons are many, but two are likely to be: (1) lack of proper education in their engineering courses and (2) inadequate time because of the need to meet a deadline.

In over 90 percent of all designs a material is selected primarily on the basis of yield strength and ability to fill space at the lowest cost. In some cases, other criteria, depending primarily on physical properties such as electrical conductivity or chemical properties such as corrosion resistance, are dictated by the end use of the design and would be the basis for the material specification.

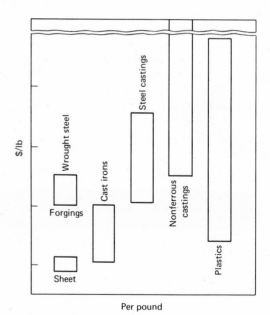

FIGURE 4-22
Comparative cost/unit volume for several materials. (*Redrawn from Foundry magazine.*)

Table 4-5   CHANGES IN THE PRICE OF METALS AND
POLYMERS IN THE DECADE 1958–1968

Polymer	Change, %	Metal	Change, %
Vinyl	−7	Gray iron	+12
Polyethylene	−7	Zinc	+14
Fiberglas	−40	Aluminum	0
ABS	−17	Copper	+30
Nylon	−13	Steel	+10

It is the designer's responsibility to match the functional and esthetic requirements of his design with the behavior of a particular material so that the total
cost after manufacture will be minimized. In a large organization this under-
taking is a team effort in which design engineers, production engineers, and materials specialists all play an important role. In a small company the designer
may handle the entire assignment.

The cost of engineering materials is in a constant state of flux. Three
components may be recognized; first, the general price changes which follow

Table 4-6   COMPARATIVE AVERAGE COSTS FOR ENGINEERING MATERIALS

Material	Specific weight, lb/in³	Cost, $/lb	Cost, $/in³
Vinyl	0.047	0.17	0.008
Polystyrene	0.033	0.18	0.006
Fiberglas	0.058	0.26	0.015
ABS	0.030	0.39	0.0115
Nylon	0.041	0.85	0.035
Gray-iron casting	0.256	0.16	0.041
Ductile-iron casting	0.256	0.32	0.082
Malleable-iron casting	0.263	0.38	0.100
Steel casting	0.285	0.40	0.114
Zinc die casting	0.240	0.42	0.101
Aluminum die casting	0.096	0.55	0.053
Brass sand casting	0.320	1.05	0.335
Manganese bronze casting	0.296	1.60	0.474
Sheet steel (SAE 1020 CR)	0.283	0.068	0.019
Bar steel (SAE 1020 HR)	0.283	0.115	0.032
Steel forging	0.283	0.350	0.099

the ebb and flow of the national economy; second, the supply-and-demand effect on the price of some metals, specifically copper and nickel; and, finally, the effect of rising production capacity and strong competition as found in the polymer industry since the 1950s. In general the polymers have dropped in price in the past decade whereas the cost of all metals but aluminum has increased (Table 4-5 and Appendix I). Inflation has led to increasing prices.

Materials cannot be compared on the basis of cost alone because they have widely varying densities and strengths. If the cost per pound is converted to cost per unit volume, a more meaningful comparison can be made (Table 4-6). When the same materials are compared on the basis of cost per unit of strength per unit of volume it is evident that not only do most materials become strongly competitive, but that in the future, if stronger polymers are created, the 4,000-year dominance of metals may be severely challenged (Figs. 4-22 and 4-23).

The inflationary increases in the cost of various materials in the decade from 1961 to 1970 is presented in Fig. 4-24. It should be noted that whereas the cost of tin bronze increased 250 percent and brass rose 75 percent, the price of steel jumped 325 percent! That tremendous increase in the cost of a basic commodity, combined with the reduced productivity of labor, is a major contributor to the economic woes of the United States. On the other hand, the price of polymers generally tended to decrease, primarily because there are few man-hours required per ton of output. The shipbuilding and textile industries are

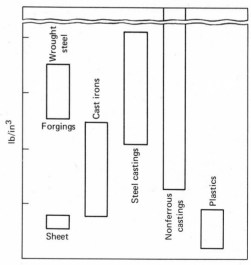

Per cubic inch

FIGURE 4-23
Strength-to-weight ratio for several materials. (*Redrawn from Foundry magazine.*)

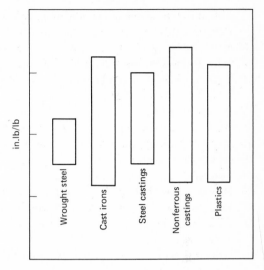

FIGURE 4-24
Strength-to-weight ratio relative to cost per unit volume for several materials.
(*Redrawn from Foundry magazine.*)

good examples of what may happen to the steel and automobile industries if progressive management and enlightened labor leaders do not seek technical innovation and greater productivity per man-hour.

QUESTIONS

1 Clearly explain the difference between the physical properties and mechanical properties of a material. Give several examples.

2 Why is a designer or production engineer interested in the properties of materials?

3 Would a designer and a production engineer be equally interested in the same properties of a material to be used, for example, in an electric motor frame? Explain your answer.

4 What is the difference between the structure-sensitive and structure-insensitive properties of a material?

5 Give several distinguishing characteristics of the four major classes of solid materials.

6 What are three general groupings which can be used in the study of the structure of materials?

7 What is the major feature of Planck's quantum hypothesis? What physical property of a material does it help us to understand?

8 What is the basic factor which affects the value of the heat capacity of a solid?

9 How would the heat capacity of two metals vary, in general?

10   What property of two materials would lead to a difference in the magnitude of their heat capacities?

11   How does the thermal conductivity of any material vary with temperature?

12   Why does the thermal conductivity of a metal exceed that of a plastic?

13   What is the relationship between true strain and engineering strain?

14   What is the advantage of the true stress–true strain relationship for a production-design engineer?

15   Why does the elastic region of the true stress–true strain graph plot as a 45° slope?

16   How is hardness determined? Give the nature of the various hardness tests.

17   What is the relationship between hardness and tensile strength for steel? Is this relationship also found for other alloys?

18   How do fatigue tests aid the production-design engineer?

19   What is the value of a creep test in a production design?

20   In a production design, what are the major advantages of thermosets and thermoplastics compared to metals?

21   What are the major advantages of ferrous alloys compared with plastics and other metallic alloys?

22   What material is strongest on a strength-to-weight basis?

23   How does cost affect the alternatives regarding which material to use for a given situation?

PROBLEMS

1   Explain the behavior of electrical and thermal insulators and conductors as a function of temperature in terms of the band theory of solids and free electrons. Show by an appropriate sketch how a semiconductor can become a conductor or an insulator.

2   A standard tensile test on a specimen 0.505 in diameter with a 2-in gage length yielded the following data: yield load, 11,700 lb; maximum load, 65,000 lb; gage length at the maximum load, 3.018 in; and reduction in area at fracture, 35 percent. Determine the yield strength, the engineering fracture stress, the natural fracture stress, the strain hardening equation, and its constants.

3   Given the strain hardening equation for an annealed material as $\sigma = 95,000\varepsilon^{0.25}$, estimate its yield and tensile strengths as annealed and after 60 percent cold work.

4   If the value for the Bhn of AISI 1040 steel as water quenched and tempered at 400°F is 560, what would be its probable tensile strength?

5   From a long-time creep test of Inconel X at 1500°F and carried out in a nonoxidizing atmosphere, the following data were obtained:

(a)   Elongation after 10 h, 1.2 percent
(b)   Elongation after 200 h, 2.3 percent
(c)   Elongation after 2,000 h, 4.5 percent
(d)   Elongation after 4,000 h, 6.8 percent

(e)   After 5,000 h, neckdown began

(f)   At 5,500 h, rupture occurred

Determine the creep rate of Inconel X at 1,500 h.

6   A specimen of 60–40 brass was cold-rolled so that it received a 63 percent reduction, what is the magnitude of the true strain in the material?

7   A cast iron bearing plate $10 \times 10 \times 2.5$ in is subjected to a uniformly compressive load. The compressive strength of gray iron is 100,000 lb/in^2 and the strain at fracture was found to be 0.002 in/in. What was

(a)   The compressive fracture load?

(b)   The total contraction at fracture?

(c)   The total strain energy corresponding to fracture?

SELECTED REFERENCES

AMERICAN SOCIETY FOR METALS: "Metals Handbook," 7th ed., Metals Park, Ohio, 1948.

———: "Metals Handbook," 8th ed., vol. 1, "Structure and Properties of Metals," Metals Park, Ohio, 1964.

———: "Metals Handbook," 8th ed., vol. 7, "Microstructures of Metals," Metals Park, Ohio, 1973.

BROPHY, J. H., R. M. ROSE, and J. WULFF: "The Structure and Properties of Materials," vol. II, "Thermodynamics of Structure," Wiley, New York, 1964.

CAHNERS, INC.: "Ceramic Data Book," Cahners, Chicago, published annually.

CHAPMAN-REINHOLD: "Materials Selector Issue," Materials Engineering Series, Chapman-Reinhold, New York, published annually.

MCCLINTOCK, F. A., and A. S. ARGON: "Mechanical Behavior of Materials," Addison-Wesley, Reading, Mass., 1966.

MCLEAN, D.: "Mechanical Properties of Metals," Wiley, New York, 1962.

PARKER, E. R.: "Materials Data Book for Engineers and Scientists," McGraw-Hill, New York, 1967.

ROSENTHAL, D.: "Introduction to Properties of Materials," Van Nostrand, Princeton, N.J., 1964.

SAMAUS, C. H.: "Metallic Materials in Engineering," Macmillan, New York, 1963.

SMITHELLS, C. J.: "Metals Reference Book," 3d ed., Butterworths, London, 1962.

STANLEY, J. K.: "Electrical and Magnetic Properties of Metals," American Society for Metals, Metals Park, Ohio, 1963.

5

ENHANCEMENT OF THE PROPERTIES OF MATERIALS

Heat treatment is commonly used to enhance the mechanical properties of a material in the solid state by first heating and then cooling at a predetermined rate. Many alloys are heat-treated, the most important be'ng ferrous and aluminum alloys. But the mechanical properties of materials can be increased by strain hardening as well. Hardening mechanisms include:

1 Strain hardening
2 Control of grain size
3 Eutectoid decomposition
    a Equilibrium
    b Nonequilibrium
    c Hardening operations
4 Alloy hardening
    a Solid solution
    b Dispersion hardening
    c Precipitation hardening
    d Diffusion reaction

STRAIN HARDENING

When a metallic alloy is plastically deformed its yield strength increase; with an increase in strain, as long as the recrystallization temperature is not exceeded. Thus controlled amounts of cold working may be used to increase the mechanical properties of a material (Fig. 5-1). The true stress–true strain curves from Chap. 4 show that on a log-log plot the strain-hardening equation is indeed a straight line and its slope is defined as the coefficient of strain hardening. Through the strain-hardening equation an engineer can predict the improvement in properties a given operation will impart to a material.

In polycrystalline alloys, the mutual interference of adjacent grains causes slip to occur on many intersecting slip planes with accompanying strain harden-

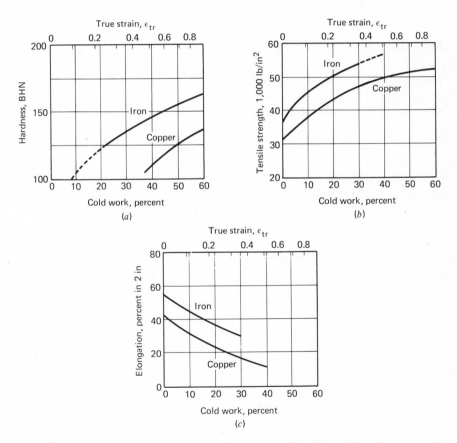

FIGURE 5-1
The effect of cold work on the mechanical properties of iron and copper. (*Redrawn from L. H. Van Vlack, "Materials Science for Engineers," Addison-Wesley, Reading, Mass., 1970.*)

ing of the metal. This progressive strengthening with increasing deformation stems from the interaction of dislocations on intersecting slip planes.

The strain-hardening behavior of a metal depends on its lattice structure. In face-centered cubic metals, the rate of strain hardening is affected by the stacking fault energy through its influence on mechanical properties. Copper, nickel, and austenitic stainless steel strain-harden more rapidly than aluminum does. Hexagonal close-packed metals are subject to twinning and strain-harden at a much higher rate than do other metals because there is only one plane of easy glide available in the close-packed hexagonal structure. Strain hardening is also affected by grain size, impurity atoms, and the presence of a second phase.

Severely strained metals have elongated grains with distorted and twisted lattices and a strong anisotropy which can be used by an astute designer if he takes advantage of directional strength in his design.

Recovery

During a cold-working process most of the energy used is dissipated as heat but a small percentage is stored within the distorted lattice structure of the alloy. That energy is the thermodynamic driving force which tends to return the metal to its original state provided there is sufficient thermal energy to enable the reactions to occur. Recovery is a gradual change in the mechanical properties of an alloy, i.e., loss in brittleness, or marked increase in toughness brought about by controlling the heat-treating times and temperatures such that there are no appreciable changes in the microstructure. However, recovery does significantly reduce the residual stresses within the distorted lattice structure.

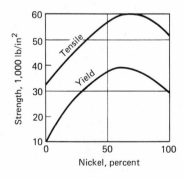 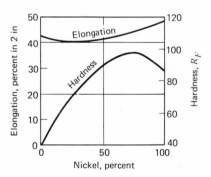

FIGURE 5-2
Effect of increasing solute on the strength of a substitutional solid-solution alloy of copper and nickel. (*Redrawn from L. H. Van Vlack, "Materials Science for Engineers," Addison-Wesley, Reading, Mass., 1970.*)

Recrystallization

All the properties of a cold-worked metal are affected to some degree by a recovery heat treatment, but the yield strength and ductility can only be restored by recrystallization. The latter may be defined as the nucleation and growth of strain-free grains out of the matrix of cold-worked metal. In general the properties of the recrystallized alloy are those of the metal before the cold-working operation. This is of commercial importance because if an alloy has become work-hardened in a drawing operation, it can be recrystallized and the drawing operation can be continued. Nucleation is encouraged by a highly cold-worked initial structure and a high annealing temperature. The presence of alloying elements in the solid solution decreases the rate of nucleation.

GRAIN SIZE

An important factor in the heat treatment of steel is the grain size, by which is meant the sizes of the microscopic grains which are established at the last temperature above the critical range to which the piece of steel has been treated.

1 Fine-grained steels show better toughness at high hardness.
2 Controlled grain-size steels will show less warpage.
3 Coarse-grained steels harden better.

Grain-size-controlled steel can be secured from the steel mill within certain limits. A high percentage of carbon steels used for heat treating is grain-size-controlled and sold to grain-size limits. Almost all the alloy steels are grain-size-controlled. If a given grain size is specified, it must be obtained by suitable control of the recrystallization process and by prevention of excessive grain growth.

SOLID STATE REACTIONS WHICH IMPROVE MECHANICAL PROPERTIES

Strain hardening (Fig. 5-1) and alloy hardening (Fig. 5-2) are widely used in industry, especially in aluminum- and copper-base alloys. In certain alloys it is possible to augment their mechanical properties by other solid state reactions which can produce much greater hardness than is possible by alloying alone, and plastic deformation is not required. However, solid state reactions are usually limited to several alloys because:

1 Relatively few alloys can be affected by a given solid state reaction.

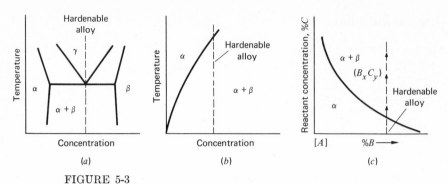

FIGURE 5-3
The form of equilibrium diagram required for three solid state reactions:
(a) eutectoid decomposition; (b) precipitation from the solid; (c) diffusion
reaction. (*Redrawn from A. G. Guy, "Elements of Physical Metallurgy," 2d ed.,
Addison-Wesley, Reading, Mass., 1959.*)

Euctectoid decomposition, for instance, is rare in alloy systems other than
the iron-carbon system.

2 To achieve significant hardening, a solid state reaction must form an
equilibrium or metastable structure such as martensite.

3 Even though an energetically favorable solid state reaction is possible
according to the equilibrium phase diagram, it may produce little or no
hardening. Therefore the occurrence of a given reaction is a necessary,
but not a sufficient, condition for strengthening.

Several solid state reactions produce gains in mechanical properties which
are important to engineers. Typical phase diagrams indicate the combination
of phases which is needed before a solid state reaction could occur, but that
alone does not mean that a significant improvement would occur. A typical
eutectoid reaction is a prerequisite to eutectoid decomposition (Fig. 5-3a).
However the iron-carbon system is the only one which is of commercial value.
When the solvus line which separates a one- and two-phase region slopes to indicate decreasing solid solubility of B in A, there is a chance that precipitation
hardening may occur (Fig. 5-3b). Several aluminum alloys have this type of
phase diagram, but although many binary alloys have similar solvus lines,
aluminum alloys are the best example of precipitation hardening. The requirements of the diffusion reaction are indicated in Fig. 5-3c. In this case the diffusion of metal C into the alloy causes the composition of the hardenable alloy
(metal B in metal A) to shift from a single-phase region to a two-phase region.
As metal C diffuses into the solid solution, the overall composition gradually
shifts into the alpha plus beta region, so that the beta phase (B_xC_y) begins to
precipitate. The nitriding process works according to this reaction because
aluminum, chromium, and vanadium form nitrides. Thus when nitrogen gas

diffuses into the steel surface which contains those elements, their nitrides form within the surface of the metal to produce an extremely hard surface. Tensile strength is only moderately improved, but creep resistance and the recrystallization temperature are much enhanced. The unusual surface hardness is thought to be a result of the fine dispersion of nitride particles rather than the inherent hardness of the nitrides alone.

SOLID STATE TRANSFORMATIONS IN FERROUS ALLOYS

Steel is unique in its ability to exist as a soft ductile material which can be easily formed or machined and then, as a result of a heat treatment, to assume the role of a hard, tough material which resists changing shape. There are two reasons for this behavior. The first is the fact that iron undergoes an allotropic change at 1330°F. Carbon has little solubility in the body-centered cubic lattice which is characteristic of iron at room temperature, but up to 2 percent carbon is soluble in the face-centered cubic lattice which is stable above 1330°F. The second is the solid state eutectoid reaction in which a solid solution at a certain temperature can react to form two new solid phases. In the case of carbon steels, the gamma phase, austenite, transforms to alpha iron (ferrite and cementite, Fe_3C) by the eutectoid reaction. The eutectoid reaction is common in many materials, but only steel exhibits such a marked change in properties.

Thus when iron with 0.8 percent carbon is heated above 1330°F, the carbon dissolves and the resulting solid solution is called austenite or gamma iron. If the austenite is suddenly quenched in water the carbon cannot escape and thus is trapped within the lattice structure as an interstitial atom which strains the lattice because of the increased volume it must occupy in the new body-centered tetragonal structure called martensite, which is characterized by a needlelike microstructure (Fig. 5-4). Martensite is a hard, brittle, metastable structure, i.e., a nonequilibrium structure, which is a supersaturated solid solution of cementite, Fe_3C, in a body-centered tetragonal iron. In the presence of a moderate temperature rise (300 to 400°F), tempering or recovery occurs and the brittle structure becomes much tougher, while still retaining its strength and hardness.

Austenitizing

Most ferrous heat treatments require that austenite be produced as the first step in a heat-treating operation. The iron-carbon phase diagram of Fig. 5-5 shows the minimum temperature at which austenite can form, but about 100°F more temperature is required if a reasonable austenitizing time is desired. As shown in Fig. 5-6, the austenitizing temperature varies with the carbon content, decreasing along line A_3 to a_1 at 0.8% C and increasing again along the A_{Cm} line. Austenitizing is a function of both time and temperature. In practice a soaking time of 1 h/in of cross section is considered to be adequate for austenitizing a

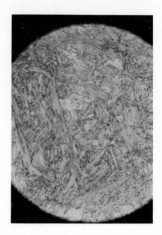

FIGURE 5-4
The microstructure of martensite.

carbon steel, although temperature and initial carbide particle size are both important factors.

HEAT TREATMENT OF FERROUS MATERIALS

All heat treatments either soften or harden a metal. The most common treatments for softening are stress relieving, annealing, and normalizing; while the major hardening processes are case hardening or surface hardening, and through hardening. The softening heat treatments for steel will be presented first.

Stress Relieving

Stress relieving is heating to a temperature below the critical temperature (1100 to 1200°F for low-carbon steel) and cooling slowly. It may be thought of as a subcritical anneal. This process is particularly applicable after straightening and cold-working operations and either prior to or after heat treatments to reduce distortion. In stress relieving, no change in microstructure is involved, but residual stresses are markedly reduced and toughness is improved.

When consideration is given to the simplicity of the stress-relieving operation and the equipment required, this important process, which can save considerable amount of straightening and reworking time, cannot be overlooked. It may be applied to relieve stresses induced by casting, quenching, normalizing, machining, cold working, or welding.

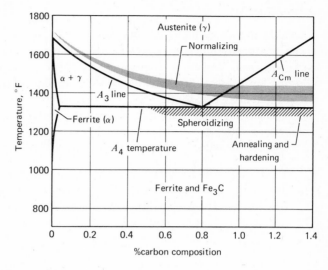

FIGURE 5-5
A portion of the iron-carbon equilibrium diagram showing the temperature ranges usually employed in various heat treatments.

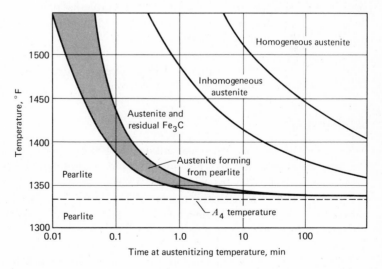

FIGURE 5-6
Approximate times necessary for the isothermal formation of austenite in a normalized eutectoid steel at various austenitizing temperatures. (*After Roberts and Mehl.*)

Annealing

Annealing is heating iron-base alloys 50 to 100°F above the critical temperature range, holding for a period of time to ensure uniform temperature throughout the part, and then allowing to cool slowly by keeping the parts in the furnace and allowing both to cool. The cooling method should not permit one portion of the part to cool more rapidly than another portion. Its purpose is to remove stresses; induce softness; alter ductility, toughness, and magnetic properties; change grain size; remove gases; and produce a definite microstructure.

Steel is annealed for one or more of the following reasons:

1 To soften it for machining or fabrication operations
2 To relieve stresses in the material after casting or welding in some cases
3 To alter the properties of the material
4 To condition the material for subsequent heat treatments or cold work
5 To refine grain size and improve ductility

Perhaps the principal reason for annealing is to give optimum machinability characteristics to the material. It is not possible to present a general rule for annealing that can be used on all types of steel subject to diversified metal-removal operations. For heavy roughing cuts the material should be as soft (Rockwell B80) as possible. This is especially true where accuracy and finish are not important; however, if a close tolerance is required, as on a broaching operation, then a finer lamellar microstructure with a hardness of approximately Rockwell B100 would be more desirable. The principal point for the production-design engineer to keep in mind when using medium carbon steel is to specify the condition of the steel as purchased.

Localized or spot annealing can be accomplished with an oxyacetylene torch and is a valuable technique when salvaging scrap work, but tool steels should not be torch-annealed; they will crack.

Normalizing

Normalizing is heating iron-base alloys to 100 to 200°F above the critical temperature range, and then cooling in still air at ordinary temperatures. Normalizing is performed in order to relieve internal stresses resulting from previous operations and to improve the mechanical properties of the material. For example, the hammer stresses developed in a forging need to be relieved prior to machining. Normalizing is closely related to annealing and produces similar results.

Gears, bolts, nuts, washers, and other parts in which low distortion is an important criterion should be made in the following general operation sequence.

1 Rough-machine
2 Normalize

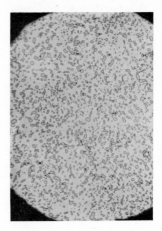

FIGURE 5-7
The microscopic structure called spheroidite.

3 Finish-machine
4 Carburize
5 Heat-treat
6 Grind

Toughness, i.e., resistance to impact or fatigue, is generally applied to medium-carbon steels where improvement in toughness over the "as-rolled" stock is desired. Normalizing refines the structure and eliminates the carbide network at the grain boundaries.

Spheroidizing

Spheroidizing produces globular carbides in the iron (Fig. 5-7). The iron-base alloy is held for a prolonged period of time (10 to 12 h) at a temperature near but slightly below the critical temperature, and then slowly cooled. A spheroidized steel has minimum hardness and maximum ductility. The structure improves machinability markedly. Normalized steels are one of the better starting spheroidized materials because their fine initial carbide size accelerates spheroidization.

Hardness and Hardenability

The maximum hardness of a steel is a function of its carbon content (Fig. 5-8). Although alloying elements such as vanadium and chromium increase the rate at which the martensite transformation occurs and thus the depth to which full

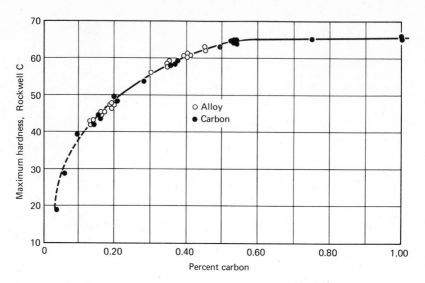

FIGURE 5-8
Maximum hardness versus carbon content. (*From "Quantitative Hardening,"* by *J. L. Burns, T. L. Moore, R. S. Archer,* Transactions ASM, *vol. XXVI,* 1938. *Courtesy American Society for Metals.*)

hardness can be achieved, no alloy steel can exceed the hardness of SAE 1055 steel.

Hardenability refers to the distance within a specimen normal to its surface that appreciable hardness can be developed. Since the quenching rate of a steel is limited by its heat diffusivity and the rate at which austenite can transform, the hardenability of steels depends on the alloy content and the grain size of the original austenite. Hot-work steel die blocks, which must be hardened throughout their mass, are alloyed to such an extent that they can transform to a bainitic structure with an air quench. The hardenability of a typical alloy steel and a plain carbon steel are compared in Fig. 5-9. Note that for all sections above $\frac{1}{2}$ in in diameter the plain carbon steel could not be quenched sufficiently rapidly to achieve full hardness even at the outside diameter.

A standard test for checking hardenability is the Jominy test. It consists of heating a 1-in bar to its austenitizing temperature and then setting it over a jet of water which hits only the bottom face. Consequently, along the length of the specimen there are various cooling rates; later, hardness values can be measured along the side of the bar, representing cooling rates which vary from that of a full water spray down to an air cool. This is a wide-range test, yet its results can be correlated with tests on fully quenched bars, as well as with work on actual parts.

Another useful fact may be mentioned. A gear or a shaft made from a

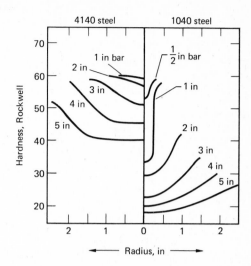

FIGURE 5-9
Comparison of the hardenabilities of 4140 and 1040 steels.

particular steel may be found to have at some point, a Rockwell hardness of C50. A Jominy hardenability test is then run, and it is found that C50 is $1\frac{9}{16}$ in from the water-cooled end. Any other steel which on a Jominy hardenability test will show C50 at $1\frac{9}{16}$ in from the quenched end will produce a similar hardness distribution, when it is used to make the same part.

Table 5-1, collected from various sources, shows the sizes of bars that are fully hardenable, that is, that will develop a hardness of at least Rockwell C50 at the center when quenched as indicated.

Table 5-1   SIZE OF BARS FULLY HARDENABLE (ROCKWELL C50
MINIMUM AT CENTER)

SAE	Water quench, in	Oil quench, in	SAE	Water quench, in	Oil quench, in
1050	$\frac{3}{4}$–1	....	3130	$1\frac{1}{8}$	$\frac{9}{16}$
1330	$1\frac{1}{4}$	$\frac{11}{16}$	3140	$1\frac{5}{8}$	1
1340	$1\frac{1}{4}$	$\frac{11}{16}$	X-3140	$2\frac{1}{4}$	$1\frac{3}{4}$
2330	1	$\frac{1}{2}$	4150	$3\frac{1}{2}$	$2\frac{1}{2}$
2340	$1\frac{3}{8}$	$\frac{7}{8}$	X-4340	6	4
5130	1	$\frac{1}{2}$	6150	$1\frac{3}{8}$	$\frac{7}{8}$
5145	$1\frac{1}{2}$	$\frac{7}{8}$	3240	4	3

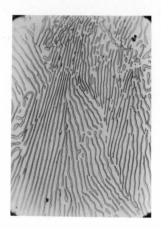

FIGURE 5-10
The microstructure of pearlite.

These are interesting figures, particularly when correlated with steel prices. You will see that SAE 2340 (more expensive than 4150) is much less hardenable. It is also interesting to note that the mechanical properties obtained for 4150 are just as good. The 1300 steels are practically as good as 2300 (3.5 percent nickel) or as 5100 chromium steels in the matter of depth of hardening; they are almost as good as nickel-chromium steels.

Microstructure of Heat-treated Steel

Today there is a broad variety of heat-treating equipment providing various ways of heating the steel, quenching it, and tempering it. Temperature control of the various stages can be assured; thus, the production engineer can depend upon accurate and reproducible control of heat-treated parts after he has established a sound heat-treating cycle.

The purpose of heat treatment is to change the form in which the carbon is distributed in the steel. Alloys present in a steel will affect the rate at which the reactions of heat treatment occur, but have little effect on the tensile properties of the steel. In ordinary carbon steel that is not hardened, the carbon is present in either flakes or rodlike particles which can be readily discerned with the microscope. The structure having the rod or parallel-plate appearance is known as pearlite (Fig. 5-10), while the flakelike structure is referred to as spheroidite (Fig. 5-7). When steel such as this is heated to a relatively high temperature (1500°F), the carbon particles become absorbed by the surrounding ferrous structure and a solution of carbon in iron known as austenite is formed

FIGURE 5-11
The microstructure of austenite.

(Fig. 5-11). As austenite is cooled, the carbon tends to separate from the solution and return to its original form. However, by controlling the rate of cooling, the return to pearlite and spheroidite may be avoided. For example, if the austenite of an 0.80 percent carbon steel is cooled from 1500°F down to 1200°F and is then allowed to remain at 1200°F until it transforms, it will take the form of pearlite, as illustrated in Fig. 5-10. This steel would be quite soft, having a Brinell hardness of about 200. However, if the austenite is cooled quickly to a lower temperature, say 600°F, it will have escaped the 1200°F transformation stage, and will change to a structure known as bainite, which has a hardness of about 550 Brinell. If the austenite is cooled to a still lower temperature, for example 250°F, before transformation takes place, the very hard structure known as martensite will be formed. Martensite gives a Brinell of about 650 (Fig. 5-4).

Figure 5-12 gives information on the hardness of the transformation products of coarse pearlite to martensite. An important rule that the production engineer should keep in mind is that, when forming any particular transformation product, the steel must thus escape transformation at a higher temperature. If this is not observed, then the desired transformation product will not be achieved. From the foregoing, it can be seen that a hardened piece of steel takes the form of martensite or bainite. The formation of pearlite must be avoided. Thus, the process of heat treating involves first heating and soaking at the correct temperature, then cooling the steel rapidly enough to avoid the formation of pearlite, and finally holding at the desired temperature to form either bainite or martensite. Figure 5-12 illustrates the transformation time in seconds required by austenite at different temperatures. This chart is known as a TTT curve (time, temperature, transformation). This isothermal transformation dia-

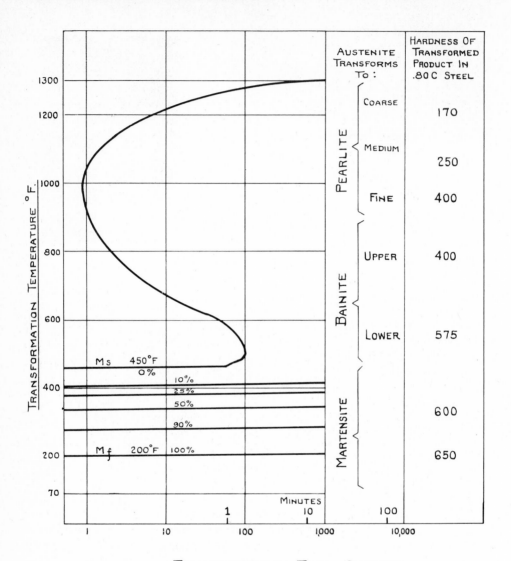

FIGURE 5-12
Time required by austenite to transform at different temperatures (TTT curve or isothermal transformation diagram). (*Courtesy United States Steel Co.*)

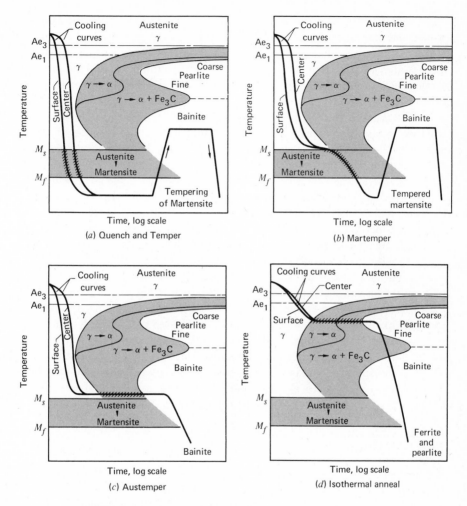

FIGURE 5-13
Various heat treatments as related to the TTT diagram.

gram is very useful, as it shows the various structures formed by a steel during its cooling period or formed while being held at a temperature to which it was cooled.

A study of this diagram will indicate the importance of cooling time in order to avoid the formation of the soft pearlite. Small parts that cool rapidly (dwell in the 1000°F zone less than 1 to 3 s) are readily transformed to either bainite or martensite and escape the pearlite formation. However, in large sections, where the cooling period is longer, it would be necessary to use a steel which had the nose of its TTT curve moved to the right so there would be a longer period of cooling time in order to avoid the formation of pearlite. When steel is alloyed with certain elements such as chromium or molybdenum, thick sections can be hardened even in the center. Such steels are referred to as having greater hardenability.

Toughness, or the ability to deform slightly under load rather than to fracture, is another reason for heat treating. Toughness is brought about by changing the shape of the individual carbide particles from platelike flakes to spheroidal shape. Tempering of steel results in this change which gives it toughness characteristics. The degree of toughness varies with the steel being treated. A high-carbon steel given the same heat treatment and tempering will result in a stronger but less tough steel than a low-carbon steel that has undergone the same treatment. In treating steel the production-design engineer must realize that a gain in one characteristic such as hardness results in a loss in another characteristic such as toughness.

From this, it can be seen that the effect of lowering the carbon content is to induce more rapid transformation rates. The effect of alloys is to shift the entire isothermal transformation diagram to the right; that is, transformation at all temperature levels starts later and is slower to go to completion. It should be recognized that although this is a general characteristic of alloying, various alloys differ substantially in both the magnitude and the nature of their effect. That is, the shape of the isothermal curve will vary considerably as well as its location with respect to the time-temperature axes. Several heat treatments are shown schematically on TTT diagrams in Fig. 5-13.

Effects of Grain Size

When steel is heated through the critical range (approximately 1350 to 1600°F) transformation to austenite takes place as mentioned. When first formed the austenite grains are small, but they grow in size as the temperature above the critical range is increased and, to a limited extent, as the time is increased.

When temperatures are raised materially above the critical range, different steels show wide variations in grain size, depending upon the chemical composition of the steel and the deoxidation practice used in making the heat. Heats are usually deoxidized with aluminum, ferrosilicon, or a combination of other deoxidizing elements.

Fine-grain steels do not harden as deeply as coarse-grain steels, and they are more resistant to cracking upon quenching after heat treatment; they tend to be tougher and have better application under impact load conditions. However, coarse-grain steels produce a discontinuous chip permitting a better machined finish and less tendency toward a built-up edge, thereby making them more applicable on designs requiring intricate machining.

Increasing the grain size of austenite lengthens the time for both the beginning and completion of transformation.

Through Hardening

Through hardening consists of heating the part to some point above the critical temperature and then quenching. This is frequently followed by tempering in order to instill toughness to the part (Fig. 5-13a). Through hardening is generally applied to medium- and high-carbon steels in order to develop a particular combination of both hardness and toughness.

Oven furnaces, pit furnaces, and liquid baths are used for through hardening, with the oven type being the most popular. The furnaces may be heated by gas, oil, or electrical resistance elements. On smaller parts, induction heating and flame heating may be used.

In order to ensure that the correct heat treatment will be achieved, the production engineer should verify that the "soaking" period in both the heat and the draw are adequate. In the heating period, it is essential that the part be at the correct hardening temperature so that full hardness throughout the section is achieved. Also it is important that the atmosphere of the furnace be controlled so that the decarburization of the surface will not be greater than anticipated.

In order to achieve the required toughness on a through-hardened piece, it is usually necessary to temper. This is accomplished by reheating the part to a temperature between 300 and 1000°F (depending on the alloy and toughness required) and then cooling.

Drawing furnaces include lead baths, salt, oils, and forced-air circulation. The latter type can be made to cover a wider range of drawing temperatures than other drawing mediums and are therefore quite popular in view of the wider variety of work that can be handled. Drawing oils are used from 300 to 600°F. Oils have the disadvantage of requiring cleaning of the parts once they are extracted from the quench tank. Drawing salts are usually used when drawing at higher temperatures (500 to 1000°F). Salts have the advantage of holding the oxidation of high-temperature draws to a minimum. The principal disadvantage of drawing salts lies in the fact that the salt must be boiled or washed off the work once it is removed from the furnace. Lead baths are also frequently used. When using a lead bath, care must be exercised in placing the part in the bath so as to avoid cracking. Lead is a rapid heating medium and the through-hardened piece may not be able to withstand the quick immersion in the molten lead. Then, too, lead tends to adhere to the parts when removed from the bath, especially if deep recesses prevail. This is frequently objectionable.

The length of time that the part should be held at drawing heat is also important. Generally speaking, the higher the alloy content, the longer should be the draw. For example, carburized SAE 1020 steel should be drawn at 350°F for 1 h; parts of 2315 or 4615 steel should be drawn at 350°F for a period of $1\frac{1}{2}$ to 2 h.

Quenching is an important factor in the through hardening of a piece of steel. The most common quenching media are water, oil, and air. Water usually is used on the plain carbon steels that require fast cooling in order to achieve the highest hardness. Water quenching is quite severe and should not be used on intricate shapes and contours where cracking at the time of quench may be induced. Oil quenching is considerably milder than water quenching and is used usually with alloy steels and on intricate shapes. Cooling in an air blast is used on the higher-alloy steels such as some of the tool steels.

Interrupted-quench treatments are used when it is necessary to keep distortion to a minimum and guard against cracking. The most common interrupted-quench treatments are martempering, austempering, and isothermal anneal (Fig. 5-13b to d).

Martempering is quenching a properly heated piece of steel in a salt bath which is maintained at a temperature between 300 and 500°F, and then holding the part in the bath until the temperature is equalized throughout the piece (Fig. 5-13b). The part is then removed from the salt bath and air-cooled. Following this, it is drawn to the desired hardness. This process, being less severe than quenching in cool oil, permits the structure of the part to change more readily with fewer internal strains. This treatment is most effective on the oil-quenching steels, although water-hardening steels are handled also in this manner. In martempering of water-hardening steels, the parts are first quenched in brine to a point somewhere between the transformation point and the lowest temperature of the critical-cooling curve. The parts are then rapidly transferred to a bath the temperature of which is slightly above the transformation point.

Austempering is similar to martempering except no draw after quenching in the salt bath is performed. In austempering, the salt bath is maintained between 400 and 850°F, depending upon the section thickness. This process is used on sections of alloy and tool steel not over 1-in thick.

Isothermal transformation is a method that gives similar characteristics to the austempering technique on larger sizes (up to 2 in). Here the part is quenched in an agitated salt bath between 450 and 600°F, and held to permit isothermal transformation. The part is then transferred to another bath at a higher temperature in order to draw the piece (Fig. 5-13d).

SURFACE HARDENING

Case Hardening

Case hardening is the hardening of a surface layer of a part made from a low-carbon steel so that this outside layer is substantially harder than the interior.

This is accomplished by one of several processes that call for addition of either carbon or nitrogen to the part being hardened. The case-hardening processes include carburizing, cyaniding, carbonitriding, and nitriding. Flame hardening and induction hardening may be used to austenitize the surface carbon prior to quenching.

Carburizing involves the heating of the steel part to between 1600 and 1850°F in the presence of either a solid carbonaceous material such as charcoal, a carbon-rich gaseous atmosphere, or a liquid salt. This process is usually applied to steels with low carbon content. Open-fired or semimuffle gas furnaces of the continuous type or electric and gas-fired bell-type furnaces have been designed for controlled-atmosphere carburizing. In all carburizing processes, the amount and depth of case is determined by temperature and time. The greatest amount of control is obtained with the gas-carburizing technique. This method also permits the greatest depth of case. Cases between 0.30 and 0.40 in are readily obtained by the gas-diffusion method. Case depths up to about 0.070 in are obtained by pack carburizing.

The salt-bath method of carburizing is particularly adapted to small parts where many components may be immersed and treated simultaneously. A depth of case up to 0.030 in is obtained by this method.

Cyaniding is performed at temperatures between 1500 and 1650°F in either salt baths, electric or radiant-tube furnaces, or in electric or gas-fired bell-type furnaces. The cyaniding technique provides a case of high hardness and good resistance to wear. The process is the result of the formation of iron nitrides in the case brought about by the release of nitrogen from the bath. Cyaniding is used effectively for case depths up to 0.025 in. It is used especially on smaller parts that can be handled in salt-bath equipment without mechanical handling equipment.

Nitriding involves subjecting the parts to be hardened to the action of ammonia gas at temperatures of 920 to 980°F, so as to introduce nitrogen to the surface of the parts. Since that temperature is below the normal or critical hardening temperature of the steels being nitrided, it is said to be a "subcritical" hardening operation. Sealed retort furnaces with close temperature control are required for the process. The time of exposure is relatively long, 72 h being required for about 0.025-in case depth. No quenching is necessary in nitriding since the hard case is formed by the nitrides developed from the combination of nitrogen, iron, and the various alloying elements in the steel. This fact, combined with the low temperature used, minimizes distortion. This case-hardening process is usually used on special Nitralloy steels. Other steels, however, are nitrided successfully. A typical nitrided steel would be a chromium-aluminum-molybdenum steel in which a 50-h cycle will give a 0.015-in case. The hardness of a nitrided surface ranges from 60 to 72 Rockwell C and the hard, tough nitrided case is from 0.002 to 0.035 in deep.

Care must be exercised in using hardness-testing devices on case-hardened parts. Tests may be misleading if they puncture the hardened surface.

Light versus heavy cases   Under most conditions, there is better wear resistance from the first few thousandths of the heavy case than there is for the light case, for this reason: When a heavy case is produced, the maximum carbon at the surface is higher; in a shallow case, the maximum carbon is lower because the migration of the carbon away from the surface is so fast that the surface carbon does not reach the maximum value. Better wear resistance is generally obtained from a carbon steel having 1.25 percent carbon than from another with 0.90 percent carbon; 1.25 percent surface carbon is to be expected after 8 h, but after 1 h, not more than 0.80 or 0.90 percent.

To summarize, the principal situations that warrant a heavy case are:

1   To withstand a heavy load, so that the case will not collapse. The danger of case collapse is a function of the load and also of the hardness of the material under the case. If the case is thin relative to the load, and if the core is soft relative to the case depth, trouble may develop.

2   To withstand severe wear. This applies only if the part can lose dimension without losing its usefulness. It is useless to put a 0.075-in case on a gage, for example, if it is discarded after wearing 0.0005 in. However, if the part is a link pin, which will wear under sandy conditions and still be good until the day it breaks, the more case put on it, the more life it will have.

3   Because of grinding to be done after heat treatment. Any case that is taken off by grinding is not only heat-treatment-wasted, but requires extra expense to remove; therefore, excessive finish grinding should be avoided. The part ground off is the most valuable part of the case as regards hardness and wear resistance. Not infrequently a heavy case is specified and then too much stock has been allowed for grinding. What is left is a very poor case.

4   To build up the mechanical properties of the surface. It should be emphasized that hardness means tensile strength; the reverse is also true—tensile strength in steel means hardness. If a high tensile strength is needed to give a high endurance limit at the surface, a deep case is required.

Properties of the core   There is still another factor in the selection of case depth—the balance between core properties and case depth. If wear resistance of the outer part of the steel is to be increased, and resistance to crushing is also required, there are two ways to do it: (1) by putting a heavy case on a poorly hardened steel, (2) by putting a light case on steel that has already hardened fairly well without a case.

The automotive industry has arrived at a compromise between the older plans of carburizing a low-carbon steel or merely oil quenching an alloy steel containing 0.50 percent carbon. The new procedure is to cyanide or lightly case-carburize a steel with perhaps 0.40 percent carbon. The plan is to get from 0.006 to 0.010 in of carburized case, and yet to have the core quench out to

Rockwell C50. A gear is economically produced in this manner because the case depth is shallow, the steel blank is relatively easy to machine because of its intermediate carbon, and the gear offers excellent resistance to crushing because the core is stronger, yet the gear teeth have desirable surface hardness. This compromise is an indication of the current trend toward use of intermediate steels case-hardened to shallow depths rather than using extremely heavy cases.

Woodvine diagram   Strength of the metal below the case is important, as indicated in Fig. 5-14 known as the Woodvine diagram. This shows the stresses on a piece which withstands bending. Line OA indicates that from the neutral axis (center) of a bar which is being stressed by bending toward the surface, the imposed stress increases from zero at the center to a maximum at the surface. In this particular example, a diameter of 0.3 in was selected and the stress on the outermost fiber assumed to be about 45,000 lb/in². If the case is on a steel part which has a stress resistance of 47,000 lb/in² (plotted as line CB), obviously it is safe at the surface. If the case extends inward to a distance equivalent to 0.045 in, the strength of this outer region is indicated as BP, still on the safe side of the requirements. At the bottom of the case a fairly rapid transition can be assumed to the unaffected core with its stress resistance of only 32,500 lb/in² (distance OG on Fig. 5-14). The metal here is barely strong enough to stand the stress, yet it probably would "get by."

Now suppose a light case were used. The effect would be to move line QP upward, since the stress resistance of the core is approached more rapidly. Point Q would then be on the wrong side, i.e., the upper side of line OA:

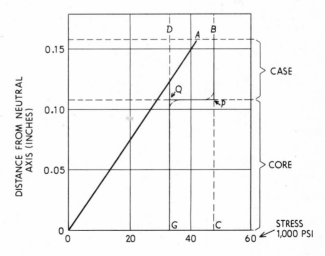

FIGURE 5-14
Woodvine diagram.

there would be less stress resistance at the junction between the case and core than the magnitude of the applied stress, and a subsurface fatigue failure could result if the bar were subjected to alternating stresses.

Figure 5-14 might also be used to explain the use of a shallow case on a toughened core. Line QP would be closer to DB, but line GQD would move over to the right because the stress resistance of the core would have a higher value. If a shallow case is used on a poor-hardening core, a subsurface failure can be expected, whereas with a stress-resistant core, such a failure will be avoided.

A higher core hardness produces a more impact-resistant gear tooth. In this connection, nickel-chromium-molybdenum alloy steels have proven to be superior to all others. In some cases it is possible to harden the case and to toughen the core in a single heat treatment. If there is sufficient carbon in the alloy, the gear teeth can be flame- or induction-heated and quenched, leaving the core with the properties produced at the steel mill.

One precaution should be noted: the harder a steel, the more notch sensitive it becomes. A case-hardened surface has maximum hardness and as a result a maximum sensitivity to nicks and tool marks. Every effort must be made to provide ground or lapped surfaces and blended radii at all surface intersections to ensure maximum impact strength.

Induction Hardening

The production engineer should be familiar with induction heating since it has so many applications. The induction-heating circuit is essentially a transformer where the conductor carrying a supply of alternating current is the primary, and the material to be heated is the secondary (Fig. 5-15). The current flowing through the inductor (copper tubing in the form of coils) sets up magnetic lines of force in a circular pattern which pass through the surface of the work. A flow of energy is therefore induced and internal molecular friction is developed in the work which dissipates itself in the form of heat. This heating effect is known as hysteresis. Heat is also generated in the work by eddy currents set up by the I^2R loss. The work is heated electrically by the alternating current, although the work itself does not come in contact with the power source. Thus, it may be seen that any conductor will become heated when placed within the immediate area of a conductor that is carrying an alternating current.

In induction hardening, electrical frequencies from 1,000 to 500,000 cycles are employed. Very high frequencies, 3,000 to 500,000, cause the current to remain on the surface of the part; therefore, only the surface is heated. Lower frequencies, from 60 to 2,000 cycles, are used for heating the part to greater depth. When a variety of parts are to be hardened, the heat-inducing coils must be changed to accommodate the particular part being hardened. The induction-heating machines on the market readily permit coil replacement. Induction

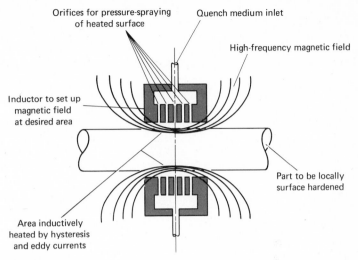

FIGURE 5-15
Schematic diagram of an induction heat-treatment device.

hardening is a fast, reliable method for surface hardening; however, since specially designed coils are necessary to accommodate different parts, it is usually economical to use on production runs only.

The equipment may be automatic or it may be operated manually. When heated to the proper temperature, the part can be quenched by a spray or stream of liquid, or it can be dropped into a tank. Little distortion results, and a uniformly hardened surface is achieved without affecting the properties of the core. To ensure good control, a temperature-controlled quenching liquid is often forced directly through the inductor elements to quench the surface as soon as it is austenized.

Induction heating is applied to radio tubes as they are evacuated. Very high frequencies are used to obtain dielectric heating of nonmetallic parts. Plywood adhesives are cured by heating in a dielectric field; thus, a thermosetting plastic can be used to give a moisture-resisting bond. Induction heating is applied to soldering and brazing. It is quick and clean. The heat can be applied where desired, and distortion or excessive losses of heat are avoided. This process is especially useful on small parts.

Flame Hardening

Flame hardening is used on small lots in view of its adaptability and freedom from special tooling and setup. This method involves the rapid heating of the

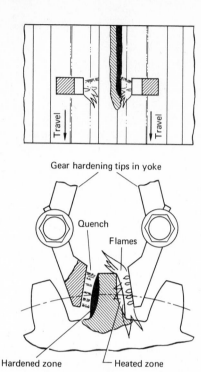

FIGURE 5-16
Flame-hardening setup for gear teeth. (*Redrawn from Doyle et al., "Manufacturing Processes and Materials for Engineers," 2d ed., Wiley, New York, 1969.*)

part by means of a torch or torches followed by a quench (Fig. 5-16). The part may be moved under the nozzles or the nozzles may move over the part. The shape and position of the nozzles are modified to suit the shape of parts, such as gear teeth or flat or V-shaped lathe bedways.

The speed of travel is 6 to 8 in/min and the penetration is $\frac{1}{8}$ to $\frac{3}{16}$ in deep. Machines are made fully automatic or semiautomatic. SAE 1040 to 1070 steels are suitable for flame hardening. Surface cracks may occur on high-carbon steels and the quenching rate must be reduced. Cast-iron lathe and machine toolways are flame-hardened to give a hard-wearing surface. Little distortion occurs, consequently little machine finishing is required. Wearing surfaces can be hardened without affecting the rest of the part. Crankshaft bearing surfaces, cams, gear teeth, and machine parts can be hardened (Fig. 5-17). Large gears, 20 or more feet in diameter, and large shafts, 20 or more feet in length, can be flame-hardened; thus, the danger of cracking is avoided and the use of large heating and quenching equipment is unnecessary.

Once a decision has been made to flame-harden a part and the most suitable

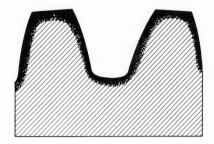

FIGURE 5-17
Section of a carburized gear tooth after hardening.

process has been selected, the next step should be the determination of the depth of case required. In the interest of economy, the case should be no deeper than necessary. The amount of finish grinding needed after heat treatment must be considered. A part that is subject to considerable abrasion, pressure, or pounding should have a relatively deep case (0.050 to 0.075 in). It should be remembered that too deep a case will result in cracking or checking on parts that are subjected to elevated temperatures. This is due to the lack of flexibility of a hard, deep case.

Where localized case hardening is required, the piece may be copperplated on all sections except where the hard case is required, provided that a carburizing grade of base metal was specified. The copperplating will prevent the part from absorbing any carbon in the plated areas. Another method would be to carburize the entire part and then machine off the unhardened case in the sections where it was not desired. In this method, allowance would have to be made for the finish-machining operations.

HEAT TREATMENT OF ALUMINUM

Aluminum is the only nonferrous material of structural importance that can be effectively heat-treated to enhance its mechanical properties. The mechanism of this heat treatment is known as precipitation hardening.

Precipitation hardening is possible wherever the phase diagram of the alloy shows a sloping solvus line, as shown in Fig. 5-18, for aluminum. If a particular aluminum alloy, as depicted by the vertical line, is cooled slowly from its molten state to room temperature, first the liquid completely solidifies to a solid phase, alpha, then upon further cooling, a different phase, beta, evolves, to some extent from the alpha phase, so that at room temperature there will be a mixture of phases alpha and beta—the resulting material being rather soft and without internal stresses.

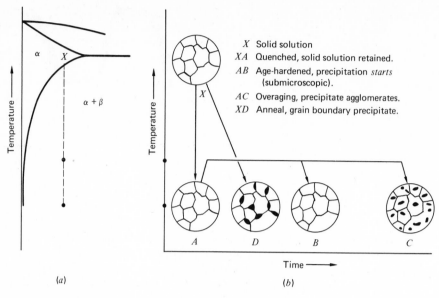

X   Solid solution

XA   Quenched, solid solution retained.

AB   Age-hardened, precipitation *starts* (submicroscopic).

AC   Overaging, precipitate agglomerates.

XD   Anneal, grain boundary precipitate.

(*a*)   (*b*)

FIGURE 5-18

Phase diagram for aluminum and copper.

On the other hand, if the alloy were cooled very fast from a high-temperature zone within the alpha phase and across the solvus line, then the beta phase would not appear, the solid solution being supersaturated at low temperature. The hardness of the material in this condition is slightly higher than in the annealed condition. However, given enough time, some of the beta phase will precipitate out of the supersaturated solution at room temperature and the

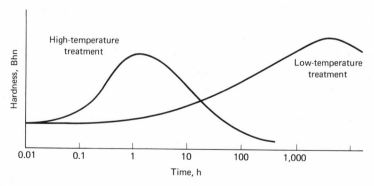

FIGURE 5-19

Variation of hardness with aging time.

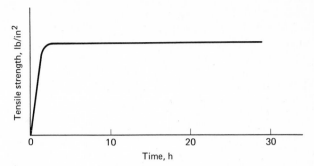

FIGURE 5-20
Increase of tensile strength of aluminum-copper alloy with aging time.

hardness of the alloy will considerably increase. The alloy is then said to be naturally aged. This precipitation can be accelerated by a slight heating to some 200°F; the alloy is then artificially aged (see Fig. 5-19).

As the cooling rates of aluminum castings can vary considerably, yielding anything from a completely annealed material to a supersaturated one, it is necessary to solution heat-treat aluminum castings before their precipitation hardening. This is done in order to establish a standard procedure. In solution heat treatment, the material is heated into the alpha-phase region, close under the solidus line, and kept there long enough to ensure that all the material is in the alpha phase. After this, the material is quenched to obtain the supersaturated solid solution. Then the aluminum is heated to a moderate temperature (about 200°F) and kept there as long as necessary to obtain the precipitation hardening desired.

The tensile strength of aluminum is also enhanced during precipitation heat treatment (Fig. 5-20). Strength, as well as hardness, depends on the aging temperature and time.

Designation of Alloy Treatment

As the properties of an alloy very often depend on its processing history, it is convenient to specify that history by means of code letters. The condition or temper of an alloy is specified by adding a letter to the alloy designation:

Letter	Condition
F	As fabricated
O	Annealed
H	(Cold)-work-hardened
T	Heat-treated

The degree of strain hardening is designated by a number affixed to the initial letter H; e.g., H1, H2, H3, ..., H9. The various tempers produced by heat treatment are denoted by adding an additional number after the T designation:

T2   Annealed (cast products only)
T3   Solution heat-treated and cold-worked
T4   Solution heat-treated and naturally aged
T5   Artificially aged only
T6   Solution heat-treated and artificially aged
T7   Solution heat-treated and stabilized (by overaging)
T8   Solution heat-treated, cold-worked, and artificially aged
T9   Solution heat-treated, artificially aged, and then strain-hardened

HEAT-TREATMENT PROCEDURES FOR ALUMINUM ALLOYS

The heat-treatable aluminum alloys are those that contain elements that have solid solubility at higher temperatures. Included in the heat-treatable group are the copper-bearing alloys, the magnesium silicide type alloys, the magnesium alloys, and the zinc-bearing alloys.

By introducing foreign atoms in the space lattice, resistance to slip will be pronounced. This increased resistance to slip along the slip planes will result in an increase in the inherent strength of the aluminum. This strengthening effect is pyramided by the precipitation of intermetallic compounds along the slip planes as the alloy is cooled. For example, when copper is alloyed with aluminum, it reprecipitates to form the compound $CuAl_2$. The particles of $CuAl_2$ precipitating out at higher temperatures (above 350°F) coalesce into large masses and, consequently, interfere little with slip. However, those particles which precipitate at lower temperatures will remain in sizes comparable to space-lattice distances and cause dislocation pileup, thus retarding slip.

In the heat treatment of aluminum, rapid cooling should take place from the solution temperature in order to maximize the precipitate that most retards slip. This is handled by subjecting the alloy to a severe quench (usually water or oil held at 100°F).

Precipitation begins as soon as the alloy is quenched. The time required for complete precipitation at room temperature may be several days. When precipitation takes place at room temperature, the alloy is said to be naturally aged and is given the designation T4.

By elevating the temperature, it is possible to shorten the period for complete precipitation (that point where the mechanical properties stabilize). This process is referred to as artificial aging. Aluminum alloys that are hardened by thermal precipitation techniques are heated to a temperature ranging from 250 to 375°F, and soaked from 6 to 24 h, depending upon the alloy. After adequate soaking, the alloy is air-cooled.

Annealing

Aluminum can be purchased in the annealed state (O temper) directly from the manufacturer; however, the production engineer will frequently find it necessary to anneal aluminum that has been strain-hardened during forming and requires still further forming. The cold working of the metal may be considered to result from using up the available "slip planes." Thus, an aluminum that is soft is thought of as having a fresh, undisturbed bank of slip planes that allow adjoining lattice planes to slide against each other. When the metal is formed through the application of force, the metal dimensions are changed as slip occurs. The planes available for subsequent slippage will not be located for the most favorable movement, consequently, greater force must be applied in order to bring about an equal amount of cold-forming.

The original workability of the aluminum can be restored by heating it above a certain temperature, known as the recrystallization temperature. This temperature is usually between 300 and 750°F. The application of heat creates fresh slip planes along which the material can move.

Homogenizing

Homogenizing is a procedure that facilitates the proper distribution of the alloying elements throughout a material so that a homogeneous structure is obtained. This is accomplished by heating the metal to a temperature just under its melting point, and then following by a slow cooling process. For most aluminum alloys it is done in the temperature range of 200 to 1000°F.

Heat-treating Equipment

The salt bath and the air furnace are the two types of furnaces used in the heat treatment of aluminum alloys.

Salts baths are heated by electricity, gas, or oil; air furnaces are heated usually either by gas or by electricity. Both types of furnaces should include automatic temperature regulators so that a control of temperature can be maintained. Accurate temperature control is a primary requisite for a furnace intended to be used for the heat treatment of aluminum alloys. The temperature should be held to a 20°F spread between the high and low points of control.

Gas-fired air furnaces should be of the muffle type so that the parts being treated will not come in contact with the products of combustion. These air-conversion-type furnaces permit accurate temperature control in low temperature ranges and consequently are quite popular.

Salt baths are usually made up of a mixture of molten sodium nitrate and potassium nitrate. This combination provides the flexibility for both heat treating and annealing. For heat treating only, a salt bath of just sodium nitrate is suitable. Because of its high melting point, it is not satisfactory for annealing operations. It might also be mentioned that sodium nitrate alone is not as stable as the mixture of the two salts, and tends to corrode the tank as well as the parts being treated.

Procedure for Heat-treating Aluminum

As with ferrous materials, the heat treatment of aluminum involves three distinct steps: heating to a predetermined temperature, soaking at the required temperature for a predetermined interval, and quenching.

Depending on the alloy, the heat-treating temperature for aluminum varies between 900 and 1000°F. It is important that the heat-treating temperature of the specific alloy be accurately maintained in order to arrive at the desired physical properties.

The soaking period will run from a few minutes to more than an hour, depending on the heating medium, the alloy being treated, and the shape and size of the part. The soaking time increases with the thickness of the part. When different-sized pieces are soaked, the cycle should be governed by the part having the heaviest section.

Quenching of aluminum alloys is usually done by immersion in oil or water. A fast quench is considered necessary in order to prevent the precipitation of $CuAl_2$ from coalescing into larger masses. This would lead to corrosion because of electrolytic reaction when the alloy is exposed to salt water. In addition to keeping the time interval of the quench as short as possible, the parts should be thoroughly washed so that all the salt is removed from the surfaces. This step is necessary as a further guard against corrosion.

If the parts being treated are of intricate design, it may not be advisable to subject them to a rapid quench because of the danger of warpage. An alternate method of cooling such parts is to use spray or fog quenching. In both of these methods, a fine spray of water from all directions is thrown on the parts. Neither of these methods is permitted on parts supplied for naval aircraft except by special permission from the Bureau of Aeronautics.

Some warpage can be corrected after hardening, and this straightening should be done immediately after the quench before age hardening takes place. If it is impractical to straighten the parts immediately after quenching, they can be kept under refrigeration, thus delaying age hardening.

SUMMARY

Heat treating of parts is often avoided because the engineer is unfamiliar with the possibilities of heat-treated materials. Tool steels, stainless steels, alloys of iron and nonferrous materials, and aluminum acquire many useful properties under proper heat treatment. The processes must be controlled carefully in the shop; proper equipment, temperatures, quenching mediums, procedures, materials, and skills are necessary for uniform and satisfactory performance. By visualizing the process and what occurs to the part during treatment, the production-design engineer quickly realizes that a quenched part can have internal stresses locked within. This is revealed by cracks occurring during treatment or breakage after a slight load has been applied. Therefore, stress concentrations caused by sharp corners, uneven sections, rapid and uneven cooling, gas pockets, and complicated designs must be avoided by the designer. Many of the principles applied for casting apply to heat-treated parts. Consider the problem of a cube. Which portion of the cube cools the slowest? The temperature gradient is affected by the shape, location, and amount of surface area. This is governed by the design of the part. The heat-treater can help alleviate some of the poor conditions of high-temperature gradients by masking or by using progressive quenching, but the problem of internal strains is present with the best possible form of part.

Typical difficulties that arise during heat-treating processes are

1   Uneven support during heating causing warpage
2   Parts having variable surface characteristics may warp when quenched
3   Warpage frequently resulting from quenching at too high a temperature, even though the parts are within the allowable temperature range
4   Low overall hardness occurring because of poor circulation in the quenching medium
5   An even penetration of case retarded by soot deposits during gas carburizing

Some general rules for quenching, tempering, and austempering, as developed by the U.S. Steel Corporation, that will provide helpful guideposts to the production design engineer are:

1   The useful structures in hardened pieces are bainite and tempered martensite. To obtain either of these structures, the piece is heated to secure adequate solution of carbides, and then quenched so rapidly that the formation of pearlite or upper bainite in the 1000°F zone is avoided, bainite or martensite being formed at lower temperatures.
2   When quenching a specific size of piece, the avoidance of pearlite is made easier by lengthening the time of possible formation. The slower rate of pearlite formation is obtained by adding alloys. If the piece is fully hardened (martensitic), the hardness is affected very little by the alloys, being governed almost exclusively by the carbon content.
3   For a given size of piece, avoidance of pearlite or upper bainite is also aided by employing a greater severity of quench. This is obtained by avoiding excess scale on the piece, by employing the quenching liquids

that generally show a greater severity of quench, and by agitating either the quenching liquid or the piece being quenched. It is well to remember that the old method of hand agitation during quenching was very effective, and it may be desirable to duplicate it in modern mechanized equipment.
4 When measuring toughness, remember that the results are affected profoundly, not only by the magnitude of the stress imposed, but also by its nature and direction and, consequently, also by the shape of the piece. A laboratory test should always simulate a service test as closely as possible.
5 Toughness in martensitic structures is obtained by tempering the quenched piece.
6 Bainite structures, obtained in austempering, are inherently tough and, when produced with 48 Rockwell C or above, offer a combination of strength and toughness superior even to tempered martensite.

QUESTIONS

1 What are the most common forms of heat treating?
2 Define the following terms: stress relieving, annealing, normalizing, spheroidizing, quenching, hardening, tempering, and carburizing.
3 For what reasons is annealing performed?
4 In finishing materials requiring a close tolerance, to what hardness would you recommend bringing the material?
5 Give the operation sequence on products requiring low distortion, close tolerance, and fine finishes.
6 To what processes may stress relieving be advantageously applied?
7 Where would you recommend the use of Nitralloy steels?
8 Explain the process of induction heating.
9 What advantages are offered by the flame-hardening process when compared to induction hardening?
10 What methods may be used to localize case hardening?
11 What is meant by "soaking time"?
12 On what type of steels is water preferred as a quenching medium?
13 Explain the martempering process.
14 Distinguish between pearlite and spheroidite.
15 Explain the importance of time in the cooling stage of heat treatment.
16 What is meant by "slip planes"?
17 How can the workability of aluminum be restored after a deep-drawing operation?
18 What is meant by precipitation when referred to aluminum alloys?
19 What is homogenizing?
20 What six general rules for quenching, tempering, and austempering of ferrous materials have been established by the U.S. Steel Corporation?

PROBLEMS

1 To provide a tough core and a hardenable surface in a carbon steel member, it is possible to diffuse carbon into a low carbon steel by a carburizing process. Since only carbon solid solution can take part in diffusion, the maximum effective carbon concentration at the surface of a steel is determined by the solubility of carbon in austenite. At 1700°F this value is approximately 1.3 percent by weight. It is desired to carburize 18-in lengths of 1 in round SAE 1010 steel. Plot the carbon penetration curve (carbon content versus distance from the surface of the work) that would be produced by carburizing for 3 h at 1700°F. Determine the time required to obtain a carbon content of 0.6 percent at a depth of 0.05 in. The diffusion constant for carbon in austenite may be estimated from

$$D = (0.07 + 0.06)(\text{weight percent C}) \exp \frac{-3,200}{RT} \quad \text{cm}^2/\text{s}$$

where R (gas constant) $= 1.987$ cal/(°K)(g mol)

T = temperature, °K

Make whatever assumptions you feel necessary.

2 A casting of plain carbon steel is specified for a wear-resistant application. Outline the proper heat treatment for the optimum carbon content to give the best microstructure for a tough core and maximum wear resistance on the surface, after the casting has been cooled to room temperature and has passed through the cleaning room. Give the temperatures, quenching media cooling times, and cooling rates.

3 In a certain corrosive atmosphere only a copper-nickel alloy has been considered suitable. Such alloys form substitutional solid solutions in all proportions. Choose the least cost alloy and its proper diameter to support a 5,000-lb tensile load without yielding.

4 An annealed copper wire which was 0.010 in OD was cold-drawn to 0.008 in OD, what is its present tensile strength after that amount of cold work?

5 A severely cold-worked metal is 50 percent recrystallized after each of the following heat treatments: 1 min at 162°C, 10 min at 138°C, and 100 min at 97°C. Estimate the temperature which would be required for 50 percent recrystallization in 10^4 min by treating recrystallization as a self-diffusion process.

SELECTED REFERENCES

AMERICAN SOCIETY FOR METALS: "Metals Handbook," Metals Park, Ohio, 1948.

————: "Metals Handbook," vol. 2, "Heat Treating, Cleaning and Finishing," Metals Park, Ohio, 1964.

BULLENS, D. K.: "Steel and Its Heat Treatment," Wiley, New York, 1948.

GROSSMAN, M. A.: "Principles of Heat Treatment," American Society for Metals, Metals Park, Ohio, 1953.

———— and E. C. BAIN: "Principles of Heat Treatment," American Society for Metals, Metals Park, Ohio, 1964.

GUY, ALBERT G.: "Elements of Physical Metallurgy," 2d ed., Addison-Wesley, Reading, Mass., 1960.

TRIGGER, K. J., and S. RAMALINGAM: "Metals Principles Treatment and Selection," Campus Book Store, Champaign, Ill., 1966.

U.S. STEEL CORP.: "Atlas of Isothermal Transformation Diagrams," Pittsburgh, Pa., 1951.

FERROUS METALS

Ferrous metals are widely used because they are abundant, they are economical on a basis of cost per unit of strength, and they have unique magnetic properties. Ferrous alloys can be easily formed in the annealed state and then can be heat-treated to be as hard as a file, tough as the hook of a 50-ton crane, or tempered properly to drill holes in other metals.

This chapter presents the types of ferrous materials available to design engineers, the properties of ferrous alloys, and how design objectives can be developed in ferrous alloys.

Ferrous metals are furnished to industry in the form of shapes and plates, bars and tool steels, sheets and strips, castings of iron and steel, and forgings (Table 6-1). Over the past 25 years the production of cast iron, weldments, and sheet products has increased markedly whereas in most other categories there has been relatively little change and the production of forgings and semi-finished steel shapes and plates has decreased. Competition from the casting process and from plastics as well as foreign imports of both steel and automotive components has been responsible for a large part of the decrease.

Ferrous materials can be broadly divided into cast irons, which contain relatively high carbon content, and steels, which usually contain 1 percent car-

Table 6-1 COMPARISON OF MATERIALS USED IN MANUFACTURE FROM 1944-1968

	1944 10⁶ × ft³	%	1948 10⁶ × ft³	%	1953 10⁶ × ft³	%	1958 10⁶ × ft³	%	1963 10⁶ × ft³	%	1968 10⁶ × ft³	%
Ferrous												
Gray-iron cast	39.44		54.38		57.99		44.12		54.29		63.95	
Ferrous forgings	8.83		3.99		6.14		3.26		3.82		5.09	
Steel cast	7.84		7.49		7.78		4.76		6.39		7.36	
Malleable iron cast	3.61		3.81		4.11		2.81		3.93		4.65	
Semifinished steel	31.97		16.20		18.90		10.31		13.42		20.50	
Steel shapes and plates	70.62		48.11		55.41		41.10		53.40		61.91	
Steel bars and tool steel	47.83		48.00		57.30		37.31		49.61		58.10	
Steel pipe and tubing	25.74		29.21		48.11		28.60		29.90		38.11	
Steel sheets and strip	51.90		81.47		112.80		94.21		131.91		138.60	
Total ferrous	287.78	98.6	292.66	95.2	368.54	94.5	266.48	92.3	346.67	90.5	398.27	87.5
Nonferrous												
Aluminum cast	3.06		2.52		3.91		3.54		7.18		9.33	
Aluminum wrought	...		9.76		13.57		15.45		24.99		42.85	
Total	3.06	1.1	12.28	3.9	17.48	4.5	18.99	6.5	32.17	8.3	52.18	11.5
Magnesium cast	0.78		0.08		0.32		0.25		0.31		0.39	
Magnesium wrought	0.11		0.05		0.18		0.17		0.23		0.25	
Total	0.89	0.3	0.13	0.04	0.50	0.14	0.42	0.10	0.54	0.11	0.64	0.14
Zinc-base cast	...		...		1.19	0.3	1.16	0.4	1.84	0.5	1.98	0.4
Copper-base cast	...		2.17	0.7	2.08	0.5	1.56	0.5	1.79	0.5	1.88	0.4
Total nonferrous	3.95	1.4	14.58	4.74	21.25	5.4	22.13	7.14	36.34	9.4	56.68	12.44
Plastics												
Phenolic	(no data)		(no data)		0.002		0.001		0.003		0.003	
PVC					0.001		0.002		0.003		0.007	
SAN, ABS											0.004	
Cellulose acetate					0.003		0.002		0.001		0.001	
Styrene									0.010		0.020	
Polypropylene									0.002		0.005	
Polyethelyne									0.008		0.030	
Total plastics					0.006	0.002	0.005	0.001	0.027	0.004	0.070	0.015
Grand total	291.73	100	307.24	100	389.80	100	288.62	100	383.06	100	455.09	100

bon or less. The latter can be further subdivided into plain carbon steels, with low, medium, and high amounts of carbon; low- and high-alloy steels; and tool steels. Each type of steel has unique properties, but low-carbon steel far exceeds all others in tonnage produced.

IRON-CARBON EQUILIBRIUM DIAGRAM

Before beginning a discussion of ferrous alloys it is well to review briefly the general relationships shown in the iron-carbon equilibrium diagram. This phase diagram shows the types of alloy structures which can be formed in a two-component system, in this case iron and carbon, under equilibrium conditions. For any binary alloy system a similar plot of change of phase from liquid or liquid plus solid to solid could be constructed from carefully measured freezing-point data. Ferrous alloys have greatly different properties and microstructures with changes in carbon content. These can be related to one another in terms of the iron-carbon diagram.

Figure 6-1 illustrates this diagram for carbon contents up to 5 percent. It will be noted that two allotropic forms of iron exist. At temperatures below 1670°F and at temperatures ranging from 2552°F to the melting point, the stable form of iron has a body-centered cubic structure. (For convenience, the low-temperature form is designated α, while the high-temperature form is designated δ.) At temperatures varying between 1670°F and 2552°F, the stable form of iron has a face-centered cubic structure which is designated as γ. An equally common practice is to refer to the body-centered forms of iron as ferrite and the face-centered form as austenite. The high-temperature body-centered cubic form is designated δ ferrite.

On heating pure iron from room temperatures, at 1670°F a change from body-centered alpha iron to face-centered gamma iron occurs. Further heating to 2552°F will cause the gamma iron to revert back to the body-centered form, delta. Since these changes are reversible, on cooling from above 2552°F, the reverse reactions will occur. At 1670°F and 2552°F,

$$Fe_{\text{bcc}} \rightleftharpoons Fe_{\text{fcc}}$$

The bcc-fcc reactions occur at critical temperatures depending upon the carbon content. The iron-carbon equilibrium diagram shows, among other things, the variation of transformation temperatures with the addition of carbon, and the regions where the phases are stable. It also shows that when carbon is added to iron a third phase may be stable under certain conditions. This phase is the iron-carbon phase, composed of Fe_3C, commonly referred to as cementite or as iron carbide.

The fact that iron-carbon alloys undergo various phase transformations largely accounts for the importance of steel as an engineering material. It is the

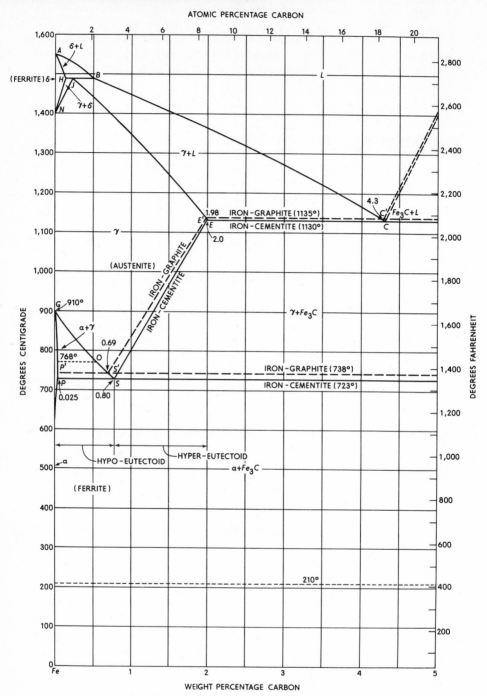

FIGURE 6-1
The iron-carbon equilibrium diagram for carbon contents up to 5 percent.

transformations which make the heat treatment (and therefore the variation and control of mechanical properties) possible. However, it must be remembered that the conditions shown by the equilibrium diagram are not necessarily those which exist in a steel after heat treatment. The phase diagram shows those phases which are thermodynamically stable (i.e., lowest energy) whereas heat treatment takes advantage of the variation in the kinetics of the reactions in order to obtain structures which are thermodynamically metastable.

The dominance of ferrous alloys in manufacturing stems from their wide range of properties with changes in carbon content. As little as one part in one thousand (0.1 percent) changes pure iron to steel. This is soft sheet steel used in drawing and forming. Mild steels have carbon contents of 2 to 3 ppt. Rail and tool steels have from 6 to 9 ppt of carbon. With somewhat above 1 percent carbon, the ability to hold a sharp edge is greatest, so such steels are used for razor blades and wood chisels.

The cast-iron range extends from 2 to 4.8 percent carbon. Actually as much as 6 percent carbon can be dissolved in molten iron, but less than 2 percent can remain in solution in the solidified alloy. It is important to recognize that cast irons also contain significant amounts of silicon. The silicon adds a third dimension to the cast irons and creates ternary alloys which are more complex than the binary alloys of steel.

CAST IRONS

There are five types of cast iron: gray, ductile, malleable, high-alloy, and white; by far the most common is gray iron. These cast irons cannot be specified or identified by chemical analysis alone. It is the form of the excess carbon that determines the type of cast iron. The mechanical properties depend on both the form of the free graphite and the matrix which surrounds it.

In *gray iron* the silicon content is high enough to cause the iron carbides to break down so that flake graphite precipitates during solidification. The simultaneous solidification of the iron and precipitation of the graphite flakes greatly reduces its volumetric shrinkage during solidification. Since gray iron is a complex alloy, it freezes in a mushy manner and therefore has little tendency to develop internal shrinkage as would a skin-forming alloy. This makes it relatively easy to obtain a sound casting even in complex shapes such as motor blocks, air brakes, or control valves.

The excess carbon is the basis for many of the good properties of gray iron, such as high fluidity, high damping capacity, low notch sensitivity, and good machinability. The amount of free carbon in a cast iron depends on the chemical composition, the rate of freezing, and the amount of silicon present. The flake graphite is in the form of a three-dimensional structure similar to the petals of a newly opened rose bud on a microscopic scale (Fig. 6-2). The space in between

FIGURE 6-2
Three-dimensional representation of the structure of flake graphite in gray iron, made from a specimen from which the matrix was dissolved by hydrochloric acid.

is filled with the matrix of either alpha iron or pearlitic iron or a mixture of both.

Alpha or ferritic iron is pure iron, with a small amount of dissolved carbon appearing as white grains in the photomicrograph. Pearlite is the eutectoid composition which occurs at 0.8 percent carbon. It has grains which are composed of alternate plates of alpha iron and combined carbon Fe_3C (Fig. 6-3).

Gray cast iron is comparatively soft, of low tensile strength, and easily machined. There are at least eight engineering grades of gray iron (Table 6-2), with tensile strengths ranging from 30,000 to 80,000 lb/in^2. One of the outstanding characteristics of gray iron is its compressive strength. Cast iron is stronger than many steels in compression, having strengths ranging from 105,000 lb/in^2 for ASTM class 30 to 225,000 lb/in^2 for ASTM class 80.

The strongest gray irons have a pearlitic matrix; those with a ferritic matrix are softer and more machinable. The presence of free carbides in a cast iron reduces its machinability markedly, so foundrymen take great care to produce irons with no carbide inclusions.

The form and size of the flake graphite also has a bearing on the properties of a cast iron. A finer, uniformly distributed type A graphite is usually specified because gray irons with that type of graphite have the best mechanical properties.

Ductile iron is a relatively new alloy. It is the fastest growing ferrous alloy because it has such a wide range of properties; it can be stronger than mild

Table 6-2 GENERAL ENGINEERING GRADES OF GRAY IRON—MECHANICAL AND PHYSICAL PROPERTIES

Class	30	35	40	45	50	60	70	80
Tensile strength, lb/in² (min)	30,000	35,000	40,000	45,000	50,000	60,000	70,000	80,000
Compressive strength, lb/in²	105,000	115,000	125,000	135,000	150,000	175,000	200,000	225,000
Torsional strength, lb/in²	40,000	45,000	54,000	60,000	67,000	76,000	85,000	90,000
Modulus of elasticity,* lb/in² × 10⁻⁶	14	15	16	17	18	19	20	21
Torsional modulus, lb/in² × 10⁻⁶	....	....	5.5	6.6	7.0	8.0	8.1	
Impact strength, Izod AB (1.2-in.-diam. unnotched), ft·lb	23	25	31	36	65	75	120+	120+
Brinell hardness	180	200	220	240	240	260	280	300
Endurance limit, lb/in² Smooth	15,500	17,500	19,500	21,500	25,500	27,500	29,500	31,500
Notched	(13,500)	(15,500)	17,500	(19,500)	21,500	23,500	25,500	27,500
Damping capacity	Excellent	Excellent	Excellent	Good	Good	Good	Fair	Fair
Machinability	Excellent	Excellent	Excellent	Good	Good	Fair	Fair	Fair
Wear resistance	Fair	Fair	Good	Good	Excellent	Excellent	Excellent	Excellent
Pressure tightness	Fair	Fair	Good	Good	Excellent	Excellent	Excellent	Excellent
Specific gravity: g/cm³	7.02	7.13	7.25	7.37	7.43	7.48	7.51	7.54
lb/in³	0.254	0.258	0.262	0.267	0.269	0.270	0.272	0.273
Thermal coefficient of linear expansion: (50–200°F) (in/in, °F) × 10⁻⁶	6.5–6.7	....	6.6–6.8	6.4–6.4				
(50–500°F)	6.9–7.2	....	7.1–7.3	6.8–7.0				
(50–800°F)	7.4–7.6		7.4–7.6	7.0–7.2				
Magnetic properties	Mag	Mag	Mag	Mag	Mag	Mag	Mag	Mag
Pattern shrinkage, in/ft	1/10 – 1/8	1/8	1/8	1/8	1/8	1/8 – 3/16	1/8 – 3/16	1/8 – 3/16
Coefficient of friction (against steel)	....	....	....	(0.19)	(0.195)	(0.20)		

* At 25% of tensile strength.
Note: Values in parentheses are estimated.

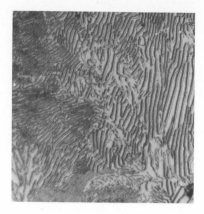

FIGURE 6-3
Pearlitic microstructures found in the matrix of pearlitic gray, malleable, and ductile cast irons. Nital etch; 1000X. (*Courtesy of John Hoke, Pennsylvania State University.*)

steel, yet poured from a low-cost melting furnace such as a cupola. It is frequently called nodular iron also because its free graphite is in the form of spheres rather than flakes.

Ductile iron is produced by adding trace amounts of elements such as magnesium to the molten alloy. The trace elements alter the surface tension of the graphite in the molten iron and cause it to condense into spheroids (Fig. 6-4). In the form of tiny balls, the graphite has no detrimental effect upon the

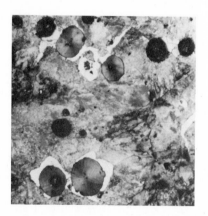

FIGURE 6-4
Microstructure of ductile iron showing the spheroidal graphite in a pearlitic matrix. Nital etch; 230X. (*Courtesy of John Hoke, Pennsylvania State University.*)

Table 6-3   PRINCIPAL TYPES OF DUCTILE IRON

Type no.*	Brinell hardness no.	Characteristics	Applications
80–60–03	200–270	Essentially pearlitic matrix, high-strength as cast. Responds readily to flame or induction hardening	Heavy-duty machinery, gears, dies, rolls for wear resistance, and strength
60–45–10	140–200	Essentially ferritic matrix, excellent machinability and good ductility	Pressure castings, valve and pump bodies, shock-resisting parts
60–40–15	140–190	Fully ferritic matrix, maximum ductility and low transition temperature (has analysis limitations)	Navy shipboard and other uses requiring shock resistance
100–70–03	240–300	Uniformly fine pearlitic matrix, normalized and tempered or alloyed. Excellent combination of strength, wear resistance, and ductility	Pinions, gears, crankshafts, cams, guides, track rollers
120–90–02	270–350	Matrix of tempered martensite. May be alloyed to provide hardenability. Maximum strength and wear resistance	Same as 100–70–03

* The type numbers indicate the minimum tensile strength, yield strength, and percent of elongation. The 80–60–03 type has a minimum of 80,000 lb/in^2 tensile, 60,000 lb/in^2 yield, and 3 percent elongation in 2 in.
SOURCE: Courtesy Gray and Ductile Iron Founder's Society.

mechanical properties of the matrix so the strength of ductile iron depends upon the type of metallic matrix. With a pearlitic matrix ductile iron can have strengths up to 120,000 lb/in^2 which is equivalent to the strength of high-carbon steel but with superior castability and machinability. The major types of ductile iron are listed in Table 6-3.

Malleable iron starts as a white-iron casting which is then heat-treated. In this case the influence of silicon is most evident because white iron, which has all its carbon combined in Fe_3C, can only be produced at low silicon and carbon levels. As the white iron is soaked at temperatures above 1600°F, the silicon causes the iron carbide to break down into iron and carbon in the form of ir-

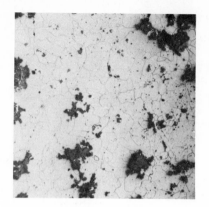 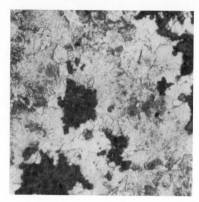

FIGURE 6-5
Microstructure of malleable iron showing the typical carbon with (left) a
ferritic matrix and (right) a pearlitic matrix. (*Courtesy of John Hoke, Pennsylvania State University.*)

regular-temper carbon nodules (Fig. 6-5). The resultant cast structure is
very ductile and is easily machined. For structural designs malleable iron is
limited to relatively thin walls (less than 1 in) because of large shrinkage values
and the need for rapid chilling to produce a white iron as cast structure.

In *high-alloy irons,* those with over 3 percent alloy, there is such a radical
change in the basic microstructure that the material can no longer be regarded
as gray, ductile, or malleable iron. High-alloy irons are more costly to produce
and are used for special applications such as those requiring extreme wear or
corrosion resistance.

Mechanical Properties of Cast Irons[1]

Tensile strength   Tensile strength is the most frequently specified property
of cast iron. Although, as mentioned, ASTM specifications list classes of gray
iron by tensile strength, a given class of gray iron will have its properties significantly affected by the cooling rate in the mold. Thus thickness of the casting
will influence both the tensile strength and hardness. Figure 6-6 illustrates these
relationships. Typical production strengths of several cast irons for both the
as-cast and the heat-treated condition are also shown (Fig. 6-7).

Since there is no definite yield point of cast iron, it is possible to use the
material at stresses approaching its maximum tensile strength under static loading. However, care must be exercised since a slight overload results in fracture.

---

[1] Much of this material was taken with permission from the "Gray Iron Castings
Handbook."

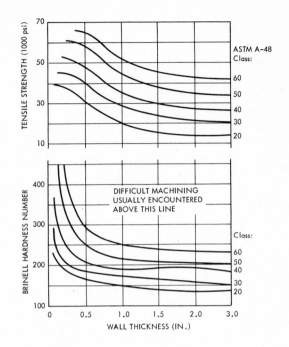

FIGURE 6-6
Tensile strength and hardness as a function of section thickness for various grades of cast iron.

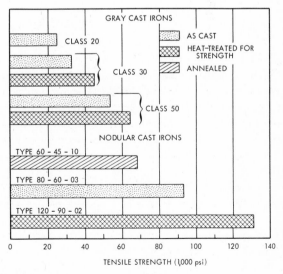

FIGURE 6-7
Tensile-strength properties of cast irons.

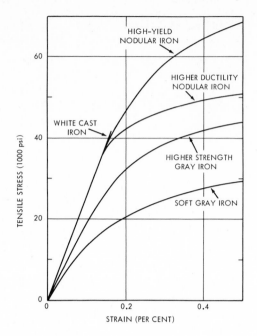

FIGURE 6-8
Stress-strain relationships of the principal cast irons in tension.

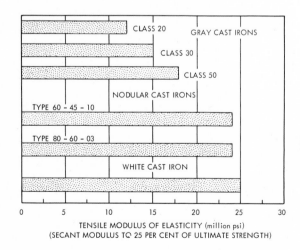

FIGURE 6-9
Tensile modulus of elasticity (million lb/in²) (secant modulus to 25 percent of ultimate strength).

Compressive strength   The compressive strength of gray iron usually is between three to five times its tensile strength and the shear strength is approximately equal to its tensile strength. Thus, where components are stressed in compression, gray iron is comparable in strength to the higher-strength steels. Consequently, gray iron has had wide application for use in machine-tool bases, die blocks, etc., and where reinforcing ribs stressed in compression may be introduced into the design.

Modulus of elasticity   Figure 6-8 illustrates the stress-strain relationships of the principal cast irons in tension. In determining the modulus of elasticity of gray cast iron, it is common to use the slope of the load deflection curve at 25 percent of the tensile strength. The designer should select cast irons with a low modulus of elasticity in applications requiring resistance to heat shock. It will be noted that high-yield nodular and white cast iron have a modulus of elasticity approaching that of steel. See Fig. 6-9.

Yield strength   Since a clearly defined yield point is not apparent during the typical tensile test, the value for yield strength of gray iron is taken at the point of 0.2 percent elongation. Figure 6-10 illustrates yield strengths of the principal cast irons.

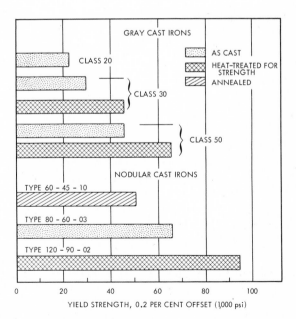

FIGURE 6-10
Yield strength, 0.2 percent offset (1,000 lb/in²).

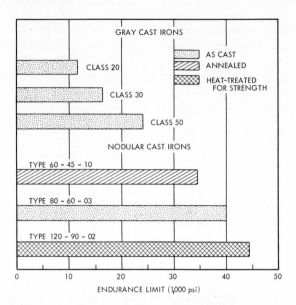

FIGURE 6-11
Endurance limit (1,000 lb/in²).

Endurance limit   The production-design engineer frequently is interested in the endurance limit, which is a measure of the resistance to fatigue of a material. The values shown in Fig. 6-11 were obtained on a rotating-beam machine where the stresses in the surfaces of the samples were alternated between tension and compression. For gray cast iron, the endurance limit usually is computed as being between 35 and 50 percent of the tensile strength.

Ductility   Cast irons have low ductility. Gray iron will give elongation ranging from 0.2 to approximately 1 percent; while the high-strength heat-treated irons show elongations of less than 0.2 percent. The nodular irons test in the range of 3 to 20 percent depending upon type. Values enumerated are based on tensile tests.

Gray cast iron is much more brittle than steel and also has lower impact value. This characteristic has given cast iron a performance edge over steel in the design of jigs and fixtures. An impact blow that will break cast iron will often deform steel to an extent that will destroy the necessary relationship between drill bushings, mastering surfaces, and so forth.

Hardness   The hardness of gray iron varies with its tensile strength. Typical Brinell hardness numbers are illustrated in Fig. 6-12. The Brinell test is pre-

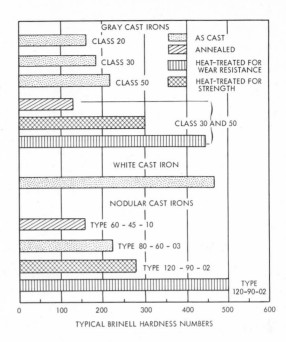

FIGURE 6-12
Typical Brinell hardness numbers.

ferred in measuring the hardness of cast iron since its ball indentor covers sufficient area to give a reading that is representative of the overall hardness.

Production and Design Characteristics of Cast Iron

Castability   To the production-design engineer, castability refers to the ease with which the material can be cast in thin and complex sections. Cast iron is a fluid casting alloy and can be considered as having good castability characteristics. Its castability is better than steel or aluminum. For example, automobile engine blocks and heads are produced readily in cast iron and would be difficult to cast with other ferrous materials.

Machinability   Machinability refers to the ease of cutting material with due regard to surface finish and tool life as well as rate of metal removal. Cast irons generally are machined easily and good finishes at low total costs are obtained. Those cast irons having higher strengths and higher hardness values obviously are not as easily machined as those with lower hardness-strength values.

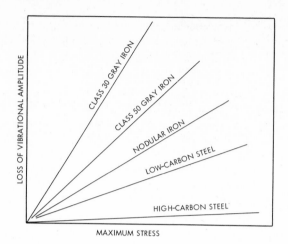

FIGURE 6-13
Relative damping capacities of several engineering materials.

Corrosion resistance   Under atmospheric conditions, cast iron will corrode readily. This corrosion, however, forms a protective surface which offers resistance to further atmospheric corrosion as well as soil corrosion. Widespread use of cast iron for water main and soil pipe is a good indication of its generally favorable corrosion resistance. Gray-cast-iron water mains and gas lines have been in service in some areas for more than 100 years.

Vibration damping   Damping capacity is an inverse function of the modulus of elasticity. Gray cast iron has good damping qualities, making it a valuable engineering material in the design of parts subject to vibration due to dynamic forces. Figure 6-13 illustrates the relative damping capacities of several engineering materials.

One way of demonstrating damping capacity is to strike the casting: steel will ring while cast iron will thud. The energy absorbed per cycle in gray iron is about 10 times that of steel, but damping capacity decreases with applied load.

Wear resistance   In sliding friction, gray iron is outstanding in its wear resistance. This is amply demonstrated by the fact that practically all engine cylinders or liners are made of gray iron. The ways of many machine tools are made of gray iron for the same reason.

Weldability   Two methods lend themselves most readily to the welding of gray-iron castings. These are oxyacetylene welding with cast-iron filler and metal arc with nickel or copper-nickel welding electrodes. Oxyacetylene welding is the fastest method of depositing metal and offers sound deposits and identical color match. When properly executed, the welds are readily machinable and free from defects. The items to be joined must be preheated.

Shrinkage rules and machining allowances Unlike most metals, gray iron shrinks very little when it solidifies. Depending upon the grade, 0.100 to 0.130 in/ft must be provided by the patternmaker to allow for solid shrinkage as the casting cools from 2100°F to room temperature.

The amount of material necessary to provide for machine finish will vary with the casting size and the materials' tendency to warp, as well as the analysis of cast iron itself. As a guide to the designer, Table 6-4 illustrates typical machine-finish allowances.

Ductile Iron

Various alloys of gray cast iron are in use today. They include alloys of nickel, chromium, molybdenum, and magnesium (often called ductile iron). These alloys have greater strength and ductility and compete well with forged steel.

Having several times greater tensile strength than ordinary cast iron plus greatly increased ductility and shock resistance, ductile iron combines the advantages of cast iron, such as availability, ease of founding, and machinability, with many of the product advantages of steel. The presence of small amounts of magnesium in ductile iron produces graphite in spheroids. The elimination of the weakening effect of flake graphite gives the magnesium-containing iron excellent engineering properties; it has particularly high tensile strength, elastic modulus, yield strength, toughness, and ductility. Under stress, ductile iron behaves elastically like steel, having proportionality of stress to strain up to high loads. It has a modulus of elasticity of about 25 million lb/in². Table 6-3 presents the principal types of ductile iron and its applications.

Table 6-4   MACHINE-FINISH ALLOWANCES
FOR GRAY CAST IRON

Casting dimension, in	Expected tolerances for as-cast dimension, in
Up to 8	$\pm \frac{1}{10}$
Up to 14	$\pm \frac{3}{32}$
Up to 18	$\pm \frac{1}{8}$
Up to 24	$\pm \frac{5}{32}$
Up to 30	$\pm \frac{3}{16}$
Up to 36	$\pm \frac{1}{4}$

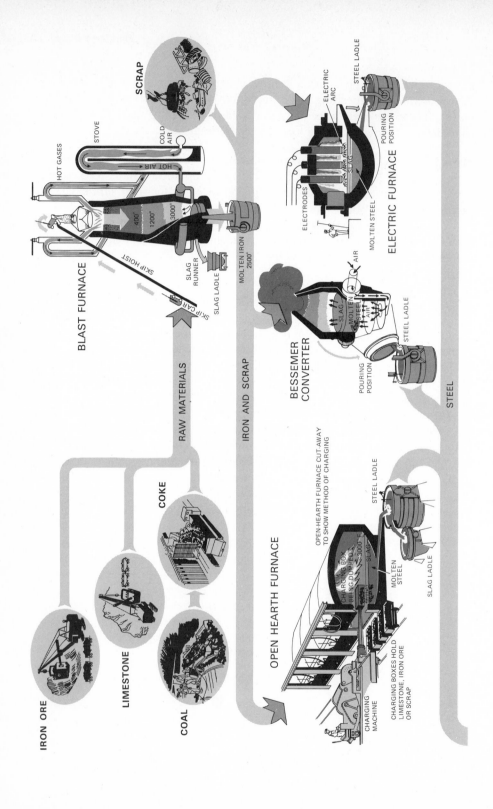

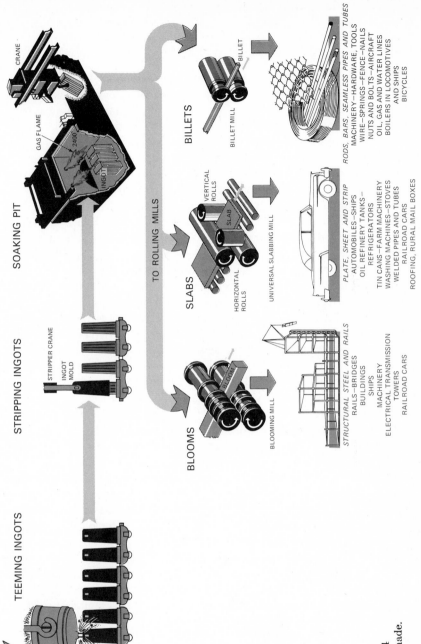

FIGURE 6-14
How steel is made.

STEELS

Steel is the most valuable metal to man; approximately 200 million tons can be produced in the United States annually. In 1900, our capacity was but 21 million tons. Although the process of steelmaking is familiar to most engineers, a review of this process would be appropriate at this time.

Iron ore, limestone, and coal are the principal raw materials used in making iron and steel. Coke is produced by heating bituminous coal in special ovens. Skip cars go up the skip hoist with loads of iron ore, coke, and limestone and dump them into the top of the blast furnace. Hot air from the stove is blown into the furnace near the bottom. This causes the coke to burn at temperatures up to 3000°F. The ore is changed into drops of molten iron which settle to the bottom of the blast furnace. The limestone which has been added joins with impurities to form a slag which floats on top of the pool of liquid iron. Periodically (approximately every 6 h), the molten iron is drained into a ladle for transporting to either the open-hearth furnace, Bessemer converter, or electric furnace. The slag is removed separately so as not to contaminate the iron.

The making of steel from iron involves a further removal of impurities. Regardless of which process is used for making steel, open-hearth, Bessemer-converter, or electric-furnace, steel scrap is added along with desired alloying elements and the impurities are burned out.

Liquid steel upon removal from the furnace is poured into ingot molds. The ingots are then removed to "soaking pits" where they are brought to a uniform rolling temperature.

At the rolling mill, the white-hot steel passes through rolls which form the plastic steel into the desired shape: blooms, slabs, or billets. These three semifinished shapes then go to the finishing mills where they are rolled into finished forms as structural steel, plates and sheets, rods, and pipes. See Fig. 6-14.

Steel is the basic and most valuable material used in apparatus manufactured today. Its application is based on years of engineering experience, which serves as a guide in choosing a particular type of steel. Each variable, such as alloy, heat treatment, and processes of fabrication (casting, forging, and welding) has its influence on the strength, ductility, machinability, and other mechanical properties, and affects the type of steel selected. The following basic concepts also assist in determining which steel should be used:

1 The modulus of elasticity in tension falls within the range of 28×10^6 to 30×10^6 lb/in^2, regardless of composition or form; therefore, sizes as determined by deflection remain the same regardless of the steel chosen.

2 Carbon content determines the maximum hardness of steel regardless of alloy content. Therefore, the strength desired, which is proportional to hardness, can determine the carbon content.

3 The ability of the steel to be uniformly hardened throughout its volume depends on the amount and kind of alloy. This is more complex, but does not necessarily change the calculation of the size of the part.

4 Ductility decreases as hardness increases.

The preliminary choice of steel for a part as well as for other factors, such as notch sensitivity, shrinkage, blowholes, corrosion, and wear, is simplified when based on the above principles. The final selection is made by matching the material with the process of manufacture used in order to obtain the shape, surface, and physical requirements of the part. The selection may be made from low-carbon steels, low-alloy steels, high-carbon steels, and high-alloy steels.

Carbon Steel

Plain carbon steels represent the major percentage of steel production and they have a wide diversity of application including castings, forgings, tubular products, plates, sheets and strips, wire and wire products, structural shapes, bars, and tools. Plain carbon steels, generally, are classified in accordance with their method of manufacture as basic open hearth, acid open hearth, or acid Bessemer steels and carbon content.

In designating carbon steels, the prefix "B" denotes acid Bessemer carbon steel and the prefix "C" denotes basic open hearth. The basic open-hearth carbon steels have a carbon range of 0.08 to 1.03 percent and a manganese range of 0.25 to 1.65 percent. The American Iron and Steel Institute (AISI) composition ranges of basic open-hearth, resulfurized steel indicate a carbon range of 0.08 to 1.65 percent and 0.30 to 1.65 percent manganese.

The principal factors affecting the properties of the plain carbon steels are the carbon content and the microstructure. The microstructure is determined by the composition of the steel (carbon, manganese, silicon, phosphorus, and sulfur which are always present and residual elements including oxygen, hydrogen, and nitrogen) and by the final rolling, forging, or heat-treating operation. However, most of the plain carbon steels are used without a final heat treatment and, consequently, the rolling and forging operations influence the microstructure. The average mechanical properties of as-rolled 1-in bars of carbon steel as a function of carbon content are shown in Fig. 6-15.

Carbon steels are predominantly pearlitic in the cast, rolled, or forged conditions. The constituents of the hypoeutectoid steels are therefore ferrite and pearlite and of the hypereutectoid steels, cementite and pearlite.

Alloy Steel

Alloy steel is an alloy of iron and carbon containing alloying elements, one or more of which exceeds the following: manganese, 1.65 percent; silicon, 0.60 percent; copper, 0.60 percent; and/or specified amounts of other alloying ele-

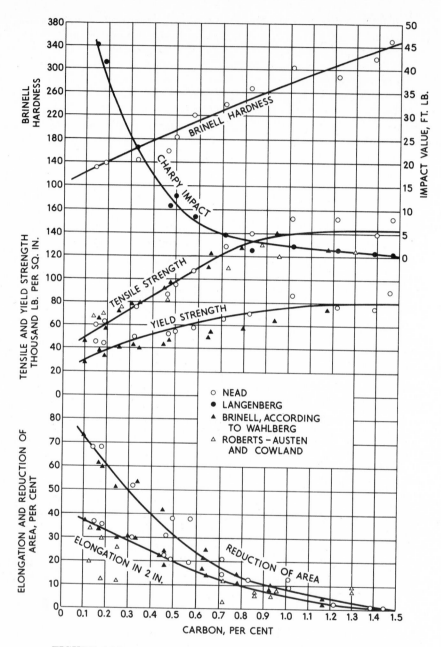

FIGURE 6-15

Variations in average mechanical properties of as-rolled 1-in-diameter bars of plain carbon steels, as a function of carbon content.

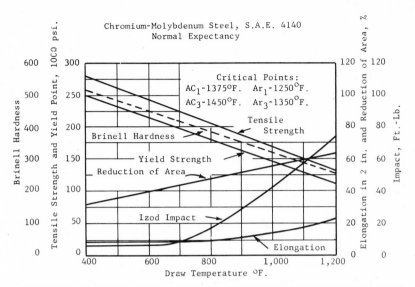

FIGURE 6-16
Typical property chart of alloy steel showing quenched from normal temperature and drawn at various ranges.

ments including aluminum, boron, and chromium up to 3.99 percent; cobalt, columbium, molybdenum, nickel, titanium, tungsten, vanadium, zirconium, or other elements added in sufficient quantity so as to give the desired properties of the steel.

Since there are more elements, some expensive, to be kept within specified ranges in alloy steel than required in carbon steel, alloy steel requires more involved techniques of quality control and, consequently, is more expensive.

Alloy steel can give better strength, ductility, and toughness properties than can be obtained in carbon steel. Consequently, the production-design engineer should consider alloy steels in designs subject to high stresses and/or impact loading (Fig. 6-16).

Almost all alloy steels are produced with fine-grain structures. A steel is considered to be fine grained if its grain size is rated 5, 6, 7, or 8. Number 1 grain size shows $1\frac{1}{2}$ grains/in² of steel area examined at 100 diameters magnification. Fine-grain steels have less tendency to crack during heat treatment but have better toughness and shock-resistance properties. Coarse-grained steels exhibit better machining properties and may be hardened more deeply than fine-grained steels.

In order to select the alloy steel that is best suited for a given design, the effects of the principal alloying elements must be taken into account. They are:

Nickel   Provides toughness, corrosion resistance, and deep hardening
Chromium   Improves corrosion resistance, toughness, and hardenability
Manganese   Deoxidizes, contributes to strength and hardness, decreases the critical-cooling rate
Silicon   Deoxidizes, promotes resistance to high-temperature oxidation, raises the critical temperature for heat treatment, increases the susceptibility of steel to decarburization and graphitization
Molybdenum   Promotes hardenability, increases tensile and creep strengths at high temperatures
Vanadium   Deoxidizes, promotes fine-grained structure
Copper   Resistance to corrosion and acts as strengthening agent
Aluminum   Deoxidizes, promotes fine-grained structure, aids nitriding
Boron   Increases hardenability

Tool Steel

Tool steels constitute a class of high-carbon alloy steels that have properties such as shock and wear resistance, hardness, strength, and toughness produced by heat treatment and fabrication. Tool steels are essentially combinations with iron of one or more of the following elements: carbon (0.80 to 1.30 percent), manganese (0.20 to 1.60 percent), silicon (0.50 to 2.00 percent), chromium (0.25 to 14.00 percent), tungsten (1.5 to 20.00 percent), vanadium (0.15 to 3.00 percent), molybdenum (0.80 to 5.00 percent), and cobalt (0.75 to 12 percent).

Tool steels are characterized by hardenability and timbre. Hardenability refers to the ability of steel to harden upon heat treatment, and timbre is that property which refers to grain size and toughness after being subjected to a given heat treatment.

The hardenability of a tool steel may be expressed as the depth to which a certain standard size will harden when quenched in a certain medium. Hardness is influenced by the speed of cooling from above the critical temperature to the necessary transformation temperature (about 1160°F) and the critical quenching speed of the steel which is influenced by the alloy additions. Plain carbon steels have a high critical quenching speed, while the addition of most alloys decreases this speed. Thus, it may be said that the critical quenching speed determines the depth to which a steel will harden in any given cooling medium.

Timbre of a tool steel is determined by inspection of its grain size. If the grain size is fine, the steel is referred to as having *tough timbre*; if the grain size is coarse, the timbre is referred to as being *brittle timbre*.

In the selection of a tool steel, the production-design engineer will encounter four requirements which will vary to a degree depending on the use of the tool. Each of these requirements may be realized in greater or lesser degrees by selecting the proper tool steel. However, improvement of one requirement will result in the diminishing of at least one of the remaining three requirements. These requirements as outlined by the Carpenter Steel Company are:

1 Wear resistance for cutting or abrasion
2 Toughness or strength
3 Hardening accuracy and increased hardenability
4 Ability to do its work at elevated temperatures

Stainless Steel

The term "stainless steel" denotes a large family of steels containing at least 11.5 percent chromium. They are not resistant to all corroding media.

Stainless steel competes with nonferrous alloys of copper and nickel on a corrosion-resistance and cost basis and with light metals such as aluminum and magnesium on the basis of cost and strength-weight ratio. Stainless steel has a number of alloy compositions and many suppliers. Information on its properties and fabrication can be obtained readily. Sound techniques have been evolved for casting, heat treating, forming, machining, welding, assembling, and finishing stainless steel. It will be found that this material usually work-hardens (which makes machining, forming, and piercing more difficult), must be welded under controlled conditions, and under inert gas. It has desirable high strength, corrosion resistance, and decorative properties.

A bright, clean surface is essential for best corrosion resistance. Traces of scale and foreign matter should be removed by machining, pickling, or polishing. Dipping in nitric acid will ensure the formation of a good oxide film on new pieces. Stainless steels may be electroplated and electropolished, anodically etched, covered with porcelain enamel, or given colored coatings through the dying of surface oxides. Highly polished sheets may be purchased directly from stainless-steel producers. A coating of plastic may be used to protect the surface during fabrication.

Stainless steel can be made very hard and its strength more than doubled by cooling to −300°F and simultaneously rolling under high pressure, then heating to 750°F for 24 h.

Corrosion resistance is the most important single characteristic of the stainless steels. This quality is due to a thin transparent film of chromium oxide which forms on the surface. It will withstand oxidizing agents such as nitric acid, but will be attacked by reducing agents such as hydrochloric acid

Table 6-5 PROPERTIES OF STAINLESS AND RUST-RESISTING STEELS[1]

Carbon %	Cr %	Ni %	1,000 Lbs/Sq. In. Ult. Str.	1,000 Lbs/Sq. In. Yield Str.	% Elong. 2	% Red.	Brin.	Machining (2)	Weldability	Rel. Corr. Res.	Scaling Temp. °F (3)	Elec. Res. (4)	Shape of Mat.	DESCRIPTION AND COST (5)		
.15	5		65	25	35	70	150	--	Fair (10)	Poor	1100	260	Sheets	H.R. annealed. Range ovens.		
.30	13		235	(6) 150	--	--	477	--	(11)	Fair	--	284	Springs	Rust resisting springs.		
.12	13		90	60	30	45	180	--			1200	286	Bars	Cold drawn, low sulphur, electrodes.		
.12	13		80	48	25	50	190	F.M.	(12)	Good	1200	293	Bars	Cold drawn, high sulphur. Screw-mach. wk.		
.15	13		65				131	--		(14)	1200	287	Strips	C.R. for electrical resistance.		
.10	13		65	35	30		140	--			1200	270	Strips	H.R. & C.R. General use.		
1.00	18		95	55	6	10	190	--	(11)		1200	306	Bars	H.R. Rollers, cam for brush arm.		
.12	17		70	40	25	--	159	--	(12)	Good	1550	307	Sheets	C.R. annealed. General use.		Non-Hardenable.
.15	17		70	40	25	50	175	--		(15)	1550	279	Bars	H.R. & C.R. annealed.		
.20	18	8	80	25	35	45	210	F.M.			1650	345	Bars	H.R. General use.	Non-Magnetic.	
.20	18	8	90 (7)	(7) 40	35 (7)	45 (7)	230 (8)	F.M.	(13)		1650	345	Bars	Cold finish. General use.		
.16	18	8	90	40	35	45	230	--			1650	345	Bars	Cold finish. General use.		
.20	18	8	90	35	35	--	--	--		Very	1650	345	Sheets	H.R. & C.R. General use.		
.08	18	8	85	30	50	--	--	--	Good	Good	1650	345	Sheets	H.R. & C.R. for deep drawing.		
.10	18	8	182	--	--	--	375	--	--	(16)	1650	345	Springs	Cold finished. Flat spring steel.		
.15	18	8	246 (9)	--	--	--	--	--	--		1650	345	Springs Wire	Cold drawn spring wire.		
.25	18	8	200 (7)	(6) 135 (7) 170	--	--	--	--	--		1650	345	Wire	Cold drawn. Tinning banding wire.		
.07	18	8	85	30	50	55	--	--	Good		1650	345	Tubes	El. welded, annealed.		
.07	18	8	85	30	50	55	--	--	Good		1650	345	Tubes	Seamless, annealed.		
.20	24	12	100	45	35	45	185	--	Good	Very	2100	378	Bars	Cold drawn.		
.20	24	12	90	40	40	50	170	--	Good	Good	2100	369	Sheets	Annealed. Furnace lining.		
.10	24	12	90	40	40	50	170	--	Good	(17)	2100	369	Tubes	Seamless, annealed.		

(1) The data given in this table are only relative for comparative purpose except the data marked (7), (8), (9).
(2) F.M.--Free machining, all other material is difficult to machine.
(3) Scale resistance is not merely a function of temperature. The atmosphere encountered is quite important.
(4) Microhms per square inch foot.
(5) Cost increases approx. with total percentage of chromium and nickel.
(6) Eleastic Limit. (7) Minimum. (8) Maximum.
(9) Min. for .0625 diam. Strength varies inversely with diam.

(10) Preheating from 300 to 500°F. essential. Anneal after welding.
(11) Not recommended for welding.
(12) Limited, preheating and annealing essential.
(13) Good; anneal after welding for corrosion resistance.
(14) Improved if heat treated.
(15) Better than the preceding specifications.
(16) Improved if annealed.
(17) Better than the preceding specifications.

or any of the halogen salts. Scaling and corrosion are accelerated in applications where the oxide layer is constantly being broken. Repeated heating and cooling with the accompanying expansion and contraction crack off the oxide layers. Since the straight-chromium grades of stainless steel have less thermal expansion than the chromium-nickel grades, they serve best where constant heating and cooling is involved. Most stainless steels show good short-time strength at 1500°F, and a few special types are good at 2000°F. Compare this with ordinary carbon steels which lose their usefulness above 900 to 950°F. The heat-conducting properties of stainless steel are poor, so copper cladding is often used in cooking utensils to distribute heat. (See Table 6-5 for physical properties of rust-resisting steels.)

Groups of Stainless Steel

Although there are many types of stainless steel available to the production-design engineer today, they may be classified in one of three groups according to their microstructures: austenitic, ferritic, and martensitic.

The austenitic group is often referred to as chromium-nickel alloy, having 18 percent chromium and 8 percent nickel, and is known as 18–8 stainless. The 8 percent nickel is generally sufficient to stabilize the high-temperature, austenitic face-centered cubic structure so that upon cooling to room temperature it does not transform. This gives the alloy high-temperature strength. There are many modifications (over 20) based on this grade in which the chromium-nickel ratio is modified, carbon content decreased, or other elements added for stabilization or increased resistance to oxidation.

In general, the austenitic stainless family may be hardened only by cold working. As a family, they are nonmagnetic, and have good resistance to atmospheric corrosion. However, after heating in the critical range, 800 to 1300°F, the steel may become prone to intergranular corrosion under corrosive environment. This intergranular failure may develop in areas that have been welded. Intergranular attack may be avoided or greatly reduced by using a grade which is stabilized with certain alloying elements (titanium, columbium), by using a grade with extra-low-carbon content (0.03 percent carbon maximum—known as the ELC grades), or by heat treating the steel in the range 1800 to 2100°F followed by quenching after it is exposed in the critical range.

The austenitic stainless grades are generally stronger than the other grades at temperatures above 1000°F. Since the austenitic steels are more ductile than the ferritic and martensitic, they are considered to have better fabrication properties. However, all the wrought stainless-steel grades can be fabricated by the methods applicable to carbon steel.

The ferritic grades differ from the austenitic in that they contain 18 to 30 percent chromium and no nickel. Ferrite is magnetic and has a body-centered cubic crystal structure. This group can be hardened to some extent by cold

Table 6-6 STEEL BAR, CARBON—MINIMUM PROPERTIES*

SAE no.	Description	Tensile strength lb/in² Yield	Tensile strength lb/in² Ultimate	% elong. in 2 in	% red. in area	Brinell hardness	Chemistry percent, and SAE equivalent	Bending recommendation	Application
1070	Hot-rolled, annealed spring steel 1 46563 gr. B type II 2 QQ-S-663, FS1070 3 51-107-WD-1070 5 A107, gr. 1070 6 1070	Ht. tr. per PS 50994				184 max.	0.65–0.75 C, 0.70–1.0 Mn, 0.35 Si max; SAE 1070	180° over pin = 2 × thk.	Springs. Requires ht. tr. following forming to develop spring properties.
1015	Hot-rolled, unpickled, unoiled 1 47S11, CLC, Fin 2 QQ-S-636, as-rolled 3 57-136-WD 1010, as-rolled 5 A 107, G 1010 6 1010 or 1015	26,000	48,000	30	55	100	0.08–0.18 C, 0.30–0.60 Mn SAE 1015	180° over pin Size $\frac{1}{2}$ × size; $\frac{1}{4}$ or less 1 × size; $\frac{3}{4}$ to 1 $1\frac{1}{2}$ × size; 1 to $1\frac{1}{2}$ $2\frac{1}{2}$ × size; $1\frac{1}{2}$ to 2 3 × size; 2	General use where a soft ductile steel is required and where fit and finish are not important. Suitable for welding but contains scale undesirable for resistance welding. Threads may be chased or cut with special dies, but this steel cannot be used to advantage on automatic screw machines. Much cheaper than cold-rolled steel.
1015	Hot-rolled, pickled and oiled 1 47S11, CLC, fin. 3 2 QQ-S-636, pickled and oiled 3 57-136-WD 1010, pickled and oiled 6 1010 or 1015	26,000	48,000	30	55	100	0.08–0.13 C, 0.30–0.60 Mn SAE 1015	Same as 1015 above	Same as 1015 except superior for resistance welding due to absence of scale.
1045	Hot-rolled, forging quality 1 4684, Gr. P 2 QQ-S-663, FS1045 3 57-107-WD 1045 4 A-107, Gr. 1045 6 1045	40,000	70,000 May be ht. tr. per PS 50765 (see S 927.8)	18	30	140	Comp. A Comp. B 0.42–0.50 C 0.42–0.50 C 0.40–0.60 Mn 0.55–0.75 Mn 0.15–0.30 Si 0.15–0.30 Si SAE 1045	180° over pin = 4 × size	Shafts, gears, cams and pinions subject to drastic heat treatment. Suitable for upsetting and forging.

Grade	Condition / Specifications	Yield, psi	Tensile, psi	Elong., %	Red., %	Brinell	Chemical composition	Bend test	Remarks
1035	Hot-rolled straightened Size: 3 in. dia. or less → Size: Over 3 to 6 dia.→ Size: Over 6 dia.→ May be ht. tr. per PS 50765 (see S 927.8) 2 QQ-S-663, FS1035 3 57-107-WD 1035 4 AN-S-4, cond. B 5 A 306, Gr. 75 6 1035 7 5080	40,000 40,000 36,000	75,000 to 95,000	18 15 15	30 25 25	150 150 150	0.30–0.40 C Approx., 0.60–0.90 Mn, 0.15–0.30 Si SAE 1035	180° over pin = 4 × dia.	Misc. shafting, bolts, studs.
1111	Cold-finished, Free machining 2 QQ-S-663, FSB 1111 3 57-107-WD 1111 5 A 108, Gr. 1111 6 B 111	45,000	60,000	8	30	120	0.13 C Max, 0.60–0.90 Mn, 0.07–0.12 P, 0.08–0.15 S SAE 1111	Not recommended	Smooth-finish high-speed automatic screw machine work. Do not use for application requiring shock resistance such as for transportation apparatus. Low temperatures reduce the impact resistance still further. Not recommended for pack carburizing.
1111	Centerless ground, round, free-machining, close plus and minus tolerances, bright smooth finish	45,000	60,000	8	30	140	0.13 C Max, 0.60–0.90 Mn, 0.07–0.12 P, 0.08–0.15 S SAE 1111	Not recommended	Close plus and minus tolerance applications. Same restrictions for use as SAE 1111.
1016	Cold-drawn, carburizing quality May be carburized per PS 50934 and PS 290519 (see S 921.1 and S 927.8) 2 QQ-S-663, FS1016 3 37-107-WD 1016 4 MIL-S-866, CIA 5 A108, Gr. 1016 6 1016 7 5060	50,000	70,000	15	45	140	0.13–0.18 C, 0.60–0.90 Mn SAE 1016	Not recommended	Same as 2084-1 except for parts to be carburized such as circuit-breaker triggers, latches, and links.
1137	Cold-drawn, relatively free-machining, stress-relief annealed May be ht. tr. per PS 50765 (see S 927.8) 5 A311, Gr. 1137 6 1137 7 5024	80,000	95,000	16	45	200	0.32–0.45 C, 1.35–1.65 Mn, 0.08–0.13 S SAE 1137	Not recommended	Shafts. Where moderate strength, negligible warping during machining and relatively free machining characteristics are required. Automatic screw machine work.

* Italicized specification numbers refer to the following: 1, Navy; 2, federal; 3, Army; 4, military or Air Force–Navy Areo; 5, ASTM; 6, SAE or AISI number. 7, SAE–AMS specification

working but cannot be hardened by heat treatment. These steels are restricted to a somewhat narrower range of corrosive conditions than the austenitic grades but may be considered to be interchangeable with austenitic types under strongly oxidizing conditions.

The martensitic grades contain chromium from 12 to 17 percent, and usually no nickel. There are exceptions where martensitic grades do contain some nickel. The martensitic stainless steels differ from the ferritic and austenitic in that they can be hardened by heat treatment. Martensitic stainless steel is magnetic and has a body-centered tetragonal crystal structure characterized by a needlelike pattern. It has excellent corrosion-resistant properties under mild conditions such as weak acids, fresh water, and the atmosphere. Severely corrosive environments will attack martensitic stainless steels.

Other Types of Steel

There are many other types of steel classed according to their alloying elements, such as boron, copper-bearing, manganese, nickel, nickel-chromium, molybdenum, chromium, chromium-vanadium, silicon-manganese, and silicon steels. There are also groups classed according to molding and forming properties and surface conditions, which include free-cutting, deep-drawing, texturized, precision-ground, polished, cold-drawn, and cold-rolled steels. Although it is not practical to describe them in detail, much of the information in the data sheets and in the chapter as a whole is applicable to all of these types of steels.

DATA SHEETS FOR STEEL

Useful reference information that can help to apply steel to a design project is included in Tables 6-6 and 6-7. These are the type of data supplied to a designer in a large corporation. Some of the nomenclature used remains in the data and is explained as follows:

1 PS numbers[1] are process specification numbers.
2 SAE numbers are Society of Automotive Engineers numbers.

The same type of information exists for other materials, such as aluminum, magnesium, copper, brass, and bronze.

The information listed is based upon the following general data.

Mechanical properties   The values listed indicate minimum mechanical properties (except where noted otherwise) that may be expected from 0.505-in-diameter tension-test specimens at room temperature.

---

[1] These numbers are used only by Westinghouse Electric Corp.

Table 6-7   STEEL SHEET, CARBON, COLD-ROLLED—THICKNESS TOLERANCES (Plus or Minus)

Thickness, in*

Width, in	0.2299 / 0.1875	0.1874 / 0.1800	0.1799 / 0.1420	0.1419 / 0.0972	0.0971 / 0.0822	0.0821 / 0.0710	0.0709 / 0.0568	0.0567 / 0.0509	0.0508 / 0.0389	0.0388 / 0.0344	0.0343 / 0.0314	0.0313 / 0.0255	0.0254 / 0.0195	0.0194 / 0.0142	0.0141 and less
To 3½ incl.													0.003	0.002	0.002
3½–6											0.004	0.003	0.003	0.002	0.002
6–12					0.006	0.006	0.006	0.005	0.005	0.004	0.004	0.003	0.003	0.002	0.002
12–15	0.008	0.007	0.007	0.007	0.006	0.006	0.006	0.005	0.005	0.004	0.004	0.003	0.003	0.002	
15–20	0.008	0.008	0.008	0.008	0.007	0.007	0.006	0.006	0.005	0.004	0.004	0.003	0.003	0.002	
20–32	0.009	0.009	0.009	0.008	0.007	0.007	0.006	0.006	0.005	0.004	0.004	0.003	0.003	0.002	
32–40	0.009	0.009	0.009	0.009	0.008	0.007	0.006	0.006	0.005	0.004	0.004	0.003	0.003	0.002	0.002
40–48	0.010	0.010	0.010	0.010	0.008	0.007	0.006	0.006	0.005	0.004	0.004	0.003	0.003	0.002	0.002
48–60			0.010	0.010	0.008	0.007	0.007	0.006	0.005	0.004	0.004				
60–70			0.011	0.011	0.009	0.008	0.007	0.007	0.006	0.005	0.005				
70–80			0.012	0.012	0.009	0.008									
80–90			0.012	0.012	0.010										
90			0.012	0.012											

* Thickness is measured at any point on the sheet not less than $\frac{3}{8}$ in from an edge.

The properties listed are obtainable only in the section sizes specified. If no section size is specified, it may be assumed that the properties apply to all sizes commercially obtainable.

Section sizes Where size ranges are indicated to which certain properties apply, the term "dia." refers to solid round bars, the term "thk." refers to the minimum dimension of a rectangular section, and the term "size" refers either to the diameter of round bars or the minimum dimension of a rectangular section.

Modulus of elasticity in tension The modulus of elasticity in tension for all specifications listed falls within the range of 28×10^6 to 30×10^6 lb/in².

Endurance limit in tension In the absence of actual data, the endurance limit in tension is approximately 0.45 times ultimate tensile strength where ultimate tensile strength does not exceed 125,000 lb/in².

Torsion In the absence of actual data, the yield point in torsion may be taken as one-half that for tension, the endurance limit in torsion as one-half that for tension where ultimate tensile strength does not exceed 125,000 lb/in², and the modulus of elasticity in torsion as 11.5×10^6 lb/in².

Machinability Steels having a hardness below 225 Brinell generally are machined easily. For average applications, the cost of machining rises rapidly above 250 Brinell. Steels may be machined at hardnesses over 300 Brinell, but with considerable difficulty and generally excessive cost. Although possible, machining of steels at hardnesses exceeding 400 Brinell is rarely attempted, but it is possible using aluminum oxide inserts.

Chemistry Where phosphorus and sulfur contents are not given, the maximum phosphorus content is in the range of 0.03 to 0.05 percent and the maximum sulfur content is in the range of 0.03 to 0.055 percent. The approximate SAE equivalent specification is listed immediately following the chemistry of each specification.

Bending recommendations Bending recommendations show maximum bendability and are based on specification values or are estimated. They are not based on actual shop practice.

Corrosion resistance The corrosion resistance of steels is complicated by many factors. In general, carbon and alloy steels do not differ greatly in their resistance to air atmospheres in rural, urban, marine, and industrial locations; they rust readily if moisture is present. Where resistance to rust is required, it is necessary to use a protective finish on carbon and alloy steel or to specify a special steel which develops an adherent oxide coat such as Corten steel.

Application

In choosing between carbon, alloy, and stainless steels, the most important considerations are usually mechanical properties, corrosion resistance, magnetic properties, and cost. Wherever applicable, carbon steel should be used because of its lower cost. Use of alloy steel is usually restricted to applications requiring high mechanical properties involving heat treatment in section sizes larger than can be hardened throughout with carbon steel. Stainless steels are used primarily to obtain corrosion resistance, although the nonmagnetic characteristics of the austenitic chromium-nickel steels are frequently of major importance. The magnetic chromium stainless steels are used generally for parts requiring corrosion resistance (significantly greater than carbon or alloy steels, but inferior to chromium-nickel stainless steel) together with high mechanical properties obtainable by heat treatment. Another group of magnetic chromium stainless steels, which are nonhardenable, are occasionally used for their magnetic and corrosion-resisting characteristics.

FACTORS ASSISTING IN THE SELECTION OF STEEL[1]

Once the production-design engineer has decided upon the properties he needs, he can then select the material that best fits the criteria he has established. Some factors that will assist in the selection of steel follow.

Useful and important properties are added to steel by heat treatment (see Chap. 5). Strength and hardness increase when steel is cold-worked. Rolled and forged steel has directional properties. Properties of steels are better with the grain than across the grain. Grain direction can be used to an advantage in forgings and should be incorporated in design. All steels have practically the same stiffness (modulus of elasticity). It is impossible to stiffen a piece of steel by heat treating. The modulus of elasticity in alloy steels ranges from 27×10^6 to 32×10^6 lb/in². Deflection is often more important in design than strength.

Temperatures less than room temperature increase tensile strength and reduce ductility.

Designs are not based on tension alone. Failures may occur because of many unknown factors, such as the overloading of equipment by an operator. The absence of failures in the field indicates that the parts are designed too con-

---

[1] Factors assisting in the selection of steel in this chapter have been taken from Gordon T. Williams, "What Steel Shall I Use," American Society for Metals, Metals Park, Ohio.

Table 6-8   STEEL SHEET, CARBON—THICKNESS.
(Preferred for Uncoated, Thin, Flat Metals
under 0.250-in Thick)*

0.004	0.014	0.040	0.112
0.005	0.016	0.045	0.125
0.006	0.018	0.050	0.140
0.007	0.020	0.056	0.160
0.008	0.022	0.063	0.180
0.009	0.025	0.071	0.200
0.010	0.028	0.080	0.224
0.011	0.032	0.090	
0.012	0.036	0.100	

* The use of the American Standard preferred thicknesses
eliminates the confusion caused by the various gage number
systems.
Note: The same composition and finish as indicated for bars
in Table 6-6, plus many more types of sheets, can be found on
the market.

Table 6-9   STRENGTH-WEIGHT FACTORS

Material	Ultimate tensile strength, lb/in²	Average specific gravity	Strength-weight ratio
Aluminum, commercial 2S	13,000	2.71	4.8
Iron, ingot (wrought)	40,000	7.87	5.1
Steel, cold-rolled	60,000	7.84	7.6
Aluminum, cold-rolled, 2S-H	24,000	2.71	8.9
Steel, SAE low-alloy, low-carbon	160,000	7.85	20.4
Aluminum alloy, 17S-T (duralumin)	58,000	2.79	20.8
Spruce for aircraft	10,000	0.435	23.0
Aluminum alloy, C17S-T	65,000	2.8	23.2
Steel, SAE medium-alloy, medium-carbon, heat-treated	150,000	7.85	24.2
Aluminum alloy, 24S-RT	68,000	2.77	24.5
Magnesium alloy, AM58S	46,000	1.85	24.9
Steel, 18–8 stainless, heavily cold-rolled	200,000	7.93	25.2
Steel, high-alloy, high-carbon, heat-treated	250,000	7.85	31.8
Steel, piano-wire, cold-drawn, very fine	400,000	7.84	51.0

servatively except where life is endangered. It is sometimes best to choose the optimum size and conditions first and modify as experience dictates. Table 6-9 gives the tensile strength, specific gravity, and strength-weight ratio of representative ferrous and nonferrous materials.

Endurance Limit of Steels

Very few machines fail through tension or shear. They usually wear out or fail because of fatigue. *Fatigue failure* occurs after repeated application of a load that can be borne safely in a single application. After all, it seems surprising that a hardened steel specimen, which will carry 215,000 lb/in^2 in a tension test without showing any plastic deformation, will break by the application of 120,000 lb/in^2 if such a stress is repeated often enough. The endurance limit (that is, the load that can be withstood for an infinite number of repetitions) is ordinarily, but not necessarily, below the yield point and is close to 50 percent of the ultimate tensile strength. The effect of the shape and size of notches on fatigue strength is considerable and should be checked by actual tests. The higher the tensile strength of the material, the greater the effect of surface corrosion. In design, notches and anything that weakens the surface, such as decarburization, should be avoided. Fatigue is affected by grooves on the surface. Corrosion severely diminishes the endurance of a material, it can be improved by surface burnishing.

"Crystallization in service" is a myth. Innumerable mechanical experts have looked at a piece of metal, broken off sharply without any apparent sign of ductility, and said, "crystallization." They believed that the metal was fibrous at the start of its life, but by the stresses and strains of its existence the material had changed over from tough fibers to large brittle crystals. They prove it by pointing to the coarse crystalline fracture (when there is one; when it is a fine porcelainic fracture, they still say "crystallization").

If the part after failure exhibits a coarse crystalline fracture, it had a coarse crystalline structure when it left the hands of the artisan who fashioned it. At least 90 percent of the failures called crystallization are fatigue failures. Service can not change the grain size of metallic alloys.

Scoring and galling *Scoring* is similar to abrasion, but the abrasive, instead of being dirt, is one or more hard particles or projections of metal, embedded in one of the surfaces. As the pieces slide past each other, a high spot on one may interlock with a slightly high spot on another piece; a fragment is torn out, and scoring results. Scoring may occur soon after a part is placed in service, and, once having made suitable tracks for the high spots on the mating part, may not go any further. All too frequently, it does go further, and then it is called by another name—*galling*. High finish is usually helpful in reducing scoring: of course, lubrication is an important factor. If galling is severe enough to stop the sliding motion of two parts, it is known as *seizure*. Some high-alloy steels are notoriously bad in the matter of galling and seizure as, for instance, 18–8 stainless steel, which is difficult to form in dies.

Pitting   Pitting is another important type of failure. A true case of pitting involves compression fatigue of the surface layers; the fatigue failures first occur below the surface in the weaker, lower-carbon material found there.

Flow and corrosion   Flow under pressure, vibration, or shock may cause a soft bearing to yield or a ball socket to expand, permitting greater movement and pounding during operation. This causes loss of dimension.

Corrosion can cause loss of dimension such as the rusting of a frame or rod until it breaks under load.

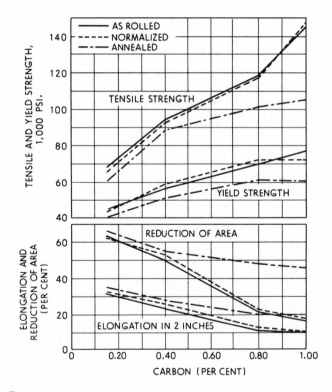

FIGURE 6-17
Average properties of 1-in commercial carbon-steel bars, as they vary with carbon content and with treatment subsequent to final rolling.

The loss of dimension due to abrasion, scoring, galling, pitting, flow, and corrosion may determine the life of a product. The loss of 5 lb of metal may make the difference between a brand new and a completely worn-out 5-ton truck; the 10,000-lb machine becomes completely worn out by the loss of 5 lb in critical places. Indefinite life can be obtained by smooth surfaces and the use of noncorrosive lubricants.

Metallurgical Factors in the Selection of Steels

Metallurgical factors that determine the useful properties of steel are not very numerous. Details of heat treatment, critical points, critical cooling rates, and similar matters will not be discussed here (information can be obtained from books on metallurgy and metals), but those things must be borne in mind in selecting steel.

General data on representative properties of dozens of much-used steels are available from any steel company. The typical chart contains the recommended heat treatment; forging temperature; quench, normalizing, annealing, or quenching temperature; size of piece treated to give the indicated results; the critical points; chemical specifications; and often the analysis of particular samples used. Curves show the Brinell hardness, tensile strength, yield point, elongation in 2 in, reduction of area, and Izod impact, all as they vary with the drawing temperature.

A group of actual test data would not fall neatly on these curves, but would occupy a band across the diagram, whose lower region is marked by the published curves. Manufacturers usually publish minimum values in order that they can be reasonably sure of meeting their own specifications (see Fig. 6-16).

Properties of Steels As Purchased

What can the production-design engineer normally expect of forging billets, hot-rolled bars, normalized bars, annealed bars, and cold-drawn bars? One alloy steel will not be praised or belittled; the primary aim is to show the similarity between steels—features that they have in common—not the differences between steels that have been talked of so much by proponents of single alloying elements. Those differences are sometimes exaggerated. As far as rolling-mill practice goes, the finishing temperature as the steel emerges from the last roll is of most significance to the buyer. The steelmaker can "roll hot," using a high temperature to the benefit of the buyer if he wants a coarse grain in the bar, as rolled. Low finishing temperatures usually induce a fine grain and an associated high degree of toughness, even after subsequent heat treatment. Grain-size differences cause a noticeable difference in machinability and hardenability. Thus, finishing temperature can be considered another variable in quality steelmaking practice that must be kept under control.

Steel can be purchased to many size tolerances. Tolerances are relatively large for hot-rolled steels; likewise, such bars may be out-of-round 0.010 or even 0.020 in. Since they have scaled surfaces, there will be handicaps for some uses, particularly in automatic screw machines where the work is held in collets. If the production-design engineer wants to use the OD (outside diameter) of a bar as a portion of a completed part without any finishing, he will then want closer tolerances. He can buy such accurate hot-rolled bars by paying extra for them, but ordinarily he would want to choose cold-finished bars.

What could fairly be expected in a steel bar of a certain carbon content, either as-rolled, normalized, or annealed (any of these conditions being purchasable from the steel mill)?

Figure 6-17 summarizes the tensile properties as affected by carbon content and final mill treatment. Note the solid lines, which show the as-rolled properties of these steels. The tensile strength goes up as the carbon content goes up. With 0.20 percent carbon steel, there may be 70,000 lb/in^2 in an as-rolled bar; when the carbon content goes up to 0.80 percent, 120,000 lb/in^2 are obtained. The yield strength also goes up, but not strictly in proportion. The reduction of area and the ductility go down, as might be expected. Between 0.20 and 0.80 percent carbon, there is a gain of perhaps 70 percent in tensile strength, but a loss of about two-thirds of the reduction of area, and almost as much of the elongation. In other words, as the carbon goes up the tensile strength increases, but the ductility goes down just as fast.

When the tensile strength passes beyond 100,000 lb/in^2, machining becomes difficult. It is, therefore, frequently necessary to buy the higher-carbon steels in the annealed or normalized condition to counteract the hardening and strengthening effect of the rather rapid cooling on the runout table from the last stand in the hot-rolling mill. Particularly is this true in small sizes. In such cases, better machinability is associated with a lower hardness number. For lower-carbon steels where normalizing does not affect the hardness much, the as-rolled bar may be expected to machine a little better, because its less-uniform microstructure would probably produce a slightly more brittle, freer-breaking chip.

Note that the tensile curves for normalized steel approximately coincide with the ones for as-rolled steels. ("Normalizing" means heating above the critical temperature and then cooling in still air.) Since the difference in tensile properties between normalized and as-rolled bars is very slight, if the final rolling temperature is low and cooling on the hot bed is at a moderate rate, normalizing would be done to ensure a more uniform microstructure and to release internal strains which would cause warpage on machining or subsequent heat treatment.

The broken lines in Fig. 6-17 show the properties of annealed bars. ("Annealing" is heating above the critical temperature range and then cooling at a slow rate.) Annealing removes all hardening that may have taken place during the more rapid cooling after rolling. It has lowered the tensile and yield strengths, more in the higher-carbon than in the lower-carbon steels. Reduction of area and elongation are likewise improved greatly.

Above 0.20 percent carbon the HR (hot-rolled) alloy steels are usually purchased in the annealed condition for the sake of machinability. Alloy steels are too expensive not to be heat-treated so as to secure full use of their properties.

Bars may be purchased with the following special surfaces:

1   Rough-turned
2   Turned and polished or centerless-ground
3   Cold-drawn
4   Cold-rolled

Effects of Cold Work

Cold work, even as little as 12.5 percent ($\frac{1}{16}$ in on 1-in bar) affects the properties of the bar to the very center, as anyone knows who has used cold-drawn steel for its good machining characteristics. This important property is often not understood; the gain is more than surface gain. Hot-rolled bars have Brinell hardness of about 125 all the way across the section. After a $\frac{1}{32}$-in draft, the surface hardness has increased to 185 and the center to 165. After a $\frac{1}{16}$-in draft, the surface increases further and the center catches up more nearly to the surface; and with a $\frac{1}{8}$-inch draft, the hardness is approximately uniform from center to surface. See Table 6-10.

The proportional limit of steels (carbon or alloy) is not raised by cold-drawing; in fact, the stress-strain curve in a tension test departs from a straight

Table 6-10   EFFECT OF COLD-DRAWING CARBON STEELS AND SCREW STOCK

SAE steel	Draft on 1-in bar	Tensile strength	Yield point	Elonga- tion	Reduc- tion of area	Charpy impact
1020	None*	64,000	48,000	36	68	53
	$\frac{1}{16}$ in.	83,000	74,000	19	61	44
	$\frac{1}{8}$ in.	91,000	82,000	15	58	34
1112	None*	71,000	54,000	34	59	30
	$\frac{1}{16}$ in.	97,000	89,000	16	48	16
	$\frac{1}{8}$ in.	105,000	98,000	13	44	12
1045	None*	109,000	58,000	25	52	—
	$\frac{1}{16}$ in.	118,000	79,000	15	44	19
	$\frac{1}{8}$ in.	130,000	85,000	12	38	14

* Hot-rolled 1-in rod.

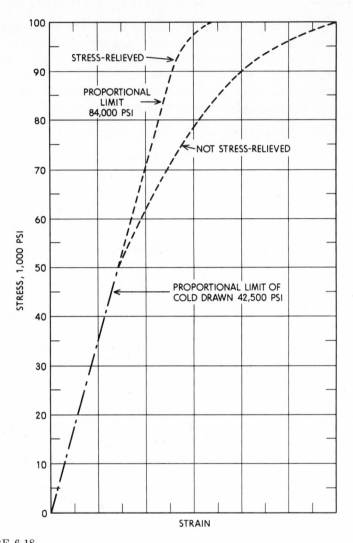

FIGURE 6-18

True proportional limit is reached by cold drawing, although the steel can be more heavily loaded without taking a permanent set. Stress relief at 900°F will restore and increase the proportional limit to a high figure.

line at a rather low figure. However, the load that causes definite yield (permanent set) is raised. This is shown in part by Fig. 6-18.

Figure 6-18 shows two stress-strain curves for the same steel plotted one on top of another; one curve was taken from a cold-drawn sample, the other from a piece of the same bar after stress relief by annealing at 900°F. The curve for cold-drawn steel deviates from true proportionality at about 42,500 lb/in^2. (It might have been even lower if the test had been performed with extensometers of higher sensitivity.) The actual yield point (beginning of plastic deformation) is much higher at a point that cannot be noted on a simple test to fracture, but the test piece is acting in an elastic manner. The process known as stress relieving (heating the steel up to a temperature of 630 to 900°F) produces a proportional limit of 84,000 lb/in^2 in the cold-drawn steel bar.

The residual stresses within cold-worked materials cause unpredictable size changes during machining. Proper stress relief is necessary for such alloys if distortion is to be avoided and proper tolerances achieved. For example, a plant made a series of large shafts from welded tubes which were swaged shut on each end for a distance of 8 to 10 in. The swaging operation was often finished below the recrystallization temperature of the steel. However the company engineers did not realize that and spent many thousands of dollars trying to solve the machining problem by calling in consultants and machining experts. In the end proper stress relief turned out to be the answer to the problem.

Selection of Suitable Steel

There is not a great difference between "this" steel and "that" steel; all are very similar in mechanical properties. Selection must be made on factors such as hardenability, price, and availability, and not with the idea that "this" steel can do something no other can do because it contains 2 percent instead of 1 percent of a certain alloying element, or it has a mysterious name. A tremendous range of properties is available in any steel after heat treatment; particularly is this true of alloy steels.

Ease of hardening (hardenability) The correct selection for a given part will often depend on hardenability, which can be defined as "the depth to which steels can be hardened in quenching." Various alloy steels will differ considerably in this respect. When they are properly hardened, all of these alloy steels with a given carbon content have similar properties. The difficulty is in hardening them. The effectiveness of heat treatment is what is bought in alloy steels. The maximum hardness attainable is a function of the carbon content and practically nothing else, but the quench necessary to get that hardness may be difficult to obtain. The effect of the alloy is to raise the hardness not at the surface, but at the center of the bar when it is quenched.

"The size of bar fully hardenable," as listed, is regarded as that bar which will show, when quenched, a hardness of not less than C-50 at the center. It is well known that the quenching rate in oil is not as fast as in water, and consequently does not have as much effect at the center of the bar. Therefore, the size that can be hardened to Rockwell C-50 minimum is smaller when quenched in oil than when quenched in water. One thing to be emphasized is that 1.5

percent manganese steels 1330 and 1340 will harden in water to Rockwell C-50 at the center of a $1\frac{1}{4}$-in bar. These are the cheapest alloy steels shown on the chart. More expensive steels have scarcely any difference: 2340, $1\frac{3}{8}$ in; 5140, $1\frac{1}{2}$ in; 3130, $1\frac{1}{8}$ in; 6150, $1\frac{3}{8}$ in.

Considerations in Fabrication

The properties of the final part (hardness, strength, and machinability), rather than properties required by forging, govern the selection of material. The properties required for forging have very little relation to the final properties of the material; therefore, not much can be done to improve its forgeability. Higher-carbon steels are more difficult to forge. Large grain size is best if subsequent heat treatment will refine the grain size.

Low-carbon, nickel-chromium steels are just about as plastic at high temperatures under a single 520-ft·lb blow as plain steels of similar carbon content. Nickel decreases forgeability of medium-carbon steels, but has little effect on low-carbon steels. Chromium seems to harden steel at forging temperatures, but vanadium has no discernible effect; neither has the method of manufacture any effect on high-carbon steel.

Formability The cold-formability of steel is a function of its tensile strength combined with ductility. The tensile strength and yield point must not be high or too much work will be required in bending; likewise, the steel must have sufficient ductility to flow to the required shape without cracking. The force required depends on the yield point, because deformation starts in the plastic range above the yield point of the steel. Work hardening also occurs here, progressively stiffening the metal and causing difficulty, particularly in the low-carbon steels.

It is quite interesting in this connection to discover that deep draws can sometimes be made in one rapid operation that could not possibly be done leisurely in two or three. If a draw is half made and then stopped, it may be necessary to anneal before proceeding, that is, if the piece is given time to work-harden. This may not be a scientific statement, but it is actually what seems to go on.

Surface A good surface on the steel is very important for any drawing operation for two reasons: (1) to get slippage through the dies, and (2) because there is a terrific amount of stress and of deformation at single points, and stress raisers will cause trouble there. For example, if a deep draw around a punched hole which is rough inside is attempted, the part may split from the punch marks around the original hole. (Better results will be obtained from reaming, followed by deep drawing.) The stress raisers combined with the work hardening that has taken place locally will interfere with satisfactory drawing. Of course the finish of the dies or forming tools is equally important from the standpoint of easy slippage.

Internal stresses   Cold forming is done above the yield point in the work-hardening range, so internal stresses can be built up easily. Evidence of this is the springback as the work leaves the forming operation, and the warpage in any subsequent heat treatment. Even a simple washer might, by virtue of the internal stresses resulting from punching and then flattening, warp severely during heat treating.

When doubt exists as to whether internal stresses will cause warpage, a piece can be checked by heating it to about 1100°F and then letting it cool. If there are internal stresses, the piece is likely to deform. Pieces that will warp severely while being heated have been seen, yet the heat-treater was expected to put them through and bring them out better than they were in the first place.

Welding   The maximum carbon content of plain carbon steel safe for welding without preheating or subsequent heat treatment is 0.30 percent. Higher-carbon steels are welded every day, but only with proper preheating. There are two important factors: (1) the amount of heat that is put in, and (2) the rate at which it is removed.

Welding at a slower rate puts in more calories and heats a large volume of metal, so the cooling rate due to loss of heat to the base metal is decreased. A preheat will do the same thing. For example, SAE 4150 steel, preheated to 600 or 800°F, can be readily welded. When the flame or arc is taken away from the weld, the cooling rate is not so great, due to the higher temperature of the surrounding metal, and slower cooling results. Even the most rapid air-hardening steels are weldable, if preheated and welded at a slow rate.

Machinability   Machinability means several things. To production men it generally means removing metal at the fastest rate, leaving the best possible finish, and obtaining the longest possible tool life. Machinability applies to the tool-work combination.

It is determined not alone by hardness, but by the toughness, micro-structure, chemical composition, and tendency of a metal to harden under cold work. In the misleading expression "too hard to machine," the word "hard" is usually meant to be synonymous with "difficult." Many times a material is actually too soft to machine readily. Softness and toughness may cause the metal to tear and flow ahead of the cutting tool rather than cut cleanly. Metals that are inherently soft and tough are sometimes alloyed to improve their machinability at some sacrifice in ductility. Examples are lead in brass and sulfur in steel.

Machinability is a term used to indicate the relative ease with which a

Table 6-11 RELATIVE MACHINABILITY RATING AS RELATED TO
MACHINING CHARACTERISTICS

Relative machinability rating	Machining characteristics
85 and above	Free machining.
70 to 80	Easily machined.
55 to 65	Difficult machining.
40 to 50	Very difficult machining. May be unmachinable in thin sections or with operations requiring extreme pressure or tools of weak section. Carbide tools generally recommended.
35 and under	Unmachinable by ordinary shop methods. Such materials require grinding or special techniques.

material is machined by sharp cutting tools in operations such as turning, drilling, milling, broaching, and reaming.

In the machining of metals, the metal being cut, the cutting tool, the coolant, the process and type of machine tool, and the cutting conditions all influence the results. By changing any one of these factors, different results will be obtained. The criterion upon which the ratings listed are based is the relative volume of various materials which may be removed by turning under fixed conditions to produce an arbitrary fixed amount of tool wear.

Although machinability ratings are relative, it must not be assumed, for example, that a steel with a rating of 40 would simply require twice as much

Table 6-12 MACHINABILITY RATINGS FOR
CERTAIN STEELS

Cold-drawn		Annealed	
SAE no.	Value	SAE no.	Value
1113	135	3140	55
1112	100	4130	65
1315	85	4150	45
1335	70	5140	45
1020	75	4640	55
1020	75	6140	40
1040	60	1330	50
3115	50	1340	40
4615	60	2350	35
2515	30	52100	30

time and cost to machine as another steel with a rating of 80. Actually, the steel with a rating of 40 may not even be machinable by ordinary methods (see Table 6-11).

The machinability of materials is a major factor in the cost of the product. The size, strength, composition, and heat treatment of the material can be selected to give the best machining conditions and the lowest costs. Machinability is an important item to be considered by the designer, as well as by the engineer in the shop.

In general practice, a Brinell hardness in the neighborhood of 180 is acceptable to the machine shop, especially if this hardness is combined with poor ductility, although, with the will and the tools to do it, materials that are heat-treated to a surprisingly high hardness can be machined. Frequently the machinability of a given steel can be helped by heat treatment; that is, it can be normalized, if it is a low-carbon material, in order to bring up the hardness to something approximating 180 and likewise reduce the ductility. SAE 1015 or 1020 steels may even be quenched to bring up the hardness.

Table 6-12 summarizes the machinability ratings for a number of steels in the cold-drawn and annealed condition. Table 6-13 lists ratings for different materials. These figures give some indication of the relative machinability, but so many things influence machining that the use of a table is not completely

Table 6-13 MACHINABILITY RATINGS FOR VARIOUS METALS

Material	Machinability rating	Brinell
B1112 screw-machine steel, high-sulfur	100	179–229
Standard malleable	120	110–145
Cast iron		
Soft	80	160–193
Medium	65	193–220
Hard	50	240–250
Cast steel (0.35 C)	70	170–212
Wrought iron	50	101–131
Carbon steel (C1010)	50	131–170
Stainless steel (18–8 FM)	45	179–212
Tool steel	30	200–218
Free-cutting brass (35.0 Zn, 3.0 Pb)	100	77
Naval brass (39.25 Zn, 0.75 Sn)	30	87
10 to 20 percent silver	20	88–94
12 percent leaded nickel-silver	60	88
5 percent aluminum-bronze	20	92

satisfactory. For the purpose of the table, the machinability of cold-drawn screw stock is taken as 100 percent. The improved 1112 will rate 135 percent. The implications of that can be realized by comparing it with the value for SAE 1020, which is similar in carbon and manganese to the 1112, but has lower sulfur and low phosphorus, and it will have a machinability somewhere in the neighborhood of 75 percent. The considerably higher percentage of sulfur and phosphorus makes 1112 less ductile in the transverse direction.

Lead has also been successfully added to steel to facilitate the cutting action. It apparently acts in essentially the same manner as sulfur, producing discontinuities in the chip transversely without wearing the tools at a high rate. The actual lubricating value of the lead is probably negligible.

Improving the machinability of a given material means that the mechanical properties must be changed, either by reducing the tensile strength or damaging the ductility. Cold drawing will help the low-carbon materials because it will increase the hardness.

From the strictly metallurgical standpoint, the material in process must be homogeneous, developing strength uniformly. If today a piece of 1045 machines with a given setup at a satisfactory rate, and tomorrow a new batch of material can no longer achieve the standard output, it may be because of nonhomogeneity of the steel or its treatment.

Design and Process Affect Selection

Variations in design and manufacturing have a greater effect on gear life, due to their influence on stress concentration, than do variations in materials and heat treatment. The effect of metallurgical change is obscured by the effect of these stress-concentration factors, which include:

1   Elastic deflection of teeth, shafts, and bearings
2   Surface cuts and scratches
3   Tooth shapes having abrupt change of section
4   Eccentric assemblies
5   Heat-treatment distortion and incipient quenching and grinding cracks
6   Nonuniform load distribution along the tooth

It is suggested that to obtain improvement in gear life the engineer should first look to improvement in design and manufacturing before resorting to metallurgical change. The sensible thing to do is to pick a steel with regard to the possibility of heat treating that steel without undue distortion.

Although steels of a wide range of laboratory properties are available, service life may frequently depend largely on nonmetallurgical factors. Selection of steel for a given job must include consideration of treatment, material cost, processing cost, and ease and accuracy of processing and manufacturing procedure (for example, stress raisers such as notches, bad machining, lack of fillets, welding damage, and the hardening due to welding).

The man who can use plain carbon steel in an application where a competitor is using alloy steel and get equally satisfactory life has a bargain, but he has earned it by virtue of a better application. There is one fundamental principle: smart engineering is to try the inexpensive way first. If the high-cost way is tried and it works, the engineer may be afraid to change. If the inexpensive and simple way does not work, a change must be made. Remember, when considering steel prices, the first cost of the raw steel (while worthy of consideration) is a small portion of the total cost of the part, and the difference in prices between any two of these alloy steels may be trivial as compared to the overall cost of making and using a particular part.

Stock material and material in steel warehouses should be used. Some of the factors in the cost of a steel part that must be kept in mind are the grade, size, quantity, special conditioning, special inspection, cost of machining, shape, and sequence of operations.

There is a possibility that the steel may be heated as the first operation; this has some advantages. On a turned shaft the advantage of no distortion and no scale on the finished part is possible. Likewise steel cut to length may be delivered adjacent to the furnace where the heat-treating operation will be performed. This will save a lot of handling in the plant, an important item of expense. Another fact that should be kept in mind is that some of the steels will occasionally come at about the maximum hardness for reasonable machinability. If such a steel is specified to be machine-cut to length, the mill must be sure that it is machinable, and in that way, annealing extras may be saved.

Heat treating first, of course, may harden and toughen the part to where it will cost a lot more in the machine shop because of the reduced tool life, the time consumed in machining, and so on. These extra costs must be added in. It might be mentioned here that one of the principal advantages of steel castings and forgings is that parts are obtained that are partially "machined" beforehand—that are approximately to shape before any work is done on them.

Another important thing that must be kept in mind is process cost. How much loss is experienced on every 100 pieces? How many pieces crack or warp? How many pieces give sufficient difficulty in machining that the pieces themselves are spoiled?

All that has been written in this chapter indicates the truth of the following statement made by one of the best-qualified production metallurgists: "Production metallurgy is 95 percent mechanical engineering." That is true. Now and then every metallurgist will get backed into a corner by a designer, experimenter, or production man who says, "See here, that part isn't heated right." The engineer may know that he has not made a mistake, but simply to be right is not enough; he must show that man that *he* made a mistake, and to convince him he must talk his language. The engineer must, therefore, be prepared to defend himself in mechanical work because the majority of problems encountered will be mechanical. Even when the problem is metallurgical, the mechanical factors may be a useful key to the solution.

SUMMARY

Ferrous materials are certainly the most widely used and most important to the majority of product design engineers. It is essential that the engineer understand the usefulness of ferrous materials, their potentialities and available properties, and how they can be used to the best advantage. He should have the knowledge to get the most out of one specific type of cast iron or steel and be certain what that "most" is. He should also understand the significant factors that interfere with the attainment of these useful properties.

Mechanical properties, as ordinarily revealed by the tensile, notched-bar, fatigue, hardness, and wear tests, are important in correct material selection. Likewise, metallurgical characteristics, such as grain size and hardenability, are important criteria in material selection. Furthermore, shop factors, such as welding and machinability, have a direct bearing on steel selection as well as the cost and availability of the ideal analysis.

The use of statistics can prove quite helpful in the selection of a material or a material treatment. For example, the student's t distribution provides information as to whether a sample whose mean value is $\bar{x}$ could have come from a population whose mean value is μ and whose standard deviation is σ.

Let us assume that a treatment of a certain steel in order to arrive at a hardness of C65 on the Rockwell scale calls for a 1-h draw after being brine-quenched from 1450°F. The soak of 1 h is specified at 350°F. The production-design engineer checks the hardness of a sample of 10, and the results are 61, 62, 64, 63, 63, 64, 62, 62, 65, 64. He would like to know if these hardness values, which appear lower than specifications, are representative of the treatment specified or if they appear to be representative of a different population and a change in treatment would be in order. The t test can provide the answer:

$$t = \frac{|\mu - \bar{x}|\sqrt{n - 1}}{s}$$

where μ = population mean
$\bar{x}$ = sample mean
n = sample size
s = standard deviation of sample
In this example:

$$t = \frac{|65 - 63|\sqrt{9}}{1.18} = 5.1$$

For 9 degrees of freedom at a value of 0.001 (one chance in 1,000), we have a tabulated value of t of 4.297. Since our computed value of t is 5.1, we reject the null hypothesis that the sample is not different from the population. Thus, with confidence, we will alter our treatment of the material in order to obtain a Rockwell reading closer to the desired hardness of C65.

QUESTIONS

1 Why is cast iron frequently used in the construction of large jigs and fixtures?
2 What properties of cast iron make it a much-used material for frames and beds of machine tools?
3 What is ductile iron? How is it produced?
4 What basic concepts assist the production-design engineer in the selection of a steel?
5 What are the principal characteristics of carbon steel? Of alloy steel? Of tool steel?
6 What does the term "stainless steel" denote?
7 What is meant by "camber" tolerances?
8 How would you recommend the finishing of the outside diameter of a piece of steel of 300 Brinell?
9 What would be the approximate tensile strength of a piece of steel with a Brinell reading of 180?
10 Differentiate between the Rockwell, Brinell, and scleroscope hardness test. What would be the equivalent value of Brinell 200 on the Rockwell and scleroscope scales?
11 What does "toughness" infer in a specimen of steel?
12 What is the most important factor in the heat treatment of steel?
13 What would be the maximum hardness of a piece of steel with 0.50 percent carbon?
14 Describe the Jominy test.
15 What would be the approximate hardness of a $1\frac{5}{8}$-in-diameter chromium-nickel bar $\frac{7}{8}$ in from the outside surface? Assume a 0.75 percent carbon content and a quench into warm brine after heat treatment.
16 Why is water usually preferred as a quenching medium?
17 How may case-hardening be accomplished?
18 How is machinability determined?
19 About what Brinell hardness is thought of as being desired for general machining work?
20 How can a piece of steel be checked to determine whether internal stresses will cause warpage?
21 Describe the two allotropic forms of iron.
22 Explain why the iron-carbon equilibrium diagram is important to the production-design engineer.
23 A class 50 gray-iron casting has a wall thickness of $\frac{1}{2}$ in. It is to be machined to a 100 root-mean-square surface finish. What recommendations can you make?
24 If the expected value of resistance to corrosion of a given steel is 200 h when sub-

jected to a 20 percent salt spray, what would be your evaluation of 10 samples of this steel that gave the following hours of resistance: 180, 190, 150, 220, 200, 175, 180, 200, 205, 170?

PROBLEMS

1   Compare the structure and mechanical properties of AISI 1020, 1050, 1080, and 1.2 percent carbon steels in the fully annealed state and as quenched.

2   Clearly differentiate between the following:

   (a)   Substitutional and interstitial solid solutions
   (b)   Eutectic and eutectoid compositions
   (c)   Peritectic and peritectoid compositions
   (d)   Ferrite and austenite
   (e)   Pearlite and cementite
   (f)   Martensite and bainite

3   A farm-machinery company received field complaints that an SAE 1020 steel lever deflected excessively when a celery picker was operated at high speed. A quick check showed that the operating load on the lever was well within the yield strength of the material but that dynamic system imposed inertial forces caused by the mass and acceleration of the lever in operation. The design engineer decided to change to an aluminum alloy for the lever to reduce its mass and thereby to decrease the inertial forces and the deflection.

   (a)   Derive an analytical relationship to test the validity of the proposed solution.
   (b)   Explain what properties of a material affect its resistance to deflection.
   (c)   Would changing from a low-carbon steel to a high-strength, heat-treated alloy steel solve the problem? Explain your answer.
   (d)   Compare the deflection of levers of the same design but made from the materials listed below:

      (1)   Aluminum
      (2)   Gray iron
      (3)   Malleable iron
      (4)   Magnesium
      (5)   Steel
      (6)   Bronze

   (e)   In what ways, other than changing the material, could the stiffness of the lever be increased?
   (f)   What material(s) could be used to provide an inherently stiffer lever of the original design? Would such materials be economically feasible?

4 Traditionally, only nonferrous alloys have been die cast, but recently a number of steel castings weighing up to a pound or more have been successfully produced by the die-casting process. Several potential die materials have been investigated, three of which are TZM, a molybdenum alloy; SiN, silicon nitride; and W-Cu, a powdered metal compact of tungsten and copper.

(a) Compare the heat absorption by each of those die materials when a flat plate 3 × 3 × ¼ in thick is die cast from 18-8 stainless steel which is injected from the shot cylinder at 2780°F (about 100°F superheat), and from aluminum alloy when injected into an identical cavity in an H-13 alloy die at 100°F superheat. The cavity is cut ⅛ in into each die half.

(b) Find the steady-state die temperature of a plane ¼ in from the flat bottom of each die cavity. Of course that temperature would be only a rough comparison between materials and not the temperature reached in practical die-casting operations. Base your calculations on a ¼ × ¼ in element at the center of the die cavity and consider the latent heat of fusion.

(c) Determine the solidification time for each alloy in each die material assuming quasi-steady-state conditions.

(d) On the basis of your calculations, discuss the relative merits of each die material from the standpoint of heat transfer compared with that of aluminum alloy in H-13 dies.

PHYSICAL PROPERTIES OF THE MATERIALS

Material	Conductivity, Btu/ (hr)(ft)(°F)	Density, lb/ft³	Specific heat, Btu/ (ft³) (°F)	Thermal diffusivity, ft²/h	Latent heat, Btu/lb	Injection temperature, °F
TZM	101.8	633.4	0.07	2.34		
W-Cu	81.4	1076	0.036	2.09		
H-13	30.2	490.8	0.12	0.52		
SiN	13.1	208.2	0.24	0.27		
304 St. St.	20.0	501.1	0.12	0.33	117.0	2780
43 Aluminum	82.2	171.6	0.23	2.09	149.5	1250

5 In high-speed mechanisms, sudden shocks should be avoided if at all possible. Frequently a cam can be provided to furnish uniform acceleration and deceleration, as in the case of a heart-shaped cam design.

(a) Design a uniform motion cam which has a throw of 2½ in.

(b) Plot a displacement-vs.-time diagram for the motion of the follower.

(c) Make detailed sketches of the follower design and select an appropriate design for operation at a speed of up to 120 Hz.

(d) Make a velocity-vs.-angular-displacement diagram for the follower.

(e) Select a material for the cam and give in detail the operation sequence, the equipment, tooling, and procedures for producing a small lot of five cams.

SELECTED REFERENCES

AMERICAN SOCIETY FOR METALS: "Metals Handbook," vol. 7, "Atlas of Microstructure of Industrial Alloys," 8th ed., Metals Park, Ohio, 1972.

————: "Tool Steels," 3d ed., Metals Park, Ohio, 1962.

BRIGGS, CHARLES W.: "Steel Castings Handbook," Steel Founders' Society of America, 1960.

DUMOND, T. C. (ed.): "Engineering Materials Manual," Reinhold, New York, 1954.

U.S. STEEL CORP.: "The Making, Shaping, and Treating of Steel," Pittsburgh, 1957.

WALTON, CHARLES F.: "Gray Iron Castings," Gray Iron Founders' Society, Inc., 1968.

7

NONFERROUS METALS

Although nonferrous metals comprise only 12 percent by volume of all metals used in commercial production, some are of major importance, such as aluminum to the aircraft industry, lead in storage batteries, or chromium for plating. A comparison based on the dollar value of the finished product would show that nonferrous metals hold a considerably higher place because they are used for components which have high finished value.

The unique properties of aluminum, copper, zinc, lead, magnesium, nickel, cobalt, chromium, and titanium make them useful for components of a large number of designs. In fact about 40 nonferrous metals are of commercial value today. Nonferrous alloys are most often used for parts in which there is considerable labor added in the form of secondary operations or where their unique properties, such as high resistance to corrosion, justifies their higher cost. Usually a combination of several properties is sufficient basis for selecting a nonferrous alloy; these include ease of fabrication, light weight, corrosion resistance, good machinability, electrical or thermal conductivity, color, ability to absorb energy, or good strength-weight ratio.

ALUMINUM

Aluminum alloys are the most important nonferrous alloys. The art and science of fabricating and applying aluminum has developed a vast store of information with which the production-design engineer must be familiar.

The selection of aluminum or an aluminum-base alloy for use in a particular product or structure is usually based on one or more of the following design considerations: (1) weight-strength ratio, that is, in applications where lightness of the final product is desirable (it has a specific gravity of 2.70); (2) when ease of machining, fabrication, or forming is of major importance; (3) where resistance to atmospheric corrosion or attack by certain chemicals is required; (4) where low electrical resistance is a requisite; (5) when high heat and/or light reflectivity or low emissivity is needed; (6) where such properties as acoustical deadness, nontoxicity (to foods), or nonsparking or nonmagnetic properties are desirable.

Aluminum alloys are produced in practically all of the forms in which metals are used: plate, sheet, rod, wire, tube, forgings, castings, and ingots, as well as rivets and screw-machine products. Shapes may be the standard structural shapes or they may be of special design that can be produced only by extrusion. Forgings may be made by pressing or by the drop-forging method. Castings may be poured into sand, in cast-iron molds, or in special plaster or other refractory material, or may be made by forcing the molten alloy into a steel die under pressure. The various forms are fabricated into finished shapes and structures by drawing, stamping, spinning, hammering, machining, welding, brazing, riveting, and, in some cases, soldering. Ease of fabrication is one of the reasons for the choice of aluminum alloys.

Notice the reference is to aluminum alloys. High-purity aluminum, while it has many desirable characteristics, has a tensile strength of only about 9,000 lb/in². Even though this strength can be slightly more than doubled by cold working, the resulting strength is still not high and it is not heat-treatable.

Pure aluminum finds little use in the production of castings, not only because of its low strength, but also because of its inferior casting qualities. The only castings practically produced from commercially pure aluminum are those requiring the higher electrical conductivity of the pure metal. Two examples are induction motor rotors and cable clamps.

Designation System for Aluminum Alloys

The wrought aluminum alloy designations follow a four-digit index system in which the first digit indicates the alloy type, the second digit indicates the alloy modifications, and the last two digits indicate the aluminum purity. The following table gives the various types of aluminum alloys and indicates the corresponding first digit in the aluminum identification.[1]

[1] Courtesy, Aluminum Corporation of America from ALCOA "Aluminum Handbook."

Aluminum—99.00% minimum	1xxx
Copper	2xxx
Manganese	3xxx
Silicon	4xxx
Magnesium	5xxx
Magnesium and silicon	6xxx
Zinc	7xxx
Other element	8xxx
Unused series	9xxx

Temper designation of an aluminum alloy employs the letter O, the letter F, and letters H and T followed by one or more numbers. The letter follows the alloy designation and is separated from it by a hyphen. The letter O indicates the annealed temper of wrought materials and T2 indicates the annealed temper of casting materials. Temper designation F, in the case of a wrought alloy, indicates the as-fabricated condition. (This indicates no control has been exercised over the temper of the alloy.)

Temper designation T followed by one or more numbers indicates a heat-treated alloy. The first digit following the T signifies the type of heat treatment given the alloy. The heat treatments available, and their symbols, are:

T3 Solution heat treatment followed by strain-hardening. Different amounts of strain hardening of the heat-treated alloy are indicated by a second digit.

T4 Solution heat treatment followed by natural aging at room temperature to a substantial stable condition.

T5 Artificial aging after an elevated-temperature, rapid-cool fabrication process such as casting or extrusion.

T6 Solution heat treatment followed by artificial aging.

T7 Solution heat treatment and then stabilization to control growth and distortion.

T8 Solution heat treatment, strain hardening, and then artificial aging.

T9 Solution heat treatment, artificial aging, and then strain hardening.

Temperature designation H followed by a number indicates a cold-worked temper of a wrought alloy. Tempers in the H series are not applicable to castings. Cold-worked alloys are designated as follows:

*H1*x Strain-hardened temper produced by cold working the metal to desired dimensions. A second digit indicates the degree of hardness. The full-hard commercial temper is designated H18 and the extra-hard temper is H19. The designation H14 indicates material cold-worked to a tensile strength approximately midway between the fully annealed temper O and the full-hard temper H18. Cold working involves putting a permanent strain in the metal below its recrystallization temperature. Different amounts of cold working result in different hardnesses.

*H2*x   Strain-hardened temper produced by cold working the metal and then partially annealing it. Numbers 2 through 8 are used for the second digit in the same manner as in the H1 series.

*H3*x   Strain-hardened and stabilized.

A temper designated by the letter H followed by a three-digit number indicates that, for the temper designated by the first two digits of the numeral, normal fabricating practices have been varied to attain certain properties in the alloy for a special application.

Castings

The improvements in the casting qualities of the aluminum alloys as compared with the pure metal are perhaps even greater than the improvements in their mechanical properties. The commercial casting alloys, containing varying percentages of one or more alloying elements, have been developed to combine good qualities from the standpoint of foundry operations, with desirable mechanical properties in the finished castings.

Sand castings are most generally used when the quantity requirement is small, or if the casting requires intricate coring, or if the castings are very large. The minimum thickness of sections that can be cast in sand depends on the size and intricacy of the casting, the pressure tightness, and the casting alloy. The minimum thickness for small and medium-sized castings is $\frac{3}{16}$ in, although small castings with $\frac{1}{8}$-in wall thickness have been made. Tolerances in the order of $\frac{1}{32}$ in are practical in small castings but should increase with increased size. If machined, a casting is normally given $\frac{1}{8}$-in allowance, but for large castings, $\frac{1}{4}$ in or more is frequently provided.

The minimum wall thickness of castings poured in permanent molds and semipermanent molds is substantially the same as sand castings ($\frac{3}{16}$ in). However, tolerances and machining allowances need be only one-fourth as large. Molds should have a minimum draft of 1 to 5°; 3 to 5° is preferable.

Relatively thin and uniform sections are desirable for die casting. These castings may weigh a fraction of an ounce or as much as 80 lb. For castings having dimensions up to about 6 in, 0.045-in sections are practical; for 15-in castings, 0.080-in sections are practical; for larger dimensions, 0.150-in sections are usually used. Tolerances of 0.002-in/in can readily be maintained.

The principal casting alloys of aluminum are copper, silicon, magnesium, zinc, and nickel. The specific gravity of the various alloys does not vary significantly and for practical purposes, all alloys may be considered to weigh about 0.10 lb/in^3. All alloys have a modulus of elasticity of approximately 10,300,000 lb/in^2, a modulus of rigidity of 3,900,000 lb/in^2 and Poisson's ratio of 0.33.

It is apparent, in view of the low modulus of elasticity as compared with ferrous materials, that deeper sections must be used when a design indicates loading as a beam. However, the lower modulus of elasticity is an asset where greater deflections are desired under a given load such as in a relay contact flexible support and where greater resilience is desired. Resilience is the amount of energy absorbed upon loading within the elastic limit, per unit weight of material. Resilience is directly proportional to the yield strength and indirectly proportional to the modulus of elasticity. Thus, aluminum with the lower modulus of elasticity can absorb more energy than steel with a higher modulus of elasticity at the same stress.

Creep in aluminum castings will take place at or near the yield strength but usually will not occur at stresses lower than half the yield-strength value.

The majority of the aluminum casting alloys are welded easily; thus, castings may be joined readily with other aluminum components.

Alloys

Since the desirable qualities associated with aluminum depend on the alloy, the next question is, "What are these alloying elements?" While a variety of alloying elements are used in the production of aluminum alloys, copper, magnesium, silicon, manganese and, more recently, zinc are the more common. In some special-purpose alloys, nickel, tin, lead, and bismuth are added, while in many alloys, both cast and wrought, titanium or chromium comprise an important part of the composition. Generally speaking, the total alloy content is greater in casting alloys than wrought alloys.

The aluminum-silicon alloys have the best castability qualities but their mechanical properties are somewhat inferior to those aluminum alloys in which copper or magnesium are the principal alloying agents.

Nickel is used mainly to aid in maintaining mechanical properties at elevated temperatures.

Heat-treatable alloys contain at least one element that is more soluble in aluminum at higher temperatures than at room temperature. Copper is the element most frequently used in the heat-treatable alloys.

Heat Treatment

An important feature of some aluminum alloys is that their strength can be varied by heat treating. The metal can be annealed to make it extremely soft and easily worked during drawing and forming. Then after it has been shaped the metal can be given another and different heat treatment to make it extremely strong and hard, so that the surface is more resistant to wear. By properly selecting the alloying elements and combining heat treatments with small amounts of cold work, strengths in excess of 80,000 lb/in^2 can be secured. Heat treatment of aluminum alloy is covered in Chap. 5.

Table 7-1 ALUMINUM NON-HEAT-TREATABLE—PROPERTIES

Strength values (TENSILE, YIELD, SHEAR, ENDURANCE) in LB/SQ. IN. IN 1,000.

TYPE	FORM	GRADE	GRADE CODE	P.D. SPEC. NO.	TENSILE STRENGTH ①	YIELD STRENGTH ②	SHEAR STRENGTH ③	ENDURANCE LIMIT ④	AVERAGE ELONGATION % IN 2 IN. ⑤	BRINELL HARDNESS	BEND TEST SPECIMEN MIN. DIAM. ON PIN	ELEC. COND. % OF COPPER	THERMAL COND. AT 100° C. C.G.S. UNITS
COMMERCIALLY PURE ALUMINUM	BAR	COLD FIN. AS FABRICATED	2 S	7601-1	17	14	11	7	15 to 25	32	180° Flat Flat	58	.52
COMMERCIALLY PURE ALUMINUM	BAR	SOFT-COLD FIN. OR ROLLED	2 S-O	7601-2	15.5 Max.	4	9.5	5	25 Min.	23	180° Flat Flat	59	.53
COMMERCIALLY PURE ALUMINUM	BAR	ROLLED AS FABRICATED	2 S	7601-3	15	13	10	6	20 to 30	28	180° Flat Flat	57	.51
COMMERCIALLY PURE ALUMINUM	FLAT STRIP	SOFT	2 S-O	7601-4	15.5 Max.	4	9.5	5	15 to 30	23	180° Flat Flat	59	.53
COMMERCIALLY PURE ALUMINUM	FLAT STRIP	HALF HARD	2S-1/2 H	7601-5	16 Min.	14	11	7	2 to 6	32	180° Flat	58	.52
COMMERCIALLY PURE ALUMINUM	FLAT STRIP	HARD	2 S-H	7601-6	22 Min.	20	13	8.5	1 to 4	44	180° 3 X Tks.	57	.51
COMMERCIALLY PURE ALUMINUM	SHEET AND PLATE	SOFT	2 S-O	7601-12	15.5 Max.	4	9.5	5	15 to 30	23	180° Flat Flat	59	.53
COMMERCIALLY PURE ALUMINUM	SHEET AND PLATE	HALF HARD	2S-1/2 H	7601-13	16 Min.	14	11	7	3 to 7	32	180° Flat	58	.52
COMMERCIALLY PURE ALUMINUM	SHEET AND PLATE	HARD	2 S-H	7601-14	22 Min.	20	13	8.5	1 to 4	44	180° 3 X Tks.	57	.51
COMMERCIALLY PURE ALUMINUM	TUBING	SOFT	2 S-O	7601-7	15.5 Max.	4	9.5	5	15 to 30	23		59	.53
COMMERCIALLY PURE ALUMINUM	TUBING	QUARTER HARD	2S-1/4 H	7601-15	14 Min.	12	10	6	4 to 10	28		58	.52
COMMERCIALLY PURE ALUMINUM	TUBING	HALF HARD	2S-1/2 H	7601-8	16 Min.	14	11	7	2 to 6	32		58	.52
COMMERCIALLY PURE ALUMINUM	TUBING	HARD	2 S-H	7601-9	22 Min.	20	13	8.5	1 to 4	44		57	.51
MANGANESE ALUMINUM ALLOY	BAR	COLD FIN. AS FABRICATED	3 S	7602-1	21	18	14	9	10 to 20	40	180° 1 X Tks.	41	.37
MANGANESE ALUMINUM ALLOY	BAR	SOFT-COLD FIN. OR ROLLED	3 S-O	7602-2	19 Max.	5	11	7	25 Min.	28	180° Flat	50	.45
MANGANESE ALUMINUM ALLOY	BAR	ROLLED AS FABRICATED	3 S	7602-3	18	15	12	8	15 to 25	35	180° Flat	42	.38
MANGANESE ALUMINUM ALLOY	FLAT STRIP SHEET AND PLATE	SOFT	3 S-O	7602-4	14.5 to 19	5	11	7	20 to 25	28	180° Flat Flat	50	.45
MANGANESE ALUMINUM ALLOY	FLAT STRIP SHEET AND PLATE	QUARTER HARD	3S-1/4 H	7602-12	17 Min.	14	12	8	4 to 9	35	180° Flat Flat	42	.39
MANGANESE ALUMINUM ALLOY	FLAT STRIP SHEET AND PLATE	HALF HARD	3S-1/2 H	7602-5	19.5 Min.	18	14	9	3 to 8	40	180° Flat Flat	41	.37
MANGANESE ALUMINUM ALLOY	FLAT STRIP SHEET AND PLATE	THREE QUARTER HARD	3S-3/4 H	7602-13	24 Min.	20	15	9.5	1 to 4	47	180° 4 X Tks.	41	.38
MANGANESE ALUMINUM ALLOY	FLAT STRIP SHEET AND PLATE	HARD	3 S-H	7602-6	27 Min.	25	16	10	1 to 4	55	180° 3 X Tks.	40	.36
MANGANESE ALUMINUM ALLOY	TUBING	SOFT	3 S-O	7602-7	19 Max.	5	11	7	20 to 25	28		50	.45
MANGANESE ALUMINUM ALLOY	TUBING	HALF HARD	3S-1/2 H	7602-8	19.5 Min.	18	14	9	3 to 8	40		41	.37
MANGANESE ALUMINUM ALLOY	TUBING	HARD	3 S-H	7602-9	27 Min.	25	16	10	1 to 4	55		40	.36

DESIGN CHARACTERISTICS		
Maximum corrosion resistance. Can be torch, arc, and spot welded without appreciable loss in properties. Used for equipment to be left with bright finish. If painting is required, prepare surface by sand blasting, scratch brushing or etching. Can be anodized for additional corrosion resistance or decoration. Electrical characteristics good. Machining characteristics fair. Application: General use bright finish aluminum.	Corrosion resistance comparable to aluminum #7601. Can be torch, arc and spot welded without appreciable loss in properties. Can be painted without preparation of surface because of gray finish. Can be anodized for additional corrosion resistance. Cost same as aluminum #7601. Machining characteristics fair. Application: General use gray finish aluminum. Chemical Properties: 1.0 to 1.5% Manganese.	

The above materials are produced in the following sizes:

7601-1: Rounds 1-1/2 diam. or less
      Squares & Hexagons 1/32 to 1-1/2 across flats
      Rectangles 1/16 x 1/8 to 1-1/2 x 4

7601-2: Cold Finish—Same as 7601-1
      Rolled—Same as 7601-3

7601-3: Rounds 1-9/16 to 8 diam.
      Hexagons 1-9/16 to 2 across flats
      Squares 1 to 4 square
      Rectangles 3/32 x 1-1/8 to 3 x 7

7601-4: Thicknesses .006 to .102 for Flat Strip

7601-5: Thicknesses .010 to .085 for Flat Strip

7601-6: Thicknesses .006 to .102

7601-12: Thicknesses .010 to .249 for Sheet
        .250 and over for Plate

7601-13: Thicknesses .010 to .249 for Sheet
        .250 and over for Plate

7601-14: Thicknesses .010 to .128 for Sheet
        .250 and over for Plate

7602-1: Same as 7601-1

7602-2: Same as 7601-2

7602-3: Same as 7601-3

7602-4: Thicknesses .006 to .102 for Flat Strip
        .010 to .249 for Sheet
        .250 and over for Plate

7602-12: Thicknesses .017 to .102 for Flat Strip
        .017 to .249 for Sheet
        .250 and over for Plate

7602-5: Thicknesses .010 to .085 for Flat Strip
        .010 to .249 for Sheet
        .250 and over for Plate

7602-13: Thicknesses .006 to .053 for Flat Strip
        .010 to .162 for Sheet
        .250 and over for Plate

7602-6: Thicknesses .004 to .102 for Flat Strip
        .006 to .128 for Sheet
        .250 and over for Plate

Values specified for tensile, yield and shear strength and endurance limit are averages, unless otherwise specified.

① Young's Modulus of Elasticity is approximately 10,300,000 lb/sq.in.

② Stress which produces a permanent set of 0.2% of initial gage length (A.S.T.M. Specification E8-40T).

③ Single shear strength values obtained by double shear test.

④ Based on withstanding 500,000,000 cycles of completely reversed stress, using the R.R. Moore type of machine and specimen.

⑤ Elongation values vary with size and shape.

Table 7-2   ALUMINUM, HEAT-TREATABLE—PROPERTIES

COPPER-MAGNESIUM-MANGANESE-ALUMINUM ALLOYS

TYPE →	7603-1 17 S-O (BAR, SOFT ANNEALED)	7603-2 17 S-T (BAR, HEAT TREATED)	7603-7 17 S-T (BAR, HEAT TREATED ROLLED)	7603-3 17 S-O (SHEET AND PLATE, SOFT ANNEALED)	7603-4 17 S-T (SHEET AND PLATE, HEAT TREATED)	7603-5 17 S-O (TUBING, SOFT ANNEALED)	7603-6 17 S-T (TUBING, HEAT TREATED)	8490-1 24 S-O (BAR, SOFT ANNEALED)	8490-2 24 S-T (BAR, HEAT TREATED)	8490-4 24 S-T (FLAT STRIP, SHEET & PLATE, SOFT ANNEALED)	24 S-T (HEAT TREATED)	8490-8 24 S-RT (ROLLED AFTER HEAT TREATING)	8490-5 24 S-O (SOFT ANNEALED)	8490-6 24 S-T (HEAT TREATED)	8490-7 24 S-RT (ROLLED AFTER HEAT TREATING)
TENSILE STRENGTH [1] (LB/SQ.IN. IN 1,000)	35 Max.	50 to 55	50 to 53	35 Max.	55 Min.	35 Max.	55 Min.	35 Max.	57 to 70	35 Max.	58 to 62	68	35 Max.	64 Min.	70
YIELD STRENGTH [2]	10	28 to 30	28 to 30	10	32 Min.	10	40 Min.	10	40 to 52	10	40 Min.	50 Min.	10	42 Min.	58 Min.
SHEAR STRENGTH [3]	18	36	36	18	35	18	35	18	41	18	41	42	18	41	42
ENDURANCE LIMIT [4]	11	15	15	11	15	11	15	12	18	12	18		12	18	
AVERAGE ELONGATION % IN 2 IN. [5]	16 Min.	16 to 18	16 Min.	12 Min.	10 to 18	10 to 12	12 to 16	16 Min.	10 to 14	12 Min.	6 to 17	10 to 12	12 Min.	10 to 16	10 Min.
BRINELL HARDNESS	45	100	100	45	100	45	100	42	105	42	105	116	42	105	116
BEND TEST SPECIMEN MIN. DIAM. ON PIN	180° 1 X Tks.	180° 4 X Tks.	180° 4 X Tks.	Flat to 6 X Tks.	180° 3 to 8 X Tks.			180° Flat to 1 X Tks.	180° 6 X Tks.	180° Flat to 6 X Tks.	4 to 10 X Tks.	4 to 9 X Tks.			
ELEC. COND. % OF COPPER	45	30	30	45	30	45	30	50	30	50	30	30	50	30	30
THERMAL COND. AT 100° C C.G.S. UNITS	.41	.28	.28	.41	.28	.41	.28	.45	.28	.45	.28	.28	.28	.28	.28

DESIGN CHARACTERISTICS

(Left group — 7603 alloys)

Corrosion resistance inferior to aluminum #7601.

Can be spot welded without appreciable loss of properties. If spot welding is not practical, riveted construction is recommended. For high efficiency joints hot steel rivets are preferred.

Can be painted without special surface preparation. Annealed material can be formed and then heat treated to obtain maximum properties.

More expensive than #7601.
Machines freely.
Application: High strength construction, frames, rotor wedges, screw machine products.

3.5 to 4.5% Cu, .20 to .75 Mg, .40 to 1.0 Mn

(Right group — 8490 alloys)

Corrosion resistance comparable to aluminum alloy #7603.

Can be spot welded without appreciable loss of properties. If spot welding is not practical, riveted construction is recommended. For high efficiency joints hot steel rivets are preferred.

Can be painted without special surface preparation. Annealed material can be formed and then heat treated to obtain maximum properties.

Slightly more expensive than #7603.
Machines freely.
Application: Similar to #7603.

3.8 to 4.9% Cu, 1.2 to 1.8 Mg, .3 to .9 Mn

The above materials are produced in the following sizes:

7603-1:	Rounds 8 diam. or less; Squares & Hexagons 1/32 to 4 across flats; Rectangles 1/16 to 1/8 to 3 x 7
7603-2:	Same as 7603-1
7603-7:	Rounds 8 diam. or less; Squares & Hexagons 1/32 to 4 across flats; Rectangles (round edge) 1/8 x 5/8 to 1/2 x 6
7603-3 and 7603-4:	Thicknesses .010 to .249 for Flat Strip & Sheet; .250 and over for Plate

8490-1:	Same as 7603-1
8490-2:	Same as 7603-2
8490-3 and 8490-4:	Same as 7603-3
8490-8:	.019 to .249 for Flat Strip and Sheet; .250 and over for Plate

---

Values specified for tensile, yield and shear strength and endurance limit are averages, unless otherwise specified.

[1] Young's Modulus of Elasticity is approximately 10,300,000 lb/sq.in.

[2] Stress which produces a permanent set of 0.2% of initial gage length (A.S.T.M Specification E-8-40T).

[3] Single shear strength values obtained by double shear test.

[4] Based on withstanding 500,000,000 cycles of completely reversed stress, using the R.R. Moore type of machine and specimen.

[5] Elongation values vary with size and shape.

Plate Sheet and Foil

Aluminum that has been rolled to a thickness of more than 0.25 in is classified as plate, while thicknesses of 0.006 to 0.25 in are known as sheet. Foil refers to aluminum that has been rolled to thicknesses less than 0.006 in. Today foil is being produced to thicknesses of less than 0.0002 in.

Sheet of 0.102 in or less is available in both flat and coiled form. Usually there is a slight price advantage by purchasing sheet in coil form. If required, sheet can be purchased (at extra cost) from the mill with a bright finish. The degree of brightness will of course depend upon the grade and the temper specified. The standard mill finish will vary from a dull gray to a relatively bright finish, depending on the process conditions. It is possible to procure sheet with a standard mill finish on one side and a bright finish on the other.

Properties

Aluminum is not affected by magnetic fields; it is nonsparking, and is considered acoustically dead (desirable for such uses as in ducts where transmission of noise is objectionable). Aluminum is about three times as flexible as steel.

The information given on the accompanying instruction sheets is for the guidance of the production design engineer when choosing materials. Such information is limited and should be supplemented by current catalogs and list prices (see Tables 7-1 to 7-5).

Table 7-3   TOLERANCES OF ROLLED ALUMINUM STRUCTURAL SHAPES

Dimensions	Tolerance
Thickness of section	Plus or minus $2\frac{1}{2}$ percent of nominal thickness—minimum tolerance: ± 0.010 in
Overall dimensions; length of leg of angles or zees	Plus or minus $2\frac{1}{2}$ percent of nominal—minimum tolerance: $\pm \frac{1}{16}$ in
Length Up to 20 ft not incl. 20 ft to 30 ft incl. Over 30 ft	 Minus 0, plus $\frac{1}{4}$ in Minus 0, plus $\frac{3}{8}$ in Minus 0, plus $\frac{1}{2}$ in
Channels, depth	Plus $\frac{3}{32}$ in, minus $\frac{1}{16}$ in
Channels, width of flange	Plus or minus 4 percent of nominal width
Weight of a lot or shipment of sizes 3 in or larger*	Plus or minus $2\frac{1}{2}$ percent of nominal weight

* For sizes under 3 in, dimension tolerances only apply.

Table 7-4 ALUMINUM AND ALUMINUM ALLOY SHEET AND PLATE—SOME THICKNESS TOLERANCES (Plus or Minus)*

Thickness, in	Width, in				
	18 or less	Over 18 to 36 (incl.)	Over 36 to 48 (incl.)	Over 48 to 54 (incl.)	Over 54 to 60 (incl.)
0.007 to 0.010	0.001	0.0015			
0.011 to 0.017	0.0015	0.0015			
0.018 to 0.028	0.0015	0.002	0.0025		
0.029 to 0.036	0.002	0.002	0.0025		
0.037 to 0.045	0.002	0.0025	0.003	0.004	0.005
0.046 to 0.068	0.0025	0.003	0.004	0.005	0.006
0.069 to 0.076	0.003	0.003	0.004	0.005	0.006
0.077 to 0.096	0.0035	0.0035	0.004	0.005	0.006
0.097 to 0.108	0.004	0.004	0.005	0.005	0.007
0.109 to 0.140	0.0045	0.0045	0.005	0.005	0.007
0.141 to 0.172	0.006	0.006	0.008	0.008	0.009
0.173 to 0.203	0.007	0.007	0.010	0.010	0.011
0.204 to 0.249	0.009	0.009	0.011	0.011	0.013
0.250 to 0.320	0.013	0.013	0.013	0.013	0.015
0.321 to 0.438	0.019	0.019	0.019	0.019	0.020
0.439 to 0.625	0.025	0.025	0.025	0.025	0.025
0.626 to 0.875	0.030	0.030	0.030	0.030	0.030
0.876 to 1.125	0.035	0.035	0.035	0.035	0.035
1.126 to 1.375	0.040	0.040	0.040	0.040	0.040
1.376 to 1.625	0.045	0.045	0.045	0.045	0.045
1.626 to 1.875	0.052	0.052	0.052	0.052	0.052
1.876 to 2.250	0.060	0.060	0.060	0.060	0.060
2.251 to 2.750	0.075	0.075	0.075	0.075	0.075
2.751 to 3.000	0.090	0.090	0.090	0.090	0.090

* Grades: Alclad 145; Alclad 245, 525, 615, and 755.

Toxicity Pure aluminum is nontoxic and will not change the flavor, purity, or color of food. Thus, it is suitable for use in processing, shipping, and cooking foods and beverages. Long and careful research has proved that there is no such thing as aluminum poisoning. As a matter of fact, traces of aluminum are found in many foods including yellow string beans, beets, lettuce, carrots, celery, onions, milk, and calves liver.

Table 7-5   OD AND WALL TOLERANCES OF ALUMINUM AND ALUMINUM ALLOY EXTRUDED TUBING
(Plus or Minus)

OD, in	Allowable deviation of mean dia. from specified dia.	Allowable deviation of dia. at any point from specified dia.	Specified wall thk., in	Allowable deviation of mean wall thk. from specified wall thk.			Allowable deviation of wall thk. at any point from mean wall thk. (eccentricity)
				Under 3 OD	3 to 5 OD	5 OD and over	
0.50–0.99	0.010	0.020	0.062 and less	0.007	0.008	0.010	Plus or minus 10% of mean wall thickness; max. ± 0.060 min. ± 0.010
1.00–1.99	0.012	0.025	0.063–0.124	0.008	0.010	0.015	
2.00–3.99	0.015	0.030	0.125–0.249	0.009	0.013	0.020	
4.00–5.99	0.025	0.050	0.250–0.374	0.011	0.016	0.025	
6.00–7.99	0.035	0.075	0.375–0.499	0.015	0.021	0.035	
8.00–9.99	0.045	0.100	0.500–0.749	0.020	0.028	0.045	
10.00–11.99	0.055	0.125	0.750–0.999		0.035	0.055	
12.00–12.25	0.065	0.150	1.000–1.490		0.045	0.065	

Heat and light reflectance, emissivity   Aluminum is a good reflector of light and heat. When brightly polished, it will reflect up to 95 percent of the light and up to 98 percent of the infrared energy falling upon it. Aluminum, even when weathered, will reflect 85 to 95 percent of the radiant heat striking it. These properties, desirable in roofing materials and other substances, make aluminum ideal for use in heat reflectors. The emissivity of aluminum is extremely low (4 to 5 percent of a perfect blackbody radiator); for this reason it is used in hot-air ducts and similar applications where low heat loss is desirable.

Weight   The advantages of lightness can be readily appreciated in a qualitative sense, by engineer and layman alike; the design engineer must also learn to place a quantitative valuation on lightness. There are a few applications where the value of "lightness" can be computed directly. An aircraft is an example of this type of product. The commercial airline operators have calculated the cost of carrying each pound of airframe at between $20 and $30 a year, the exact cost depending on the type of service in which the particular aircraft is used. In most products the value of lightness cannot be measured directly in terms of dollars, but nonetheless the engineer must evaluate the intangible benefits in cases where weight of product can be reduced at a cost. Weight is related to wear on bearings and surfaces in sliding contact in machines, for the forces resulting from the motion of the parts take the form of added pressure on these surfaces.

In bridge spans of extreme length, the truss weight is greater than the weight of useful load supported by the bridge; the structural members are, in effect, mainly supporting themselves. Here reduced weight of members would permit reduced loads on other members and the reduction would have a multiplying effect. Aluminum structural members offer interesting possibilities in this area, but little work has been done in this field thus far.

The designer should recognize two types of weight problems in design, and should use two sets of criteria in selecting light alloys. First, there is a so-called *strength-weight criteria*. In this type of problem, the strength requirements are set and the problem is to find the material which will meet these requirements without adding any more weight than necessary. In this case the designer will consult a strength-to-weight ratio table and select the material with the highest strength-to-weight ratio which meets all requirements in other respects.

The other type of design problem is one in which the interest is in weight per unit volume. For many parts almost any of the common aluminum alloys will meet the strength requirements because other design considerations necessitate the use of a large volume of metal. Cylinder blocks and aircraft wheels are examples of this type product. In this case, cost and other design considerations will determine the choice of material.

This distinction is quite important to the designer. The high-strength aluminum alloys are generally more expensive, are harder to machine and weld, and are often less corrosion-resistant than the low-strength alloys. The high-strength alloys will also require heat treatment if extensive forming is to be done.

Machining   Aluminum is about the cheapest of all common metals to machine (magnesium can be machined slightly faster). The following table gives a comparison of relative cutting speeds for aluminum and various other metals.

Material	Relative cutting speed (drilling and turning)
Aluminum alloys	0.5–1.0
Steel, mild (up to 0.3 percent)	0.25–0.35
Brass or bronze	0.6–1.0
Magnesium and its alloys	0.9–1.3

Actually, in most instances, the equipment controls the cutting speed as the material can be machined faster than the speed at which the facility is capable of running. The process designer for general machining of aluminum should employ high machining speeds and moderate feeds and depths of cut.

Where extensive forming and machining operations are required to produce a part, manufacturing costs can be substantially lowered if aluminum can be substituted for a ferrous metal. The reduction in manufacturing costs may more than offset the higher cost of the raw material.

Forming   Most aluminum alloys can be drawn, forged, or extruded, but the high-strength, heat-treatable alloys must be annealed before extensive cold working is done. This will in turn necessitate heat treating the alloy after forming to restore strength properties.

The substantial savings in both cost and weight which may result from the use of extruded or forged parts instead of machined or fabricated members should not be overlooked by the designer. Aluminum extrusions in a wide variety of shapes are stocked by distributors; for volume production, specially made extrusions and forgings can be ordered from aluminum producers. Extruded parts are almost always stronger and lighter than comparable machined or fabricated members; forgings are likewise stronger than castings or sections machined from solid stock.

Fabricating   Aluminum can be joined in numerous ways, but riveting is probably the most common process. It is usually preferable to use rivets made of the same alloy as the material being joined. When designing for riveted joints, the holes drilled, pierced, or subpunched and reamed to size to take the rivets should provide a clearance between the inside diameter and the rivet. The best clearance (approximately 0.010 in) is the smallest one that will allow the rivet to be inserted readily. Hot rivets require more clearance than cold so that they can be inserted without delay.

All of the aluminum alloys can be joined by some form of welding. Spot welding is widely used to join aluminum sheet. When this process is selected it is necessary to remove any protective oxide from the surfaces of the metal being joined so as to prevent electrode pickup and produce sound consistent welds. Since spot welding is a form of resistance welding and aluminum has high electrical conductivity, extremely high welding amperage is required to produce ample heat for fusion. Current requirements are between 13,000 and 36,000 A (amperes), depending upon the alloy and the thickness of the material. Electrode pressures of 58,000 lb/in^2 immediately before and after and 23,000 lb/in^2 during the welding process are typical.

Aluminum is also satisfactorily joined by the gas, arc, and inert gas-shielded arc processes; however, these techniques should not be selected for sections less than 0.032 inch in thickness. When welding thick sections, it is wise to preheat the material to around 600°F so as to prevent cracking. This procedure will also simplify bringing the base material to a molten condition at the time the molten filler material is introduced.

Corrosion resistance Aluminum is highly resistant to progressive atmospheric oxidation. Pure aluminum readily forms an oxide, but this oxide is tough and protects the base metal from further oxidation. For this reason it is not necessary to give aluminum or the near-pure aluminum alloys any protective coating for service under ordinary atmospheric conditions. If the service conditions of the design result in erosion or abrasion, the protective oxide will be removed and corrosion will be accelerated.

Several aluminum alloys (mainly, high-strength alloys) will corrode even under normal conditions. To overcome this difficulty, such alloys are sometimes coated with pure aluminum. The alloys sold under the trade name Alclad fall in this category.

Substantial savings in painting costs may be realized from the use of aluminum instead of the more easily corroded ferrous metals. Aluminum roofing has been in use for some time and prefabricated aluminum wall panels have recently come into limited use. In this latter application, speed and ease of erection have been the chief selling points.

Aluminum is far from being chemically inert, but there are numerous compounds which will attack steel and not aluminum. Aluminum thus has found many applications in the chemical industry; piping in chemical plants, containers for shipping chemicals, and tank cars for chemical shipment serve as common examples of this use.

Care must be exercised to avoid direct contact of aluminum with other metals in the presence of moisture so as to avoid galvanic corrosion. Cadmium or zinc plating on ferrous materials will reduce the corrosive effect.

Electrical properties Aluminum is one of the best electrical conductors known. It is inferior in this respect only to silver and copper. The relative conductivity of aluminum is 61 percent that of copper, but conductivity is measured on a unit-area basis, and copper weighs 3.3 times as much as aluminum. An aluminum conductor with the same resistance as a comparable copper conductor has only about 0.4 of the weight of the copper conductor. Both metals are sold by the pound and aluminum is the cheaper of the two metals. For power-line wire, aluminum conductors are wound on steel wire, the latter providing the strength required for long unsupported spans. The long spans, made possible by this combination of a light conductor and strong steel wire, permit the use of approximately 30 percent fewer supporting poles and towers on transmission lines.

Aluminum conductors should also be considered for use on aircraft and other applications where weight is important. It must be remembered, though, that aluminum is more difficult to solder than copper, so its use will make it more difficult to get good electrical connections.

MAGNESIUM

Magnesium, with a specific gravity of only 1.74, is the lightest metal available for use by the production-design engineer. The chief source is sea water, which contains about 0.13 percent magnesium.

Like other pure metals, magnesium must be alloyed to give maximum strength and usefulness. Although lightness is its outstanding characteristic, it has other properties that are equally significant. Magnesium is extremely easy to machine and is adaptable to practically all the usual methods of metal working. It can be joined by gas, arc, and electric-resistance welding and by riveting. It can be cast by the sand, permanent mold, and die-casting methods. It can be painted, plated, or anodized. Other properties include good stability to inland atmospheric exposure and resistance to attack by alkalies and most oils. The alloys of magnesium have relatively low electrical resistivity and high thermal conductivity. Magnesium is also nonmagnetic.

Magnesium as an alloy is more expensive than aluminum. It is more difficult to produce both in wrought and cast forms. Its tensile strength is less and it must be heat-treated to get good strength characteristics. If weight represents a critical item, or if much machining is required, then the selection of magnesium should be considered.

Uses

Sand castings of magnesium find widespread use in aircraft engines and aircraft landing wheels where light weight, high strength, and shock resistance are of prime importance. In high-speed rotating and reciprocating parts, its light weight is used to effect smoother operations. Every principal American aircraft power plant uses numerous magnesium sand castings for parts such as the blower and supercharger housings, gear cases, carburetor bodies, ignition harnesses, oil sumps, air-induction systems, and numerous small parts and covers. Of particular current interest is the extensive use of magnesium parts in turbojet engines; parts of the air-compression system are built almost entirely of magnesium castings. An example of the use of magnesium castings is the cast magnesium aircraft wing section made by the Northrop Aircraft Corporation. The casting is 16 ft long and adheres to exacting tolerance.

The excellent surface finish, reduction in machining, and low cost have caused magnesium die castings to be widely used in such applications as instrument parts, various kinds of housings, vacuum cleaner parts, and innumerable other parts where light weight is essential.

Characteristics

The outstanding characteristic of magnesium is, of course, its light weight, which is only one quarter that of iron and two thirds that of aluminum. Electric conductivity on a volume basis is about 38 percent that of standard copper and 60 percent that of pure aluminum. Like aluminum, magnesium can be fabricated by any of the industrial methods, although its strong tendency to oxidize rapidly makes precautions necessary in foundry, welding, and machining processes. Magnesium and its alloys are the most readily machinable of all industrial metals, and deep cuts can be made at high speeds. The amount of cold work that can be done on magnesium alloys is limited, but they are readily hot-worked. Commercially pure magnesium has so little tensile strength that it is of practically no value as a material of construction, but the tensile strength of some of its alloys is in excess of 50,000 lb/in².

Tensile strengths and yield strengths decrease at elevated temperatures to the extent that magnesium alloys usually are unsatisfactory for use at temperatures exceeding 400°F. The strength of magnesium alloys is not changed significantly at subnormal temperatures. The modulus of elasticity of magnesium alloys is 6,500,000 lb/in². In view of this low modulus of elasticity, magnesium alloys have a good capacity for energy absorption.

Magnesium alloys have lower thermal conductivity than aluminum alloys but greater than that of ferrous alloys. The coefficient of thermal expansion is approximately twice that of ferrous alloys, being about 0.0000145 per °F. Thus allowances must be made in designs for the thermal-expansion differential between magnesium and other members if an assembly involves dissimilar metals and temperature variations will be encountered.

Magnesium and its alloys have satisfactory resistance to corrosion by solutions of most alkalies and by many organic chemicals. They are attacked by practically all common acids except chromic and hydrofluoric. They are also attacked by salt solutions. Protective coatings are essential for any magnesium alloy used in the vicinity of salt water and are recommended in all cases for protection against atmospheric corrosion when the installation is not inland. Small amounts of iron or nickel in magnesium-aluminum alloys increase susceptibility to corrosion, so these impurities are eradicated as far as possible in the extraction processes. Because magnesium is higher in the electrochemical series of the metals than are copper and iron, magnesium in contact with either of those metals or their alloys, especially in the presence of moisture, is subject to severe galvanic corrosion.

Available Alloys

Magnesium is alloyed readily with most elements except iron and chromium. However, aluminum is the most commonly used alloying element and is used in amounts varying from 3 to 10 percent. Magnesium alloyed with aluminum results in alloys with increased strength and hardness. The greater the percentage of aluminum added, the harder the alloy.

Zinc and manganese are used as alloying materials in quantities to 3 and 0.3 percent respectively in order to provide greater resistance to corrosion.

Forms

Magnesium alloys are available as forgings, extruded shapes, sheets, plates, strips, and castings including sand, permanent mold, and die. When design requirements specify lightness, yet the components are subjected to high or repeated impact stresses, a magnesium forging may be the most appropriate fabricated form. As with other metals, forging results in a fine-grained, homogeneous structure. The strongest forging alloy is AZ80X, while the alloys AT35 and AZ61X provide good mechanical properties. If the design necessitates weldability as well as forgeability, the alloy M1 is suitable. In the forging of magnesium, hydraulic presses are usually used since more dependable forgings are obtained when forging is done at slow speeds.

Magnesium can easily be extruded and is available in standard shapes including round, square, rectangular, and hexagonal. The maximum width of extrusions is approximately 12 in. Extruded tubing is available with ODs from $\frac{3}{16}$ to 12 in and with walls ranging from 0.022 to 1 in.

M1 is the most widely used alloy in the form of sheet, plate, or strip although AZ31X and AZ61X are also available. All three of these alloys are furnished in the annealed, as-rolled, and hard conditions. Sheet magnesium alloy is available in thickness ranging from 0.016 to 0.219 in and in widths from 24 to 48 in. Magnesium sheets can be welded (either arc, gas, or spot), formed, or riveted.

Magnesium alloys AZ92 and AZ63 are most commonly used in sand casting and permanent-mold casting, while AZ90 is most frequently used in die casting.

The characteristics of good design of magnesium-alloy castings are similar to those of aluminum castings. To prevent stress concentration leading to fatigue failure, sharp internal corners, gouges in surfaces, and other surface irregularities in stressed areas should be avoided. Ample fillets should be used around bosses, but not to the extent that the mass is such as to cause draws and cracks. Bosses should be located so that they can be risered or chilled to secure maximum soundness if it is desired. The thin and heavy sections must be so blended as to avoid a sharp transition line. Wall thickness in sand and permanent-mold castings preferably should not be less than $\frac{5}{16}$ in but may be $\frac{3}{16}$ in in limited areas. In die castings, walls may go down to 0.050 in.

Machining

Magnesium alloys may usually be machined at very high speeds. In view of its low flash point, a fire hazard exists in the grinding and machining of magnesium alloys where a finely divided magnesium dust is produced. In order to guard against this danger, coolants of neutral, high-flash-point mineral oil are recommended.

Cutting tools should be ground with large clearances (10 to 12°) and extra-smooth tool faces. It is not difficult to maintain a good smooth finish while machining magnesium alloys. Magnesium machines easily, even though it has an hexagonal close-packed structure, because the temperature in the cutting zone is above its recrystallization temperature.

Joining

Magnesium alloys can be joined by most of the common joining techniques including welding, riveting, adhesives, and soldering. Of these, riveting is the most commonly used method. Here aluminum-alloy rivets are used since magnesium rivets would become exceedingly hard after being driven because of the work-hardening characteristics of magnesium. In this connection, it is advisable to drill the rivet holes in the magnesium rather than pierce them so as to avoid work hardening of the sheet adjacent to the rivet. If it is desirable to pierce the rivet holes, the sheet should be heated to at least 400°F so as to minimize cold work hardening.

M1 alloy is most suited for gas welding and can be joined by means of oxyacetylene, oxyhydrogen, or oxycarbohydrogen gas. Butt joints are recommended so as to avoid trapping of fluxes that may enhance subsequent corrosion. It should be mentioned that magnesium alloys are not satisfactorily welded to other metals and should not be cut by torch methods.

The inert-gas-shielded arc is the most widely used welding method for joining magnesium alloys. Under this method, no flux is used so any type of joint is satisfactory.

Spot, butt, seam, or flash welding may be employed on magnesium alloys. However, these resistance welding methods are only satisfactory when joining magnesium to magnesium.

COPPER AND COPPER-BASE ALLOYS

Copper, one of the first metals used by man, has become one of the most useful metals to the production-designer engineer. The major reasons for its importance are: (1) its high electrical conductivity, surpassed by only one other substance, silver; and (2) the alloys which it forms, the most important of which are brasses and bronzes. Brass is the second most commonly used nonferrous alloy. (Table 7-6 shows some of the common coppers and their alloys.)

Properties

Copper is a yellow-red metal of specific gravity 8.93 and a melting point of 1083°C. The metal is very malleable, as it can be rolled to sheets 0.0026 mm thick. It is very ductile and this property improves with purification. Its heat conductivity is next to those of gold and silver and its electrical conductivity is second only to silver.

The principal reasons for the widespread industrial use of metallic copper are its ease of working, its resistance to corrosion, the pleasing color of the metal itself and of its alloys, the wide variety of alloys possible, the ability for the metal to be hardened by alloying or by cold working, and its high electrical and thermal conductivity.

Copper is not used extensively in the pure state because of its mechanical properties: pure copper is soft and relatively weak. The majority of all pure copper used today is in the form of electrical conductors.

Uses

Copper bus bars or buses are used for carrying heavy currents over short distances while copper wire is used principally for conductivity purposes over longer distances. Copper rod and wire is not restricted to electrical work but is used for bars for locomotive staybolts, woven wire screen or cloth, and welding electrodes. Aluminum is the principal competitor of copper for certain types of electrical conductors.

Copper sheet is used for producing fabricated objects by such processes as cold working, stamping, spinning, welding, soldering, and brazing. Sheet may

Table 7-6 GENERAL DATA FOR COPPER-BASE ROD ALLOYS

ALLOY GROUP	ALLOY	NOMINAL COMPOSITION Percent					DENSITY Pounds per Cubic Inch at 68° F	SPECIFIC GRAVITY
		COPPER	LEAD	ZINC	TIN	OTHER		
1	ELECTROLYTIC TOUGH PITCH COPPER	99.90 Min				0 0.04	0.321-.323	8.89-8.94
	TELLURIUM COPPER	99.5				Te 0.5	0.323	8.94
	SELENIUM COPPER	99.4				Se 0.6	0.322	8.91
	LEADED COPPER	99.0	1.0				0.323	8.94
	TELLURIUM-NICKEL COPPER	98.2				Te 0.5 Ni 1.1 P 0.2	0.323	8.94
2	COMMERCIAL BRONZE, 90%	90.0		10.0			0.318	8.80
	CARTRIDGE BRASS, 70%	70.0		30.0			0.308	8.53
3	LEADED COMMERCIAL BRONZE	89.0	1.75	9.25			0.319	8.83
	MEDIUM-LEADED BRASS	64.5	1.0	34.5			0.306	8.47
	HIGH-LEADED BRASS	62.5	1.75	35.75			0.306	8.47
	FREE-CUTTING BRASS	**61.5**	**3.0**	**35.5**			**0.307**	**8.50**
	FORGING BRASS	60.0	2.0	38.0			0.305	8.44
	ARCHITECTURAL BRONZE	57.0	3.0	40.0			0.306	8.47
4	LOW-SILICON BRONZE	96.0 Min	0.05 Max	1.50 Max	1.60 Max	Si 1.5 Max Fe 0.80 Mn 0.75	0.316	8.75
	PHOSPHOR BRONZE, 5%	95.0			5.0		0.320	8.86
	LEADED PHOSPHOR BRONZE, 5%	94.0	1.0		5.0		0.322	8.92
	ALUMINUM-SILICON BRONZE	91.0				Al 7.0 Si 2.0	0.278	7.69
	FREE-CUTTING PHOSPHOR BRONZE	88.0	4.0	4.0	4.0		0.320	8.86
	ALUMINUM BRONZE, 9%	87.0 Min				Al 7.0-10.0 Fe 1.5 Max Mn Ni 2.0 Max Sn	0.274	7.58
	ALUMINUM BRONZE, 10%	82.0				Al 9.5 Fe 2.5 Ni 5.0 Mn 1.0	0.274	7.58
	NAVAL BRASS	60.0		39.25	0.75		0.304	8.41
	LEADED NAVAL BRASS	60.0	1.75	37.5	0.75		0.305	8.44
	MANGANESE BRONZE, (A)	58.5		39.2	1.0	Fe 1.0 Mn 0.3	0.302	8.36
5	NICKEL SILVER, 65 - 18	65.0		17.0		Ni 18.0	0.316	8.73
	LEADED NICKEL SILVER, 61.5 - 10	61.5	1.0	27.5		Ni 10.0	0.314	8.70
	LEADED NICKEL SILVER, 61.5 - 12	61.5	1.0	25.5		Ni 12.0	0.315	8.73
	LEADED NICKEL SILVER, 61.5 - 15	61.5	1.0	22.5		Ni 15.0	0.316	8.75
	LEADED NICKEL SILVER, 61.5 - 18	61.5	1.0	19.5		Ni 18.0	0.317	8.78
	EXTRUDED LEADED NICKEL SILVER	46.5	2.75	40.75		Ni 10.0	0.306	8.47

Chemical symbols used are as follows:
Al Aluminum   Fe Iron   Mn Manganese   Ni Nickel   O Oxygen
P Phosphorus   Se Selenium   Si Silicon   Sn Tin   Te Tellurium

PHYSICAL PROPERTIES				FABRICATION PROPERTIES					
COEFFICIENT OF THERMAL EXPANSION Per 0°F from 68°F to 572°F	ELECTRICAL CONDUCTIVITY (Volume Basis) Percent IACS at 68°F (Annealed)	MELTING POINT (Liquidus) °F	MODULUS OF ELASTICITY (Tension) psi	CAPACITY FOR BEING COLD WORKED	CAPACITY FOR BEING HOT WORKED	RELATIVE POWER REQUIRED FOR BEING HOT WORKED	HOT WORKING TEMPERATURE °F	ANNEALING TEMPERATURE °F	MACHINABILITY RATING Free-Cutting Brass = 100
0.0000098	101.	1981	17,000,000	Excellent	Excellent	Moderate	1400-1600	700-1200	20
0.0000079	90.0	1980	17,000,000	Good	Excellent	Moderate	1400-1600	700-1200	90
0.0000098	99.0	1958	17,000,000	Good	Excellent	Moderate	1400-1600	700-1200	90
0.0000098	99.0	1974	15,000,000	—	—	—	—	—	80
0.0000098	50.0¹	1980	17,000,000	Good	Good	Moderate	1400-1600	1150-1450	80
0.0000102	44.0	1910	17,000,000	Excellent	Good	Moderate	1400-1600	800-1450	20
0.0000111	28.0	1750	16,000,000	Excellent	Fair	Moderate	1350-1550	800-1400	30
0.0000102	42.0	1900	17,000,000	Good	Poor	—	—	800-1200	80
0.0000113	26.0	1700	15,000,000	Good	Poor	—	—	800-1200	70
0.0000113	26.0	1670	15,000,000	Fair	Poor	—	—	800-1100	90
0.0000114	**26.0**	**1650**	**14,000,000**	**Poor**	**Fair**	**Low**	**1300-1450**	**800-1100**	**100**
0.0000115	27.0	1640	15,000,000	Fair	Excellent	Low	1200-1500	800-1100	80
0.0000116	28.0	1630	14,000,000	Poor	Excellent	Low	1150-1350	800-1100	90
0.0000099	12.0	1940	17,000,000	Excellent	Excellent	Moderate	1300-1600	900-1250	30
0.0000099	18.0	1920	16,000,000	Excellent	Poor	—	—	900-1250	20
0.0000099	16.0	1920	15,000,000	Fair	Poor	—	—	900-1250	50
0.0000100	9.2	1841		Poor	Excellent	—	1300-1600	—	60
—	12.0	1830	15,000,000	Fair	Poor	—	—	900-1250	90
0.0000094	12.8	1908		—	Excellent	Low	1650-1750	1450-1650	20
0.0000094	7.5	1931	—	Poor	Excellent	Low	1650-1750	1450-1650	20
0.0000118	26.0	1650	15,000,000	Fair	Excellent	Low	1200-1500	800-1100	30
0.0000118	26.0	1650	15,000,000	Poor	Good	Low	1200-1400	800-1100	70
0.0000118	24.0	1630	15,000,000	Poor	Excellent	Low	1150-1450	800-1100	30
0.0000090	6.0	2030	18,000,000	Excellent	Poor	—	—	1100-1500	20
—	—	1830	17,500,000	Fair	Poor	—	—	1100-1300	50
—	7.0	1880	18,000,000	Fair	Poor	—	—	1100-1300	50
—	6.0	1970	18,000,000	Fair	Poor	—	—	1100-1300	50
—	6.0	2010	18,000,000	Fair	Poor	—	—	1100-1300	50
—	8.5	1715	—	Poor	—	—	—	—	80

¹Age Hardened.   NOTE: The values listed above represent reasonable approximations suitable for general engineering use. Due to commercial variations in composition and manufacturing limitations, they should not be used for specification purposes.

Table 7-7 MECHANICAL PROPERTIES OF ROD ALLOYS

Temper	Diameter (inches)	Tensile Strength psi — Specification Minimum Value	Tensile Strength psi — Average	Yield Strength psi — Specification Minimum Value	Yield Strength psi — Average	Elongation % — in 4×D Gage Length Spec. Min.	Elongation % — in 2" Gage Length Average	Rockwell Hardness F Scale	Rockwell Hardness B Scale	Contraction of Area % Average	Shear Strength psi Average	ASTM Specification	Alloy	Alloy Group
As Hot Rolled			32000		10000		55	40		70	22000	B133	ELECTROLYTIC TOUGH PITCH COPPER	1
Soft	1/8 to 1/2 incl.	37000*	32000		10000	25	45	40		70	22000			
Soft	Over 1/2	37000*	32000		10000	20	55	40			22000			
Hard	1/8 to 1/4 incl.	50000	55000		50000	8	10	94	60		29000			
Hard	Over 1/4 to 3/8 incl.	50000	55000		50000	10	12	94	60		29000			
Hard	Over 3/8 to 1 incl.	45000	45000		44000	12	16	87	47	55	27000			
Hard	Over 1 to 2 incl.	40000	38000		36000	15	20	85	45		26000			
Hard	Over 2	35000					22	85	45		25000			
Half Hard	1/8 to 1/4 incl.	38000	42000	30000	39000	15	20		40		26000		TELLURIUM COPPER	1
Half Hard	Over 1/4 to 1/2 incl.	38000	42000	30000	39000	15	22		40		26000			
Half Hard	Over 1/2 to 1 incl.	38000	42000	30000	39000	15	25		45		26000			
Half Hard	Over 1 to 2 incl.	38000	42000				35		45		26000			
Hard	1/8 to 1/4 incl.	48000	54000	40000	50000	10	11		50		27000			
Hard	Over 1/4 to 1/2 incl.	44000	48000	38000	46000	10	18		45		27000			
Hard	Over 1/2 to 2 incl.						20		50		27000			
Soft Anneal			32000		10000		45	35			22000		SELENIUM COPPER and LEADED COPPER	1
Hard			45000		40000		12		50		26000			
Hard	1/8 to 1/2 incl.	70000	80000	60000	70000	12	20		85		43000		TELLURIUM-NICKEL COPPER	1
Hard	Over 1/2 to 1 incl.	68000	77000	57000	67000	12	27		84		40000			
Hard	Over 1 to 2 incl.	68000	73000	57000	62000	15	38		84		39000			
Soft Anneal	1/2 to 1 incl.	35000	40000	10000	47000		50	55			32000	B134 Alloy 2	COMMERCIAL BRONZE, 90%	2
Half Hard	1/2 and under	45000	52000	27000	45000		25		60		35000			
Half Hard	Over 1/2 to 1 incl.	40000	50000	25000			20		58		34000			
Hard	1/4 to 1 incl.	48000		30000										
Quarter Hard	1 and under		58000		41000		40		65	70	36000	B134 Alloy 6	CARTRIDGE BRASS, 70%	2
Half Hard	1 and under	55000	65000	25000	48000	20	30		75	68	42000			
Soft	All sizes	35000	37000	10000	12000	15	45	55			24000	B140 Alloy 8	LEADED COMMERCIAL BRONZE	2
Half Hard	1/2 and under	50000	55000	30000	50000	10	14		61		31000			
Half Hard	Over 1/2 to 1 incl.	45000	52000	27000	47000	10	18		58		30000			
Half Hard	Over 1	40000	50000	25000	45000	12	25		58		30000			

0.25 mm.	1		50000		19000		60	70			34000	B121 Alloy 3	MEDIUM LEADED BRASS	3
Quarter Hard	1/4 to 1 incl.	50000	55000	20000	42000		35		60		36000			
Half Hard	1/4 to 1 incl.	55000	65000	25000	48000		25		75		42000			
Half Hard	1/2 to 1 incl.	55000	60000	25000	42000	10	23		70	45	34000		HIGH-LEADED BRASS	
Soft	**1 and under**	**48000**	**50000**	**20000**	**22000**	**15**	**36**	**68**	**48**	**54**	**30000**	**B16**	**FREE-CUTTING BRASS**	
Soft	**Over 1 to 2 incl.**	**44000**	**48000**	**18000**	**21000**	**20**	**32**		**45**	**40**	**29000**			
Soft	**Over 2**	**40000**	**45000**	**15000**	**19000**	**25**	**36**		**40**	**35**	**28000**			
Half Hard	**1/2 and under**	**57000**	**65000**	**25000**	**42000**	**7**	**15**		**73**	**44**	**38000**			
Half Hard	**Over 1/2 to 1 incl.**	**55000**	**60000**	**25000**	**42000**	**10**	**23**		**70**	**45**	**34000**			
Half Hard	**Over 1 to 2 incl.**	**50000**	**54000**	**20000**	**37000**	**15**	**31**		**66**	**45**	**32000**			
Half Hard	**Over 2**	**45000**	**50000**	**15000**	**28000**	**20**	**35**		**62**	**35**	**30000**			
Hard	**1/8 to 3/16 incl.**	**80000**	**85000**	**45000**	**58000**		**7**		**85**	**50**	**47000**			
Hard	**Over 3/16 to 5/16 incl.**	**70000**	**75000**	**35000**	**52000**	**4**	**9**		**81**	**45**	**42000**			
As Extruded	1		52000		20000		45	78		65	—	B124 Alloy 2	FORGING BRASS	
As Forged			42-65000	—			25-60							
Half Hard	1/2		75000				14		85					
As Extruded			60000		20000		30		65		35000		ARCHITECTURAL BRONZE	
Soft	All sizes	40000	45000	12000	16000	30	40	55			35000	B98 Alloy B	LOW-SILICON BRONZE	4
Half Hard	Up to 1/2 incl.	55000	60000	20000	48000	11	16		—	—	41000			
Half Hard	Over 1/2 to 1 incl.	55000	60000	20000	48000	12	19		—	—	41000			
Half Hard	Over 1 to 2 incl.	55000	60000	20000	48000	12	22		—	—	41000			
Hard	Up to 1/2 incl.	65000	70000	35000	55000	8	13		—	—	45000			
Hard	Over 1/2 to 1 incl.	65000	70000	35000	55000	10	15		80	—	45000			
Hard	Over 1 to 2 incl.	65000	70000	35000	55000	10	17		—	—	45000			
Extra Hard (Bolt)	Up to 1/2 incl.	85000	90000	55000	67000	6	11		90	—	50000			
Extra Hard (Bolt)	Over 1/2 to 1 incl.	75000	80000	45000	62000	8	13		—	—	48000			
Extra Hard (Bolt)	Over 1 to 1-1/2 incl.	75000	80000	40000	62000	8	15		—	—	48000			
Soft	Under 1/4	40000	45000		—		—		—	—	—	B139 Alloy A	PHOSPHOR BRONZE, 5%	
Half Hard	1/2 to 1 incl.		75-70000		65-58000		25		80-78	—	—			
Hard	Under 1/4	80000	105000		—		—		—	—	—			
Hard	1/4 to 1/2 incl.	70000	92000	—	—	13	23		—	—	—			
Hard	Over 1/2 to 1 incl.	60000	68000	—	61000	15	32		—	—	—			
Hard	1 and over	55000	60000	—	57000	18	34		—	—	—			
Spring	Under .026	125000		—	—		—		—	—	—			
Spring	.026 to 1/16 incl.	115000	127000	—	—		—		—	—	—			
Spring	Over 1/16 to 1/8 incl.	110000	123000	—	—		—		—	—	—			
Spring	Over 1/8 to 1/4 incl.	105000	122000	—	—	3.5	—		—	—	—			
Spring	Over 1/4 to 3/8 incl.	100000	121000	—	—	5.0	—		—	—	—			
Spring	Over 3/8 to 1/2 incl.	90000	100000	—	—	9.0	—		—	—	—			

be used directly for roofing, sheathing, flashing drains, and numerous other purposes.

Copper pipe can be made by rolling copper sheets into a cylinder and welding the seam; or seamless tubing can be made by piercing billets and rolling them over a mandrel. This tube is widely used in radiators, refrigerators, air-conditioning equipment, and similar equipment where maximum heat transfer is desired.

Copper Alloys

Copper can be alloyed with many elements including zinc, tin, lead, iron, silver, phosphorus, silicon, tellurium, and arsenic (see Table 7-7). Copper and most of its alloys act like a pure metal. Its alloys are not heat-treatable and cannot be hardened by application of heat followed by a quench. Hardness is produced by cold working, and softening is produced by heating above the recrystallization temperature. Copper-base alloys have good ductility, elongation, electrical conductivity, thermal conductivity, and corrosion resistance. Most alloys are readily drawn, stamped, coined, forged, formed, spun, soldered, brazed, and welded. Also many of the alloys can be obtained in various tempers and finishes. Copper and its alloys are nonmagnetic and nonsparking. In some instances, these two properties are quite important.

The tensile strength is also good, approaching 80,000 lb/in² for several of the alloys (aluminum-bronze, manganese-bronze, hard bronzes).

BRASS

When copper is alloyed with zinc in various proportions, the resulting material is known as "brass," which is stronger than either of the materials from which it is made. In general, the physical properties of brass depend largely upon manipulation, that is, whether it has been cast, rolled, extruded, drawn, or annealed, to all of which it lends itself readily. Properties such as ductility, springiness, toughness, strength, and stiffness are controlled either by annealing or by the amount of reduction by cold rolling or drawing after the last anneal. The variation of these properties relative to the amount of manipulation can be read from charts issued by the various brass companies.

Machinability, corrosion resistance, hot and cold workability, and other similar properties are controlled largely by modifying the proportions of copper and zinc, and by the addition of small amounts of elements such as lead, tin, silicon, aluminum, nickel, phosphorus, and arsenic. Thus, a wide range of desired physical characteristics can be imparted to brass by variations in composition, heat treatment, and manipulation. The largest measure of satisfaction can be secured when the brass maker works hand in hand with the designer and is cognizant of the exact purpose for which the material is to be employed.

On a basis of overall or final cost, the brasses offer a combination of unusual properties for the production of a wide variety of cast, forged, machined, stamped, drawn, or spun articles. The comparatively high initial per-pound price should not influence the choice of material because finishing costs may be lower, especially where buffed or plated surfaces are required. Some of the unique properties that should be taken advantage of are:

1 High ductility
2 Rapid drawing speeds in draw presses and dies
3 Lower maintenance cost on dies
4 Easily plated or used for plating
5 Can be alloyed with many metals to give desirable characteristics
6 Easily machined, formed, and cast
7 High value of scrap

. Brass is one of the oldest alloys used by man. Every handbook and textbook on materials and alloys covers the essential information. New materials have displaced brass in many fields, but it still has an increasing economic use.

The information in Table 7-7 is for guidance of the production-design engineer when choosing materials. Such information is limited and should be supplemented by current catalogs.

Machining

In general, the machinability of most of the brasses is good. With the free-cutting alloys and with some of the moderately machinable alloys, the chip breaks up at a rapid rate and consequently is only in momentary contact with the cutting tool. Then the principal function of the cutting medium is to serve as a coolant and not a lubricant.

Table 7-7 gives the approximate machinability and tool-wear ratings of the principal brasses and bronzes. Excellent finishes can be obtained by keeping the cutting tools sharp and by using high speeds, slower feeds, and minimum rake angles. Addition of a few percent of lead improves machinability of cast copper-base alloys.

BRONZE

Alloys of copper with material other than zinc are known as bronze. The major alloying element in bronze is tin although other elements such as silicon, aluminum, manganese, phosphorus, and nickel may be included. Similar to the brasses, the various bronzes can be hardened by cold working. The most widely used bronzes are beryllium bronze, aluminum bronze, phosphor bronze, and silicon bronze.

Beryllium bronze (beryllium copper) is an alloy of copper and beryllium which came into active use during World War II. It is used where a nonferrous, nonsparking, or good electrical conductor (45 percent when hardened) with high strength (100,000 to 190,000 lb/in²) is required. It has a modulus of elasticity of approximately 19,000,000 lb/in². It is one of the best corrosion-resistant spring materials. Springs of all forms, x-ray windows, molds for plastics, and diaphragms illustrate its versatile use. This copper alloy can be heat-treated and shaped at the same time by holding in steel molds.

Aluminum bronze has wide applications since its physical properties can be controlled over a broad range by varying the amount of aluminum in the alloy and also by heat treating the bronze. This alloy has excellent resistance to corrosion and has superior strength.

Phosphor bronze is an alloy of copper and tin deoxidized with phosphorus. This alloy is known for its strength and hardness, which increase as the percentage of tin increases. These bronzes are produced with varying amounts of tin although usually the tin percentage is somewhere between 2 and 10 percent.

Silicon bronze includes between 2 and 4 percent silicon and usually 1 percent zinc or manganese. Silicon bronze can be readily worked by the common techniques. It is easily cast, forged, welded, stamped, rolled, and spun. Silicon bronze finds its principal uses where resistance to corrosion plus high strength are the criteria to be met. Cast bronzes frequently have lead additions to improve their pressure tightness. Bronze alloys freeze over a long temperature range so leakage is a problem.

NICKEL AND NICKEL ALLOYS

Nickel ranks ninth among all metals from the standpoint of world consumption. Its principal use is as an alloying element of both steel and nonferrous metals although large amounts of nickel are also used in electrodeposition.

Nickel is magnetic up to 680°F, is resistant to corrosion and the majority of the nonoxidizing acids (with the exception of nitric acid), and melts at 2635°F. Pure nickel is a silvery-white, tough metal having approximately the same density as copper; it costs about three times as much as copper. There are many uses for practically pure nickel in cast and wrought form, and there are over 3,000 active alloys of nickel.

The production-design engineer will find nickel and its alloys useful materials for meeting the requirements of his design. Often when a cast-iron part has blowholes or chills and is difficult to machine due to hard spots, the addition of a small percent of nickel will improve the quality.

The properties of several alloys containing more than 50 percent nickel are summarized in this text. These alloys are stronger, tougher, and harder than copper and aluminum alloys, and are as strong although more costly than steel alloys. Nickel alloys are highly resistant to corrosion and exhibit good heat resistance. In general, they can be cold-worked, although frequent anneals may be required in order to remove the rapid work-hardening effect characteristic of these alloys.

In view of their high heat resistance, ability to maintain strength at elevated temperatures, and high corrosion-resisting characteristics, these alloys are finding increased uses in furthering the space age.

The kind and quantity of elements added to the nickel form a convenient method of grouping the alloys. The following summary gives the principal classifications according to composition:

Group 1 Nickel: 93.5 to 99.5% nickel (and a maximum of 4.5% manganese)

Group 2 Nickel-copper: 63 to 70% nickel, 29 to 30% copper

Group 3 Nickel-silicon: 83% nickel; 10% silicon

Group 4 Nickel-chromium-iron: 54 to 78.5% nickel, 12 to 18% chromium, 6 to 28% iron

Group 5 Nickel-molybdenum-iron: 55 to 62% nickel, 17 to 32% molybdenum, 6 to 22% iron

Group 6 Nickel-chromium-molybdenum-iron: 51 to 62% nickel, 15 to 22% chromium, 5 to 19% molybdenum, and 6 to 8% iron

In the nickel group of alloys, there are five grades of commercial nickel: A, D, E, L, and Z. Grade A nickel contains an average of 99.4 percent nickel and is distinguished from electrolytic nickel which is 99.95 percent pure. Grade A nickel is in the wrought form and combines high mechanical properties with excellent resistance to corrosion. It can be fabricated readily and joined by welding, brazing, or soldering.

In composition, grades D and E nickel are similar to A, the principal difference being the inclusion of small percentages of manganese (usually between 2 and 5 percent) which replaces a like percentage of nickel. The manganese increases the resistance of the alloy to atmospheric corrosion and to sulfur compounds at elevated temperatures.

In L nickel, the carbon content is kept low (between 0.01 and 0.02 percent); this alloy is softer than the others, having a lower yield strength and elastic limit. This alloy is fabricated easily and for this reason frequently is specified where drawing and forming operations are required.

Z nickel contains about 93.5 percent nickel. It is harder and stronger than the other group 1 alloys. This alloy can be heat-treated, and tensile strengths up to 240,000 lb/in^2 can be obtained.

The group 2 alloys, referred to as the nickel-copper alloys, are perhaps the best known of all the nickel alloys and are commonly known as the Monel type. Monel is a trade name of the International Nickel Company, and it contains approximately 67 percent nickel and 33 percent copper.

Six grades of Monel metal are available. These are known as Monel, K Monel, R Monel, H Monel, S Monel, and KR Monel.

Monel

Monel metal is the most important nickel alloy of this group. It has an average composition of 67 percent nickel, 28 percent copper, and 5 percent iron, manganese, and silicon combined. Monel resists corrosion of distilled water, salt water, foods, and acids, except hydrochloric acid above 120°F. Its strength and its thermal expansion are about the same as steel; therefore, it can be used in machinery in parts that must resist corrosion. It is used extensively in the chemical, marine, power-equipment, laundry, food-service, pickling, roofing, petroleum, household-equipment, and pulp and paper industries. Steam valves, pipes, turbine blades, and shafting are normal applications. At 750°F it retains 75 percent of its strength at room temperature. It resists oxidation up to 1350°F.

The color of Monel is close to that of nickel. This has brought about its general use in kitchen equipment and utensils. It is produced in several alloys giving flexibility in strength and machining qualities without reducing its corrosion-resistance properties. Although its cost per pound is relatively high, it has an equally high scrap or chip value. This factor makes it competitive with stainless steel in many installations.

K Monel   K Monel is a nonmagnetic, age-hardening alloy containing approximately 3 percent aluminum. Its strength and hardness, particularly in large sections, are comparable with those of heat-treated alloy steels.

R Monel   R Monel is a free-machining alloy that is particularly adapted for use in automatic screw machines. Its relatively free-machining properties are a result of a 0.025 to 0.060 percent sulfur content.

H Monel   H Monel includes 2.75 to 3.25 percent silicon. The silicon provides increased hardness with good ductility.

S Monel   S Monel includes approximately 4.0 percent silicon, which induces age-hardening characteristics. This alloy in the cast state will give a hardness reading of about 320 Brinell, which can be increased to 350 with suitable heat treatment.

KR Monel   KR Monel is similar in composition to K Monel but it has a higher carbon content. This alloy has better machinability characteristics than its companion K.

Group 3 alloys are extremely hard (Brinell of approximately 360) and consequently are quite difficult to machine. These alloys are resistant to corrosion in sulfuric acid, acetic acid, formic acid, and phosphoric acid.

The best known of this group are manufactured by the Union Carbide Corp. under the trade name Hastelloy O.

Group 4 alloys are characterized by large percentages of chromium and iron. In this group, the alloy Inconel as produced by the International Nickel Company is one of the best known. These alloys have excellent resistance to corrosion, good strength and toughness properties, and can withstand repeated heating and cooling in the temperature range of 0 to 1700°F.

Group 5 alloys represent the nickel-molybdenum-iron alloys. This group, in addition to having good corrosion-resistance properties (in particular to hydrochloric acid), has strength and ductility properties comparable to alloy steel. The best known of this group are Hastelloy A and Hastelloy B as produced by the Union Carbide Corp. These alloys can be work-hardened but do not respond to age-hardening heat treatment.

The sixth group of alloys are those containing large amounts of chromium, molybdenum, and iron with nickel. These alloys find application where high mechanical properties combined with resistance to oxidizing acids and agents are required. Hastelloy C is one of the better known of this group.

MOLYBDENUM

Molybdenum has unusual properties that make it useful in alloys of steel and cast iron and high-temperature alloys.

Machining and welding of high-purity molybdenum is difficult due to its high melting temperature of 4750°F and hardness of 150 to 250 Brinell at high temperatures. The high melting temperature makes it a basic material in high-temperature alloys and an alloy agent for steel, cast iron, cast steel, nickel steel, chromium-nickel steel, and tool steel. The general physical properties of these alloys are increased toughness, tensile strength, uniformity, corrosion resistance, and hardenability.

When used in the foundry, molybdenum does not oxidize and permits close control of the alloy. It can be added at any time during melting or casting. During heat treatment and die forging, it promotes free scaling. When the part is cast, increased uniformity and decreased section sensitivity are obtained and a thicker and more complex casting may be designed.

The steel and cast-iron alloys at elevated temperatures have increased hardness and tensile strength.

Molybdenum is sold in the form of pure molybdenum, molybdenum oxide, and molybdenum sulfide. Economy is possible because of high recovery from scrap steel and because it does not oxidize in foundry practice and permits close control of the alloy.

Typical applications are in the production of gas-turbine rotor and stationary blades, high-temperature steam-power-plant equipment, exhaust valves, and tool steel.

COBALT

Cobalt was first introduced as an alloy in tool steel. Cobalt cutting tools were superior to high-speed steel tools. Later, cobalt was introduced as a binder in cemented carbide tools and it improved their toughness.

Cobalt is an important constituent (along with other alloys) of high-temperature, high-strength steels having dampening characteristics. These steels are used in turbine blades subject to vibration.

Cobalt and its alloys can be spot-welded, brazed, and fusion-welded.

Cobalt and cobalt alloys have the properties and applications shown in Table 7-8.

Cobalt and its alloys have been developed in the past 40 years. This is an illustration of how research has extended the use of materials.

TITANIUM

Titanium is classed as a light metal (0.16 lb/in^3). It is 60 percent heavier than aluminum but only 56 percent as heavy as alloy steel. Titanium-base alloys are extremely strong, far more so than aluminum alloys. Because of its high strength-weight ratio, which is superior to both steel and aluminum, and its resistance to corrosion, titanium competes primarily with aluminum and to a lesser extent with various alloy steels.

Titanium in the ore is available on the open market at less than \$0.02/lb; however, alloyed titanium in sheet form costs today between \$6 and \$10/lb. Because of the difficulty of removing the metal from its contaminants and because the metal is so very active chemically when molten (absorbing oxygen or nitrogen from the air with ruinous rapidity), it has been extremely difficult and costly to produce.

Commercially pure titanium has the following approximate mechanical properties:

1   Hardness, 85–95 Rockwell B   Sp.ht.: 0.13 Btu/(lb)(°F)
2   Ultimate strength, 80,000 lb/in^2 Therm. cond.: 105 Btu/(ft^2)(h)(in)
3   Yield strength, 70,000 lb/in^2 Coef. exp.: 5×10^{-6}/°F
4   Elongation, 20%   Stress rupture: @750°F; 35,000
                                         lb/in^2 for 1,000 h

Table 7-8 PROPERTIES AND APPLICATIONS OF COBALT AND COBALT ALLOYS

Properties	Alloy name	Application
Low expansion	Invar	Thermostat controls
	Kovar	Glass to metal seals
		Ceramic to metal seals
Constant modulus (E) during temperature change (has low elastic limit)	Elinvar	Hair springs in watches
		Electric-current-conducting springs
		Springs working in high temperatures
Good castability, casts to high dimensional accuracy and surface smoothness	Vitallium	Replaces precious metals in dental work
Resistance to tarnish and abrasion		
Compatible with mouth tissues		Bone surgery
High strength and stiffness, lower weight and low cost in comparison with precious metals		Prosthetic devices
High-temperature strength		Electronic-tube parts
Corrosion resistant		Springs
		Cemented carbide tools
Alloys with and hardens copper, gold, silver, platinum		Jewelry
Colorant and decolorizer		Glass
		Ceramics
Reduced resistance to electric current at increased temperatures		
Catalysts		Oil refineries
Radioactive		Radioisotopes
Nutrition element		Animal nutrition (Vitamin B12)

Commercially pure titanium does not respond to heat treatment, but may be cold-worked to well above 120,000 lb/in² tensile and 100,000 lb/in² yield, with a resultant drop in elongation to about 10 percent.

The good strength and weight characteristics result in a higher strength-weight ratio for titanium alloys than can be obtained from alloy steel at temperatures up to 800°F. Above 800°F its strength-weight ratio falls below that of alloy steels.

Many titanium-base alloys that have superior physical properties are available. For example, the Titanium Metals Corporation type T1-155A is a high-strength forging alloy, beta-stabilized by iron, chromium, and molybdenum, with sufficient aluminum to maintain strength at temperatures to 1000°F. It has the following mechanical properties:

1 Hardness, 300–320 Brinell
2 Tensile strength, 155,000 lb/in^2
3 Yield strength, 140,000 lb/in^2
4 Elongation, 12%

Titanium is being successfully clad on carbon steel by rolling; and investigation continues on the use of titanium as an electrolytic plating material.

Several alloys of titanium are commercially available and are used as flat products, bars, and forgings. They contain combinations of iron, aluminum, chromium, manganese, molybdenum, tin, and vanadium. The elements iron, chromium, vanadium, molybdenum, and manganese stabilize the high-temperature modification and produce alloys that can be hardened by heat treatment. Aluminum, tin, nitrogen, and oxygen facilitate the strengthening of the low-temperature modification.

The melting point of titanium (3135°F) is higher than the majority of metals. This property led to the belief that the alloy would be suitable for high-temperature application. However, because of its affinity for both oxygen and nitrogen at elevated temperatures it has been found impractical for sustained use at temperatures greater than 1000°F.

Titanium is very resistant to corrosion by salts and to oxidizing acids. However, it is not resistant to hydrofluoric, sulfuric, oxalic, and formic acids. Although it is attacked by concentrated alkalies, it is resistant to dilute alkalies. Its adherent oxide film resists moist air and O_2 up to 600°F.

Today titanium is fabricated by sheet-metal forming, forging, casting, and machining. In these processing methods, titanium behaves much as stainless steel. A word of caution should be stated, however, relative to titanium castings. In producing castings, both the mold and furnace must be housed where either a vacuum or inert gas atmosphere is maintained since liquid titanium reacts with all elements except the inert gases. Titanium alloys forge well at 1600 to 1800°F in inert atmosphere. Strip, wire, and rod may be cold-worked up to 50 percent reduction. The tensile strength increases to 135,000 lb/in^2 and elongation drops to 10 percent.

On the basis of response to heat treatment, titanium alloys have for the most part not been hardened by heat treatment. However, both the martensitic alloys and the metastable beta alloys are potentially capable of hardening by heat treatment. Surface hardening does take place when titanium-base materials are heated in air because of the absorption of oxygen and nitrogen. Commercial titanium can be annealed and is usually done in the range 1100 to 1350°F.

Titanium can be joined by inert-gas-shielded–arc-welding procedures as well as brazing. Adhesion and riveting find application as joining methods in addition to welding and brazing.

LOW-MELTING ALLOYS

The field of low-melting-point alloys includes all combinations of metals that have melting temperatures below 1000°F. This classification can be further broken down into three categories: alloys that become fluid between 300 and 1000°F, those which melt between 200 and 300°F, and finally the ultra-low-melting alloys which melt below 200°F.

Alloys Becoming Fluid between 300 and 1000° F

The first of these three categories embraces, primarily, the zinc-, tin-, and lead-base alloys. These are by far the most important of all of the low-melting alloys.

Zinc-base alloys Zinc-base alloys are used widely in present-day industry in all types of products. The auto industry uses many zinc-base alloys in the manufacture of carburetors, fuel pumps, door handles, and numerous other items. They are also used in the construction of household appliances, business-machine parts, lathe gears, blower rotors, and even the lowly zipper.

Zinc-base alloys can be easily cast; in fact, more die castings are made of zinc-base alloys than any other metal. The advantages of using this type of alloy are lower production costs and lower die-maintenance costs due to the low temperatures inherent in the process.

A widely used zinc alloy is ASTM XXI, which corresponds to SAE 921 and to New Jersey Zinc Company's alloy Zamak 2. This alloy consists of 4.1 percent aluminum, 2.7 percent copper, 0.03 percent magnesium, and 93.17 percent pure zinc. The melting point is 733.6°F and the shrinkage is 0.0124 in/in. Average tensile strengths of 47,300 lb/in^2 are obtained from 6-month-old cast specimens. Another important alloy, ASTM XXIII (SAE 903 or Zamak 3), contains 4.1 percent aluminum and 0.04 percent magnesium, and the remainder is pure zinc. This alloy has slightly lower tensile strength and hardness but it has excellent retention of impact strength and dimensions.

Tin-base alloys Tin-base alloys, more commonly called *Babbitt metals*, are used principally for bearings. The main alloying elements are copper, antimony, and lead. The melting point of pure tin is 450°F; however, for complete lique-faction of the alloys, a temperature of 700 to 800°F must be used. The effect of the antimony is to harden these alloys and increase their antifriction properties.

The alloy ASTM B23-26, grade 1 (SAE 10), contains 91 percent tin, 4.5 percent copper, and 4.5 percent antimony. Grade 2 is composed of 89 percent tin, 3.5 percent copper, and 7.5 percent antimony. Both of these alloys are used extensively in automotive and aircraft bearings. The latter grade is slightly harder and stronger. Alloy SAE 12, which has 60 percent tin, 3.5 percent copper,

10.5 percent antimony, and 26 percent lead, is much cheaper than the other two alloys because of the high lead content. This alloy is used chiefly for light-duty bearings.

All of these tin-base alloys are corrosion-resistant and consequently can be used in food-handling equipment and for soda fountain hardware.

Lead-base alloys Lead-base alloys are composed principally of lead and antimony. Pure lead melts at 621°F; however, the addition of 17 percent antimony reduces the melting point to 570°F. The antimony in these alloys hardens the lead and reduces shrinkage. The chief lead-base alloy used in bearings is ASTM B23-26, grades 12 and 17, which have respectively 10 and 15 percent antimony, copper held to 0.50 percent maximum, and arsenic held to 0.25 percent maximum.

These alloys have low-strength properties, but are cheap and easy to cast. They are used for light-duty bearings, weights, battery parts, x-ray shields, and for noncorrosive applications.

Shrinkage of both the lead- and tin-base alloys is exceptionally low, ranging from 0.002 to 0.003 in/in. This characteristic facilitates the casting of parts to accurate size and also lessens the danger of shrinkage cracks in casting.

Solder The final alloys of importance in this heat range are the tin-lead alloys more commonly called "solder." The solders contain small percentages of antimony and sometimes silver. The physical properties are affected most significantly by varying the lead-tin ratio. Solders have a wide possible range of composition from 100 percent tin to 100 percent lead. The most widely used solders are those which contain 38.1 percent lead and 61.9 percent tin, and those which are composed of 68 percent lead and 32 percent tin. The first of these is known as "tin solder." It is completely molten at 362°F and, after losing its latent heat, is completely solid at 362°F. From this it can be seen that it is quick setting. The latter composition is termed "plumbers' solder." It is completely molten at 488°F and totally solid at 362°F. This range of temperature between the molten and solid states makes possible the plumbers' wiped joint.

Alloys Becoming Fluid between 200 and 300° F

The second category, those alloys which melt between 200 and 300°F, are among the most interesting and the most misunderstood of the low-melting alloys. There are more than 2,000 commercial and experimental types but all of them are relatively new and as yet their full potentialities have not been appreciated.

Bismuth is a common constituent of all of the alloys in this and the ultra-low-melting groups. The usual alloying elements are tin, lead, and cadmium.

Certain peculiar characteristics of bismuth and bismuth alloys are responsible for their wide range of usefulness.

Bismuth is one of the few elements that does not shrink when it solidifies. Water and antimony are two other substances that expand on solidification, but bismuth expands more than the former, namely, 3.3 percent of its volume. This expansion is somewhat modified when bismuth is alloyed. Some of the alloys shrink slightly upon solidification but expand again after a few hours of aging at room temperature.

This characteristic of bismuth alloys makes them excellent for casting because, when they expand into the mold, they pick up every detail. This expansion characteristic is also the basis of the alloys' widest use, anchoring. The growth enables them to grip keyed, notched, or bossed surfaces on which they are cast, and to fill annular spaces tightly, despite the absence of fusion or bond with the materials with which they come in contact. Examples of this are the anchoring of bearings, bushings, and stationary parts in machinery. This is done by providing oversize holes to receive the bearings or bushings and then by accurately aligning the parts in jigs. Babbitting clay dams are formed around the part and the molten alloy is poured in. Upon solidification, it holds the parts rigidly. This method of locating bearings and parts eliminates all costly machining to press-fit parts, makes possible easy field servicing, and safeguards against overheating through failure of oil supply, since the inexpensive alloy melts before the bearing is damaged.

Dies and punches for blanking, piercing, and trimming are secured in a steel shell by use of these alloys. This method of construction eliminates the three-stage dies formerly required. Chucks are also made of low-temperature alloys to hold odd-shaped parts while they are being machined, ground, and inspected.

Cores for electroforming are also made from these alloys. The core is easily molded into intricate shapes and is then placed in the electrolytic bath. The plating metal is then deposited upon the core, and as the final step, the low-melting alloy is melted out.

Alloys Becoming Fluid at Less Than 200° F

In alloys of the 100 to 200°F range, tensile strengths up to 1,500 lb/in² and compressive strengths up to 20,000 lb/in² can be obtained. Bismuth is the principal metal used in this class of alloys. These permit the use of low-melt alloy casts as form blocks or dies for the fabrication of sheet-metal products with hydropress and drop-hammer equipment.

The ultra-low-melting alloys have only limited application in present-day industry. As materials for finished products, they have been used for such purposes as fire-extinguishing systems, the production of special molds and patterns for dentures, and molds for plaster casting. Their greatest and most important use is in the bending of thin-walled tubing. This tubing requires internal support during bending to prevent wrinkling, buckling, and flattening. The tubes are filled with alloy and the operations completed. The alloy is then melted out by immersing in boiling water.

Low-melting alloys, although relatively new, are finding increasingly more important roles in industry. It would be wise for every engineer to keep abreast of all new developments in this field for the many moneysaving applications they provide.

SUMMARY

Only nonferrous metals that are principally used by the production-design engineer have been discussed in this chapter. Many other metals such as gold, silver, zirconium, vanadium, thorium, thallium, sodium, tantalum, rhodium, ruthenium, osmium, palladium, platinum, rhenium, manganese, mercury, indium, iridium, lithium, germanium, gallium, columbium, cerium, and calcium have application, but to a lesser degree; consequently, they will not be discussed in this text.

When the production-design engineer selects a nonferrous material, it is usually because he is looking for certain characteristics discussed in this chapter. Familiarity with the metal's physical properties and ability to be worked will facilitate the selection of the material that will meet the design requirements.

QUESTIONS

1 What are the principal properties of aluminum that lead to its selection by the production-design engineer?
2 What are some of the uses of aluminum foil?
3 Why has aluminum been widely used as a roofing material? Why is it advisable to use aluminum nails with aluminum roofing?
4 Describe the characteristics of the single-point cutting tool when used to cut aluminum.
5 Why is it not advisable to join thin sections (less than 0.032 in) of aluminum by arc-welding methods?
6 What advantages does magnesium have over aluminum in product design?
7 What precautions should be observed when grinding magnesium?
8 What are the principal reasons for the widespread use of metallic copper?
9 How is hardness produced in copper alloys?
10 What is the relative machinability between free-cutting brass, manganese bronze, and aluminum bronze?

11 Make a sketch of the recommended tool shape using high-speed and carbide single-point tools for turning brass.

12 What is the relationship in ductility between free-cutting brass and leaded commercial bronze?

13 When would the production-design engineer specify bronze in preference to brass?

14 What is Monel metal? What is its strongest competitor? Why?

15 Briefly describe the application of the low-melting alloys.

16 Why has titanium received so much publicity in the last few years?

PROBLEMS

1 Select material and determine the thickness of the contact springs of a control relay. Show all calculations, formulas, and reasons for selecting the material on a right-wrong form. Choose between at least two materials. Please list all references from which you obtain material data. The design of the relay is limited by the size of the contact spring and the contact spring limits the size of the relay. Every design is a compromise or a balance between its various components. Some parts have more material than required from an operating viewpoint in order to accommodate another part which in turn is limited by other restrictions.

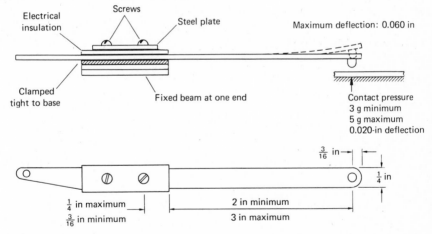

2 In boring operations, there can be a considerable deflection of the tool tip because of the cantilever support required owing to the nature of the operation. For this reason the boring bar is made with as heavy a cross section as possible. The deflecting force from a certain boring operation has been found to be 228 lb when the overhanging distance from the toolpost to the tool tip was 8 in.

(a) Determine the taper in a 4-in-diameter hole, 6 in long if the bar is 3 in diameter and made from AISI 1035 steel.

(b) How much would the taper be reduced if the material of the boring bar were changed from steel to tungsten carbide?

(c) What is the cost differential between the two bars if each is 12 in long, and the cost of steel is $0.15/lb and the cost of tungsten carbide is about $7.00/lb.

3 Find the resistance of a copper wire which is 0.06 in diameter and 250 ft long if it has a resistivity of 1.7 $\mu\Omega \cdot$ cm.

4 What diameter copper wire should be specified if 50 ft of wire can have a maximum resistance of 2.5 Ω?

5 Explain the difference between ductility and malleability and correlate these properties with the crystal structure of five particularly ductile metals. Repeat the correlation but choose particularly brittle metals such as titanium and magnesium. What structure do they have? In light of the foregoing facts why does zinc have particularly good ductility? Why are titanium and its alloys difficult to machine? Explain.

SELECTED REFERENCES

ALUMINUM COMPANY OF AMERICA: "Structural Handbook: A Design Manual for Aluminum," Pittsburgh, Pa., 1958.

AMERICAN SOCIETY FOR METALS: "Metals Handbook," 8th ed., vol. 1, "Properties and Selection of Metals," Metals Park, Ohio, 1972.

————: "Metals Handbook," 8th ed., vol. 7, "Atlas of Microstructure of Industrial Alloys," Metals Park, Ohio, 1972.

8
PLASTICS

INTRODUCTION

The term "plastics" has been applied to those synthetic nonmetallic materials that can be made sufficiently fluid to be shaped readily by casting, molding, or extruding, and which may be hardened subsequently to preserve the desired shape. Synthetic rubber, ceramics, and glass, which may seem to be in this category, are often classified separately. Both natural and synthetic rubber are included in this chapter, while ceramics, including glass, are discussed in Chap. 11.

Plastics as engineering materials are constantly proving themselves useful in present-day designs. In spite of the continued introduction of new materials and processes on an extensive scale, the industry has achieved maturity. The use of plastics is following the same course of development and application that was characteristic of the introduction of light metals and stainless steel. Costs have been reduced, uniformity and reliability of materials have been improved, and possible applications have been determined. Plastics are attractive materials and offer advantages in weight, cost, moisture and chemical resistance, toughness, abrasive resistance, strength, appearance, insulation (both thermal and electrical), formability, and machinability.

One of the earliest and largest users of plastics was the electrical industry, which is a leader in the use of thermosetting and laminated plastics. Electrical companies entered the field of plastics to capitalize on benefits to electrical products. Their products spurred, in turn, the development and application of plastics for many parts that had no electrical function.

For example, high-density polyethylene milk containers produced in 1971 approximate 1 billion units. In 1973 approximately 25 percent of dairy products were packaged in plastic containers, whereas only 6 percent used plastic in 1968. Today most major stadiums in the United States are covered by synthetic turf.

In 1972 the total plastics consumption by the automotive industry exceeded 700 million lb, whereas only 625 million lb were used in 1968. The 1972 automobile averaged about 85 lb of plastic. The market for the new rigid urethane materials for furniture and decoration is expanding rapidly. As many as 400,000 new homes utilized ABS (acrylonitrile butadiene styrene) drainage plumbing in 1970, which is an increase of over 30 percent from 1968.

By virtue of their thermal characteristics, plastics usually are divided into two groups, thermoplastic or thermosetting. Those that undergo no chemical change in the molding operation may be softened again by heating to the temperature at which they originally became plastic, and therefore are termed *thermoplastic*. Since they become increasingly softer with increase in temperature, certain members of the thermoplastic family are liable to permanent distortion under mechanical strain at relatively low temperature (140°F). They may flow to an appreciable extent under load at room temperatures.

The basic structural units of thermoplastics are referred to as monomers. Monomers are molecules consisting of carbon atoms with attached ribs of other atoms such as hydrogen, chlorine, and fluorine. These ribs determine to a large extent the intrinsic properties of the plastic. For example, thermoplastics composed of the ethylene monomer

have both good flexibility and toughness. By replacing one of the hydrogen atoms with chlorine, the monomer vinyl chloride is formed:

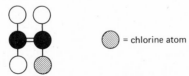

This provides a thermoplastic with more rigid characteristics than ethylene.

These structural units, monomers, are joined end to end to produce long chainlike molecules known as polymers. The process by which polymers are formed is a chemical reaction known as *polymerization*, which is defined as the "chemical reaction by which single molecules are linked to form large molecules."

Plastics that are capable of being changed into a substantially infusible or insoluble product when cured by application of heat or chemical means are called *thermosetting* plastics. Here reactive portions of the molecules may form cross-links, that are attached to other molecules, during polymerization.

These materials, once molded, will distort under stress at approximately 250 to 500°F, but they will not become soft or fusible. Thermosetting materials will char and burn at high temperatures. They are inclined to have greater tensile strength and hardness, and, in some cases, are lower in raw-material cost than thermoplastics. Thermoplastics, on the other hand, have generally higher impact strength, pleasing appearance, and can be converted into a finished product at lower manufacturing cost.

All plastic articles are initially derived from molding compounds. These molding compounds consist of a resin or binder and one or more of the following components: fillers, plasticizers, dyes and pigments, and lubricants. The "resin," as the principal component, gives the compound its name and classification, and imparts the primary properties to it. It is the cohesive and adhesive agent which provides rigidity and binds together the filler particles. The "filler" is usually an inert, fibrous material which modifies the properties of the resin or imparts special properties to it. "Plasticizers" are added to the compound if the flow or softness of the compound must be regulated, while "dyes and pigments" are added to impart color to the molded part. Lubricants of wax or stearates are added occasionally to a molding compound to facilitate its removal from the mold. (Its degree of granulation also has a major bearing on molding qualities.)

Natural and synthetic rubber are plastics which are very important to industry. These incompressible elastic plastics present problems very similar to those encountered in the manufacture and application of other types of plastics. Some specialized equipment, such as callenders, is required for the preparation of raw material. Rubberlike plastics may be extruded into shapes, filaments, or sheets, and formed by the extrusion, transfer, and compression types of molds.

Some thermoplastic and thermosetting plastics are used as adhesives. They are tough, strong, and reliable, and can be applied between almost any combination of materials. One of the first and most successful applications was the use of transparent plastic sheets between two plates of glass to form our present-day safety glass. The tough plastic adheres to the glass and prevents splinters from flying. Plywood made with plastic adhesives withstands weathering and water and now can be used for concrete forms and outside sheathing for homes. Large wooden columns and thick panels can be built by curing the adhesive by dielectric heating. The combination of plastic and wood makes a strong structural member. The C process of sand molding, as described in Chap.

10, uses a phenolic resin for binding the sand in cores and molds. Wood furniture and metal cabinets are held together by adhesives which simplify their design and reduce cost of manufacture.

Fabrics are made of many plastics in pleasing colors. They are durable, tough, and easy to clean. Natural fabrics such as cotton and wool are facing stiff competition from these new plastic filaments.

Reinforcing of plastics by metal and glass fibers has produced strong, flexible, and light materials, such as that used in bullet-proof vests for the armed services.

Plastics are manufactured under controlled conditions to give uniform raw material and finished products. Color, surface, strength, and size variations are minor. Failures usually are due to misapplication by the engineer. There is no such thing as a bad plastic; all plastics are good if compounded properly for a particular use. Sufficient information for making a proper choice is available from material suppliers and plastic-molding companies.

The cost of plastics is being reduced constantly as improvements are made and demand increases. In general they are more expensive than metals on a per-pound basis. It will probably be some time before automobile bodies, kitchen cabinets, or refrigerator housings are predominantly made of plastics. By coating metals with plastics the benefits of both materials can be obtained. Plastics must do the same job as another material at less cost, or a better job for the same money, or be in such a position that a plastic is the only material that will do the job, before they are used more extensively.

The cost of metal molds for plastics is about the same as that of die-cast dies. A superior polish in a mold for plastic is transferred authentically to the molded piece. The molding operation in some cases may be slower than metal-die and permanent-mold castings; however, the trimming of flash associated with die casting is not usually as time consuming with plastics.

GENERAL PROPERTIES

The problem of selecting plastic materials is that of finding the material with suitable properties from the standpoint of intended service, methods of forming and fabricating, and cost. New and improved plastic materials possessing almost any desired characteristic are being introduced continually. There are plastics that do not require plasticizers, that have greater flexibility under lower temperatures, and are stable under higher temperatures. Some resist water, acids, oils, and other destructive agents. The wide use of plastics testifies to their value; however, fundamental limitations should be considered when applying a new material or adapting an old material to new applications.

Effects of Temperature*

LOW TEMP.
&
HIGH TEMP
} TO BE AVOIDED

Plastics are inclined toward rigidity and brittleness at low temperatures, and softness and flexibility at high temperatures. They are fundamentally unstable dimensionally with respect to temperature, and are susceptible to distortion and flow when subjected to elevated temperatures. The thermoplastics are particularly susceptible, while the thermosetting plastics are much more resistant, differing, however, only in degree. The distinction between the thermal stability of the thermosetting and thermoplastic resins is not well defined. A true distinction can be drawn only between individual plastics, rather than between classes of plastics. High temperatures not only seriously reduce the mechanical properties of plastics, but accelerate the destructive action of external agents to which they are sensitive. Continuous heating also may induce brittleness and shrinkage in heavily plasticized materials by volatilization of plasticizers. The use of one plastic in contact with a dissimilar plastic in a proposed application should be checked first in the light of possible "migration of plasticizer," sometimes resulting in discoloration or hardening of one of the plastics.

In general, moderate temperatures are required for storage of plastics over long periods; low temperatures are to be avoided because of the low-temperature brittleness of most of the plastics, and high temperatures should be avoided because of the rapid loss of mechanical properties, volatilization of plasticizers, and the susceptibility of a large number to distortion.

Effects of Humidity*

Plastics, with only a few exceptions, are extremely sensitive to the effects of water. High-humidity atmospheres induce water absorption and varied resulting effects, depending upon the composition and formulation of the plastics. Increased water content plasticizes some materials, and there is a general lowering of the mechanical properties. Water absorption is responsible for swelling in certain plastics and the ultimate decomposition of a few. Moist or wet atmospheres may extract plasticizers from heavily plasticized materials and also provide conditions favorable to fungal growth. In recent years, however, new plastics have come into use that have first-class moisture resistance and may contain water indefinitely while resisting other influences at the same time.

Extremely dry environments may cause brittleness in certain plastics as a result of loss of water that normally contributes to their plasticity. Cyclic wet and dry atmospheres are more destructive to plastics than continuous exposure at constant humidity because of the mechanical stresses induced in the plastics by swelling and shrinking with moisture absorption and moisture

* Starred sections are taken from a report prepared for the Office of the Chief of Ordnance. This is from an abridged account published in "Modern Plastics Encyclopedia."

emission. Relatively constant, moderate to low humidities are preferred for plastic storage because of the adverse effects of water on the structure and properties of these materials, and the possibility of plasticizer loss by extraction and fungal attack in moist atmospheres.

Effects of Light*

Prolonged exposure to sunlight will affect adversely all plastics with the exception of tetrafluoroethylene (Teflon). The change induced by the ultraviolet components may vary in kind and severity from slight yellowing to complete disintegration as a result of the chemical degradation of the polymeric compound or plasticizers. Loss of strength, reduced ductility, and increased fragility usually accompany such action. Many plastics are offered in special formulations containing "ultraviolet inhibitors" which should be utilized when this influence is present. Exposure of plastics to sunlight during storage should be avoided, especially when the transparency of clear materials is to be preserved.

Weight

As a family, plastics are light when compared to metals. Most plastics have a specific gravity between 1.35 and 1.45, which is less than that of magnesium.

Electrical Resistivity

Plastics have excellent electrical resistivity making them have wide application as an insulating material. In the high-frequency applications, plastics are particularly advantageous and, consequently, are being used to a large extent in the fields of radar and television.

Heat Insulation

Plastics have low heat conduction and, consequently, have application as an insulating material. In particular, they are used as handles for appliances and tools subjected to heat.

Fabrication

The principal characteristic of plastics from the fabrication standpoint is ease of molding. Both thermosetting and thermoplastic materials lend themselves to molding irregular and complex shapes with relatively short curing cycles.

Plastics may be joined by using various cements, chemical solvents, and mechanical fasteners. Heat sealing, which parallels somewhat the welding process of metals, is used extensively in joining light thermoplastic films. In such cases, dielectric heating is the technique usually used. Friction adhesion has had moderate application also in the joining of small thermoplastic parts.

Plastics can be machined with conventional machine tools. However, certain cautions should be exercised. In order to maintain a good finish, a heavy flow of coolant should be used so as to avoid temperatures that will distort the work. In some thermosetting laminates (glass, for example), the customary high-speed steel tool will not stand up in view of the abrasive action of the laminating material. Here, either tungsten carbide or ceramic cutting tools must be used.

Effects of Oxygen*

Organic plastics are nearly all subject to oxidation when exposed to the atmosphere. The process is accelerated by high temperatures and light; but, over long periods of time, oxidative deterioration may take place at room temperature. Oxidation susceptibility depends largely upon the chemical nature of the plastic and its compounding. Materials with the greatest number of double bonds in their molecular structure will generally be the most sensitive to oxidation. Yellowing and a gradual loss of strength and ductility are the principal results of oxidative processes.

Oxidation is not a problem of great magnitude in storage, since the rigid plastics are rather resistant to oxidative deterioration under moderate conditions.

Effects of Loading*

Under moderate conditions the common thermoplastic materials are subject to distortion and flow when significantly loaded. Such plastics cannot be expected to maintain a high degree of mechanical stability over extended periods when subjected to stress; especially is this true when they are also exposed to relatively high temperatures. The thermoplastics should, however, maintain themselves fairly well when not subject to load or when subjected to only moderate load. Recently, fillers, such as glass wool, have been added to thermoplastics to further improve this property.

The thermosetting plastics are much more load-stable than the thermoplastics because of their structure and the inclusion of fillers in their formulation. In the laminated form they provide a rather high order of distortion and creep resistance. When not subjected to mechanical stress they may be considered to be highly stable. These materials, however, may suffer creep over long periods, especially when maintained at elevated temperatures.

The thermoplastic types should not be subjected to load when stored; and, whenever possible, the loading of stress-bearing thermosetting moldings or laminates should be removed or reduced.

Chemical Stability

Plastics, in general, possess a high degree of inherent stability with respect to chemical deterioration. In many instances, this stability may be fortified by the addition of the proper stabilizers during compounding. While there is vast difference from one plastic to another, the general statement may be made that there is a plastic available to resist virtually any commercial chemical.

THERMOPLASTICS

Thermoplastics can be reshaped upon being reheated, or more significantly, can be reground and remolded. Many uses have been found for the various types of materials, and processes have been developed to produce economical finished and semifinished products. Thermoplastics can be molded, rolled into sheets, used for strip coatings, and extruded into shapes. The material can be reused; therefore, there is little waste. New materials that will offer additional advantages constantly are being developed. Thermoplastic materials can be obtained in any shade of any desired color or mixture of colors as well as in any desired degree of transparency, including crystal clear and opaque. This is a definite advantage in product development and eliminates costly painting or finishing operations.

Acrylic Resins (Transparent)

Polymethyl methacrylates (Plexiglas, Lucite) are almost perfectly clear and transparent. Sheets and shapes are stable under normal temperature (140 to 190°F) and light loading (4,000 to 10,000 lb/in²), but are subject to creep under heavy loads. They have low water absorption and are affected very little by weather conditions. They have the best optical properties of the transparent plastics and are known for their ability to pipe light, their use in windows and optical lenses, and their beauty of color.

Cellulose Plastics

Cellulose acetates (Tenite I, Plastacele, Lumarith) are tough transparencies with high impact strength and are pleasant to touch because they are low conductors of heat. They can be varied from rigid types to elastic types. Good molding qualities make them suitable for compression and injection molding. All colors can be provided. Water absorption is high, tensile strength is 5,000 to 11,000 lb/in², and flexural strength is 4,000 to 8,000 lb/in². Their heat resistance is higher (140 to 250°F) than that of many thermoplastics. Their physical properties are quite sensitive to temperature. Strength, weather, and chemical resistance are not good because of crazing, embrittlement, and loss of transparency. They are used for general-purpose parts.

Cellulose Acetate Butyrate

Cellulose acetate butyrate (Tenite II) is tougher and has higher impact strength and weather resistance than cellulose acetates. Otherwise it is similar in many respects. It is used for fuel lines, tool handles, and general-purpose parts. This plastic has a characteristic odor and its use in confined places is thereby limited.

Cellulose Nitrate

Cellulose nitrate (Pyralin, Nitron) is tough, water resistant, and the most stable of the cellulosic plastics. Otherwise the general characteristics listed previously apply. It is commonly known as "celluloid" and is extremely combustible.

Ethyl Cellulose

Ethyl cellulose (Ethocel, Colon, Nixon) has the general characteristics of the other cellulose types and is used for parts requiring good strength and electrical characteristics at low temperatures, such as camera cases, flashlights, and some military equipment.

Aniline-Formaldehyde Resins

Aniline-formaldehyde (Cibanite), unlike other formaldehydes, is thermoplastic. It has good mechanical and electrical properties (8,500 to 10,000 lb/in² tensile strength), in addition to resistance to moisture and ultraviolet light. Unfilled types resist distortion, cold flow, and creep at moderate temperatures (180 to 190°F). They absorb very little moisture and maintain mechanical and electrical characteristics under humid conditions. They are used for terminal blocks and parts on which electrical items are mounted.

Polyamide Resins

Polyamide resin (nylon), a thermoplastic, has physical stability at high temperatures and a high, rather sharp melting point (450 to 505°F). This resin excels in toughness and strength (see table below), chemical resistance, and low friction value.

Characteristic	Temperature, °F	Type FM1, lb/in²	Type FM3, lb/in²
Strength	70	15,700	12,900
Strength	77	10,530	7,600
Strength	170	7,600	6,760

Molded nylon materials have a high tensile strength, large elongation, and comparatively good impact strength at normal temperatures, but like other thermoplastics, the tensile strength decreases with rising temperature, and the elongation increases. Because of its crystalline structure, nylon preserves its physical properties at high temperatures to a much greater extent than most thermoplastics.

Nylon is water sensitive and will absorb large amounts of water when immersed or exposed to atmospheres of high humidity. Variation in the equilibrium moisture content causes a corresponding variation in the mechanical properties of nylon. With an increased moisture content the stiffness and tensile and flexural strengths decrease, and the impact strength increases.

Fibers and fabrics of nylon possess the same general characteristics as the molded materials. They are tough, extremely strong, stable over wide temperature variation, and more resistant to weathering and fungal attack than natural materials. Nylon fabrics are not completely weather resistant, and they may support fungal growth when fabricated with susceptible lubricants or sized.

Nylon gears recently tested in a kitchen mixer showed practically no wear after a 24-h run with the beaters turning in a bowl of sand. Bronze gears on a similar test were worn out at the end of that time. Three months later, after 2,400 h of continuous operation, the nylon gears were still running and showed very little wear. Another application of nylon is for bearings. A thin, $\frac{9}{64}$-in-thick section of nylon was molded to a steel shell to form the bearing. A $\frac{1}{2}$-in-diameter shaft of SAE 1112 steel operating at 350 r/min was loaded at 200 lb/in² with a clearance of 0.005 in between shaft and bearing. The nylon showed no visible wear after 2,500,000 r without lubrication.

Since molding requires special experience, the supplier should be selected with care. Many parts are made by machining because nylon works like metal. Guides for the best way to machine can be obtained from the E. I. DuPont de Nemours Co.

Polystyrenes

Polystyrene (Dylene, Lustrex, Styron) is usually molded without a plasticizer, as it has very good molding qualities. The resins have good electrical properties, clarity, and low sensitivity to water (5 percent in 48 h). They are rigid to 150 to 205°F (in various formulations), resist distortion, and do not creep or flow under their load-bearing capacities. Films of the material are good protection, as they resist water and weathering. They are affected very little by many chemicals, including alcohols and gasoline. Polystyrenes are low in cost, averaging today about $0.25 per pound. They have unlimited colorability.

Vinyl Resins and Vinylidene Chloride Resins

Polyvinyl alcohol (Elvanol, Resistoflex) is neither a thermoplastic nor a thermosetting plastic. It is resistant to petroleum hydrocarbons and is quite unstable mechanically. The vinyl chloride, vinyl butyral resins (Vinylite, Saran) are thermoplastic, and, depending on the use of plasticizers, range from strong, rigid products to flexible or elastomers. In general, they are insensitive to water at low temperature. They are tough, have high inherent strength, abrasion resistance, and unlimited colorability. This family has good dielectric characteristics, does not support flame, is not toxic, and resists moisture. Saran, for example, is molded into pump parts, handles, plumbing pipe, and elbows; extruded into tubing, tape, and wire coating; and formed into sheets used to enclose and seal packages for storage under dehydrated conditions. It is made into filaments that are woven into warm fabrics, window screen, and shoe fabrics. The resins are used to coat and finish paper and as components of varnishes and paint.

Shellac Compounds

Shellac resin is a natural product of the lac bug *Tacchardia lacca*. The resin is thermoplastic and is generally employed as an impregnant and binder in heavily filled moldings or in laminates. Shellac compounds are characterized by their good electrical properties, scratch hardness, and resilience.

The mechanical properties of shellac compounds depend upon the type and characteristics of the filler or lamina. Wood-flour-filled moldings have low tensile strength, while that of the laminates may be quite high. Shellac resins soften at rather low temperatures, and the compounds are not suitable for use much above 150°F.

Polyethylene

Polyethylene is produced in three types: low, intermediate, and high density. All three types have in common the properties of toughness; resistivity to solvents, alkalies, and mold acids; excellent dielectric properties; good colorability; very low moisture absorption; and relatively low cost.

Type I (low-density polyethylene) is quite flexible, having a stiffness modulus of approximately 20,000 lb/in². It has high impact strength and low heat resistance. 175°F represents the upper temperature limit for sustained use.

Type II (intermediate density) is less flexible than type I. It has a stiffness modulus of about 45,000 lb/in² and can withstand temperatures up to 200°F.

The high-density polyethylene materials known as type III are fairly rigid, having a stiffness in flexure up to 140,000 lb/in². Type III materials are able to withstand temperatures similar to type II, that is, temperatures up to 200°F.

The polyethylenes are blow- and injection-molded, extruded, and calendered in sheets. Typical products include housewares such as bowls, plates, dishpans, paint-brush handles, flexible tubing, bags for packaging vegetables, hardware, squeeze bottles, etc.

Polypropylene

This thermoplastic is similar in properties to type III polyethylene. It is slightly lighter, somewhat more rigid, and capable of withstanding higher temperatures (up to 230°F). Polypropylene has good resistance to creep. It has low water absorption and can be extruded, blow-molded, and injection-molded. It has application as a packaging material where high strength and heat resistance is necessary. It also is used as a material for appliance components such as small gears, cams, bearing surfaces, housings, etc.

ABS Materials

This group of thermoplastics includes those containing acrylonitrile, butadiene, and styrene. The most important property of the ABS family is toughness. Tensile strength is about average for thermoplastics, running about 7,000 lb/in². These materials can be injection-molded, extruded, and calendered into sheets.

Typical applications include housings for household appliances, football helmets, utensil handles, etc.

Fluorocarbons

This family includes tetrafluoroethylene, commonly known as Teflon, and chlorotrifluoroethylene, usually referred to as Kel-F. Both of these fluorocarbons have the disadvantage of comparatively low strength and high cost. Teflon costs about $4.50/lb and Kel-F approximately $7.00/lb.

The principal advantages of this family of thermoplastics include high-temperature resistivity (500°F for Teflon, 390°F for Kel-F) and inertness to most chemicals. They have low coefficients of friction and good dielectric properties.

Acetal

This is one of the newer plastics developed by the DuPont Company. It has been developed as a material for mechanical parts including sprinkler nozzles, handles, gears, housings, etc. It has good tensile strength (approximately 10,000 lb/in²), resistance to temperature (240°F), low friction characteristics, and resistance to most solvents.

THERMOSETTING PLASTICS

Thermosetting resins are used with or without modifying agents or fillers. The base materials are compounded and then formed in molds. Heating the compound changes the resin to a binder. The molded product usually is hard and dense as a result of forming under pressure.

Thermosetting Resin

The resin imparts certain properties to the finished product and functions principally as a binder for the filler. Thermosetting plastics are classified according to the resin used. The main classes are: (1) phenol formaldehyde, (2) urea formaldehyde, (3) melamine formaldehyde, (4) polyesters, and (5) epoxies.

Modifying Agents and Fillers

Modifying agents include dyes, pigments, lubricants, plasticizers, accelerators, graphite, and so forth.

Fillers, together with the resin, affect the properties of the final product. Wood flour, for example, reduces the cost and improves the mechanical properties. Cotton flock improves impact still further. Fabric cuttings, cord, and string are used for high impact, paper pulp for improved impact of special shapes. Mica filler will improve the dielectric strength and reduce the electric losses. Asbestos improves the heat and moisture resistance. Slate, silica, soapstone, or clay may be selected as filler for improved appearance, for electrical resistance, and for thin sections of the final product.

The mineral fillers increase the wear of the mold and the wear of tools such as cutters, saws, drills, and taps when further machining after molding is required. Short fiber fillers keep the cost down; the long fiber materials, having improved impact, are more expensive. It is not economical to select an impact material better than that required for the application under consideration.

Phenolic Resins

Phenolic resins have been a popular engineering material since the early part of this century in view of their excellent heat resistance and, more recently, because it has been discovered that they can be injection-molded. Today there are as many different parts produced from phenolic resins as any other single engineering plastic; however, the principal uses are in electrical equipment and automotive components.

Because of the relatively low initial cost of phenolics (it runs on the average less than $0.02/in³) and the fact that they can be injection-molded, phenolics are assured use as a competitive plastic.

Phenolic resins are formed from phenol and formaldehyde; they are of three types: general-purpose, impact, and heat-resistant. The general-purpose phenolics are molded with wood-flour fillers and cost the least, about $0.0125/in³. The impact grades incorporate either glass fibers or cotton flock as filler material, with costs approaching $0.015/in³. The heat-resistant phenolics usually use asbestos as a filler and cost approximately $0.0135/in³.

The main advantages of the phenolic family are heat resistance, rigidity, dimensional stability, wear resistance, good dielectric properties, resistance to many chemicals, and, as mentioned, low cost. The main disadvantages are its impact strength and colorability. Even the high-impact division of phenolic resins are capable of impact strength of only 10 to 33 ft·lb/in based on the Izod notched test. The natural color of phenolics is black. If black or the darker colors of brown and blue are not satisfactory, the only alternative is to paint the phenolic part, which will, of course, increase the cost.

As mentioned above phenolic resins today are being molded successfully by the screw-injection process in addition to transfer- and compression-molding methods. The more rapid injection process allows phenolics to compete economically with thermoplastic materials.

In the screw-injection process the temperature relationships between the barrel and the mold when molding phenolics is just reversed from what is characteristic when molding thermoplastics. With phenolics barrel temperatures are kept at 200 to 230°F to plasticize the plastic, whereas thermoplastics need much higher temperatures, usually in the range of 500 to 600°F. With phenolics the mold temperatures are raised to 340 to 390°F to cure the material, while with thermoplastics the mold temperature is reduced to around 200°F to cool the molded part.

Laminated-type Phenolics

Another group of plastic products, known as laminated phenolic plastics, uses sheets of cloth or paper to give greater mechanical and electrical strength. The term "laminations" applies to the layers of plastic-treated paper, cloth, and wood veneer which are piled up to form (by pressure and temperature) the thickness of the molded sheet or plate. The use of this material requires another set of instructions, a portion of which will be given so they may be compared with other molded materials. The greater proportion of laminated materials is used in the form of sheets, plates, channels, angles, bars, rounds, and tubes (round and rectangular). It is molded under greater limitations than are regular molding materials.

Plates are furnished in sizes of 36×36, 36×72, 48×48, and 48×96 in. The thickness starts with $\frac{1}{32}$ in.

From $\frac{1}{32}$ to $\frac{1}{16}$:   varies in $\frac{1}{64}$-in steps
From $\frac{1}{16}$ to $\frac{3}{16}$:   varies in $\frac{1}{32}$-in steps
From $\frac{1}{4}$ to $\frac{3}{8}$:   varies in $\frac{1}{16}$-in steps
From $\frac{1}{2}$ to $\frac{3}{4}$:   varies in $\frac{1}{8}$-in steps
From 1 to $\frac{1}{2}$:   varies in $\frac{1}{4}$-in steps

Thicker plates can be molded on order. Thickness tolerances vary with type of material, thickness, and size (0.030 to 0.060 per inch of thickness).

Information on the use of rods, bars, angles, Z bars, and channels is omitted. The tubes and rods are made by wrapping the continuous sheet around a mandrel of the desired shape, and baking. Angles and channels are made by molding laminated sheets with an inside radius at least equal to the thickness of material and a corresponding outside radius at the corners.

The dielectric strength of the laminated type is best at right angles to the laminations and, therefore, in applying the laminated type electrically, advantage should be taken of this characteristic wherever possible.

Thermosetting phenolic resin should be applied mechanically with the thought in mind that it is not a ductile material like common metals. The laminated-type products, plates, angles, channels, tubing (also bars to some extent), and other laminated-molded-type shapes have the inherent characteristic of splitting relatively easily (as compared with metals) between the laminations as a result of incorrect machining or application of mechanical loads that place the bond or binder of these laminations under adverse stresses. The higher the bond-strength value, the more resistance to splitting.

The laminated type (also the laminated-molded type generally) has better mechanical properties (except resistance to splitting) than the chopped-molded types when manufactured of the same paper, cloth, or other material in relatively simple shapes or molds. The chopped-molded type may be produced in fairly intricate molds similar to those of certain nonlaminated compositions. The chopped-molded type generally has better resistance to splitting as compared with the previously mentioned laminated type.

This material, especially in plate form, is subject to a small, but indeterminate warping, and is therefore not adaptable to parts requiring refined accuracy of alignment. Wherever one surface of laminated plate is subjected to a different treatment from the other (such as machining, sanding, or varnishing), warping is almost certain to result. Heat treatments of laminated plate are also likely to cause warping.

Subjection of the laminated type to temperatures above 140°C may result in splitting or blistering. Exposure of this material to temperatures above 115°C for more than 4 h will also cause blistering.

Urea Formaldehyde

The ureas (Beetle, Plaskon) are valuable primarily because pastel shades and translucent color can be produced readily. Although comparatively low in impact strength, they are remarkable for their hardness, color stability, and color fastness, and have excellent electrical properties, notably, very high arc resistance. Their resistance to organic commercial solvents, weak acids, and weak alkalies is excellent. The ureas are odorless, tasteless, and have high heat resistance which makes them suitable for use in contact with foodstuffs.

As with the phenolics, the properties of the urea resins can be modified by varying the composition. The resin is used with water or in solvent solution for adhesives of several kinds. As an adhesive, it is used for the bonding of plywoods and fine furniture veneers where it combines good bonding properties with ease of application. It also is used for paperbacked abrasives (belt sanding). Urea resins are compounded with various alkyd derivatives as a low-cost surface on metal or wood.

Melamine Formaldehyde

Melamine resins (Melmac) maintain their properties over a substantially greater range of temperatures and offer higher electrical resistance than do either the phenol or urea resins. The melamines have all the advantages of the ureas and phenolics with the added feature of better water, acid, and alkali resistance.

Because of their special properties, melamine resins can be used with all fillers, including glass fibers. When melamine is compounded with alpha-cellulose fiber, a hard, tough plastic results. For general-purpose molding, a wood-filled material shows good flexural strength. High-heat-resistant plastics are made with asbestos as the filler, and these combine high arc resistance with their elevated-temperature characteristics, making them valuable materials in the electrical field. Melamine-type plastics have gained extensive use as commercial and domestic tableware in view of their durability and pleasing appearance.

Polyesters

The polyester resins are copolymers of a polyester and usually styrene. They have good strength, toughness, and resistance to chemical attack; low water absorption; and ability to cure at low pressures and temperatures.

These resins have had wide application in low-pressure laminates. Here the unpolymerized resin can be ladled on the laminating fabric and subsequent cure can be completed at temperatures as low as 160°F for the majority of formulations, while some will cure at room temperature. Curing pressures range from ordinary contact pressure to 30 lb/in². One widespread application of this family of thermosetting plastics is the manufacture of boat hulls using glass fibers as the laminating material and a polyester as the resin.

Epoxies

The epoxies are one of the most versatile plastics. Their laminates have been used extensively as a tool material since they can reproduce contours and dimensions readily and because they have good physical and mechanical properties. Dies for stamping and forming have made use of this material at considerable savings. For example, one manufacturer experienced a $16,000 savings in cost of dies used in producing station-wagon roofs by converting to epoxy from steel. Because of epoxy's lighter weight, it also has considerable application as a jig and fixture material where much material handling is involved.

As a casting resin, it has been used extensively for encapsulating capacitors, resistors, and other electrical equipment.

Epoxies require two separate components which react chemically and harden soon after mixing. The hardening is accelerated by application of moderate heat.

The epoxy resins give good chemical resistance, excellent bonding properties, and exceptionally high strength when reinforced with Fiberglas or similar laminates, and the epoxy resins are resistant to temperatures up to 350°F. Their principal drawback is their relative high resin cost, which today runs about $0.65/lb.

Silicone Resins

Silicone resins have excellent resistance to heat, water, and certain chemicals. The resins are usually filled or used in laminated parts. Good physical properties are maintained from 450 to 500°F.

DESIGN SUGGESTIONS

Selecting Materials

The following step-by-step procedure will help in selecting the most suitable material for an application, that is, the material giving the best service at the lowest cost:

1 Analyze carefully the requirements of a new application and the conditions under which the apparatus has to operate in service. Determine whether such factors as the mechanical (impact-resistance), electrical, or heat-resistance properties are of primary importance for selecting a specific material.

2 With the above data on hand, check the properties of several plastics. Compare the specific values of those materials selected in order to make the choice of the material that comes closest to the requirement.

3 Estimate the working stresses within the material which may be prevalent in the assembly under consideration.

4 Determine the required minimum sizes of the important cross sections of the part to be designed, using the property values of the material selected. Calculate the strength of the cross sections, using a minimum safety factor of 4 for the mechanical properties and a minimum safety factor of 6 for the electrical properties listed on the property sheet. These minimum safety factors must be increased by the designer, depending upon:

a The importance of the molded part in the functioning of the assembly

b The accuracy of the estimates of the working stresses involved

c The deterioration or decrease of the test values given due to conditions prevalent at the place of use or service

5 Before releasing the design, consult the molding department.

The following information is given as a guide for designing molded parts of plastic.

Minimum wall thickness   The minimum thickness of any wall or rib should be between $\frac{3}{32}$ and $\frac{5}{32}$ in depending upon the moldability of the plastic employed.

The above ruling will not affect the minimum wall thicknesses around holes and inserts. Deep draws ($\frac{1}{2}$ in and above) require separate consideration. Consult the molding department.

Maximum wall thickness   It is advantageous not to exceed the above minimum wall thicknesses. Heavier walls waste material and increase the time for molding and curing. Any wall thickness of urea resin should not exceed $\frac{1}{8}$ in. Design suggestions for the size of holes apply to urea resins only as long as a $\frac{1}{8}$-in-thick wall is not exceeded. Heavier walls will be porous or undercured. Consult the molding department in case deeper holes or inserts are required.

Draft on side walls   A draft or taper of 1 to 2° is desirable (see Fig. 8-1b) on the vertical surfaces or walls parallel with the direction of mold pressure. A minimum draft of 0.5° may be permissible in order to facilitate removal of

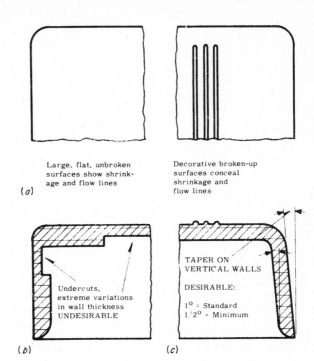

FIGURE 8-1

(a) Using decorative design to conceal shrinkage and flow lines. (b) Illustration of undesirable undercuts. (c) Application of taper on vertical walls.

molded parts from the mold cavity. However, this amount of draft may necessitate the use of a more substantial knockout mechanism. Short outside surfaces approximately ½-in high formed vertically by the matrix of a compression mold in some cases may have the taper omitted. Thin walls or barriers require a minimum taper of 1° on both sides (Fig. 8-1c). A minimum surface of $\frac{3}{16}$ in should be provided at the end for the stripper or ejector pin (Fig. 8-2).

Fillets   Fillets (Fig. 8-3) must be added to facilitate molding with minimum distortion and breakage. The use of fillets and generous radii facilitates the flow of plastic material into the mold cavity, minimizes stress concentrations, and lessens material warpage after molding. Figure 8-3 shows the effect of fillet radius for a given section thickness.

Ribs and bosses   Ribs (Fig. 8-4) can be used to increase part strength without increasing wall thickness and bosses can provide reinforcement around holes. Ribs and bosses must have 5° taper and adequate fillets.

Pronounced markings (flow lines or shadows) may appear opposite ribbed surfaces. Specify decorations (Fig. 8-1a) to conceal this undesirable effect. Ribs should not be higher than three times the thickness. Bosses should not be higher than twice the diameter.

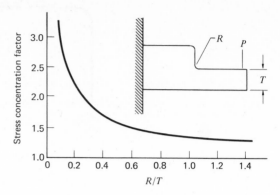

FIGURE 8-2
Fillet, rib, and surface for stripper pin design.

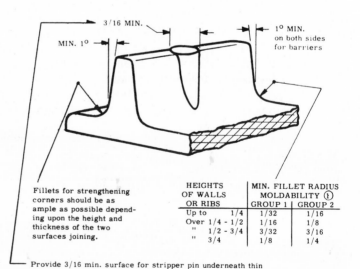

Fillets for strengthening
corners should be as
ample as possible depend-
ing upon the height and
thickness of the two
surfaces joining.

HEIGHTS OF WALLS OR RIBS	MIN. FILLET RADIUS MOLDABILITY ①	
	GROUP 1	GROUP 2
Up to 1/4	1/32	1/16
Over 1/4 - 1/2	1/16	1/8
" 1/2 - 3/4	3/32	3/16
" 3/4	1/8	1/4

Provide 3/16 min. surface for stripper pin underneath thin
ribs, high barriers, or inside walls.

FIGURE 8-3
Effect of fillet radii on stress concentration.

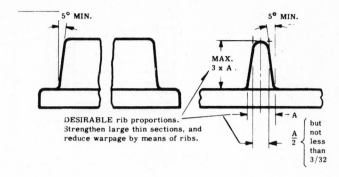

FIGURE 8-4
Design of rib proportions.

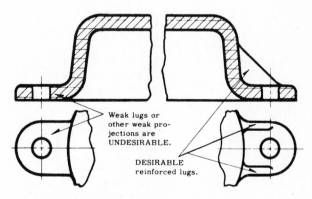

FIGURE 8-5
Design of lugs.

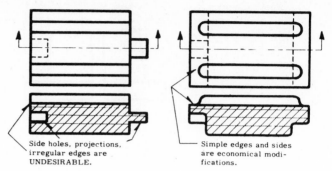

Side holes, projections,
irregular edges are
UNDESIRABLE.

Simple edges and sides
are economical modi-
fications.

FIGURE 8-6
Design should avoid side holes and side projections.

Lugs and projections Lugs (Fig. 8-5) and projections on molded parts should be reinforced by ribs and fillets to reduce warpage and distortion and to increase the strength without increasing the section of lugs or projections.

Irregular parting lines, surfaces, and projections Irregular surfaces at the parting line for the mold and projections requiring featheredge members of the mold should be avoided as they increase the cost of the mold, of the molded piece, and of the mold maintenance (Figs. 8-6 to 8-8).

Location of cavity number, parting line, and ejector pins Cavity num-

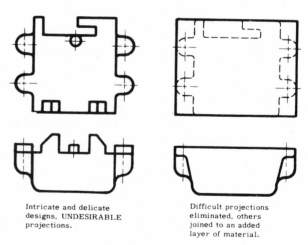

Intricate and delicate
designs, UNDESIRABLE
projections.

Difficult projections
eliminated, others
joined to an added
layer of material.

FIGURE 8-7
Design should avoid intricate and delicate sections.

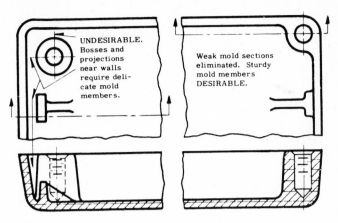

FIGURE 8-8
Design for sturdy mold members.

bers on parts are required for identification purposes during manufacture. Such numerals will be engraved in the mold 0.094-in high and 0.006 to 0.008 in deep, which will make it raised on the molded piece. The location of these numerals should be indicated on the drawing at a place on the finished piece where it will not affect the appearance, design, or function of the assembly. The location of the parting line and ejector (stripper) pins should also be discussed with the molding department because they leave a mark on the molded surface of the part thus affecting appearance.

Threads   Threading (Fig. 8-9) below $\frac{5}{16}$ in in diameter should be cut after

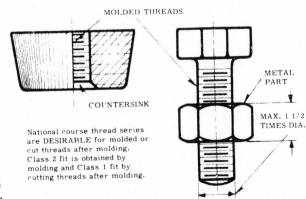

FIGURE 8-9
Thread design.

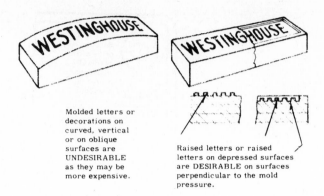

Molded letters or decorations on curved, vertical or on oblique surfaces are UNDESIRABLE as they may be more expensive.

Raised letters or raised letters on depressed surfaces are DESIRABLE on surfaces perpendicular to the mold pressure.

FIGURE 8-10
Letter design.

molding. Threads of $\frac{5}{16}$ in in diameter and above should be molded. Threaded through holes, tapped or molded, should start with a countersink in order to eliminate chipping. Due to shrinkage, long-molded thread will not assemble with standard metal thread, but will fit mating molded thread or mating short-metal thread of approximately $1\frac{1}{2}$ times diameter. A suitable grip or straight knurls should be provided for unscrewing mating thermoplastic parts.

Letters Molded letters (Fig. 8-10), raised 0.006 to 0.008 in above the basic surface of the molded part, are reproductions of engraved characters in the mold except on hobbed molds for high activities. These characters may be engraved on pins or blocks inserted in deep molds or cavities if the chosen location is inaccessible for the engraving tool. These engraving pins or blocks permit raised letters on depressed surfaces of the molded piece by inserting the pins or blocks to 0.010 in above the level of the mold cavity if molded raised letters are objectionable.

Standard characters are recommended in order to avoid costly engraving. Filled characters engraved in the surface of the molded piece are expensive and therefore should be avoided. They may be obtained by embossing or branding characters into the surface of dark-colored pieces. A heated die is pressed into the surface of the finished part and is coated by means of a colored foil for better contrast.

Labels and printings do not adhere sufficiently to a glossy surface although the adhesion may be improved by sand blasting the mold or the molded surface of the finished piece. However, special paints and inks that etch and adhere very satisfactorily to glossy surfaces are now available for most plastics.

Baselines for tolerances and oblong holes In order to eliminate loose or

Table 8-1   HOLE SPECIFICATIONS

Type of detail	Molding method	Max. depth of holes
Blind hole and slot with straight wall and recess (Fig. 8-11), taper 5° min.	Compression Transfer	$H \leq 2D_1$ $H \leq 3D_1$
Through hole with recessed wall (Fig. 8-11)	Compression Transfer	$T \leq 2(D_2 + D_3)$ $T \leq 3(D_2 + D_3)$
Through hole with offset wall (Fig. 8-11)	Compression Transfer	$T \leq 4D_2$ $T \leq 6D_2$
Through hole and slot with straight wall (Fig. 8-14)	Compression Transfer	$T \leq 3D_6$ $T \leq 3D_6$

careless fits, it is well to establish baselines for dimensioning and for specifying tolerances.

Holes—slots and recesses   Holes smaller than $\frac{1}{16}$ in in diameter must be drilled or formed after molding. The maximum depth of molded vertical holes, slots, or recesses depends upon the type of hole, method of molding, and the moldability of the material used as indicated in Table 8-1. There is a problem of varying diameters due to draft that must be considered. This is, of course, especially prevalent on deep holes. The minimum thickness of walls or solid material required around or between molded vertical holes or recesses depends principally upon the depth of hole as shown in Table 8-1.

The molded recessed through hole in Fig. 8-11 should be given preference because this type will result in the sturdiest mold members — and slots. Therefore, ample fillets on holes are desired for economical production and mold maintenance.

If any of the previously mentioned rounded edges or corners are critical for the proper functioning of an assembly and if the rounding must not exceed a permissible radius, the maximum allowable radius should be specified on the drawing. A recommended minimum radius of 0.020 applies to most designs. The elimination of sharp corners reduces the stress concentration at these points and consequently the molded parts will have greater structural strength.

Straight-wall through holes are formed by molding pins, mounted usually in the bottom half of the cavity and entering into holes on the top half. The vertical holes formed by this type of pin have a radius at the lower end only.

The four extreme edges of a narrow slot may be improved as shown in

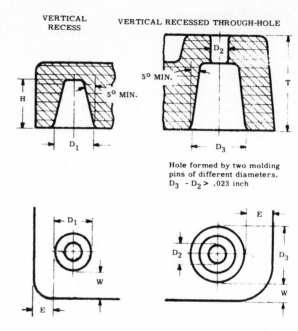

FIGURE 8-11
Design for vertical recesses and recessed through holes.

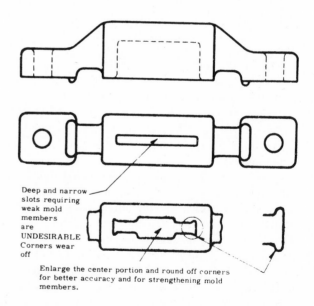

FIGURE 8-12
Design for slots.

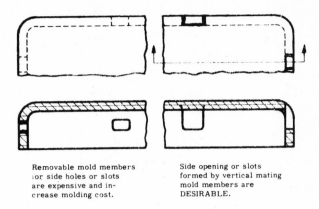

Removable mold members
ior side holes or slots
are expensive and in-
crease molding cost.

Side opening or slots
formed by vertical mating
mold members are
DESIRABLE.

FIGURE 8-13
Design for side openings.

Fig. 8-12. By increasing the center section of the mold pin, it is strengthened to assure greater service life.

Molded horizontal holes Molded horizontal holes are uneconomical, as they require removable mold members. Short side openings or slots may be formed by vertical mating mold members as indicated in Fig. 8-13.

Long-molded horizontal holes require a support at the free end of their molding pins and a removable mold section, increasing the mold cost and the cost of production. It will be more economical to drill such round horizontal holes or other holes which do not have their longitudinal axes parallel with the direction of the molding pressure, in smaller quantity requirements. If volume is substantial, then horizontal holes can be cored by cam action. This will increase the tool cost but usually does not alter the production cost.

Inserts—general Avoid inserts if a satisfactory assembly may be made with speed nuts, speed clips, spring clips, threads, self-tapping screws, drive screws, or tubular rivets. Inserts in urea resins are impractical and are in general more suitable in plastics with higher "elongation" property. When it is necessary to use inserts, it is important that the insert be surrounded with enough plastic so as to absorb shock so that the insert does not break loose when the part is in service.

Molded-in inserts Use standard inserts of brass. Metal insert surface must be cleaned for good contact or alignment after molding.

Use round end inserts instead of square or other shapes (Fig. 8-14). Shapes other than round will be expensive to anchor and seal in recesses required in the mold. Use closed end inserts to prevent flow of molding material inside.

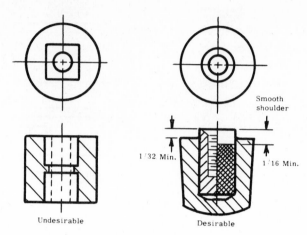

FIGURE 8-14
Design for inserts.

Standard molded-in inserts are held by the diamond knurl on the imbedded metal surfaces. A straight knurl would not prevent the inserts being pulled out. Provide a smooth shoulder of $\frac{1}{16}$-in minimum width on the projecting end of the insert. This will allow a better seal in the recess of the mold to eliminate flow of material inside.

The imbedded end of the insert should not extend close to the surface. The recommended minimum thickness of material around molded-in inserts should be $\frac{3}{32}$ in for inserts up to $\frac{1}{4}$-in diameter, $\frac{5}{32}$ for inserts above $\frac{1}{4}$- to $\frac{1}{2}$-in diameter, and $\frac{1}{4}$ for inserts above $\frac{1}{2}$- to 1-in diameter.

Molded-in inserts extending through the part should be avoided for compression-molded parts.

Standard inserts in a horizontal position while molding must be supported on both ends. Short inserts having an imbedded length shorter than 120 percent of its diameter may be drilled deeper and retapped. Whenever possible, avoid inserts in a horizontal position while molding, because they require special mold construction and molding procedure, increasing the cost. Use pressed-in inserts in case the pull-out load is light.

Avoid inserts if they weaken the maximum cross section permissible for the design. Molded or tapped threads, drilled holes together with self-tapping screws, or drive screws may be used for light loads and permanently assembled parts.

Special inserts anchored in holes of the mold, instead of on pins, should be designed with the imbedded section (diameter and length) conforming with standard inserts; that is, the imbedded length should be limited to 150 percent of the diameter. They should have round smooth shoulders for good seal in the

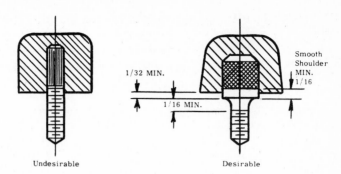

FIGURE 8-15
Do not extend threaded portion of the insert into the molding.

recesses of the mold. The threads should end $\frac{1}{16}$-in minimum from the shoulder, as shown in Fig. 8-15.

Long, slender inserts may be bent during molding. Inserts having a long, knurled surface or anchoring shoulders far apart ($\frac{5}{8}$ in or further, depending upon the shrinkage) may cause cracking. Specify short knurls and short distance between outer shoulders to permit the shrinkage, or specify compounds with small differences of shrinkage between metal and molding material, in critical cases.

Inserts of other materials, like asbestos, are feasible for such purposes as improving the arc resistance or the mechanical properties.

Pressed-in metal inserts   These inserts should be considered only if the previously mentioned assembly methods and the molded-in metal inserts are not practical or are not economical. These inserts are advisable only in case the pull-out load is light, as for terminals. The resulting press fits of the pressed-in inserts are variable due to the necessary working tolerances in the diameters of the holes and across the knurled inserts, as shown in Table 8-2. Heavy walls and impact-resisting materials are most suitable for pressed-in inserts. Specify 30 straight knurl on this type of insert when it has an OD of $\frac{3}{16}$ in and below and 46 straight knurl above $\frac{3}{16}$-in OD.

Tolerance Specifications

Several factors must be taken into account in establishing the tolerance specifications that can be expected in molded plastic compounds. The principal factor is the tolerance capability of the toolmaker in the construction of the mold. The magnitude of the material shrinkage also plays a factor in tolerance capabilities of molded parts, and distortion of the molded piece due to internal stresses resulting from high mold pressures causes a dimensional variation.

Table 8-2   PRESS FIT FOR PRESSED-IN INSERTS

Hole dia., in	Wall thickness, in	Recommended interference, in
$\frac{1}{8}$	$\frac{3}{16}$ to $\frac{1}{4}$	0.0005–0.0035
$\frac{1}{8}$	Over $\frac{1}{4}$	0.001–0.004
$\frac{3}{16}$	$\frac{3}{16}$ and over	0.002–0.005
$\frac{1}{4}$	$\frac{1}{4}$ and over	0.002–0.005
$\frac{3}{8}$	$\frac{3}{8}$ and over	0.002–0.005
$\frac{1}{2}$	$\frac{3}{8}$ and over	0.003–0.006

Table 8-3 provides a summary of tolerances that can be held during the molding process for several of the important plastics.

Coating of Plastics

Both painting and metal-vapor coating are practical and economical ways of coating part or all of plastic components in order to impart certain characteristics to the product. Painting includes the use of either lacquers or enamels. Enamels are more frequently used with thermosetting plastics in view of the high baking temperatures required.

Most metal-vapor coating involves the deposition of a thin film of aluminum on the plastic surface to be coated. The thickness of this coat is only two to three millionths of an inch and is applied primarily for reasons of appearance. One lb of aluminum will cover over 20,000 ft². Aluminum is usually used because it is inexpensive and is easily evaporated in the vacuum chamber.

Copper alloys, silver, and gold may also be used for vacuum metallizing.

Table 8-3

Plastic compound	Typical molding tolerance, in/in
Methyl methacrylate	0.0015
Polystyrene	0.0020
Cellulose acetate butyrate	0.0025
Vinyl	0.0100
Polyethylene	0.0150

Electroplating is also an important coating process applied to plastics. There are several distinct advantages in the use of plated plastics over plated metals. Some of the more important of these are the following:

1 Plated plastics are lighter, making them easier to handle and less expensive to ship. For example, a recent model of Chevrolet used a plated ABS grille which weighs slightly over 6 lb. This is about one-third the weight of a comparable zinc die-cast grille.

2 Plated plastics can be handled in higher-temperature environments because of their lower thermal conductivity.

3 Complex shapes may be more readily achieved with plastics.

4 Cost per unit may be less in view of a lower cost of resin on a volume basis.

To electroplate, a conductive substrate must be applied to the plastic. This is accomplished by chemical means and is referred to as *electroless metal plating.*

Although all plastics can be plated, there are only three types today which provide good adhesion supplemented with a high-gloss surface and can be plated economically on a production basis. These are ABS, polysulfone, and polypropylene.

Interior applications for plated plastics have proved quite successful during the past 10 years. However, when plated plastics are subjected to severe extremes of outdoor weathering, blistering, cracking, and peeling are more likely to occur. For this reason the automotive industry has been cautious in converting zinc die-cast exterior parts such as grilles, trim, headlamp bezels, etc.

NATURAL AND SYNTHETIC RUBBER

Rubber is the term commonly applied to substances, either natural or synthetic, that are characterized by exceptional elastic deformability. That is, properly prepared rubbers may be stretched without rupture far beyond the limits of any other engineering material, and upon release of the deforming stress they will return almost to their original shape. Thus, in addition to high elasticity, rubbers may possess resilience to a useful though not perfect degree. Even if rubbers possessed no other useful properties, these two would make them unique among engineering materials.

As its name implies, natural rubber occurs in nature as a latex, a milky fluid contained in specialized tissues which grow between the bark or protective sheath and the main body of certain plants. A much larger number of plants than is generally known yield rubber latex. Several, such as milkweed and dandelions, grow in temperate climates. The rubbery component known as rubber hydrocarbon is identical in all these plants. The nonrubber components

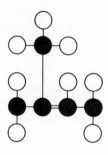

FIGURE 8-16
Isoprene monomer composed of five carbon atoms and eight hydrogen atoms.
This is the monomer that makes up natural rubber.

of the latex, however, vary widely from plant to plant, and are often injurious
to the quality of the finished rubber. It has not been economically practicable
to remove these impurities. As a result, commercial natural rubber is produced
from less than a dozen species of rubber-bearing plants. Of these, by far the
largest source is the tree *Hevea brasiliensis*, a native of the Amazon valley,
which has been transplanted to other tropical regions.

Rubber hydrocarbon is more technically polyisoprene. Isoprene is a volatile
organic liquid with the chemical formula C_5H_8 (see Fig. 8-16). Like many
organic substances, isoprene molecules have the ability to combine with themselves to form very large molecules called polymers. Rubberlike polymers are
unique in that the individual molecules are believed to join together to form
long chains which coil around somewhat like a spring. This is believed to account
for their unusual elasticity, and has led to the creation of the special term
"elastomer" to describe rubbery polymers.

Because of its unique properties, rubber is vital to the operation of many
types of machinery, both military and civilian. It thus has an importance as a
strategic material that far surpasses its annual dollar volume. The so-called
"synthetic rubbers" that are now commercially available have done much to
free the rest of the world, and the United States in particular, from the grip of
a monopoly that at times has seriously threatened the whole economy.

During the search for synthetic rubber a number of polymers have been
developed that possess to some degree the properties of natural rubber. These
have become generally known as synthetic rubbers. It should be noted however
that this is a misnomer. They are synthetic elastomers, but none of them is
real rubber. Some of them possess one or more properties which far surpass
natural rubber. For example, polysulfide elastomers, which are marketed under
the trade name of Thiokol, greatly exceed natural rubber in resistance to petroleum oils, greases, and fuels. Silicone elastomers retain useful rubberlike

properties at temperatures both higher and lower than rubber can withstand. By the proper choice of a synthetic elastomer the design engineer can achieve many results which would not be possible through the use of natural rubber. But one or more desirable results will be obtained often at the sacrifice of others.

The search for improved elastomers has been made more difficult because of the dual nature of the rubbery state. All elastomers act in part like solids and in part like fluids. As a consequence, elastomers completely follow the laws neither of solids nor of fluids. This dual nature accounts for many of the virtues as well as the defects of rubberlike materials. It has also made a systematic and rigorous study of the rubbery state very difficult. As a pragmatic art, rubber chemistry has grown to be a highly developed and most useful industrial tool. But the essential nature of an elastomer is still only imperfectly understood. As a result, most of the research and development in this field is by the "cut and try" method, and relatively few fundamental laws have yet been discovered.

In the raw state none of the elastomers possesses much engineering or practical utility. Raw natural rubber, for example, is quite temperature-sensitive, being hard and hornlike in the winter, and soft and tacky in the summer. In general, the fluid properties of raw elastomers are objectionably prominent.

The first great technical advance, and the one that really established the rubber industry, was the discovery of vulcanization by Charles Goodyear. Vulcanization is the establishment of chemical cross-links between the elastomer molecules. This acts in much the same way as a reinforcing strut in a mechanical system, and in effect ties the whole mass of rubber together into a single molecule. By this means the slippage between molecules is restrained, which minimizes the fluid properties and reinforces the solid properties. Sulfur is the principal vulcanizing agent for natural rubber, and will cross-link several of the synthetic elastomers also. Subsequent advances in the art of vulcanization have been largely the development of catalysts or accelerators to speed up and control the reaction.

If very finely subdivided solids, ranging roughly from 20 to 100 nanometers (nm) in diameter, are mixed into elastomers, they appear to wedge into the molecular interstices. Again this interferes with plastic or fluid properties and reinforces the solid properties. For this reason, such materials are sometimes known as reinforcing fillers. Carbon has been relatively easy to prepare in this small particle size, and has long been the principal reinforcing filler. More recently finely divided silica, magnesia, and other minerals have become available. Light-colored stocks of good physical properties have thus become possible.

The sulfur-accelerator system and the reinforcing fillers contribute most to the improvement of elastomers as engineering materials. In addition, ingredients are added to minimize attack by oxygen, ozone, and sunlight. All in all, a dozen or more ingredients may be incorporated into a modern rubber compound, and these may be widely varied to meet specific needs.

PROPERTIES OF VULCANIZED RUBBER

In speaking of rubber as an engineering material, we thus do not refer to a single species, but to a complex family of compounds. Each individual elastomer possesses its own specific virtues, and each may be widely modified by compounding to meet a myriad of specific engineering needs. No single compound excels in all virtues. By wise selection, elastomers may be used by engineers to do things that no other engineering materials can do. Elastomers are useful engineering materials by virtue of their mechanical, physical, and chemical properties. The most useful properties to the engineer will be discussed.

Mechanical Properties

Elasticity The ability of elastomers to undergo large deformations and yet to return to substantially their original shape makes them very useful for connecting structures that undergo large relative motions. No other engineering material will accommodate as much relative motion in such a small space. Several elastomeric materials can be elongated as much as five to ten times their original length.

Resilience Because of their ability to restore the energy by which they are deformed, elastomers may be used as springs. Natural rubber can be compounded to have a resilience approaching 95 percent. Synthetic elastomers, in general, will have lower resilience. Because of their large elasticity, elastomeric springs can be made very compact. Moreover, when rubber is used in systems apt to become resonant, high resilience is undesirable. Metal springs for such uses require dash pots or other dampers. These are bulky and often expensive. Elastomers, on the other hand, can be compounded for low resilience. With no increase in size, elastomer springs can be built with inherent damping.

Tensile strength Natural rubber can be compounded to a tensile strength in excess of 4,000 lb/in². Synthetic elastomers will, in general, have tensile strengths of the order of 3,000 lb/in², while silicone rubbers have much lower tensile strengths, of the order of 800 to 1,000 lb/in². Although high tensile strength is a measure of a high-quality rubber compound, it should be noted that a rubber stock loaded in tension to this value would be extended to seven or more times its original length. Hence, tensile strength does not have the significance to the design engineer for rubber that it has for steel.

Fatigue resistance Because elastomers serve some of their most useful functions interposed between structures in relative motion, ability to withstand repeated and often severe flexure is of great engineering importance. The nature of the principal function to be served may determine the type of compound to be used. Elastomers are relatively poor conductors of heat. Therefore the heat generated through hysteresis loss (the inverse function of resilience) may be significant. If ability to withstand flexure is the major consideration, as in a flexible coupling which must accommodate shaft misalignment, low-hysteresis stocks would be desirable. In a vibration isolator under steady-state vibration, some hysteresis might be desirable to damp out high resonant amplitudes. In a shock absorber, it may be desirable to absorb the energy of large-magnitude impacts. In such cases high-hysteresis stocks may be desirable, provided the impacts are sufficiently infrequent to permit the energy absorbed to be dissipated as heat.

Cut-growth resistance As with most structural materials, cuts, sharp notches, and other surface defects can cause local stress concentrations, and thus serve as focal points for fatigue failure. The designer of rubber parts for dynamic service should provide fillets or radii at sharp corners to minimize this effect. This situation illustrates a point frequently encountered in designing elastomer compounds for specific uses: strengthening of one property of a compound will be at the sacrifice of another. High-resilience stocks, which may be desirable for resistance to fatigue, are often not especially resistant to cut growth. This is quite comparable to spring steels, which are often poor in notch-impact resistance.

Abrasion resistance Perhaps because of their ability to yield in the path of a sharp object which would otherwise shear off some of their surfaces, elastomers in general possess unusual resistance to abrasion. This property has been very important in the development of tires, the largest segment of the rubber industry. High abrasion resistance is dependent in large measure upon skillful compounding, although proper design, especially to provide for the proper dissipation of heat, is also of great importance.

Friction properties It is well known that elastomers possess very high coefficients of friction when in contact with dry surfaces. This property has been of great importance in the tire industry, for shoe soles, and in many mechanical applications. In the presence of films of water or of other substances, the coefficient of friction is greatly reduced. Because of this property, propeller-shaft bearings for ships are frequently lined with rubber. Not only are the friction properties adequate for this service, but the high abrasion resistance and ability to yield to stresses helps prevent scratching of the shaft in the presence of sand or other abrasive substances. Friction properties may be modified through choice of elastomer, through compounding, and through design. The process for making the butadiene-styrene elastomer known as Buna-S or GR-S had to be modified before this elastomer could be safely used for tire construction. Many motorists will remember the poor performance on wet pavements of the first synthetic tires brought out during the early days of World War II. By proper choice of compounding ingredients, the coefficient of friction can be modified over a fairly wide range. In particular, it is often possible to include lubricants in the compound to reduce the coefficient of friction. Finally, design may be very important. In tire design, for example, proper tread design will permit the wiping away of the water film on a wet pavement through squeegee action. Again, the design engineer must provide for the dissipation of heat, since the elastomer surface may otherwise melt and provide a fluid film at the friction interface.

Moldability With varying degrees of ease, depending on the elastomer chosen and the type of compounding employed, elastomers may be molded under heat and pressure during the vulcanizing or curing operation. Molded articles can be made to conform accurately to very intricate mold configurations, as is demonstrated in many tire treads. If the mold surfaces are polished, high-gloss surfaces may be imparted to the molded object. Since the coefficient of thermal expansion of most elastomers is about 10 times as great as for steel, proper allowance must be made in the mold design for the shrinkage that will occur as the article cools to room temperature from the curing temperature.

Extrudability By pressure applied either by a hydraulic piston or by a worm screw, elastomers can be forced through orifices to form extruded shapes of considerable complexity. Hose and gaskets are examples of items conveniently and economically formed by this means. The unvulcanized stocks must be fairly heavily loaded with fillers to give them enough stiffness to prevent their collapsing under their own weight; therefore, high-quality mechanical goods of complex shape will not ordinarily be formed by extrusion. However, simple, solid shapes of high mechanical quality, such as tire-tread camelback, are often extruded. Because of the extreme elastic deformation that occurs as elastomers pass through the extrusion orifice, the design of these orifices for complex shapes is a highly skilled and expensive operation, applicable, in general, only to large-volume mass production.

Bondability Most of the elastomers can be made to adhere to metals, glass, plastics, and fabrics by a process known as bonding. By proper attention to details and by using high-quality stocks, the bond so attained is usually stronger than the elastomer itself so that rupture will occur within the body of the elastomer. The first great advance in the art of bonding was the discovery that excellent bonds can be formed to brass plate of closely controlled composition. While brass plating for bond formation is still used extensively, within recent years bonding adhesives have been developed which will form excellent bonds to the thoroughly cleaned surfaces of most commercial metals or other structural materials without brass plating. Because of the great difficulty of controlling brass plating, and the expense of the installation, this process is now becoming obsolete. A variety of bonding adhesives are available to meet the individual needs of the engineer.

Electrical Properties

Electrical resistance   Uncompounded elastomers are, in general, good nonconductors of electric currents. Most of the compounding ingredients used are also nonconductors; therefore, the majority of elastomer products are excellent insulators and are widely used for that purpose. Carbon, however, is a moderately good conductor of electricity, and its use as a filler in rubber compounding can be taken advantage of to make the stock itself an electric conductor. A number of types of carbon are available for rubber compounding, some of which are better conductors than others. In order for the stock to conduct electric current, a large proportion of carbon must be added so that the particles are in contact with each other. In effect, therefore, the rubber remains nonconductive, and merely acts as the matrix for a continuous chain of conducting particles embedded within it. Because of the large volumes of filler required only relatively inelastic compounds are conductive. Fairly low ohmic resistance can be achieved, but because of the poor heat conductivity of elastomers it is difficult to dissipate the heat generated as I^2R loss. Conducting stocks are thus used primarily to conduct static electricity or where the power is low.

Electrodeposition   In addition to natural rubber, several of the synthetic elastomers either occur at some step in their synthesis or can be otherwise prepared as latices. These are colloidal suspensions in an aqueous medium. The particles are electrically charged. If a direct current field is created in the latex, the charged particles of elastomer will migrate to the pole of opposite charge and be deposited upon it. Thereupon their charge is neutralized and a layer of the elastomer will be built up. By the use of compounding ingredients, also in colloidal suspension, it is possible to deposit compounded elastomers requiring only drying and vulcanization to complete. This process is used to line vessels with rubber or to coat wire mesh for abrasion-resistant screens Because of the problem of removing residual water from massive sections, the process is usually restricted to the deposition of relatively thin coatings for chemical or abrasion resistance.

Chemical Properties

An examination of Table 8-4 will disclose that natural rubber, in general, will surpass the synthetic elastomers in useful physical and mechanical properties.

Table 8-4   COMPARISON OF VARIOUS PROPERTIES OF TYPES OF RUBBER

Compound Type / Property	Natural Rubber NR	Synthetic Rubber				Reclaimed Rubber RR
		GR-S Butadiene-Styrene Copolymers (Buna S)	GR-M Chloro-butadiene Polymers (Neoprene)	GR-N Butadiene-Acryloni-trile (Buna N)	GR-I Iso-butylene Copolymers (Butyl)	
Cost	3	2	4	6	5	1
Hardness, Durometer "A"	20-100	30-95	25-100	25-100	—	40-90
Tensile Strength	1	5	2	3	4	6
Elongation	1	3	2	4	5	6
Low Compression Set	1	4	3	2	5	6
Resilience	1	4	2	3	6	5
Color Adaptability	1	2	3	4	—	5
Flexibility	1	4	2	3	—	5
Insulation	1	2	5	4	3	—
Tear Resistance	2	5	4	3	1	6
Low Staining Effect on Enamel	2 (Pale Crepe)	1	4	5	3	6
Minimum Odor	1 (Pale Crepe)	2	3	4	5	6
Adherence to Metals	1	4	2	3	5	6
Adherence to Fabric	1	2	3	4	6	5
Resistance to:						
Abrasion	1	4	2	3	5	6
Aging	5	4	1	3	2	6
Alcohol	2	4	3	1	—	5
Animal Fats	3	4	2	1	—	—
Acids (Weak)	1	5	2	4	3	6
Benzene	—	—	2	1	—	—
Cold (to 40 F)	1	3	5	4	2	6
Cold (to 70 F)	1	3	5	4	2	—
Carbon Dioxide	3	4	2	1	—	—
Carbon Tetrachloride	—	—	—	1	—	—
Ethylene Glycol	—	—	2	1	—	—
Flame	—	—	1	—	—	—
Freon	—	—	2	1	—	—
Gas (Natural)	3	4	2	5	1	6
Gasoline	—	3	2	1	—	—
Heat (to 180 F)	1	4	3	2	—	5
Heat (to 250 F)	4	3	2	1	—	—
Kerosene	—	3	2	1	—	—
Methyl Chloride	—	3	2	1	—	—
Moisture	2	4	5	3	1	—
Naphtha (Coal Tar)	—	—	—	1	—	—
Naphtha (Petroleum)	—	3	2	1	—	—
Oil (Vegetable)	1	4	3	2	—	—
Oil (Lubricating)	4	3	2	1	—	—
Prestone	—	3	2	1	—	—
Sunlight	5	4	1	3	2	6
Turpentine	—	—	—	1	—	—
Weather Aging	5	4	1	3	2	6

SOURCE: Adapted from data furnished by Lavelle Rubber Co., Chicago.

Note: No. 1 indicates the compound most favorable for each property, etc.

In the area of useful chemical properties, the synthetic elastomers will often surpass natural rubber. It should be noted that the pattern varies, and that each elastomer excels in a limited area but may be relatively inadequate in other areas. Even the useful physical properties may, in time, be impaired through undesirable chemical reactivity of the elastomer. Therefore the design engineer must carefully balance physical and chemical properties to select the elastomer most useful for his purposes. Again, it will be frequently necessary to accept a compromise. For this reason it will be of great importance to recognize and provide for those properties that can least afford to be impaired and to sacrifice in other areas of relatively less importance. From the point of view of the engineer, the useful chemical property of an elastomer may be defined as its resistance to serious and irreversible damage through chemical interaction with atmospheric or other agents to which it must be exposed throughout its proposed service life. It should be remembered that wherever a system is created in which materials of different chemical nature are associated, each can react in a manner different from the way it might behave if its neighbor were not present. This is well known in the field of metals where the galvanic interaction of dissimilar metals in contact in the presence of a corrosive atmosphere is of recognized importance. It is less well known but equally true that a similar situation can also exist where nonmetallic materials are involved in the system. The design engineer must therefore keep in mind all materials that can enter into a proposed system and allow for their possible interaction. If a bonded-rubber system is to be employed, the bond interface should not be ignored.

Atmospheric effects   Except in very special cases, of course, almost all elastomer structures are exposed to the air. The oxygen in the air slowly attacks natural rubber at room temperatures, and more rapidly at elevated temperatures. This reaction constitutes part, at least, of the phenomenon known as "aging." Usually the first manifestation is a hardening of the surface due to the formation of a resinlike film somewhat similar to hard rubber. At this stage rubber may crack at the surface when flexed, thus accelerating fatigue failure. Later the surface may become gummy and tacky. Because aging is practically universal, antioxidants are regularly used in compounding practically all natural rubber, and to the degree justified by the expected service life of the compound.

Ozone, which is normally present in air at sea level in only microscopic proportions, is extremely active in its attack. Other oxidizing substances, such as the oxides of nitrogen, will have a similar effect. Certain ground-level areas appear to be particularly bad in this respect, the Los Angeles area being a notorious offender. Electric sparks, corona discharges, and ultraviolet radiation will all generate ozone. Synthetic elastomers should be chosen for use in close proximity to such ozone generators whenever their other properties will permit. If not, excellent antiozonant materials may be used in compounding, or coatings may be applied to protect the rubber. It should be noted, however, that most resistant coatings tend to crack or flake off if the elastomer is subject to much flexure. Elastomers exposed to sunlight are also attacked in a similar fashion. The effect appears to be indirect, since those surfaces subject to direct exposure may be less attacked than shaded surfaces in a generally sunlit area. Normal compounding of high-quality mechanical rubber stocks will include ingredients to protect against this "sunchecking."

Temperature effects   Two types of temperature effect must be considered. The first is irreversible but indirect. In general, the rate at which any chemical reaction proceeds increases rapidly with increasing temperature; therefore, it may be expected that any of the chemical effects here discussed will take place at a faster rate if the system is at elevated temperatures. This does not necessarily mean that the effect will proceed to a greater extent. However, the service life of any elastomer subject to degradation through chemical means may be expected to be shorter at elevated temperatures.

The second effect is purely physical and reversible. As the temperature of any elastomer is lowered, the elastomer becomes gradually less resilient. After going through a stage where it is still pliable but leathery, it eventually becomes brittle and will fracture easily upon impact. By the use of plasticizers, which are, in effect, internal lubricants, the useful low-temperature ranges of elastomers can be extended somewhat below their inherent value. The silicone elastomers have the broadest low-temperature range, and are useful below $-100°F$. Specially compounded natural rubber is useful to $-70°F$, but the low limit for most of the synthetic elastomers will not exceed $-40°F$.

Raw rubber undergoes what is known as a second-order transition as it is cooled below about 40°F. It loses elasticity, becomes more opaque, and undergoes a volume change. Vulcanized rubber either loses this transition or has it shifted below the temperature where it will impair its useful properties. Some of the synthetic elastomers, notably some of the neoprenes, undergo this transition even in the cured state. It is important that although it is a reversible phenomenon, a higher temperature is required for reversal; it may be necessary to heat above 100°F. If much work is done on the elastomer in a dynamic system, the heat generated by hysteresis may bring about reversal. Otherwise care may be required in selection of elastomers for service or storage in cold climates.

With increase in temperature above room temperature, both reversible and irreversible changes may take place. Reversibly, as the temperature rises, all elastomers lose in tensile strength. Relatively, the silicone rubbers suffer least in this respect. The poor initial tensile strength must be remembered, however. At a given temperature, other elastomers which have lost a much greater percentage of their initial tensile strength may still have higher absolute tensile strength.

The acceleration of aging reactions at elevated temperature has been discussed previously. This effect is irreversible. Most reports in the literature to date have dealt only with this phase of high-temperature effects. Very little has been published to date concerning tests run at the actual temperature of operation. Obviously, both phases should be carefully considered by the design engineer before an actual installation is made.

Solvent effects   The degrading effect of mineral oils, gasoline, benzene, and similar lubricants, fuels, and solvents, is well known. This is a *solvent effect*. In unvulcanized natural rubber, the molecules are actually separated to form a colloidal suspension. This is not a true solution, but is similar to it. After the molecules have been cross-bridged by vulcanization, complete separation does not occur; however, a more or less pronounced swelling takes place, and the elastomer becomes weak and flabby. The degree of swelling will depend on the nature of the solvent. Aromatic substances of the benzene family will in general be more severe than the straight-chain compounds of the aliphatic family. The size of the solvent molecules may affect the rate at which swelling occurs, but may not greatly affect the degree of swelling which will ultimately occur. Heavy lubricating oil, for example, may swell a rubber stock quite slowly. If only slight or occasional exposure is to be met, and if the elastomeric part can be fairly easily replaced, advantage may be taken of other desirable properties of that elastomer, even though it may not have the highest oil swell resistance. On the other hand, if constant exposure is expected, or if the part is inaccessible for maintenance, another elastomer may be necessary, even at the sacrifice of other desirable properties.

Swelling due to organic solvents is to some degree reversible. Volatile fluids may evaporate in time, leaving little permanent swelling or damage. The process is slow, however, since the solvent must diffuse to the surface first. Moreover, the work done in swelling shows that some active forces of attraction are at work. It is doubtful, therefore, if swelling is ever completely reversed, so it is preferable to choose an elastomer more resistant to swelling wherever possible.

Corrosive chemical effects   Many elastomers possess excellent resistance to inorganic acids, bases, and salts. Most of them are not highly resistant to strong oxidizing chemicals such as nitric acid, concentrated sulfuric acid, and peroxides. With these exceptions, elastomer containers or linings are widely used in the chemical industry. Since the compounding ingredients may be less resistant, however, special compounding is necessary for such uses.

Although the elastomers themselves may be resistant, bonded assemblies or other combinations of elastomers with nonelastomeric materials must often be considered for such service. In such cases consideration must be given not only to each of the materials involved, but also to the interaction each may have upon the other. A bonded-rubber assembly, for example, contains at least three major components: an elastomer, usually (though not always) a metal,

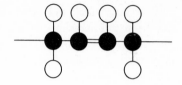

FIGURE 8-17
The monomer butadiene consists of four carbon atoms linked with six hydrogen atoms.

and the thin but very real bonding system. Such an assembly is usually attached to another structure by screws or rivets or by other mechanical means. The entire system now consists of many parts, all assumed to be exposed to the corroding medium. In such cases, for example, galvanic corrosions may occur between the bolt and the supporting structure. Even though the entire bonded assembly has been proved to be satisfactorily resistant to this corroding medium, the electrochemical effects of the bolt-supporting structure may be in some degree related to it, with consequent early failure of the bond. The design engineer should bear in mind that all marine atmospheres contain appreciable amounts of salt which can deposit on any exposed equipment. Moreover, inorganic salts are frequently used on highways either for dust or ice control. Finally, rains wash appreciable amounts of acids and other corroding materials out of smoke-filled or industrial atmospheres. There are, therefore, few places where elastomers can be designed into equipment where some problem of corrosion may not potentially exist.

SYNTHETIC RUBBERS

None of the synthetic rubbers is equal to natural rubber in resilience, yet each synthetic has some physical or chemical property that is superior to the same property in natural rubber. Since World War II, the synthetics have opened up new fields in the design of such items as oil seals, gasoline hose, gasoline-pump diaphragms, couplings used under oil exposure conditions, inner tubes, and linings impermeable to air, gases, and chemical solutions.

SBR

SBR, frequently referred to as Buna-S, is a synthetic copolymer of styrene and butadiene (see Figs. 8-17 and 8-18). It is one of the earliest synthetics and even today represents the largest variety being made. It is about 70 percent as resilient as natural rubber at room temperature, and has a lower tensile strength than natural rubber. At 40 Durometer hardness, SBR is approximately only

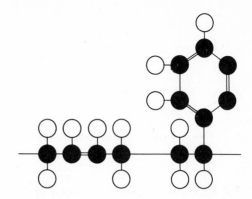

FIGURE 8-18
One butadiene monomer linked with a styrene monomer forming a chain
produces the styrene-butadiene copolymer.

10 percent as strong as natural rubber. Natural rubber has a tensile strength of
2,000 lb/in² and SBR will run only about 200 lb/in², with compounds resulting
in 40 Durometer or softer.

The tear resistance of SBR is poorer than natural rubber. As with natural
rubbers, SBR should not be used in products where resistance to oil and solvents
is needed. SBR compounds have fair resistance to set, and resistance to abrasion
is also satisfactory. Certain types give good wear resistance and consequently,
SBR has had application as a tire material. It has superior water resistance as
compared with natural rubber. Resistance to sunlight and ozone is about the
same as natural rubber. It is less expensive than natural rubber.

NBR

NBR is a copolymer of acrylonitrile and butadiene (see Fig. 8-19). NBR has
a low tensile strength in comparison with natural rubber and is approximately

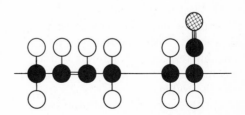

FIGURE 8-19
One butadiene monomer linked with an acrylonitrile monomer forming a chain
produces the acrylonitrile butadiene copolymer.

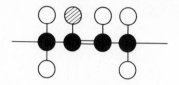

FIGURE 8-20
Replacing one of the inner hydrogen atoms of the butadiene monomer with a
chloroprene atom gives the chloroprene monomer characteristic of Neoprene.

two-thirds as resilient as natural rubber; however, it can be compounded to
give tensile strengths approaching natural rubber. It loses its resiliency rapidly
as the temperature declines, and at $-40°F$ it becomes brittle. Its resiliency
improves at higher temperatures. For this reason, it was used extensively during
World War II years as a substitute for natural rubber on installations where
temperatures were higher than normal room temperatures.

NBR has good compression set resistance and consequently has had wide
use as a gasket or seal material. The important quality of NBR is its resistance
to swelling and deteriorations when exposed to gasoline and oils. For this reason,
it is being used in quantity for such purposes as gasoline hose, oil-pump seals,
and carburetor and fuel-pump diaphragms.

Neoprene

Neoprene is a polymer of chloroprene (see Fig. 8-20). It approximates natural
rubber in resiliency at room and elevated temperatures. In this respect, it is
superior to the other synthetic materials. The general physical properties of
Neoprene, including resistance to set and tensile strength are good; however,
because of its cost and the fact that it stiffens appreciably at lower temperatures,
it has not been seriously considered as a replacement for rubber.

Because of certain chemical properties which Neoprene possesses, it does
have wide application. Although Neoprene swells when subject to gasoline and
oils, it will not disintegrate as does natural rubber. Also, at elevated temperatures
(150 to 250°F), it will not soften as does natural rubber. In general, Neoprene
resists the effect of oxygen, ozone, and aging better than the other rubberlike
materials. Neoprene tends to be impermeable to air and many gases, although
in this respect it is inferior to butyl rubber. It is flame resistant and will not
support combustion. It has good resistance to the corrosion action of chemicals
and to water.

Because of its desirable characteristics, Neoprene is being used for such
items as garden hose, insulation for wire and cable, gasoline-pump hose, packing
rings, motor mounting, and oil seals.

Butyl Rubber

Isobutyl-isoprene rubber is produced by copolymerizing isobutylene with small amounts of isoprene. The general physical properties of butyl rubber, including tensile strength, resistance to tear, and resiliency, are somewhat poorer than those of natural rubber. Butyl rubber is not resistant to gasoline and oils, and has only fair resistance to set under compression and tension. Also, its compounds stiffen considerably when subject to low temperatures.

Butyl rubber is highly resistant to ozones and oxidation and thus is used when resistance to aging is a requisite. It has excellent dielectric properties, and is resistant to sunlight and all forms of weathering. It is also practically impermeable to air and many gases; thus it is useful for products such as inner tubes, tubeless tires, dairy hose, and gas masks. In addition, since it is resistant to acids, it is useful for many applications in the chemical industry.

Polysulfide

Linear polysulfide's mechanical properties, including tensile strength and resistance to tear and abrasion, are poor as compared to natural rubber. Abrasion resistance is about one-half that of natural rubber and tensile strengths average about 1,300 lb/in².

Polysulfide polymers are resistant to swelling and deterioration in the presence of gasoline and oils. They are also quite resistant to oxidation, ozone, and aging. They are not permeable to liquids and are more impermeable than natural rubber to most of the gases.

Polysulfide rubber has been used for solvent-resistant molded parts and coated fabrics. It is used also for such items as gasoline, paint and lacquer hose, printers' rolls, and newspaper blankets. It has not proved very useful for mechanical goods.

Silicone Rubber

Silicone rubber has a different type of structure from other elastomers. It is not made up of a chain of carbon and hydrogen atoms but is made up of an arrangement of silicon, oxygen, carbon, and hydrogen atoms (see Fig. 8-21).

The mechanical properties of the silicone rubbers, including tensile strength, tear and abrasion resistance, resistance to set, and resiliency, are inferior to those of natural rubber. Today, silica-reinforced elastomers have tensile strengths of about 2,000 lb/in², and elongations of 600 percent can be realized.

This rubber's most important property is resistance to deterioration at elevated temperatures. Intermittent temperatures as high as 550°F and continuous temperature of 450°F produce negligible changes in flexibility and surface

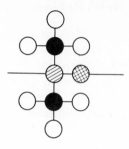

FIGURE 8-21
The silicone elastomer is based on arrangement of silicon, oxygen, carbon, and
hydrogen atoms known as the dimethyl siloxane polymer.

hardness of silicone rubber, whereas natural rubber and the other rubberlike
materials would quickly deteriorate or harden.

Silicone rubber also has excellent resistance to oxidation, ozone, and aging.
It is fairly resistant to oils, although it does deteriorate in gasoline.

Because of the high cost of silicone rubber (about 20 times the cost of
natural and most synthetic rubbers), it has had limited industrial application.

SUMMARY

During the past 20 years, the number of plastic materials that have become
available to the production engineer has greatly increased. Today there are
more than 30 chemically distinct plastic families. The selection and design of
elastomers, thermoplastic and thermosetting plastics can be quite complex.

The alert production-design engineer should become familiar with fundamentals of good design as related to plastics in general and should have an
understanding of the mechanical and physical properties as well as methods of
fabrication and manufacturing limitations of the principal thermosetting, thermoplastic, and elastomer materials. To meet both static and dynamic chemical,
electrical, or mechanical needs, plastics are widely used in many structures and
potentially could be used to advantage in many more. By the use of properly
selected plastic systems, longer useful life, more comfort, convenience, and
utility can be built into many engineering structures.

Today plastics cost an average of about $0.50/lb, although there is con-
siderable variation in the cost of different plastics. For example, in hopper-car
lots high-density polyethylene runs $0.15/lb, while ABS resins are approximately
$0.60/lb and nylon is $1.00/lb. Aluminum and magnesium in similar shapes cost
between $0.35 and $0.40/lb, and steel costs $0.12/lb. This is a significant differ-
ence in cost and is the principal reason why plastic automobiles are not in full-
scale production. It must be remembered that for a given section, plastics

weigh approximately 15 percent as much as steel and 45 percent as much as aluminum; so from a volume standpoint, they are reasonably competitive.

QUESTIONS

1 Why was the electrical industry one of the earliest users of plastics?
2 What is meant by polymerization?
3 Of what do molding compounds consist?
4 What is the effect of plasticizers in the molding compound?
5 When would it be advisable to add a lubricant to a molding compound?
6 What are the effects of temperature on thermoplastic materials?
7 Why is "migration of plasticizers" an important consideration in designing assemblies?
8 Outline the effects of humidity on plastics.
9 How does prolonged exposure to sunlight affect plastics?
10 Is oxidation a serious objection to designing components to be made of plastics? Why?
11 What are the effects of loading on plastic materials?
12 When would you specify the plastic Lucite?
13 In what type of parts would it be advisable to call for the use of cellulose-acetate plastics?
14 What are the principal applications of the aniline-formaldehyde resins?
15 What are the desirable characteristics of nylon?
16 When would it be desirable to specify the use of phenol-formaldehyde resins? Of urea-formaldehyde resins? Of melamine-formaldehyde?
17 What five-step procedure will help in selecting the most suitable plastic material?
18 Why will a long male thread cut in metal not be readily assembled in a female threaded plastic member?
19 What thought should be given to molded horizontal holes?
20 What are the relative costs of plastics, ferrous, and nonferrous materials? Explain the relationship of these costs to their specific gravities.
21 Design a plastic part with inserts and the mold for making the part according to good molding practice. Select materials and give reasons for selection.
22 What is the chemical constitution of isoprene?
23 How are rubberlike polymers unique?
24 What is "Coral" rubber?
25 What does the vulcanization process entail?
26 Define the mechanical property "resilience." Where would materials of high resilience be used?
27 What is the approximate tensile strength of natural rubber? Why is the tensile strength of rubber of little concern to the production-design engineer?
28 How can a compounded elastomer be made an electrical conductor?
29 In the design of what type of products would you recommend the use of Buna-S?
30 Why is Neoprene widely used as a material in the manufacture of oil seals?

31 Why has Thiokol not proved very useful for mechanical goods?

32 What is silicone rubber's most important property?

PROBLEMS

1 Design a plastic cover for an oil reservoir for an automatic forming machine. There is a filter element inside the reservoir which may become clogged in use. In that event the cover will be subjected to an internal pressure of up to 200 lb/in². The cover must have eight threaded bolt holes with the heads of the bolts on the other side of the flange to which the cover is attached. The cover must also have the company name on the outside in raised or indented letters whichever you recommend. Specify the type of plastic, the size of the bolts, the size of the enlargements for the bolt holes, the minimum wall thickness, the type and size of the lettering and how the holes are to be threaded or if inserts are to be used. Also specify the molding process to be used if the production quantity is expected to be 100,000 per year.

SELECTED REFERENCES

KAUFMAN, MORRIS: "Giant Molecules, The Technology of Plastics, Fibers, and Rubber," Doubleday Science, New York, 1968.

ROSEN, STEPHEN L.: "Fundamental Principles of Polymeric Materials for Practicing Engineers," Barnes and Noble, New York, 1971.

THE SOCIETY OF THE PLASTICS INDUSTRY, INC.: "Plastics Engineering Handbook," Reinhold, New York, 1960.

CERAMICS AND POWDERED METALS

CERAMICS

The American Ceramic Society has defined ceramic products as those manufactured "by the action of heat on raw materials, most of which are of an earthy nature (as distinct from metallic, organic, etc.) while of the constituents of these raw materials, the chemical element silicon, together with its oxide and the compounds thereof (the silicates), occupies a predominant position." With the introduction of such materials as silicon, carbide, fused cordierite, pure alumina titania, and others, the ceramic industry, which is one of the oldest, has made rapid strides in the development of new materials. Almost every plant has a chance to garner extra profits by taking advantage of recent developments in this field. For example, ceramic cutting tools are beginning to show promise. Ceramic tool tips are reported to allow less friction loss between the tool face and the cut chips and to be more resistant to cratering than conventional tools.

The American Ceramic Society has classified ceramics into the following groups.

1 Whitewares
2 Glass

3 Refractories
4 Structural clay products
5 Enamels

Of these, whitewares, glass, and refractories will be discussed, as well as carbon, which rightfully can be classed as a ceramic.

Whitewares

Ceramic whiteware is a ceramic body that is usually white and may be glazed. It includes such families of products as earthenwares, china, and porcelain. The typical ceramic whiteware body is composed of a nonplastic, a plastic, and a flux. The nonplastic comprises the structural skeleton which is held together by the bond developed from the flux during the forming process. The plastic gives the ceramic the workability required in the forming process.

Production methods Once the body has been prepared, there are four possible ways to form the required shape: jiggering, casting, extruding, and pressing.

Jiggering Jiggering, which is usually automatic is a forming process patterned after the old potter's wheel. Here blanks of the material are placed in a heavy mold made of plaster of paris. The inside of the mold has the shape desired of the outside of the ceramic piece being made. The inner surface of the blank is formed by having a shaped tool forced into the plastic material while the blank is rotated.

Casting In casting, a suspension of the body composition is poured in a gypsum mold. The mold, which absorbs moisture from the suspension, causes the body to cast against the plaster face of the mold. In the event the part is to be hollow, the excess body may be poured off once the correct wall thickness has been realized. This is known as *drain casting*.

Extruding In extruding, the raw material is mixed to a plastic state in the extrusion press. Frequently the plastic material, prior to the actual extrusion, is deaired in what is known as a *pug mill*. The nozzle of the extrusion press allows mandrels to be placed in position so that the extruded rod can have a variety of internal openings as well as shapes.

Pressing In pressing, the base aggregate (which has a reduced amount of water as compared with extrusion) is placed in a steel die and subjected to pressure. Pressing, which is done anywhere from several hundred to several thousand pounds per square inch with hydraulic or automatic presses, has gained extended

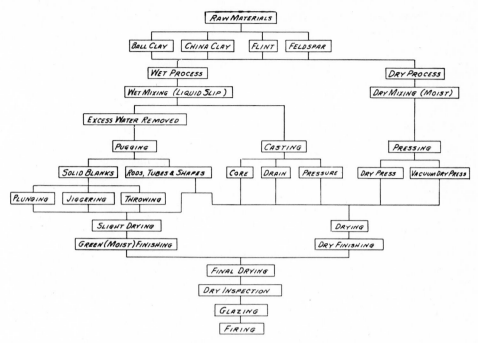

FIGURE 9-1
Manufacturing process commonly used in making porcelain insulators.

use where adaptable. However, since the granular material is limited in its plasticity, complex shapes in the lateral plane cannot be formed successfully.

The formed ceramic is in the *green* state; after drying it is said to be in the *leather-hard* state.

If a glaze is to be applied, it is sprayed on and the body and the glaze are matured in a single firing. Sometimes the body is matured first with a bisque fire; then the glaze is applied, which is matured with a lower temperature firing known as a ghost fire. Figure 9-1 illustrates the manufacturing processes commonly used in making porcelain insulators.

Joining In joining ceramic whiteware to metals, five techniques are employed: mechanical seals, brazed seals, soldered seals, matched seals, and unmatched seals. Mechanical seals are the most widely used; although with the expanding field of electronics, soldered and brazed seals are popular in the miniaturization of components.

Properties of whitewares Almost all whiteware ceramics are resistant to all chemicals except hydrofluoric acid and, to some extent, caustic solutions. They will withstand prolonged heating to temperatures over 1800°F. Indeed, many will bear temperatures considerably higher. Ceramics, in general, are weak in tensile strength and are much stronger under compressive loading. There is a wide range of strength in the different ceramics depending upon their composition and vitrification.

Whitewares have good thermal endurance and high dielectric strength, making them valuable materials for many electrical applications.

The required characteristics of the spark-plug insulator illustrate those properties attainable with high-tension porcelain. Here we must have: (1) good dielectric properties at elevated temperatures, (2) high dielectric strength, (3) resistance to heat shock, (4) good mechanical strength, (5) resistance to lead compounds, and (6) resistance to attack by carbon. A typical composition of such a vitrified porcelain would be 30 percent kaolin, 20 percent ball clay, 30 percent feldspar, and 20 percent silica.

Types of whitewares   The production-design engineer today inevitably will find application for a ceramic whiteware at some time. The principal types of these ceramics today, exclusive of earthenwares and chinas, include: alumina-type bodies, beryllia-type bodies, steatite-type bodies, zirconium bodies, and titania ceramics.

The alumina-type bodies are typified by the spark-plug insulator. These ceramics may be defined as those which contain at least 80 percent of alumina in the alpha form. Most of the remaining composition is silica plus the alkaline earths. The alumina, silica, and alkaline earths when fired form a stable, hard, dense, chemically inert, gas-tight mass.

The typical characteristic of a high alumina ceramic is its relatively good strength. Transverse strengths of 15,000 lb/in^2 can be achieved at 2500°F in some alumina ceramic materials. Dimensional stability and chemical inertness are other important attributes of alumina ceramics.

High alumina ceramics have made possible construction of high-powered klystrons which have been used to bounce radio signals from the moon. The very low dielectric loss-factor property of alumina ceramics has been instrumental in allowing these klystrons to be built. High alumina ceramics are being used to a great extent in microwave communications.

Beryllia-type ceramics usually have beryllia contents which range from 95 to 100 percent BeO. As with alumina ceramics, the beryllia ceramics are hard, dense, gas-tight materials but they have a much higher thermal conductivity. Thus, with the beryllia ceramics, we are able to incorporate the thermal conductivity of a metal-like aluminum with electrical resistivity. This ceramic promises to be one of the important forthcoming nuclear materials since it is capable not only of shielding a reactor, but of withstanding temperatures which are approximately double those experienced in reactors today.

Steatitic porcelain is produced largely from talc (3MgO·4SiO$_2$H$_2$O). This family of ceramics has its greatest use as an insulating material, and has such applications as crystal cases, coaxial cable insulators, coupling insulators, heater wire supports in appliances, etc. These ceramics have high mechanical strength (20,000 lb/in² flexural strength and 85,000 lb/in² compressive strength), high hardness values (7.5 Moh hardness), and good resistance to abrasion and dielectric strength (240 V/mil).

Where large temperature gradients are encountered, an improvement in the thermal endurance of steatite may be realized by adding appreciable (up to 20 percent) barium zirconium silicate in place of an equivalent amount of talc for zirconium types of ceramics. Other ceramics recently developed for improvement of thermal-shock characteristics include: zircon-calcium, zirconium, and zircon-alumina.

Cordierite-type ceramics have application where low thermal expansion and/or thermal endurance is a requisite. The theoretical composition of this type is $2MgO \cdot 2Al_2O_3 \cdot 5SiO_2$. In general, the cordierite ceramics are characterized by having fair mechanical properties (8,000 lb/in² flexural strength and 40,000 lb/in² compressive strength), high dielectric strength (220 V/mil), wide firing range with low thermal expansion, and excellent thermal endurance.

The most important characteristic of titania ceramics is the high dielectric constant. Representative dielectric-constant values of the more important titanias are:

Barium-strontium titanate	10,000
Barium titanate	1,200–1,500
Strontium titanate	225–250
Calcium titanate	150–175

High dielectric values have given the titania ceramics much application in resistors, high-capacity capacitors, waveguides, photoelectric cells, and substitutes for mica capacitors.

Design considerations Since most whiteware parts shrink during the firing process, which takes place at about 2500°F, shrinkage must be taken into consideration in the design. In order to finish-machine the formed material in the dried state, a certain amount of material must be allowed; however, it usually is not necessary to finish-machine formed ceramics, as fairly close tolerances can be achieved when the material is formed. These are ±0.012 in for casting and ±0.007 in for dry pressing.

Where finishing operations are required, they are usually achieved by wet grinding at high speeds. In cylindrical grinding, ±0.004 in can be held, and parallel grinding will give an accuracy of ±0.0012 per 4-in length.

The minimum thickness of flat surfaces should be $\frac{1}{8}$ to $\frac{1}{4}$ in, depending on the method of production and the size of the finished part. As the surface area increases, the minimum thickness will of course increase.

Table 9-1 COMPARISON OF PRINCIPAL TYPES OF GLASSES

	Silica	96% Silica	Boro-silicate	Lead	Lime
Cost	Highest	High	Moderate	Low	Lowest
Electrical Resistance	High	High	High	Highest	Moderate
Thermal Shock Resistance	Highest	High	Good	Low	Low
Strength	Highest	High	Good	Low	Low
Hot Workability	Poorest	Poor	Fair	Best	Good
Chemical Resistance	Highest	High	Good	Fair	Poor
Impact Abrasion Resistance	Best	Good	Good	Poor	Fair
Heat Strengthening Possibilities	None	None	Poor	Good	Good
Ultraviolet Light Transmission	Good	Good	Fair	Poor	Poor
Weight	Lightest	Light	Medium	Heaviest	Heavy

Glass

About 95 percent of the glass produced today is made from a mixture of oxides of silicon and certain metals. Modern refinements in the method of manufacture, together with the general properties of this material, have made possible the development of one of the most versatile of all materials. Today, glass is woven into cloth; made into doors, cookware, and self-defrosting windshields; used as a glazing material for buildings; made into filters, prisms, and other light-separating devices; and made into bottles, jars, and many other products.

Types of glass The principal types of glass, classified according to composition, are silica, borosilicate, lead, and lime. Table 9-1 shows a comparison of these types of glass.

Silica Silicon oxide, when fused to a clear glass, produces properties that have considerable application under thermal conditions. This type of glass, frequently referred to as *quartz glass*, has very low thermal expansion. It can be heated red hot and then immersed into cold water without cracking or failure. Silica glass also has high strength characteristics, chemical resistance, electrical resistance, and good ultraviolet light transmission. If it were not for the difficulty in producing it, and its consequent high cost, it would be an excellent general-purpose glass.

Borosilicate The borosilicate glasses represent a class to which oxides of boron, sodium, and sometimes potassium and aluminum are added. Boron is the principal additive, usually comprising 14 percent of the glass. A representative borosilicate glass would be made up as follows: silica, 80 percent; boron oxide, 14 percent; sodium oxide, 4 percent; aluminum oxide, 2 percent. Borosilicate glasses are readily worked and still have high strength, high chemical resistance, high electrical resistance, and low thermal expansion. Because of these characteristics and a lower cost than the silica group, the borosilicate glasses have wide industrial usage. Typical applications include sight glasses, gage glasses, electrical insulators, laboratory glassware, and glass cookware.

Lead Lead glasses are those to which lead oxide has been added to the sodium oxide–silicon oxide combination. The amount of lead oxide added varies considerably with the intent of the final product. For example, some glasses used for protective purpose against x-rays have as much as 90 percent lead oxide. A representative lead glass for utility purposes could be: silica, 70 percent; lead oxide, 15 percent; sodium oxide, 10 percent; potassium oxide, 5 percent.

Lead glasses have excellent electrical-resistance properties and ability to be readily hot-worked. Consequently, they are used for such products as thermometer tubing, fluorescent lamps, lamp tubing, and television tubes. Lead glasses also have a high refractive index, ranging from 1.50 to 2.2, making them ideal for optical glasses. Since these glasses have a high dispersion of light, they are especially desirable for cut glassware and jewelry.

Lead glasses, the heaviest of the glasses, are soft and, consequently, easily scratched. These inferior characteristics should be considered before selecting lead glass as an engineering material.

Forms of glass Glass is available in numerous forms including sheet, rod, tube, and various finished forms. When special finished forms are desired, the production-design engineer should consult the glass manufacturer, who usually produces the glass in the finished form required. Costs can frequently be lowered by following the suggestions of the glass manufacturer.

Window Glass Window glass is usually a soda-lime sheet glass produced by extruding a thin strip of the material vertically from the melting tank. It cools as it descends and is cut to length from the bottom of the extruded sheet. Window glass is produced in thicknesses up to $\frac{1}{4}$ in, in standard size sheets of 76 × 120 in. Window glass is not acceptable for automotive or aircraft glazing but is widely used for domestic and commercial buildings. Sheet window glass also finds wide application in such items as mirrors, table tops, and photographic plates.

Sheet window glass can be heat-treated so as to increase its tensile strength from two to five times. With increased strength the sheet glass can be converted to many additional uses, including fire screens, safety mirrors, office-building glazing, and gage shields. Wire mesh can be imbedded in sheet glass in the molten state so as to give a glass that is stronger from the standpoint of penetration of missiles and less vulnerable to the danger of fragmentation.

Plate Glass   Plate glass refers to sheet glass that has been ground. It is produced by rolling the plastic glass to thicknesses ranging from $\frac{3}{16}$ to $1\frac{1}{4}$ in. This rough-rolled stock is then ground and polished with special-purpose equipment to obtain a surface that is optically flat. This quality glass has been found useful for automotive glazing, storefront windows, tracing tables, and surface plates. Plate glass can be heat-treated to give extra strength; when this is done, it is sold under the name of tempered plate glass.

Laminated Glass   Laminated glass, frequently referred to as safety glass, is produced by placing a transparent vinyl plastic between two layers of plate glass. The plastic layer prevents splintering of the outside glass layers if broken. It is used for such purposes as automotive and aircraft glazing, protection shields, and storefront windows.

Formed Glassware   Formed glassware may be classified as pressed ware, blown ware, and drawn ware. Pressed ware is produced by forcing molten glass into a mold. This is usually done automatically, although it may be done by hand. Typical products by this method are household appliances, eyeglasses, glass gages, and decorative and ornamental pieces.

Blown ware includes those hollow forms produced by blowing a jet of air into a glob of the molten material so that it takes the form of the closed mold upon solidification. Typical blown ware includes bottles, jars, vases, and bulbs.

Drawn ware includes rod and tubing and is used in gage glasses, chemical pipe, and insulation.

Properties of glass   When designing for the use of glass, structurally a maximum load of 1,000 lb/in² is considered suitable for annealed glass and a load of 2,000 to 4,000 lb/in² (depending upon its composition) for heat-treated glass. Glass usually fails in tension, as it is about 10 times as strong in compression as it is in tension. Glass is subject to fatigue: under load, it will lose strength in time. Temperature affects the strength of glass. At low temperatures, glass is the strongest; it tends to reach its lowest strength at around 350°F; thereafter, its strength increases with additional temperature rise. Generally speaking, glass is a hard material, being harder than soft steel, aluminum, and brass. The thermal conductivity of glasses at room temperature ranges from 0.0016 to 0.0029 cal/(cm)(s)(°C). Glass makes an excellent material for cookware since about 96 percent of radiated heat is absorbed by glass. When in loosely compacted batts of glass fibers, it is a poor conductor of heat and so an excellent thermal insulator. Because of its electrical resistivity, glass is widely used in the electrical industry as a standard resistance material.

Design considerations Since glass is weak in tension, designs should avoid stressing it in this manner. Large radii should be provided on molded shapes, as glass is quite notch sensitive. Designs that require heat-treated glass should be symmetrical and have constant section thickness. With formed glass, sharp changes in section, long thin necks, and heavy sections near the top of the mold should be avoided.

Pressed glass necessitates sections heavy enough to permit adequate fluidity to fill the mold. Ample draft and liberal radii should be provided so that the molded part may readily be removed from the mold. Block molding is the lowest in cost of the pressing methods and should be utilized when possible. Tables 9-2 to 9-5 provide the production-design engineer with pertinent information as to the various types of glass.

Refractory Ceramics

A refractory ceramic is a nonmetallic material resistant to a state of fluidity or breakdown due to heat. Refractories are primarily used to provide linings for furnaces where they must resist high temperatures, slag corrosion, and abrasive action of the charge. In the form of slurries, refractories are forced through a hose and nozzle and sprayed over the inside of the cupola and open-hearth furnace or the fire box of a boiler. In this manner, spalled areas can be repaired and a sealed furnace can be obtained. Almost all refractory materials are available as brick in a wide range of standard sizes and shapes. Most refractories are also available as mortars for forming joints between the brick or for patching or coating brickwork.

Refractory types Refractories are classified according to chemical composition into four major groups: alumina-silica, silica, basic, and special. The alumina-silica group includes the refractories composed practically entirely of silica and alumina. Fire-clay brick is the most widely used type of refractory, comprising about 75 percent of all refractories produced in the United States.

High-temperature ceramics are being used in paints and coatings. In paints, they are being used to coat metal surfaces exposed to temperatures of 1800°F. Ceramic paints may be applied without firing. They provide protection against oxidation and radiant-energy transmission. Ceramic coatings, including graphites, oxides, and endothermic materials, are fired on the base metal which they protect within the plastic range of the metal.

Design of refractory and high-temperature ceramics Refractory ceramics

Table 9-2   GLASS AS AN INDUSTRIAL MATERIAL

Type		Preparation	Composition, Sizes, Etc.	Uses
Sheet Glass				
Window		Drawn vertically from the molten bath as a continuous sheet, firepolished and annealed. Shows only slight wave or distortion.	Soda-lime glass. Thicknesses from very thin to ¼ in. max. standard sheets 76 in. x 120 in.	For general purpose glazing of building. Also for table tops, induction heating jigs, shields, electrical insulators, small mirrors, etc. Thin sheets for photographic and microscopical uses.
Tempered		Glass is quickly heated to about 1150 F. and chilled quickly. Strength increased 2-5 times.	Same as above—must be cut to size, drilled, etc., before tempering.	For gage glasses, safety mirrors, business machine windows, fire screens, hospital glazing, etc.
Wire		Wire mesh is embedded in the molten glass.	Sometimes uses heat-absorbing types of glass.	For skylights, roofing, air raid precautionary glazing, etc.
Heat-Resisting		Special compositions, usually of borosilicate type.		For oven sight glasses, heat protection shields, electrical insulators, furnace door glasses, etc.
Ultraviolet		Glass of a special composition to transmit about 50% of the ultraviolet rays in solar radiation.		For hospital sun porches, greenhouses, poultry houses, etc.
Water White		An exceptionally "white" glass with higher light transmission.		For greenhouses, picture glazing, photographic uses, etc.
Light Reducing		Blue-tinted glass for cutting down amount of sunlight transmitted.		For glazing in sunny climates.
Colored		Obtainable in a wide range of colors, transparent and translucent. Plain and varigated colors.		For ornamental glazing, decorative panels, modern furniture.
Figured		A figure is rolled into the glass to diffuse or otherwise reduce lighting.		For obscured windows, skylights, glare-reducing windows, etc.
Polarizing		Glass is coated with a chemical that polarizes the light transmitted.		For anti-glare applications, as sunglasses.
Plate Glass				
Rough Rolled		Rolled to sheet form with a knurled surface in a variety of patterns.	Soda-lime glass. Standard thicknesses 3/16 in. to 1¼ in. in usual grades.	The rough stock for manufacture of plate glass. Also used for ornamental glazing, obscured windows, skylights, etc.
Polished		Rough rolled stock ground and polished to substantially optical flatness of surface.	Standard thicknesses 7/64 in. to 1¼ in.	For tracing tables, surface plates, blueprint machines, and show windows and counters, illuminated signs, mirrors, windows, doors, stairways, walls, etc.
Tempered		Plate glass is rapidly heated to about 1150 F. and cooled suddenly in a blast of air. Strength is increased to 4 or 5 times that of ordinary plate.	Same as above—must be cut to size before tempering.	For pickling tanks, gage glasses, pressure tanks, oven sight glasses, store front panels, advertising signs, etc.
Special Purpose	X-Ray	Specially prepared with lead content of about 61%, and lead equivalent of about 0.32. About ¼ in. thickness. About twice as heavy as ordinary plate glass.		For X-ray protection.
	Water White	Special composition to provide higher transmission of all light.		For photographic and blueprint purposes, refrigerated showcases, display windows, etc.
	Document	Designed to cut down the amount of ultra-violet light transmitted. About ¼ in. thickness.		For protection of old documents and collections in museums.
	Heat Absorbing	A blue-tinted glass transmitting 70% of solar light but only 45% of the heat.		For double-glaze units on trains, skylights, airport control towers.
	Colored	Blue, flesh-tinted, and other colors are produced in transparent and opaque glasses.		For decorative mirrors, panels, chalkboards, bulletin boards, etc.
Laminated Glass				
Safety Plate		Two or more lights of glass bonded by interlayers of transparent plastic.	Plate glass—composite is usually ¼ in. thick.	For automobile windshields and windows, aircraft and bus glazing, railroad car windows, protective shields, jewelry store windows, safety lenses, etc.
Safety Sheet			Sizes from about 3/32 in. to ½ in.	

SOURCE: From "Materials and Methods Portfolio of Engineering File Fact," 5th ed., no. 81. May, 1945. pp. 65–68.

Table 9-3

Type	Preparation	Composition, Sizes, Etc.	Uses
Bullet Resisting Plate Bullet Resisting Sheet	Several lights of glass, usually three, laminated as for safety plate.	Standard thicknesses from ½ in. to 1½ in.	Primarily for protection against firearms, as in banks, money collecting trucks, police cars. Also for protective shields in laboratories.
Double Glass	Not strictly a laminated glass—2 lights are separated by cemented glass spacer around edges, leaving air space between.	Must be made to order as to sizes—usual thickness is two ⅛-in. lights with ⅛-in. space.	For glazing air conditioned buildings, railway cars, etc.
Formed Glassware			
Pressed Ware	Molten glass is pressed into a mold, by hand or automatically. Metal inserts or attachments may be used. Parts may be tempered.	Compositions to suit requirements. Borosilicate glasses where thermal resistance or extra strength is required. Lead glass or soda-lime-magnesia-alumina for household ware.	For electrical insulators, centrifugal pump parts, glass gages, godet wheels, rotary valves, sight glasses, etc. for industry; household utensils, tableware, decorative glass articles, etc., for commercial fields.
Blown Ware	Molten glass is blown into a mold to produce hollow forms. Work may be done by hand or automatically. Parts may be tempered.	Frequently, a soda-lime-magnesia-alumina glass is used because of its easy workability. Heat resisting glasses or other special compositions where desired.	For bottles and similar containers, battery jars, floats, laboratory glassware, reaction columns, lantern globes, domestic utensils, etc.
Drawn Ware	Glass is drawn directly from the melting furnace in some standard form.	A variety of compositions can be used, usually soda-lime-magnesia-alumina or a heat-resisting variety.	For tubing, glass pipe, rods, gage glasses, and similar forms.
Multiformed	Glass is cold molded to close tolerances and sintered. Wide range of shapes.	A variety of compositions can be used. The glass usually is white and translucent.	For electrical insulators, pump seal rings, etc.
Glass Fiber Materials			
Batts	Very fine glass filaments drawn out from molten glass and collected as a woolly mass—may be treated with a resin binder and compacted to any desired degree, or faced with wire mesh. Forms a non-absorptive, non-combustible, chemically resistant material having good sound-, heat-, and electrical-insulating properties.		For insulation of military aircraft, domestic stoves, roofs of industrial buildings, railroad tank cars; as tower packing; insulation of furnaces; sound deadening in test rooms, as a dust filter in air conditioning.
Blocks	Mats of glass fiber compacted to blocks or boards under pressure and held with a binder. Available in thicknesses to 2 in. for use to 600 F. for one grade, to 1200 F. for another.		For insulation of boilers, ovens, breeching, pipes, valves, tanks, low temperature grade for refrigerated spaces.
Textiles Cord Sleeving Tape or Cloth	Glass fiber filaments may be spun to threads and woven into fabric on standard textile equipment. The textile materials so produced have exceptionally high strength, exceeding that of the natural and of most other synthetic fibers, weight for weight.		As heat insulating material—for lagging pipe, aircraft engine exhausts, etc. As electrical insulation—for winding wire, for covering motor coils, ignition cable. As a decorative or fireproof, or rotproof fabric—for draperies, military fabrics.
Composites	*Glass Fiber-Plastic Laminates*—an extremely high-strength plastic laminate. *Glass Fiber-Asbestos*—for high-temperature insulation. *Glass Fiber-Neoprene*—for conveyor belts to operate at elevated temperatures. *Glass Fiber-Mica*—slot insulation in motors and generators, etc. *Glass Fiber-Rubber*—for military and naval tarpaulins, etc.		
Glass Structural Forms			
Block	Made by fusing together two halves of pressed glass to form a hollow, partially evacuated block.	In 3 standard sizes—5¾ in. sq., 7¾ in. sq., 11¾ in. sq., all 3⅞ in. thick. Standard radial and corner blocks available.	For light-transmitting masonry walls, interior partitions, permanent windows or panels in air-conditioned structures, etc.
Cellulated	Gas is trapped in molten glass to give a lightweight material containing numerous closed cells.	As usually made, is lighter than water. Available as blocks in sizes to 12 in. x 18 in. x 6 in.	For building blocks, thermal insulation; also for life rafts, fishing floats, etc.
Cast Ornamental Panels	Sculptured designs cast in glass, and back face frosted, polished, mirrored, etc.	In several standard patterns and modeled designs. Strip or square panels of various sizes.	For door or fireplace trim, indirect lighted panels, screens and partitions, strip decoration, fountains, etc.

have little ductility and points of stress concentration, such as holes, notches, and sharp corners, should be avoided. Tolerances should be generous so as to avoid machining, which frequently is difficult. On coated ceramics, the thickness of coat will thin out on sharp corners, so they should be avoided if uniform thickness is required.

Table 9-4

Type	Fused Silica	96% Silica	Soda-Lime Glasses		Alumino-Silicate
			Plate	General Purpose	
PHYSICAL PROPERTIES					
Density, Lb/Cu In.	0.079	0.078	0.09	0.089	0.091
Softening Point, F (Approx)	3050	2800	1330	1285	1675
Thermal Cond, Btu/Hr/Sq Ft/Ft/F @ 212 F	0.80	0.80	0.53	0.53	—
Coeff of Exp per ° F, 32–570 F	3.0×10^{-7}	4.5×10^{-7}	48×10^{-7}	51×10^{-7}	23×10^{-7}
Spec Ht, Btu/Lb/F	0.185	0.185	0.20	0.20	—
Elect Res, Ohm-Cm @ 212 F	$>10^{15}$	$>10^{15}$	—	4×10^9	$>10^{15}$
Power Factor, 68 F 1 MC, %	0.025	0.02–0.04	0.80	0.90	0.37
Dielectric Constant, 68 F 1 MC	3.75	3.8	7.4	7.2	6.3
Thermal Stress Resistance[1]	—	390	—	65	85
Thermal Shock Resistance[2]	Very high	Very high	135	125	240
MECHANICAL PROPERTIES					
Mod of Elasticity, Psi	10×10^6	9.7×10^6	$9–10 \times 10^6$	$9–10 \times 10^6$	12.7×10^6
Normal Work Stress (Annealed) Psi	1000	1000	1000	1000	1000
THERMAL TREATMENT					
Annealing Temp, F (Stress Relieving)	2080	1670	1010	950	1315
FABRICATING PROPERTIES	Glass can be ground or polished without great difficulty. Machining is limited to sawing or drilling. Sawing is readily accomplished with an impregnated wheel. Drilling is difficult, but is done with special carbide-tipped drills using kerosene as a lubricant.				
Joining	Glass can be joined by heat-sealing using a blast-lamp, gas torch or by electric heating. Parts must have closely similar coefficients of expansion or intermediate pieces having intermediate coefficients must be used. Glass-to-metal seals are made by this latter process, also.				
Highest Service Temp F, Annealed	1650	1500	900	840	1200
Tempered	—	—	550	480	840
CORROSION RESISTANCE	Glass is attacked by hydrofluoric acid, hot concentrated phosphoric acid and strong alkalies. It is resistant to most other chemicals.				
USES	Ultra-violet energy transmission, chemical apparatus, thermocouple protection tubes, laboratory apparatus.	Ultra-violet energy transmission, chemical reaction vessels, thermocouple protection tubes, laboratory apparatus.	Sheet and plate glass, molded glassware, bulbs for electric lamps, bottles, vials, fluorescent lamp tubing.		

Carbon

Carbon graphite is coming into more and more prominence as an engineering material. Its value as an automotive water-pump seal nose is widely recognized and it is employed considerably throughout the industry. In the aeronautical field, especially in fuel pumps, carbon in the form of seals and bearings is one of the few successful materials. In similar fashion, general industry, textile bleacheries, oil refineries, chemical processing plants, and oven-conveyor users find that carbon is frequently a solution in cases where bearings cannot be lu-

Table 9-5

Type	Borosilicate Glasses			
	Low Expansion Chem. Resistant	Baking Ware	"Kovar" Sealing	Low Electrical Loss
PHYSICAL PROPERTIES				
Density, Lb/Cu In.	0.080	0.081	0.082	0.77
Softening Point, F (Approx)	1500	1425	1300	—
Thermal Cond, Btu/Hr/Sq Ft/Ft/F @ 212 F	0.67	—	—	0.67
Coeff of Exp per ° F, 32–570 F	18.5×10^{-7}	20×10^{-7}	25×10^{-7}	18×10^{-7}
Spec Ht, Btu/Lb/F	0.195	0.195	—	—
Elect Res, Ohm-Cm @ 212 F	10^{12}	—	3×10^{13}	$>10^{15}$
Power Factor, 68 F 1 MC, %	0.46	0.28	0.26	0.06
Dielectric Constant, 68 F 1 MC	4.6	4.7	5.1	4.0
Thermal Stress Resistance[1]	—	—	—	—
Thermal Shock Resistance[2]	320	—	—	—
MECHANICAL PROPERTIES				
Mod of Elasticity, Psi	9.8×10^6	—	—	6.8×10^6
Normal Work Stress (Annealed), Psi	1000	1000	1000	1000
THERMAL TREATMENT				
Annealing Temp, F (Stress Relieving)	1020	975	890	910
FABRICATING PROPERTIES	Glass can be ground or polished without great difficulty. Machining is limited to sawing or drilling. Sawing is readily accomplished with an impregnated wheel. Drilling is difficult, but is done with special carbide-tipped drills using kerosene as a lubricant.			
Joining	Glass can be joined by heat-sealing using a blast-lamp, gas torch or by electric heating. Parts must have closely similar coefficients of expansion or intermediate pieces having intermediate coefficients must be used. Glass-to-metal seals also are made by this latter process.			
Highest Service Temp F, Annealed	900	840	800	810
Tempered	550	500	420	450
CORROSION RESISTANCE	Glass is attacked by hydrofluoric acid, hot concentrated phosphoric acid and strong alkalies. It is resistant to most other chemicals.			
USES	Heat exchanger tubes, chemical apparatus, cooking-ware, electrical insulators, sight and gage glasses, containers for chemicals and medicinals, metal sealing, industrial piping, industrial glassware requiring thermal resistance, special lighting ware, heat resisting lenses.			

bricated due to high temperatures or inaccessibility, or because they are immersed in corrosive mixtures.

The most important properties of carbon and graphite materials are inertness to chemical action, good high-temperature strength, excellent resistance to thermal shock, high sublimating or boiling points, good heat and electrical conductivity, high heat of vaporization, low friction losses, and freedom from swelling or warping.

Fabrication of carbon   Carbon powder is fabricated by combining the powder with a tar which acts as a binder material and shaping either by extrusion or pressure molding. The green parts are then baked at around 1800°F in order to remove the volatile compounds in the binder and give a stronger finished product.

Designing carbon parts   In the baking process, the material shrinks and distorts somewhat, so that shrinkage should be provided for when designing the mold. Relatively close tolerances can be achieved in the molding process. Small parts are frequently held to dimensional tolerances as close as 0.0005 in. On most dimensions, a tolerance of 2 percent can readily be maintained; this amount of variation represents standard molding tolerance for carbon parts.

Some general design suggestions for molded parts are:

1   Avoid close tolerances (0.001 in/in or closer).
2   Avoid intricate external and internal shapes.
3   Avoid thin-sectioned flanged parts. Better to use two parts.
4   Avoid press-fit of carbon sleeve over shafts.
5   Do not make seals (washers) from extruded tubing. Better to let manufacturer produce finished seals to specifications.
6   Do not specify wall thicknesses of less than $\frac{3}{16}$ in.
7   Do not specify threads in graphite.
8   Specify chamfered edges rather than radius edges.

For special application, the pores of molded carbon parts can be impregnated with thermosetting resins to make the product impervious to liquids and gases. Also greases, oils, and metals such as copper and babbitt can be used as impregnants.

Carbon parts can be vulcanized to rubber and can be attached to metal parts through the use of adhesives. Carbon parts are readily machined after molding and are usually processed dry. Parts can be lapped and polished to accuracies of a few millionths of an inch.

POWDER METALLURGY

Powder metallurgy is the production of metallic components by feeding metal powder into a die of the desired shape, pressing the loose metal powder into a compact briquette which is of the approximate geometry of the desired finished product, and then heating the briquette in a furnace so as to bond the powder particles together to produce parts with the desired physical and mechanical properties. If necessary, final sizing or coining of the product to meet specified dimensional tolerances may take place.

Today parts weighing as much as 30 lb are being produced in quantity from iron powders. Powdered-metal components are advantageously competing with machined castings, forgings, and stampings. They are finding increasing usage in the automotive and appliance industries for producing gears, bearings, splined parts, and the like. The use of iron powders in 1973 was more than six times what it had been 10 years earlier. Various combinations of metals and mixtures of metals and nonmetals are possible.

Advantages of Designing for Powdered-Metallurgy Components

The two main advantages of powdered-metallurgy components are: (1) it allows production of complex parts such as gears, eccentrics, splines, and irregularly shaped holes with a minimum of machining and scrap, and (2) it can produce parts, such as porous bearings, cemented carbide cutting tools, and refractory metals, with special properties not otherwise attainable.

Today not only are iron, carbon steel, alloy steel, copper, brass, bronze, and nickel available in powder form, but also stainless steels, refractory metals, and the precious metals. Thus, parts of complex geometry can be produced with a choice of a variety of materials giving unique properties of the specific material employed. Where high-density parts are needed with increased physical and mechanical properties a re-pressing and second sintering operation will provide beneficial results. Tensile strengths up to 180,000 lb/in^2 are achieved in heat-treated high-density low-porosity powdered-metal components.

Special properties such as self-lubrication in porous bearings are an important advantage of powder metallurgy. The porosity characteristic of powdered-metallurgy construction allows between 10 to 40 percent of oil by volume to be contained within the component. As the part heats, because of friction, the contained oil will expand and emerge on the bearing surface providing a film of lubricant. Subsequent cooling results in reabsorption of the oil into the pores of the part.

Powder-metallurgy components have damping characteristics resulting in quieter operation of mating parts.

Powder metallurgy can be used to produce high-melting (refractory) metal such as tungsten, molybdenum, and tantalum. An example of the type of product is heavy metal, a tungsten alloy containing 6 percent nickel and 4 percent copper. It can be used for radium containers, balancing weights, and similar applications where high density is desirable.

Copper and tungsten can be joined together by pressing and sintering a mixture of the metal powders to produce special electrical contacts and resistance welding electrodes. With such a mixture it is possible to retain much of the relatively high electric conductivity of the copper while realizing high strength from the tungsten.

One of the important applications of the powder-metallurgy process is the production of cemented carbide products. Mixtures of hard metal carbides, tungsten carbide, molybdenum carbide, and tantalum carbide are made with a small amount of cobalt. After pressing, sintering is carried out at a temperature above the eutectic, the cobalt cementing the hard carbide particles into a strong solid mass.

Combinations of metals and nonmetals are used for special purposes. Examples are clutch plates and brake linings, in which nonmetallic abrasive powder is included. Copper alloys containing graphite are widely used as bearings.

With controllable density, the production-design engineer can select predetermined mass-weight ratios. Density can be distributed within a structure to achieve high strength in one area and self-lubrication in another. Metals can be oil-impregnated for self-lubrication or made to be completely porous for use as a filter.

Limitations of Powder Metallurgy

The principal limitations are those imposed on the size and the shape that can be economically produced. Relatively high pressure is required to compact the powder into the desired shape, and the size of a part is consequently limited by the available press capacity. Seldom is a part weighing more than 30 lb produced by this process.

Since the compacted part must be ejected from the die without fracture, the shapes that may be made by the powder method are limited. Designs with undercuts or reentrant angles cannot be made. Thin sections must be avoided, as well as featheredges and narrow or deep splines.

Another limitation is the die cost. Precision dies for forming the powder are expensive, and unless the high cost can be distributed over large production runs, the die cost per piece produced may be excessive. Usually 10,000 pieces represent a minimum quantity before this method of fabrication would be economically practical.

A final limitation is that protection against corrosion during the processing cycle must be provided.

Designing for Powdered Metallurgy

Before specifying a powdered-metallurgy component, the production-design engineer should consider four constraints. These are:

1 Is the quantity large enough to justify the tooling? As mentioned earlier, this figure in most cases is approximately 10,000 pieces of the same part.

2 Can the shape of the part be produced to the required tolerances by powdered metallurgy? With relation to tolerances, the expected tolerance that is achieved on diameters and dimensions at right angles to the direc-

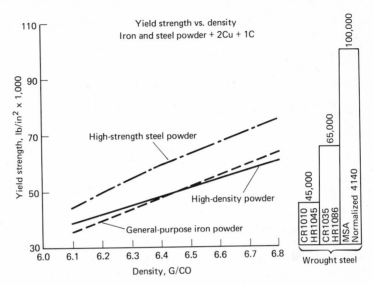

FIGURE 9-2
Yield strength versus density—iron and steel powder with 2% copper and 1% carbon added.

tion of compacting is $\pm$ 0.001 in/in. Tolerances of $\pm$ 0.004 in/in are characteristic of tolerances held in the direction of compacting.

Secondary sizing operations such as shaving can, of course, be performed at added cost to improve tolerance accuracy.

The production-design engineer should recognize that undercuts cannot be molded, since it would not be possible to eject the briquette from the die. Likewise, complete spheres cannot be molded since the point of contact of the mating punches would involve a sharp edge which would fracture as the compressing pressure built up.

Reentrant angles cannot be produced. Thin walls (less than 0.030 in) should be avoided as well as abrupt changes in cross section and large thin forms.

Five design rules relative to shape should be borne in mind. First and most important is the shape must be such that the briquette can be ejected from the die. Second, do not design a part that requires the metal powder to flow into a thin wall or flange or sharp corner. Third, the geometry of the part must permit the design of strong die sections. Fourth, the length-diameter ratio of 2.5:1 should not be exceeded for most thin-walled parts. Fifth, to assure uniform density and consequent high strength changes in section thickness should be minimized.

3 Are the required physical and mechanical properties within the capabilities of powdered metals? There is wide range of physical properties that can be achieved. Low-density iron-product parts have tensile strengths as low as 10,000 lb/in^2, whereas high-density high-strength steel powders have tensile strengths over 70,000 lb/in^2. Heat-treated high-density alloy powders have tensile strengths of 180,000 lb/in^2 or more. Figure 9-2 provides a graphic illustration of the relationship between density and yield strength of three ferrous powders.

Other properties, such as hardness, yield strength, etc., should be carefully checked to ensure that the attainment of the required specifications is within the realm of powdered metallurgy.

4 Economic considerations enter into every design decision. It should be established that powdered-metallurgy design is the most economical possible, assuming other factors are at least equal to alternative methods. Increased physical properties can be obtained by additional sintering and infiltration. However, this increases cost. The various common processes for the manufacture of sintered structural parts and their relation to increased cost are illustrated in Fig. 9-3.

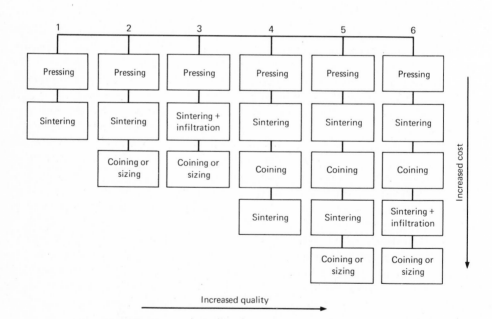

FIGURE 9-3
Processes for the manufacture of sintered structural parts and their relation to increased cost.

Steps in Producing Powdered-Metallurgy Parts

Three steps are involved in producing powdered-metallurgy parts: mixing and blending; compacting; and sintering.

Mixing and blending This involves the mixing of the metal powder with any alloying elements and lubricants required. Precise control of the volume of the raw powders and lubricants dispensed is essential. The best mixing is obtained when all particles have similar size, shape, and density.

Compacting Compacting the powder is an important step in the powder-metallurgy process. In addition to producing the required shape, the compacting operation also influences the subsequent sintering operation and governs, to a large extent, the properties of the final product. Several of the most important effects of compacting are:

1 Reduction of voids between powder particles and increased density of the compact
2 Adhesion and cold welding of the powder and sufficient green strength
3 Plastic deformation of the powder to alloy recrystallization during subsequent sintering
4 Plastic deformation of the powder to increase the contact areas between the powder particles, increasing green strength and facilitating subsequent sintering

Compacting is performed by pouring a measured amount of the appropriate powder into the die cavity and then introducing one or more plungers to press the metal powder into a coherent mass. Pressures up to 100,000 lb/in² are used. A measured volume of powder is usually used, although in some cases a definite weight of powder is more suitable.

An ideal pressed compact should have the desired shape and uniform density. However, metal powders do not flow uniformly, nor do they uniformly transmit pressure. Friction along the die walls and between powder particles, and the inability of powder to transmit uniform pressure at right angles to the direction of the applied force, result in nonuniform compacting force within the die. The powder is not compressed to uniform density throughout the compact, and does not react uniformly to subsequent sintering. However, this inequality may be minimized in several ways. Thin compacts are more evenly pressed throughout their thickness than are thick compacts; therefore the design thickness of powder-metallurgy parts should not be excessive. Double-end pressing, in which the part is simultaneously pressed from both top and bottom, helps enhance the pressure gradient in the compact and provides more even density. When irregularly shaped parts are made, multiple punches are often necessary.

Lubrication of the die walls and the powder is helpful in reducing friction and minimizing pressure gradients within the compact. Metallic stearates, graphite, and other lubricants may be mixed with the powder or coated on the die walls to provide lubrication.

For production runs the punches and dies are rigidly mounted in a suitable press and the die cavity filled with a measured amount of properly mixed metal powder. The volume of the loose powder is approximately 2 to 10 times the volume of the pressed compact.

The speed of compacting, which is proportional to the rate of punch travel, is relatively unimportant. Compacting pressure may be applied rapidly or slowly, the only limitation on speed being due to entrapment of air within the pressed compact. When the pressing speed is excessive, air between the loose powder particles does not have sufficient time to escape, and is trapped within the compact. Subsequent sintering may then produce considerable expansion of the compact, rather than the usual shrinkage.

Following the compacting operation, the relatively fragile but coherent mass of metal powder must be ejected from the die without injury, and transferred to a furnace for sintering.

Sintering   Sintering entails heating the pressed compact to below the melting temperature of any constituent of the compact, or at least below the melting temperature of all principal constituents of the compact. The purpose of such heating is to facilitate a bonding action between the individual powder particles that increases the strength of the compact.

Sintering is generally carried out in a controlled-atmosphere furnace because oxidation of individual metal particles weakens the sinter bond or even entirely prevents sintering.

The mechanism by which individual metal particles are joined into a coherent mass having increased density and strength has been the subject of many investigations and much discussion. The mechanisms involved in sintering appear to be principally the following: (1) diffusion, (2) recrystallization, (3) grain growth, and (4) densification.

Increase of strength developed in the powder compact during sintering is principally due to a disappearance of the individual particle boundaries through diffusion and recrystallization. The powder particles are usually deformed by greater than critical strain during pressing, so that recrystallization and coalescence are common during sintering.

The important factors controlling sintering are temperature, time, and furnace atmosphere. An increase in sintering temperature and time increases the sintering effect. Strength, hardness, and elongation are increased by longer sintering time or higher sintering temperature.

Ordinarily, the density of the compact is increased by sintering, and shrinkage will occur. Shrinkage entails the filling of holes within the compact, a process which necessitates closer contact between individual particles, increase of areas of contact, and reduction of the size of holes in the compact.

In some instances sintering is done at a temperature at which a liquid phase exists. When a liquid phase exists, the compacting conditions are less critical and high density can be attained quickly. In products of this type the nature of the alloy formed between the solid and liquid metals, the ability of the liquid metal to wet the solid particles, the capillary action of the liquid in the solid compact, and other factors play important roles in sintering.

Preparation of Metal Powders

A great variety of mechanical, physical, and chemical methods are employed for the production of metal powders; and powders widely differing from each other in particle size and shape, chemical purity, and microstructure are on the market. The powders for use in powder metallurgy should be nearly spherical in shape, with a minimum of fine dust and no large particles. The properties of the final product and the techniques used for its production depend upon the characteristics of the powder, which are principally influenced by the method of manufacturing the powder.

Most powders intended for use in powder-metallurgical processes are produced by one of two methods, reduction of metal oxides or electrolysis, although atomization, mechanical comminution, thermal decomposition of carbonyls, and intergranular corrosion are also sometimes employed.

Reduction of metal oxides   Very fine metallic oxides are reduced in a gaseous medium to the metallic state, producing a spongy metallic mass which is later pulverized to the desired particle size. Gases such as H_2, CO, coal gas, and carbon are employed as reducing agents. The powders produced by this process are in most cases spongy, and of angular and granular spherical shape.

Electrolysis   Metal powders can be produced by electrodeposition from solutions, as well as from fused salts. The electrolysis of solutions is employed for the commercial production of metals such as iron, copper, nickel, cadmium, tin, antimony, silver, and lead. The electrolytic process exhibits a control of powder characteristics, particularly particle size. Means of control are the regulation of current density, temperature, composition and circulation of the bath, size and arrangements of the electrodes.

Atomization   This process consists of forcing molten metal through a small orifice and breaking up the thin stream of liquid metal by a jet of compressed air, steam, or inert gases. The cooling effect produced by the sudden expansion of the metals leaving the nozzle, combined with the cooling effect of the gases, instantaneously solidifies the metal. The controlling factors for particle size are nozzle design, temperature and rate of flow of the molten metal, temperature and pressure of the gas.

Mechanical comminution Crushing and milling of the brittle metals generally produce angular particles which are suitable for powder metallurgy. Ductile metals may also be milled to powder form, but the resulting powder is flakelike and not desirable for most powder-metallurgy work.

Thermal decomposition of carbonyls Iron and nickel powders are often made by the decomposition of iron carbonyl and nickel carbonyl. The metallic carbonyls $Fe(CO)_5$ and $Ni(CO)_4$, which are gaseous at the operating temperature maintained during the process, are decomposed to iron or nickel. These powders are usually spherical.

Intergranular corrosion This method is of particular interest since it permits the preparation of powders of 18:8 stainless steel which cannot be prepared by other methods. In this process, the steel is sensitized and after cooling is disintegrated at the grain boundaries by a corrosive solution—for example, by a boiling aqueous solution of 11% $CuSO_4$ and 10% H_2SO_4. The particle size of the resulting powder is determined by grain size of the corroded material after sensitization.

Forgings from Powder Preforms

By combining preform sintering with subsequent forging it is possible to produce parts made of powdered metal having little flash and of such dimensional accuracy so as to minimize secondary operations while providing parts with strength and toughness properties much superior to the typical sintered powder-metal component.

Furthermore, the preform is usually forged in but one die cavity and normally given only one blow to attain the desired component. In those cases where the composition of the preform is similar to a conventional forged grade of material and the density is approximately 100 percent of its theoretical value, the powdered-metallurgy forgings will have about the same properties as the conventional material. For example, die-forged preforms of 1050 carbon steel have shown yield strengths of 76,000 lb/in^2 when the density is 7.86 g/cm^3. Tensile strength, yield strength, and ductility fall off rapidly as the percent porosity increases. For example, yield strengths of 70,000 lb/in^2 at 0.005 percent porosity fall off to 40,000 lb/in^2 at 0.05 percent porosity using low-alloy steel powders. Mechanical properties appear to be a function of final pore shape rather than particle size prior to sintering.

SUMMARY

New ceramic materials are being developed continually. In the future, we may well see this engineering material used in structural, load-bearing applications as well as electrical and electronic uses, and furnace- and crucible-lining applications. Ceramic materials that have been developed in recent years include high-alumina and -magnesia ceramics with density approaching the theoretical, recrystallized graphites, foamed ceramics, and ceramic fibers.

Today, many metals are challenged by the ceramic glass. Thus we have glass furniture, automobiles, fishing rods, and many other products that for a long time were produced exclusively from metallic materials.

The alert production-design engineer will continually follow the development of ceramic materials. A material that seems to have little application in a given design today may be the ideal material in that design tomorrow.

Today powder metallurgy is one of the lowest-cost mass-production methods of manufacturing complex parts to close tolerances. Ferrous- and nonferrous-based powders are finding an ever-widening market in the design of metal parts in almost every consumer industry. As a result of improved control and tooling, the powdered-metallurgy process is being advantageously chosen for producing complex and critical application parts in the automotive, appliance, farm-equipment, and other industries.

QUESTIONS

1 What four processes are available to the production-design engineer for forming ceramic whitewares? Describe each.
2 Where does the property of high dielectric strength find application?
3 What type of glass is best suited to high electrical resistance?
4 What is the difference between laminated glass and plate glass?
5 Would you recommend glass as a suitable material for the manufacture of Go-No-Go plug gages? Why?
6 What are the properties of carbon that make it a valuable engineering material?
7 Would you recommend a threaded graphite bushing? Why?
8 What are the three main steps involved in the powdered-metallurgy process?
9 Why are powdered-metallurgy components frequently used in the manufacture of bearings?
10 What are the principal limitations of powdered-metallurgy parts?
11 What finishing operations, if any, would be required on a 2-in powdered-metal gear having a tolerance of ± 0.008 in on the major diameter?
12 What yield strength can be expected from a general-purpose iron powder having a density of 6.3 to 6.5 g/cm^3?

PROBLEMS

1   A general-purpose iron powder is compressed to a density of 6.5 g/cm³. The expected yield strength of the resulting sintered product is 50,000 lb/in² with a standard deviation of 2,000 lb/in². With further compaction the production-design engineer can expect a density of 6.8 g/cm³ at an added cost of $0.11 per product unit. This added compaction will give an expected yield strength of 65,000 lb/in² with a standard deviation of 4,000 lb/in². The product is stressed at an average tensile load of 42,000 lb/in² with a standard deviation of 5,000 lb/in². If the company computes the cost of product failure at $10 each and they expect to produce 500,000 parts, what level of compaction would you recommend?

2   Calculate the volume of metal required to compact a T-shaped body to 85 percent of its theoretical density. Give the size of the dies and height of the compact if the sintering operation results in 5 percent linear shrinkage in all directions. The finished parts need to be 1 in $\pm \, {}^{0.005}_{0.000}$ OD on the body, $1\frac{1}{4}$ in $\pm \, {}^{0.005}_{0.000}$ on the head diameter. The cored hole wants to be 0.625 in $\pm \, {}^{0.003}_{0.000}$. The head is $\frac{3}{8} \pm 0.015$ in thick and the overall length is $1\frac{1}{4} \pm 0.015$ in.

SELECTED REFERENCES

HAUSNER, HENRY H.: "Powder Metal Processes," Plenum, New York, 1967.

KINGERY, W.: "Ceramics," Wiley, New York, 1960.

SCHWARZKOPF, PAUL: "Powder Metallurgy," Macmillan, New York, 1947.

VAN VLACK, LAWRENCE H.: "Ceramic Materials," Addison-Wesley, New York, 1964.

10
BASIC MANUFACTURING PROCESSES: LIQUID STATE

INTRODUCTION

Bronze arrowheads were cast some 6,000 years ago in open-faced clay molds; bronze statues were poured soon afterward. The art of founding is fundamental to civilization and was practiced throughout the ancient world—in Europe, Central and South America, India, the Orient, and Northern Africa.

Today, foundries are our fifth largest industry, producing 15 to 20 million tons of castings each year (Table 10-1). Without castings, automobiles, household appliances, and machine tools would become much more costly. Therefore engineers and designers need to understand the capabilities and limitations of the casting process, to develop consumer products which are safe, are of high quality, and which can be competitive in the world market. The modern engineer has a responsibility to the customer to build in long-lasting quality despite his employer's insistance on reducing material to save cost.

When an engineer considers whether or not a part should be cast, he must first decide which casting alloy and process can most nearly meet the required dimensional tolerance, mechanical properties, and production rate. If he finds that a casting may be suitable, he must estimate the product cost of fabrication,

Table 10-1   ANNUAL TONNAGE OF CASTINGS SHIPPED, 1960–1973*

Year	Gray iron	Cast steel	Malle- able iron	Ductile iron	Alumi- num alloys	Brass- bronze	Zinc alloys	Mag- nesium alloys
1960	12.70	1.80	0.80	0.17	0.380	0.50	0.380	0.300
1961	10.82	1.22	0.72	0.18	0.381	0.365	0.301	0.012
1962	11.55	1.42	0.87	0.25	0.467	0.380	0.346	0.015
1963	12.76	1.50	0.93	0.37	0.476	0.406	0.353	0.016
1964	14.32	1.84	1.00	0.50	0.627	0.496	0.457	0.016
1965	15.71	1.96	1.14	0.60	0.714	0.445	0.532	0.022
1966	15.72	2.16	1.13	0.80	0.816	0.503	0.487	0.020
1967	14.31	1.86	1.04	0.86	0.762	0.483	0.419	0.020
1968	14.08	1.73	1.09	0.96	0.794	0.396	0.426	0.020
1969	14.65	1.90	1.15	1.29	0.845	0.426	0.439	0.020
1970	12.34	1.72	0.85	1.61	0.753	0.375	0.348	0.017
1971	11.73	1.59	0.88	2.11	0.788	0.353	0.368	0.022
1972	13.49	1.61	0.96	1.83	0.929	0.382	0.400	0.021
1973	14.51	1.78	1.03	2.18	1.024	0.380	0.450	0.020

* Tabulated values are millions of tons.

forging, or machining. Once it appears that a particular casting process offers the least cost, the engineer proceeds with a specific casting design. In summary, if a casting will be economical, design for the casting process.

Casting has several distinct assets: the ability of a liquid to fill a complex shape; economy when a number of similar pieces are required; and a wide choice of alloys suitable for use in highly stressed parts, where light weight is important or where corrosion may be a problem. Casting permits the engineer to place the metal where it is needed and remove it where it is excess, for example, cast crankshafts and connecting rods or parts for farm machinery. Parts as thin as $\frac{1}{8}$ in can be sand-cast in gray iron. Gray iron also has a unique damping quality which is important in piano frames, machine-tool bases, and engine blocks.

There are inherent problems, too, including internal porosity, dimensional variations caused by shrinkage, and solid or gaseous inclusions which stem from the melting operation. But these problems can be minimized or avoided by proper design and good foundry practice.

The design and production of a successful casting requires the simultaneous application of a number of basic sciences—metallurgy, fluid flow, heat transfer, and materials engineering to name a few. Defects such as metal penetration into the mold wall or pinholes have occurred through the ages, and they still occasionally plague even the most sophisticated foundryman, despite years of experience and greatly expanded technological know-how.

To be of acceptable quality a sand casting requires concurrent success in molding, alloy analysis, pouring, and feeding. For example, an anxious customer finds little comfort in the fact that a foundry has met the exact alloy specification for a casting but has left too little stock for machining!

To establish the relationship between the casting processes and design this chapter is divided into three major areas: (1) casting fundamentals, (2) casting processes, and (3) design considerations.

FUNDAMENTALS OF CASTING

Castings have a number of general features and specific characteristics which arise from the casting process itself. This section discusses casting alloys; weight and dimensional limitations; shrinkage; and economic considerations.

Materials

Most parts are cast from alloys which have good casting properties such as fluidity; chemical stability with respect to air, refractories, and mold materials; nonwetting of mold materials and refractories; and machinability. Pure metals

Table 10-2   COMMONLY CAST ALLOYS

Ferrous	Nonferrous	Heat and corrosion-resistant
Cast Irons	Aluminum	Cobalt-base
Gray iron	Al–Si	(stellite, vitallium)
Ductile iron	Al–Cu	
White iron	Al–Mg	
Malleable iron	Al–other	
Alloyed cast irons (chromium,	Copper	
nickel, molybdenum, others)	Brass (red, yellow)	
Steel	Bronze	
Carbon (low, medium, high)	Monel	
Alloy (low, medium, high)	Other	
Stainless steel (iron-chromium,	Magnesium	
iron-chromium, nickel)	Nickel (nickel-	
	molybdenum-iron,	
	nickel-molybdenum-	
	chromium-iron)	
	Lead	
	Titanium	
	Zinc	

Table 10-3   MECHANICAL PROPERTIES OF THE PRINCIPAL TYPES OF CAST FERROUS ALLOYS*

Property	Gray cast iron	Ferritic and pearlitic malleable iron	Nodular iron	Carbon and low-alloy steel
Tensile strength, lb/in^2	20,000 to 80,000	48,000 to 120,000 +	60,000 to 160,000 +	60,000 to 200,000 +
Yield strength, lb/in^2	Essentially same as tensile strength	30,000 to 95,000	40,000 to 135,000	30,000 to 170,000 +
Compressive strength, lb/in^2	3 to 5 × t.s.	About same as in tension	About same as tensile yield str.	60,000 to 200,000
Shear strength, lb/in^2	1.0 to 1.6 × t.s.	About 0.90 × t.s.	0.90 × t.s.	
Elongation in 2 in, %	3 to 0	26.0 to 1.0	26.0 to 1.0	35 to 5
Reduction of area, %	0	23.0 to 0	30.0 to 0	65 to 5
Brinell hardness no.	135 to 350 + (t.s./Brin. no. = 0.16 to 0.21)	125 to 285 + ‡	140 to 330 +	130 to 750† (t.s./ Brin. no. = 0.47 to 0.58)
Max. Rockwell C hardness	60 to 64	60 to 64	60 to 64	65 to 66
Modulus of elasticity, lb/in^2	12,000,000 to 22,000,000	25,000,000	24,000,000 to 26,000,000	30,000,000
Endurance limit, lb/in^2	0.4 to 0.6 × t.s.	0.40 to 0.6 × t.s.	0.4 to 0.55 × t.s.	0.4 to 0.5 × t.s.
Impact resistance, ft·lb, Charpy V	Low§	1 to 20.0	1 to 20.0	3 to 65

* The information covers the average range of commercially obtainable properties of the ferrous casting alloys; mechanical properties are given for room temperature.
† Surface hardnesses up to 900 Vickers.
‡ t.s. = 1.82 (Brinell hardness no.)$^{1.85}$ also used.
§ 20 to 80 ft·lb on 1.125-in-diameter round bars, unnotched, machined from halves of 1.20-in-diameter transverse bars and broken on 6-in supports.
SOURCE: Courtesy Hans J. Heine, Technical Director, Malleable Founders Society.

Table 10-4 MECHANICAL PROPERTIES OF THE PRINCIPAL TYPES OF CAST NONFERROUS ALLOYS[a]

Property	Aluminum-base alloys	Copper-base alloys	Magnesium-base alloys	Nickel-base alloys		Zinc-base alloys
				Cast Ni and Ni-Cu	Other Ni alloys	
Tensile strength, lb/in²	19,000–55,000	21,000–125,000	23,000–45,000	50,000–145,000	68,000–118,000	25,000–52,100
Yield strength,[b] lb/in²	8,000–45,000	11,000–100,000[c]	14,000–30,000	25,000–115,000	40,000–118,000	
Compressive strength,[b] lb/in²	About same as yield strength	. . . .	Same as tensile yield strength	18,000–80,000[d]	. . . .	55,000–93,000
Shear strength, lb/in²	14,000–36,000	. . . .	14,000–21,000	. . . .	. . . .	31,000–46,000
Elongation in 2 in, %	22–0	52–0	12–1	45–1	30–0	10–0.5
Reduction of area, %	. . . .	40–4	. . . .	35–1	47–0	
Brinell hardness[e]	40–140	47–425	50–85	100–375[e]	155–390[e]	75–100[f]

Property					
Rockwell E hardness	......	......	60–90		
Modulus of elasticity, lb/in²	10,300,000	9,100,000–20,000,000	6,500,000	21,500,000–24,000,000	22,500,000–28,000,000
Endurance limit, lb/in²	6,500–23,000	4,000–15,000	9,000–13,000ᵍ	......	6,875–8,500
Impact resistance, ft·lb					
Charpy unnotched bar	......	......	......	......	1–48ʰ
Charpy keyhole notched bar	......	......	0.5–10	4–70	3–60ⁱ
Izod bar	......	½–40	......	3–80	25–85ⁱ

ᵃ General information covers the average range of commercially obtainable properties of the principal nonferrous casting alloys; mechanical properties are given for room temperature.

ᵇ 0.2% offset.

ᶜ 0.5 extension under load.

ᵈ Few data available; compressive yield strength.

ᵉ 10 mm. ball: 500-kg load, except nickel alloys tested with 3,000-kg load.

ᶠ Load held 30s.

ᵍ Based on 500,000,000 cycles on R. R. Moore type machined and specimen.

ʰ ¼-in square cast bar.

ⁱ Data not available for all alloys. Some have low impact resistance.

SOURCE: Courtesy Hans J. Heine, Technical Director, Malleable Founders Society.

Table 10-5 OTHER PROPERTIES OF CAST FERROUS AND NONFERROUS ALLOYS—RELATIVE

Property	Gray cast iron	Ferritic and pearlitic malleable iron	Nodular iron	Carbon and low-alloy steel	Aluminum-base alloys	Copper-base alloys	Magnesium-base alloys	Nickel-base alloys	Zinc-base alloys
Machinability[a]	Good	Good	Good	Less machinable than other ferrous alloys	Good to excellent	Fair to good (2.3 for leaded yellow brass)	Excellent	Comparable to steel	Excellent
Damping capacity	Approximately 10 × that of steel	Roughly related inversely to the modulus of elasticity							
Wear resistance, lubricated sliding friction	Excellent	Pearlite—excellent; Ferrite—good	Good to excellent	Good, improved by heat treatment	Poor to excellent	Good to excellent	Poor to excellent	Comparable to steel(?)	Poor
Suitability as a bearing material[b]	Poor to excellent	Pearlitic—poor to excellent; Ferritic—poor	Poor to excellent	Inferior to cast iron	Poor except for special bearing alloy	Good to excellent	Poor	Not normally used as a bearing	Poor
Abrasive wear[c]	Excellent for special alloys ↓ Depends on hardness	Pearlitic malleable[c] Depends on hardness	Pearlitic nodular[c] ↑	Excellent for special alloys ↑	Poor	Poor to good depending on hardness	Poor ↓	Poor to good Depends on hardness ↑	Poor
Notched sensitivity[d]	Ferrous alloys fairly comparable on basis of unnotched to notched bar in fatigue test. Cast steel less sensitive than wrought steel.				Yes	Few data available	↓	Few data available	↑
Section sensitivity[e]	Yes	No	As cast—Yes Annealed—No	To a limited extent	Yes	To a limited extent	To a limited extent	To a limited extent	No, when used in die casting

Suitability for joining by:	1	2	3	4	5	6	7	8	9
Brazing	Yes	Yes	Yes	Yes	Yes (few alloys)	Yes	No[f]	Yes	No[f]
Soldering	Yes	Yes, with special precaution	Yes	Yes	Yes	Yes	Yes	Yes	Yes
Welding	—	Yes	Yes	→Yes, with ease	Yes[g]	Yes	Yes	Yes	Yes
Fluidity	Excellent	Good	Excellent	Inferior to cast iron	Excellent	Fair to good	Good to excellent	Comparable to steel	Excellent
Susceptibility to hot tearing	No	Yes	No	Yes	Depends on composition	Depends on composition	Depends on composition	Comparable to steel	No
Pressure tightness	Yes	Yes	Yes	Yes	Depends on composition	Depends on composition	Depends on composition	Yes	Yes
Properties improved by heat treatment	Yes	Yes; 100% heat-treated	Yes	Yes	Yes, for most alloys	In a few alloys	Yes, for most alloys	In a few alloys	No, except dimensional stability
Subject to property control through inoculation, ladle treatment, etc.	Yes	Mildly so	Treatment a necessity	Mildly so	For a few alloys	Mildly so	Yes	No	No

a Figures give relative power requirement to machine.
b Excellence as a bearing alloy, among other things, requires proper metal structure; for instance, ferrite as a constituent in ferrous castings is generally avoided. In cast irons, proper size and distribution of the graphite flakes is also important.
c Good abrasion resistance in ferrous alloys can also be obtained by surface hardening treatments.
d Limited data does not justify a more critical comparison.
e Variation in properties of a casting depending on section size or cooling rate.
f Low melting point of these alloys precludes use of this method.
g Preparation of surface and selection of welding rod and flux important.
SOURCE: Courtesy Hans J. Heine, Technical Director, Malleable Founders Society.

are seldom cast because they have high shrinkage and are too soft and weak for economic use. Alloying strengthens a metal and reduces its melting point which are both helpful to an engineer. The commonly cast alloys and their properties are given in Tables 10-2 to 10-5. Steels are the strongest alloys in the as-cast condition. Stainless steels and certain copper-base alloys are the most corrosion-resistant alloys, and magnesium or aluminum alloys have the least weight.

Weight and Dimensional Limitations

Depending on the process used, castings vary in weight and dimensional tolerances from a few tenths of an ounce and thousandths of an inch for investment castings to 200 or more tons in weight and a tolerance of an inch for pit-molded sand castings of gray iron or steel. For each process there is also a limitation on the size and depth of cored holes; thickness, length, and spacing of fine dimensional tolerance; and surface finish. General limits for the various processes are shown in Table 10-6. These limits are not absolute, but correspond to normal production practice with an acceptable number of rejects. The limits of weight and size may be exceeded, for instance, by casting two separate parts and joining them by Electroslag welding or in the case of tolerance by a willingness to accept a larger number of rejects or the expense of using a more precise casting process.

The specific details of the casting process are not needed to determine whether or not a casting could comply with design specifications, but in determination of costs such details are frequently significant.

The Casting Process

Casting can be done in several ways. The two major ones are sand casting, in which the molds used are disposable after each cycle, and die casting, or permanent molding, in which the same metallic die is used thousands or even millions of times. Both types of molds have three common features. They both have a "plumbing" system to channel molten alloy into the mold cavity. These channels are called sprues, runners, and gates (Fig. 10-1). Molds may be modified by cores which form holes and undercuts (Fig. 10-1) or inserts that become an integral part of the casting. Inserts strengthen and reduce friction, and they may be more machinable than the surrounding metal. For example, a steel shaft when properly inserted into a die cavity results in an assembled aluminum step gear after the shot.

After pouring or injection, the resulting castings require subsequent operations such as trimming, inspection, grinding, and repairs to a greater or lesser extent prior to shipping. Premium-quality castings from alloys of aluminum or steel require x-ray soundness that will be acceptable by the customer.

Certain special casting processes are precision-investment casting, low-

Table 10-6 COMPARISON OF TYPICAL FEATURES OF THE MAJOR CASTING PROCESSES

Casting feature	Sand casting	Precision casting	Permanent-mold	Low-pressure	Die-cast	Centrifugal casting
Alloy cast	All	All	Nonferrous gray iron	All	All ferrous	All
Size range						
Min., lb	Few oz	0.1 oz	$\frac{1}{2}$	0.1 oz	0.1	Over
Max, lb	200,000	1,500	300	1,000	100	10,000
Surface finish (μin)	300-600	20-125	150-1,000	20-125	20-125	40-300
Tolerance 1st in (in/in)*	$\pm\frac{1}{32}$	±0.003	$\pm\frac{1}{64}$	±0.010	±0.006	±0.010
Porosity	5	1	1	2	2-3	2
Min. section, in	$\frac{1}{8}$ to $\frac{3}{16}$	$\frac{1}{64}$	$\frac{1}{8}$ to $\frac{3}{16}$	$\frac{1}{16}$ to $\frac{1}{8}$	$\frac{1}{32}$ to $\frac{1}{16}$	$\frac{1}{16}$
Draft	1-3°	$\frac{1}{2}$°	3-5°	2-3°	1-7°	
Tensile strength, lb/in² (in thousands)	19	17	23	25	30	25
Prod. rate (pc./h)	10-15	....	40-60	50-80	120-150	30-50
Typical pattern cost	$300	....	$2,000	$3,000	$5,000	$1,500
Scrap loss	5	3	4	3	2	1

* Over 3 in add 50% to tolerance given per inch.

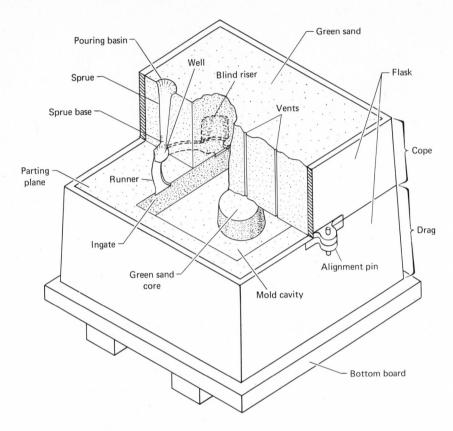

FIGURE 10-1
A typical mold partially sectioned to show detail.

pressure casting, and centrifugal casting. A comparison of the major features of different casting processes is given in Table 10-6.

Behavior of Alloys During Solidification

Alloys have three solidification ranges: (1) short, (2) medium, and (3) long (Fig. 10-2). Low-carbon steel and pure metals are typical of the first; aluminum-silicon alloys are representative of the second; and certain tin-bronzes have long freezing ranges. As a cast alloy loses its heat, islands of crystallites begin to form within the melt. The crystallites migrate toward the mold wall which is cooler and there they begin to align themselves with the chill grains. Then long columnar grains develop and grow into the melt, with a mushy zone between the dendrites as they thrust forward toward the thermal center of the casting,

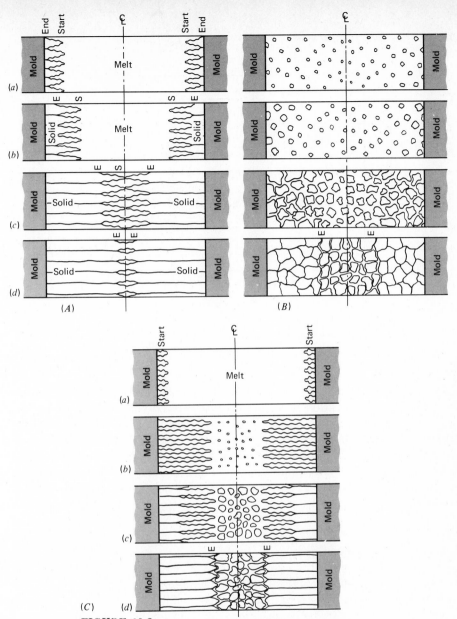

FIGURE 10-2

Typical freezing modes for alloys poured in sand molds. (*A*) Diagrammatic representation of freezing in a casting of low-carbon steel: (*a*) early stage; (*b*) intermediate stage; (*c*) intermediate stage—impingement of start-of-freeze waves; (*d*) late stage—near impingement of end-of-freeze waves. (*After Bishop and Pellini.*) (*B*) Diagrammatic representation of freezing in a sand casting of an alloy of long-freezing range: (*a*) very early stage; (*b*) early stage; (*c*) intermediate stage; (*d*) late stage. (*C*) Diagrammatic representation of freezing in a sand casting in an alloy which yields mixed structures: (*a*) very early stage; (*b*) early stage—impingement of start-of-freeze waves; (*c*) intermediate stage; (*d*) late stage.

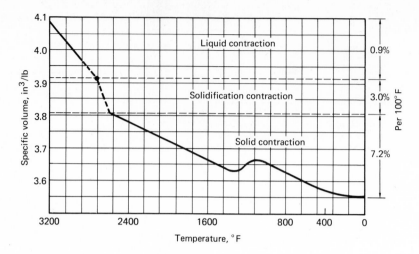

FIGURE 10-3
Thermal zones in a solidifying steel casting.

where in most cases equiaxed grains evolve (Fig. 10-2). As time passes the casting solidifies with the liberation of the heat of fusion and a volumetric shrinkage on the order of 3 percent in the case of steel. Subsequent cooling to room temperature has further shrinkage of about 1 percent for every 400°F temperature drop (Fig. 10-3). A typical freezing curve for a binary alloy is shown in Fig. 10-4, which also depicts how an equilibrium phase diagram can be determined from a series of freezing curves (Fig. 10-5). Cast metals do not freeze under equilibrium conditions and therefore the extent of the mushy zone is increased (Fig. 10-6).

Dendrites and Properties

A major aspect of alloy solidification is the simultaneous formation of treelike structures called dendrites throughout the melt (Fig. 10-7). Dendrites have a layered structure of varying composition—like a candle which has been successively dipped in waxes of different colors. The outer layers are most heavily alloyed. The structure, called coring, is evidence of microsegregation.

As solidification proceeds the dendrites grow until they impinge upon each other with the result that impurities, low-melting constituents, and microporosity occur at the interstices between the interlocking dendrites. The properties of cast structures depend upon the proper control of the chemical and structural heterogenities that occur in dendrite solidification. Recently metallurgists have dramatically improved the strength of cast aluminum alloys to 70,000 lb/in². The outlook for ferrous alloys is equally promising.

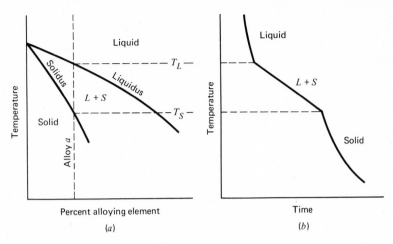

FIGURE 10-4
Effect of alloying on the solidification temperature of a metal. (a) Plot of solidification temperature versus alloy composition; (b) ideal cooling curve (equilibrium cooling) of alloy.

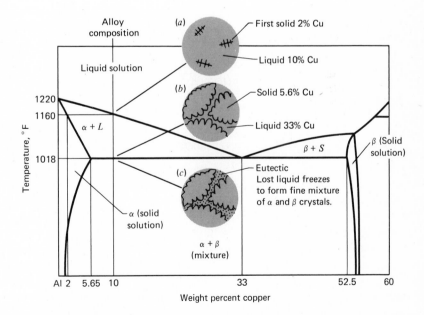

FIGURE 10-5
Phase diagram of the aluminum-rich end of the aluminum-copper phase diagram. Microstructures shown are obtained by very slow solidification. (Redrawn from Taylor et al., "Foundry Engineering," Wiley, New York, 1959.)

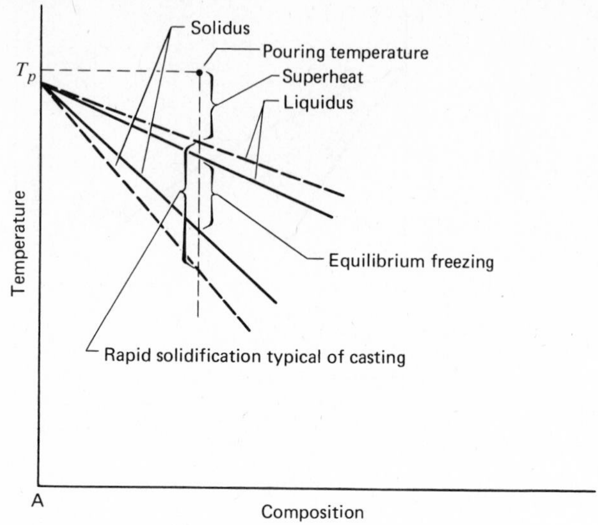

T_p

Solidus
Pouring temperature
Superheat
Liquidus

Temperature

Equilibrium freezing

Rapid solidification typical of casting

A

Composition

FIGURE 10-6
Partial phase diagram to show how rapid chilling extends the mushy region of a cast alloy.

Risers Compensate for Shrinkage

Properly placed risers can compensate for solidification shrinkage, but compensation for solid contraction is furnished by the oversize pattern. The proper riser size can be calculated rather readily for steel castings, but simple calculation for risers are not known for nonferrous alloys.

In general risers are reservoirs of heat and feed metal and are needed for sound castings—either to prevent centerline shrinkage or to compensate for

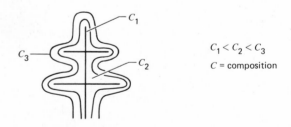

C_1

C_3

C_2

$C_1 < C_2 < C_3$

C = composition

FIGURE 10-7
Diagram of the compositional gradients within a dendrite of a binary alloy after casting. This structure is called coring and can be reduced by a lengthy annealing cycle.

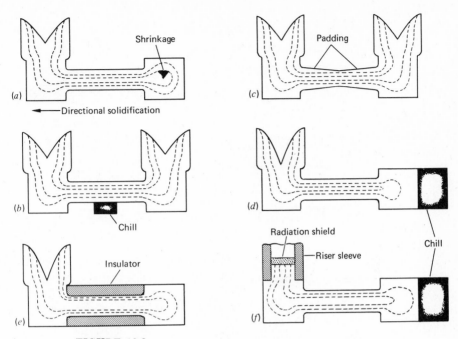

FIGURE 10-8
Several ways to feed an isolated heavy section.

shrinkage at the thermal center. Risers should be located to promote directional solidification, i.e., solidification from a chilled small section by a gradually tapered wall to a larger section which is risered (Fig. 10-8). The simplest example of multiple risers is a dumbbell which has a riser on each large end with a tapered section from the center to each end (Fig. 10-8). Any sizable casting can be divided into a number of such sections each of which has its own riser and gate.

Shrinkage Defects

There are a number of shrinkage problems which can occur in a casting but good design and skillful foundry practice can overcome them. The major problems are hot tears, or excessive residual stress, centerline shrinkage, and dispersed microporosity.

Hot tears   Hot tears and excessive residual stress occur because sections of a casting are restrained from shrinking by either massive cores or hard rammed molds. Such problems can be overcome by designing castings with gradual changes in shape and by making the molds and cores more collapsible. Generous use of fillets and large radii at joints also help to reduce residual stress levels.

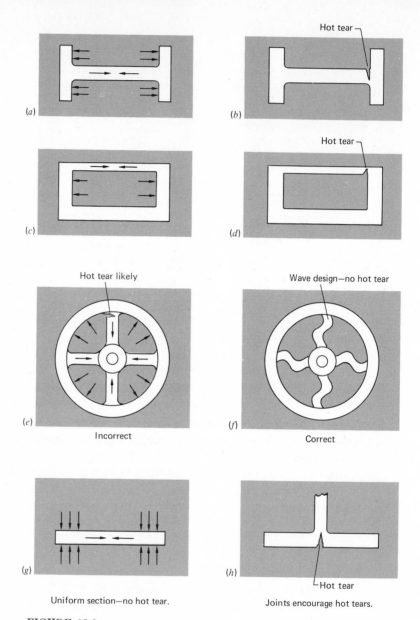

FIGURE 10-9
Mold constraints are the source of hot tears in castings. (*Redrawn from Taylor et al., "Foundry Engineering," Wiley, New York, 1959.*)

Heavy sections can be made to cool faster by adding chills or reducing the mass by coring. Hot tears are particularly troublesome in steel castings (Fig. 10-9).

Centerline shrinkage   Centerline shrinkage occurs in short-freezing-range alloys such as steel. In this case there is too little feed metal to fill the entire center of the casting at the time the last metal solidified. Thus a series of voids appear at the thermal center of the casting. Although these defects appear on x-ray negatives of the casting, they usually have little or no effect on its mechanical properties. Fatigue tests of steel castings have demonstrated that centerline shrinkage has no effect on the failure of steel castings subjected to fatigue loading.

Dispersed microporosity   Castings which have a long freezing range such as high-carbon steel, aluminum alloys, and copper alloys are subject to dispersed microporosity. This defect occurs at the interstices between interlocking dendrite arms and if uncontrolled does lead to a marked decrease in mechanical properties. The defect can be controlled by using a large enough thermal gradient to supply feed metal at the moment of final solidification. Grain size may also be reduced by supplying sufficient grain refiners such as titanium or boron powders to solidifying aluminum alloys, or by supplying mechanical agitation such as ultrasonic vibration to the freezing melt. By whatever means grain size and the dendrite arm spacing within the grain are reduced, mechanical properties soar. Dendrite arm spacing now appears to be a function of the alloy composition and cooling rate at that particular section of the casting and is independent of grain size itself. At high chilling rates dendrite arm spacing can be reduced by an order of magnitude, and the time to homogenize a cast structure depends on the square of the dendrite arm spacing, and thus the homogenizing heat-treatment cycle time would be reduced by 2 orders of magnitude. The improvements for an aluminum alloy are shown in Fig. 10-10, where the strength of the chilled metal is double that of an unchilled casting and its ductility is increased by a factor of 4. Thus castings can be designed to over twice the strength previously possible.

Fluidity and Minimum Section Thickness

The fluidity of all cast alloys is closely related to superheat, i.e., the difference between the pouring temperature and the liquidus of the alloy. Fluidity is important because it is an index of the filling of detail in thin sections. With the maximum practical superheat, cast iron can be cast as thin as $\frac{1}{8}$ in, aluminum and steel down to $\frac{5}{32}$ in and $\frac{3}{16}$ in, respectively. Where too thin or too extensive a section is designed, a casting is prone to develop "cold shuts" or holes during pouring.

Thin sections display enhanced mechanical properties because of the fine

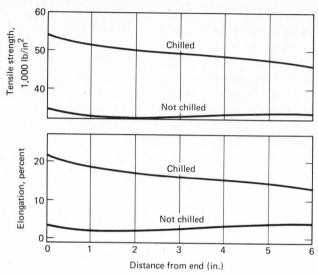

FIGURE 10-10

Tensile strength and ductility of a chilled aluminum alloy as a function of the distance from the chilled end. *Note:* Tensile strength doubled; elongation quintupled.

grain size *caused* by the rapid cooling. A similar improvement can occur in thicker sections of nonferrous alloys by ramming up chills of cast iron or copper in the sand mold so that the metal lies against them in the mold. If gray iron of the proper composition is cast against a heavy chill, white iron forms in that area. Formerly plowshares and the rims of the wheels for mining cars were made that way.

The product designer should work closely with a foundryman to achieve proper directional solidification, the optimum parting line, proper draft, and compensation for shrinkage. Venting is usually no problem in sand molds, which are permeable, but in metallic molds proper gating and venting are real challenges to any designer.

Major Casting Processes

Casting processes include sand casting, refractory mold casting, and metallic mold casting. Other ways to classify casting methods might be by alloy poured, such as malleable-iron practice, steel practice, etc.; by type of tolerance, such as precision investment casting; or by the process itself, such as centrifugal casting.

SAND CASTING

A sand casting is produced when molten metal is poured at atmospheric pressure into a cavity formed in a compacted molding sand. In typical ferrous casting the mold would be bonded with 6 to 8 percent western bentonite mulled with 2 to 3 percent water. Patterns would be reusable, but the sand molds would be broken from the casting after the solidified casting had cooled sufficiently. The sand would be returned to the muller for recycling by mulling with some new clay and water to reactivate the old clay. Sand molds define the external geometry of a casting, whereas cores define the interior or undercuts.

Sand casting accounts for more than 90 percent of all metal poured on a tonnage basis. Sand casting is an extremely high-production process—for instance, 125-lb gray-iron bathtubs are molded and poured at the rate of 72/h using a crew of five men. Isn't that amazing! Automotive engine blocks are poured at the rate of one every 15 s! For comparison a die-cast engine block requires at least 2 min floor-to-floor time for one block.

On the other hand, it requires a high investment in plant facilities and sand-handling equipment to make such production rates possible. Sand casting is versatile and is capable of making parts ranging from small knobs to huge radiotelescope bases weighing more than 200,000 lb.

Sand Molds

Sand molds are made by ramming sand around a pattern in a flask (Fig. 10-11). The sand is either shoveled by the molder, thrown by a sand slinger, or dropped in from an overhead chute. The molding sand is compacted by jolting several times and squeezing or by jolting alone for larger deep flasks. The pattern is withdrawn, cores are added, and the mold is reassembled, weighted, and sent on to the pouring station. Sand molds include green sand, skin dried, carbon dioxide, shell molded, and cement bonded.

The texture of the mold surface is imparted to the surface of the casting. If the pouring temperature is too high, sand grains may adhere to the cast metal surface. Such sand inclusions cause problems in the machine shop and are to be avoided. Proper selection of sand grain size and distribution, and proper clay content and compaction, can result in excellent surface appearance and detail in sand molds for aluminum, brass, bronze, and cast iron.

In high-pressure molding the compaction force must exceed 100 lb/in^2 at the parting plane, which is about 50 percent more than conventional practice during the 1960s. The greater compaction pressures yield harder, more uniform molds which permit much less mold wall movement and therefore result in closer dimensional control of the final casting.

Sand molds are made in many sizes, and machines have been developed to

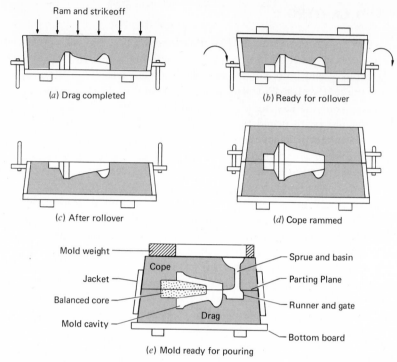

FIGURE 10-11
Steps in the construction of a mold by hand ramming.

mechanize molding of all but the largest sizes. Thus sand molds are produced by bench molding, jolt-squeeze molding, automatic molding, flaskless molding, high-pressure molding, floor molding, and pit molding.

In brief, green-sand molds must be strong enough to withstand the poured weight of the molten metal, must resist erosion by the flowing metal, must be sufficiently permeable to liberate mold gases, and must be refractory and yet strip easily from the casting after cooling. *Cores must burn out and collapse* to permit the cast metal to contract freely during solidification and cooling.

Chemically Bonded Sands

Carbon dioxide molds and cores In this process the molding aggregate is bonded with 3 to 5 percent sodium silicate (water glass). After the aggregate is packed around the pattern or into the core box, it is cured by passing carbon dioxide gas at about 10 lb/in^2 through the mix for about 10 s using a probe or a gasing head. The sand immediately becomes extremely strongly bonded as the sodium silicate becomes a stiff gel:

$$Na_2O \cdot x\ SiO_2 + nH_2O + CO_2 \rightarrow Na_2CO_3 + \underbrace{xSiO_2 \cdot n(H_2O)}_{gel}$$

where x is 1.6 to 4, most often $x = 2$.

Silicone parting agents should be sprayed on the patterns and core boxes before each use to prevent sticking because the cured molds and cores are very strong and have little yield.

The process is more expensive than molding with clay-bonded sand because the binder is more costly and carbon dioxide–bonded cores have very poor collapsibility, yet the process offers a number of advantages which are more widely accepted abroad than in the United States. Notable advantages are:

1 Uses conventional equipment for molding and sand mixing.
2 Eliminates the need for internal support for cores and for pouring jackets.
3 Molds and cores can be used immediately after processing.
4 Eliminates baking ovens and core driers.
5 Improved dimensional accuracy is available because of greater fidelity of pattern or core-box detail in the cured mold.

Shell-mold Casting

Shell molding uses a phenolic resin bond mixed with sand for casting steel, iron, or nonferrous alloys. The pattern is made of metal, preferably cast iron. It should be accurate because the process duplicates parts to close tolerances. The pattern is heated to 400 to 500°F; then, after a silicone parting agent is sprayed on the surface, the resin and sand mixture is deposited on the pattern by blowing or dumping (Fig. 10-12). The excess material is shaken off for reuse and the shell or crust ($\frac{1}{8}$ to $\frac{1}{4}$ in thick) is removed from the pattern after curing is complete. Halves are matched and located by integral bosses and matching recesses and glued or clamped together. Finally they are placed in a metal case, and surrounded by $\sim$1.5 in of steel shot, sand, or other backup material to support them during pouring. The gates, sprues, and risers are usually a part of the mold.

The shell-molding process illustrates how one industry (plastics) affects another industry (sand casting). This process today has proved practical for parts ranging from malleable iron chain hooks ($\frac{1}{2} \times 1 \times \frac{1}{2}$ in) to automotive crankshafts.

Fully automatic shell-molding machines are available which can produce a thin shell of resin-bonded sand every 30 s. This equipment, in the production shop, is coupled with a shell-closing machine which seals the cope and drag halves of the mold together. Close tolerance between the cope and drag is maintained, thus minimizing objectional fins in the finished casting. After the

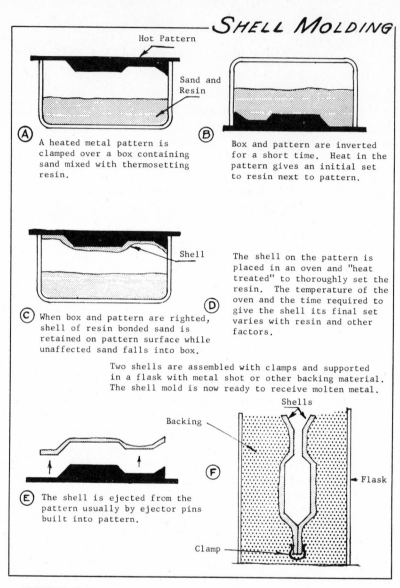

SHELL MOLDING

Hot Pattern

Sand and Resin

A A heated metal pattern is clamped over a box containing sand mixed with thermosetting resin.

B Box and pattern are inverted for a short time. Heat in the pattern gives an initial set to resin next to pattern.

Shell

C When box and pattern are righted, shell of resin bonded sand is retained on pattern surface while unaffected sand falls into box.

D The shell on the pattern is placed in an oven and "heat treated" to thoroughly set the resin. The temperature of the oven and the time required to give the shell its final set varies with resin and other factors.

Two shells are assembled with clamps and supported in a flask with metal shot or other backing material. The shell mold is now ready to receive molten metal.

Shells

Backing

Flask

E The shell is ejected from the pattern usually by ejector pins built into pattern.

F

Clamp

FIGURE 10-12
The shell-molding process.

shell-closing operation, the molds can be placed in a mold-storage area until needed. The mold shown has a vertical parting line, but most molds have a horizontal parting if the casting is small. High production rates, good surface finish, and close dimensional tolerances can be achieved in gray iron and small steel castings.

No backing is used, just a bed of dry sand and a weight. When the metal is poured, the smooth shell promotes easy flow and the metal is not held back by gas pressure; the porous mold permits gases to escape easily. Heat is removed quickly through the thin shell and backup material (Fig. 10-12). Some resin formulations are slightly flexible after setting; therefore, patterns do not need as much draft and molds can be ejected readily. The molds can be stored and do not contain moisture. Thus they are suitable for pouring at any time.

This process will reduce the dust present in the sand foundry and reduce the amount of sand to be handled. Since the sand cannot be reused, without pneumatic or thermal reclamation, shell molding is relatively expensive, which is the major disadvantage of the process.

The shell-molding process has the following advantages over typical green-sand molding:

1 Productivity can exceed that of conventional sand-casting practice.
2 Thin sections (down to 0.010 in) can be cast.
3 Machining of the castings is reduced and, in some cases, eliminated.
4 Cleaning is considerably reduced; shot blasting is frequently eliminated.
5 Saving of metal through the use of smaller gates, sprues, and risers in the casting process results in a higher yield from the metal.
6 Savings in work space, material handling, and storage are also realized through use of 90 percent less molding materials.
7 The cured resins are not hygroscopic, thus permitting prolonged storage of the molds.
8 Closer dimensional tolerances can be obtained (0.002 in/in) in one-half of a mold but 0.010 in across the parting line.
9 Better surface finishes are realized (100 μin).
10 Shell molding is particularly useful for small steel castings (up to 20 lb).
11 Shell core production is particularly attractive because it reduces core weight and eliminates core ovens.

As mentioned, the principal disadvantage of the shell-molding process is that in view of the cost of pattern and curing equipment, the process is not economically advantageous in small quantities. A representative phenolic resin will cost about $0.30/lb and cannot be reused. The typical clay binder in green-sand molds costs only about $0.01/lb and may be used many times. Secondly, the strong phenolic odor bothers some people and may be enough to cause disagreeable working conditions.

Catalyzed-resin Bonding (Hot- and Cold-box Cores)

Sands may be bonded with any ceramic or organic material which forms a gel with a 1- to 2-h shelf life and sets hard and strong in a reasonable length of time.

A number of such binders have been developed and are known as hot-box, no-bake, or air-setting binders. Certain organic resins, usually various blends of formaldehyde, urea, and furfural alcohol tend to cross-link into highly immobile long-chain polymers in the presence of a catalyst such as phosphoric acid. The furfural alcohol gives a relatively weak cross-linking tendency, whereas in the presence of the phosphoric acid catalyst the curing speed is greatly enhanced.

The urea component of the mix must be carefully controlled or even eliminated in the case of steel foundry sands for cores or molds because it promotes pinhole porosity.

A typical core sand mix contains 2 percent resin and 0.6 to 0.8 percent catalyst by weight. The setting time is a function of the amount of catalyst, the temperature of the mulled sand, and the mulling time. Since the curing cycle begins as soon as the materials are combined in the muller, the bench life is short. It is possible to mix the ingredients and run them directly into a slinger for ramming up the core boxes.

The hot-box process uses basically the same materials but the mix must have good flowability. In this case a weak catalyst is used. When the mix is blown into core boxes at temperatures of 400 to 500°F an exothermic curing action occurs and the core is completely cured in 30 s. The water content of the sand mix is critical—too little moisture results in friable cores whereas excess water increases the baking time. A major drawback of the process is caused by the fact that the core must cure before another can be made.

In the cold-box process a two-component liquid organic binder is used. Part 1 is a phenolic resin dissolved in a solvent, and part 2 is a polyisocyanate which is also in solution. In the presence of an airborne catalyst, triethylamine, the hydroxyl group of the liquid phenolic resin combines with the isocyanate group to form a stiff urethane resin. The two binder components are usually used in a 1:1 ratio with from 1 to 2 percent by weight of the resin mixture added to the sand.

The process has a 10-s curing time, so it is a simple, high-production core-making process. Care is needed to ensure that the triethylamine does not leak into the work area. Once it has been used it must be catalytically decomposed and then vented to comply with the regulations of the Occupational Safety Health Act. However, in the cured state, the nitrogen gas content is only 3 percent of the binder rather than 10 percent as is found in the hot-box process. Therefore cold-box cores are used in steel castings to try to minimize pinhole porosity. A quick comparison of the two processes is given in Table 10-7.

Flaskless Molding

Flaskless molds with a vertical parting line are made at a rate of up to 750 molds per hour by high compaction of sand into blocks which are about 14 × 18 in in

Table 10-7 COMPARISON OF HOT- AND COLD-BOX CORES

Process	Resin	Bench life, h	Catalyst	Nitrogen content, %
Hot box	Phenolic	8	Furfural alcohol	10
Cold box	Phenolic Polyisocyanate	2–3	Triethylamine	3

cross section and up to 8 in thick. The sand is blown into the machine shuttle at 90 to 100 lb/in^2 (Fig. 10-13a), then the entrance is stopped off and a squeeze ram moves forward and the block is compacted at up to 2,000 lb/in^2. The shuttle moves to one side, bringing a second shuttle to the molding station. While the second mold is compacted, a hydraulic ram strips the first compact and adds it to the previous blocks which are already on the pouring conveyor (Fig. 10-14). Note that the conventional cope and drag patterns are on opposite sides of the compacted block. Only when two blocks have been completed and pushed together can a mold be poured.

To meet the high pouring rates which are required by the fast molding cycle, a multilipped pouring ladle can be used to pour as many as five molds at one time. The shakeout and sand-return details are indicated also. Of course all elements of the operation must be fully balanced, and the length of the cooling conveyor is dependent upon the section thickness of the casting and its cooling rate.

A specially blended high-clay, synthetically bonded molding sand is required to make the process feasible in production. With the proper sand and a balanced cooling rate, high production rates can be achieved on such diverse parts as gas stove grills, brass valve bodies, and malleable iron pipe fittings. The process is competitive with ferrous die casting on small brass and ferrous parts. In fact it may make the latter process desirable only when a superior surface finish is needed as cast. Flaskless molding can achieve a tolerance as small as ±0.005 in across the parting line on a routine basis.

Full-mold Process

This is also called the foam vaporization process. The polystyrene foam pattern, complete with sprues, bottom gates, runners, and risers, is rammed up in a suitable flask. The pattern is expendable, and no parting line is needed. During pouring the heat from the molten metal vaporizes the pattern which then moves through the capillary-porous molding media to a point where it condenses. Cores

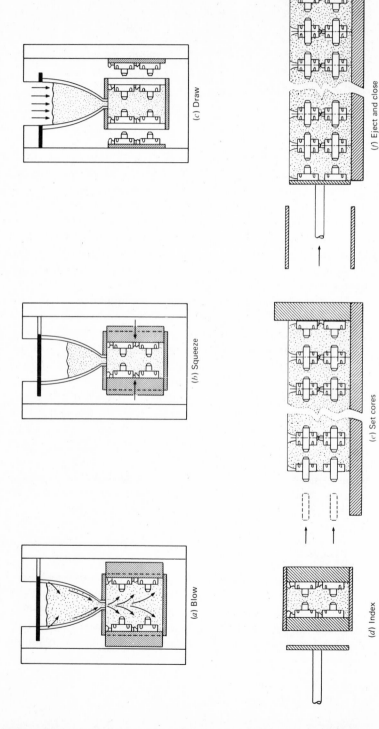

(a) Blow

(b) Squeeze

(c) Draw

(d) Index

(e) Set cores

(f) Eject and close

FIGURE 10-13
Diagram of operation of flaskless molding machine.

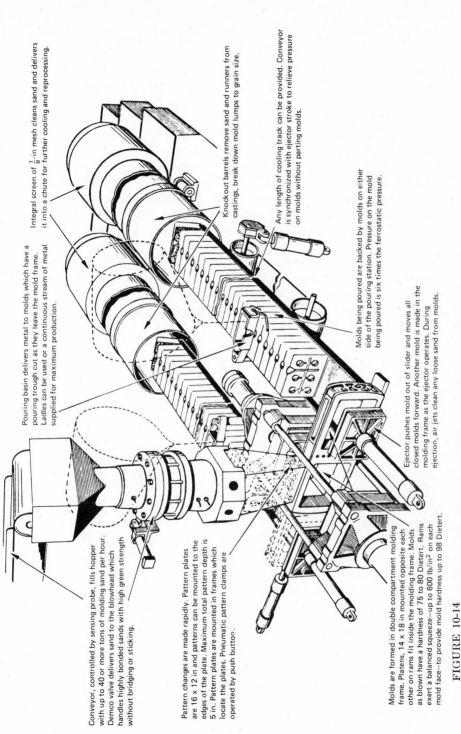

Conveyor, controlled by sensing probe, fills hopper with up to 40 or more tons of molding sand per hour. Demco valve delivers sand to the blowhead which handles highly bonded sands with high green strength without bridging or sticking.

Pattern changes are made rapidly. Pattern plates are 16 x 12 in and patterns can be mounted to the edges of the plate. Maximum total pattern depth is 5 in. Pattern plates are mounted in frames which locate the plates. Pneumatic pattern clamps are operated by push button.

Molds are formed in double compartment molding frame. Platens, 14 x 18 in mounted opposite each other on rams fit inside the molding frame. Molds as blown have a hardness of 75 to 80 Dietert. Rams exert a balanced squeeze—up to 600 lb/in² on each mold face—to provide mold hardness up to 98 Dietert.

Pouring basin delivers metal to molds which have a pouring trough cut as they leave the mold frame. Ladles can be used or a continuous stream of metal supplied for maximum production.

Integral screen of $\frac{1}{8}$-in mesh cleans sand and delivers it into a chute for further cooling and reprocessing.

Knockout barrels remove sand and runners from castings, break down mold lumps to grain size.

Any length of cooling track can be provided. Conveyor is synchronized with ejector stroke to relieve pressure on molds without parting molds.

Molds being poured are backed by molds on either side of the pouring station. Pressure on the mold being poured is six times the ferrostatic pressure.

Ejector pushes mold out of slider and moves all closed molds forward. Another mold is made in the molding frame as the ejector operates. During ejection, air jets clean any loose sand from molds.

FIGURE 10-14
Flaskless molding machine. (*Courtesy The Herman Corp; Zelienople, Pa.*)

are not needed if holes are properly located and sized because the molding medium itself can be used as a green-sand core, or an air-setting sand may be rammed into the cored areas. Thus the pattern itself can serve as a core box!

Suitably formed polystyrene foam can also be used as an extension to or modification of an existing pattern or as loose pieces which remain in the mold. In addition spherical risers can be rammed up in a conventional cope and drag mold wherever desired for best feeding without regard to a parting line because they will stay in the mold when the pattern is drawn.

The process is especially useful for large castings of which one or two pieces are required. Most automotive forming dies for quarter panels, roofs, or hoods are made by the full-mold process because the polystyrene foam can be formed so easily. Pattern shops have determined that they can make three polystyrene patterns before they reach the break-even point with a wood pattern.

The future of the process is particularly bright if one considers that a foam of the proper density could be made right in the foundry in a proper steel pattern box. A nonbonded molding media could be poured in, and vibrated around the pattern and core assembly. Then the mold could be poured, the casting shaken out, and the sand passed through a pneumatic reclaimer. Then the process would be repeated.

PRECISION-CASTING PROCESSES

Plaster-mold Casting

Plaster molding as an art has been known for years. It is now competing successfully for castings because engineers have improved accuracy, quality, range of applications, and costs. Costs have been reduced by standardization of methods, flasks, materials, and material-handling facilities. The process is similar to sand casting, since cope and drag molds are used, and has many of the features of permanent molding.

Plaster molds cannot be used for ferrous castings because the plaster ($CaSO_4 \cdot \frac{1}{2}H_2O$) is not sufficiently refractory. However, nonferrous castings can be made with smooth, accurate surfaces and fine details. Their cost is about three times as great as that of sand castings, but the elimination of machining and finishing operations frequently compensates for the additional cost. This process, due to the recent improvements in foundry mechanization and techniques, should receive consideration when design and process are selected.

Mold material Gypsum or calcium sulfate (plaster of paris), talc, asbestos, silica flour, and others, and a controlled amount of water, are used to form a slurry.

Process steps   The steps in processing are as follows:

1   Parting compound is sprayed on flask and pattern.
2   Slurry is poured from hose around pattern and flask is filled.
3   Pattern is removed after setting.
4   Mold is dried and baked in conveyor oven.
5   Mold is separated from flask and flask is returned to step 1.
6   Inserts and cores are placed and guide pins are placed in holes in four corners of mold.
7   Cope and drag are matched or assembled and guided by pins.
8   Metal is poured.
9   Casting is cooled on conveyor in mold.
10   Casting is shaken out and the mold is destroyed. (The mold material is not salvaged.)
11   Castings are trimmed of gates, sprues, and flash, and are inspected.

Antioch Process

In the Antioch process a mold material is used that includes the advantages of sand and gypsum plaster. Cast surfaces and details are comparable to those produced by plaster molds, but, like plaster molding, the Antioch process is confined to nonferrous casting. Silica sand is the bulk material and gypsum the binder. Talc, terra alba, sodium silicate, and asbestos are used to control some characteristics. Water and dry material produce a slurry that is piped to the flask and pattern setup. The material develops initial set in 5 to 7 min, is air-dried for about 5 h, and then placed in an autoclave at about 2 atm of steam pressure. The molds are again air-dried for about 12 h and then reheated in an oven at 450°F. This autoclave operation and drying procedure helps develop excellent permeability. Tolerances of ±0.005 in can be kept on small castings, and ±0.015 in on automotive tire molds.

Ceramic-mold Casting

This process, which is also known as cope and drag investment casting has two proprietary variations: (1) Shaw process and (2) Unicast process. The major distinction between ceramic-mold casting and investment casting is that the former relies on precision-machined metal patterns rather than the expendable patterns used in conventional investment casting. Ceramic-mold casting is similar to plaster molding except that the mold materials are more refractory, require higher preheat, and are suitable for most castable alloys, particularly ferrous alloys. The refractory slurry consists of fine-grained zircon and calcined high-alumina mullites or in some cases fused silica.

In ceramic molding, a thick slurry of the mold material is poured over the

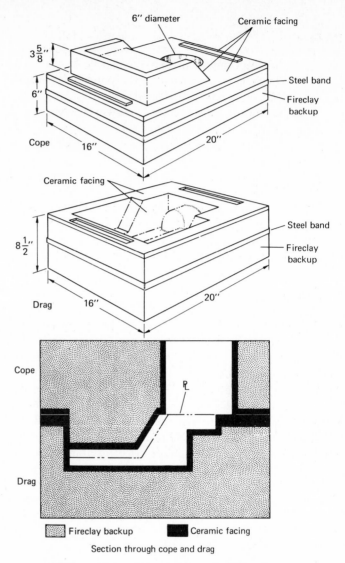

This Shaw-process mold was used for casting H13 tool steel dies for use in the hot upset forging of steel axles.

FIGURE 10-15
Typical cope and drag setup for ceramic molding. (*Redrawn from "American Society for Metals," Metals Handbook, vol. 5, Metals Park, Ohio, 1970.*)

reusable split and gated metal pattern which is usually mounted on a match plate (Fig. 10-15). A flask on the match plate contains the slurry which gels before setting completely. The mold is removed during the time when the gel is firm to prevent bonding to the pattern. The basic difference between the Shaw and Unicast processes is that in the Shaw process stabilization results from the burn off of an alcohol binder whereas in the Unicast process the mold is stabilized by immersion in a liquid bath or a gaseous atmosphere which cures the gelled slurry. Before pouring, the molds are usually preheated in a furnace to reduce the temperature difference between the mold and the molten metal and to maximize the permeability available through the microcrazed mold structure.

Ceramic-mold casting is preferred when: (1) the parts are too large for conventional precision-investment casting or (2) when parting-line defects are not objectionable. The process replaces sand casting if a superior surface finish is required or improved dimensional accuracy is needed, or if sand inclusions or hot tears can not be tolerated in a casting of complex geometry. Ceramic molding can handle parts weighing 1,500 lb or more. Even on such large castings the process has excellent accuracy and reproducibility.

It is impractical to control dimensional tolerance across the parting line to the same tolerance as within one-half of the mold. Also, mechanical properties suffer the loss of the chill effect because the mold materials are such good insulators that a coarse-grained structure results. The designer can compensate for such a reduction in strength by making critically stressed sections somewhat heavier. The process is expensive because the mold materials are high in cost and expendable. Ferrous alloys are the most commonly cast; they include die-casting dies, large-trim dies, components for food machinery, milling cutters, structural components for aircraft, and hardware for aerospace vehicles and atomic reactors.

An as-cast surface finish of 125 μin or better can readily be achieved. Dimensional tolerances are ±0.003 in/in for the first inch, with incremental tolerances of ±0.002 in/in for larger dimensions. Across the parting line an additional tolerance of ±0.001 in should be provided.

Investment Castings

Investment casting is of interest to the production-design engineer because this process offers greater freedom of design than any other metal-forming operation. Accurate and intricate castings can be made from alloys that melt at high temperatures. Parts such as gas-turbine blades and latches can be cast to such close tolerance that little or no machining is required.

Investment casting is like sand casting in that the mold is broken to release the casting; it is like permanent molding and plaster casting in that good surfaces, accuracy, and consistency can be obtained; and like plaster mold-

PREPARING A MOLD FOR INVESTMENT CASTING
The "Lost Wax" or precision casting process

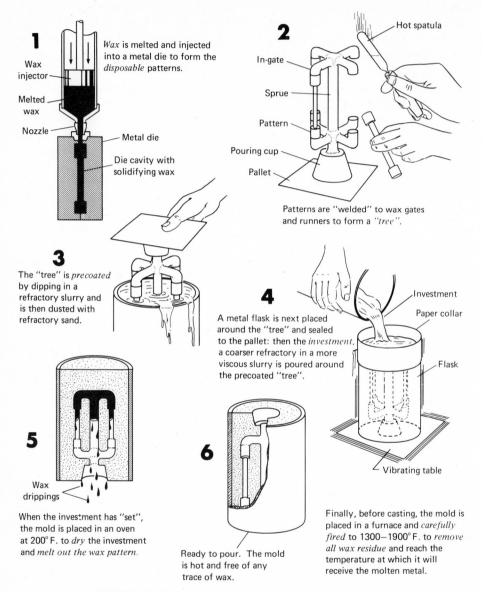

1 *Wax* is melted and injected into a metal die to form the *disposable* patterns.

Wax injector

Melted wax

Nozzle

Metal die

Die cavity with solidifying wax

2

Hot spatula

In-gate

Sprue

Pattern

Pouring cup

Pallet

Patterns are "welded" to wax gates and runners to form a *"tree"*.

3 The "tree" is *precoated* by dipping in a refractory slurry and is then dusted with refractory sand.

4 A metal flask is next placed around the "tree" and sealed to the pallet; then the *investment*, a coarser refractory in a more viscous slurry is poured around the precoated "tree".

Investment

Paper collar

Flask

Vibrating table

5 Wax drippings

When the investment has "set", the mold is placed in an oven at 200° F. to *dry* the investment and *melt out the wax pattern*.

6

Ready to pour. The mold is hot and free of any trace of wax.

Finally, before casting, the mold is placed in a furnace and *carefully fired* to 1300–1900° F. to *remove all wax residue* and reach the temperature at which it will receive the molten metal.

FIGURE 10-16
Preparing a mold for investment casting.

ing in that the mold is made from a slurry and baked. Investment casting, precision casting, and lost-wax processes are all essentially the same.

Investment casting is unlike other casting processes in that an expendable pattern of wax, plastic, or frozen mercury is used (Fig. 10-16). Intricate castings can be made by forming expendable patterns separately in dies and assembling them to make up the final and more complicated pattern unit. Gates and risers are attached for molding. The whole expendable pattern unit is coated with a fine colloidal silica wash and is dusted with a refractory sand. A slurry is then poured around it and allowed to set. By this means no parting-line inaccuracies prevail and thus close tolerances can be maintained in all directions. The flask with its contents is vibrated to pack the slurry around the pattern and separate the solid material from the water, which can be poured or withdrawn from the top. The flask is then placed upside down in a furnace and baked. The pattern material melts and runs out, leaving the mold ready for final baking at high temperatures. The mold is in one piece and cannot be inspected before pouring. After it is baked at high temperature, it is ready for casting a high-melting-point alloy while the mold is still hot. When sufficient cooling time has elapsed, the mold is broken and the casting removed. This process can be used for any type of material. Low-temperature alloys such as brass, silver, gold, and bronze were cast in this manner in ancient times.

The various controls used for materials, temperatures, time of baking, and pouring ensure uniform products. The wax is of a known composition and has definite shrinkage factors. The ingredients can be adjusted to vary the shrinkage; thus compensation can be made for mold shrinkage, and the difficulty of adjusting mold sizes for shrinkage is reduced.

Dies for the wax pattern can be made of rubber, plaster, soft metals, or steel. Their composition is determined by the accuracy requirements, quantity, and ease of removing the wax. Plastic and mercury require metal molds. Units made in one mold are joined to gates to make multiple patterns so that many parts can be cast at the same time. The accuracy of the original mold influences the final size of the part.

Castings are usually poured by gravity with parts arranged in a vertical fashion. Centrifugal, vacuum, or air pressure is sometimes used to pour metal into the hot mold and to obtain accurate, homogeneous castings.

Ceramic-shell Casting

Ceramic-shell casting has proved to be a boon to investment casters because it has permitted the automation of the ancient art of *cira perdue* or lost-wax process. Ceramic-shell molds may cost less than 50 percent as much as conventional investment molds for the same pattern. As in the investment process, patterns are wax-welded to a central sprue (Fig. 10-17). The assembly is care-

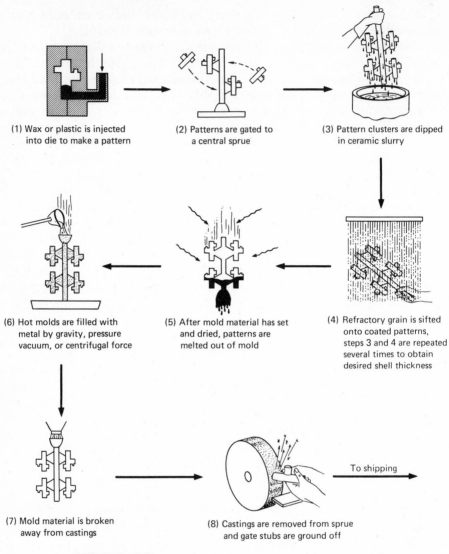

(1) Wax or plastic is injected into die to make a pattern

(2) Patterns are gated to a central sprue

(3) Pattern clusters are dipped in ceramic slurry

(6) Hot molds are filled with metal by gravity, pressure vacuum, or centrifugal force

(5) After mold material has set and dried, patterns are melted out of mold

(4) Refractory grain is sifted onto coated patterns, steps 3 and 4 are repeated several times to obtain desired shell thickness

(7) Mold material is broken away from castings

(8) Castings are removed from sprue and gate stubs are ground off

To shipping

FIGURE 10-17
Sequence of operations in the ceramic-shell process.

fully cleaned in a suitable solvent, dipped in a colloidal ethyl silicate gel, and drained for several minutes. When dripping stops it is stuccoed by dipping into a fluidized bed of a fine-grained fused silica. The process is repeated. On the third and subsequent stuccoing operations a coarser grade of fused silica is used; usually five coats are sufficient. Over 85 percent of all precision casting is carried out in ceramic-shell molds.

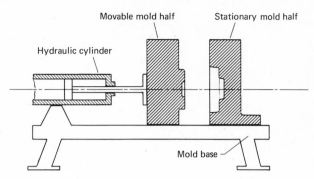

FIGURE 10-18
Schematic drawing of straight-line permanent-mold machine equipped with
two-piece deep-cavity die.

Permanent-mold Casting

When sand castings are made, the sand mold is destroyed. Therefore, it has been
an ambition of engineers to develop a mold that can be reused thousands of
times. Permanent molds which are filled by gravity pouring have come into
general use because they are economical, and because they are particularly useful
in casting low-melting-point alloys (Fig. 10-18). Metal-mold castings have some
distinct advantages over the typical sand-mold casting. These include closer
dimensional tolerances, better surface finish, greater strength, and more economical production in larger quantities. Some disadvantages of metal molds are
their lack of permeability, the high cost of the mold, the inability of the metallic
mold to yield to the contraction forces of the solidifying metal, and difficulty in
removing the casting from the mold since the mold cannot be broken up. Recently, bronze, cast iron, and even steel have been cast in metal molds and have
extended the use of permanent molds.

In general, castings to be produced by permanent-mold methods should
be relatively simple in design with fairly uniform wall sections and without
undercuts or complicated coring. Undercuts on the exterior of the casting com-
plicate the mold design, resulting in additional mold parts and increased cost.

If the design requires undercuts or relatively complicated coring, semipermanent molds should be considered. The semipermanent molds are like
permanent molds, except that they use sand cores. The metal is poured into the
permanent mold under the force of gravity and therefore there are no high
pressures.

It should be emphasized that all permanent molds have a thick (0.030-in)
coating of sodium silicate and clay or other insulating materials over the cast-
iron surface. The coating causes a poorer surface finish and wider tolerance than
is found in a typical die casting. Generally speaking permanent-mold castings

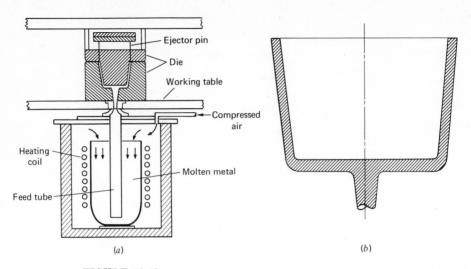

FIGURE 10-19
Low-pressure permanent-mold casting. (a) General equipment; (b) enlarged
view of casting showing typical "carrot" ingate.

are sounder than equivalent die castings. For this reason automotive pistons are
cast in permanent molds rather than die-cast.

Permanent-mold dies are preheated to 300 to 400°F before pouring and
are given a graphite dusting every three or four shots. Thermal balance is very
important, and auxiliary water cooling or radiation pins are used to cool heavy
sections. Proper venting of the cavity is very important in order to avoid misruns.

Low-pressure permanent-mold casting This casting method is gaining
acceptance as a production process. Low-pressure permanent molding depends
on directional solidification and proper die design. Other factors include the position of the casting within the dies, the location of the ejection pins, and die
coating and venting. Finally, heat balance in the die, ingate location, and gate
removal are important factors in the economical production of sound castings.

In this type of casting, molten aluminum is forced by low air pressure from
a silicon carbide crucible up the delivery tube and into the die cavity. The solidified casting is removed from the upper die by ejector pins, which are activated
when the upper platen is raised (Fig. 10-19). The casting is located so that directional solidification takes place toward the mouth of the tube. If a proper
heat balance is maintained, risers are eliminated and there is a high casting
yield. As soon as solidification occurs at the end of the tube, the air pressure is

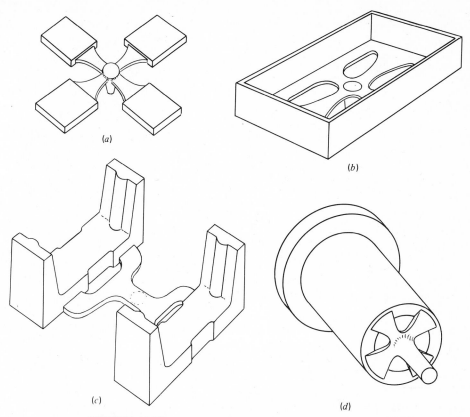

FIGURE 10-20
Typical parts cast of aluminum alloys by the low-pressure permanent-mold process. (a) Four pieces on a central sprue; (b) box-shaped part with a gate system which is sheared all around; (c) twin pieces to keep heat flow in balance; (d) cylinder with chill at far end.

released and the metal returns to the crucible where it is reheated for the next shot. Typical parts cast by this process are shown in Fig. 10-20.

The process has a number of useful features:

1   It makes possible thin-walled castings.
2   Aluminum alloys can be cast by this process.
3   The process can use expendable cores.
4   Heavy sections and castings of large projected areas may be cast because there is no need for heavy locking forces.
5   Low-pressure permanent-mold castings usually require no risers to ensure adequate feeding, so the average yield approaches 90 percent.

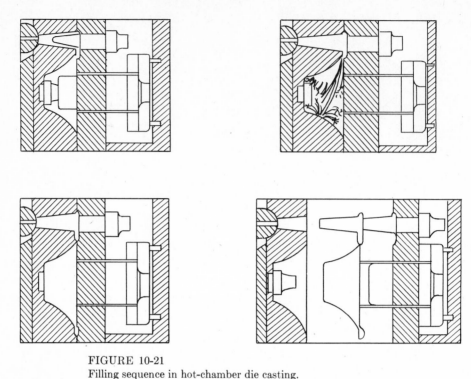

FIGURE 10-21
Filling sequence in hot-chamber die casting.

6 Die materials and die coatings are similar to those used for permanent-mold castings.

7 Accurate control and reproducible results can be virtually assured.

Die Casting

Die casting is the art of rapidly producing accurately dimensioned parts by forcing molten metal under pressure into metal dies. The term also applies to the resultant casting. Die castings can be used economically in designs having moderate to large activity because the completed piece has a good surface, requires relatively little machining, and can be held to close tolerances. The principles of die casting follow those of good practice in any casting operation. The steel dies are permanent and should not be affected by the metal introduced into them, except for normal abrasion or wear. Die-casting dies are usually more expensive than those used in plastic or permanent molding of a part of similar size and shape.

The rapidity of operation depends upon the speed with which the metal can be forced into the die, cooled, and ejected; the casting removed; and the die prepared for the next shot (Fig. 10-21).

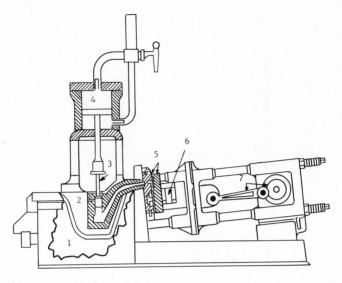

FIGURE 10-22
A typical hot-chamber die-casting machine on which are indicated the seven
basic elements. (1) Furnace for keeping the alloy molten. (2) Alloy-holding
pot. (3) Plunger that forces the alloy from the cylinder through the passage
into the die cavity. (4) Air cylinder that operates the plunger. (5) Die with
its cavity. (6) Mechanisms for ejecting the casting. (7) Connecting rod and
crank for opening and closing the dies. The transition from molten alloy to
die-cast part is accomplished in a fraction of a minute with each cycle of the
machine.

Hot-chamber type (Fig. 10-22)   The hot chamber refers to the pot in which
the metal is melted and from which it is led to the die by what is known as a
gooseneck. The hot metal is forced into the die by two methods:

1 By maximum air pressure of 600 lb/in² over the molten metal. With
this method, 100 shots per hour can be made. This is obsolete for zinc and
aluminum alloys, but can be used for ferrous alloys.
2 By a cylinder and plunger which are able to exert a pressure of 1,500
to 6,000 lb/in². The cylinder is submerged in the pot from which the
molten metal flows; up to 700 shots per hour can be made.

These machines are used for metals having a maximum melting point of
800°F.

Cold-chamber type (Fig. 10-23)   The second group of machines is called
the cold-chamber type because the metal is heated in a separate pot suitable
for high temperatures. The metal is poured into the cylinder by a hand ladle
or automatic pour device, and the plunger forces the metal through the orifice

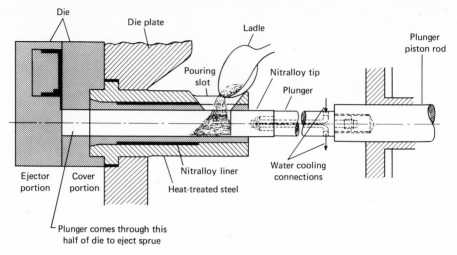

FIGURE 10-23
Schematic view of the cold-chamber die-casting process.

into the die. Pressures of 2,000 to 6,000 lb/in² are normal, but some machines offer much higher pressures for castings with a small projected area. The metal is usually hand-ladled into the cylinder, but automatic ladles have been developed to deliver the correct amount directly into the cylinder.

In both the hot-chamber and cold-chamber machines, the metal may pass through the orifice and gates into the die as a spray. The metal quickly covers over the surface and fills in the voids. Thus all corners and shapes of the die are completely filled, giving remarkable detail. The surface of the part has no folds and reproduces the surface in the die. Of course, the usual troubles of trapped air and shrinkage defects can result from casting design, die design, variations in operation, and metal conditions. Generally, all die castings have trapped mold gases, but a solid-front fill through larger gates results in better-quality castings. Locking pressure is the limiting factor. If a casting had a total projected area of 400 in² including the runners and biscuit, then at 3,000 lb/in² injection pressure, $400(3,000) = 12 \times 10^5$ lb locking force would be required to contain the metal after the shot. Therefore an 800-ton machine would be needed to provide about a 30 percent safety factor.

Cold-chamber machines can be used to cast copper alloys because the die can receive metals at high temperatures and pressures. The shot cylinder is placed close to the die, thus cutting down the length of travel of the molten metal through gates and orifices. The injection cylinder may be horizontal or vertical (Figs. 10-23 and 10-24).

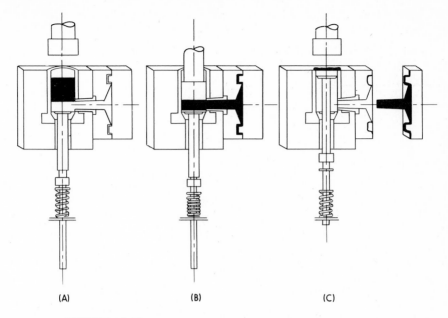

FIGURE 10-24
Cycle in a vertical cold-chamber die-casting machine. (*a*) Cold chamber filled with metal for injection. (*b*) Upper piston descends, lower piston is pushed down, and metal flows into die under pressure. (*c*) Ascending lower piston shears off excess metal, die opens, and casting is ejected.

Centrifugal Casting

For more than a quarter of a century, centrifugal-casting techniques have been used in routine production operations in this country. Up to the advent of World War II, centrifugal casting was employed mostly in the production of cast iron pipe in sand and metal molds. Necessity developed its use as a method of producing tubular steel shapes, which, until then, had been produced by rolling and welding.

Theory of centrifugal casting Figure 10-25 indicates the chilling action of sand castings as compared with that of centrifugal castings. The sketch at the

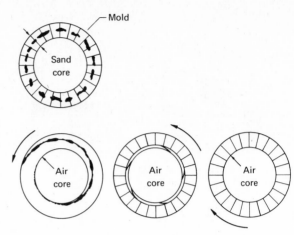

FIGURE 10-25
Sand-cast versus centrifugally-cast cylinder.

upper left shows the direction of grain flow during the cooling of a static or sand casting. Note that the chilling action originates at both the outside and inside surfaces and progresses to the center of the casting, thus developing an area of weakness in the center of the wall. This is caused by the meeting of the grain boundaries at final solidification and the entrapment of impurities in this central section.

The three lower diagrams indicate the progressive formation of the grain structure in a centrifugal casting. The sketch at the lower left illustrates the layering effect present immediately after pouring, with minute impurities distributed throughout the mass or casting.

The center sketch shows the casting in the state of partial solidification from the outside surface, inward. Note that the impurities are now concentrated in the still molten metal near the inner diameter.

The lower right sketch shows the casting in the state of complete solidification—dense, sound, and free of impurities, with no weakened sections caused by grain growth from outer and inner surfaces at the same time. All impurities, being lighter than the molten metal, have been forced to the inner diameter of the casting by centrifugal force and lie deposited in the inner bore, from which they may be removed by a rough machining operation if required.

Centrifugal casting can be broken down into three general types: centrifuged, semicentrifugal, and true centrifugal (Fig. 10-26).

True centrifugal casting True centrifugal casting involves not only feeding the metal by centrifugal force, but also using no core or riser. An example of this is the production of steel tubing by centrifugal methods.

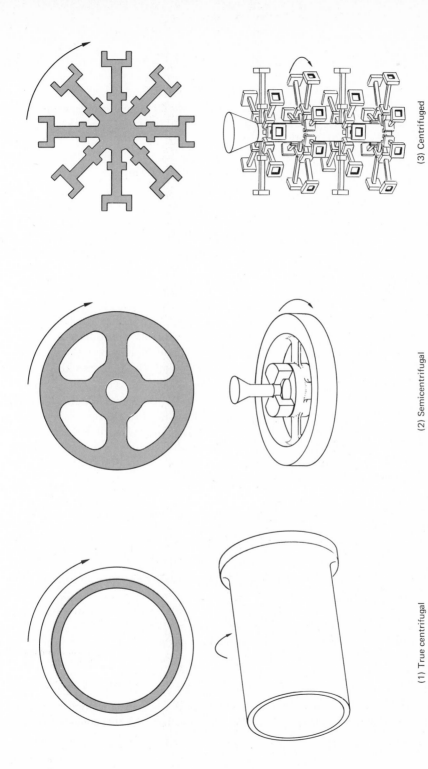

(1) True centrifugal

(2) Semicentrifugal

(3) Centrifuged

FIGURE 10-26
Three types of centrifugal castings. *(From Ekey and Winter, McGraw-Hill, New York, 1958.)*

A tubular flask with a baked- or green-sand lining is placed horizontally on a casting machine, where it is rotated by a system of rollers actuated by an electric motor. Molten metal is fed into one end of the mold cavity and is carried to the walls of the cavity by centrifugal force. Rotative speeds sufficient to produce a force of 75 g's are feasible. The wall thicknesses of the tubes are controlled by carefully weighing each charge of metal and knowing the volume, and hence the weight, of metal needed to produce a given wall thickness. Tolerances of $\frac{1}{64}$ in have been adhered to in this type of casting.

Semicentrifugal casting Semicentrifugal casting is a means for forming symmetrical shapes about the rotative axis, which is usually placed vertically for small parts. The molten metal is introduced through a gate, which is placed on the axis, and flows outward to the extremities of the mold cavity. During World War II, nozzles for aircraft turbo-superchargers were made by this method. Wheels are often cast by this method with the gate placed in the hub, but risers and cores are needed.

Centrifuged casting Centrifuging provides a means for obtaining greater pouring pressures in casting. In this process, several molds are located radially about a vertically arranged central riser or sprue, and the entire mold is rotated with the central sprue acting as the axis of rotation. The centrifugal force provides an acceleration, and consequently a pressure, at every point within the mold cavity. This pressure is directly proportional to the distance from the axis of rotation and the square of the rotative speed. This type of centrifugal casting is best used for small, intricate parts where feeding problems are encountered. This method can be used to advantage for the stack molding of six or more molds mounted one above the other (Fig. 10-26).

Applications and problems Centrifugally cast steel tubes in small diameters, and made of the ordinary carbon steels, are not competitive with welded or rolled tubes. Economy will result when thick-walled tubes are cast, and when mechanical properties are a major consideration and high-alloy grades of steel are required.

Iron and steel centrifugally cast tubes and cylinders are produced commercially with diameters ranging from $1\frac{1}{8}$ to 50 in, wall thicknesses of $\frac{1}{4}$ to 4 in, and lengths up to 50 ft. It is impractical to produce castings with the outside diameter to inside diameter ratio greater than about 4:1.

The as-cast tolerances for centrifugal castings are about the same as those for static castings. On cast gray-iron pipe, for example, tolerances of 0.06 in for 3-in diameter and 0.12-in for 48-in diameter are typical.

Centrifugally cast tubes can be produced in various external shapes, such as square, elliptical, hexagonal, or fluted, simply by constructing the proper pattern. In large quantities, production of these shapes will prove to be very economical.

A designer or other user may secure long-length, centrifugally cast steel tubes in a large range of diameters and of almost any analysis he may require. The foundry can produce centrifugally cast steel tubes from minimum heats of 2,000 lb. The designer may experience difficulty in securing the correct sizes and alloys from the high-production rolling mills, which only produce and stock standard sizes of the most popular alloys.

Centrifugally cast steel tubes are used as propeller shafts for naval and coast guard vessels. Hollow shafts, which are stronger than solid shafts of the same weight, have been used successfully and have started a new trend in propeller-shaft design.

In the De Lavaud method, gray and ductile iron is centrifugally cast in water-cooled metal molds to produce high-grade cast-iron pipe; 3- to 48-in diameter pipe in standard lengths up to 20 ft is easily cast. Sand cores are used to form the bell end of the pipe. The pipe is poured from a long-lipped ladle which moves along the axis of rotation at a given rate while the pipe spins. The pouring temperature and fluidity of the iron is very critical.

Several different problems have arisen in the production of centrifugally cast shapes, but solutions have been found. One of these problems was shrinking and cracking due to pouring directly into bare metal molds. The resultant chilling and contraction of the outer surface caused cracks and tears to appear. This was cured by spraying the mold cavity with a refractory coating which retarded the heat transfer to the mold and allowed a uniform outer wall to form. Heat transfer to the inner air space was also encountered. This caused a thin wall to form on the inside diameter, with the result that feeding of the outer section caused shrinkage within the casting itself. To overcome this it was necessary to minimize the heat transfer to the inner air pocket. This was accomplished by using an exothermic and insulating material of a lower specific gravity than the metal, which stopped the heat transfer and permitted the metal to solidify at 100 percent density.

One of the most interesting and significant applications of centrifugal casting is the production of dual metal tubes which combine the properties of two metals in one application. One example of this is the production of dual-metal cylinder liners for diesel engines with a mild steel outer shell for strength and gray iron on the inside for wear resistance.

In producing these dual-metal cylinders, the workman prepares each metal in a separate furnace. The metal that is to form the outer wall is poured first and allowed to freeze partially; then a flux is added; and finally the inside metal is poured. This process must be properly timed if the metals are to fuse properly at the line between the two metals. With proper timing, this bond is as strong as the metals themselves because all of the basic requirements of a weld are met. The following data are a summary of process-design considerations regarding centrifugal casting.

Dimensions handled   Length, up to 347 inches; diameters, 2 to 50 inches.

Tolerances $\frac{1}{64}$-in minimum—about the same as other casting techniques.
Surfaces Same as static casting processes.
Material handled Steel (carbon steel, alloy steel, stainless steel), copper and its alloys, aluminum alloys, aluminum and manganese bronzes, Monel, Everdur, and plain and alloy cast irons.
Economical quantities Foundry heats as low as 2,000 lb can be economically cast.
Supplementary operation Metal molds sprayed with refractory material.
Design factors Odd sizes and uncommon alloys can be economically used.
Process control Heat transfer to walls of metal molds and to center air space must be controlled.

Slush and Pressed Castings

Slush casting, a form of permanent-mold casting, is limited to some tin-, zinc-, or lead-base alloys. The process involves filling the metal molds with molten metal. Then, after a brief solidification period, the mold is turned over and the liquid metal in the center is allowed to run out. The casting is in effect a shell, the outside having the same appearance as the mold contour and the inside being an irregular cavity. The thickness of the casting can be controlled by the length of the solidifying period before dump out.

Another minor variation of permanent-mold castings is Corthias or pressed casting. In this procedure, a core is inserted into the partially filled permanent mold causing the fluid metal to fill the remaining space. Corthias castings are also shell-like but with a controlled inside cavity.

These types of castings find application in the production of such products as toys, ornaments, and lighting fixtures, where strength is not of prime importance and good appearance is an absolute necessity. Considerable quantities of slush and pressed castings are made in britannia metal, pewter, and zinc.

Continuous Castings

Savings in equipment, fuel, and processing time have attended the development of continuous casting for metals. The process is different from extrusion in that the raw materials are melted and transferred to equipment that transforms the molten material directly into semifinished mill products. This eliminates the expensive equipment of soaking pits, blooming mills, and reheating cycles before the finishing operations for mill products. The process eliminates the casting of ingots that go through a steel mill and reduces floor space and maintenance costs.

Material cast continuously has uniform grain structure and composition; little slag, impurities, or porosity; and is more suitable for future processing operations than noncontinuously cast material. On nonferrous castings, close

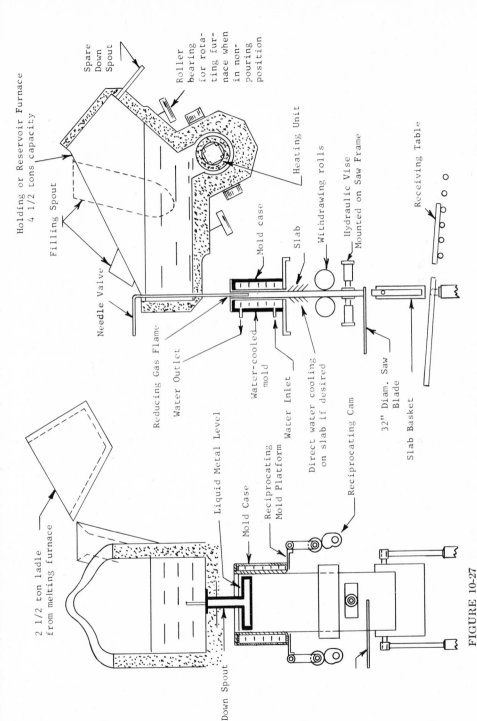

FIGURE 10-27
Sketch of a continuous-slab casting unit which illustrates particularly the twin-spout arrangement of the holding furnace.

tolerances can be met, and different shapes and sizes can be made with moderate changeover expense.

The process (Fig. 10-27) involves a source of molten alloy, which is prepared in batches, poured into a holding reservoir, kept at the proper temperature, and controlled to remove impurities and prevent contamination. From the reservoir, the metal is continuously fed through launders into the mold which has the ability to remove heat rapidly and solidify the molten metal. The casting exit is regulated by rollers, which travel at the proper speed and contact the cooled alloy. The alloy is water-cooled in the mold for only a short distance (on the order of 1 ft) because contraction prevents contact. The internal portion is still molten and, as the material contracts, the molten metal feeds in from the top. After passing through the die, the metal is cooled by direct-contact sprays or muffled, as the requirements dictate. The material is then cut to suitable lengths by saws or burning equipment that travel at the speed of the casting. These lengths are processed in mills for various applications. Alternatively, large-diameter casting may be made 12 or 20 ft long. As soon as the proper length is reached, pouring stops, the hot casting is removed, and another pour is begun.

Nonferrous equipment can make tubes and other shapes in a manner similar to the extrusion processes.

Steel is much more difficult to handle because of its high melting temperatures and the resulting problem of heat transfer at the mold. Mold sections are very thin and are made of steel, copper, and brass. Large quantities of water are forced around the mold to cool it and the molten material.

Since this is a basic process, the production-design engineer will not use this method in normal manufacture unless it is in connection with designing equipment for the continuous-casting processes. A description has been given in order to point out how ideas of long standing are being developed into practical processes by engineers.

Vacuum Casting of Metals

The need for metals and alloys that are free from contaminants has led to the development of furnaces for melting, reducing, and pouring metals under a vacuum at very low pressures. Typical vacuum-melting equipment operates at temperatures up to 3000°F and under 10 μm pressure. Undesirable gases are abstracted from the molten metal, and little contamination can take place in view of the controlled environment.

Vacuum casting has progressed from the small electric furnace in the laboratory to equipment that can pour the largest ingot made. In the typical vacuum-molding installation, an electric furnace melts the alloys (usually through induction heating) in a vacuum furnace. The ingot mold is poured within the vacuum furnace, and the vacuum is maintained until the ingot is completely solid.

Turbine spindles are produced today from vacuum-poured steel ingots. The steel ingots are forged to shape and the resulting forgings are more uniform and free from occlusions. Metals such as titanium, aluminum, and steel are affected by the oxygen, hydrogen, and nitrogen of the air, and titanium, for example, can be cast only in a vacuum in view of its great affinity for nitrogen and other elements.

DESIGN FOR CASTING

The present-day engineer is indeed fortunate, for, according to Dr. Flemings of MIT, "As the secrets of alloy solidification unfold, our most versatile metal working process comes of age. The fruits are vital to engineer and artist alike— castings of unparalleled quality and economy." Premium-quality castings are in routine production in aluminum foundries and the development of ferrous die casting may herald cast steels of untold strength and economy. But major breakthroughs in die materials, die coatings, and metal input are still needed to make the process a competitive success.

Yet to be effective and economic, a cast component must be especially designed for the casting process or converted to a casting design. In several cases quality products made by other processes have been successfully converted to the casting process. Some said that it couldn't be done! But engineers from General Motors did. Cast crankshafts for automobiles and trucks are here to stay as are steering knuckles and universal joints. In the late 1940s no engineer would consider casting such parts even with a 4:1 factor of safety. Today engineers often think of castings first, even for dynamically stressed parts. Careful engineering design and testing are required for successful conversion to the casting process.

Designers must realize that cost estimates are based on the normal production of good castings. If a casting design leads to greater than normal scrap rates, for example, the original estimate will be invalid. To avoid that pitfall, the engineer must be aware of casting-theory design and practice.

Since all castings begin with patterns and cores, we should logically look at them first.

Patterns

A pattern forms a mold cavity such that after the cast metal reaches room temperature the product will be the expected size. Shrinkage allowances (Tables 10-8 and 10-9) are at best approximations; the size and shape of a casting and the mold wall's resistance to deformation by the liquid metal affect the shrinkage

Table 10-8 SHRINKAGE, MACHINING, OUTSIDE-DIMENSION ALLOWANCE, MINIMUM SECTION THICKNESS, AND TYPICAL TOLERANCES FOR SAND-CAST FERROUS ALLOYS

Pattern dimension, in	Shrinkage, in		Machining allowance,* in			Minimum section size, in	Typical tolerance, in
	Solid	Cored	Bore	Outside dimension	Cope side		
Gray iron							
Up to 6	$\frac{1}{8}$	$\frac{1}{8}$	$\frac{1}{8}$	$\frac{3}{32}$	$\frac{3}{16}$	$\frac{1}{8}$	$\pm\frac{1}{32}$
6–12	$\frac{1}{8}$	$\frac{1}{8}$	$\frac{1}{8}$	$\frac{1}{8}$	$\frac{1}{4}$	...	$\pm\frac{1}{16}$
13–24	$\frac{1}{8}$	$\frac{1}{8}$	$\frac{3}{16}$	$\frac{5}{32}$	$\frac{1}{4}$		
25–36	$\frac{1}{10}$	$\frac{1}{10}$	$\frac{1}{4}$	$\frac{3}{16}$	$\frac{1}{4}$		
37–48	$\frac{1}{10}$	$\frac{1}{12}$	$\frac{5}{16}$	$\frac{1}{4}$	$\frac{5}{16}$		
49–60	$\frac{1}{12}$	$\frac{1}{12}$	$\frac{5}{16}$	$\frac{1}{4}$	$\frac{5}{16}$		
61–80	$\frac{1}{12}$	$\frac{1}{12}$	$\frac{3}{8}$	$\frac{5}{16}$	$\frac{3}{8}$		
81–120	$\frac{1}{12}$	$\frac{1}{12}$	$\frac{7}{16}$	$\frac{3}{8}$	$\frac{7}{16}$		
Cast steel							
Up to 1	...	...	Cast solid	...	...	$\frac{3}{16}$	...
Up to 6	$\frac{1}{4}$	$\frac{1}{4}$	$\frac{1}{4}$	$\frac{1}{8}$	$\frac{1}{4}$	...	$\frac{1}{4}$
6–12	$\frac{1}{4}$	$\frac{1}{4}$	$\frac{1}{4}$	$\frac{3}{16}$	$\frac{1}{4}$	...	$\frac{1}{4}$
13–18	$\frac{1}{4}$	$\frac{1}{4}$	$\frac{9}{32}$	$\frac{1}{4}$	$\frac{5}{16}$	...	$\frac{1}{4}$
19–24	$\frac{1}{4}$	$\frac{3}{16}$	$\frac{9}{32}$	$\frac{5}{16}$	$\frac{3}{8}$	...	$\frac{5}{16}$
25–48	$\frac{3}{16}$	$\frac{3}{16}$	$\frac{5}{16}$	$\frac{3}{8}$	$\frac{1}{2}$	...	$\frac{3}{8}$
49–66	$\frac{3}{16}$	$\frac{5}{32}$	$\frac{3}{8}$	$\frac{3}{8}$	$\frac{1}{2}$	...	$\frac{1}{2}$
67–72	$\frac{3}{16}$	$\frac{1}{8}$	$\frac{1}{2}$	$\frac{7}{16}$	$\frac{9}{16}$	...	$\frac{5}{8}$
Over 72	$\frac{5}{32}$	$\frac{1}{8}$	$\frac{5}{8}$	$\frac{1}{2}$	$\frac{5}{8}$	...	$\frac{3}{4}$

Ductile iron

Section thickness, in	Shrinkage,† in
Up to 24	$\frac{1}{10}$ $\frac{1}{8}$
	$\frac{1}{10}$
	$\frac{3}{16}$
	$\frac{5}{32}$
	$\frac{1}{4}$

Malleable iron

Section thickness, in	Shrinkage,† in	Section thickness, in	Shrinkage,† in	Minimum sec. size, in	Typical tol., in
$\frac{1}{16}$	$\frac{11}{64}$	$\frac{1}{2}$	$\frac{7}{64}$	$\frac{1}{16}$	Up to 5 $\pm\ \frac{1}{32}$
$\frac{1}{8}$	$\frac{3}{32}$	$\frac{5}{8}$	$\frac{3}{32}$	...	5–8 $\pm\ \frac{3}{64}$
$\frac{3}{16}$	$\frac{19}{128}$	$\frac{3}{4}$	$\frac{5}{64}$	...	9–12 $\pm\ \frac{1}{16}$
$\frac{1}{4}$	$\frac{9}{64}$	$\frac{7}{8}$	$\frac{3}{64}$	...	13–24 $\pm\ \frac{1}{8}$
$\frac{3}{8}$	$\frac{1}{8}$	1	$\frac{1}{32}$		

* Allowance on bore given for radius.
† No core.

Table 10-9 SHRINKAGE AND MACHINING ALLOWANCES FOR NONFERROUS CASTINGS POURED IN SAND MOLDS

	Pattern dimensions, in	Section thickness, in	Shrinkage allowances, in	Machining allowances, in	Min. section size, in	Typical tolerance, in
Aluminum	Up to 24	$\frac{5}{32}$	$\frac{5}{32}$	$\frac{3}{32}$	$\frac{3}{16}$	$\pm\frac{1}{32}$
	25–48	$\frac{5}{32}$	$\frac{9}{64} - \frac{1}{8}$	$\frac{1}{8}$		
	49–72	$\frac{9}{64}$	$\frac{1}{8} - \frac{3}{16}$	$\frac{1}{8}$		
	Over 72	$\frac{1}{8}$	$\frac{1}{8} - \frac{3}{16}$			
Magnesium	Up to 24	$\frac{11}{32}$	$\frac{5}{32}$	$\frac{3}{32}$	$\frac{5}{32}$	$\pm\frac{1}{32}$
	25–48	$\frac{11}{32}$	$\frac{5}{32} - \frac{3}{16}$	$\frac{1}{8}$	...	$\pm\frac{1}{32}$
	Over 48	$\frac{5}{32}$	$\frac{5}{32} - \frac{3}{16}$	$\frac{1}{8}$	...	$\pm\frac{1}{32}$
Admiralty metal	Up to 24	...	$\frac{1}{8}$	$\frac{1}{4} - \frac{3}{8}$		
Copper	....	...	$\frac{3}{16} - \frac{7}{32}$			
Brass	....	...	$\frac{3}{16}$	...	$\frac{3}{32}$	$\pm\frac{3}{32}$
Bronze	....	...	$\frac{1}{8} - \frac{1}{4}$	...	$\frac{3}{32}$	$\pm\frac{3}{32}$
Beryllium-copper	....	...	$\frac{1}{8}$	...	...	$\pm\frac{1}{16}$
Everdur	....	...	$\frac{3}{16} - \frac{1}{4}$			
Harteloy	....	...	$\frac{1}{4}$			
Nickel and nickel alloys	....	...	$\frac{1}{4}$			

allowance significantly. For example, a round steel bar shrank $\frac{9}{32}$ in/ft; with a large knob on each end, it shrank $\frac{3}{16}$ in/ft; when the knobs were replaced by flanges, it shrank $\frac{7}{64}$ in/ft. Thus, if tolerances are important, an engineer must work closely with qualified foundrymen. In addition, the geometry of a casting may cause it to shrink differently from the allowances in the tables. For example, in a large bull gear the shrinkage across the diameter will be much different from the shrinkage across the rim.

Material for patterns must resist the moisture and abrasion of green sand. Which material is chosen depends on the quantity of castings to be made and on the casting process to be used. The most common pattern materials are wood and metals, but epoxy resins and plaster are used on occasion.

Softwood patterns (white pine) are suitable for short runs (1 to 50 pieces) of medium castings (up to 6 ft long) but they wear rapidly; hardwoods (mahogany, cherry, birch) are used for runs of 50 to 200 pieces. At the latter production requirement, gates should be attached to the pattern; flasks and core boxes should have metal wear strips. In some cases, additional equipment may be needed (e.g., core setting and pasting fixtures, drier patterns or core driers). At still higher production requirements (200 to 5,000 pieces), patterns and accessories must be metal with hardened-steel wear plates, and core boxes should be constructed with venting devices to permit blowing.

To decide what pattern material to use, make an engineering cost estimate considering the average annual investment in tooling, the annual maintenance cost, the production usage, and the risk of obsolescence. Wood is inexpensive and easily worked; therefore, it is commonly used for low-production needs and large patterns. But even when painted, humidity causes wood patterns to vary in size and shape, and they are easily damaged on the molding floor or on their way to and from storage. Metal patterns are more accurate and durable. They are used for all match plates (aluminum for manual operations and gray iron or brass for machine-lifted plates).

The factors to consider in designing a pattern are:

1 Have gating and risering been provided?
2 Can the pattern be removed from the mold?
3 Has the minimum castable section thickness been considered?
4 Is there proper draft to permit the removal of the pattern?
5 Will draft be permissible on the final casting?
6 Can cores be anchored?
7 Can loose pieces, multiple partings, and irregular partings be eliminated?
8 Is size of pattern adjusted to allow for shrinkage of casting and expansion of mold?
9 Has machining allowance been provided?
10 Has a warping or distortion allowance been provided?
11 Has the foundry been consulted?

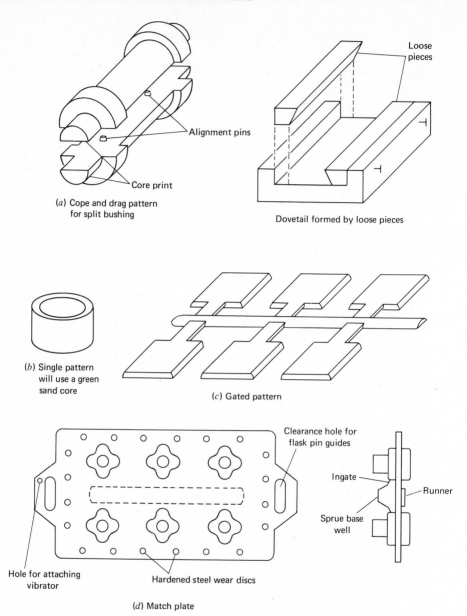

(a) Cope and drag pattern
   for split bushing

Dovetail formed by loose pieces

Loose
pieces

Alignment pins

Core print

(b) Single pattern
    will use a green
    sand core

(c) Gated pattern

Clearance hole for
flask pin guides

Ingate

Runner

Sprue base
well

Hole for attaching
vibrator

Hardened steel wear discs

(d) Match plate

FIGURE 10-28
Typical small pattern equipment.

Classification of patterns *Single patterns* Single patterns are usually solid with no gates attached. They are single copies of the casting to be produced with adequate allowances for making the casting. A skilled molder is required to gate them properly and determine a parting line (Fig. 10-28b). They are used mostly in casting limited quantities of a part.

Gated patterns Gated patterns are one or more single patterns fastened to a gate (Fig. 10-28c). They require the same skill as the single pattern in providing a parting line; however, the hand cutting of the gates is eliminated. They are used when the production requirements are medium.

Match-plate patterns Match-plate patterns are mounted on a board or plate with gates and runners provided (Fig. 10-28d). The pattern may be split along the parting line formed by the board. The cope pattern is on one side and the drag pattern is on the opposite side. Match-plate patterns are economical only for quantity production, handled by unskilled operators on rollover, jolt, and squeeze-type molding equipment. In very high production or for large castings there may be two plates—one for the cope and one for the drag—which are made at two separate places on the production line.

Cope and drag patterns Cope and drag patterns are of great variety (Fig. 10-28a). They are made so they will split and can be drawn from the cope and drag. Most patterns are of this type and are used for the largest castings. At times the cope or drag may be separated into one or more parts and the patterns are divided accordingly.

Temporary and permanent patterns Patterns or core boxes are called "temporary" or "permanent," depending upon the material used in their construction. Temporary patterns, of soft wood, are easily made. They soon wear, warp, or crack and have a short life. Hardwood patterns and core boxes are used more than any other type of pattern. The portions that wear may be protected by sheet metal. Permanent patterns are made of metal—usually aluminum or brass —or plastics which are easily cast and machined.

Pattern colors In 1958, a tentative pattern color code for new patterns was adopted by the Pattern Division of the American Foundrymen's Society:

1 Unfinished casting surfaces, the face of core boxes and pattern or core-box parting faces: clear coating.
2 Machined surfaces: red.
3 Seats of and for loose pieces: aluminum.

4 Core prints: black. In the case of split patterns and where cores are used, paint the core area black.

5 Stop-offs:[1] green.

The clear coating on most surfaces will disclose the quality of material and workmanship of the pattern. Likewise important construction details and layout or centerlines used by the patternmaker will not be obliterated and will thus be available for use during repair or modification of the equipment.

There are numerous types of pattern-coating materials on the market. Nitrocellulose lacquers cure rapidly but have relatively poor moisture resistance. Present-day modified lacquers have much better moisture resistance. Shellac modified with various synthetic resins is superior to pure shellac. Wood patterns coated with synthetic resin (plastic) have been used under the most adverse conditions, such as being rammed up with hot sand or left in the mold for many hours, without noticeable damage. Plastic coatings require considerable time and care for application, but prolong the pattern life up to several times. In addition, there is less tendency for the sand to cling to the pattern surface, and so such patterns draw more easily.

Draft Allowance

Once a mold has been rammed around a pattern, the latter must be removed from the mold cavity. This is aided by rapping or vibrating the pattern and by a taper called *draft* on all vertical surfaces of the pattern. Usually draft will vary between $\frac{1}{16}$ to $\frac{1}{8}$ in/ft (1 to $1\frac{1}{2}°$). Inside surfaces such as cored holes require greater draft than outside surfaces, and manual molding equipment needs greater draft than that required if mechanical drawing equipment is used.

Shrinkage and Machining Allowance

Solidification shrinkage must be compensated by the riser, but shrinkage which occurs during the cooling of a casting from the solidification temperature to room temperature is compensated for by making the pattern oversize. The allowances given in Tables 10-8 and 10-9 are general averages. Closer dimensional tolerances can be achieved in green-sand molding provided that there is sufficient production to warrant pattern modification and that a controlled foundry practice is followed. Machining allowances also can be reduced if, for instance, high-pressure molding is used with good sand control or if shell molding is used in the foundry production. For instance automotive foundries consistently use

---

[1] Stop-offs are portions of a pattern that form a cavity in a mold which is refilled with sand before pouring. This might be desirable to prevent breakage of a frail pattern member. The stop-off may be filled by a core later. This procedure may simplify the location of the pattern parting line by making it unnecessary to carry it out of a plane position (to provide for lugs or certain cored holes).

smaller-allowance figures than those published here. This is where a detailed knowledge of the practice of a particular plant is important because the saving in machining can be significant.

Venting of Molds

Molds have other functions besides forming the cast material. They provide passages or gates through which the metal flows into the casting cavity. In steel castings the mold must be designed with parting line, gates, and vents, so that the molten metal can enter at the bottom, without turbulence and without creating hot spots in the mold by impinging on certain points. In sand molds aluminum is directed through traps or strainer_cores or glass-fiber screens that help eliminate sand and other foreign material that may have been picked up by the molten metal.

The molten metal generates steam and gases as it comes in contact with the molding material. These gases must escape and cannot be confined in pockets. In synthetic sand molds there is usually sufficient permeability in the rammed mold to vent any trapped gases. However, if the mold has been compacted by a high-pressure molding machine, it may require the drilling of additional vents. Vents should be so arranged that gas in the mold, as it is being pushed ahead of the gradually rising level of molten metal, is not trapped in crevices or other branches not directly in line with risers or overflows. If these gases did not escape, the casting would be defective and the mold might explode. The molds are vented by small holes through the mold material. Definite vent passages are provided in metal molds and, when required, in large sand molds. Overflow of excess material is provided for by passages to pockets.

Cores

When it is necessary to leave an opening in a casting that cannot be made by the external mold, a core made to the shape of the opening is provided and located in position by anchoring it to the external mold or supporting it by chaplets. Chaplets are fusible metal supports placed between the cores and the mold wall for the purpose of separation. They melt and are absorbed by the liquid metal. If the chaplet is not absorbed, it may cause porosity.

Cores can be made by any of the sand processes described under sand molding, i.e., the traditional oil-bonded cores (usually a linseed-oil derivative), the carbon dioxide, synthetic resin, or inorganic bonded ceramic cores. Die casting and permanent mold-casting dies usually use metallic cores of H13 die steel or a molybdenum alloy such as TZM. Metal cores are actually a part of the die and will not be discussed further.

Sand cores are usually made from specially bonded and cured silica sands rammed into a core box. Although hand-ramming methods are used, blowing of the bonded mix is by far the most common production method. Curing may be by baking, but that is fast losing out to curing in the core box by gassing or by stripping from the box in a semicured condition and allowing the core to set through the use of catalysts.

Components of cores may be made separately and then pasted together to make complicated cores. The cores used in an automobile engine block for water jackets and gas passages are an example of intricate coring. The cores for a railroad airbrake cylinder and valve are complex and are held to very close tolerances. Each individual core part is molded in a separate core box, baked, put in place, and held there by the core paste.

Cores are used externally to form flat and perpendicular surfaces that will not permit draft, on vertical surfaces to form undercuts or bosses that cannot be drawn, to form surfaces in pit molding where there is no cope to mold the top part of the casting, and to prevent mold erosion at critical points. An example of a completely cored part is the crankshaft casting of an automobile, which is made of pancakelike layers stacked one on top of the other and keyed in place.

Metals are heavy in comparison with nonmetallic mold materials; if they flow too rapidly, they will wash the molds and pick up foreign material. The heavy liquid can float the cores, as well as the cope, as the metal enters the mold. This causes shifts in cores and distortion of the mold. Weights on the molds and clamps are used to counteract this liquid pressure. A feeder 1 ft high for a typical ferrous material will exert 4 lb/in² on a core. If the core has a face area of 5×8 in, the force necessary to shift it out of place is 160 lb. The heavy, hot fluid material must be guided to its place without damage to the mold or contamination of the material. This mass of hot molten or mushy material has no strength at temperatures above and near the melting point. Therefore, the mold and cores must support the material until it cools to the point where it is strong enough to carry its own weight.

Cores are made on automatic machines, on production lines, where a combination of machine and hand operations produce them in great quantities, and at benchwork stations. In considering cores, it must be remembered that the labor for making a core is almost equivalent to the labor required to make a casting of the same size. They must be formed in a core box, baked or cured (unless a green-sand core is used), placed in the mold, and removed after the casting is made. The removal of a core should be provided for in the design of the part. A green-sand core, formed in a similar manner but not baked is less expensive. Green-sand cores are often not possible, however, due to low strength.

Core sand bonds disintegrate under heat, and thus the sand can be easily removed after the casting is made.

Cores are often anchored solidly by core prints in the mold so they will not wash out. The correct location of the core depends upon the accurate size of the cores and molds and the care taken by the molder in placing the cores. At times the molder is supplied with measuring gages to check that the core location is correct.

Cores are expensive and should be kept to the minimum when designing parts. Holes that can be made by a separate core can sometimes be drilled at a cost less than that of the core.

Core-making equipment Core boxes contain cavities to form sand cores to the desired shape. Patternmakers construct core boxes from wood, metal, synthetic resin, or other suitable materials. Generally, the core-box material would be the same as that chosen for the pattern. There are several types of core boxes and accessories:

1 Two- or three-piece split core boxes (hand- or machine-rammed)
2 Dump boxes for making a half core (later the two halves are pasted together)
3 Blow boxes for the high production of relatively small cores (a few ounces to 400 lb)
4 Multiple piece, loose piece, or special-purpose core boxes
5 Auxiliary equipment such as core driers and pasting jigs
6 Shell core boxes, equipped with integral heating devices
7 Hot-box core equipment for curing furan resin-bonded sand

The design and construction of most core boxes and patterns are often left largely to the discretion of the patternmaker. But a tool engineer should understand the principles of core-box design, vent location, vent areas, and the general break-even point at which it is better to move from wooden core boxes to metal core boxes or from hand ramming to core blowing.

Core blowing The core blower rapidly produces small- and medium-sized cores. It clamps the core box shut, seals the sand reservoir tube to the box, then fills and rams the core by the kinetic energy of a sand-laden air stream. Filling can only be achieved if the sand is free to enter the box through the blow-holes and to exit from the box cavity through suitable vents which will impound the sand (Fig. 10-29). Proper orientation of the blow and vent holes will promote uniform filling and ramming of the core in less than 2 s—even for the largest core boxes.

Core blowers require large volumes of compressed air—up to 12 to 30 ft³/min of free (14.7 lb/in²) air per cycle. Consequently high-capacity air compressors are required for large core-blower installations. Supply piping must also be of sufficient diameter to permit large flow (2 to 4 in diameter).

Suitable air cleaners such as centrifugal filters and oil traps must be provided to eliminate oil and water from the air stream. Also, adequate clamping pressure must be developed to assure tight sealing of the box sections in both horizontal and vertical planes. All joined edges must be machined square and parallel to assure correct seating and to eliminate blow-by. Core boxes are usually aluminum with added wear plates of steel at points of high attrition.

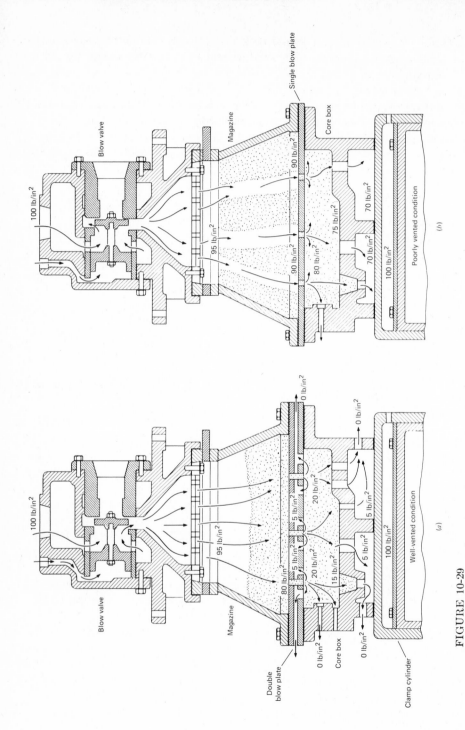

FIGURE 10-29
Pressure distribution in a core-box blowing system. (Good venting versus poor.)

Blowholes   A steel blow plate fits the bottom of the sand reservoir and provides holes from $\frac{3}{16}$ to $\frac{1}{2}$ in diameter to direct the sand to the proper location within the core box (Fig. 10-29). The number and size of the blowholes are at present largely a matter of experimentation. An insufficient number of blowholes prevents the box from filling completely and promotes channeling. On the other hand, too many holes cause the box to fill before the remote corners are rammed sufficiently hard.

The blow plates and sand reservoirs should be designed to be interchangeable for as many core boxes as possible to obtain maximum returns from the original expense and to minimize storage space.

Core-box venting   All the air which enters the core box through the blowholes must be vented. The air enters the core box, expands, moves at a lower velocity, deposits the entrained sand, and flows out through the vents. The impact of one grain upon another rams the core.

Venting must be designed in proportion to the blowhole area to achieve adequate ramming. Less venting is required for more flowable sands because a smaller volume of ramming air is required to ram cores to suitable strength.

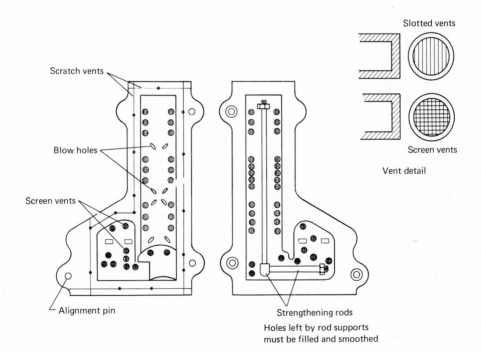

FIGURE 10-30
Vents for core boxes.

Conversely, more vent area is required for core sands with high green properties because larger volumes of air are needed for equivalent rammed properties.

Venting is obtained by using vent screens or slotted vents (Fig. 10-30). If screens are used, the venting area should be about twice the blowhole area; slotted vent plugs require greater area. Vent plugs do not leave surface imperfections on a core as screens do because the former can be contoured to the cavity. If surface finish is important either vent plugs must be used or screens should be placed on core prints and at the other spots where blemishes are not critical.

Venting of core gases Cores must be permeable to permit the hot gases which are generated by the burning of bonding materials to escape to the atmosphere. Such vents are usually aided by placing wires or wax rods in the core box prior to blowing the core. Subsequently the wires are removed or the wax melts during the core baking. If possible, the vents should pass through the locating prints to permit gases to escape through the back of the core prints.

Core support Larger cores require internal reinforcement with embedded wires or rods to prevent breakage caused by handling or premature sagging in the mold after the casting is poured.

Irregularly contoured cores require special core driers for support during the curing process. In high production the drier may be the lower half of the core box itself. Such a practice would require a considerable investment in tooling which would have to be balanced by a cost saving in core production and scrap loss.

Inserts

Inserts are separate parts made of a metal that is generally different from the metal of the casting, which are "cast in" to provide locally some special properties such as hardness, wear resistance, strength, bearing qualities, electrical characteristics, corrosion resistance, resilience, or special decoration not obtainable from the cast metal. Typical inserts are the heating units cast in aluminum flatirons and hot plates, tubing for the passage of a liquid, bushings, bearings, anchorage for soldering connections, and studs. Inserts are knurled, grooved, or surfaced in such a manner that the material, in freezing around the insert, grips it firmly so that it cannot turn or pull out in tension.

Attention must be paid to the danger of electrolytic or galvanic corrosion when the assembly is exposed to any kind of humidity. The corrosion is caused by the galvanic potential difference between the base metal of the casting and the dissimilar metal of the insert. Aluminum- or magnesium-base alloys in intimate contact with copper-base, tin-base, lead-base, or nickel-base alloys, or steel and iron, are apt to corrode when covered by moisture, especially when the area of the aluminum- or magnesium-base casting surrounding the insert is smaller than that of the inserted metal. The same holds true for aluminum joined to magnesium. When joints of dissimilar metals are exposed to moisture, it is necessary to protect the joints against the entry of moisture by painting, dipping, or plating.

Furthermore, in the design of inserts it is advisable to avoid sharp corners, projections surrounded by thin sections of material, or other factors that might lead to stress concentration with its injurious effect on the mechanical properties of the casting, particularly in fatigue and shock. Yet inserts should have a knurled or grooved surface at the casting-insert interface if a tight mechanical fit is desired.

When inserts are used, the cost of the casting is usually increased because of the cost of the insert and the cost involved in placing it in the mold. Also, allowances must be made for damaged inserts if they are not reclaimable; however, inserts may reduce the overall cost of the product.

Gating Systems for Sand Molds

Design   The gating system for a casting is a series of channels which lead molten metal into the mold cavity. It may include any or all of the following: pouring basin, sprue, sprue base, runners, and ingates. A well-designed gating system should:

1   Minimize turbulence within the molten metal as it flows through the gating system. The use of tapered sprues and proper streamlining will reduce excessive erosion and gas entrainment.

2   Reduce the velocity of the molten metal in order to attain minimum turbulence.

3   Deliver the molten metal at the best location to achieve proper directional solidification and optimum feeding of shrinkage cavities.

4   Provide a built-in metering device to permit uniform, standardized pouring times regardless of variations in pouring techniques.

Figure 10-31 is a typical nonpressurized gating system for an aluminum alloy. Note that each gate has a riser.

Turbulence within a gating system   Extensive research has shown that molten metal and water flow similarly and that gating systems can be designed using the principles of fluid mechanics.

Several limitations are apparent. The high density of metals (up to 10 times that of water) makes it difficult to force molten alloys to turn a corner as from a runner to an ingate. But once Newton's law of inertia is applied to flowing metal, proper gating systems can be readily designed. The density of a metal does not affect its flow characteristics because the rate and nature of fluid flow

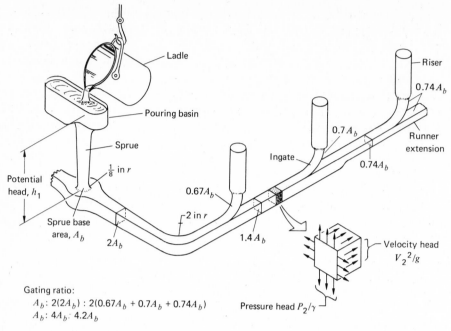

FIGURE 10-31

Design of a nonpressured gating system using a 1:4:4 gating ratio.

depend on the inertia of the fluid and the forces applied to it. Both factors depend on density in the same way, so it has no effect on the fluid flow. But impact depends upon density alone; hence mold erosion increases directly with density.

Although water has a surface tension approximately one-tenth that of molten metals, recent experiments have shown that Wood's metal and mercury have nearly equal ability to entrain air. However, the greater surface tension and the natural oxide coating which envelopes a stream of molten metal seem to permit it to flow in a nondisruptive fashion at a greater velocity than that suitable for water in a given channel.

Velocities within a gating system   The flow of molten metal in a gating system is a function of a number of other variables. Bernoulli's theorem states that the sum of the potential, pressure, kinetic, and friction energies at any point in a flowing liquid is a constant, or:

$$Wh_1 + \frac{WP_1}{\gamma} + \frac{WV_1^2}{2g} + WF_1 = Wh_2 + \frac{WP_2}{\gamma} + \frac{WV_2^2}{2g} + WF_2 \qquad (1)$$

Dividing by W the total weight of liquid flow per unit time, one obtains the usual form:

$$h_1 + \frac{P_1}{\gamma} + \frac{V_1^2}{2g} + F_1 = h_2 + \frac{P_2}{\gamma} + \frac{V_2^2}{2g} + F_2 \qquad (2)$$

where h_1, h_2 = respective heads at stations 1 and 2, in
P_1, P_2 = respective pressures on liquid, lb/in²
V_1, V_2 = respective liquid velocities, in/s
γ = specific weight of liquid, lb/in³
g = gravitational constant on earth, 386 in/s²
F_1, F_2 = respective head losses from friction, in

The velocity of the molten metal at any point in a gating system can be evaluated by the use of Bernoulli's theorem. Proper streamlining will permit a significant increase in the flow rate of a gating system (Fig. 10-32).

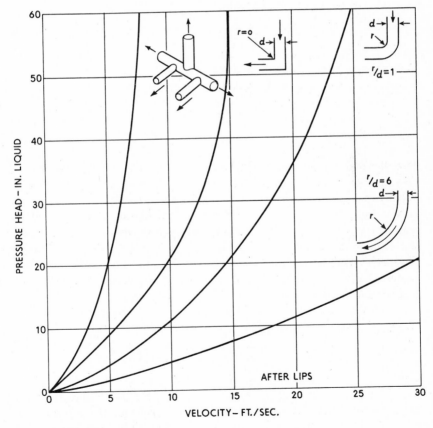

FIGURE 10-32
Effect of streamlining gating on velocity.

Effect of streamlining a gating system   Round-cross-section gates are more efficient than those of any other shape because they have the smallest surface-area-to-volume ratio and, consequently, can pass a greater volume of metal with the least heat loss. The gating system should be streamlined and of correct magnitude so as to control the velocity of the flowing metal. Too high a velocity will cause disruptive turbulent flow, resulting in sand inclusions and erosion of the mold cavity wall. Streamlining can increase effectively the volumetric capacity of a gating system and thereby allow smaller-size gates and runners which will consequently increase effective melt utilization. The effect of stream-lining on metal velocity is shown in Fig. 10-32. A method for improving stream-lining at the base of a T section from the sprue to the runner is shown in Fig. 10-31.

The law of continuity   A second fundamental relationship in fluid flow is the *law of continuity*, which states that the flow rate of a fluid is a constant at any point in a continuous stream:

$$q = A_1 v_1 = A_2 v_2 \quad \text{or} \quad Q = Avt \tag{3}$$

where q = flow rate, in^3/s
$\quad Q$ = volume of flow in a given time, in^3
A_1, A_2 = respective cross-sectional areas of the flow channel at points 1 and 2, in^2
$\quad v_1, v_2$ = respective velocities of flow at points 1 and 2, in/s
$\quad t$ = time, s

The vertical elements of a gating system   The law of continuity requires that the same quantity (flow rate) of material must exist at all points in a flow-ing stream. In the vertical part of a gating system (sprue), the acceleration of gravity increases the velocity of flow. If a straight-sided sprue is used, the cross-sectional area of the flowing stream at the sprue base will be less than that of the sprue. Consequently, air will be aspirated from the surrounding mold until the sprue volume is completely filled. However, if a tapered sprue is designed to conform to the dimensions of the descending stream, such a condition can no longer exist and the metal quality will improve.

If we neglect friction and take a horizontal plane through the ingate as a reference, Bernoulli's equation becomes

$$h_t + \frac{P_t}{\gamma} + \frac{v_t^2}{2g} = h_b + \frac{P_b}{\gamma} + \frac{v_b^2}{2g} \tag{4}$$

where t refers to the top and b to the base. Then h_b = zero because it is on the reference plane, $v_t = 0$ because there is no velocity at the top, and P_t/γ and $P_b/\gamma = 14.7$ lb/in^2 because the system is at atmospheric pressure at both ends.

Then we have

$$h_t = \frac{v_b^2}{2g} \quad \text{or} \quad v_b = \sqrt{2gh_t} = 27.8 \sqrt{h_t} \tag{5}$$

To design a sprue of suitable proportions let the area at the top of the sprue be A_t and the velocity there be v_t at a flow rate q_t; then $q_t = A_t v_t$ from continuity. Similarly, at the sprue base $q_b = A_b v_b = q_t = A_t v_t$. Then $A_t = A_b(v_b/v_t)$. From Bernoulli's equation, $v_t = \sqrt{2gh_t}$ and $v_b = \sqrt{2gh_b}$, where h_t and h_b are the heads, in inches of metal, at the top and bottom of the sprue. Then

$$A_t = A_b \frac{\sqrt{2gh_b}}{\sqrt{2gh_t}} \quad \text{or} \quad A_t = A_b \sqrt{\frac{h_b}{h_t}} \tag{6}$$

Thus once the area of the sprue base is known, the vertical gating system can be designed. Equation (6) indicates that the sprue should have parabolic sides but experience shows that a straight-sided sprue having the calculated diameters at the top and bottom is satisfactory (when solved for a series of heights).

When the height h_t of the pouring basin above the sprue base is known, if a tapered sprue is used, and if the pouring basin is kept full throughout the pour, Bernoulli's equation can give an approximate answer, but much of the flow is transient and the cavity must be in the drag and nonpressurized.

To find the diameter of the sprue base, it is convenient to use the result from previous research in which it was found that for an unpressurized system, poured with aluminum alloy, an average of 5.75 lb of alloy per min passed through each square inch of the sprue area. This is equivalent to 60 in^3/(min)/(in^2) of cross-sectional area.

In the case where $h_t = 9$, the following calculation may be made:

$$A_b = \frac{60 \text{ in}^3/(\text{min})(\text{in}^2)}{27.8 \sqrt{h_t}} = 0.72 \text{ in}^2$$

or

$$A_b = \pi r^2 = 0.72 \text{ in}^2 \tag{7}$$

$$\therefore r = \sqrt{\frac{0.72}{\pi}} = \sqrt{0.229} = 0.479$$

$$\therefore d = 0.95 \text{ in}$$

If a bottom gate is used, then the filling time is longer because the metal is subject to a decreasing head during filling. Therefore in an increment of time dt, the height will increase dh and the volume of the metal will increase by the area of the mold $A_m \, dh$, while the flow through the ingate in time dt will be $A_g v \, dt$ and the velocity of the metal through the gate will be $\sqrt{2g(h_t - h)}$ at any instant. If we equate the increase in casting volume in time dt to the flow through the ingate in that same time interval, we find

$$A_m \, dh = A_g \sqrt{2g(h_t - h)} \, dt \quad \text{or} \quad \frac{A_g}{A_m} \, dt = \frac{dh}{\sqrt{2g(h_t - h)}} \quad (8)$$

Let t_p be the time to fill the mold and h_m be the height of the mold cavity. Then

$$\frac{1}{2g} \int_0^{h_m} \frac{dh}{\sqrt{h_t - h}} = \frac{A_g}{A_m} \int_0^{t_p} dt \quad (9)$$

$$\therefore t_p = \frac{2A_m}{A_g \sqrt{2g}} \left(\sqrt{h_t} - \sqrt{h_t - h_m} \right) \quad (10)$$

It is found that bottom gating takes twice the pouring time of a top gating system, obviously. If parting-line gating is used, the calculation is made in two parts: (1) top gating until the drag is filled, and (2) bottom until the mold is filled. If a top riser is used, a third calculation is required. Of course we have considered the simple case of a mold of constant cross section without a core. Appropriate corrections must be made for the more complex shapes.

Ratio gating   There are two types of gating systems: *nonpressurized*, or free flowing like a sewer system (Fig. 10-31), and *pressurized*. The latter has less total cross-sectional area at the ingates to the mold cavity than at the sprue base. The gating ratio relates the cross-sectional areas of each component of the gating system taking the sprue base area as unity, followed by the total runner area and finally the total ingate area. Thus a pressurized system would have a ratio of 1:0.75:0.5, whereas a nonpressurized system might be 1:1.5:2 or 1:4:4 as in Fig. 10-31.

Unpressurized gating systems reduce velocity, turbulence, and aspiration but must use tapered sprues, enlarged sprue base wells, and pouring basins to achieve proper flow control. In addition, they can deliver metal uniformly to each ingate only if the runners are in the drag with the ingates in the cope and if the runner area is reduced by the area of each ingate after the junction in a manner similar to that used for ducts in a heating system (Fig. 10-31).

Risers for Sand Molds

Risers serve a dual function: they compensate for solidification shrinkage and are a heat source so that they freeze last and promote directional solidification (Fig. 10-8). Risers provide thermal gradients from a remote chilled area to the riser. Gating systems and risers are closely interrelated; in some cases the ingate is through a riser (Fig. 10-31).

Riser design includes supplying feed metal for shrinkage and any mold enlargement, riser location, spacing of risers for casting soundness, adequate sizing, proper connection to the casting, and the use of chills or insulation. Risers designed according to these concepts can result in improved casting quality and reduced cost.

Solidification shrinkage   Gray iron with a carbon equivalent of 4.3 percent actually expands up to 2.5 percent because of graphite precipitation, but other ferrous alloys contract 2.5 to 4 percent during freezing. Nonferrous metals contract even more: pure aluminum contracts 6.6 percent and copper 4.9 percent. Their alloys usually shrink somewhat less, with near eutectic compositions contracting least. Lead has a 7.7 percent reduction in volume at its phase change.

When an alloy has a short solidification range as in a eutectic, pure metal, or low-carbon steel, a solid skin freezes at the mold-metal interface. Then solidification proceeds slowly toward the thermal center of the casting. Alloys which solidify over a long freezing range, such as aluminum or bronzes, are subject to dispersed microporosity more or less uniformly distributed throughout the cast structure. Skin-forming alloys are likely to have centerline shrinkage. Eutectic alloys require the least feed metal therefore most commercial alloys are of near eutectic composition. The risering concepts given are for low-carbon steel.

Riser location   No matter how complex, any casting can be reduced to a series of geometrical shapes which consist of two heavier sections joined by a thinner one (Fig. 10-8). Then each heavier section needs its own riser. If the thinner section is not tapered to become larger toward the heavier sections, centerline shrinkage is probable. Chills at the thinner section may prevent such shrinkage and may promote directional solidification from the chill to the riser.

Feeding distance   Past research has shown that an adequate riser can provide soundness for a distance of $4.5t$ for a plate casting; $2t$ is the riser contribution and $2.5t$ is from the edge effect. The maximum distance between risers is $4t$ for plates but is only $1t$ to $4t$ for bars. Chills increase the feeding distance for plates to a total of $4.5t + 2$ in for a plate and to $6\sqrt{t} + t$ for a bar. Thus the maximum spacing between risers if chills are used midway between them is $9t + 4$ in for plates and $12\sqrt{t} + 2t$ for bars. Note that the distances are from the outside edge, not the centerline, of the riser (Fig. 10-33).

Riser size   The riser size for a given application depends primarily on the alloy poured and the volume-to-surface-area ratio of riser relative to that of the casting section which is to be fed. Obviously to be effective a riser must freeze more slowly than the casting.

Chvorinov's rule is the basis of most methods now used to calculate the proper size for short freezing range alloys such as steel or for pure metals. There

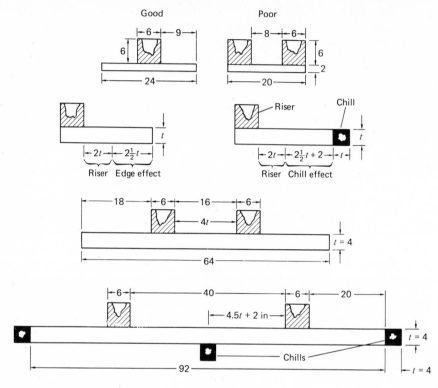

FIGURE 10-33
Feeding distance of risers for steel plates.

is no satisfactory method for calculating the riser size for nonferrous alloys. Chvorinov's rule states that the solidification time for an alloy is

$$ t = k \left(\frac{v}{sa} \right)^2 $$

where t = solidification time, min
   v = volume of the casting section, in³
   sa = cooling surface of the casting section, in²

There are two types of risers—top risers and side risers. Top risers are placed above the volume to be fed and extend to the top of the cope. Side or blind risers are located at the parting line to feed locally or in the case of skin-forming alloys they can feed many times their own height provided that a cured or green sand core connects the mold to the thermal center of the riser. This permits atmospheric pressure to be exerted on the molten metal to force it to rise to a height equal to atmospheric pressure.

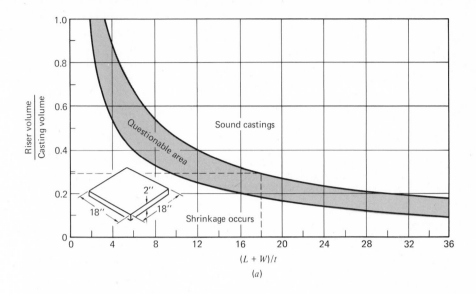

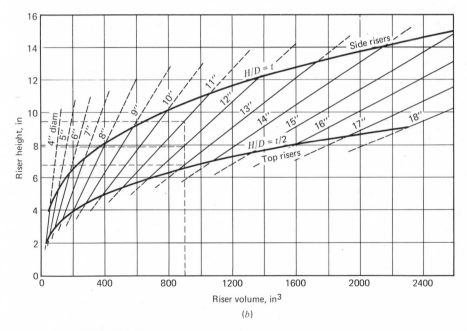

FIGURE 10-34

Riser size by the shape factor method. (a) Riser volume to casting volume ratio as a function of the shape factor. (b) Chart for determining riser diameter. (*From E. T. Myskowski, H. F. Bishop, and W. S.Pellini.*)

There are several alternative procedures for determining an adequate riser size for an alloy which freezes in a skin-forming manner. All give approximately the same result. One of the most direct depends on the *shape factor* which is defined as the sum of the length and width of the section in question divided by its average thickness. In the 1950s the Naval Research Laboratory devised the shape factor chart (Fig. 10-34a) and riser height and volume chart (Fig. 10-34b). With these charts the proper risers for steel castings can easily be selected, once the casting has been divided into the proper sections and their volume-to-surface-area ratios have been calculated. This method gives conservative answers, but that is good when only one casting is to be poured and it must be correct the first time. In high-volume production some additional experiments would be in order before final pattern construction.

Riser connections   The riser connection to the casting warrants considerable care because it determines how well the riser feeds and secondly how readily a riser can be removed. To aid in top riser removal, it is wise to consider placing an annular core similar to a large washer in the riser neck. The thin section of core sand soon becomes hot, so that little chilling occurs and feeding takes place through the center hole. The connection length for a riser should not exceed $D/3$ to $D/2$, where D is the diameter of the riser and the inner hole should be about 1.2 times the connection length.

Riser topping and insulation   Riser size can be reduced by about 50 percent if insulated sleeves and riser topping or antipiping compound are placed on top of open risers as soon as pouring is complete. Sleeves are most important for aluminum and copper alloys where the radiant energy loss from the riser is a smaller proportion of the total heat loss from the riser. Whereas for steel castings the antipiping compounds are most important because they reduce radiant energy loss and leave a flat-topped riser. The proper selection of these materials and procedures can increase casting yield significantly. Not only is remelt cost reduced but the labor for cleaning and cutting off risers is much reduced.

Design for Plaster-mold Casting

Flasks   Frames of the flask are made of solid metal with internal drafts so that the plaster mold can be pushed out. They are accurately machined so they fit snugly on locating points of match plates. Conveyors carry flasks from station to station and are used over again after cleaning. Flask sizes are as follows:

$12 \times 18 \times 3\frac{1}{2}$ to 4 in depth (standard)
$10 \times 18 \times 2\frac{1}{2}$ in
$12 \times 21 \times 6$ in
$24 \times 36 \times 12$ in

Patterns   Patterns are of split type mounted on metal plates. There is one for drag and one for cope molds. They are made of engraver's brass, are accurately machined, and easily changed. More than one part can be grouped on the standard pattern plates so that they can be poured at one time. There is no wear from slurry. The average cost is $300 to $600.

Size   Size is determined by standard flasks. Maximum weight cast is 200 lb and 35 in in diameter. (The size capable of being cast is continually increasing, and these limitations will probably soon be exceeded.)

Material   Yellow brass, aluminum bronze, manganese bronze, silicon-aluminum bronze, nickel brass, and aluminum alloy all can be cast in plaster molds. The material of the mold can stand a maximum pouring temperature of 2200°F. Further development of mold materials may raise this temperature.

Quantity   Small or large quantities can be run by merely placing the pattern in the line when proper metal is being poured. One pattern can produce 150 to 250 pieces per week. Multiple patterns or plates will increase production accordingly.

Surface   The surface of the pattern can be reproduced. It is satin in texture, but can be plated easily without special preparation and can be buffed and polished easily (30–70 to 150 μin).

Tolerances   Tolerances are as follows:

±0.005 in/in
±0.010 in/in across parting line
Cores same as above
±0.010-in horizontal shift between cope and drag
Flatness—1 in² area held within 0.005 in

Markings   Markings can be clearly reproduced.

Inserts   Inserts can be used with due regard to corrosion problems and ability to locate properly.

Wall thickness   This process can produce thin and sound walls:

Maximum of 2 in²—0.040-in minimum thickness
Maximum of 6 in²—0.062-in minimum thickness
Maximum of 30 in²—0.093-in minimum thickness

Tapers with knife edges can be easily cast.

Stock allowances   Stock allowances are $\frac{1}{32}$ in for machining. On small parts, they are less than $\frac{1}{32}$ in for holes and broaching. (Note this can be a great saving over other processes.)

Cored holes   Reasonable holes are cored easily with good accuracy and finish. Holes less than $\frac{1}{2}$-in diameter are more economically drilled. Drill spots are accurate and can eliminate jigs.

Draft   Draft for outside surface is $\frac{1}{2}°$; for inside surface, $3°$.

Advantages and disadvantages   Plaster-mold casting has the following advantages:

1   Ability to vary mold material in thermal capacity from a heat insulator to a chill
2   Slow cooling of metal
3   Mold yield is minimal
4   Distortion and warping at a minimum
5   Smooth surfaces
6   Easy venting
7   No metal agitation
8   Uniform density; no gas pockets
9   Close tolerances

Disadvantages of this process are:

1   It can be used for nonferrous materials only.
2   The mold material is expendable and cannot be reused.
3   Large castings can be made, but they are expensive compared to sand castings.

Investment Castings

Design   The fundamentals of good sand-casting design apply to the production of investment castings. Three of the most important elements of good casting design would include: (1) minimize nonfunctional mass, (2) provide uniform wall sections, and (3) provide a suitable gating system to ensure complete filling of the mold cavity.

By minimizing nonfunctional mass, material is saved and a sound casting free of shrinkage porosity is more likely. The larger the mass, the greater the difficulty of providing sufficient gating to "feed" the part. By taking advantage of the correct placement of ribs, strengthening webs, etc., it usually is possible to design a part so as to minimize heavy sections.

As with all castings, the wall thickness should be as uniform as possible.

If abrupt section changes are necessary, generous radii and fillets should be incorporated to reduce the turbulence of metal flow and to minimize stress concentration and porosity in these areas.

Tolerances and quality Investment castings have been called precision castings because they are much more consistent in size and composition than sand castings. The precision does not approach that of precision machining: ± 0.002 in/in for low-temperature alloys and ± 0.005 in for ferrous and high-temperature alloys are far from precision-machined tolerances. Nothing has as profound an effect on the shape and size of the desired casting as the master pattern or pattern die used to make the disposable patterns from which the part will ultimately be formed. Master patterns are required when the disposable pattern is to be made from a castable soft-metal alloy.

The size of a casting depends on many factors:

1 The master pattern must be made to allow for the four shrinkages and other adjustments listed below. If no master pattern is used, the machined mold eliminates the first shrinkage factor.

a Shrinkage of the die alloy during solidification and cooling
b Shrinkage of the wax or plastic pattern
c Shrinkage of the investment during setting and firing
d The solidification and cooling contraction of the metal in the casting

2 Experience in the art of investment casting governs the quality of castings. Variations are affected by design details such as:

a Presence or absence of cores
b Location of cored passages
c Ratio of core mass to metal mass
d Conjunction of light and heavy sections
e Location and design of angular projections
f Position and mass of gates
g Size and location of risers and vents
h Orientation of the pattern in the flask
i Position of casting in the flask relative to the gate and the sprue
j Method of forcing metal into the mold
k The position of the gates, which may distort the pattern and casting when they solidify
l The inability of the pattern material to shrink due to restrictions of the die
m The pouring temperature of the alloy and mold temperature when poured

3 Wax patterns must not be distorted when removed from mold or when stored. Plastic patterns do not require racks for storing in order to maintain size.

4   Parting lines and the location of cores influence the removal of the pattern from the die and the final tolerance of the investment casting.

5   Wax, plastic, and low-melting metals and alloys may be used for disposable pattern materials, but the low-melting alloys need more development because they do not leave a clean mold surface when they are melted out. Frozen mercury has good accuracy, but has other limitations.

6   In producing a pattern die it is usually necessary to hold the deviation from design tolerance to one-tenth of total permissible deviation. For a dimension to be held to ± 0.005 in, the toolmaker must work to ± 0.0005 in. For ± 0.002 in in the casting, the toolmakers must attempt to work to ± 0.0002 in. It is desirable that all nonfunctional dimensions never be specified closer than 0.010 in to simplify the die or master construction problem and thereby minimize tool costs. It is comparatively easy for a competent toolmaker to work to tolerances of ± 0.001 in in a die or on a master pattern. The fact that dies can be held to tolerances of ± 0.0002 in is no justification for requiring such expensive work unless the function of the casting can justify it.

7   If the refractory material next to the casting is different from that in the rest of the mold, it must be of such composition that it will not crack or spall when it is fired and thus affect the cleanliness and soundness of the casting. The ferrous and refractory metals are usually cast into molds which were formed from two different investments. The first material is a dip coat. This is applied to the gated patterns to impart a fine finish to the castings and to prevent contact between the metal and the coarse, porous, and usually chemically active material used to form the bulk of the mold. The expansion characteristics of the coating and backup investments must be closely matched.

8   The greatest variation in size comes from cooling the metal after pouring, rather than from the steps preceding.

There are many other factors, such as water content of refractory, uniformity of the materials, and pouring temperatures of wax or metals. All have an effect on the size of the final casting. Therefore, the engineer must not emphasize accuracy, but should design for more allowances.

The usual cautions concerning fillets, large flat areas, and threads (internal threads are very difficult to cast) that are considered in other casting processes apply to investment casting.

Cost of parts   Investment casting has received the attention of engineers because it can be used to cast high-temperature and very hard alloys to a size that can be ground to final dimensions economically, and because it can be used to combine several parts into one cast unit and eliminate machining operations.

Table 10-10 OPERATION FOR SHUTTLE LIFTER

Number	Part	Operations	Cycle hours	Setup hours
		Old method of machining subassemblies		
1	Lifter body	7	0.255	0.148
1	Upper plate	20	0.390	0.388
1	Lower plate	20	0.300	0.316
1	Plate	9	0.235	0.168
1	Pad	8	0.145	0.124
4	Lock washer			
4	Cap screw			
	Subassembly	1	0.060	
13		65	1.385	1.144
		Grand total		2.529 h
		New method of machining single casting		
1	Lifter	8	0.260	0.164
		Total		0.424 h
		Savings		
12 parts		57 operations		2.105 h

SOURCE: Courtesy *Metal Progress*.

For example, L. G. Daniels, in *Metal Progress*, discussed an operation which combined 13 parts into one casting and saved 57 machining operations (see Table 10-10).

An example of the potentialities of intelligent application of investment castings is a radar waveguide mixer body cast in B195 aluminum. Part cost is $350 when fabricated; casting cost is $75. Production rate as investment casting is 2/h. Interior rectangular passages are held within ±0.0002 in parallel and from center to center.

Size of parts The size of parts depends upon the size of equipment available in the foundry. Average flask size is 10 in in diameter and 20 in long. Box flasks are 9 × 12 × 15 inches. The weight of steel castings made by investment is 5 lb maximum. The sizes can be increased by further development and installation of larger production equipment.

Tooling for Casting in Metal Molds

Casting in metallic molds often successfully challenges highly mechanized match-plate molding or automatic shell molding. A typical gray-iron "permanent" mold yields 50,000 or more pieces of aluminum or magnesium before failure. Such mold life is achieved by coating the interior mold surfaces periodically with a refractory mold wash.

Permanent molding is most likely to be successful if the

1 Production volume is high (at least several thousand pours)
2 Strength of a nonferrous product (or gray iron) is sufficient
3 Shape is basically simple and suitable to the process

Compared with casting in sand, tooling and auxiliary equipment is costlier; but the advantages include lower cost and weight per unit of product, less entrained gas, finer grain structure, smoother surfaces, closer dimensional tolerances, and lower machining costs. Tooling for permanent molds is more challenging; the designer must thoroughly understand metal flow, solidification, and shrinkage. Certain casting problems demand more careful attention. Fillet design is more critical; mold-wall thickness becomes significant in terms of the cavity volume enclosed, the alloy cast, proximity to the sprue, and other variables. New problems such as cyclical thermal balance and casting removal appear.

Most of these comparisons are also true for die castings as compared to sand casting but in greater degree. For example, tooling cost is still higher but is often more than offset by less operator time per unit of product, more production relative to die life, smoother surfaces, and closer dimensional fidelity.

Tooling for permanent molding, which utilizes gravity feeding, is treated here. Tooling for die casting, which is distinguished by pressurized injection, is the subject of the following section. Parts can be more economically made by die casting up to a certain production rate; then permanent-mold casting becomes more attractive (Fig. 10-35).

In practice, a permanent-mold casting usually weighs less than 20 lb and not more than 300 lb. Die castings now weigh less than 100 lb, and sand castings frequently weigh many tons. Shapes too complex for metallic molds can be cast in sand. The strength of permanent-mold castings exceeds that of die castings because correctly designed gating systems in the former entrain little or no gas (Table 10-11). Tooling costs depend on production volume (Fig. 10-35).

Design for Permanent Molding

A product which is suitable for permanent-mold casting must be carefully studied, and modified if needed, before the permanent mold can be considered

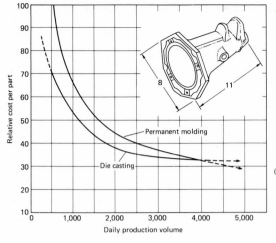

(a) Cost comparison of a transmission extension housing showing breakeven point for permanent molding and die casting. For a production rate of 4,000 the permanent mold casting method was chosen.

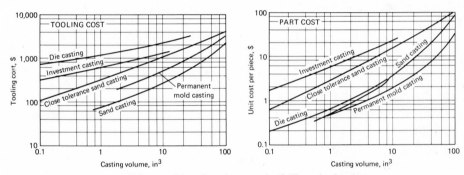

(b) Summary of the tooling and parts costs for 49 different aircraft castings

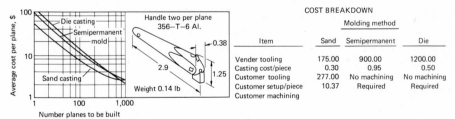

COST BREAKDOWN

Item	Molding method		
	Sand	Semipermanent	Die
Vender tooling	175.00	900.00	1200.00
Casting cost/piece	0.30	0.95	0.50
Customer tooling	277.00	No machining	No machining
Customer setup/piece	10.37	Required	Required
Customer machining			

(c) Cost comparison of a valve handle cast by three different methods

FIGURE 10-35

Cost comparisons for aircraft castings. (*Redrawn from American Society of Metals, "Metals Handbook," Metals Park, Ohio, 1970.*)

Table 10-11 COMPARISON OF SAND- AND METAL-MOLD CASTING PROCESSES

	Sand	Permanent mold	Die casting
Minimum section thickness, in	$\frac{3}{16}$	$\frac{1}{8}$	$\frac{1}{16}$
Dimensional tolerance, in	$\pm\frac{1}{32}-\frac{1}{8}$	$\pm0.020-0.050$	$\pm0.004-0.020$
Machining allowance, in	$\frac{1}{8}-\frac{1}{4}$	$\frac{1}{32}-\frac{1}{16}$	$0.010-0.020$
Minimum core diameter, in	$\frac{3}{8}$	$\frac{1}{4}-\frac{1}{8}$	$\frac{3}{32}$
Maximum weight, lb	Almost unlimited	Up to 300	Up to 100

in detail. The basic features of such a mold design include:

1 Simplicity—to minimize the cost of the mold, turbulence, and the need of machining
2 Liberal tolerances
3 Foresight in the choice of the parting plane which largely establishes the location of the gating and risers
4 Progressive solidification toward the riser from thinner sections remote from it

Dimensional standards and tolerances for permanent molding (Table 10-12) must also be included in the production design of the casting.

Gating system design Success of the casting once depended largely on the operator's skill; today the designer uses fundamental fluid-flow principles for gating design as shown in gating for sand casting.

Using a gating ratio of 1:2:2 or 1:2:1.5 with a pouring basin, conical sprue, and sprue base well which are proportioned as in sand casting, will give faster flow with least heat loss and turbulence. Most molds are gated to fill from the bottom to the top (Fig. 10-36). The metal flows from the sprue base well along the runner bottom, feeds the riser, and passes through a slot gate along the vertical side of the casting into the cavity. Note that a horizontal extension of the base runner receives the initial dross.

Risers For proper feeding the casting must have directional solidification from thin remote sections toward the riser. The riser in turn is fed last with hot metal and has a surface-area-to-volume ratio such that it will freeze more slowly than the casting.

Table 10-12 TYPICAL DIMENSIONAL STANDARDS FOR PERMANENT
MOLDS AND CORES

Casting wall thickness, average,

Casting size, in	Under 3	3–6	Over 6
Minimum wall thickness, in	$\frac{1}{8}$	$\frac{5}{32}$	$\frac{3}{16}$

Fillet radii
Inner radius = t (t = average wall thickness)
Outer radius = $3t$ blend tangent to walls

Unsupported length of cantilever for cores (must be over $\frac{1}{4}$ in of diameter, d):
10 d maximum; 8 d preferable; 4 d if $d < \frac{5}{8}$ in

Draft (cores and cavity), degrees

	Short cores	Average cores	Outside surfaces	Recesses
Minimum	2	2	1	2
Preferable	3	3	3	5

Typical tolerances, in

	Up to 1 in	Additional over 1 in
Basic tolerance (in one mold section)	$\pm\frac{1}{64}$	±0.001 in/in
Between points produced by core or slide and mold	$\pm\frac{1}{64}$	±0.002 in/in
Dimensions across parting plane	±0.020	±0.002 in/in

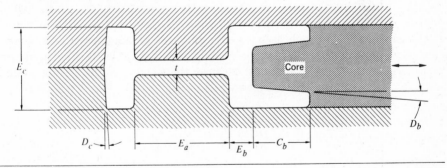

Machining allowance, in

	<10	>10	Sand-cored
Minimum	$\frac{1}{32}$	$\frac{3}{64}$	$\frac{1}{16}$
Preferred	$\frac{3}{64}$	$\frac{1}{16}$	$\frac{5}{64}$

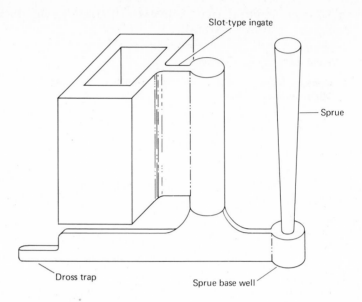

Slot-type ingate

Sprue

Dross trap

Sprue base well

FIGURE 10-36
A typical gating system for a permanent mold.

There exists at present no detailed study of risering for aluminum alloys. In order to determine a reasonable estimate of the riser size for permanent molds the concept of casting yield may be used. Yield refers to the ratio of the casting weight divided by the total weight of metal poured. Metals which exhibit greater shrinkage characteristics require larger risers to make them sound. Low shrinkage alloys such as a eutectic alloy of aluminum with 12 percent silicon, require little feed metal for a sound casting (Table 10-13). The figures given are at best approximations because the weight of the gating system must be estimated and the surface-area-to-volume ratio is not considered.

Table 10-13   EMPIRICAL YIELD DATA
FOR ALUMINUM ALLOYS

Alloy	Yield, %
380 (4% copper, 7.5% silicon)	80
Aluminum–12% silicon	80
Aluminum-zinc	50
Alcoa 142 or 750	40

Table 10-14 TYPICAL GRAY-IRON COMPOSITION FOR PERMANENT
MOLDS AND CORES

| Elements | Composition, % | For longer life add: | |
		Element	Composition, %
Carbon	3–3.5 (with 0.4–0.5% combined C)		
Silicon	1.6–2	Chromium	0.6
Manganese	0.5–0.8	Molybdenum	0.4
Phosphorus	0.2 max.	Nickel	0.2
Sulfur	0.05 max.		

Vents Vents must discharge the gas in the gating system and mold cavity as fast as the metal enters the mold, but natural venting along sliding members and at the parting line is usually inadequate. Additional venting may be as follows:

1 Cut slots as deep as 0.010 in and of suitable width across the parting seal surface.
2 Drill small clusters of holes 0.008 to 0.010 in in diameter in the mold wall at a point where venting is needed.
3 Drill one or more $\frac{1}{4}$-in-diameter holes into the area requiring additional venting. Drive into them square pins $\frac{1}{4}$ in across the corners (pin vent).
4 Drill holes and install slotted plugs (plug vent).

Mold material The mold material is chosen on the basis of three criteria: material cost, the expected total number of pours required, and the casting alloy. Most common and suitable for a permanent mold is high-quality pearlitic gray iron, inoculated at the ladle to achieve uniform grain size and highly dispersed fine graphite. This material, frequently referred to as meehanite, has the composition given in Table 10-14.

Large castings or high pouring temperatures may require cores of alloy cast iron or H11 die steel. Undercuts or complex internal features can be formed with sand cores.

Venting Details for Low-Pressure Permanent Molding

In low-pressure permanent molding, ejection of the casting and venting are the major problems. The use of ejector pins which bear on overflow wells can reduce

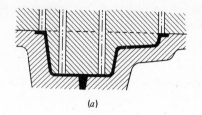

(a)

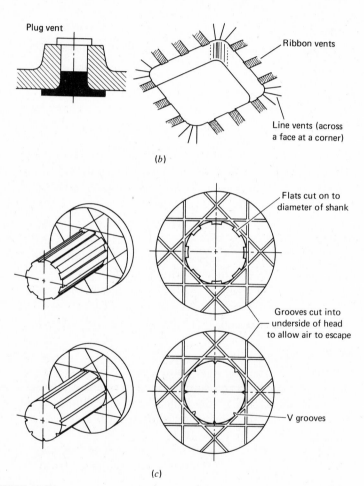

(b)

(c)

FIGURE 10-37
Ejection and venting details for dies to make low-pressure permanent-molded
parts. (a) Ejector pins must be balanced to provide for uniform part removal;
(b) typical plug and ribbon vents; (c) plug vent details.

the damage to a casting in any metallic-mold process, as is shown in Fig. 10-37a. Several means of venting large flat areas include the use of ribbon vents (Fig. 10-37b) and plug vents (Fig. 10-37c).

Die Casting

Design considerations   Die castings have the advantage of reproducing accurate casting in large quantities and with good die life. Subsequent machining and finishing operations can be reduced or eliminated. The extent of incorporating machining operations, such as holes and external and internal threads, in the casting depends on the ingenuity of the designer and the additional die and operating expense.

If inserts are used, the time for placing them in the die will increase the molding-machine operation time. If it is necessary to use slides and cores operated by separate mechanisms other than those opening the die, these slow down the operation and increase die and machine expense.

Ejection of the casting without distortion and unfavorable ejection pin marks requires the attention of the production-design engineer. When the die-casting operation is observed, it often is discovered that more labor is expended in punching, machining, and cleaning the casting of fins and burrs than in the casting operation itself. Therefore, the engineer must consider all operations for making the part, whether by the toolmaker or in his own shop. For example, a cored hole may be made by a two-part core which leaves a fin in the hole. This fin must then be punched out. Since this punching operation is required, the punching die can be designed to pierce the hole and eliminate the metal core in the casting die, making the die less expensive and possibly speeding up the operation.

The cost of tools can be reduced by using drill prints and pilot holes for drilled holes. This eliminates the need for a drill jig. The cores for small-diameter holes are expensive to maintain in die-casting molds, and often the holes can be drilled at less expense. It is the prerogative of the process-design engineer to determine which is the most economical method for a given design.

Die-cast parts made of ductile material can be formed, bent, spun, or joined to another part after casting. They can be pierced, twisted, embossed, swaged, and staked; and lugs can be used as rivets. Such operations make it practical for the production-design engineer to consider the use of die casting in a broad variety of assembly components.

The cleaning of flash depends largely on the design of the part and die. The polishing and finishing costs also can be reduced when the designer eliminates reentrant corners, avoids flat surfaces, and places ejector pins where marks can be removed or hidden.

Cores are an asset in die casting; nevertheless, they should be carefully

Table 10-15   MINIMUM WALL THICKNESS FOR DIE CASTINGS

Surface area, in^2	Tin, lead, zinc	Base casting alloy		
		Aluminum, magnesium		Copper
Up to 3.9	0.0236–0.0394	0.0315–0.0471		0.0589–0.0787
4.0–15.5	0.0395–0.0589	0.0472–0.0707		0.0788–0.0982
15.6–77.5	0.0590–0.0787	0.0708–0.0982		0.0983–0.118
Over 77.6	0.0788–0.0982	0.0983–0.118		0.119 –0.157

questioned in the light of subsequent operations and die maintenance costs. Cores save weight, keep sections more uniform, and thus secure sounder castings. They help carry the heat away and reduce the casting cycle of the machine. Often the dies can be vented through the core. Cored holes will reduce machining due to the accuracy of their size and location.

Minimum section thickness   Good casting design dictates the use of as uniform a wall thickness as possible or one which tapers slightly from the thinnest section remote from the gate to the heaviest section at the gate. In addition, alloy fluidity dictates the minimum practical wall thickness for each die-castable metal (Table 10-15). Walls must be thick enough to permit proper filling but sufficiently thin for rapid chilling to obtain maximum mechanical properties.

Undercuts and inserts   Wherever possible redesign a part to eliminate undercuts. In most cases such part modification is more economical than the labor required to handle a loose piece each casting cycle. In one case, redesigning to reduce undercuts saved $1.50 per casting on a part that is still in production 15 years later. Over 4.5 million castings have been made during that time.

Table 10-16   DEPTHS OF CORED HOLES IN DIE CASTINGS

Alloy	Min. diameter castable	Hole diameter, in.								
		$\frac{1}{8}$	$\frac{5}{32}$	$\frac{3}{16}$	$\frac{1}{4}$	$\frac{3}{8}$	$\frac{1}{2}$	$\frac{5}{8}$	$\frac{3}{4}$	1
Zinc	0.039	$\frac{3}{8}$	$\frac{9}{16}$	$\frac{3}{4}$	1	$1\frac{1}{2}$	2	$3\frac{1}{8}$	$4\frac{1}{2}$	6
Aluminum	0.098	$\frac{5}{16}$	$\frac{1}{2}$	$\frac{5}{8}$	1	$1\frac{1}{2}$	2	$3\frac{1}{8}$	$4\frac{1}{2}$	6
Magnesium	0.078	$\frac{5}{16}$	$\frac{1}{2}$	$\frac{5}{8}$	1	$1\frac{1}{2}$	2	$3\frac{1}{8}$	$4\frac{1}{2}$	6
Copper-base	0.118	...	...	...	$\frac{1}{2}$	1	$1\frac{1}{4}$	2	$3\frac{1}{2}$	5

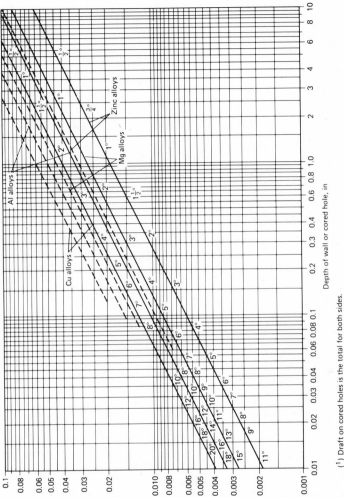

(1) Draft on cored holes is the total for both sides.

(2) Draft for outside walls is 50 percent of inside wall value.

(3) Values given herein represent normal production practice.
Greater accuracy should be specified only when absolutely
required because it usually involves extra cost.

FIGURE 10-38
Draft allowance for die-casting dies (wall: solid lines; cored holes: dashed lines). Note that (1) draft on cored holes is the total for both sides; (2) draft for outside walls is 50% of inside wall value; (3) values given represent normal production practice.

Inserts such as bearings, wear plates, bushings, and shafts can be incorporated in die castings, but they must be easily and precisely located in the die. If they are not securely placed, they may slip between the dies during the casting cycle and cause great damage. Inserts must be provided with properly knurled, crimped, or grooved surfaces to ensure a good mechanical bond between the insert and casting. If the insert is large relative to the casting, better results will be obtained if the insert is preheated, e.g., cylinder liners for automotive engines.

Cored holes   Holes can be readily cast in all alloys according to the data given in Table 10-16.

Draft requirements   The amount and location of draft in a die-cast part depends upon how it is located in the die and whether it is an external surface or cored hole. Draft on the die surfaces normal to the parting line permits the parts to be ejected without galling or excessive wear of the die impression (Fig. 10-38). The values shown represent normal production practice at the most economic level. Greater accuracy involving extra close work or care in production should be specified only when necessary because it may involve extra cost.

Table 10-17   TOLERANCE FOR DIMENSIONS OF DIE CASTINGS

	Zinc	Aluminum	Magnesium	Copper
Basic tolerances (up to 1 in)	±0.003	±0.004	±0.004	±0.007
Additional tol. (1–12 in)	±0.001	±0.0015	±0.0015	±0.302
Additional tol. (over 12 in)	±0.001	±0.001	±0.001	
Add. tol. across movable die section, projected area:				
Up to 10 in²	±0.004	±0.005	±0.005	±0.010
11–20 in²	±0.006	±0.008	±0.008	
21–50 in²	±0.008	±0.012	±0.012	
51–100 in²	±0.012	±0.015	±0.015	
Add. tol. across parting line, projected area:				
Up to 50 in²	±0.004	±0.005	±0.005	±0.005
51–100 in²	±0.006	±0.008	±0.008	
101–200 in²	±0.008	±0.012	±0.012	
201–300 in²	±0.012	±0.015	±0.015	

Table 10-18   RECOMMENDED MATERIALS FOR DIE-CASTING DIES

Alloy	Total castings produced		
	50,000 or less	250,000	1,000,000
Zinc (1-in cavity)	P20	P20	P20 or H13
Zinc (4-in cavity)	P20	P20 or 4150	4150 or H13
Aluminum or magnesium	H11 or H13	H13 or H11	H13 or H11
Copper	H12, H20, H23		
Copper	Molybdenum		

Dimensional tolerance   The dimensional tolerance which can be achieved in die casting depends on several factors:

1 The accuracy to which the die cavity and cores are machined
2 The thermal expansion of the die during operation
3 The injection temperature and shrinkage of the alloy being cast
4 Normal wear and erosion of the die cavity and cores
5 Position of the movable parts relative to each other during casting
6 Surface finish—50 to 125 μin is common, a better finish can be achieved if needed

The tolerances for basic linear dimensions in one-half of the die are summarized in Table 10-17. Another tolerance must be added to the basic tolerance if the linear dimension is for a part which extends across a movable die section. More tolerance must be added to dimensions which extend across the parting line.

Die materials   Die-casting die materials must be resistant to thermal shock, softening, and erosion at elevated temperatures. Of lesser importance are heat treatability, machinability, weldability, and resistance to heat checking. The performance of die materials is directly related to the injection temperature of the molten alloy, the thermal gradients within the die, and the production cycle. Tool steels of increasing alloy content are required as the injection temperature of the molten alloy, the thermal gradients within the die, and the production cycle increase. Dies for use with zinc can be prehardened by the manufacturer in a range of R_c 29 to 34. The higher-melting alloys require hot-work tool steels (Tables 10-18 to 10-20).

Table 10-19 NORMAL COMPOSITION OF DIE-CASTING DIE STEELS

Steel	Carbon	Chromium	Tungsten	Molybdenum	Vanadium	Manganese
H11	0.35	5.00	. . . .	1.50	0.40	
H12	0.35	5.00	1.50	1.50	0.40	
H13	0.35	5.00	. . . .	1.50	1.00	
H20	0.35	2.00	9.00			
H21	0.35	3.50	9.00			
H22	0.35	2.00	11.00			
P4	0.05	5.00				
P20	0.30	0.75	. . . .	0.25		
4150	0.50	0.95	. . . .	0.20	. . . .	0.87

Table 10-20   MATERIALS FOR CORES, SLIDES, AND EJECTOR PINS

Casting alloy	Cores and slides	Ejector pins*
Zinc	440B,† H11, H12, H13	H12, 7140 nitrided, H11, H13
Aluminum or manganese	TZM, H11, H12, H13	H12, 7140 nitrided, H11, H13
Copper	TZM, H11, H12, H13	H21, H20, H22
Ferrous	TZM	H13

* Nitride all steels except for copper.
† Use for cores only.

Wear of cores, slides, and pins

Cores, slides, and pins must provide abrasion resistance in addition to heat resistance. Wear can be reduced by the following.

1   Use contacting materials of differing hardness.
2   Use nitride on one or both surfaces in contact.
3   Use a lubricant on the areas of contact (avoid contamination of the molten metal).
4   Establish and maintain proper clearance between mating parts.
5   Polish the wearing surfaces of the mating parts.

Materials for such moving parts are listed in Table 10-20.

OTHER CONSIDERATIONS IN CASTING AND MOLDING

Quality-Accuracy-Uniformity

Castings are fairly accurate and uniform in size in any of the casting processes. Accuracy of casting depends upon:

1   The accuracy of the pattern and mold and its construction to prevent warpage and shrinkage
2   The uniform temperatures of molten or molded material and mold, and the moisture content of the mold
3   The kind of surface produced
4   The construction and accuracy of the flasks and molding machines

5 The skill of the operator in placing cores and inserts in the mold, and his skill in withdrawing the casting from the mold
6 The uniformity of the cast material as to composition and quantity
7 The wear of the pattern or die

For example, the accuracy of precision casting depends upon the size of the wax pattern, the uniformity of the wax material, wax-expansion characteristics, the care and assembly of wax parts, the mixing and baking of the slurry material for the mold, the wax residue, and the pouring of the metal. When these factors are uniform and under control, the casting will stay within reasonable limits of accuracy and the finish will be uniform (usually ± 0.005 in/in, but ± 0.0005 in/in is possible). The dentist casts crowns, dentures, and bridges to exact sizes by this process. Dentists have been pioneers in adapting the lost-wax process and other precision-casting techniques for modern use.

Good casting practice now produces homogeneous castings that are strong and reliable. X-ray, sonic, penetrant, gamma-ray, and holographic inspection techniques detect faults and assist the molder or foundryman in correcting the defects. Castings are not subject to grain flow and are of nearly equal strength in all directions. Castings of the same chemical composition and heat treatment are competitive, with respect to mechanical properties, with rolled and forged products. Cast crankshafts for automobiles are an example of a product that is now improved because of the inherent characteristics of the cast material.

The quality and uniformity of plastic, die-cast, permanent-mold, and sand castings are generally accepted and can be relied upon. These high standards, along with low costs, have kept casting in competition with other processes.

Blueprint Information

All the requirements of draft, risers, gating, parting lines, cores, locating points, finish allowance, size of sections, and pressure-tight castings should be placed on the drawing to guide machinists and foundrymen. If possible, draft, finish, and core supports and prints should be indicated on the drawing. All such specifications should have the approval of the foundry. This vital information is supplied by the process designer as he works and collaborates with the product designer.

As an engineer works closely with these casting industry representatives, he learns the limitations and possibilities of the particular foundry or mold shop he is using. If another supplier is used, definite and detailed information about that shop must be obtained, because each shop has distinct characteristics and offers specific equipment.

The design drawing should indicate the locating points and the chucking or holding procedure desired for the machining of the casting, so that the foundry, mold shop, and machine shop will use the same checking routine.

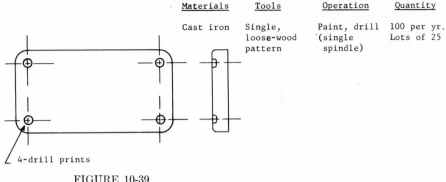

Materials	Tools	Operation	Quantity
Cast iron	Single, loose-wood pattern	Paint, drill (single spindle)	100 per yr. Lots of 25

4-drill prints

FIGURE 10-39
Cast iron cover plate.

Quantity and Costs

The quantity of castings to be made within a period of time determines to a large extent the cost of the casting. The greater the quantity, the more that can be invested in pattern or die equipment and the less the casting will cost. Multiple patterns or dies produce two or more castings in practically the same time it takes to produce one. Increased quantities resulting in better equipment improve the quality of the casting and permit the spreading of the cost of molds or patterns over more castings with less development cost per unit.

Close competition exists between the various casting processes, as well as with other types of processes. Use of each process is determined by the following factors: quantity to be produced; material required to meet strength, corrosion, weight, and appearance conditions; accuracy required; complexity; and cost of subsequent operations such as machining and finishing.

For example, say a cover plate for a piece of equipment is required. As the market for it develops, improvements are made, quantities increase, and functions change. In Figs. 10-39 to 10-43, note the influence of quantity and function on the type of material and operation used. For a brief period, castings lost to rolled steel. Note how considerations of appearance and accuracy played an important part in the choice of process. The metal molds (Fig. 10-43) could not be used until the activity of the part justified the expensive mold.

In looking at the figures, assume that a gasket is necessary to prevent oil leakage. Assume also that the part comes from the foundry or molder with all operations—such as removal of fins and marks due to parting lines, gates, push-out pins, and cores—performed to give a piece which is finished except for machining and final finishing operations.

A permanent mold aluminum cover plate in Fig. 10-43 is being applied to equipment that requires a better seal and a gasket which must be held in a gasket groove. Better finish and greater accuracy are required. Bolt holes can be cast and better finish and accuracy can be had by permanent metal molds.

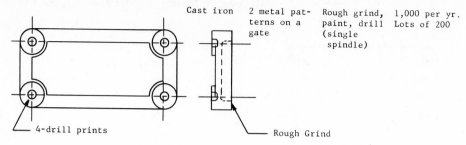

FIGURE 10-40

Cast-iron cover plate molded two at a time. Quantity increases; tools and dies can be paid for in 1 year.

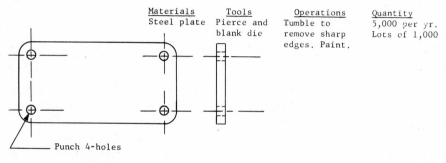

FIGURE 10-41

Stamped-steel cover plate. Quality of apparatus demands better appearance, and aluminum is chosen, No other finish is required.

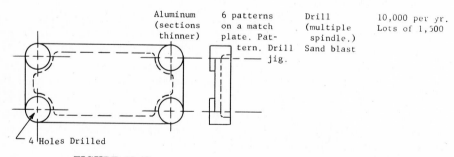

FIGURE 10-42

Sand-cast aluminum cover plate. Cover plate is applied to other equipment that required a better seal and gasket which must be held in a gasket groove. Better finish and greater accuracy are required. (Bolt holes can be cast, and better finish and accuracy can be had using permanent metal molds.)

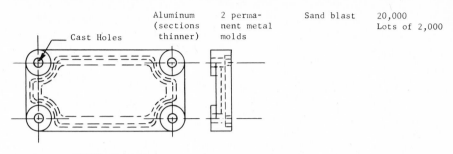

FIGURE 10-43
Die-cast aluminum cover plate.

Figure 10-44 shows a die-cast zinc cover plate. A lacquer finish was applied to harmonize with equipment. Countersink head screws were necessary, and an insignia on the cover was added. Here the gasket is cheaper, but sand blasting is required. This cover plate is used when activity has increased and the accuracy of the part must be improved. The quantity required has also increased. By using a zinc die-casting alloy, cost of material is reduced and less material is required in die casting the part. Greater accuracy is obtained. The die-cast part requires no finishing operations other than cleaning and application of finish, as the surface is better than with permanent-molded parts.

Material and Overhead Costs

Usually the material cost is the least portion of the cost of casting. Material costs can be reduced by using a standard material which is used frequently in the foundry or molding shop. A special material may not be uniform. It may be expensive to prepare or to adapt equipment to handle it in small quantities.

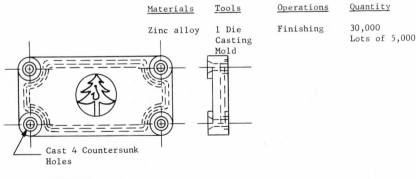

FIGURE 10-44
Die-cast zinc cover plate.

Table 10-21 ANNUAL COST AT $0.10/LB REGARDLESS OF VOLUME

Weight classification, lb	Volume, lb/yr	Actual labor cost, ¢/lb	Raw material cost, ¢/lb	Actual cost, ¢/lb	Selling price per year, @ $0.10/lb	Actual cost per year
1-9	20,000	30	2	32	$ 2,000	$ 6,400
10-49	20,000	20	2	22	2,000	4,400
50-199	10,000	15	2	17	1,000	1,700
200-499	50,000	5.25	1.75	7	5,000	3,500
500-2,000	100,000	2.5	1.5	4	10,000	4,000
					$20,000	$20,000

With the mechanization of foundries to handle sand, and raw and finished material, by conveyors, special equipment, and molding machines, the overhead costing rates have increased considerably over the days when sand was hand-shoved. The maintenance of such equipment is high because of the sand and dust hazards. In spite of high overhead costs, the foundries have reduced the cost per pound of castings through advanced mechanization.

Molding shops have high costs of maintaining molding equipment—with air, steam, electric power, and water service. "Permanent" molds are also subject to wear and are replaced after a given service life. This requires expensive machine equipment and high-priced toolmakers to maintain the quality of the product.

An example of competition between materials and processes might be the same automobile company die casting an engine block out of aluminum and at the same time perfecting the sand molding and casting of thin-walled accurate castings for a similar engine block.

Auxiliary Operations

The cost of molding is sometimes less than the auxiliary operations of removing gates, risers, and parting-line material; trimming, cleaning, grinding, and polishing surfaces; and inspection. Costs are reduced by eliminating these auxiliary operations. Heat-treating and straightening operations are expensive and their cost can sometimes be reduced by proper design and specification of material.

Purchasing Castings

Purchasing castings on the straight cost-per-pound basis is simple from the accounting standpoint, but purchasing castings on a classified weight basis is less misleading. The most economical arrangement is to deal with an organization that knows the detailed costs of each operation required to make the casting, pattern, and tools. Then these costs can be compared with detailed costs of other processes and the correct decisions can be made.

For example, company A purchases castings at a flat price of $0.10/lb; the weights of the castings vary from 1 to 2,000 lb; and the volume of castings is distributed as shown in Table 10-21. It can be seen that inexpensive small castings are being purchased due to the excessive price of the heavy castings. If the true costs were allowed for, better pattern equipment would be ordered for the production of the smaller castings. For example, small parts may be more economically made as weldments than as castings. The possible savings resulting from using the best process, on an *overall* cost basis, are shown in Table 10-22.

Table 10-22 ACTUAL COST CAST VERSUS FABRICATED AT SEVERAL PRODUCTION LEVELS

Weight classification, lb	Actual cost of casting	Correct process		Savings
		Cost per lb	Total cost	
1–9	$ 6,400	Fabricated, 25¢	$ 5,000	$1,400
10–49	4,400	Fabricated, 18¢	3,600	800
50–199	1,700	Fabricated, 15¢	1,500	200
200–499	3,500	Cast, 7¢	3,500	
500–2,000	4,000	Cast, 4¢	4,000	
	$20,000		$17,600	$2,400

Design Factors for Castings

The part to be made from a casting may be intricate or simple. Intricate castings can be economical because they may combine many parts into one piece and thus save the cost of the fabrication and joining of several separate pieces. Joints must be machined and bolted together. Misalignment and loosening during vibration may result. Sand castings weighing several tons (such as a locomotive frame) have been made to advantage because of their ability to combine several parts into one piece. The more complicated the casting, the more ingenuity and control required. The simpler the part, the less the cost of the mold and pattern equipment and hence the less expensive the part. Variations in size and strength may be more difficult to control in more complex parts; thus a more highly skilled molder may be required.

The design of a part to be cast depends upon the behavior of the material as it cools, the construction of the mold, and the functions of the part in service. The art of casting and molding has progressed to such an extent that practically anything can be cast that is within the size range of the equipment available. It may not be economical to cast the part today, but in a few years the process may be improved so that it may be economical to redesign the part from a sheet metal or welded design into a casting. To make a casting simple and easy to cast demands the highest skill and the best judgment on the part of the designer and foundryman.

Some design factors to be considered are:[1]

   1   For maximum strength and stiffness, material should be kept away from the neutral axis. This is important to all designs and is related to

[1] Much of this information is taken from "The Steel Castings Handbook."

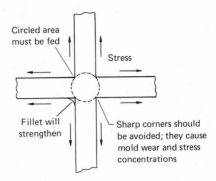

FIGURE 10-45
Stress conditions around hot spot.

the moment of inertia I which is a measure of the ability of a cross section to resist rotation or bending about the axis passing through its center.

The modulus of rupture, which designates the load in pounds per square inch imposed on a beam, is given by the well-known equation

$$S = \frac{MC}{I}$$

where S = modulus of rupture or maximum fiber stress, lb/in^2, imposed on metal at greatest distance from the neutral axis

M = bending moment, or load in pounds times distance from concentrated load to point under study

C = distance from neutral axis to outer surface

2 Keep plates in tension and ribs in compression. This is desirable since the compressive strength of some materials, particularly cast iron, is much greater than the corresponding material tensile strength. The plate distributes the load over its entire surface and is more effective when placed in tension rather than compression. Ribs, on the other hand, are most effective when placed in compression.

3 Assure that the pattern or part can be easily removed from the mold. This factor is probably violated more than any other and causes the use of loose pieces and complicated cores, gates, and risers.

4 Ensure that cores can be removed.

5 Use smoothly tapered sections to eliminate high stress concentrations.

6 Sharp corners and abrupt section changes at adjoining sections should be eliminated by employing fillets and blending radii (see Figs. 10-45 to 10-47).

7 Determine the best location (locations) where material should be fed into the part.

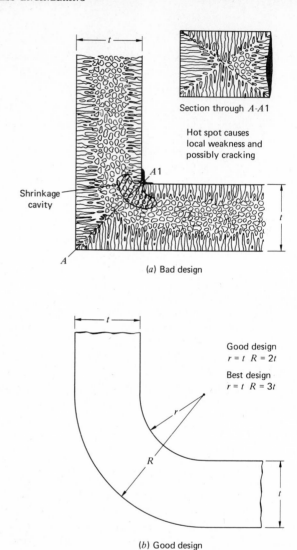

Section through A-A1

Hot spot causes
local weakness and
possibly cracking

Shrinkage
cavity

(a) Bad design

Good design
r = t   R = 2t

Best design
r = t   R = 3t

(b) Good design

FIGURE 10-46
Design of a corner joint to produce soundness.

8   Does the design avoid the development of tears and spongy sections?
9   Does the necessary draft on the pattern interfere with the part design?
10   Can parts be clamped and located easily for machine operations without interference from fins or excess material at parting lines and junctions of gates and feeders?

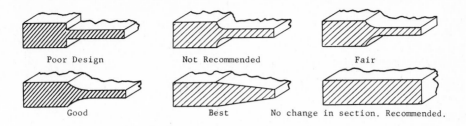

Poor Design   Not Recommended   Fair

Good   Best   No change in section. Recommended.

FIGURE 10-47
Design for changes in section thickness.

11 Are locating points for chucking or holding indicated on the drawing
for foundry or molding-shop and machine-shop information?

Effect of Grain Size

A relatively large dendrite grain size may be developed in the process of solidification as a result of the freezing rate or section size of the casting. Large castings
and ingots freeze with coarse grains. Thin-sectioned castings, or castings made
in metal molds, develop a fine grain size due to rapid freezing. Normally, a
fine grain size is desirable since higher ductility and impact strength values are
obtained at a given tensile strength level.

Hot Spots

Hot spots are the last portions of the casting to solidify. They usually occur at
points where one section joins another, or where a section is heavier than that
adjoining it—at a square corner for instance. The outside of the corner should
be rounded to reduce the section change (Fig. 10-46).

Hot spots are weak and the metal may tear where they occur. Figure 10-45
shows stress conditions around the hot spot within the circle. In casting steel
and other high-temperature metals, the spot enclosed by the circle must be fed
to avoid cavities.

The shape of the part should be such as to make possible directional solidification of the molten metal in unbroken sequence from the farthest part of the
casting to the point of entry. If the casting should "freeze" somewhere along
the line ahead of its turn, the sections between it and the farthest end, which
are still liquid, would be cut off from the supply of feeding liquid metal; cracks,
or at least dangerous internal stresses, would result. In view of this, the casting
should be designed to avoid abrupt changes from a heavy to a thin section (Fig.

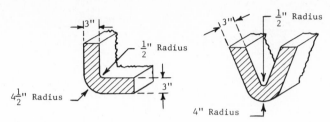

FIGURE 10-48
Unfed joining section.

10-47). Where light and heavy sections join, the thickness of the lighter section should be increased gradually as it approaches the heavier section.

Steel is difficult to cast. Most steel castings should be stress-relieved because of the high shrinkage of the metal and resulting internal stress. The design rules for steel castings can be modified for other ferrous materials. If they are followed in principle, better castings will be obtained regardless of the material being cast. The following six design rules for casting provide an important supplement to the 11 general principles mentioned previously.

1   In designing unfed joining sections in L or V shapes, it is suggested that all sharp corners at the junctions be replaced by larger radius corners, so that these sections become slightly smaller than those of the arms (Fig. 10-48).

Reason: Outside corners of joining sections are positions of extra mass and result in hot spots. If these positions are not fed from outside sources (by risers) they will be the location of shrinkage cavities. Cutting off the

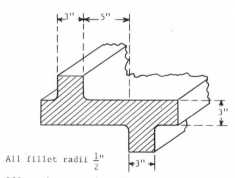

Offset the arms of an X section to produce two T sections

FIGURE 10-49
Offset the arms of an X section to produce two T sections thereby reducing the hot spot.

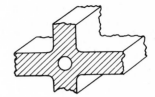

FIGURE 10-50
Reduction of a hot spot by coring.

outside corner is similar to producing a uniform section and is a fundamental of good design.

2   In designing members that join in an X section, it is suggested that two of the arms be offset considerably (Fig. 10-49).

Reason: The center of an X section which cannot be fed by a riser is a hot spot and will result in the formation of a shrinkage cavity. Offsetting the arms permits the use of external chills by the foundryman to produce a section free from cavities.

If the design does not permit offsetting the arms at the X section, then the possibility of placing a core in the middle of the junction should be considered. This will permit uniform sections with small stresses through the center (Fig. 10-50).

3   Isolated masses not fed directly by risers are details of poor design. They should be hollowed out with a core construction or constructed of a lighter section. If this cannot be done, then internal chills should be used.

Reason: Areas of heavy metal attached on all sides to members of much smaller thicknesses are considered isolated masses. When these areas are so located that the foundryman has no opportunity to feed the heavy portions properly by means of conveniently placed risers, shrinkage cavities occur within the section.

4   The use of webs, brackets, and ribs at joining sections should be kept

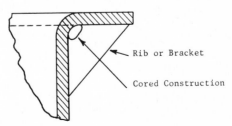

Rib or Bracket

Cored Construction

FIGURE 10-51
Cored construction in stiffening members.

to a minimum. If they are required as stiffeners for a casting, the poor design effect can be remedied by the extensive coring of the web, bracket, or rib in the region of the adjoining section (Fig. 10-51). Small brackets will help prevent hot tears and consequently are frequently used.

Reason: Stiffening members at joining sections are sources of considerable trouble with regard to the formation both of hot tears and of shrinkage cavities. They increase the mass at the intersection of the adjoining members. Coring will not impair the stiffening features of these webs, brackets, or ribs.

5 When the design of a one-piece cast steel structure is very complicated or intricate or when section thickness variations are larger, it can sometimes be broken up into parts so that they may be cast separately and then assembled by welding, riveting, or bolting.

Reason: Very complicated designs often lead to enclosed stress-active systems and any mold relieving that may be employed by the foundry would not be sufficient to produce the casting without cracking or failure. In such cases, appendages, cast-in baffles, and other extraneous members should be removed from the design and cast or prepared separately, and then welded into place.

6 Cast members that are parts of an enclosed stress-active system should be designed slightly waved or curved (Fig. 10-52).

Reason: Curve construction or wave design reduces distortion resulting from contraction of the steel during cooling. A good example of this type of design is the use of curved spokes in many classes of wheels (Fig. 10-52). In such castings the rim, hub, and spokes may cool at different rates, thus subjecting the casting to considerable internal stress. Spokes designed with a wave in them will, under stress, tend to flex, thus preventing tearing or distortion. Residual stresses can be relieved through heat treatment.

Details of Design for Steel Castings

The greater the shrinkage and the higher the temperature of the metal, the more difficult the problem of casting. While iron castings are common, and iron is relatively easy to cast, steel casting is much more difficult, and the principles established for steel castings are described as a review and as an illustration of the more severe conditions. The practices described may be modified for other kinds of materials.

1 In order to prevent cavities in the casting resulting from shrinkage, thought must be given to the casting design, so that the volumetric contraction is compensated for by a supply of liquid steel.

2 The casting and mold should be so designed that stresses are as small as possible when the casting is at a temperature just below the solidifica-

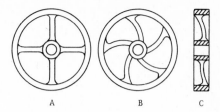

FIGURE 10-52
Wave construction. Design (A) produced cracked spokes. It is corrected by the use of five curved spokes as in (B) or the alternative design as in (C).

tion temperature. It is at this point that steel has poor strength and ductility characteristics so that it is likely to hot-tear.

3 Steel has low fluidity as compared with other metallic alloys. Thus, thin sections should be avoided, especially when remotely located from the orifice feeding the casting.

4 A poured steel casting will solidify from the outside of the casting inward and the rate of solidification is about the same regardless of section size; therefore, heavier sections will take proportionally longer to solidify than lighter sections. The principal defects in steel castings that result from poor design are hot tears, cold shuts, shrinkage cavities, misruns, and sand inclusions.

Hot tears Hot tears are cracks at various points in a casting brought about by internal stresses resulting from restricted contraction. Sharp angles and abrupt changes in cross section contribute to large temperature differences within a casting, which may result in hot tears.

Shrinkage cavities Shrinkage cavities are voids in the casting brought about by insufficient metal to compensate for the volumetric contraction during the solidification of the casting. Shrinkage cavities are more pronounced in areas fed by thin sections. This results because the thin feeding section will solidify too rapidly to allow the introduction of additional metal to the casting which has diminished in size because of volumetric contraction.

Misruns When a section of a casting is incompletely filled with metal, it is known as a misrun. This is usually brought about by solidification prior to the complete filling of a mold cavity. Thus thin casting sections should be avoided.

Sand inclusions When a portion of the mold breaks away and is wholly or partially enclosed by the molten metal, a sand inclusion occurs. This may be

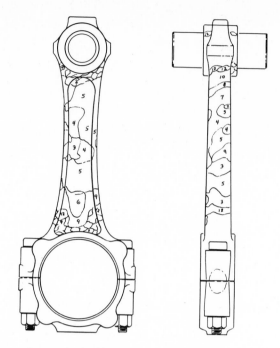

Compression load indicating compressive stresses

FIGURE 10-53
Brittle-lacquer stress analysis.

brought about by abrupt turns and complicated passages of flow of the metal
as it is poured into the casting.

Comparison of Cast and Forged Connecting Rods

In 1962 the Central Foundry Division of General Motors introduced the cast
connecting rod. Since then it has become the standard of the industry because
of its economy and reliability. This development is a triumph of engineering
perseverance and casting design. The success of this design depends on the use
of the inherent strength of *pearlitic malleable iron* and its machinability. The
presence of temper carbon in the matrix reduces the cost of cutting operations
by acting as an internal lubricant and eliminating deburring.

Why cast a connecting rod?   The casting process permitted greater freedom
of design because the engineer could put the metal where it was needed. But
stress analyses were needed to determine where the greatest stresses occurred.
In their original work, General Motors engineers took conventional forged

connecting rods and ran exhaustive stress analyses. First static stress loads were calculated as follows:

$$\text{Compressive load} = P_{\max} \times A$$

where $P_{\max}$ = Max. gas pressure in cylinder, lb/in²
    A = area of cylinder bore, in²

$$\text{Tensile load} = KW\left(\frac{S}{2}\right)N^2\left(1 + \frac{S}{2L}\right)$$

where K = constant 2.84×10^{-5}
    W = reciprocating weight, lb
    S = stroke, in
    N = Engine speed, rev/min
    L = Length, pin center to crank center, in

Analysis of stresses in a forged rod   Brittle-lacquer-stress-coat patterns showed the location of highly stressed areas in the connecting rod (Fig. 10–53). Strain gages attached to those areas gave the quantitative values of stresses in the rod. The largest stress was found to be in the center of the column of all rods studied.

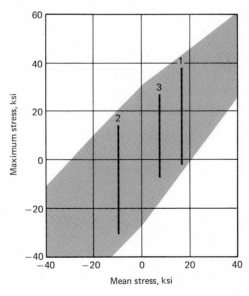

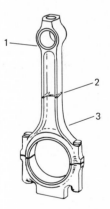

Stress ranges in three areas
of a connecting rod

FIGURE 10-54
Goodman diagram of fatigue stress levels at three sections of a forged connecting rod.

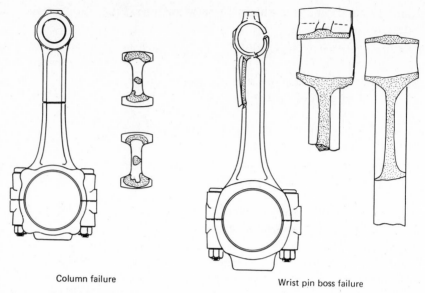

Column failure

Wrist pin boss failure

FIGURE 10-55
Typical connecting-rod fatigue-test failures.

Next, fatigue testing was carried out with alternating axial loading to produce a Goodman diagram (Fig. 10-54) which represents fatigue strength under combined alternating loading and a static load. If no failure occurred in 10×10^6 cycles, the rod was considered to be acceptable.

On the basis of the Goodman diagram the stresses in three particular areas were plotted (Fig. 10-54).

Table 10-23   ROD LOADING AS A FUNCTION OF
ENGINE SPEED

Engine speed, rev/min	Tensile load, lb	Compressive load, lb
500	27	8,156
1,000	109	8,077
2,000	437	7,755
3,000	983	7,224
4,000	1,747	6,477
5,000	2,728	5,519
6,000	3,925	4,372

Area 1   Compressive loading should result in high tensile stresses normal to the long axis of the rod on the underside of the wrist pin.

Area 2   Compressive and tensile loads each produce moderate stresses in the column in the direction of the long axis, but combined they produce a high alternating stress.

Area 3   The tensile load should induce high tensile stresses in the rail at the crankpin end, but, in fact, relatively low stress was induced there because of the compressive loading.

The validity of the fatigue tests were revealed when actual failure occurred as predicted when the rods were subjected to engine tests. Failures took the form of transverse fractures in the center of the column and longitudinal fractures below the wrist-pin bosses (Fig. 10-55). Although the fatigue tests were based on the combined maximum compression and tension loads, in actual practice the rod never reached such a load combination at any speed (Table 10-23).

The design must be such that the alternating stress in all areas falls within the limits of the Goodman diagram.

Design of a cast connecting rod of pearlitic malleable iron   After the basic design parameters had been determined, the design of the pearlitic malle-

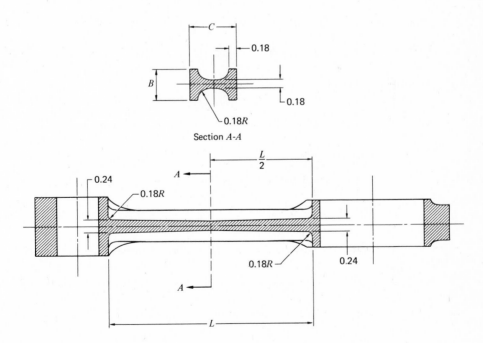

FIGURE 10-56
Typical cross sections of a cast connecting rod.

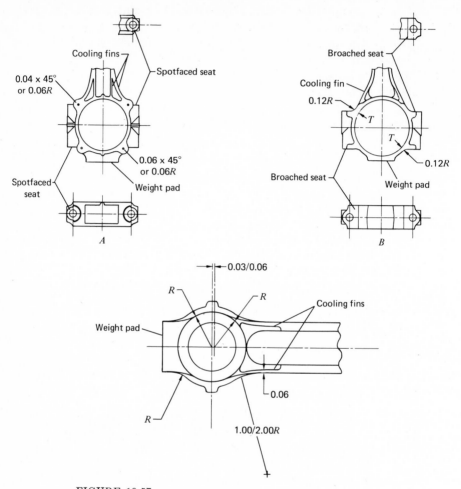

FIGURE 10-57
Typical design details for the crankpin and wrist-pin ends of the cast connecting rod.

able-iron connecting rod was tailored to the casting process. The column was specified to have a minimum and uniform cross section consistent with good foundry practice (Fig. 10-56). Wherever the stress level was low, metal was removed and added where more stress was found. The ability to redistribute the metal permitted the weights of the cast and forged rods to be equal (Fig. 10-57).

Design of gates and risers To determine the best foundry practice, gating studies were made using high speed photography and Lucite models. On the basis of the tests the casting was gated at only one end because the meeting of the streams from dual end gates could cause defects in the highly stressed center of the rod arm.

Risering was designed to achieve directional solidification by tapering the central web so that solidification began at a point remote from the risers, in this case the center of the thin arm. Thermocouples placed in test castings at critical points indicated the sequence of freezing and helped achieve proper riser sizes for both ends of the connecting rod. Directional solidification was achieved from the center of the arm to each end despite gating from the small end. The *I*-beam section of the arm provided additional chilling so that solidification began at the arm center where it was critically stressed and travelled to the risers. Cooling fins were added to the boss and column rail to provide better local solidification.

A match plate showing the gates, risers, and other arrangement details reveals the care which was taken in maximizing the product yield when using the casting process (Fig. 10-58).

Match-plate features   Use of rounded corners, generous fillets, and uniform section thickness with increasing taper toward the direction of solidification provided for sound castings. Other general requirements include:

1   Draft, 4°
2   Stock for machining, 0.060 in

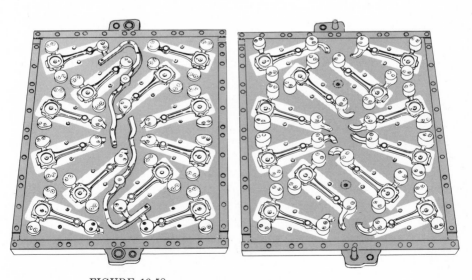

FIGURE 10-58
Match plate for the cast connecting rod. Note 12 are cast in one mold, gating is from one end, and three risers are used on each rod.

Table 10-24 COMMON FACTORS IN CASTING PROCESSES

Process	Permanent molds	Expendable molds	Gravity	Vacuum	Air pressure	Plunger	Centrifugal	Squeeze of mold	Rough	Smooth	Expendable	Metal	Mech. removal	Mech. removal	Annealing	Shakeout	Expendable inserts	Permanent inserts	Chills	Risers	Draft	Expendable patterns	Permanent patterns
			(Feed type)						*(Surface)*		*(Cores)*			*(Castings)*									
Investment (high-temperature alloys)		X		X			X			X	X					X	X		X		X	X	X
Steel sand casting		X	X						X		X				X	X	X			X	X	X	X
Carbon-iron sand casting		X	X						X		X				X	X	X			X	X		X
Nonferrous sand casting		X	X						X		X				X	X	X			X	X		X
Plaster casting (nonferrous only)		X	X							X	X					X	X			X	X		X
C process			X							X	X				X	X	X		X	X	X		
Permanent molding	X		X							X	X							X		X	X		
Slush and Corthias	X		X				X			X								X		X	X		
Die casting	X									X		X									X		
Bronze	X					X				X		X	X	X				X			X		
Aluminum	X				X					X		X	X	X				X			X		
Zinc alloys	X				X					X		X	X	X				X			X		
Plastic casting	X							X				X	X	X				X					
Compression	X							X		X		X	X	X				X					
Transfer	X					X		X		X		X	X	X									
Rubber	X							X		X		X	X	X				X			X		
Ceramic	X									X		X	X					X			X		
Continuous	X		X							X					X			X			X		
Centrifugal	X	X	X				X		X						X	X					X		X

3 Fillet radius, 0.12 in; corner radii, 0.06 in

4 Dimensional tolerances: connecting rod length $\pm$ 0.01 in; other dimensions, $\pm$ 0.02 in; gate removal, $+0.06$, to -0.03 in

Inspection features Of course not only must the proper melting procedures and foundry practice be followed but also hot trimming and mechanized inspection procedures must be provided. All connecting rods pass through conveyorized sonic, ultrasonic, and Magnaglow inspection stations where defective castings are immediately removed from the line. Provision for such elaborate inspection procedures is needed in order to guarantee to the customer that the cast product is equal to or exceeds the quality of rods produced by any other process.

SUMMARY

The casting processes provide a versatility and flexibility that is second to no other broad process classification. In small and medium-sized production runs, sand casting with expendable patterns represents the most economical method for producing a vast variety of parts. Intricate designs frequently can be cast, at an overall savings over other methods, using the investment-casting technique. Likewise, each of the other casting processes offer the production-design engineer definite advantages under certain conditions.

Table 10-24 indicates some of the principal characteristics of the various casting processes.

In spite of competition from welded structures, sheet-metal parts, and forgings, the foundry and molding industries have expanded. Larger equipment has been developed. The frame of the automobile door has been die-cast. These same industries cast in large quantities many smaller items—as small as the die-cast lugs used in zippers for clothing. All of the casting industries have become highly mechanized, especially the foundries. It is difficult to get skilled labor because of relatively poor working conditions; therefore, dust removal, material handling, and reduction of heavy labor receive considerable attention from engineers and management.

Each new material (alloy, plastic, ceramic) opens new opportunities for developing processes and making parts at less cost. As accuracy and quality of castings are improved, the amount of machining is reduced. Castings can join several parts into one, and as mechanization of processes improves, more parts will be cast.

QUESTIONS

1 How do casting and molding processes differ from forging or extrusion processes?

2 What factors control the rate of solidification?

3   How can riser size be calculated in order to feed a given casting satisfactorily?

4   What is meant by directional solidification?

5   Give a sketch of the ideal shape of a casting to obtain directional solidification.

6   What design factors should the production-design engineer observe when specifying castings?

7   Why is steel difficult to cast?

8   How are patterns classified?

9   What factors should be considered in pattern design?

10   What is the function of chaplets? Make a sketch of chaplets in use.

11   When would it be advisable to specify green-sand cores instead of dry-sand cores?

12   Of what materials are permanent molds made?

13   What dimensional tolerances are usually held in precision casting? What tolerances are possible?

14   Upon what does the accuracy of any casting depend?

15   What information is the process designer expected to provide in the engineering drawing?

16   What are the principal metal-casting processes?

17   In what quantity requirements are sand castings usually advantageous?

18   What precautions should be observed in the design of a steel casting so that internal stresses are as small as possible?

19   Why is it especially important not to have thin sections when designing steel castings?

20   What is meant by hot tears? How can they be eliminated?

21   What is a misrun? How is it caused?

22   What are sand inclusions and how can they be minimized?

23   What finish allowance would you specify for a gray-iron casting 5 ft in length?

24   What are the principal advantages of permanent-mold castings?

25   What is the strength-weight ratio of aluminum versus bronze permanent-mold castings?

26   Why is it difficult to core holes less than $\frac{1}{4}$ in in diameter in permanent-mold castings?

27   Why is the hot-chamber type of die-casting machine not applicable to alloys having melting temperatures in the neighborhood of 1400°F?

28   If porosity were causing trouble in the die-casting department of your plant, what steps would you take to minimize this complaint?

29   How could die life be prolonged for zinc-base die-cast parts?

30   Why are the weight and area limitations of zinc-base die castings greater than aluminum-base die castings?

31   When would you recommend incorporating internal threads on die-cast parts?

32   Why is the plaster-mold casting process not adaptable to ferrous materials?

33   Why can investment castings be produced to close tolerances?

34   When would you advocate the investment-casting process?

35   What economies are gained by the shell-mold casting process?

36   Explain the theory of centrifugal casting.

37   Why is slush casting limited to the lower-melting-temperature alloys?

PROBLEMS

1 A foundry is producing gray-iron blocks 12 × 6 × 4 in. The parting line is at the midpoint of the height of the block so that there the mold cavity extends 2 in into both the cope and the drag. The flasks are 20 × 12 × 5 in over 5 in so the combined height of the cope and drag is 10 in. The mold is poured with molten gray iron which weighs 0.22 lb/in³ and the compacted sand weighs 100 lb/ft³. What would be the total lifting force in pounds tending to separate the cope from the drag as a result of the metallostatic pressure within the mold?

2 The volume of a sand core is 100 in³. Find the buoyant force on the core if poured with the following alloys: (a) cast iron, (b) cast steel, (c) aluminum. Note that the density of molten metal must be obtained by taking the density at room temperature and correcting for expansion by using the volumetric coefficient of expansion applied to the solid. The error in neglecting the liquid expansion is not large because most foundry alloys are poured at only 100 to 200 degrees superheat.

3 Below are given the solidification times for top cylindrical risers which are 4 in diameter × 4 in high for steel, copper, and aluminum.

Solidification times, min

Metal	No treatment	Insulating sleeve	Radiation shield	Insulation + shield
Steel	5	7.5	13.4	43.0
Copper	8.2	15.1	14.0	45.0
Aluminum	12.3	31.1	14.3	45.6

(a) Determine the effective constant for Chvorinov's rule from the data for each of the metals listed above.

(b) Discuss the relative effect of radiation shielding for each metal listed, in terms of specific heat, latent heat of fusion, and pouring temperature.

(c) Repeat item b in terms of using an insulating sleeve only.

(d) Discuss why the solidification times for the steel and the aluminum risers are about the same when they have both insulating sleeves and radiation shielding.

4 Explain the use of blind risers and the type of alloy which can be fed by them. Calculate the height in inches to which a blind riser can feed a casting which has a specific weight in the molten state of 0.264 lb/in³ if it freezes like low carbon steel.

5 Derive an equation for the injection velocity of an alloy in a die-casting operation as a function of the maximum plunger pressure p. Assume that the shot chamber is large in cross section compared with the ingate, the cavity is vented through the entire injection process, the energy at any point in the molten metal stream is a constant, and all friction and orifice coefficients are neglected.

6 A casting as shown in Fig. P10-6 is produced in a green sand mold which measures

30 × 30 in with a 6-in cope over an 8-in drag. The casting is specified to be class 40 gray iron. Determine the total weight of metal required to fill the mold, the gating ratio, whether the gating system is pressurized or unpressurized, and the casting yield, i.e., the ratio of good casting to the total weight poured.

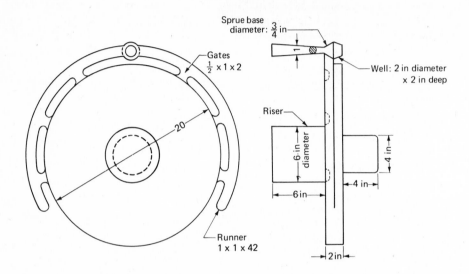

7 Two castings are molded in green sand. They differ in weight by a factor of 3.8 but they are both cubes. An experiment has shown that the lighter casting solidifies in 8.7 min. How much time would you estimate that it would take for the larger casting to solidify?

8 In true centrifugal casting, i.e., spun about a horizontal axis, too low a speed will permit slipping or raining of the metal and too high a speed will result in hot tears on the periphery of the casting. A force on the metal rim of about 75 g works well for sand molds, whereas about 60 g is sufficient for metal molds because of their greater chilling power. As much as 100 g's are needed for molds spun about a vertical axis.

(a) Derive a relationship to express the minimum force in terms of g to make a true centrifugal casting D inches, diameter and spinning at n, rev/min. Use the centrifugal force on the annular ring and the angular velocity to obtain the relationship.

(b) Find the centrifugal force f at any radius r for a molten metal of specific weight d.

(c) Use the relationship obtained above to calculate the rotational speed required to cast ductile iron centrifugally in the form of a pipe with a $\frac{3}{8}$-in wall thickness. What speed would be needed if the casting were produced in a sand mold? What speed if cast in a metal mold?

(d)   If an unbonded sand layer is first distributed uniformly throughout the inner surface of the centrifugal mold, explain why molten metal can be poured into the mold without dislodging the sand layer.

(e)   Why is a greater force needed to make a centrifugal casting with a vertical axis than is required to make the same casting about a horizontal axis. Consider the wall thickness and contour in each case.

9   Sketch a match-plate layout for the gate valve casting shown in Fig. P10-9. Determine the parting line, the number of castings on a plate if the production rate calls for delivery of 800 valves per month excluding scrap which has run at an average of 8 percent of castings for porosity at the leak test which follows the machining operations at the customer's plant. The casting is a globe valve for domestic water systems, so choose the alloy accordingly. Specify the alloy to be used, the casting process, and the type of melting equipment. Would you modify the design of the part to make a better casting? If so, how? What gating ratio would you use? Show the calculations on your drawing.

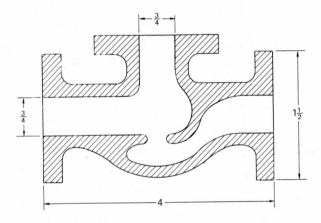

10   Design a cantilever load carrying member for a machine tool which must withstand 2,500 lb at its outer end. Use a flange design with a supporting rib. What is the best design in terms of strength to weight ratio. Compare this design to that of a steel weldment using an I beam to carry the same load. Use a class 40 gray iron for the cast beam.

11   A small aluminum casting is required to be cast to close dimensional tolerances including the location of its two cored holes. The casting weighs 5 lb. Specify the tolerances and pattern equipment and materials for production rates of 5, 100, and 20,000 castings. The part has an average wall thickness of 0.200 in and is the cover half of an enclosure for electronic equipment which will be exposed to outdoor environments throughout the world.

12   The part in Fig. P10-12 is to be made as a permanent-mold casting. Suggest what

changes in design should be made and explain why. Make a careful sketch to scale
to show the details of the part redesigned for permanent molding.

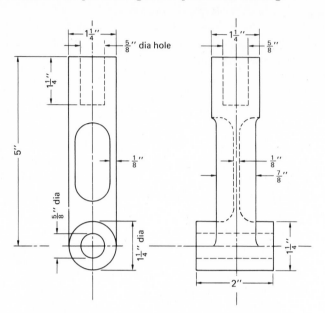

SELECTED REFERENCES

AMERICAN FOUNDRYMEN'S SOCIETY: "Cast Metals Handbook," Des Plaines, Ill., 1953.
———: "Design of Ferrous Castings," Des Plaines, Ill., 1963.
———: "Molding Methods and Materials," 1st ed., Des Plaines, Ill., 1962.
AMERICAN SOCIETY FOR METALS: "Casting of Brass and Bronze," Metals Park, Ohio, 1950.
———: "Casting Design Handbook," Metals Park, Ohio, 1962.
———: "Metals Handbook—Forging and Casting Handbook", vol. 5, Metals Park, Ohio, 1970.
FLEMINGS, M. C.: "Casting Metals," *Science and Technology*, December, 1968, p. 13.
GRAY AND DUCTILE IRON FOUNDERS' SOCIETY: "Gray and Ductile Iron Castings Handbook," Rocky River, Ohio, 1971.
MALLEABLE FOUNDERS' SOCIETY: "Malleable Iron Castings," Rocky River, Ohio, 1960.
STEEL FOUNDERS' SOCIETY OF AMERICA: "Steel Casting Design Engineering Data File," Rocky River, Ohio, 1963.
———: "Steel Castings Handbook," 3d ed., Rocky River, Ohio, 1960.

11

BASIC MANUFACTURING PROCESSES: SOLID STATE

When metals are formed at temperatures which exceed their recrystallization temperature, which is about 50 percent of their absolute melting temperature (Fig. 11-1), they are being hot-worked and behave as perfectly plastic materials (Fig. 11-2). The products of such hot-working operations are called wrought metals, which are important engineering materials because they are worked under pressure to:

1 Obtain the desired size and shape from the original ingot, thereby saving time, material, and machining costs

2 Improve the mechanical properties of the metal through: refinement of the grain structure; development of directional "flow lines" (Fig. 11-3); and breakup and distribution of unavoidable inclusions, particularly in steel

3 Permit large changes in shape at low power input per unit volume

Examples of wrought metals are structural shapes such as I beams, channels, and angles; railroad rails; round, hexagonal, and square bar stocks; tubes and pipes; extrusions and forgings. Most wrought products become the raw material

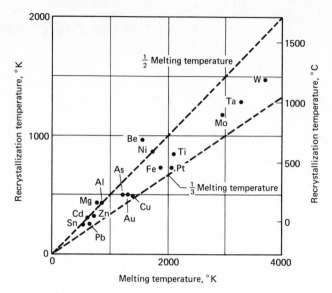

FIGURE 11-1

Recrystallization temperature of a number of metals as a function of their melting temperature.

for secondary processes which produce finished items by cutting, forming, joining, and machining.

FUNDAMENTALS OF HOT WORKING

When a metal is hot-worked, as by rolling, it is passed through opposing rollers and reduced in section thickness. The original grain structure of the initial

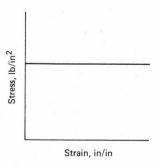

FIGURE 11-2

Typical stress-strain diagram for a hot-worked alloy.

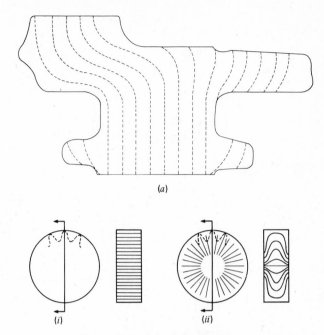

Direction of grain flow in a gear blank;
(*i*) bar stock and (*ii*) forged stock.

(*b*)

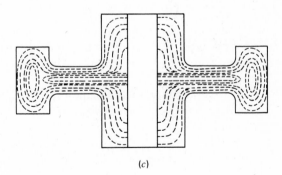

(*c*)

FIGURE 11-3
Typical fiber flow lines of a forging. (*a*) Flow lines in a forged crankshaft; (*b*) flow lines in a gear blank [*i*, bar stock, *ii*, forged blank]; (*c*) flow lines in a typical flanged gear blank.

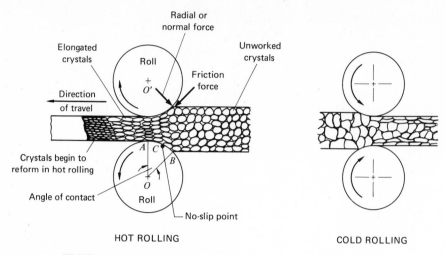

FIGURE 11-4
Hot rolling refines the grain structure, whereas cold rolling distorts it.

material is elongated and broken up in the deformation zone, and the fragments of crystals become nuclei for new smaller crystals so that a fine-grained structure is produced at the exit end (Fig. 11-4). The higher the metal temperature and the longer the time at temperature the more rapidly and larger the new crystals will grow. The larger the grains the less the hardness and strength. Hot working gives the following properties to an alloy:

1 It makes little change in the hardness or ductility. Recrystallization is spontaneous and the resultant fine-grained structure is stronger than before deformation, whereas cold working increases the strength and hardness of an alloy (Table 11-1).

2 The worked metal has enhanced properties in the direction of working because impurities segregate into stringers which lie parallel to the direction of metal flow.

3 Metals can receive large reductions when they are hot-worked compared to when they are cold-worked—and there will be no ruptures or hot tears.

There are disadvantages too. Hot working of steels requires expensive heat-resistant tools, but the hot working of aluminum alloys is not so severe on tools because the temperatures are less and there is little scale and oxidation. In the forging of steel close tolerances cannot be maintained, although with aluminum alloys this is not a problem.

Forging may be defined as shaping a metal under impact or pressure and improving its mechanical properties through controlled plastic deformation.

Table 11-1   COMPARISON BETWEEN PROPERTIES OF HOT- AND
COLD-WORKED METALS

Alloy	Condition	Ultimate tensile strength, lb/in²	Yield strength, lb/in²	Hardness, Brinell No.
Aluminum 1100	O	16,000	6,000	28
	H18	29,000	25,000	55
Electrolytic tough pitch copper	Hot-rolled	34,000	10,000	42
	Extra spring	57,000	53,000	99
Steel, SAE 1010	Hot-rolled	62,000	32,000	94
	Cold-rolled	81,000	50,000	174
Brass, yellow	Annealed	49,000	17,000	57
	Spring hard	91,000	62,000	174

Forged products can be given the optimum strength properties with a minimum variation of properties from piece to piece. Typical structures of a casting, bar stock, and forging are given in Fig. 11-5.

Forgings have a history of structural integrity which is not excelled by any other metalworking process. Internal voids and gas pockets are dispersed and welded shut. Forging is also foremost in uniformity and resistance to impact and fatigue. Yet when all factors are considered, forged connecting rods, crankshafts, and steering knuckles have had to yield to castings in a number of cases on the basis of economics despite the fact that these parts are critically stressed. The ability of the casting to put the metal where the stress is located has surpassed the advantage of enhanced strength because of the fiber flow pattern. In other cases the total cost of the casting is less because of the improved machinability of the equivalent cast alloy; or the ductility, strength, and economy of a

Grain

Casting   Bar stock   Forging

FIGURE 11-5
Schematic diagram of the grain flow in a casting, bar stock, and forging.

properly designed ductile iron casting has proved to be more economical than a competitive forging. Welding and cast-weld design also give forgings severe competition. But in nonferrous precision forgings, the improvement in properties and the low strength-weight ratio frequently give large aluminum forgings a competitive edge.

Materials

The predominant characteristic of a forging material is its tough, fibrous structure. When a heterogeneous material is formed or rolled to shape at a suitable temperature range, the distortion and subsequent recrystallization of the material results in a fibrous structure. In steel forging, this fibrous structure imparts directional properties which are an advantage when the lines of flow of the fibers are parallel to the applied forces, and a disadvantage when forces are applied at an angle to the lines of flow. There is less resistance to shearing forces parallel to the lines of flow than to those applied across the lines of flow.

The special properties of forged material are:

1 A homogeneous structure, free from voids, blowholes, and porosity.

2 Greater strength per unit of cross-sectional area under static loads, and better resistance to shock.

3 Superior machining qualities. The uniform structure permits higher machining speeds; the freedom from imbedded impurities allows longer tool life and fewer grindings; and homogeneity reduces machining scrap.

Depending upon the initial quality of the ingot and the processing history, the wrought product may be either superior to the cast product in every respect or only partly superior in some respects and distinctly inferior in others. The directional properties can be reduced by working the material in all directions. Grain size has very little effect on working properties of material that has high ductility. Grain size is controlled by heat treatment.

Forgeable Metals[1]

An almost unlimited variety of forging metals is available in ferrous and nonferrous alloys. The following are general classifications of forgeable metals. An exhaustive treatment may be found in books on forging and metallurgy.

1 Carbon steels:

a Low-carbon (up to 0.25 percent)—forgings for moderate conditions and for carburized parts where resistance to abrasion is important.

b Medium-carbon (0.30 to 0.50 percent)—forgings for more severe service. Some heat treatment is generally desirable.

---

[1] "Standard Practices and Tolerances for Impression Die Forgings," Drop Forging Association.

c   High-carbon (above 0.50 percent)—forgings for hard surfaces and for springs. Heat treatment is essential.

2   Alloy steels (manganese, nickel, nickel-chromium, molybdenum, chromium, vanadium, chromium-vanadium, tungsten, silicon-manganese).

3   Corrosion- and heat-resisting and stainless steels: Generally, but not necessarily, forged surfaces should be polished to obtain full benefit of corrosion-resisting properties.

4   Iron: Either wrought iron or ingot iron is forged for special applications where ductility is required. Wrought iron furnishes a moderate degree of corrosion resistance. The copper-bearing irons and low-carbon steels are in this class.

5   Copper, brasses, and bronzes.

6   Nickel and nickel-copper alloys: Pure nickel is forgeable. The alloy of nickel and copper known as Monel metal offers a desirable combination of strength, toughness, and corrosion resistance.

7   Light alloys (aluminum, magnesium).

8   Titanium alloys.

Note that any ductile metal can be forged. The kind of material is selected primarily for its ultimate properties in the part, such as corrosion resistance, strength, durability, and machinability. The forging process is a secondary consideration. It can be used as long as the high tooling cost can be spread over a large number of pieces.

Effect of Temperature on Material

The forming properties of any metal or alloy depend on the temperature of the material:

1   The hot-working range is near the melting point. The working temperature cannot exceed the melting point of any one of the elements of the alloy. Many such alloys can only be cold-worked. Most hot-worked metals do not acquire a permanent hardening.

2   The cold-working range usually is near room temperature. Most metals strain-harden when cold-worked, except zinc, lead, and tin.

3   At low temperatures, steel becomes brittle. Face-centered cubic metals like aluminum, copper, nickel, gold, and platinum remain ductile at all temperatures.

The forging processes work material cold or hot, depending upon the nature of the material and its size. The ability to work cold material depends upon its ductility and malleability as indicated by its stress-strain curve. The ability to work hot material depends upon its range of plasticity at higher temperatures. The greatest ductility is near the melting point. In general, most materials become more plastic as the temperature increases. The characteristics of each material at higher temperatures should be studied. For example, steel has a blue-brittle range between 450 and 700°F. Also, materials may oxidize rapidly and objectionable scale may form at high temperatures. The region of plasticity before the material becomes liquid may be very short and close temperature control may be required.

Stress-strain curves cannot predict the behavior of materials formed under various speeds. A study of equilibrium diagrams and phase changes used in metallurgy indicates little about the forming properties of a material. Experiments are required to determine the best temperatures and speeds of forming. Here again, statistical tools are quite helpful. Let us look at a typical example. A four-impression die involving drawing, edging, prefinal, and final form is designed. The part has intricate geometry and the material being forged is alloy steel. The production design engineer notes that there is considerable variation in the total number of blows required to shape the part. He suspects that the supplier of the steel is permitting undue variation in the material being furnished since the other parameters including forging temperature and forging speed are being closely controlled. He takes four samples of five forgings. Obviously, he would not expect each sample to have the same mean. He would, however, expect the within sample variation to be not significantly different from the between sample variation, unless the samples were drawn from different populations.

The results of these four samples are:

	Sample 1	Sample 2	Sample 3	Sample 4
	7 blows	7 blows	4 blows	10 blows
	6 "	8 "	4 "	9 "
	8 "	8 "	5 "	9 "
	5 "	7 "	4 "	9 "
	4 "	5 "	3 "	8 "
Totals	30	35	20	45
	$\bar{\bar{x}} = 6$	7	4	9
	$N = 20$	$\bar{\bar{x}} = 6.5$ blows		

Using simple analysis of variance, we get:

Source of variation	Sum of squares	Degrees of freedom	Variance estimate
Between samples	65	3	$\frac{65}{3} = 21.67$
Within samples	20	16	$\frac{20}{16} = 1.25$
Total	85	19	

Using an F test, we get a computed value of F equal to:

$$F = \frac{\text{greater estimate of the variance of the population}}{\text{lesser estimate of the variance of the population}}$$

$$= \frac{21.67}{1.25}$$

$$= 17.4$$

Comparing this value of F with the tabulated values for $F_{.99}$ (ninety-ninth percentile of the F distribution), which is 5.29, we can conclude that it is highly probable that the samples were not drawn from the same population. Thus, either a change in the steel furnished, a variation in forging temperature, or some other parameter was altered.

Failures

Failures of material as it is worked or formed depend on a combination of many factors. Material, as it is forced into the various types of tools and processes, may fail in shear, compression, buckle, or neck; or it may break under tension. If the friction between material and tool is great and the forces insufficient, the material may not fill the die and conform to the proper shape. Overlaps and folds may occur because of friction between die and material, or as a result of buckling. The cause of split sheets and cracked or buckled edges can be traced by analyzing the forces applied to the material.

The combination of forces shown in Fig. 11-6 can be applied to a small cube of material. The principle can be illustrated by cutting a cube from a rubber eraser and applying forces according to Fig. 11-6.

External forces always cause other forces in other planes. For example, in Fig. 11-7 a compressive force S_2 on a cube of material will result in tension force on the periphery at S_1. Thus we can have failure of material in the form

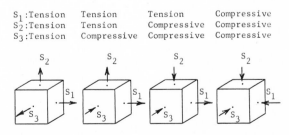

S_1:Tension	Tension	Tension	Compressive
S_2:Tension	Tension	Compressive	Compressive
S_3:Tension	Compressive	Compressive	Compressive

FIGURE 11-6
Four possible combinations of principal stresses.

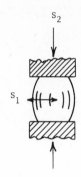

FIGURE 11-7
Cracking occurring in direct-compression-type working processes due to the presence of secondary tension (S_1).

of cracks around the middle due to tension forces (S_1) at right angles to the compression forces (S_2), which force the material into the center and outward faster than the outer material can stretch. This same phenomenon occurs when edges crack as sheets are rolled without proper annealing and the right amounts of reduction at each roll. Failures of materials due to splits, tears, and cracks can be understood by the action of the applied forces. A modification of these forces by annealing, reduction of the speed of the material passing through the rolls, or reduction of working pressures will prevent these failures. The engineer learns from these failures to improve his designs and tools.

Forgeability

The basic lattice structure of metals and their alloys seems to be a good index to their relative forgeability. The face-centered cubic metals are the most forgeable, followed by body-centered cubic and hexagonal close-packed. In alloys other metallurgical factors may affect their forgeability, such as: (1) the

Table 11-2 RELATIVE FORGEABILITY OF VARIOUS ALLOYS
(In Order of Decreasing Forgeability)

Good	Somewhat difficult	Difficult	Very difficult
Aluminum alloys	Martensitic stainless steel	Titanium alloys	Nickel-base alloys
Magnesium alloys	Maraging steel	Iron-base super alloys	Tungsten alloys
Copper alloys	Austenitic stainless steel	Cobalt-base super alloys	Beryllium
Carbon and alloy	Nickel alloys	Columbium alloys	Molybdenum
steels	Semiaustenitic PH stainless	Tantalum alloys	alloys

composition and purity of the alloy, (2) the number of phases present, and (3) the grain size. Forgeability increases with billet temperature up to the point at which a second phase appears or where there is incipient melting at grain boundaries or if grain growth becomes excessive. Certain mechanical properties also influence forgeability. Metals which have low ductility have reduced forgeability at higher strain rates, whereas highly ductile metals are not so strongly affected by increasing strain rates.

The relative forgeability of various alloys is presented in Table 11-2.

HOT-WORK PROCESSES

Hot-working processes, i.e., rolling, forging, upsetting, and extruding (Fig. 11-8) change the shape of metals by pressure. The metal is brought to the viscous or plastic state by subjecting it to elevated temperature, and by the use of pressure it flows without fracture, in contrast to the casting process in which the metal flows as a liquid and is chilled and solidified in the mold or die. In forging and other metalworking processes, the metal may be cold or it may be heated in order to bring it to a more plastic condition. The hot working of metals is accomplished at temperatures from 1700 to 2500°F (Fig. 11-9) for ferrous materials, from 1100 to 1700°F for copper, brasses, and bronzes, and from 650 to 900°F for aluminum and magnesium alloys.

In forging, the die is often heated slowly to working temperature before

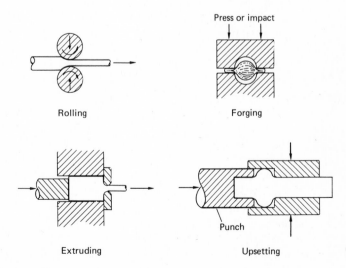

FIGURE 11-8
Schematic diagram of the four major hot-working processes.

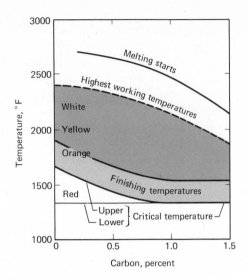

FIGURE 11-9
Range of rolling and forging temperatures for carbon steel.

starting the operation so that stresses due to differential temperatures will be avoided. Also cold dies in hot-forging operations may chill the blank sufficiently to cause considerable resistance to metal flow. The hot material remains in the die long enough to take the form of the die and is not chilled to any degree. Materials, even when they are in a plastic condition, resist change of shape with considerable force. Most operations are rapid, and therefore large amounts of power must be applied by the machines that operate the dies. This quick application of power heats up the material and aids the plastic flow. The art in forging, pressing, and working this viscous, doughy material is obtaining the final shape with the least number of tools, operations, and loss of material. The skills of the designer and the forger join in the art of creating the shape of the part and the dies to obtain the best flow of metals.

Metal is practically incompressible; therefore, excess material must escape or it will prevent the dies from closing or will break the rolls. With the more complex shapes and when less ductile and malleable materials are used, it is necessary to go through a series of stages before the final shape can be produced from the raw material. Each stage requires a different set of tools and sometimes requires different equipment and skills. Reheating of the material is usually required between these stages.

The forging and metalworking equipment described in this chapter requires hydraulic, steam, compressed-air, and mechanical systems, and all forms of mechanical actuating devices and controls. The heat is supplied by coal, oil, gas, and electrically heated furnaces. Sometimes resistance heating or induction heating is applied directly to the material in the machine.

HAMMER AND PRESS FORGING

The original forging of metals was done with a hammer held in the artisan's hand. The blacksmith with his forge and anvil shaped metal into useful objects; up until 1920 (and even now in a few places) he could be observed by the boys who were to become the future engineers.

Today only a few engineering schools teach hand forging in their laboratories because the hand process has been mechanized by industry.

When steam activates a vertical piston attached to the hammer, we have the steam drop hammer that can be controlled by a skilled operator as easily as the hammer in the hands of a blacksmith. Handling the raw material are mechanical and electrical manipulators that perform the same operations as the arm, wrist, and the hand holding the tongs. The board-type drop hammer is a drop hammer in the true sense, because the hammer is attached to and lifted up by a board clamped between two counterrotating rollers. When the hammer is at various heights, the pressure between the rollers is released and the hammer drops on the material and anvil. Drop hammers are very noisy and the shock of impact is difficult to isolate. The size of the drop hammer is indicated by pounds of falling weight. Board drop hammers range in size from 200- to approximately 8,000-lb falling weight and steam hammers range from 600 to about 50,000 lb.

The Chambersburg impactor has two hammers placed in horizontal position so that the two die parts held by impellers or hammerheads meet in the center. The impellers are actuated by air cylinders which drive them and return them to their resting positions. Special electronic controls regulate the entrance

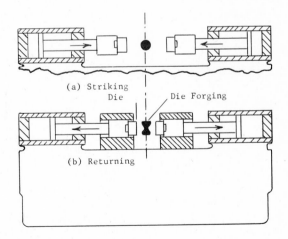

(a) Striking Die

Die Forging

(b) Returning

FIGURE 11-10
Operation of the impactor.

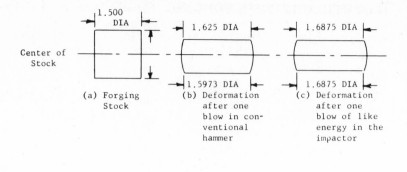

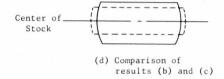

(d) Comparison of
results (b) and (c)

FIGURE 11-11
Comparison of results of a single blow of a given intensity on like forging
stock in conventional hammers and in the impactor.

of air into the cylinders and govern the speed of impact and the location of the
point of striking on the centerline of the machine (Fig. 11-10).

The impactor is a new application of a long-established principle—which
states that when two inelastic bodies of equal mass traveling at alike but opposite
speeds collide, both bodies come to rest with a complete absorption of energy. If
the bodies are elastic, the rebound of both is equal and opposite. The work done

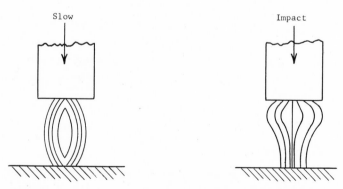

FIGURE 11-12
Flow of material under slow pressure or squeeze and under impact.

in both cases in bringing the two bodies to rest is concentrated at the impacting faces of the two bodies. Figure 11-11 compares conventional forging with the forging results of the impactor. No shock is felt when the impactor strikes. The equipment does not require special foundation or buildings. Multiple operations can be performed rapidly by placing additional units side by side so that no reheating is necessary between operations. The part is passed by a conveyor from one impactor to the other.

Hydraulic-actuated Forging Presses

Hydraulic-actuated forging presses are much slower than drop-hammer presses. In press forging, pressure or squeeze is applied to the raw material and the intensity of this pressure increases as the plastic metal resists deformation (Fig. 11-12). Due to the great pressures available, these presses can be made to have very large capacities. In World War II, Germany used a 30,000-ton press; presses up to 75,000 tons have been built. Such presses, the equipment to operate them, and the furnaces to heat the billets or blanks cost over a million dollars. The presses are used to develop materials and processes for the cold extrusion of steel and extrusion molding. Of course, hydraulic-actuated presses also are made in small sizes.

Operator skill is an important factor in the production of hammer and press forgings. The press may be entirely operated by one man, although more frequently, depending upon the size and nature of the work, two or more operators may work as a team. Hammer forging has low setup and tool costs and therefore is economical for low-quantity items. Tolerances should be liberal and in line with skills and customary practices; otherwise costs will be excessive. The cost of heating the material is a major item, as well as the loss of material due to scaling and removal of crop ends.

Mechanically actuated Hammers

For the most part, mechanically actuated hammers are made in small sizes. They impart a series of blows regulated by the speed and power applied to the hammer beam pivoted at one end. The power goes through a spring system which permits greater amplitude of the beam, and thus a heavier blow is struck when full power is applied.

DIE FORGING

Die forgings are made in steel dies constructed of hard, tough, and wear- and heat-resisting tool steel. As the mating halves of the die close, the material is shaped into the form of the die. When the die halves are separated, the forged part may be removed easily. In forming the part, several successive operations may have to be performed. The material may be shaped by extrusion or rolling and then cut to length. For example, preliminary to making a four-pronged handle for a water faucet, the shape may be cruciform before forging. This material may be partially formed into shape one or more times and then given the final form. The operations may be done in the same press and die, and the operations may be so close together that it is unnecessary to heat between them. As the part takes its final shape, a flash begins to form between the two halves of the die. This will be trimmed off either before or after the final forming operation. The amount of material within the die determines the thickness of the flash. The production-design engineer must exercise care in determining the amount of material to be inserted in the die. If there is not enough material, the part will not fill up the die; if there is too much material, the flash will be excessive, resulting in wasted material and greater die wear.

Parting lines usually are straight, although they may have a contour shape. In order to remove the forging from the die without distortion, the drafts in the die are made larger than in the casting and molding processes.

Die forgings usually are formed hot. The hot material cannot stay in contact with the die too long, for the forging then will not be hot enough for the next operation. Also, if the hot material is in contact with the die too long, it will overheat the die and so cause excessive wear, softening, and breakage. The rapidity with which the part can be formed into uniform shapes and with uniform properties permits the die-forging operation to compete with other processes on high-quality parts.

The use of closed-impression dies improves both the strength and toughness of the metal in all directions. The fiber structure characteristic of metals can be formed so as to improve the mechanical properties in areas where most needed to meet specific service conditions.

After the forging operation, the part must be trimmed to remove the flash. If the carbon content is low and the forging small, it is usually trimmed cold. Most medium-sized and large forgings are trimmed hot. Subsequent operations to remove the scale or oxide include shot blasting, tumbling, or pickling.

ROLL DIE FORGING

In roll die forging, often referred to as die rolling, the die configuration is placed on the periphery of the rolls and the material is formed as it passes between the rolls. The die cavities in the upper and lower rolls match and are held in position by a gear train on the ends of the rolls. The rolls are large in diameter and the frames and bearings are strong enough to withstand the forging pressures as the material is squeezed into shape. A series of dies can be placed side by side on one roll, so that the part can be passed through each die until the final shape is achieved. Such multiple operations are usually performed on single pieces. Roll forming is generally used for shaping parts that can be formed in a continuous chain from long bars heated to proper temperature and then cut to length. Such parts as automobile axle blanks, chain-sprocket rim segments, and shaker bolts are made by this method.

In many respects roll die forging is similar to the regular mill technique of hot rolling, the principal difference being that in hot rolling several passes are made to bring the billet to the desired section, whereas in roll die forging only one pass is made.

The most important use of this process is in the preparation of preformed blanks for forging.

Die rolling is a fast, mass-production process that requires considerable volume to justify the investment in dies and equipment. On light articles, a minimum requirement of about 150 tons would be required, whereas on heavier parts a volume of 600 tons should be available.

It should be recognized that roll die forging has limited application to parts that can be made continuously in chain form. Any material that can be forged can be die-rolled. In fact, even materials that present difficulties when subject to forging operations, such as AISI 6412 steel, can be die rolled at temperatures from 1850 to 2300°F. Another form of rotating die consists of two disks, set at an angle, which rotate while pressed together. Dies are mounted on each disk and the forging is squeezed into one die by the disk which is set at an angle. The pressure is applied at only one point. Gear blanks are made in this manner.

Roll die forming permits close tolerances. Outside diameters may be held from $\pm \frac{1}{32}$ in on small work to $\pm \frac{1}{16}$ in on large work.

UPSETTING

Upsetting is similar to die forging in that the shaping of the part is done by dies. The term *upsetting* refers to the process of increasing the size of a portion of the part by forcing material from the rest of the part. The forming of the head of a bolt from a small rod is an example. Upsetting was performed by the blacksmith by heating one end of the rod and then hammering it on the end until the desired amount of material was available to produce the final shape. Today automatic machines clamp the hot rod and a plunger strikes the end, forcing the material into a die that shapes the formed head. The depth of the plunger travel determines the height of the head. Other machines can perform more than one forming operation, as well as cut the material to proper length.

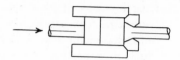

FIGURE 11-13
Direct extrusion.

HOT EXTRUSION

Extrusion (Fig. 11-13) is the process of shaping material by forcing it, under pressure, to flow through a die. In hot extrusion, the material is heated to a temperature that will cause it to become plastic without becoming liquid as it is forced through the die. In order to maintain its shape, the material is cooled rapidly on leaving the die. The hot-extrusion process is applied to both ferrous and nonferrous metals, plastics, ceramics, rubber, and other materials that have a plastic or doughy consistency.

Material can be extruded with internal holes running lengthwise. The sizes and shapes extruded are limited by the strength of the die and the capacity of the press that forces the material through the die. Metals can be formed into a final bar shape with good surface and workable limits. Dies are not expensive and they are maintained by the supplier of the extruded material. Due to the materials' plastic condition on leaving the die, the bar may twist or bend and have to be straightened after it cools. This distortion depends on the symmetry of the section. It is corrected by passing the section through drawing dies to obtain the final shape and straightness.

EXTRUSION MOLDING

Extrusion molding combines the hot-extrusion process with molding (Fig. 11-14). The dies are closed firmly so that the enormous internal pressures will not separate them. The plastic material is then forced in under extreme pressures. The metal remains in a plastic state—no portion of it is liquid—and therefore it does not have the shrinkage characteristics of die-cast materials, which cool

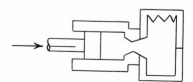

FIGURE 11-14
Extrusion molding.

from the liquid to the solid state. The raw metal is in billet form, prepared by the usual casting and forging methods. It is heated so that it can be extruded continuously. The heat and pressure change the solid material into a mushy plastic. Great pressures and a shock force are used to push the material into every portion of the die. There are no vents. The air or gas is trapped in overflow cavities so that the material is under pressure in all parts of the mold. Speed of entry and the shock are controlled by a hydraulic accumulator, which is permitted to fall at a given rate. Cores that can be pulled after the material is molded can be used. Bosses and flanges can be produced without setting up internal stresses, slip planes, or concentrated grain flow. The physical and mechanical properties of the resulting shape are comparable to those of the same material when forged, and are better than when the material is cast. The parts are uniform and accurate. The material breaks down or plasticizes as it passes through the restricting orifice, which reduces wear because the material does not touch the orifice except at the entrance. This orifice action is the same as that in the flow of fluids and as that described under impact extrusion. Because of the pressure and compression of the metal, heat is generated, which causes the metal to be uniformly refined and to have a fine-grained, mushy, plastic consistency. Too great a reduction in the orifice, the wrong orifice contour, or too fast an extrusion speed may cause the metal to overheat and become partly liquid, a most undesirable condition. This causes segregation and actual burning, which results in a cavity full of oxidized or burned metal.

This process is relatively new, so there have been many experiments to determine its potentialities. Empirical data sufficient for developing proper molds and orifices have been obtained. The process is controlled through electrical and hydraulic apparatus, and with properly regulated temperatures and skilled operators, uniform results that exceed those of other processes are obtained. The extrusion-molded parts are designed similar to other metal-molded parts in reference to uniform cross sections, radii, and entrance of material. Many materials that cannot be molded or cast by other processes can be successfully extrusion-molded. The molding of high-temperature alloys still presents the same problem of obtaining suitable materials and producing the molds, cylinders, and plungers that can withstand the high temperatures.

Since material molded by the extrusion-molding process is so uniform and possesses the best characteristics, the designer should base his factor of safety on the high proportional limits and the uniform performance of the actual part under test. In many cases, the weight of airplane parts can be cut one-third by using extrusion molding.

HOT ROLLING AND TUBE FORMING FROM SOLID BARS

Unlike the previously described processes, hot rolling and tube forming from solid bars are mill processes usually performed by the supplier of raw material.

The end product of the processes is standardized as to size limitations, tolerances, finishes, and materials available. Occasionally an engineer requires special mill shapes, such as elevator rails and conveyor tracks, for which a mill is willing to make special rolls and run an order of considerable tonnage. Since these processes are seldom used by the production-design engineer, they will not be described in this text. They are important processes used by the raw material supplier who furnishes special shapes for secondary operations.

HIGH-VELOCITY FORMING

A number of machines have been designed which can achieve high rates of energy release for forging operations. Their major contribution to conventional forging operations is their high impact velocity which is 2 to 10 times larger than conventional velocities. For example, a large steam hammer with a total die, ram, and piston assembly weight of 50,000 lb can deliver 850,000 ft·lb at an impact velocity of 30 ft/s. In contrast, a high-velocity forming machine with a 2,500-lb ram, piston, and die-block assembly at a velocity of 100 ft/s can deliver 1,250,000 ft·lb of energy at the moment of impact as determined by:

$$ KE = \frac{1}{2} \left(\frac{w}{g} \right) v^2 $$

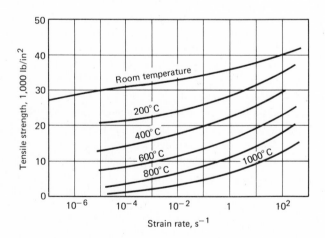

FIGURE 11-15
Effect of strain rate on the tensile strength of copper for tests at various temperatures.

where w = weight of moving ram and die, lb

g = acceleration of gravity, ft/s^2

v = velocity at impact, ft/s

In the conventional practice, the mass term has been made ever larger to obtain heavier forging forces, whereas in high-velocity forming the velocity term has been increased.

Although a great amount of engineering and development effort has been devoted to implementing high-velocity forming, there are two severe limitations. First, most metallic alloys have an inherent resistance to increased strain rates. The faster they are deformed, the greater their flow stress (Fig. 11-15). Unfortunately, that behavior is diametrically opposed to the reduction of the flow stress which is achieved by heating the alloy. Secondly, during deformation at high velocity considerable internal energy is released adiabatically. This heat has no where to escape, except into the die surfaces where it causes tremendous instantaneous expansion of the surface layers of the die. Thus an extremely steep thermal gradient is developed at the die surface which in turn causes thermal fatigue at an earlier time than would be the case in conventional forging.

The thermal fatigue is further complicated by the fact that the dies must overcome the augmented flow stress which results from the high rate of energy input to the dies. The two factors lead to a rapid deterioration of the die surface and to reduced die life compared with die life for the same part made by conventional forging practice. Despite the fact that certain more complex parts may be made by high-velocity forming, the total forging cycle time is reduced very little. Short die life has proven to be a major hurdle which has not yet been solved.

DESIGN FACTORS FOR HOT-WORKING PROCESSES

The processes and types of equipment that forge or work metals are based on many factors which apply in some degree to each process and forgeable material.

Some design factors for hot-metalworking processes are listed below.

1 The various sections of the forging should be balanced.

2 Generous fillets and radii should be allowed (see Fig. 11-16).

3 Sufficient draft (preferably 7 degrees) should be allowed.

4 Deep holes and high projections are not desirable.

5 Holes in two planes will make removal of the forging from the dies impossible.

6 Flash thickness variation should be specified (not less than $\frac{1}{32}$ in).

7 Raised letters or numbers for marking should be used.

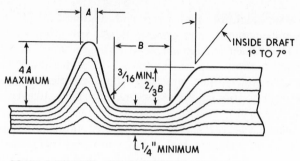

DRAFT, THICKNESS, AND PROJECTION LIMITATIONS
FOR NONFERROUS FORGINGS

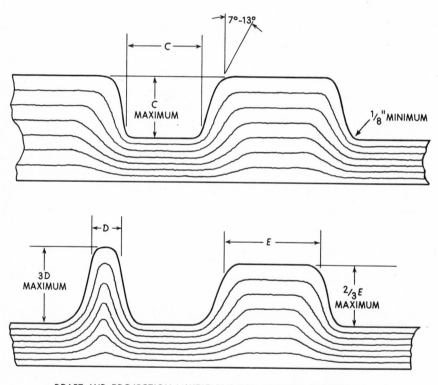

DRAFT AND PROJECTION LIMITATIONS FOR FERROUS FORGINGS

FIGURE 11-16
Forging design factors.

Advantages and Limitations

Designing parts for the various processes of forging should result in the following advantages:

1 Uniformity of quality for parts subject to high stress or unpredictable loads
2 Weight saving
3 Close tolerances
4 Less machining (none in some cases)
5 Smooth surface
6 Speed of production
7 Incorporation in welded structures

Some disadvantages in using the forging processes are:

1 High tool cost
2 High tool maintenance
3 No cored holes
4 Limitations in size and shape

Heat treating is used to increase the physical properties of the material, but it also increases the cost of the part.

Standard Specifications for Forgings

Impression die forgings are sold by the piece and not by the pound. It is understood without specific mention that the excess metal or flash of forgings shall be removed by trimming and that forgings shall be free from injurious defects.

Quantity   The quantity specified permits standard practice limits on overruns and underruns.

Size   Forgings within commercial size limits will be furnished unless closer tolerances are specified.

Coining or sizing   Closer tolerances may be obtained by additional hot- or cold-sizing operations.

Surface conditions   Forgings are generally, but not always, furnished in a cleaned condition by either tumbling, pickling, or blast cleaning.

Special requirements   Any special requirements, such as heat treatment or special tests, should be stated clearly.

Table 11-3   THICKNESS TOLERANCES, in*

Net weights, max., lb	Minus	Plus
0.2	0.004	0.012
0.4	0.005	0.015
0.6	0.005	0.015
0.8	0.006	0.018
1	0.006	0.018
2	0.008	0.024
3	0.009	0.027
4	0.009	0.027
5	0.010	0.030
10	0.011	0.033
20	0.013	0.039
30	0.015	0.045
40	0.017	0.051
50	0.019	0.057
60	0.021	0.063
70	0.023	0.069
80	0.025	0.075
90	0.027	0.081
100	0.029	0.087

* These apply to a thickness perpendicular to the plane of the parting line of the die.

Dies   Impression die forgings require special dies and tools for their production. The original charge for dies and tools conveys exclusive use but does not permit removal without additional payment. Dies and tools are maintained without additional charge.

Tolerances   It is important that the production-design engineer be aware of the tolerances that can be held by the forging process. Actually five tolerance areas should be considered. These are:

1   Thickness (applicable to thickness perpendicular to the plane of the parting line of the die)
2   Widths and lengths
3   Draft angle
4   Quantity
5   Fillets and corner

Width and length tolerances include shrinkage and die wear, mismatching, and trimmed size.

Table 11-4 SHRINKAGE PLUS DIE WEAR

Lengths or widths, max., in	Plus or minus	New weight, max., lb	Plus or minus
1	0.002	1	0.016
2	0.003	3	0.018
3	0.005	5	0.019
4	0.006	7	0.021
5	0.008	9	0.022
6	0.009	11	0.024
For each additional inch add	0.0015	For each additional 2 lb add	0.0015

The tolerances shown in Tables 11-3 to 11-8 may be used as a guide by the production design engineer when designing a part to be produced by the forging process.

Design for Upset Forging

1 Design for smallest diameter or section of stock.
2 Use the minimum of upset material and shape it into conventional styles—round, hexagon, and square.
3 Avoid square corners. Use as large a radius as possible at inside corners.
4 Avoid using head diameter greater than four times stock diameter. A maximum of $2\frac{1}{2}$ in diameter can be upset at one time.

Table 11-5 MISMATCHING TOLERANCES*

Net weight, max., lb	in
1	0.010
7	0.012
13	0.014
19	0.016
For each additional 6 lb add	0.002

* Mismatching is due to the displacement of one die block in relation to the other. This tolerance is independent of, and in addition to, any other tolerances.

Table 11-6   DRAFT-ANGLE TOLERANCES

	Normal angle	Close limits
Outside	$5\frac{1}{2}°$	$0-7\frac{1}{2}°$
Inside holes and depressions	$7°$	$0-7\frac{1}{2}°$

5 For hot forging follow these general rules when part is to be made by one stroke of the press without injurious buckling.

a Maximum length of unsupported stock is three times diameter of bar.
b When material is gathered in a recess $1\frac{1}{2}$ times (maximum) the diameter of the bar, any length can be gathered that is within the limits of the stroke of the machine without buckling.

Table 11-7   QUANTITY TOLERANCES

No. of pieces on order	Overrun pieces	Underrun pieces
1–2	1	0
3–5	2	1
6–19	3	1
20–29	4	2
30–39	5	2
40–49	6	3
50–59	7	3
60–69	8	4
70–79	9	4
80–99	10	5
	Percent	**Percent**
100–199	10	5.0
200–299	9	4.5
300–599	8	4.0
600–1,249	7	3.5
1,250–2,999	6	3.0
3,000–9,999	5	2.5
10,000–39,999	4	2.0
40,000–299,999	3	1.5
300,000 and up	2	1.0

Table 11-8 FILLET AND
CORNER
TOLERANCES

Net weight, max., lb	in
0.3	$\frac{3}{64}$
1	$\frac{1}{16}$
3	$\frac{5}{64}$
10	$\frac{3}{32}$
30	$\frac{7}{64}$
100	$\frac{1}{8}$

Ball-bearing blanks, small cups, and nuts, as well as bars with upset portions within their lengths, are made by the upsetting process.

Hot electric upsetting consists of passing a high current through the portion to be heated and pressing it into shape as it is heated. Longer lengths of material can be gathered by means of electric upsetting than are possible when the material is upset after furnace heating.

Die Design

The following considerations should be observed so as to minimize forging die cost per piece:

1 Forge two (or more) pieces as one and, subsequently, separate them in either a cutting or trimming operation.

2 Design forgings to incorporate straight lines and circular arcs so as to minimize toolroom time.

3 If possible, have the lower die a flat surface; thus only one-half (upper) will require machining to the required geometric contour.

4 In order to eliminate draft, consider locating the parting line so that the forging geometry provides a natural draft.

5 Consider the possibility of subsequently hot (or cold) twisting or bending of a simply designed forging rather than developing a complex die to forge completely the more complicated forging.

SUMMARY

When the production-design engineer specifies a forging or metalworking process, he is usually trying to design a product with more attractive mechanical characteristics, such as greater strength, or else to economize on the weight of the finished part, thus saving on material cost.

If a forging is selected, the designer must keep in mind the following principles of sound manufacturing design:

1   Plan carefully all draft angles so that the forging can readily be removed from the die and machining costs kept low.

2   Forgings should be designed to have a flat plane parting if possible. Equal volumes of metal should be on either side of the parting line.

3   Use generous fillets at all times.

4   Avoid abrupt changes in section thickness.

5   Avoid sections thinner than $\frac{1}{8}$ in.

6   Consider where it will be necessary to hold the forging for machining. Avoid using the parting plane as a location point.

7   Allow sufficient stock on areas that are to be machined. Usually this will vary from $\frac{1}{16}$ to $\frac{1}{8}$ in, depending on the size of the forging.

8   Design forgings so that they are supplied with indentations to spot holes that are to be subsequently drilled.

9   Avoid the design of forgings with deep recesses and pockets. Especially avoid complex contoured deep pockets.

10   Locate identification marks on surfaces that do not require machining.

QUESTIONS

1   What is the temperature range for the hot forging of ferrous metals? Nonferrous metals?

2   What is the difference between hammer and press forging?

3   Explain how a board hammer operates.

4   What is the range of capacity of the steam hammer?

5   Upon what basic principle is the Chambersburg impactor based?

6   Why is it important that in the final forging operation the dies should close?

7   When would you advocate the continuous heating of a forging die during a production run of forgings?

8   What is roll die forging? Why is the process limited to parts that can be made continuously in chain form?

9   What seven design rules should be observed by the production-design engineer when designing for hot upsetting?

10   To what materials is the hot-extrusion process applied?

11   When would you advocate extrusion molding?

12   What design factors should be considered when specifying a forging process?

13   What is meant by the "blue-brittle" range?

14   Why are forgings sold by the piece rather than by the pound?

15   What five tolerance areas should be considered in the forging process?

16   What property of metals makes justification of high-velocity forming difficult?

PROBLEMS

1   Design the upsetting dies needed to produce a blank for this pinion. Sketch the die
features to scale on quadrille paper and include dimensions. Include for each pair
of dies the shape at that stage of the preform with dimensions of the pinion form.
Give supporting calculations as needed.

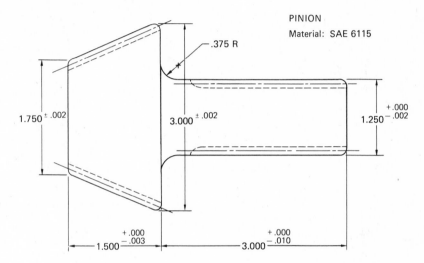

PINION

Material: SAE 6115

.375 R

$1.750^{\pm .002}$

$3.000^{\pm .002}$

$1.250^{+.000}_{-.002}$

$1.500^{+.000}_{-.003}$

$3.000^{+.000}_{-.010}$

2   A drop hammer weighing 1,200 lb has a free fall of 30 in. What is (*a*) the energy of
the blow and (*b*) the velocity at the instant of impact? What is the average force
exerted by the hammer if it deforms the forging $\frac{3}{32}$ in in one blow? What would be
the projected area of the workpiece if it were made of steel which has an average
yield strength of 12,000 lb/in² at 2000°F?

SELECTED REFERENCES

ALEXANDER, J. M., and R. C. BREWER: "Manufacturing Properties of Materials," Van
Nostrand, London, 1963.

AMERICAN SOCIETY FOR METALS: "Metals Handbook," 8th ed., vol. 5, "Forging and
Casting," Metals Park, Ohio, 1970.

AMERICAN SOCIETY OF TOOL AND MANUFACTURING ENGINEERS: "Fundamentals of Tool
Design," Dearborn, Mich., 1962.

CHASE, HERBERT: "Handbook on Designing for Quantity Production," 2d ed., McGraw-Hill, New York, 1950.

JENSON, J. E.: "Forging Industry Handbook," Forging Industry Association, Cleveland, Ohio, 1970.

JOHNSON, C. G.: "Forging Handbook," American Technical Society, Chicago, 1938.

KYLE, P. E.: "The Closed Die Forging Process," Macmillan, New York, 1954.

NAUJOKS, W., and D. C. FABEL: "Forging Handbook," American Society for Metals, Metals Park, Ohio, 1939.

PARKINS, R. N.: "Mechanical Treatment of Metals," American Elsevier, New York, 1968.

RUSINOFF, S. E.: "Forging and Forming of Metals," American Technical Society, Chicago, 1952.

SABROFF, A. M., F. W. BOULGER, H. J. HENNING, and J. W. SPRETNOK: "A Manual on Fundamentals of Forging Practice," Battelle Memorial Institute, Columbus, Ohio, 1964.

UNITED STATES STEEL CORPORATION: *The Making, Shaping and Treating of Steel*, Pittsburgh, 1964.

BASIC MANUFACTURING PROCESSES: PLASTICS

INTRODUCTION

In the fabrication of plastics, either thermoplastic or thermosetting, only a limited number of basic processes are available. Each basic process is able to impart some geometrical form to the plastic. If we know the geometry requirements of our design, the tolerance requirements of the design, the plastic that is to be used, and the quantity requirements, we can select the most advantageous basic process to produce the design. The selection of this "best" process can be done in advance of design completion if we are able to identify in advance those parameters that affect process selection. By having this information early, we can design for the most appropriate process and thus develop a functional design that is truly designed for production.

The following basic processes represent those that can be used in the fabrication of plastics.

COMPRESSION MOLDING

Compression molding is that basic forming process where an appropriate amount of material is introduced into a heated mold, which is subsequently closed under

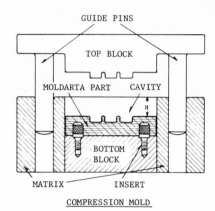

FIGURE 12-1
Typical compression mold.

pressure. The molding material, softened by heat, is formed into a continuous mass having the geometrical configuration of the mold cavity. Further heating (thermosetting plastics) results in hardening of the molding material. If thermoplastics are the molding material, hardening is accomplished by cooling the mold.

Figure 12-1 illustrates a typical compression mold. Here the molding compound is placed in the heated mold with inserts in position. After the plastic compound softens and becomes plastic, the top block moves down and compresses the material to the required density by a pressure of $\frac{1}{2}$ to 3 tons per sq. in. Some excess material will flow (vertical flash) from the mold as the mold closes to its final position.

Continued heat and pressure produce the chemical reaction which hardens the compound. The time required for polymerization or curing depends principally upon the largest cross section of the product and the type of molding compound. The time may be less than a minute, or it may take several minutes before the part is ejected from the cavity.

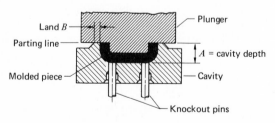

FIGURE 12-2
Flash compression mold.

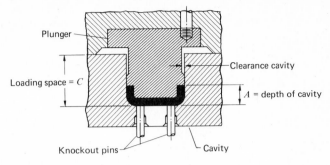

FIGURE 12-3
Positive compression mold.

The dimension H (Fig. 12-1) depends upon the nature of the charge. Bulky, nonpreformed materials require more depth than the more dense types: a mold made for dense materials (molding compound) frequently cannot be used for the bulky types. The difference in shrinkage of the materials will further reduce the number of cases in which it is feasible to change from one material to another after the mold has been made. The difference in the flow characteristics of the molding compounds is another factor limiting the change from one material to another type if the same mold without expensive changes is to be used.

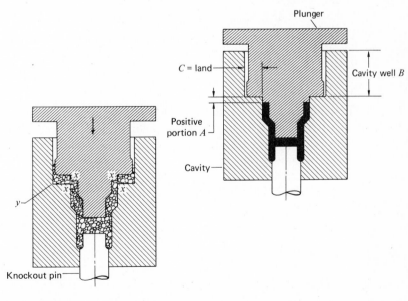

FIGURE 12-4
Semipositive compression mold.

Compression molds are of three types: flash, positive, and semipositive.

An illustration of a flash compression mold is shown in Fig. 12-2. Here, any excess material is squeezed out between the plunger and cavity across the land area. The land is usually $\frac{1}{8}$ in in width. Figure 12-3 illustrates a typical positive mold. Here the excess material moves vertically in the clearance between the plunger and cavity into a clearance cavity. The maximum clearance between the plunger and the cavity is 0.006 in across the entire part or diameter. Positive molds find application for long-draw parts or parts involving fabric-filled or impact-type thermosetting materials. In this type of compression mold it is important that exact weight be used when introducing the molding powder or preform.

In the semipositive mold shown in Fig. 12-4, we have a design that is partly positive and partly flash. When the semipositive mold is partially closed, some of the excess material escapes across the land as shown at y. When the corners at x pass into the mold cavity, the mold becomes positive and the remaining entrapped compound is compressed.

This type of compression mold finds particular application when the design involves sections of various thicknesses. As in the case of positive molds it yields a vertical flash which is usually easier to remove by belt sanding equipment.

Compression molding of thermosetting plastics (it is seldom used for production quantities of thermoplastic materials) offers the following advantages:

1 It permits molding of thin-walled parts (less than 0.060 in) with a minimum of warpage and dimensional deviation.
2 The absence of gate markings is advantageous. This is especially true on small parts.
3 Lower and more uniform shrinkage is a characteristic of compression molding.
4 Compression molding is usually more economical for large parts (parts weighing over 3 lb).
5 High-impact materials are difficult to find and consequently may be more easily processed by compression molding.
6 First costs are usually less since compression molds are usually more simple to design and construct.
7 In molding materials with reinforcing fibers, maximum impact strength is permitted since the reinforcing fibers are not broken up, which is the case with closed-mold methods.

TRANSFER MOLDING

Transfer molding is the process of forming articles in a closed mold, where the plastic material is conveyed into the mold cavity under pressure from an auxiliary chamber. The molding compound is placed in this hot auxiliary chamber and

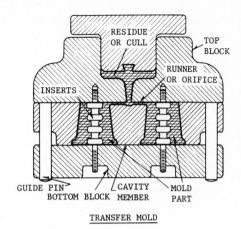

TRANSFER MOLD

FIGURE 12-5
Transfer mold.

subsequently is forced in a plastic state through an orifice into the cavities by the application of pressure. The heat and pressure must be held for a definite time for the chemical reaction (polymerizing or curing) to take place, depending upon the cross sections of the piece and the material used for molding. These same conditions are found in compression molding. The molded part and the residue (cull) are ejected upon opening the mold, and the need for flash trimming is not usually present, in contrast with conventional compression molding. Figure 12-5 illustrates a transfer mold used to produce a plastic component having inserts.

Transfer molds cost more than compression molds. In transfer molding some sections of the finished part may be weak due to incompletely filled sections. The pattern of flow may introduce sectionalizing and internal stresses that will result in weak or warped sections. Changing the material without producing a new mold is limited to approximately the same few cases as mentioned under compression molding.

The production-design engineer frequently will need to compute the maximum number of mold cavities that can be placed on a mold plate, giving consideration to pressure limitations. This can be determined through use of the relationship of mold-clamping pressure and plunger pressure. It has been determined[1] that the mold-clamping pressure should be at least 15 percent more than the plunger pressure in order to ensure safe operation and to avoid flashing. Since the mold-clamping pressure also equals the mold-clamping force divided by the area of the mold, we can express the projected area of the mold in terms of the plunger pressure and the mold-clamping force.

[1] Society of the Plastics Industry, Inc., "Plastics Engineering Handbook," Reinhold, New York, 1960.

For example, let us assume the following conditions applied and we wished to determine the maximum number of cavities that we would be able to attach to the mold plate in order to produce satisfactory molded parts.

Projected area of one molded article = 4 in²
Projected runner area per mold cavity = 1 in²
Total area per cavity = 5 in²
Diameter of clamping ram = 16 in
Diameter of injection ram = 6 in
Plunger diameter = 4 in
Line pressure = 2,000 lb/in²
Clamping ram force = $(\pi)\,(8^2)\,(2,000) = 128,000\pi$ lb
Injection ram force (preform) = $(\pi)\,(3^2)\,(2,000) = 18,000\pi$ lb

$$\text{Plunger pressure} = \frac{18,000}{(2^2)} = 4{,}500 \text{ lb/in}^2$$

$$\text{Mold clamping pressure} = (1.15)(4{,}500) = 5{,}175 \text{ lb/in}^2$$

$$\text{Mold clamping pressure} = \frac{\text{clamping ram force}}{\text{mold area}} = \frac{128{,}000\pi}{5N}$$

Then the number N of mold cavities would be

$$N = \frac{128{,}000\pi}{(5)(5{,}175)} = 15.6 \text{ or } 15 \text{ cavities}$$

INJECTION MOLDING

Injection molding is the process of forming articles by placing raw materials (granular, pellets, etc.) into one end of a heated cylinder, heating the material

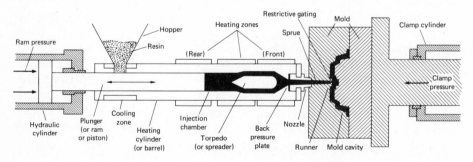

FIGURE 12-6
Schematic cross section of a typical plunger injection-molding machine.

FIGURE 12-7
Multiple-cavity mold used in conjunction with injection-molding machine.

in the heating chamber, and pushing it out the other end of the cylinder through a nozzle into a closed mold, where the molding material hardens to the configuration of the mold cavity. Figure 12-6 illustrates the injection-molding process and Fig. 12-7 shows a multiple-cavity mold used in conjunction with an injection machine for molding thermoplastic materials.

So as to allow adequate time for the heating of successive charges of the plastic molding material, the heating chamber is designed to accommodate several charges based upon the size of the molded piece. Unheated compound is metered to the cylinder every cycle so as to replenish the system for what has been forced into the mold. The pressure required to force the plastic molding compound through the heating cylinder and into the mold cavities varies between 10,000 and 25,000 lb/in². Heating temperatures in the cylinder vary with the plastic being injected, but are usually in the range of 200 to 600°F. Thermo-

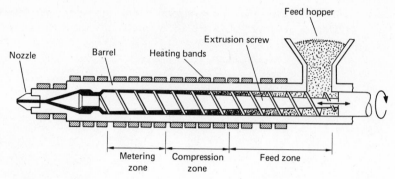

FIGURE 12-8
A typical reciprocating-screw injection-molding machine.

plastics need high barrel temperatures, usually between 350 and 600°F, and those thermosetting plastics that can be injection-molded, such as phenolic resins, use low barrel temperatures such as 150 to 250°F.

The number of variables that need be controlled in plastic molding is larger than most industrial processes. Here the equipment should control the material temperature, mold temperature, injection pressure, mold fill rate, clamping force on the mold, machine cycle time, shot volume, and the viscosity of the plastic at injection. Today the most commonly used injection system is the in-line reciprocating screw type (Fig. 12-8). It should be noted that the depth of the screw flights in this equipment varies from a maximum amount at the feed zone to a minimum at the metering zone. Also, the screw drive is designed to cause the screw to reciprocate as an injection plunger. Thus, the screw acts as a combination injection and plasticizing unit. As the plastic material passes from the hopper to the nozzle it encounters three zones: the feed zone, a compression zone, and a metering zone. At the metering zone the melted plastic is conveyed through an anti-flow-back valve to the nozzle. When the chamber in back of the nozzle becomes full, the resulting pressure forces the rotating screw back to a point where a switch is tripped causing the screw to be forced forward by hydraulic pressure. This forward thrust of the screw injects plastic into the closed mold. A nozzle shutoff valve is incorporated to prevent plastic from seeping out of the nozzle when the mold is open.

There are several parameters that determine the maximum number of cavities that may be attached to the mold plate. These include:

1   Shot capacity of the machine
2   Clamping capacity of the machine
3   Plastifying capacity of the machine
4   Cost of mold per cavity and cost of the total operation

Usually, it is wise to assign not more than two-thirds the shot capacity of the machine to the mold. Thus, if Q_1 = number of cavities based on the shot capacity of the machine, W_r = weight of the sprue and runner in ounces, and W_p = weight of the molded piece in ounces, we have this relationship:

$$Q_1 = \frac{2/3S - W_r}{W_p}$$

where S is the shot capacity of the machine in ounces.

Shot capacity provides a satisfactory estimate on heavy-sectioned designs. However, when thin flat articles are molded, the clamping capacity of the machine provides a better estimate of the number of cavities. Here we use 5 tons of clamping force for each square inch of projected cavity area. Thus, if

Q_2 = estimated cavities based on clamping capacity of machine
C = clamping capacity of machine, tons
A_r = projected area of sprue and runner, in²
A_p = projected area of molded piece, in²

we have the relationship

$$Q_2 = \frac{0.2C - A_r}{A_p}$$

It is also desirable to compute the number of cavities (Q_3) based on the plastifying capacity of the heating cylinder.

If we denote the plastifying capacity of the heating cylinder by P (pounds per hour) and the overall cycle in seconds by T, we have

$$Q_3 = \frac{0.00445\,PT - W_r}{W_p}$$

Of course, final decisions are based ultimately on cost, and the break-even chart can be used to determine the optimum number of cavities from a cost standpoint. Here fixed cost will be plotted as the cost of the mold (whether it be four, five, six, etc., cavities) and the variable cost will be the cost of production using that particular mold. Crossover points of various plottings will reflect the optimum number of cavities for various quantity requirements.

EXTRUSION

Extrusion is the process of continuous forming of plastic articles by softening the plastic material through heat and mechanical working or solvents, and subsequently forcing the soft plastic through a die orifice which has approxi-

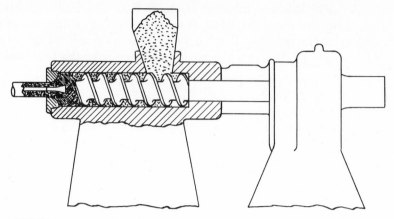

FIGURE 12-9
Continuous extrusion-molding machine.

mately the geometrical contour of the desired profile. The extruded form is
hardened by carrying it through cooling media. The operation of an extrusion
machine requires exacting heat and feed control accurately regulated for the
requirements of the material to be molded, as each plastic has its own individual
characteristics. The material, passing from hopper through machine body and
head, is forced through a die (Fig. 12-9). The extruded material passes onto a
conveyor operated at a speed regulated to avoid or control deformation of the
extruded shape, and is then wound into coils or cut to desired lengths.

The continuous-extrusion process of plastic fabrication makes possible the
use of plastics for many articles that it has previously been impractical or too
costly to produce by other methods of plastic fabrication. Rods, tubes, strips,
and forms of uniform cross section can be extruded in the wide range of colors
offered by thermoplastics, and, in many cases, with the elimination of costly
finishing operations. Extrusion molding almost always precedes blow molding.

CASTING

Casting of plastics is the process in which plastic materials (usually in the liquid
form) are introduced into a mold shaped to the contour of the piece to be formed.
Depending on the molding material, polymerization in the mold may occur at
room temperature or by controlled heating. The most important families of the
castable plastics are the ethoxylene, or, more commonly, the epoxy, family,
the phenolic family, and the polyester family. The mold that is used is usually
made of lead, plaster, or a flexible material such as rubber latex.

COLD MOLDING

Cold molding is that process in which mixed plastic compounds are introduced into room-temperature molds which are closed under pressure. The formed component is then removed from the mold, transferred to a heating oven, and baked until it becomes hard.

Conventional presses with pressures ranging from 2,000 to 12,000 lb/in² are used in this process. The molding cycle is relatively short. It consists of filling the mold, closing the press, and then removing the formed article. No curing takes place in the press, since the molding is done cold.

The molds are made of abrasive resistant tool steel with wall thicknesses sufficiently heavy to accommodate the pressures required to form the molding material.

THERMOFORMING

Thermoforming is the shaping of hot sheets or strips of thermoplastic materials into a desired geometrical contour through the utilization of either mechanical or pneumatic methods. Mechanical methods involve the use of a solid mold either moving or stationary. The pneumatic process involves the use of a solid mold and a differential of air pressure, created either by vacuum or compression.

Strips of plastic may also be thermoformed by passing the stock between a sequence of rolls that produce the desired contour. Heat is applied locally to the areas where bending takes place. As the formed shape emerges from the series of rolls, it is cut off to the desired length.

The sheets of plastic used in the thermoforming process are produced by either extrusion, calendering, pressing, or casting. Thicknesses of sheets that are processed range from 0.003 in to as heavy as 0.50 in or even greater.

The temperature range for forming varies significantly with the material being processed. Most thermoplastic materials become soft enough for thermoforming somewhere between 275 and 425°F. They are brought to these temperatures by infrared radiant heat, electricity, or forced-air ovens heated by gas or oil. The heating of the plastic sheet should be accomplished as rapidly and as evenly as possible over the whole area to be thermoformed.

Those thermoplastics that cannot tolerate intense heat, such as the acrylics, are brought to forming temperature in gas- or electric-heated ovens. The temperatures involved here usually range between 275 and 350°F.

The production-design engineer must realize that in thermoforming, the wall thickness in the finished product will be reduced proportionally to the increase in area of the formed piece over the area of the original sheet. For example, Fig. 12-10 illustrates a cycle in thermoforming a thermoplastic dish-shaped housing. It can be seen that the wall thickness of the finished part is considerably less than the original stock.

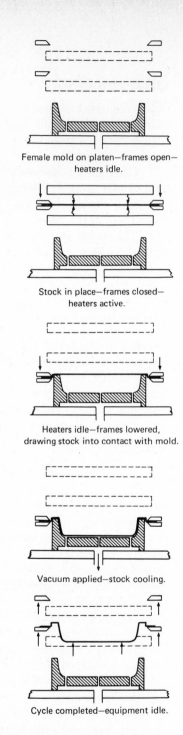

Female mold on platen—frames open—
heaters idle.

Stock in place—frames closed—
heaters active.

Heaters idle—frames lowered,
drawing stock into contact with mold.

Vacuum applied—stock cooling.

Cycle completed—equipment idle.

FIGURE 12-10
Cavity forming. (*Courtesy Society of the Plastics Industry, Inc., "Plastics Engineering Handbook," 1960.*)

BLOW MOLDING

In blow molding a tube of molten plastic material is enclosed in a split mold and then this tube is blown out to match the shape of the mold. The tube of plastic material is the end product of the extruder. Typical materials that are extruded for subsequent blow molding include high- and low-density polyethylene (e.g., squeeze bottles), nylon, PVC, polypropylene, polystyrene, and polycarbonates.

Today, blow molding can be integrated with the extrusion process so as to provide a continuous operation. In this arrangement, continuous extrusion of preforms take place by having the extruded stock cut to length and then drop into open molds mounted on a turntable. After the hot section of extruded stock, known as a *parison*, enters the open mold, the mold closes so as to seal and pinch off the open ends of the deposited extruded stock. Air is now injected, and the tube of stock is blown out to match the contour of the mold. The part is now cooled and then ejected automatically from the opened mold.

In blow molding the air pressure required is between 40 and 100 lb/in^2. Usually, the higher the pressure the better the surface finish of the blown item. In order to control the wall thickness of the finished part a low pressure (usually between 5 and 20 lb/in^2) is used to blow the molten parison to its initial molded form; subsequently, a higher pressure (usually between 60 and 100 lb/in^2) is used to firmly hold the material against the mold cavity wall while it cures, thus minimizing wrinkling and distortion due to shrinkage.

In very thick sections, carbon dioxide or liquid nitrogen may be used to hasten the internal cooling while the geometry is being formed.

MACHINING FROM STOCK

When only one or a few of the finished parts are needed, it may be more economical to rough and finish machine standard stock to the form required rather than provide the tooling to utilize one of the aforementioned basic processes. Most plastic materials can be purchased in standard sheets, rounds, or flats. These standard shapes can be transformed to the desired geometry by typical machining operations (sawing, drilling, tapping, turning, grinding, etc.).

In turning and milling plastics diamond tools provide the best accuracy, surface finish, and uniformity of finish. Surface speeds of 500 to 600 ft/min with feeds of 0.002 to 0.005 in are typical.

SUMMARY

With the exception of machining from stock, a mold, or sequence of rolls, must be made for all plastic forming processes; each process involves a different mold type. The design of a part governs whether simple or complex and expensive molds will be used.

Molding may be done in molds mounted in automatic presses, in semiautomatic presses, or in simple hand molds. The proper method of molding is determined by the plastic material being processed; the number of parts to be produced within a given time; the size, geometry, and tolerance requirements of the design; and the inserts to be molded into the piece.

Some general guides to mold design are:

1 Specify how many cavities the mold should contain. Indicate each cavity identification and location in the mold.

2 Specify make and model of press that will accommodate the mold. Mold length and width and minimum and maximum shut height need be considered.

3 For handling purposes specify eye bolts on two sides 90 degrees apart. A $\frac{5}{8}$-in-diameter eye bolt is typical.

4 Consider parting-line location and the location and size of ejector pins in both the mold cavities and the runners.

5 Provide for venting if a gas-trapping problem is predictable.

6 In designing the mold consider the shrinkage rate of the plastic material being processed.

7 Specify runner size. In case of doubt, start small—it can always be enlarged.

8 Specify gate position and gate size.

9 Specify draft.

10 Specify mold finish. Never underspecify finish.

11 Specify metal and hardness from which mold cavities and cores should be made. In the case of molds for vinyl plastics, stainless steel or hard chrome plating is recommended to avoid corrosion.

12 A working tolerance of $\pm$ 0.005 in is practical in mold construction. Tolerances of $\pm$ 0.0002 in can readily be held but at an increase of 15 to 20 percent in cost.

QUESTIONS

1 What parameters should be considered in the selection of the most appropriate basic process for producing plastic components?

2 Is the compression-molding process capable of producing thermoplastic parts containing inserts? If 1,000 pieces of such a design were required, would you recommend compression molding? What other considerations should be investigated? If flash minimization were critical what process would you recommend?

3 If the total projected area of a mold cavity and its runner is 7 in², how many cavities can be made on a mold for a transfer press having a clamping ram of 20 in diameter, an injection ram of 6 in diameter, a plunger of 5 in diameter, and a line pressure of 2,000 lb/in²?

4 What is the function of the torpedo in the injection-molding machine?

5 What can you say relative to the cost of molds used in the cold-molding process?

6 Would you say the casting process is a high-volume process for producing epoxy components? Explain.

7 Describe the blow-molding process.

8 Explain the relation between vinyl plastics and mold corrosion.

PROBLEMS

1 An injection press has a shot capacity of 18 oz, a clamping force of 180 tons, and a plastifying capacity of 120 lb/h. The following information has been estimated by the manufacturing engineer in conjunction with a new plastic component that is to be produced on the injection press: Weight of molded piece, 3 oz; weight of sprue, 3.5 oz; weight of runners, 1 oz $+ 0.5N$, where N = number of cavities; projected area of piece, 4 in²; projected area of sprue, 1 in²; projected area of runners, $1.5N$; estimated cycle time, 1 min; expected shrinkage (losses and rejects), 10 percent; production requirements, 800 pieces per 8-h shift. How many cavities should the mold design include?

2 The case for a volt-ohm meter can be produced from cellulose acetate by injection molding or from a urea formaldehyde compound by compression molding. The case requires 6.8 in³ of material. Cellulose acetate has a cost of $0.032/in³, and can be molded in 15 s in a die with an initial cost of $6,800. In contrast urea resin has a cost of $0.019/in³. It takes 90 s to make one piece in a mold which costs $1,300. Determine the break-even point analytically and graphically if the operator rate is $15/h including the overhead allowance.

SELECTED REFERENCES

DOYLE, ET AL.: "Manufacturing Processes and Materials for Engineers," 2d ed., Prentice-Hall, Englewood Cliffs, N.J., 1969.

LINDBERG, ROY A.: "Materials and Manufacturing Technology," Allyn and Bacon, Boston, 1968.

THE SOCIETY OF THE PLASTICS INDUSTRY, INC.: "Plastics Engineering Handbook," 1960.

13

SECONDARY MANUFACTURING PROCESSES: MATERIAL REMOVAL

INTRODUCTION

In many cases products from the primary forming processes must undergo further refinements in size and surface finish to meet their design specifications. To meet such precise tolerances the removal of small amounts of material is needed. Usually machine tools are used for such operations, and metals are chosen for the product because of their long life and ease of shaping.

In the United States material removal is big business—in excess of 36×10^9 per year, including materials, labor, overhead, and machine-tool shipments, is spent. Since about 60 percent of the mechanical and industrial engineering and technology graduates have some connection with the machining industry either through sales, design, or operation of machine shops, or working in related industries, it is wise for an engineering student to devote some time in his curriculum to studying material removal and machine tools.

It is evident that the advances of our technological civilization have been achieved largely because man has developed measuring devices and machine tools, along with an ability to use them. Without machining, many of America's cherished technological devices would soon disappear and life would revert to a simpler plane. The precise sizing and smooth surfaces which are produced by machining underlie all the technological developments of our day—rapid transportation, hydrogen bombs, jet engines, and nuclear power plants, as well as kitchen appliances and sanitary facilities. Machine tools have justly been called the master tools of industry.

Although James Watt designed his steam engine and separate condenser in 1775, it took Wilkinson and Watt 25 more years to build it. They considered it to be a major triumph when their first cylinder was bored so true that "when a piston was tightly fitted at one end, a clearance no greater than a worn shilling was present at the other." Today machinists routinely bore cylinders to a tolerance of ± 0.001 in and on request can achieve ± 0.0001 in. The difference between a new car engine and one which is worn out is only a few ounces of metal in critical areas such as in crankshaft bearings or cylinder walls. Yet this small amount of wear can cause an engine block weighing a hundred or more pounds to be scrapped.

In the early 1800s Eli Whitney developed the concept of mass production in his design of tools and gages for the manufacture of muskets for the U.S. government. Until that time all guns were custom-made by hand, so that no two muskets were exactly alike. After a number of years of development, under a government grant, Whitney finally traveled to Washington to show the Army and congressional officials his new concept. From a table covered with 10 dismantled muskets, Whitney picked parts at random and quickly assembled 10 muskets, all of which worked perfectly. But even more astonishing was the fact that the jigs, fixtures, and machine tools at Whitney's factory could produce thousands more with relatively little effort—a significant advantage for the United States in the War of 1812. Interchangeable manufacture and mass production are two concepts which foreshadowed America's production supremacy. The final concept emerged in 1914 when Henry Ford introduced assembly-line production. The simultaneous use of these three concepts—mass production, interchangeable manufacture, and assembly-line production—meant an unprecedented increase in the productivity of American labor relative to that of the rest of the world. However, since World War II we have lost much of our advantage in productivity because of our ever-increasing labor cost and the growing use of automation abroad. Other countries have been developing rapidly, and the United States is no longer assured of perennial first place. Japan is first in shipbuilding and is vying for leadership in high-quality cameras, radios, televisions, and automobiles.

GENERAL PRINCIPLES OF MACHINING

For ductile materials, machining can be defined as the removal of material in the form of chips. Hard or brittle materials can also be machined by the process of material removal, but in such cases the residue is in the form of finely fractured fragments or has been chemically dissolved from the surface of the main material.

We may define a tool as a device which enables man to perform a given task more easily or efficiently. Hand tools include screwdrivers, pocket knives, hammers, vises, etc. Machine tools are power driven and remove material by means of cutting tools from a workpiece in the form of chips. Machine tools must rigidly support the cutting tool and workpiece and supply relative motion between them. Consider a drill press. The drill is held by a chuck mounted in the machine spindle and the workpiece is held in a vise which in turn may be clamped to the machine table after the work is correctly located with respect to the drill. Metal removal occurs at the drill point where the revolving cutting edges form chips which travel up the drill flutes to make way for other chips. A twist drill is an efficient cutting device because the hole itself provides considerable support during the cutting operation.

In the case of hard materials, such as tungsten carbide, chips cannot be formed, so material removal is accomplished by chemical dissolution or by controlled localized disintegration. These methods include electrolytic grinding, chemical machining, electric discharge, or ultrasonic machining.

Speeds and Feeds in Machining

Speeds, feeds, and depth of cut are the three major variables for economical machining. Other variables are the work and tool materials, coolant, geometry of the cutting tool, and tool life. The rate of metal removal and power required for machining depend upon these variables.

The depth of cut, feed, and cutting speed are machine settings which must be established in any metal-cutting operation. They all affect the forces, power, and the rate of metal removal. They can be defined by comparing them to the needle and record of a phonograph. The *cutting speed* is represented by the velocity of the record surface relative to the needle in the tone arm at any instant. *Feed* is represented by the advance of the needle radially inward per revolution, or it is the differences between two adjacent grooves. The *depth of cut* is the penetration of the needle into the record or the depth of the grooves. A simplified formula for converting cutting speed CS, usually expressed in feet per minute, to revolutions per minute (r/min) of the machine spindle is derived as follows:

$$\text{r/min} = \text{CS} \times \frac{12 \text{ in}}{\pi D} \quad \text{or} \quad \text{r/min} \approx 4 \times \frac{\text{CS}}{D}$$

where D = diameter of work, in.

Suitable cutting speeds for each workpiece material and tool material have been derived through extensive experimentation which in turn has evolved into

Table 13-1   TOOL SHAPES AND CUTTING SPEEDS FOR TURNING STANDARD MATERIALS*

Work material	Tool material	Back rake, °	Side rake, °	Relief angles, °	Nose rad., in	Cutting speed, ft/min	Tool life, min	Cutting fluid
Aluminum	HSS	15	15–40	8–10	$\frac{1}{16}$	400–1,000		PO
Aluminum	WC	10–20	10–20	8–10	$\frac{1}{16}$	1,000–3,000		SO
Brass	HSS	0	0	6	$\frac{1}{16}$	300		Dry or SO
Brass	WC	0	4–14	6	$\frac{1}{32}$	700		Dry
Cast iron (med.)	HSS	8	14	6	$\frac{1}{16}-\frac{1}{2}$	80		Dry
Cast iron (med.)	CA	0	6	6	$\frac{1}{16}-\frac{1}{8}$	200	180	Dry
Cast iron (med.)	WC	0–4	0–4	6–8	$\frac{1}{16}-\frac{1}{8}$	240–350	660	Dry
Steel SAE B112	HSS	8	18	6	$\frac{1}{16}$	180		SO
Steel SAE B112	CA	0	10	6	$\frac{1}{16}$	350		SO
Steel SAE B112	WC	0	6	6	$\frac{1}{16}$	300–600		Dry
Steel SAE 1020	HSS	8	22	6	$\frac{1}{16}-\frac{1}{2}$	90		SO
Steel SAE 1020	CA	0	10–12	6	$\frac{1}{16}-\frac{1}{8}$	150		SO
Steel SAE 1020	WC	0	10–23	6	$\frac{1}{16}$	260		Dry
Steel SAE 1045	HSS	8	20	6	$\frac{1}{16}-\frac{1}{2}$	75		SO
Steel SAE 1045	WC	0	10–14	6–8	$\frac{1}{32}$	200–400		Dry
Steel SAE 1090	HSS	8	14	6	$\frac{1}{16}-\frac{1}{2}$	50		SO
Steel SAE 1090	WC	0	5–10	6	$\frac{1}{16}$	200		Dry

HSS = High-speed steel
 CA = Cast alloy
 WC = Tungsten carbide

SO = Soluble oil emulsion
PO = Paraffin oil and 5% oleic acid

* Depth of cut, $\frac{3}{32}$ in; feed, $\frac{1}{32}$ in.
SOURCE: Adapted from O. W. Boston, "Metal Processing," J. Wiley, New York, 1951, p. 144.

Table 13-2   TYPICAL FEEDS AND DEPTH OF CUT FOR LIGHT
MACHINING

General feed designation	Typical feeds, in/r, or in/tooth	Depth of cut designation	Depth of cut, in $\times$ 10^{-3}
Fine	0.001–0.003	Light	3 $\times$ (feed)
Medium	0.005–0.015	Medium	5 $\times$ (feed)
Coarse	0.018–0.060	Heavy	10 $\times$ (feed)

relative machinability ratings for most engineering materials. Such data are available in many machining and engineering handbooks. Whereas such standardized data may vary somewhat from optimum machine settings because of slight differences in material specifications, they are a good basis for original machine-tool setups (Table 13-1).

Feed is the advance of the cutting tool (in thousandths of an inch per revolution of the spindle) through the workpiece in rotary-motion machine tools. The feed on a milling machine is usually given in inches of table travel per minute, while cutter feed is expressed on the operation sheet as the amount each tooth of the milling cutter advances into the workpiece. Table travel is calculated as feed per tooth $\times$ the number of teeth in the cutter $\times$ the r/min of the spindle. For example,

$$0.002 \text{ in} \times 20 \text{ cutter teeth} \times 100 = 4 \text{ in/min table feed}$$

On a reciprocating machine tool, the feed is the amount the tool advances normal to the direction of cutting (planer) or the amount the workpiece advances normal to the direction of cutting (shaper) at each reciprocation of the table or ram. The proper feed can only be selected after careful consideration of the combined effect of depth of cut, rigidity of the cutting tool and the workpiece, the method by which the workpiece is held in the machine, the microstructure of the material being cut, the geometry of the cutting tool, and the surface finish specified (Table 13-2). The depth of cut is a function of the machine setup, capacity and rigidity of both the machine and the workpiece, and the horsepower of the machine tool.

On rough material or workpieces where a large amount of material must be removed, three cuts are required. The roughening cut should be of sufficient depth to remove all traces of sand, scale, and eccentricity on cast metals and forgings. The semifinish cut should remove from $\frac{1}{32}$ to $\frac{1}{16}$ in of metal. The purpose of the semifinish cut is to remove any traces of out-of-roundness and also to serve as an index for sizing the finishing cut. The allowance for the finishing cut should be approximately 0.015 to 0.030 in.

MACHINE TOOLS

Cutting-Tool Geometry

Before a systematic study of metal cutting can be undertaken the standard nomenclature for the cutting-tool angles must be understood (Fig. 13-1). There are six single-point tool angles which are important to the machinist and processing engineer. These can be divided into three groups:

Rake angles   These affect the direction of chip flow, the characteristics of chip formation, and tool life. Rake angles are on the top of the tool bit.
Relief angles   These avoid excessive friction between the tool and the workpiece and allow better access of coolant to the tool-work interface.
Cutting-edge angles   The side cutting-edge angle allows the full load of the cut to be built up gradually, thus reducing the initial shock to the cutting tool. The end cutting-edge angle allows sufficient clearance so that the surface of the tool behind the cutting point will not rub over the work surface and cause increased frictional heat. In finish machining the end cutting edge might be flattened for about 1.5 times the feed increment to permit a burnishing action between the tool and the work surface. Thereby the ridges caused by the feed increment would be eliminated.

Tool Signature

The elements of a single-point tool are written in the following order, which is the tool signature: back rake angle, side rake angle, end relief angle, side relief angle, end cutting-edge angle, side cutting-edge angle, and nose radius (Fig. 13-1). For a carbide cutting-tool insert in a standard holder the signature might be $-5°$, $-5°$, $5°$, $5°$, $15°$, $15°$, $\frac{3}{64}$ in. Note that the second angle of each pair refers to the side designation; i.e., first comes the back rake angle and then the side rake angle; etc.

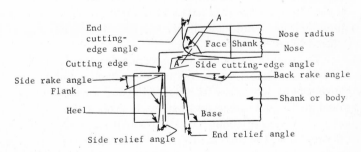

FIGURE 13-1
The geometry of a single-point cutting tool.

In general, the cutting force decreases as the rake angle increases. An economic balance must be struck between reduced cutting force and tool life because larger rake angles mean reduced efficiency of heat removal, less tool rigidity, and shorter tool life.

When using cemented carbide tools and especially cemented oxide (ceramic) tools, the solid angle at the cutting edge must be as large as possible to prevent tool failure by chipping of the cutting edge. In these cases, negative rake angles are commonly used because, despite greater horsepower requirements per unit volume of metal removed, the tool life is much improved and the total cost is significantly reduced.

Note   When a mechanical tool holder is used in conjunction with tool bits of high-speed steel, several tool angles must be corrected by an amount equal to the angle of inclination of the slot for the tool bit. Be sure to determine which angles are involved and make appropriate corrections. The usual angle of inclination of the tool slot is 15° above a horizontal plane. It should always be checked in a specific case.

[handwritten margin note: BE SURE AND CHECK]

Twist Drill

[handwritten note: FEED RATE IS PROPORTIONAL TO DIAMETER.]

The twist drill is designed to originate holes in metal parts. It must have a reasonably long tool life when used at a high rate of penetration. Often drilling is a preliminary operation to reaming, boring, or grinding during which final finishing and sizing are accomplished. Holes as small as 0.005 in can be drilled using high-speed and special techniques, but about 0.015 in is considered to be a small hole. On the other hand, holes more than 2 to $2\frac{1}{2}$ in diameter are seldom drilled because coring or burning and boring to size is frequently less expensive.

Twist drills are related to lathe tools (Fig. 13-2). Note the important features of the twist drill shown in the figure: the helix angle s, the point angle p, the web thickness w, and the drill diameter d. A typical drill has a helix angle of 30°, point angle of 118°, a web thickness of approximately $0.015d$, and a clearance angle of 10°. A twist drill must be sharpened so that the point angle lies symmetrically about the drill axis and the two lips are equal. When hard materials are drilled, the point angle is increased to about 140° and the clearance angle is reduced to about 5°.

The feed rate of a drill is proportional to its diameter because it depends on the volume of chips which the flutes can handle, but feed is independent of the cutting speed, which is a function of the tool work combination. A rule of thumb would give a feed rate as approximately $d/65$, so that a $\frac{3}{4}$-in-diameter drill would have a feed rate of about 0.012 in/rev. Although the hole wall tends to support the drill when the hole depth exceeds three times the drill diameter, there is a tendency for buckling to occur and the feed rate should be reduced.

Most drills are made from high-speed steel because of its relatively low

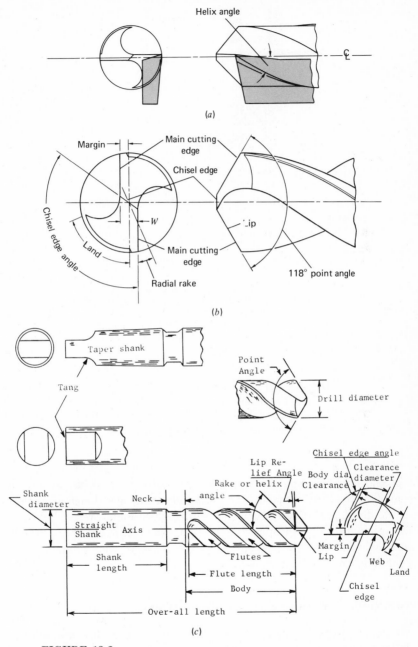

FIGURE 13-2
Geometry of the twist drill. (*a*) Comparison of twist drill and single-point
tool; (*b*) standard designation of drill point features; (*c*) standard designation
of twist-drill body and shank.

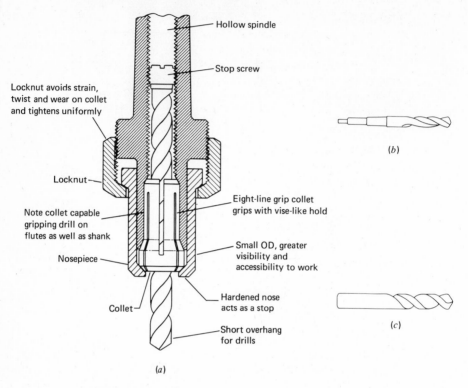

FIGURE 13-3
Specialized twist drills for numerically controlled operations. (*a*) Collet for twist drill to permit short stickout; (*b*) double margin drill to increase accuracy of drilled hole; (*c*) short stubby drill with heavy web and special spiral point.

cost and ease of manufacture. Some types of carbide drills are now available commercially. The demands of numerically controlled machine tools have led to the development of drills which will produce more precise holes and which will originate a hole in line with the centerline of the drill-press spindle. Drills with heavier webs, less stickout, double margins, and ground with a spiral point help meet these new demands (Fig. 13-3).

Multipoint Cutting Tools—Milling Cutters

Milling chips are relatively short and, when produced by slab milling, the undeformed chip thickness, i.e., original thickness of the layer removed by machining, varies along the length of the chip. There are two major types of milling operations: (1) plain milling and (2) face milling. A plain milling cutter rotates about a horizontal axis (Fig. 13-4) has cutting teeth only on its periphery, and

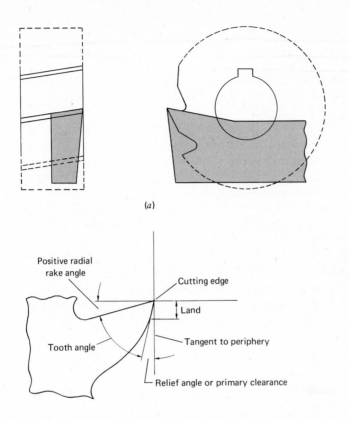

(a)

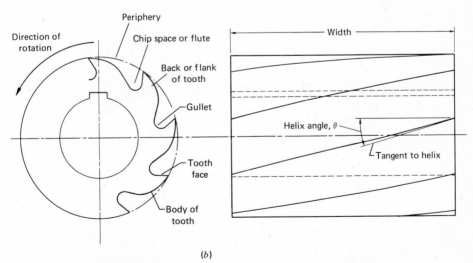

(b)

FIGURE 13-4

Principal elements of a helical milling cutter. (a) Comparison with a single-point tool; (b) nomenclature of a typical milling cutter.

is designed to produce flat surfaces. A face mill has cutting teeth on the end as well as on the periphery, and is designed to be used with its axis of rotation normal to the surface which is machined.

Each tooth of a milling cutter may be compared to a single-point cutting tool (Fig. 13-4a). In fact, a milling cutter may be likened to a series of single-point cutting tools designed to rotate about a common axis (Fig. 13-4b). The nomenclature for a typical helical milling cutter is given in the same figure.

Cutters may be solid or they may be designed to be supplied with inserted blades of cemented carbide. A solid cutter becomes smaller with repeated sharpening and the chip accommodation space becomes reduced. Inserted blade cutters usually are supplied with cemented carbides because typical cast alloy and ceramic materials have too little shock resistance for interrupted cutting.

Inserted blades permit more economical maintenance of large production cutters, because small inserts can be sharpened individually and the original diameter can be maintained by resetting the insert farther out to compensate for the material loss resulting from the regrinding operation. Step milling or gang milling, which produces two or more surfaces simultaneously, can be economically feasible because a constant differential in diameter or width can be maintained regardless of the number of regrinds.

The Grinding Wheel

The grinding wheel approaches a cutting tool with an infinite number of cutting edges. In grinding operations, if a cutting crystal is poorly oriented or dull, it pulls away from the wheel and a new crystal is exposed. Since the hard, sharp-

FIGURE 13-5
Relationship between the abrasive grains in a grinding wheel and the workpiece. Note the microchips at *a* and *b*.

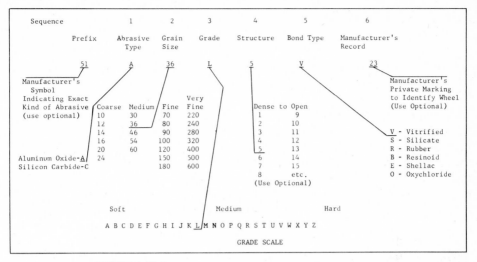

FIGURE 13-6
Standard marking system chart for grinding wheels. (Approved by the Grinding Wheel Manufacturers Association.)

edged crystals are randomly located in the matrix of the grinding wheel, all types of cutting geometry are found. A conceptual view of a grinding wheel and the tiny chips produced is found in Fig. 13-5. Note that ideally the wheel is much larger than the workpiece so that the effect of wear is negligible when grinding any one piece.

By removing a succession of small chips, the grinding wheel can produce surfaces to very close dimensions and a high degree of smoothness. Hard abrasives can cut hard materials; often grinding is the only way in which some materials, such as tungsten carbides, may be accurately shaped to final size. To obtain various rates of cutting and different kinds of surfaces on the many types of materials, the engineer uses many combinations of abrasive crystals and bonding materials.

Designation of grinding wheels   In the past it was the custom of a manufacturer to accept the recommendations of a supplier for a particular grinding wheel for a special operation. It is good practice to obtain the advice of grinding experts on difficult operations; but now, when wheels are specified, the designations are standardized. Variations can be made, and more than one supplier can furnish comparable wheels. The standard marking system is given in Fig. 13-6. The markings are in six parts:

1   Abrasive type
2   Grain size

3 Grade
4 Structure
5 Bond type
6 Manufacturer's record

Abrasive types Abrasive types are aluminum oxide and silicon carbide. Aluminum oxide crystals are tough and resist fracture to a high degree and consequently are used principally in grinding ferrous and other materials having a high tensile strength. Silicon carbide is harder than aluminum oxide, but its crystals are not as tough. The crystals break easily, leaving particles angular in form. Since silicon carbide crystals fracture easily, it is especially adapted to cutting materials with low tensile strength such as brass, aluminum, copper, cast iron, rubber, and plastics. It is also used in grinding hard, brittle materials such as carbide, stone, and ceramics.

Grain size The number designating size represents the number of openings per linear inch in the screen used to size the abrasive grains. Selection of size of grain will depend on the amount of material to be removed, the finish desired, and the mechanical properties of the material to be ground.

The larger the grains, the faster material will be removed; however, hard, brittle materials will not permit the large grains to penetrate to their full depth, so that any advantage in using grains of this size is lost. In this case, smaller grains are used, as in this way more cutting edges will be brought into contact with the work per revolution and the net result will be the same. Coarse grains are better adapted to grinding soft, ductile materials; fine grains are best when fine finishes and close accuracy are required.

Grade The letters designating grade indicate the relative strength (holding power) of the bond that holds the abrasive in place. In general, with a given type of bond it is the amount of bond that determines the hardness (grade). When the amount of bond is increased, the size of the bond posts connecting each abrasive grain to its neighbors is increased. The larger bond posts are naturally stronger, thereby increasing the hardness of the wheel. The strength or grade depends upon the forces applied to the abrasive grains. An abrasive grain should remain in a wheel as long as it cuts. When it dulls and does not cut the surface of the material, the increased forces should tear it out of the bond so that fresh, sharp abrasives can cut. Thus:

1 When the area of contact is small (OD of a small cylinder) the forces on each grain are high and a hard wheel is required to hold the abrasives before they dull. The forces are high on each grain because there are fewer grains cutting on the line of contact. When a flat surface or internal surface is ground, the number of grains in contact is greater and the forces are less; therefore, a softer-grade wheel can be used.

2 When material is hard, a soft grade is required because the cutting edges of the abrasive dull quickly and must be removed so fresh grains can cut.

3 When machine, wheel, or work vibrates, the pounding action causes the grains to break out before they are dull; therefore, a harder wheel is required.

4 When the speed of the cut surface is increased in relation to the speed of the wheel surface, a thicker chip is cut by the abrasive and the forces increase; this tends to break the grains from the wheel. This increased wear of the wheel is called *soft action* because it is similar to the action of a soft wheel. It can also result from increased traverse or infeed.

Structure The structure number indicates the relative spacing of the abrasive grains. When the grains are close together relative to the grain size, the wheels have low structure numbers such as 1, 2, 3, 4, 5. Wider spacing relative to grain size is designated by higher numbers. Selection of the proper structure, or spacing between the grains, is governed by the finish required, the nature of the operation, and the mechanical properties of the material to be ground.

Wide spacing A wheel with wide spacing between the grains is best for grinding soft, ductile materials. The grains are thus allowed to penetrate to their maximum depth and clearance is provided between the grains for the large chips. Wide spacing is also best when grinding exceedingly hard materials like cemented carbides, because the grains are released more promptly and new, sharp grains are exposed as needed. In snagging and other operations where a variable application of pressure is involved, as well as in surface grinding, wide spacing is essential so that the pressure on the work will be distributed evenly over all the grains.

Close spacings A closely spaced structure should be selected when grinding hard, brittle materials in order to utilize the cutting properties of the maximum number of small grains. If heavy pressures are used, there will be a tendency to break down the form of shaped wheels and close grain spacing will be required. From the standpoint of finish, smoothness will vary directly with the closeness of the spacing.

Medium spacing A medium-spaced structure should usually be selected for center and centerless cylindrical, tool, and cutter grinding. The best policy to follow is to experiment somewhat with work speeds to find the best combination before making any definite selections as to wheel grade. This will give better results than arbitrarily selecting a grade and then fitting the work speed to it.

Effect of grade on wheel life It might appear that the harder the wheel, the longer its life will be; but this is not always true. If the grade is too hard, the wheel will load and frequent dressing will be necessary. A hard wheel is often worn down more by the dressing than by the grinding it does. This fact merits consideration, especially since dressing-tool cost is high. Another factor to consider in this respect is that while a hard wheel may last longer on a given job, it will cut more slowly, with the result that labor and overhead costs per piece will be higher. If these increased costs are not more than offset by the saving in wheel cost, the use of the hard wheel will be uneconomical.

Bond   Bond is the medium that holds the grains together in the form of a wheel, or on a belt or disk. The bond functions in the same way as a tool post and holds the grains or cutting tools in position until they become dull and are torn out and fresh grains exposed. The bond itself ordinarily has little or no cutting action. Great advancement has been made in recent years due to resinoid and rubber bonds that permit flexibility of the grinding medium, such as thin wheels and belts. They are not affected by water, which can be used to keep them cool. The features of these new bonds have extended the application of abrasives and have reduced costs.

Vitrified bond   Vitrified bond, which is essentially glass, is used in over 75 percent of the grinding wheels manufactured. Wheels made with this type of bond have porosity and strength, and are used when a high rate of stock removal is required. Vitrified bond is unaffected by water, acid, oils, or ordinary temperature conditions.

Silicate bond   Silicate bond is composed principally of silicate of soda or water glass. This bond does not hold the grains as tightly as vitrified bond and, for this reason, less grinding stress is required to tear them out. The principal application of silicate-type wheels is for grinding edged tools that cannot be subjected to excessive heat while grinding.

Shellac bond   Shellac-bonded wheels are elastic and are used to produce high finishes. Other applications are the cool cutting of hardened tool steels and for cutting-off operations. They are affected by coolants and heat.

Resinoid bond   Resinoid-bonded wheels are held together by a synthetic resin or plastic. They can be made in various structures and have the characteristics of cutting cool and removing material rapidly. Resinoid wheels are very strong and can be run at high speeds.

Rubber bond   Rubber-bonded wheels are very strong and tough; for this reason, thin wheels are usually of this type. The principal application of rubber-bonded wheels is in jobs where a good finish is required and in cutting-off operations.

Cutting-Tool Materials

All machine tools must be provided with an easily attached and rigidly supported cutting-tool bit or cutter. It must be able to withstand severe shock, and erosive

Table 13-3 COMPARISON OF CUTTING-TOOL MATERIALS

Cutting-tool material	Approx. carbide volume, %	Chemical composition	When first avail.	Red hardness R_a	Red hardness R_c	Typical cutting speed, ft/min	Max. operating temperature, °F	Relative cutting cost, $/in³	Cost relative to high-speed steel (h-s steel = 1)
Carbon steel	15	0.9–1.2% C, balance Fe	Only material up to 1900	...	...	40	350–400	0.42	0.3
High-speed steel	10–20	18% W, 4% Cr, 1% Va; 0.6% C, 8% Mo; 4% Cr, 1.5% W, 1% Va (Also many variants)	1900	76	57	90	900–1000	0.22	1
Cast cobalt alloy	25–35	43–48% Co, 17–19% W 30–35% Cr; 2% C	1915	74	47	150	1200	0.10	2.5
Sintered tungsten carbide	75–95	Cast iron grade: 94% WC, 6% Co Steel grade: WC; TiC; TaC; 8% Co	~1930 ~1945	82	62	500	1200–1400	0.06	10
Ceramic (aluminum oxide)	...	Al₂O₃ plus binders	1955	84	65	800	1500	0.03	15
Diamond	...	Crystalline C	...	...	...	400	1100		

and abrasive conditions for 10 to 30 min of cutting, with no more than 0.030 in wear in the flank area. Cutting-tool materials must have *red hardness*; i.e., they must retain their hardness at high temperatures because excessive temperature is the major cause of tool breakdown. Toughness is also a decided asset because cutting tools are subject to shock loading during interrupted cuts or at the beginning of a cutting operation. There are a number of cutting-tool materials available but high-speed steel and tungsten carbide tools do the bulk of the metal cutting. High-speed steel tools are preferred for use on older machine tools or for interrupted cuts because of their toughness and high tensile strength. Carbide tools have higher hardness at red heat; if properly supported, they will give long life at about $2\frac{1}{2}$ times the cutting speed of high-speed steel. The hardness of the principal cutting-tool materials as a function of operating temperature is given in Table 13-3. The first cutting-tool material was high-carbon steel (SAE 1090), but it could not withstand temperatures above 400°F.

High-speed steels   Although high-carbon steels were the only tool material until 1900, Taylor and White's discovery of high-speed steel and the demands of World War I relegated carbon steel to use in wood chisels and small drills or form tools. The 18% W–4% Cr–1% Va type of high-speed steel (Table 13-3) revolutionized the metal-cutting industry and soon machine tools were redesigned to incorporate heavier bearings, a greater range of feeds and speeds, and individual motor drives which now can provide up to 50 hp or more for a single machine tool.

To save tungsten during World War II, a molybdenum type of high-speed steel was developed. This has proved to be a good, but not necessarily less expensive, substitute. When up to 12 percent cobalt is added to certain types of high-speed steel, super high-speed steel is produced. These alloys have superior red hardness and resistance to abrasion with some loss in toughness, but at 50 percent more cost. However, users find that cobalt high-speed steels are less easily ground and that their heat treatment is considerably more difficult than that used for conventional high-speed steel.

Infusion of carbides into high-speed steels   A limited amount of work has demonstrated that the tool life of certain materials, such as high-speed steel, could be markedly improved by infusing materials such as tungsten carbide into the surface layers only. Both laser and electron-beam energy sources can be used to produce such carbide-enriched surface layers.

Yet, typical sputtered coatings of many types seem to spall off the tool materials in erratic fashion. Research along these lines is underway, but it may be a number of years before the results are available for everyday metal cutting.

Cast nonferrous tools   This tool material was first introduced in 1915 just in time for use in the production of material for World War I. A typical cobalt-base alloy, Stellite, has a composition of 43–48% Co, 17–19% W, 30–35% Cr, and 1.85–2.15% C. This nonforgeable alloy, which must be cast to shape and ground to give it precise size and shape, cannot be hardened by heat treatment. It is heat-treated after casting to break up the carbide network, although its hardness results from metallic carbides in a softer cobalt matrix. Cobalt alloys are not affected by heat up to 1500°F and are actually tougher at a dull red heat than they are cold. The cobalt alloys take a high polish, which helps to reduce initial edge buildup, and their high abrasion resistance makes them especially good for cutting cast iron at speeds 50 to 100 percent higher than those suitable for 18-4-1 high-speed steel, which places them in competition with carbides at their lower ranges of cutting speed. Cobalt alloys are expensive, so they too are used in the form of inserts.

Cemented carbides   Cemented carbides were first developed in Germany in 1928. Early grades were extremely brittle and could not be used for machining steel because of their tendency to form a built-up edge. In less than 1 min the built-up edge was so large that it put the nose of the tool in tension and the tool tip failed catastrophically. During World War II it was discovered that the addition of up to 30 percent titanium carbide and tantalum carbide to the tungsten carbide, and more cobalt, prevented the formation of the built-up edge by destroying the weldability of the steel to the tool material. The cobalt increased the toughness of the material at the expense of hardness. Although cemented carbides have higher initial cost, they can be used in the form of mechanically clamped inserts. In this way one rectangular tool bit can be used up to eight times before disposal and requires no resharpening cost. Thus cutting cost is in the order of $0.10 to $0.15 per edge. The tool holders have special locating pockets which provide a standard negative-rake angle into which the inserts fit; the base or seat is also carbide for good support and reproducible alignment.

Carbides have now been developed into a wide range of cutting-tool materials varying from hard, brittle, wear-resistant carbides to soft, tough carbides. Their properties depend upon their binder content and the size, distribution, and chemical composition of the carbide particles.

Certainly the development of the disposable insert and electrolytic grinding techniques have made cemented carbides the workhorse of industry. However, even though carbides are easily available, each successful installation of carbide tooling must be studied carefully to determine the availability of sufficient power, the tightness of all spindle bearings, and carriage and cross-feed gibs. Only a rigid setup can be used to machine with carbide tools.

Ceramic or metallic oxides   Ceramic inserts were first developed in the late 1930s, but their extreme brittleness made them unusable in the machine tools of that day. During World War II, the shortage of tungsten made the Germans search for other types of cutting tools, and ceramic tools were intensely investigated. Captured documents led to further development work at Watertown Arsenal in the United States. These materials are now commercially available and are particularly useful for high-speed machining operations on cast iron using light depths of cut, as in the finishing of brake drums.

Ceramic tools are primarily aluminum oxide compacts with or without additional binders. They must be used in negative-rake tool holders and frequently a 45° chamfer on the cutting edge has been proven to improve tool life. The low coefficient of heat transfer results in most of the heat produced by the machining operation passing to the chips. Thus the inserts are cool to the touch even after a heavy cut.

It has also been found that compressive residual stresses can be induced into the ceramic compact, which in turn may lead to tools which can be used under impact conditions and which can sustain larger rates of stock removal for longer times before failure. Ceramic tools can cut at top speeds, of 1,500 to 2,000 ft/min, but their tool life as seen in Table 13-3, is much less than that of tools made of any other cutting tool material. Their mode of failure is brittle fracture at random locations of point defects. Tool life is less predictable than it is in sintered carbides. One tool may last several times longer than another, even though both appeared to be identical and were subjected to the same cutting conditions.

Harder and even more brittle ceramic tools have been developed, such as other metallic oxides and zirconium boride or silicon nitride. Although these materials are extremely hard and brittle, they have poor mechanical shock resistance and, at the present time, are merely laboratory curiosities, which they will remain until suitable binders, coating films, or other technological advances are developed to improve their characteristic and unpredictable brittle mode of failure.

Ceramic tools have tremendous potential because they are composed of materials which are abundant in the earth's crust. This is one important consideration in view of the limited amount of tungsten available. As our knowledge increases in the areas of fracture mechanics, the role of surface compressive stresses, the control of point defects, and the nature and application of surface films, the use of ceramic tools will increase markedly. Thus there is a need for continuing long-range research in this area.

Diamonds Diamonds are the hardest known material, but they can be used only at temperatures of about 1100 to 1200°F because they oxidize readily at higher temperatures. They are suitable for cutting very hard materials such as cemented carbides or ceramic tools and can produce a very fine surface finish if a high speed and low feed are used in conjunction with a good point geometry. They are extremely brittle and are a relatively poor conductor of heat, so they are limited to depths of cut of only a few thousandths of an inch. Typical applications include the precision boring of holes and the finishing of plastics or other abrasive materials, although ceramic tools are finding uses in similar applications.

Cutting-tool diamonds are off-grade chips of natural diamonds which are not of gem quality or they are synthetic diamonds. The present-day demand is so great that both natural and synthetic diamonds are needed. The growth of inserted carbides has created a tremendous new market for carbide cutting tools, which in turn has caused an increasing demand for industrial diamonds.

Machinability

Machinability refers to a system for rating various alloys on the basis of their relative ability to be machined easily, with a good surface appearance. Good machinability means:

{
 1 Better than average tool life is obtained.
 2 A smooth untorn surface is produced.
 3 Less than average power is required per unit volume of metal removal.

The rating system was established using B1112 steel as 100 percent because when the original research was carried out, that free-cutting steel was the most machinable ferrous alloy.

The machinability of B1112 steel was improved to B1113 steel by adding small amounts of lead (0.3 percent) in addition to the manganese sulfide stringers which are found elongated in the direction of rolling. Sodium sulfide may also be added. Lead additives of about 0.25 percent can improve the machinability of stainless steel and copper-base alloys such as brass and bronze. In the latter alloys, up to 1 to 3 percent lead is added, especially if cast, in order to make those long-freezing-range alloys pressure tight. Free machining grades of both ferrous and nonferrous alloys are widely used in automatic screw machines and bar-feed machines where rolled or extruded shapes are machined in high volume and where good surface finish and tolerances are needed. In the ferritic grades of cast irons, the free carbon in the form of graphite flakes, temper carbon, or nodules is a distinct aid to machinability. In fact, even pearlitic grades of cast irons have better machinability than forged steels of equivalent strengths. In aluminum alloys, additions up to 1 to 3 percent of zinc and magnesium improve their machinability.

Effect of cold work   In general, face-centered cubic metals which are soft and have high ductility, particularly pure aluminum or copper and the body-centered-cubic very low carbon steel (SAE 1010 or less), are difficult to machine because they are too soft to form chips of uniform thickness. The metal piles up ahead of the tool, like a plow pushing wet snow, until a chunk is removed. In such cases cold working of the surface layers of the bars of those metals produces a marked improvement in their machinability because the cold-worked structure promotes the formation of continuous chips.

Effect of microstructure   The machinability of the cast irons is definitely a function of their microstructure. In all cases ferritic cast irons have better machinability than do pearlitic grades regardless of whether the alloy is malleable, ductile, or gray iron. Machinability also is a function of the size and shape of the free graphite; the coarser the graphite structure, the more machinable is the iron. In fact, microstructure is a better index of the machinability of cast irons than Brinell hardness, although that is also a rough index. In the case of cast irons, surface finish improves with finer graphite particles, and so does overall machinability. The machinability of steel decreases as the carbon content increases, low-carbon steel has a rating of 80 percent but SAE 1080 steel has a rating of 40 percent. Spheroidization of a high-carbon (0.7 to 0.9 percent) steel improves its machinability markedly, but for lower carbon steels the coarse pearlitic structure is more machinable. It should be noted that it is frequently economic to heat-treat steels to the most machinable microstructures and then retreat them for the service requirements after machining.

The effect of atomic structure   The machinability of some alloys is definitely related to their crystallographic structure. In the case of hexagonal close-packed metals such as zinc, cadmium, magnesium, beryllium, and titanium there should be machining problems because they all have only one plane of easy slip. However, zinc, magnesium, and cadmium have low recrystallization temperatures so that the single plane of easy glide presents no problem. As soon as a significant amount of cold working appears as a result of machining, it is spontaneously recrystallized by the heat of deformation which is generated within the shear zone.

But this is not true in the case of titanium and beryllium. In these metals the one plane of easy glide is soon used up, and the alloys are said to work-harden readily and to have very poor machinability. For those alloys, hot machining might be of definite value because a high-intensity torch is used to heat the surface layers of metal immediately in advance of the cutting tool to a temperature well above the recrystallization range. Thus, there should be a marked improvement in machinability. When machining titanium there also appears to occur a loss in machinability because the tool material diffuses into the chip and thus exhibits a rapid wear phenomenon. This propensity for microwelding between the tool and chips causes a severe built-up edge particularly at lower cutting speeds.

The machining of beryllium is particularly difficult because of its brittleness and toxicity. Cutting fluids help in machining by improving surface finish, but they interfere with the collection and evacuation of chips so they are little used.

Analysis of Metal-Cutting Forces

The mechanism of chip formation in the cutting of ductile metals has received much study in the last 30 years, during which the mechanics of metal cutting have been well developed. An analysis of the machining or cutting process can most easily be made when based on orthogonal or right-angle cutting. Here the cutting edge is perpendicular to the tool-work motion, or the cutting edge inclination is zero (Fig. 13-7), thus enabling the analysis to be made by means of a two-dimensional view which can be used to derive the basic relationships which apply to the mechanics of chip formation. The analysis of orthogonal metal cutting is truly international. It has been independently discovered three times, beginning with a Russian named Zvorykin in 1893, continuing with a Finnish engineer named Piispannen in 1939, and finally being widely recognized after Merchant published it in 1944. We shall designate this last analysis as Merchant's model of orthogonal metal cutting. A host of other more sophisticated models have been proposed over the years but none is consistently better over the wide range of materials and cutting conditions found in metal machining. All models suffer from the fact that they are primarily based on geometry and trigonometry, and after-the-fact, room-temperature measurements; whereas metal cutting is a

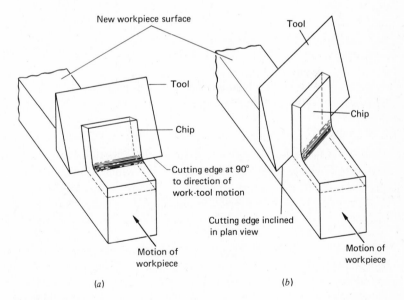

FIGURE 13-7
Two types of single-point machining: (a) orthogonal; (b) oblique. (*After G. Boothroyd, "Fundamentals of Metal Machining," Edward Arnold, London, 1965.*)

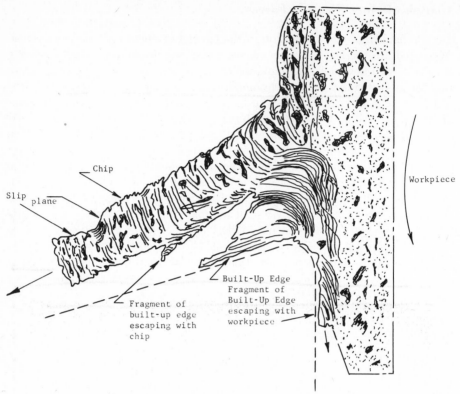

FIGURE 13-8
Details of a built-up edge on a single-point tool. (*Courtesy Cincinnati Milling Machine Company.*)

dynamic process which may be carried out on metals with many crystal structures and widely varying properties, and which occurs at high hydrostatic pressures and temperatures. The Merchant model is useful because it is simple, it is up to 10 percent conservative, and no other model has proved to be a consistently better estimator of the magnitude of the cutting forces in production practice.

Although most industrial machining takes place in three dimensions and is called oblique cutting (Fig. 13-7), orthogonal cutting is more generally used for the derivation of the approximate force relationships. With the advent of the computer, oblique cutting analyses could be used if the increased cost would warrant it.

The shear angle can be determined from the chip-thickness ratio (cutting ratio) and the tool geometry, provided that a type II chip (continuous chip without a built-up edge) is formed. The chip thickness can be measured with point micrometer calipers or by using a measuring stage in conjunction with a toolmaker's microscope. In other cases, if the cutting speed is too low to avoid a built-up edge, type III chips are cut; or if there are microconstituents such as flake graphite within the workpiece, all chips become discontinuous (type I). In either case, good determination of the chip-thickness ratio would be impossible. The chip-thickness ratio is of major importance in the analysis of metal cutting because it is the only place where the properties of the work material or the friction coefficient are brought into the solution of the problem.

Built-up Edge

If there is a large amount of built-up edge (Fig. 13-8), the surface of the workpiece has a rough and torn appearance. This is because the continuity of the work surfaces is broken at frequent intervals by fragments torn from the built-up edge and squeezed between the tool and work. These fragments stick to the tool and the cut surface to cause a scored or pitted surface. Hence, for a high-quality surface finish, all conditions should be chosen to reduce the magnitude of the built-up edge.

The following factors act to decrease the amount of the built-up edge:

1 A decrease in chip thickness
2 An increase in rake angle
3 An increase in tool sharpness
4 An increase in cutting speed
5 A reduction of adhesion between the chip and tool
6 Use of ceramic tools
7 Use of a chlorinated or sulfurized cutting fluid

Chip-thickness ratio (Fig. 13-9)   Chip-thickness ratio r is always less than 1 and is defined as

$$r = \frac{t_0}{t_c} = \frac{l_c}{l_0}$$

As the tool advances into the workpiece, the metal in front of it is severely stressed and plastic deformation takes place in a localized shear zone which begins at the tip of the cutting tool and extends obliquely forward to the work surface (Fig. 13-9). The depth of cut (planing) or the feed increment (turning) is designated by t_0. The chip-thickness ratio, which is an index of the amount of plastic deformation which the metal in the chip has undergone, depends on the relative forces on the chip and the geometry of the cutting tool.

The back rake angle, α, is of major significance in the cutting process;

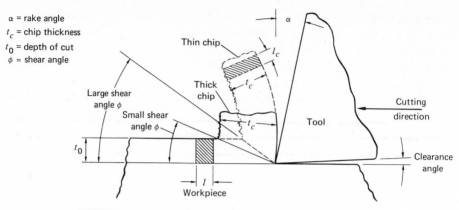

α = rake angle
t_c = chip thickness
t_0 = depth of cut
ϕ = shear angle

FIGURE 13-9
Basic model for orthogonal metal cutting showing the effect of a small and large shear-plane angle on the chip thickness.

however, the shear angle is also a function of the cutting speed through the chip-thickness ratio, r, so that even tools with negative rake angles can yield a high shear angle if the cutting speed is sufficiently high. The friction between the tool and workpiece also affects the shear angle, so that the use of proper lubricants can increase the angle especially at low cutting speeds.

It is possible to determine r by any one of three methods.

1   Compare feed increment to the actual chip thickness, which can be measured by a micrometer (preferably point or ball-point type).

2   Compare actual chip length l_c to the corresponding distance the tool has traveled through the work l_0:

$$r = \frac{l_c}{l_0}$$

3   Compare chip weight per unit length to parent metal weight per unit length:

$$r = \frac{t_0 b \rho}{m}$$

where t_0 = depth of cut or feed increment
b = chip width
m = chip weight per unit length
ρ = density of parent metal

The relationship between chip-thickness ratio r and the shear angle ϕ is illustrated in Fig. 13-9.

Thus we have

$$r = \frac{t_0}{t_c} = \frac{AB \sin \phi}{AB \cos(\phi - \alpha)}$$

$$r = \frac{\sin \phi}{\cos \phi \cos \alpha + \sin \phi \sin \alpha} \dagger$$

or

$$r \cos \phi \cos \alpha + r \sin \phi \sin \alpha = \sin \phi$$

Dividing by $\cos \phi$:

$$r \cos \alpha + \frac{r \sin \phi \sin \alpha}{\cos \phi} = \frac{\sin \phi}{\cos \phi}$$

$$r \cos \alpha + r \tan \phi \sin \alpha = \tan \phi$$

$$\tan \phi - r \tan \phi \sin \alpha = r \cos \alpha$$

$$\tan \phi (1 - r \sin \alpha) = r \cos \alpha$$

$$\tan \phi = \frac{r \cos \alpha}{1 - r \sin \alpha}$$

When ϕ is small, t_c is large compared with t_0, which causes a low chip-thickness ratio, a relatively long shear plane, and inefficient metal removal. When ϕ is large, t_c is small and the chip-thickness ratio approaches unity. For a given depth of cut and shear strength of the workpiece, a reduction in the shear-plane area reduces the forces required to produce sufficient shearing stress to cut the work material.

Assumptions Required in the Analysis of Metal Cutting

In order to analyze metal cutting according to the principles of applied mechanics, the following assumptions must be made:

1   The shear surface is a plane extending upward from the cutting edge.
2   The tool is perfectly sharp and there is no contact along the clearance face.
3   The cutting edge is a straight line extending perpendicular to the direction of motion and generates a plane surface as the work moves past it.
4   The chip does not flow to either side.
5   The depth of cut remains constant.
6   The width of tool is greater than that of the workpiece.

† By function of sums of angles $\cos(x + y) = \cos x \cos y + \sin x \sin y$.

7   The work moves with uniform velocity relative to the tool.

8   A continuous chip is produced with no built-up edge.

In practical metal-cutting operations many of these assumptions cannot be met, as for instance, (1) and (2). Real metal cutting takes place in a zone, and certainly a tool soon dulls to some extent.

Force Relations

By treating the chip as an isolated free body in equilibrium, the force between the tool face and the chip must be equal to the force between the workpiece and the chip along the shear plane. We can denote the force between the tool face and the chip as R (see Fig. 13-10), which is the resultant of a friction force F and its normal component N. The force between the workpiece and the chip along the shear plane may be denoted by R' and is the resultant of two pairs of force components. One of these pairs of force components is the shearing force F_s acting along the shear plane and the normal compressive force F_n. The other pair of force components whose resultant is R' includes F_c, which is the force required to cause movement between the tool and work, and F_v, which is the thrust force required to hold the tool against the machined surface.

Analysis of forces   If the forces R and R' are plotted at the tool point instead of at their actual points of application along the shear plane and tool face, the relationship illustrated in Fig. 13-10 is obtained. Here R and R' (which are equal and parallel) are coincident, and are made the diameter of the reference circle. Thus, analytical relationships may be obtained for the shear and friction components of force in terms of the horizontal and vertical components (F_c and F_v), which are the components normally determined experimentally by means of a strain gage dynamometer.

From Fig. 13-10 it is evident that:

$$F_s = F_c \cos \phi - F_v \sin \phi$$
$$F_n = F_v \cos \phi + F_c \sin \phi$$
$$F = F_c \sin \alpha + F_v \cos \alpha$$
$$N = F_c \cos \alpha - F_v \sin \alpha$$

Once the tool-face components have been determined, they permit the calculation of the coefficient of friction μ for the tool face:

$$\mu = \tan \beta = \frac{F_c \sin \alpha + F_v \cos \alpha}{F_c \cos \alpha - F_v \sin \alpha}$$

$$\mu = \frac{F_v + F_h \tan \alpha}{F_c - F_v \tan \alpha}$$

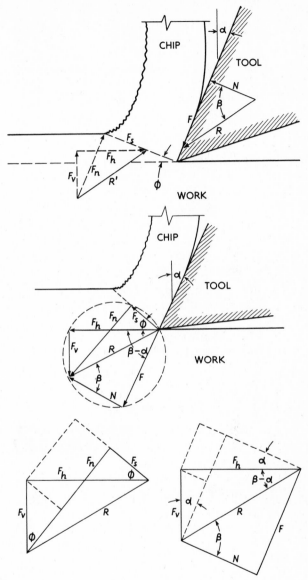

FIGURE 13-10
Force relationships. (*a*) Tool plane and shear plane forces; (*b*) Merchant's circle of forces; (*c*) triangles for calculating (1) F_s, and (2) F.

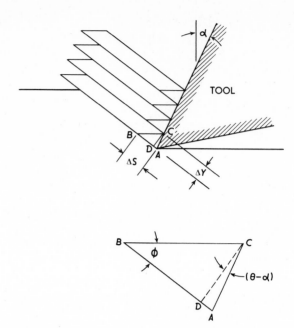

FIGURE 13-11
Shear strain in metal cutting.

Or, for negative rake angles, we have

$$\mu = \frac{F_v - F_c \tan \alpha}{F_c + F_v \tan \alpha}$$

Stress Analysis

Knowing the shear-plane components of force, we are able to determine the mean shear and normal stresses on the shear plane. Letting

$$S_s = \text{shear stress}$$

$$S_n = \text{normal compressive stress}$$

we know that:

$$S_s = \frac{F_s}{A_s}$$

and

$$S_n = \frac{F_n}{A_s}$$

where F_s = shearing force

F_n = normal compressive force

A_s = shear plane area = $\dfrac{A_0}{\sin \phi}$

A_0 = cross-sectional area of chip before removal

= $t_0 \times b$ where b = chip width before removal

Therefore:

$$S_s = \frac{(F_c \cos \phi - F_v \sin \phi) \sin \phi}{bt_0}$$

$$S_n = \frac{(F_c \sin \phi + F_v \cos \phi) \sin \phi}{bt_0}$$

Shear Strain

The shear strain in cutting may be thought of as the ratio of the distance the material being cut travels along the shear plane to the spacing of successive shear planes. Referring to Fig. 13-11, the shear strain γ is

$$\gamma = \frac{\Delta s}{\Delta y} = \frac{AB}{CD}$$

$$= \frac{AD}{CD} + \frac{DB}{CD}$$

$$= \tan (\phi - \alpha) + \cot \phi$$

$$= \frac{\cos \alpha}{\sin \phi \cos (\phi - \alpha)}$$

The rate of strain in cutting is

$$\gamma' = \frac{\Delta s}{\Delta y \, \Delta t} = \frac{V_s}{\Delta y}$$

where V_s is the speed of the chip relative to the workpiece along the shear plane and Δt is the time elapsed for the metal to travel a distance Δs along the shear plane. Δy is the spacing of successive shear planes.

Velocity Relations

In the cutting process, there are three velocities that are of interest to the production-design engineer. These are the cutting velocity, the chip velocity,

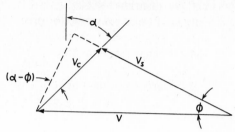

FIGURE 13-12
Velocity relationships in orthogonal metal cutting.

and the shear velocity. The cutting velocity V is the speed of the tool relative to the work directed parallel to the cutting force F_c. The chip velocity V_c represents the speed of the chip relative to the cutting tool and directed along the tool face. The shear velocity V_s is the speed of the chip relative to the workpiece and directed along the shear plane. Figure 13-12 illustrates these three velocity vectors. Here, the vector sum of the cutting velocity and the chip velocity is equal to the shear velocity vector.

Thus we have:

$$V_c = \frac{\sin \phi}{\cos(\phi - \alpha)} V = rV$$

$$V_s = \frac{\cos \alpha}{\cos(\phi - \alpha)} V = \gamma V \sin \phi$$

Energy Considerations

The total energy consumed per unit time in the cutting process may be expressed as

$$E = F_h \cdot V$$

and the total energy per unit volume of metal removed per unit time is

$$U = \frac{E}{V \cdot b \cdot t_0} = \frac{F_c}{bt_0}$$

This total energy is consumed in several ways:

1 As shear energy per unit volume u_s on the shear plane
2 As friction energy per unit volume u_f on the tool face
3 As surface per unit volume u_a due to the formation of new surface area in cutting

4   As momentum energy per unit volume u_m due to the momentum change associated with the metal as it crosses the shear plane

The shear energy per unit volume is obtained by the expression

$$u_s = \frac{F_s V_s}{V \cdot b \cdot t_0} = \frac{S_s V_s}{V \sin \phi}$$

since:

$$V_s = \gamma \sin \phi \; V$$

$$u_s = S_s \gamma$$

In a similar way, the friction energy can be stated as

$$u_f = \frac{F V_c}{V \cdot b \cdot t_0}$$

The surface energy of a solid is analogous to the surface tension of a liquid. The units of surface energy are inch-pounds per square inch. The surface energy per unit volume in cutting is:

$$u_a = \frac{T \cdot 2 V b}{V \cdot b \cdot t_0}$$

$$= \frac{2T}{t_0}$$

where T is the surface energy in pounds per square inch of the material being cut. For most metals T has a value of 0.006 in·lb/in². The momentum energy per unit volume may be expressed as

$$u_m = \frac{F_m V_s}{V \cdot b \cdot t_0}$$

where F_m = momentum force = $\rho V^2 b t_0 \; \gamma \sin \phi$
ρ = density of material being cut

Power Requirements

Power tests may be made by measuring the forces acting on the cutting tool or by measuring the power input to the machine. There are four methods frequently used in estimating the power consumed in machining. These are:

1   The gross or total horsepower: H_{pt}
2   The net horsepower: $H_{pn} = H_{pt} \times$ efficiency of machine tool

3   The specific or unit horsepower: $U_{hp} = \dfrac{H_{pn}}{Q}$, where Q = cubic inches of metal removed per unit time

4   The metal removal factor $K_n = \dfrac{1}{U_{hp}}$

The horsepower required for cutting may be resolved into three components: (1) the power required for tangential cutting, (2) the power required to feed in a longitudinal direction, and (3) the power required to feed in a radial direction. Thus the net horsepower may be expressed as:

$$H_{pn} = \frac{F_c \cdot V}{33,000} + \frac{F_v f_1 \cdot N}{(12)(33,000)} + \frac{F_r f_r N}{(12)(33,000)}$$

For turning cuts, it is recognized that the radial feed is zero, the longitudinal power is negligible so the net horsepower consumed becomes

$$H_{pn} = \frac{F_c \cdot V}{33,000}$$

In facing cuts the longitudinal feed is zero, and we have

$$H_{pn} = \frac{F_c \cdot V}{33,000} + \frac{F_r f_r N}{396,000}$$

Of the three cutting forces F_c, F_v, and F_r, the tangential or cutting force, F_c, is the greatest and consequently is of the most importance in determining the power required for cutting.

Tool Wear and Tool Life

Economic decision making in metal cutting is primarily based upon the criterion of tool life, although other criteria, such as: metal-cutting force, power requirements, surface finish, and dimensional accuracy, may also play a part. The optimum output from a machining operation is achieved when the specified surface finish and dimensional tolerance are obtained at the lowest unit cost. A number of machining factors may be varied, each of which interacts with the others and thus affects the final cost. These variables include: (1) speed, (2) feed, (3) depth of cut, (4) work material, (5) tool material, (6) tool geometry, (7) cutting fluid, (8) rigidity of the machine tool, workpiece, and tool support.

If one considers that the tool-work combination is fixed and a uniformly rigid setup is available, then relative machinability ratings can be established for the common alloys. Such a ranking permits one to make a good estimate of what cutting parameters to start with for estimating a new setup. In the classical system B1112 steel is taken as 100 percent and all other materials are compared to it (Table 13-4). It is easily seen that nonferrous metals are up to 10 times more machinable than steel, and that stainless steels and some gear alloys have low machinability ratings. Gray cast iron and C1120 steel are close to an 80 percent rating.

Thus the machinability rating describes in a general way the relative difficulty encountered in cutting various alloys. Hardness, shear strength, microstructure, rate of strain hardening, formation of a built-up edge, and other properties of a material determine the permissible cutting speed for a given tool-work combination.

A more accurate estimate of the optimum cutting speed can be obtained from the empirical Taylor equation:

$$VT^n = C$$

where V = cutting speed, ft/min
T = tool life, min
n = slope of the equation on a log-log plot
C = constant, the intercept of the equation at 1 min tool life

Typical values for n and C for various tool-work combinations are given in

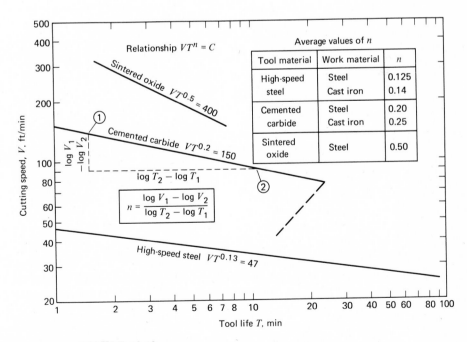

FIGURE 13-13
Log-log plot of tool-life test data.

Table 13-4 MACHINABILITY RATINGS OF AISI STEELS AND VARIOUS FERROUS AND NONFERROUS METALS TO THE NEAREST 5 PERCENT, BASED ON 100 PERCENT RATING FOR BESSEMER SCREWSTOCK AISI B112, COLD-DRAWN WHEN TURNED WITH A SUITABLE CUTTING FLUID AT 180 FT/MIN UNDER NORMAL SCREW-MACHINE CONDITIONS

AISI no.	Machinability rating, %	Brinell hardness	AISI no.	Machinability rating, %	Brinell hardness
C1008	50	126–163	6120	50	179–217
C1010	50	131–170	6145*	50	179–235
C1015	50	131–170	6152†	45	183–241
C1016	70	137–174	NE8620 or 8720	60	170–217
C1020	65	137–174	NE8630* or 8730*	65	179–229
C1022	70	159–192	NE8640* or 8740*	60	179–229
C1030	65	170–212	NE8645* or 8745*	55	183–235
C1035	65	174–217	NE8650* or 8750*	50	183–241
C1040	60	179–229	9260*	45	187–225
C1045	60	179–229	E9315	40	179–229
C1050	50	179–229	NE9440*	60	179–229
C1070	45	183–241			
C1109	85	137–166	Type	Machinability rating, %	Brinell hardness
B1111	95	179–229	Aluminum 2-S	300–1,500	
B1112	100	179–229	Aluminum 11-S	500–2,000	
B1113	135	179–229	Aluminum 17-S	300–1,500	
C1115	85	143–179	Brass—leaded	300–600	
C1117	85	143–179	Brass—red	200	
C1118	80	143–179	Brass—yellow	200	
C1120	80	143–179	Bronze, lead bearing	200–500	
C1132	75	187–229	Bronze, manganese	150	
C1137	70	187–229	Copper—cast	70	
C1141	65	183–241	Copper—rolled	60	
1320	50	170–229			
1330*	50	179–235			

Material		
1335*	50	187–241
1340*	45	187–241
2317	55	174–217
2330*	50	179–229
2340*	45	187–241
2515*	30	179–229
3120	60	163–207
3130*	55	179–217
3140*	55	187–229
3145*	50	187–235
E3310*	40	170–229
4023	70	156–207
4027*	70	166–212
4032*	65	170–229
4037*	65	179–229
4042*	60	183–235
4047*	55	183–235
4130*	65	187–229
4137*	60	187–229
4145*	55	187–229
4150*	50	187–235
4320	55	179–228
4340	45	187–241
4615	65	174–217
4640*	55	187–235
4815	50	187–229
5120	65	170–212
5140*	60	174–229
5150*	55	179–235
E52101†	30	183–229

Material		
Dowmetal	500–2,000	
Everdur	120	
Gun metal	60	
Inconel	45	
Iron—cast (hard)	50	220–240
Iron—cast (medium)	65	193–220
Iron—cast (soft)	80	160–193
Iron—ingot	50	101–131
Iron—malleable (standard)	120	110–145
Iron—malleable (pearlitic)	90	180–200
Iron—malleable (pearlitic)	80	200–240
Iron—stainless (12% Cr.F.M.)	70	163–207
Iron—wrought	50	101–131
Magnesium alloys	500–2,000	
Monel metal—cast	35	
Monel Metal—"K"	50	
Monel metal—rolled	45	
Nickel	20	
Ni-Resist*	30	
Steel—cast (0.35 carbon)	70	170–212
Steel—high-speed*	30	
Steel—manganese	55	
(oil hardening)†	30	
Steel—stainless	65	
(18–8 austenitic)*	25	150–160
Steel—stainless (18–8 F.M.)	45	179–212
Steel—tool (high-carbon, high-chromium)†	25	
Steel—tool (low-tungsten, chromium and carbon)†	30	200–218
Zinc	200	

·†| = "annealed" and † = "spheroidized anneal." These terms refer specifically to the commercial practice in steel mills, before cold drawing or cold rolling, in the production of the steels specifically mentioned.

SOURCE: E. M. Slaughter and O. W. Boston, Jan. 27, 1947.

Table 13-5 THE RELATION BETWEEN CUTTING SPEED AND TOOL LIFE FOR VARIOUS TOOL MATERIALS AND CONDITIONS

No.	Material	Shape	Material cut	Size of cut		Cutting fluid	$VT^n = C$	
				Depth	Feed		n	C
1	High-carbon steel	8, 14, 6, 6, 15, $\frac{3}{64}$	Yellow brass (0.60% Cu, 0.40%	0.050	0.0255	Dry	0.081	242
2	High-carbon steel	8, 14, 6, 6, 15, $\frac{3}{64}$	Zn, 0.8% Sn, 0.006% Pb)	0.100	0.0127	Dry	0.096	299
3	High-carbon steel	8, 14, 6, 6, 15, $\frac{3}{64}$	Bronze (0.90% Cu, 0.10% Sn)	0.050	0.0255	Dry	0.086	190
4	High-carbon steel	8, 14, 6, 6, 15, $\frac{3}{64}$	Bronze (0.90% Cu, 0.10% Sn)	0.100	0.0127	Dry	0.111	232
5	High-speed steel (18–4–1)	8, 14, 6, 6, 15, $\frac{3}{64}$	Cast iron (160 B)	0.050	0.0255	Dry	0.101	172
6	High-speed steel (18–4–1)	8, 14, 6, 6, 15, $\frac{3}{64}$	Nickel (164 B)	0.050	0.0255	Dry	0.111	186
7	High-speed steel (18–4–1)	8, 14, 6, 6, 15, $\frac{3}{64}$	Ni-Cr (207 B)	0.050	0.0255	Dry	0.088	102
			Steel					
8	High-speed steel (18–4–1)	6, 14, 6, 6, 6, 0, 0	SAE B1113 (C.D.)	0.050	0.0127	Dry	0.08	260
9	High-speed steel (18–4–1)	6, 14, 6, 6, 6, 0, 0	SAE B1112 (C.D.)	0.050	0.0127	Dry	0.105	225
10	High-speed steel (18–4–1)	6, 14, 6, 6, 6, 0, 0	SAE B1120 (C.D.)	0.050	0.0127	Dry	0.100	270*
11	High-speed steel (18–4–1)	6, 14, 6, 6, 6, 0, 0	SAE B1120 + Pb (C.D.)	0.050	0.0127	Dry	0.060	290
12	High-speed steel (18–4–1)	6, 14, 6, 6, 6, 0, 0	SAE 1035 (C.D.)	0.050	0.0127	Dry	0.110	130
13	High-speed steel (18–4–1)	6, 14, 6, 6, 6, 0, 0	SAE 1035 + Pb (C.D.)	0.050	0.0127	Dry	0.110	147
14	High-speed steel (18–4–1)	8, 14, 6, 6, 15, $\frac{3}{64}$	SAE 1045 (C.D.)	0.100	0.0127	Dry	0.110	192
15	High-speed steel (18–4–1)	8, 22, 6, 6, 15, $\frac{3}{64}$	SAE 2340 (185 B)	0.100	0.0125	Dry	0.147	143
16	High-speed steel (18–4–1)	8, 14, 6, 6, 15, $\frac{3}{64}$	SAE 2345 (198 B)	0.050	0.0255	Dry	0.105	126†
17	High-speed steel (18–4–1)	8, 14, 6, 6, 15, $\frac{3}{64}$	SAE 3140 (190 B)	0.100	0.0125	Dry	0.160	178
18	High-speed steel (18–4–1)	8, 14, 6, 6, 15, $\frac{3}{64}$	SAE 4350 (363 B)	0.0125	0.0127	Dry	0.080	181
19	High-speed steel (18–4–1)	8, 14, 6, 6, 15, $\frac{3}{64}$	SAE 4350 (363 B)	0.0125	0.0255	Dry	0.125	146
20	High-speed steel (18–4–1)	8, 14, 6, 6, 15, $\frac{3}{64}$	SAE 4350 (363 B)	0.025	0.0255	Dry	0.125	95
21	High-speed steel (18–4–1)	8, 14, 6, 6, 15, $\frac{3}{64}$	SAE 4350 (363 B)	0.100	0.0127	Dry	0.110	78
22	High-speed-steel (18–4–1)	8, 14, 6, 6, 15, $\frac{3}{64}$	SAE 4350 (363 B)	0.100	0.0255	Dry	0.110	46

No.	Tool material	Tool angles	Work material	(Hardness)	Feed	Depth	Fluid		
23	High-speed steel (18-4-1)	8, 14, 6, 6, 6, 15, $\frac{3}{64}$	SAE 4140	(230 B)	0.050	0.0127	Dry	0.180	190
24	High-speed steel (18-4-1)	8, 14, 6, 6, 6, 15, $\frac{3}{64}$	SAE 4140	(271 B)	0.050	0.0127	Dry	0.180	159
25	High-speed steel (18-4-1)	8, 14, 6, 6, 6, 15, $\frac{3}{64}$	SAE 6140	(240 B)	0.050	0.0127	Dry	0.150	197
26	High-speed steel (18-4-1)	8, 22, 6, 6, 6, 15, $\frac{3}{64}$	Monel metal	(215 B)	0.100	0.0127	Dry	0.080	170
27	High-speed steel (18-4-1)	8, 22, 6, 6, 6, 15, $\frac{3}{64}$	Monel metal	(215 B)	0.050	0.0255	Dry	0.074	127
28	High-speed steel (18-4-1)	8, 22, 6, 6, 6, 15, $\frac{3}{64}$	Monel metal	(215 B)	0.100	0.0127	Em	0.080	185
29	High-speed steel (18-4-1)	8, 22, 6, 6, 6, 15, $\frac{3}{64}$	Monel metal	(215 B)	0.100	0.0127	SMO	0.105	189
			Steel, ann.						
30	Stellite 2400	0, 0, 6, 6, 0, $\frac{3}{32}$	SAE 3240		0.187	0.031	Dry	0.190	215
31	Stellite 2400	0, 0, 6, 6, 0, $\frac{3}{32}$	SAE 3240		0.125	0.031	Dry	0.190	240
32	Stellite 2400	0, 0, 6, 6, 0, $\frac{3}{32}$	SAE 3240		0.062	0.031	Dry	0.190	270
33	Stellite 2400	0, 0, 6, 6, 0, $\frac{3}{32}$	SAE 3240		0.031	0.031	Dry	0.190	310
34	Stell, No. 3	0, 0, 6, 6, 0, $\frac{3}{32}$	Cast iron	(200 B)	0.052	0.031	Dry	0.150	205
			Steel, ann.						
35	WC (T64)	6, 12, 5, 5, 10, 45, 0	SAE 1040		0.052	0.025	Dry	0.156	800
36	WC (T64)	6, 12, 5, 5, 10, 45, 0	SAE 1060		0.125	0.025	Dry	0.167	660
37	WC (T64)	6, 12, 5, 5, 10, 45, 0	SAE 1060		0.187	0.025	Dry	0.167	615
38	WC (T64)	6, 12, 5, 5, 10, 45, 0	SAE 1060		0.250	0.025	Dry	0.167	560
39	WC (T64)	6, 12, 5, 5, 10, 45, 0	SAE 1060		0.062	0.021	Dry	0.167	880
40	WC (T64)	6, 12, 5, 5, 10, 45, 0	SAE 1060		0.062	0.042	Dry	0.164	510
41	WC (T64)	6, 12, 5, 5, 10, 45, 0	SAE 1060		0.062	0.062	Dry	0.162	400
42	WC (T64)	6, 12, 5, 5, 10, 45, 0	SAE 2340		0.062	0.025	Dry	0.162	630

* 180 ft/min is more normal.

† For values for many cutting fluids, see *Trans. ASME*, May, 1937, p. 343.

Table 13-5. Since the Taylor equation plots as a straight line on logarithmic coordinates, considerable predictive information can be obtained by fitting a least squares trend line to relatively few (5 or 6) points, each of which is replicated at least 3 or 4 times in a random order. This test is valid for a given tool-work combination and a standard feed, depth of cut, and tool geometry. Typical Taylor plots for the major cutting-tool materials are given in Fig. 13-13.

Effect of feed and depth of cut on tool life   In general, as either the feed or depth of cut is decreased from values of 0.010 in, the force per unit area rises rapidly. However, changes in the feed f and depth of cut d result in a family of curves which are parallel to one another when they are plotted on logarithmic coordinates, in the manner of Taylor equations. A typical equation would take the form of:

$$V = \frac{C_1}{f^x d^y}$$

where C_1 = constant depending on the tool-work combination.

According to Boston, for low-alloy steel, $x = 0.77$ and $y = 0.37$.

Research has shown that the maximum tool life depends on the slope of the Taylor equation, which does not change significantly with changes in either feed or depth of cut. For example, a change in feed of 2:1 can almost double the rate of metal removal at little change in cost because the cutting speed must be reduced only slightly if tool life is kept constant in both cases. That fact represents a major discovery which is borne out by production experience. The easiest way to earn a bonus in a machine shop which is on the incentive plan is to increase the feed to the point where the workpiece still passes inspection and then determine the appropriate cutting speed for the tool life desired.

Thus from both theory and practice it is found that the amount of metal removed during a unit of production time increases as f and d increase, provided that the cutting speed is decreased sufficiently to maintain the same tool life. Furthermore, since the exponent of d is larger, an increase in the depth of cut is less detrimental to tool life than is a change in feed at the same tool life. Therefore, low feeds and deeper cuts are more efficient than heavy feeds and shallow depths of cut. Practically, the depth of cut cannot exceed the stock available for machining. Heavier cuts require more power and frequently result in poor surface finish, which may be of little importance if a finishing operation is required.

The power required in machining is primarily a function of the cutting force if there is no large nose radius or cutting-edge angle. The power at the cutting point can be calculated as follows:

$$P = \frac{F_c V}{33,000} \qquad \text{hp}$$

The power of a machine tool required for a given operation can be estimated from the metal removal rate, the specific horsepower, and the overall machine efficiency, η, as follows:

$$P = \frac{\text{metal removal rate} \times \text{specific hp}}{0.7} \quad \text{hp}$$

where the metal removal rate in turning is: $12(CS)f(d/c)$ and representative values for specific horsepower are given in Table 13-6; CS = cutting speed, ft/min; f = feed increment, in/rev; d/c = depth of cut, in.

Cutting Fluids

Excess temperature is the most serious limitation to tool life because cutting-tool materials soften markedly at sufficiently high temperatures. Thus as industry strives for faster metal-removal rates and ever-shorter cycle times, the increased speeds and feeds magnify the thermal problems of the tool and the surface finish of the workpiece. Cutting fluids have long been used to reduce the average temperature within the cutting zone. The high specific heat, ready availability, and low cost would seem to make water an ideal cutting fluid, but its low lubricity and tendency to form rust on machine-tool components and ferrous work materials preclude its use.

Several types of fluids are used to reduce the heat from friction and the severe plastic deformation which occurs in metal cutting. Solids may also be used in the form of graphite in gray iron, or manganese sulfide in free machining grades of ferrous alloys, or lead in copper-base alloys. Certain low-shear-strength solids may be formed on the tool face by chemical reaction with chlorinated or sulfurized cutting fluids at low cutting speeds.

There are two major types of cutting fluids: water base and oil base. Each may have a number of additives to provide germicidal properties, to increase the wettability of the fluids, and to combine with the tool or work material to produce low-shear-strength solids at the tool-work interface. Gaseous cutting fluids include oxygen and water vapor in the air or a foggy mist which may be directed onto the cutting tool.

Cutting fluids perform several functions; they:

1 Reduce the weldability between the chip and the tool surface, which results in less friction, heat, wear, and built-up edge
2 Cool the workpiece, chip, and tool and thereby markedly reduce the average temperature in the cutting zone
3 Clear chips from the cutting region
4 Leave a residual film on the work surface to reduce corrosions
5 Improve the surface finish of the workpiece, which is especially important in the case of workpieces which are made on screw machines, gear teeth, or the surfaces of tapped holes

Table 13-6 TYPICAL VALUES OF SPECIFIC HORSEPOWER

Material	Feed, in/rev	Unit power, hp·min/in³	Specific energy, in·lb/in³
Steel (120 B*)	0.001	1.12	443,000
Steel (120 B)	0.003	0.86	347,000
Steel (120 B)	0.005	0.76	301,000
Steel (120 B)	0.010	0.64	254,000
Steel (120 B)	0.020	0.54	214,000
Steel (160 B)	0.001	1.25	495,000
Steel (160 B)	0.020	0.59	234,000
Steel (200 B)	0.001	1.50	594,000
Steel (200 B)	0.020	0.73	290,000
Steel (300 B)	0.001	1.87	740,000
Steel (300 B)	0.020	0.92	364,000
SAE 302	0.003–0.011	0.72	285,000
SAE 350	0.006–0.009	1.20	475,000
SAE 410	0.003–0.013	0.75	297,000
Gray CI (130 B)	0.006–0.012	0.29–0.35	127,000
Meehanite	0.006–0.012	0.55–0.76	262,000
K-Monel	0.004–0.010	0.80	317,000
Inconel 700	0.003–0.007	1.40	554,000
High-temperature alloy A286	0.006–0.011	1.20	475,000
High-temperature alloy S816	0.004–0.009	1.25	495,000
Titanium A55	0.010	0.65–0.76	281,000
Titanium C130	0.010	0.81–0.93	345,000
Aluminum 2014-T6	0.003–0.008	0.24	95,100
Aluminum 2017-T4	0.003–0.008	0.21	83,200
Aluminum 3003-O	0.003–0.008	0.16	63,400
Aluminum 108 (55 B)	0.003–0.009	0.15	59,400
Muntz metal	0.007–0.012	0.55	218,000
Phosphor bronze	0.002–0.006	0.33	131,000
Cartridge brass	0.003–0.009	0.48	190,000

* Brinell hardness number.
SOURCE: Courtesy of Monarch Machine Tool Co.

Chemical nature of cutting fluids There are two general types of cutting fluid actions; first, there are gaseous boundary reactions which occur during cutting in air. The oxygen in the air reacts instantaneously with the hot nascent surface and produces a very thin boundary-layer film of oxide coating on the work surface and on that of the tool. That layer reduces the tendency for a built-up edge to occur. The second type is an extreme boundary-layer film which is provided by chemical reaction to produce a low-shear-strength solid at the asperities on the interface between the tool and workpiece. In this case special additives to dipolar lard oils provide an easily adsorbed film of low-shear-strength sulfides or chlorides directly on the tool surface. These materials are adsorbed in the liquid or vapor phase by chemical adsorption. Such boundary-layer lubricants may be physically adsorbed at lower temperatures in multi-molecular layers. They are particularly effective at low cutting speeds, i.e., 40 ft/min and below, and they are most useful in tapping and reaming operations or for automatic screw machines where surface finish is most important. When lard oils or less expensive mineral lard oils are used, a special fungicide or anti-bacterial additive must also be used to prevent the cutting fluid from deteriorating in the storage tank in the machine base. Also operators must use a special skin cream to protect their arms and hands from rashes and blackheads. Even with the best of care some people are allergic to cutting fluids with a high oil content. They must be switched to other areas.

Cooling action of cutting fluids   In high-production operations, the type of cutting which has reactive additives may be of little value because the cutting speed exceeds the reaction time required to develop a low-shear-strength solid at the tool-work interface. In these cases the major role of the cutting fluid is that of removal of the heat of friction and deformation. The former is about 25 percent and the latter is about 75 percent of the total heat in metal cutting. On the average, 80 percent of the heat goes to the chip and 10 percent each to the tool and workpiece. In this case a cutting fluid with a maximum specific heat is required. Thus soluble oil emulsions, which have a milky appearance, are commonly used at reduction ratios, i.e., water to oil ratios, of 80:1 to 20:1. Whereas the cutting oils give improved surface finish and reductions in edge buildup and power, soluble oil emulsions give reduced average temperature at

Table 13-7   APPROXIMATE THERMAL PROPERTIES OF THE THREE MAJOR TYPES OF CUTTING FLUIDS

Cutting fluid	Spec. wt, lb/gal	Spec. ht., Btu/(lb)(°F)	Thermal conductivity, Btu/(hr) (ft)(°F)	Boiling point, °F
Water	8.34	0.998	0.349	212
Lard oil	7.35	0.500	0.075	~350
Air	0.010	0.240	0.015	−318

the tool-work interface and thereby permit a sufficiently high cutting speed to inhibit the formation of the built-up edge. The latter is not stable at speeds of 150 to 300 ft/min. The higher cutting speed also leads to a higher shear-plane angle and a reduced coefficient of friction, both advantageous with respect to optimum cutting conditions.

Cutting fluids for heat removal should have good thermal conductivity, high specific heat, and a high heat of vaporization. The properties of the three major types of cutting fluids are listed in Table 13-7.

From the table it is obvious that water would be the most suitable base material, but it must be and is commercially modified to enable it to achieve its potential. Thus to water are added special emulsifiable oils, rust inhibitors, antifoaming compounds, wetting agents, and bactericides.

BASIC MACHINING OPERATIONS

Machine tools have evolved from the early foot-powered lathes of the Egyptians and John Wilkinson's boring mill. They are designed to provide rigid support for both the workpiece and the cutting tool and can precisely control their relative positions and the velocity of the tool with respect to the workpiece. Basically in metal cutting a sharpened wedge-shaped tool removes a rather narrow strip of metal from the surface of a ductile workpiece in the form of a severely deformed chip. The chip is a waste product which is considerably shorter than the workpiece from which it came but with a corresponding increase in thickness of the uncut chip. The geometrical shape of the machined surface depends on the shape of the tool and its path during the machining operation.

Most machining operations produce parts of differing geometry. If a rough cylindrical workpiece revolves about a central axis and the tool penetrates be-

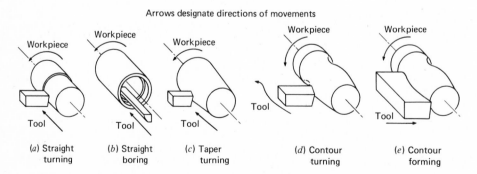

Arrows designate directions of movements

(a) Straight turning   (b) Straight boring   (c) Taper turning   (d) Contour turning   (e) Contour forming

FIGURE 13-14
Diagrams showing how surfaces of revolution are generated and formed. (*Redrawn from Doyle et al.*)

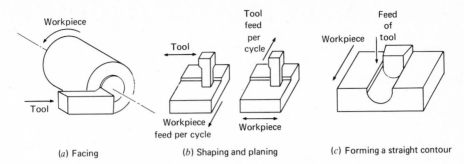

(a) Facing   (b) Shaping and planing   (c) Forming a straight contour

FIGURE 13-15
Diagram showing how plane surfaces are generated and formed. (*Redrawn from Doyle et al.*)

neath its surface and travels parallel to the center of rotation, a surface of revolution is produced (Fig. 13-14a) and the operation is called turning. If a hollow tube is machined on the inside in a similar manner, the operation is called boring (Fig. 13-14b). Producing an external conical surface of uniformly varying diameter is called taper turning (Fig. 13-14c). If the tool point travels in a path of varying radius, a contoured surface like that of a bowling pin can be produced (Fig. 13-14d), or, if short enough (approximately 1 in) and the support is sufficiently rigid, a contoured surface could be produced by feeding a shaped tool normal to the axis of rotation (Fig. 13-14e). Short tapered or cylindrical surfaces could also be contour formed.

Flat or plane surfaces are frequently required. They can be generated by radial turning or facing, in which the tool point moves normal to the axis of rotation (Fig. 13-15a). In other cases, it is more convenient to hold the workpiece steady and reciprocate the tool across it in a series of straight-line cuts with a crosswise feed increment before each cutting stroke (Fig. 13-15b). This operation is called planing and is carried out on a shaper. For larger pieces it is easier to keep the tool stationary and draw the workpiece under it as in planing. The tool is fed at each reciprocation. Contoured surfaces can be produced by using shaped tools (Fig. 13-15c).

Multiple-edged tools can also be used. Drilling uses a twin-edged fluted tool for holes with depths up to 5 to 10 times the drill diameter. Whether the drill turns or the workpiece rotates, relative motion between the cutting edge and the workpiece is the important factor. In milling operations a rotary cutter with a number of cutting edges engages the workpiece, which moves slowly with respect to the cutter. Plane or contoured surfaces may be produced depending on the geometry of the cutter and the type of feed. Horizontal or vertical axes of rotation may be used, and the feed of the workpiece may be in any of the three coordinate directions (Fig. 13-16).

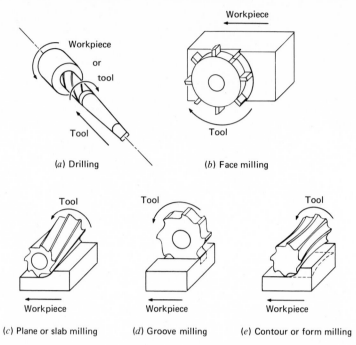

Arrows designate directions of movements

(a) Drilling   (b) Face milling

(c) Plane or slab milling   (d) Groove milling   (e) Contour or form milling

FIGURE 13-16
Typical machining operations performed by multipoint tools. (*Redrawn from Doyle et al.*)

BASIC MACHINE TOOLS

Machine tools are used to produce a part of a specified geometrical shape and precise size by removing metal from a ductile material in the form of chips. The latter are a waste product and vary from long continuous ribbons of a ductile material such as steel, which are undesirable from a disposal point of view, to easily handled well-broken chips resulting from cast iron. Machine tools perform five basic metal-removal processes: turning, planing, drilling, milling, and grinding (Table 13-8). All other metal-removal processes are modifications of these five basic processes. For example, boring is internal turning; reaming, tapping, and counterboring modify drilled holes and are related to drilling; hobbing and gear cutting are fundamentally milling operations; hack sawing and broaching are a form of planing and honing; lapping, superfinishing, polishing, and buffing are variants of grinding or abrasive removal operations. Therefore there are only four types of basic machine tools, which use a cutting tool of a specific controllable geometry: (1) lathes, (2) planers, (3) drilling machines,

Table 13-8 COMPARISON OF BASIC MACHINING OPERATIONS FOR DUCTILE MATERIALS

Operation	Shape produced	Machine tool	Cutting tool	Relative motion		Surface roughness, µin	Min. prod. tolerance, in
				Tool	Work		
Turning (external)	Surface of revolution (cylindrical)	Lathe, boring machine	Single point	(transverse)	(rotation)	32–500	±0.001
Boring (internal)	Cylindrical (enlarges holes)	Boring machine	Single point	(rotation)	(rotation + linear)	16–250	±0.0001 ±0.001
Shaping and planing	Flat surfaces or slots	Shaper, planer	Single point	(transverse)	(linear)	32–500	±0.001
Drilling	Cylindrical (originates holes 0.010 to 4 in dia.)	Drill press	Drill: twin edges	(rotation + linear)	Fixed	125–250	±0.002
Milling End, form Face, slab	Flat and contoured surfaces and slots	Milling machine	Multiple points (cutter teeth)	(rotation)	(multi-directional)	32–500	±0.001
Grinding Cylindrical Surface Plunge	Cylindrical and flat	Grinding machine	Multiple points (grind wheel)	(rotation + transverse)	(rotation)	8–125	±0.0001

and (4) milling machines. The grinding process forms chips but the geometry of the abrasive grain is uncontrollable.

The amount and rate of material removed by the various machining processes may be large, as in heavy turning operations, extremely small, as in lapping or superfinishing operations where only the high spots of a surface are removed.

A machine tool performs three major functions: (1) it rigidly supports the workpiece or its holder and the cutting tool; (2) it provides relative motion between the workpiece and the cutting tool; (3) it provides a range of feeds and speeds usually ranging from 4 to 32 choices in each case.

Range of Speeds and Feeds

In production tools the feeds and speeds are frequently provided by a gear-driven transmission in which the various steps are arranged in such a way that no desired cutting speed deviates more than a given percentage from one available through the machine-tool transmission. Thus if the design criterion was that no more than a 10 percent deviation should be allowed, the successive steps could be no more than ± 10 percent apart. The ratio between steps would then be 1.2, and the transmission speeds would vary in a geometric progression from a base of b in the following manner: $b, br, br^2, br^3, \ldots, br^{(n-1)}$. If the last term, $br^{(n-1)}$, is defined as a and n is the number of steps in the progression, the value of r is:

$$r = \sqrt[n-1]{\frac{a}{b}}$$

Many machine-tool drives provide a choice of 8, 12, 16, or 24 speeds with corresponding values of r of 1.58, 1.36, 1.26, or 1.12. As the number of choices rises, the cost of the machine tool also rises; but in most cases a ratio of 1.26 is adequate for a production machine. In a research lathe an infinitely variable cutting speed is desirable, but for a 30-hp lathe the drive to provide such variability is almost as large as the lathe and is at least as costly as the lathe alone.

Block Diagrams of the Basic Machine Tools

The major features of the four basic machine tools can most easily be described by block diagrams. In this case the principal components of a machine tool are represented by blocks and the relative motion of each component is represented by appropriate arrows.

The four basic chip-producing machine tools are: lathes, planers, drilling machines, and milling machines. Grinding machines are specifically removed from this list because the grinding wheel merely replaces the cutting tool used in the other types. The basic feed motions are the same. For instance lathes and cylindrical grinders have identical feed motions, but the necessity of providing for a large rotating wheel has modified the machine design.

Engine lathe The engine lathe produces surfaces of revolution by a combination of a single-point tool moving parallel to the axis of work rotation and a revolving workpiece (Fig. 13-17). In an engine lathe the workpiece is (1) clamped in a chuck which is mounted on the spindle or is (2) mounted between centers in the spindle and the tailstock and then driven by a dog which connects the workpiece to the work driver. The headstock contains the drive gears for the spindle speeds, and through suitable gearing and the feed rod, drives the carriage and cross-slide assembly. The carriage provides motion parallel to the axis of rotation, and the cross slide provides motion normal to it. The leadscrew is used only for chasing threads and thus is not required on a modern production lathe because thread milling or rolling would be used for threads made in quantity.

The tailstock is adjustable longitudinally along the lathe bed. If the lathe must accommodate long pieces the bed must be longer, and for slender pieces special supports (called back or steady rests) must be provided to prevent workpiece deflection away from the tool during the machining process.

The tool is mounted in the toolpost and must be located on the vertical centerline of the workpiece. The proper height is provided by packing strips and shims in a production lathe. The heavy-duty toolpost is provided with two

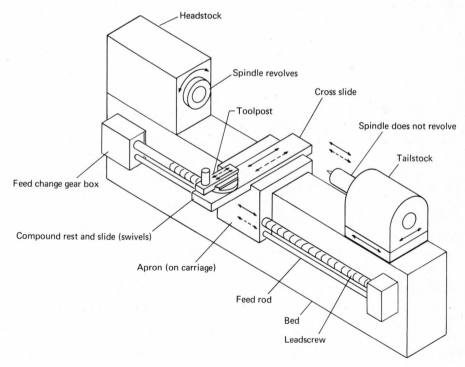

FIGURE 13-17
Block diagram of an engine lathe. (*Redrawn from Doyle et al.*)

clamping bolts to support the shank of the tool firmly during the machining operation.

Planing machines The planing operation generates a plane surface with a single-point tool by a combination of a reciprocating motion along one axis and a feed motion normal to that axis. Slots and limited vertical surfaces can also be produced. Planers handle large heavy workpieces and shapers handle smaller parts. In the planer the workpiece reciprocates and the feed increment is provided by moving the tool at each reciprocation (Fig. 13-18). In shaping the tool is mounted on a reciprocating ram and the knee upon which the workpiece is fastened is fed at each stroke of the ram. To reduce the lost time on the return stroke both planers and shapers are provided with a quick-return mechanism.

The planer drive is furnished by a large 50- to 75-hp ac-dc motor generator set. The large amount of dc power is needed to provide for the huge acceleration and deceleration braking needed because of the inertia of the heavy table and its associated work load. The distinguishing features of a planer are its large size and single-point cutting tool, and the fact that the table reciprocates. The feed is provided by motion of the tool slide along the crossrail at each cycle of the table. Two heavy housings are located at the center of the long bed. They support the crossrail which in turn carries one or more tool heads complete with tool slides and clapper boxes for the cutting-tool holders. The clapper box permits the tool to be raised sufficiently to clear the work on the return cycle of the planer table.

Drilling machine A simple drilling machine consists of a vertical column, on which is located the table for supporting the work, and at a higher level, the variable-speed drill head (Fig. 13-19). Small, straight-shank drills are mounted in a small three jaw chuck. Larger drills have tapered shanks which can fit directly in the spindle. Smaller drills use hand feed only; larger-production machines are equipped with a power feed mechanism. More versatile heavy-duty drilling machines, called radial drills, have a horizontal arm which can be adjusted vertically on a cylindrical column for supporting the drill head. In that case the arm can be swung radially and the drill head can move along the arm to bring the drill to large cumbersome workpieces.

Milling machine A milling machine uses a multitoothed rotary cutting tool to remove metal chips from a workpiece. The original cutters used by Eli Whitney were similar to overgrown dentist's burrs. Since that time milling cutters have been developed to produce a multitude of contours in a finished part. There are two broad classifications of milling operations: (1) plain milling and (2) end or face milling. The basic plain milling cutter produces a flat surface

PLANER

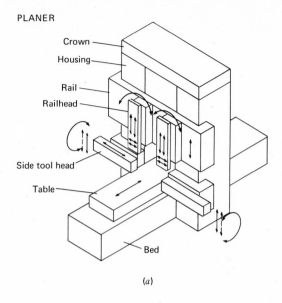

(a)

SHAPER

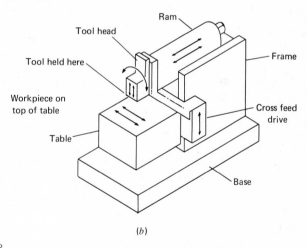

(b)

FIGURE 13-18
Block diagrams for two types of planing machines: (a) planer for large work;
(b) shaper for light work. (*Redrawn from Doyle et al.*)

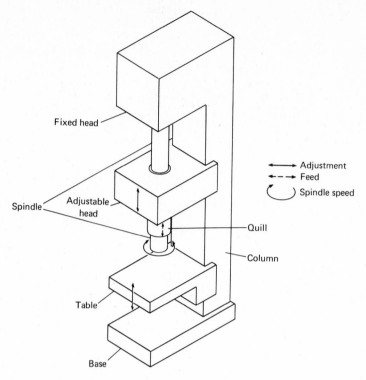

Adjustment
Feed
Spindle speed

Fixed head

Spindle

Adjustable head

Quill

Column

Table

Base

FIGURE 13-19
Block diagram of a typical drilling machine. (*Redrawn from Doyle et al.*)

through the use of cutting teeth on its periphery which are parallel to the axis of rotation.

Plain and form milling are carried out primarily on horizontal milling machines (Fig. 13-20a), whereas face milling and end milling operations are carried out on vertical spindle machines (Fig. 13-20b).

An end milling cutter has its cutting teeth located at the end as well as on its periphery and rotates around an axis which is normal to the surface being cut. It also produces a flat surface. A face milling cutter is large in diameter (>6 in) and produces flat surfaces such as the top of a six-cylinder block for an internal combustion engine. In face and end milling the cutting occurs primarily by the peripheral section of the teeth not the end of the cutter itself.

Light- and medium-duty machines are of the column and knee type. That is, the workpiece can be located in space within 0.001 in in any of the three coordinate directions which lie within the travel range of the machine. This flexibility of location and feed combined with the diversity of milling cutter geometry make the milling machine a tremendously versatile machine tool for

MILLING MACHINE—COLUMN AND KNEE TYPE

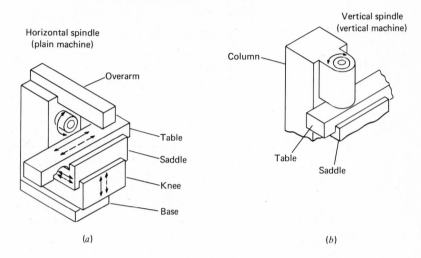

(a)

(b)

MILLING MACHINE—BED TYPE

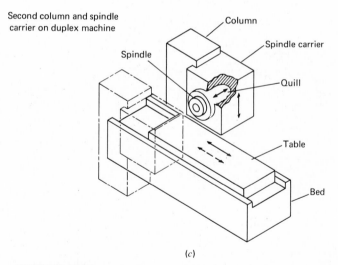

(c)

FIGURE 13-20
Block diagrams of typical milling machines: (a) horizontal milling machine;
(b) vertical milling machine; (c) bed-type milling machine. (*Redrawn from Doyle et al.*)

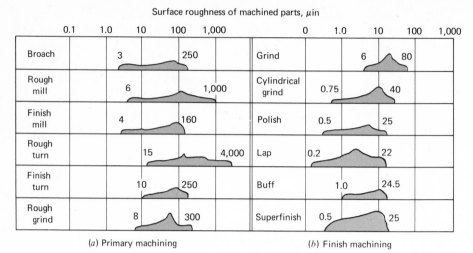

(a) Primary machining   (b) Finish machining

FIGURE 13-21

Machined surface finishes. Height of curve is proportional to occurrence. (*Physicists Research Co., Ann Arbor, Michigan.*)

toolroom and production use. In the latter case, several operations are frequently performed at once and the versatility of the machine is increased by the use of proper holding fixtures.

High-production milling operations are carried out on bed-type milling machines (Fig. 13-20c) which have less versatility but more rigidity. In such cases, machine utilization is increased by placing a work fixture at either end so that one fixture is loaded while the work in the other is being machined. In horizontal milling machines of both types, greater support is provided for the cutter arbor by adding overarms or a ram to provide an outer bearing for the cutter-arbor support. This feature is mandatory to keep the cutter running true and to provide support for heavy loads.

Surface Finish of Machined Parts

Machining processes can be divided into two groups: *primary* machining methods, which use single- or multiple-edged tools and remove large amounts of metal in heavy roughing cuts; and (2) *secondary* methods, which usually follow primary machining and impart greater dimensional accuracy, improved surface qualities, and leave less residual stress in the finished surface. Secondary machining processes are especially useful for removing dimensional inaccuracy caused by heat-treatment procedures (Fig. 13-21), but they add greatly to the total cost of the manufactured item, and should be used only when absolutely necessary to meet engineering requirements. The allowable surface roughness depends upon many factors, such as the functions and size of the piece, the fit and dimensional accuracy required, the load-carrying requirement, the uniformity of motion, or the wear characteristics of sliding or rolling surfaces.

Surface roughness can be evaluated in three ways. First, a profilometer can be used; this utilizes a diamond tracer similar to a phonographic pickup. The tracer reciprocates over the surface at a known speed and generates a current which is proportional to the surface irregularities. Second, evaluation can be done by tactual and visual standards which are normally arranged with average roughness values of 4, 8, 16, 32, 63, 125, 250, and 500 μin. When a fingernail is moved slowly across the surface of the standard, it can detect a minimum roughness of about 4 μin. Finally, evaluation can be made by direct microscopic examination of the surface, or by studying a cellulose (or other plastic) replica produced by pressing the film against a metal surface coated with a thin film of acetone or other solvent. Greater magnification of the surface profile can be obtained by cutting through the surface on a shallow taper (18°), polishing the resultant wedge, and observing the surface profile under magnification.

Formerly, the roughness of a surface was reported as a root-mean-square value; however, this statistical value yields an answer which is consistently approximately 11 percent greater than the arithmetic average based on a true sinusoidal model. The difference is so small that it is negligible for most purposes. Some countries use the maximum peak-to-valley roughness, which gives values considerably greater than the average roughness used in the United States.

In most cases the character of a machined surface depends upon the process used to produce it. For example, there are several sources of roughness when machining with a single-point tool: (1) feed marks left by the cutting tool; (2) built-up edge fragments embedded in the surface during the process of chip formation; (3) chatter marks from vibration of the tool, workpiece, or machine tool itself. When a surface is turned at high speed without chatter, the primary surface roughness lies in an axial direction and may be computed quite accurately from the feed and the tool geometry.[1] The built-up edge causes cyclic gouging and smoothing of the surface as the built-up edge sloughs off, whereas chatter marks are cyclic with the revolving of the workpiece. Both of these latter surface marks are superimposed on the feed marks. An analytical investigation of the characteristics of lathe feed marks as a function of tool geometry and feed is presented in Figs. 13-22 and 13-23.

The surface roughness caused by a built-up edge is primarily affected by cutting speed. As the latter increases, up to a point, the surface finish improves. The critical speed at which the built-up edge becomes insignificant varies from 250 to 500 ft/min, depending on the cutting conditions, tool-workpiece combination, tool geometry, cutting fluid, etc.

For example, from the partial trace of a particular surface (Fig. 13-24)

---

[1] According to the American National Standards Institute, the average roughness for a turned surface is approximately equal to the feed divided by 60.

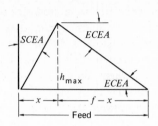

$$\text{Feed} = x + (f - x)$$
$$x = h_{max}/\cot SCEA$$
$$(f - x) = h_{max}/\tan ECEA$$
$$h_{max} = \text{feed}/(\tan SCEA + \cot ECEA)$$
$$h_{ave} = 2/3\, h_{max}$$

FIGURE 13-22
Analysis of surface profile produced by a pointed tool.

and the tabular data (Table 13-9) from nine particular deviations from the mean value, compute the arithmetic and root-mean-square values of surface roughness and compare the answers. The arithmetic average value is

$$\text{AA} = \frac{1}{L}\int_0^L |y|\, dx = \frac{\sum\limits_0^L |y|}{N}$$

and the root-mean-square value is

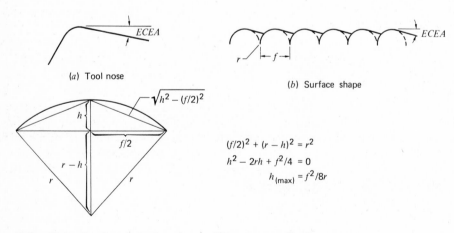

(a) Tool nose

(b) Surface shape

$$(f/2)^2 + (r - h)^2 = r^2$$
$$h^2 - 2rh + f^2/4 = 0$$
$$h_{(max)} = f^2/8r$$

(c) Magnified surface shape (approximate)

Where r = nose radius, in
f = feed, in/r
$h_{(max)}$ = surface roughness, peak-to-valley

FIGURE 13-23
Analysis of a surface profile produced by nose radius.

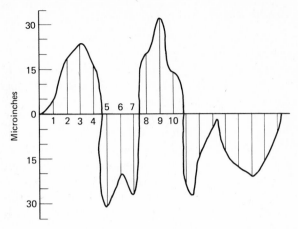

FIGURE 13-24
Partial trace of a surface.

$$\text{rms} = \sqrt{\frac{1}{L} \int_0^L y^2 \, dx} = \sqrt{\frac{\Sigma y^2}{N}}$$

where L = cutoff length
N = number of equally spaced discrete points in length L

The latter is the same as the standard deviation as defined in statistics.

Table 13-9 PEAK AND VALLEY VALUES OF THE
SURFACE TRACE TAKEN AT DISCRETE
POINTS (See Fig. 13-24)

Point	Values of y at Δx intervals, μin	Point	Values of y at Δx intervals, μin
1	4	10	13
2	19	11	23
3	23	12	15
4	16	13	6
5	31	14	12
6	20	15	18
7	27	16	21
8	20	17	17
9	31	18	9

Production Turning Operations

The era of modern machine tools is relatively recent, dating back to the Industrial Revolution in England. The first screw cutting lathe was invented by Maudsley in 1798. The modern engine lathe takes its name from its gear-driven headstock which typically provides 16 spindle speeds. Although an engine lathe can bring one to four tools mounted in its one toolpost to bear on a workpiece, it can provide only a single feed motion at one time. A typical turned part frequently requires several operations, each of which needs a different tool and feed motion. The replacement and resetting of the tooling and handling of the work in and out of the machine adds a significant cost to the finished part. Moreover, the cost per piece may be three to six times the cost of using a more specialized machine tool. When a number of identical parts are needed, positioning of the tools for each operation not only involves time but may also result in workpieces which will not pass inspection. Thus an engine lathe is suitable for low production only.

From the simple lathe, more complex machines have evolved for producing duplicate parts efficiently and in quantity. A basic feature of such machines is their ability to apply several tools to the work simultaneously using different motions. In addition, once the tools are set, they can be used repeatedly in the same sequence without being reset for each operation. Semiautomatic machines like turret lathes (Fig. 13-25) are fitted with two indexing tool holders called turrets. A square turret is mounted on a cross slide, and a hexagonal turret replaces the tailstock. They require constant attention by a machine operator, but a high degree of skill is only needed when the machine is being set up for the first piece. Such features as multitool machining, overlapping or simultaneous operations, turning lengths to positive stops, rapid indexing of the square and hexagonal turrets at the end of each cutting operation, and automatic stock feeding and clamping result in increased production. The use of a turret lathe enables each cut to be consistently within normal manufacturing tolerances of $+0.002$ to -0.003 in over a production run of several hundred pieces.

The turret lathe can bring at least 11 tools into play in any one sequence of turning operations: four from the square turret, six from the hexagonal turret, and one from the rear toolpost (Fig. 13-26). Special tooling such as a combined stop and center drill or a drill and an external turning tool can provide even more production possibilities through the mounting of two or more cutting tools at a single position.

The ram-type turret lathe (Fig. 13-25, left) is used for light duty operations, whereas the saddle type (Fig. 13-25, right) with an overhead pilot bar is used for pieces over 2 in in diameter and for heavy chucking work. The saddle type has power assists so that although it is slower, it requires no more manual effort in operation. In fact, this type of turret lathe was one of the first to be numerically controlled. In this case, the operator was replaced by a punched

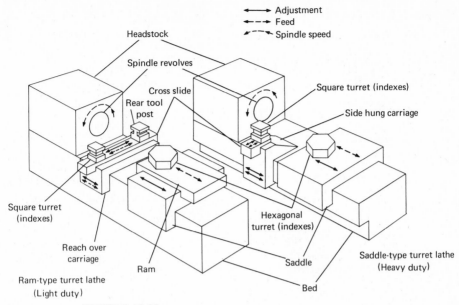

FIGURE 13-25
Block diagrams for turret lathes: ram type for smaller work; saddle type for heavy work. (*Redrawn from Doyle et al.*)

tape programmed with a sequence of instructions which performed all the functions which the operator normally carried out.

The automatic screw machine can make small parts from bar stock on a completely automatic basis. In one type of single-spindle automatic screw machine, two cams control the cross slides and a third cam controls the motion of the six-station turret which revolves on a horizontal axis (Fig. 13-27). The three cams must be designed and machined for each new part. However, the tooling is so standardized and the procedure for designing and cutting the cams is so well developed that in less than a day cams can be designed and cut for a standard job. Details of how the cams operate are given in Fig. 13-28.

Multiple-Spindle Automatic Screw Machines

The highest production rates of all automatic screw machines can be achieved by four-, five-, six-, and eight-spindle machines. Bar-type machines are rated by the largest diameter stock which can be passed through the spindles. Chucking machines are rated by the diameter of the part which can clear the tool slides.

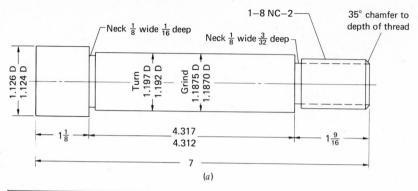

(a)

Operation	Hexagon turret	Square turret	Rear toolpost
I	Feed to stock stop, (center withdrawn)		
II	Turn diam (4)		
III	Turn diam (2)	Turn diam (6)	
IV	Face (1)		
V	Center drill (1)		
Double index			
VI	Support on center	Neck diam (3) and (5)	
VII	Thread diam (2)		
VIII	....	....	Cutoff and chamfer (7)

(b)

(c)

FIGURE 13-26
Turret-lathe production: (a) typical piece; (b) operation sequence; (c) plan view of the required tooling for the square and hexagonal turrets.

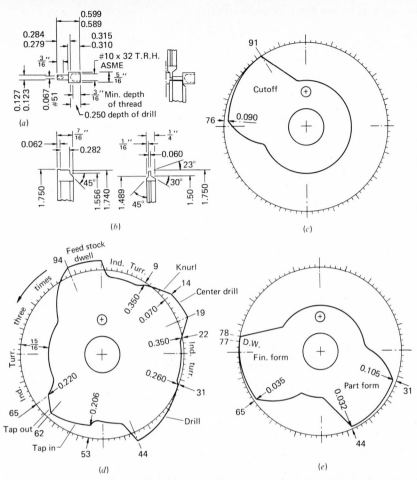

Round brass stock cold-drawn to 5/16-in diameter is used with a No. 00 spring collet and feeding finger. The spindle speed is 1,500 rpm forward and 5,000 rpm reverse. The spindle runs reverse for the carbon-tool-steel cross-slide tools. The driving shaft rotates at 240 rpm with change gears having teeth as follows; on the driving shaft, 40; first on the stud, 60; second on the stud, 52; and on the worm, 30. Production time per piece is 3.25 s, equivalent to 1,108 pieces per hour gross.

Spindle revolutions	Order of operations and tools	Feed per revolution, in
24	Index turret	0.020
37	Knurl and center drill L.H. 0.250 in diameter	0.040 and 0.005
24	Index turret	
35	Part form with front slide	0.0019
24	Index turret and reverse spindle	
8	Tap in, 10–32 thread	
	Reverse spindle	
8	Tap out, then index turret three times	
35	Finish-form with front slide	0.001
35	Cutoff with back slide	0.0025
24	Feed stock to stop	

(f)

FIGURE 13-27

Production tooling for a knurled brass insert made on an OOG Brown and Sharp automatic screw machine. (a) Workpiece and position of cross-slide tools; (b) high-speed steel cross-slide tools; (c) cam for rear-slide cutoff tool; (d) cam for turret slide; (e) cam for front slide; (f) order of operations.

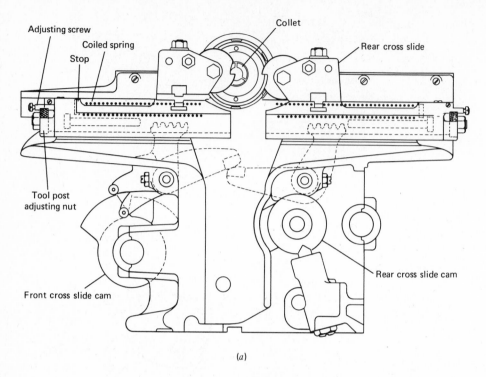

(a)

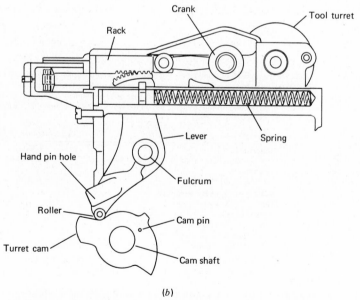

(b)

FIGURE 13-28

Details of cam operation. (a) Section through cross slides; (b) section through turret slide.

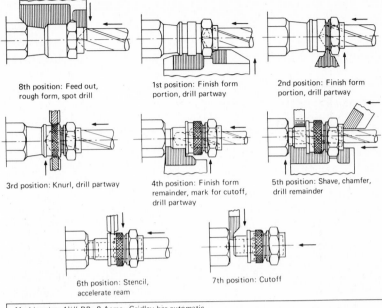

8th position: Feed out, rough form, spot drill

1st position: Finish form portion, drill partway

2nd position: Finish form portion, drill partway

3rd position: Knurl, drill partway

4th position: Finish form remainder, mark for cutoff, drill partway

5th position: Shave, chamfer, drill remainder

6th position: Stencil, accelerate ream

7th position: Cutoff

Machine size: 1¼" RB–8 Acme–Gridley bar automatic
Name of piece: Sparkplug shell; = machine time: min, 4.5 s
Material steel: Open-hearth grade A leaded,   (SFM 416);   gross production: 800 pieces per hour
Overall dimensions: 13/16 hex x 1¼" long;   spindle speed; 1692 rpm;   toolside cam: 11/32 @ 0.0042

(a)

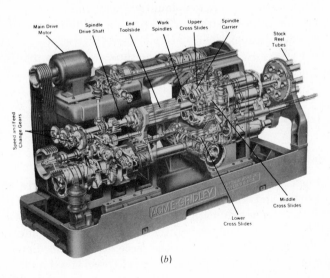

(b)

FIGURE 13-29
Multiple-spindle automatic-screw-machine operations for making spark-plug shells at the rate of 800 per hour. (a) Details of the operations; (b) typical eight-spindle automatic screw machine. (*Courtesy of Acme Gridley*, Cleveland, Ohio.)

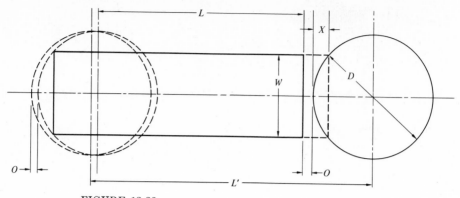

FIGURE 13-30
Minimum travel of the workpiece for face milling operations.

End working tools are mounted on the end tool slide which corresponds to the hexagonal turret of a turret lathe. The end tools do not index but slide forward and back on the bedways. The cross slides for forming and the cutoff tools are mounted next to the spindle stations and move radially toward and away from the work.

In operation, the spindle carrier indexes after each operation one station at a time. At each rotation of the spindle carrier one piece is completed; that is all operations are completed simultaneously and the cycle time for one part is the time for the longest operation plus an indexing time of 1 to 2 s. A typical part is the spark-plug shell which is made at a gross production rate of 800 pieces per hour (Fig. 13-29a). Note that the through holes in the body are being drilled in seven of the eight stations. The machine on which the spark-plug bodies are made is shown in Fig. 13-29b. Even at the production of 800 parts per hour some 90 machines are required on a three-shift basis, 7 days a week, to meet the daily demand of one manufacturer.

Production Milling Operations

Fixed-bed milling machines are more rigid and have greater accuracy than do the column and knee milling machines. The distinctive feature of fixed-bed machines is their automatic cycle. Once a piece is clamped in the fixture the start button is pushed, the cutter rotates, the table moves at rapid travel to the work, the feed begins, and the piece is cut; immediately the table reverses to the end-of-cycle position at the rapid travel speed and the cycle stops. Thus the operator merely unloads and loads the fixture and pushes a button to start the cycle. Dual fixtures can be used holding two or more parts at each end of the table. This operation is called continuous reciprocal milling and is described more fully under climb milling.

Special Milling Machines

There are a number of special types of production milling machines. Planer-type milling machines are like planers but with heavy-duty vertical milling heads replacing the planer tools. Another type of heavy-duty mill has large vertical milling heads and a rotary table which can be loaded on one side while cutting along the other side.

Duplicating mills, tracer mills, pantographs, engraving machines, or die sinkers can accurately reproduce one or more cavities by tracing a standard form or master pattern. Larger tracing units are found in die shops where large forging or die-casting dies are produced.

Planning for Milling

Standard parts may be easily supported between the jaws of a standard milling machine vise, but in many cases irregularly shaped castings or forgings require special holding devices called fixtures to properly support the workpiece during machining. Such fixtures and any special cutters need to be designed and built before any parts can be machined.

If a part is to be face-milled, the approach distance must include the travel until the diameter of the cutter fully engages the workpiece (Fig. 13-30). Thus it can be seen that the total distance traveled to face the workpiece is

$$L' = L + 2(O) + x$$

where L = length of cut, in
$\quad x$ = cutter approach, in
$\quad O$ = overtravel, in

Also
$$x = \frac{D}{2}\left[1 - \sqrt{1 - \left(\frac{W}{D}\right)^2}\right]$$

where D = diameter of cutter, in
$\quad W$ = width of workpiece, in

The power required in milling can most easily be computed from the metal-removal rate. Although milling takes a slightly greater amount of power than does turning, for estimating purposes the two can be considered to be the same and the values of specific horsepower can be used. Thus the metal-removal rate is:

$$\text{Mrr} = W \times d \times f$$

where f = in/min feed
$\quad d$ = depth of cut, in

Thus
$$\text{Hp} = \frac{\text{mrr (specific hp)}}{0.7}$$

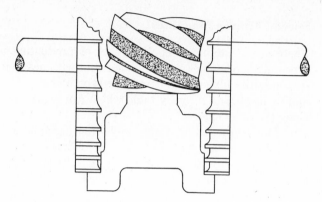

FIGURE 13-31
Gang milling a 20-mm bolt body. Three separate operations are combined.
The outer cutters are straddle milling the body; the helical milling cutter is
finishing the top surface.

where 0.7 is a factor to correct for friction in the machine drive.

Economies in manufacture can be made by milling more than one surface
at a time, or by cutting the separate surfaces without removing the part from
the machine. More than one surface may be cut by grouping the milling cutters
on an arbor, called gang milling (Fig. 13-31). Different diameters in various
positions can mill surfaces parallel or vertical to the axis of the milling arbor.
When slots or broad, flat surfaces are milled, interlocking blade cutters can be
used to assure a continuously milled surface or the correct width of a slot. The
interlocked cutters are spaced apart by washer shims at the hub of the cutter
to compensate for the grinding of the sides of the cutter teeth.

Parts are often designed with milled surfaces at an angle to other milled
surfaces, or on a plane higher or lower than another. These surfaces can often

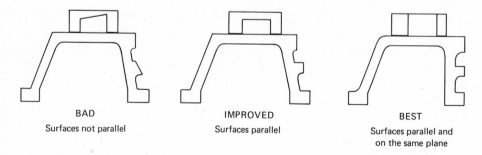

BAD
Surfaces not parallel

IMPROVED
Surfaces parallel

BEST
Surfaces parallel and
on the same plane

FIGURE 13-32
Design for ease of milling. Keep surfaces parallel and if possible in the same
plane.

be located in the same plane by altering the design of the part attached to the milled surface. In this way they can be milled in one operation (Fig. 13-32).

Climb Milling Versus Conventional Milling

When the cutter enters the material in the direction of feed, it is known as climb milling or down milling; when it enters the material opposite the direction of feed, it is known as conventional milling or up milling (Fig. 13-33). Climb milling requires a well-built and well-maintained machine, with no backlash in the feed screws that will permit the cutter to take too large a bite. On the other hand, it does not have a tendency to lift the part off the table. There are advantages and disadvantages to each method. After some experience, the engineer can determine which is more suitable for a given job.

Climb milling permits simpler fixtures because an end stop is the major requirement for resisting the cutting forces. There is less tendency for chatter so higher speeds and feeds may be used. But the machine must have a backlash

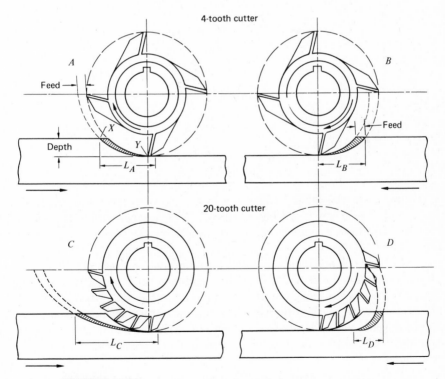

FIGURE 13-33
Comparison of climb milling and conventional milling, A and B using a 4-tooth cutter and C and D using a 20-tooth cutter.

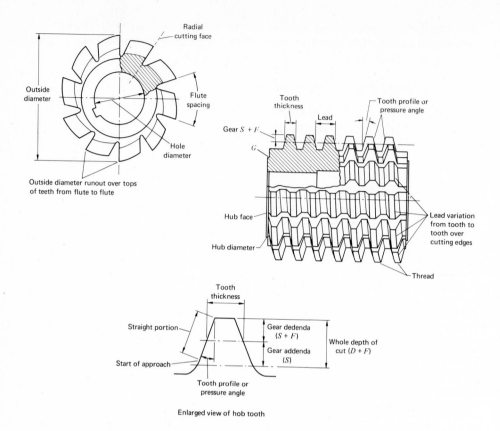

FIGURE 13-34
A hob and its elements.

eliminator. On the other hand, conventional milling gives less wear and tear on the table feed screw and nut assembly. Conventional milling is used on most column and knee machines because it keeps the screw engaged securely with the nut at all times.

In bed-type machines equipped with a proper backlash eliminator, double-fixture milling is possible. The machine can be set up to run almost continually; i.e., when a piece is being cut at one end of the table, the operator is loading an identical fixture at the other end. In this case, the machine both climb-mills and conventional-mills alternately with a rapid travel at about 300 in/min between the fixtures. If the milling time is long, an operator can handle two machines.

If the feed per tooth is kept constant and the same tool-work combination is used, the chip made by conventional milling will be longer and thinner than that produced by climb milling (Fig. 13-33A and B). This phenomenon is even more apparent when a 20-tooth cutter is used because then the feed per tooth is the same but the table feed has increased by a factor of 5 (Fig. 13-33C and D).

In conventional milling the tooth comes in contact with the work at point X and pushes until the force between the tooth and the work is sufficient to cause the cutting edge to dig in. If the tool is dull, the force reaches a high value before cutting begins. The chip starts at zero and reaches a maximum equal to the feed per tooth at point Y. The cross-sectional area of the chip removed can be calculated and is found to be equal to the feed per tooth times the depth of cut.

Gear Cutting

Gear production is really a specialized milling process which can be carried out as a repair job on a milling machine or on special gear-cutting machines. Gears can be produced either by using a series of special form cutters or by using the generation process. In the latter case, a special straight-sided tooth form is used to generate the proper gear-tooth shape. In the milling type of operation the cutter is called a hob (Fig. 13-34). The hob rotates about an axis which is set so that the helix angle of the hob is parallel with the axis of the gear being cut. Then as the gear blank revolves, the hob revolves in mesh with it. The process is called generation because the straight-sided tooth contacts the side of the gear tooth in a series of lines and actually forms a true involute shape as the hob rotates with the gear being cut (Fig. 13-35).

A gear shaper uses a vertically reciprocating cutter which has straight-sided teeth. The cutter slowly revolves while geared to be in mesh with the gear

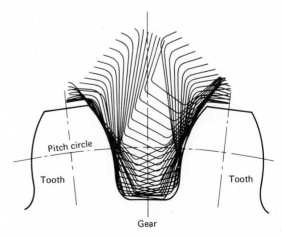

FIGURE 13-35
The action of a hob as its teeth progress through and cut or generate the tooth space in a gear.

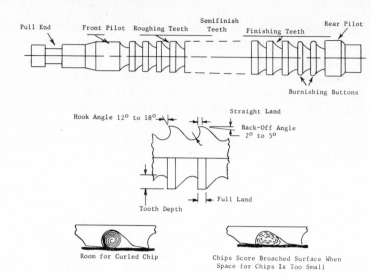

FIGURE 13-36
Broaching tool details.

blank being cut. It also generates a true involute shape in much the same manner as the hob does.

Broaching

Broaching may be thought of as a production planing process. It is a machining method wherein one or more cutters with a series of teeth are pushed or pulled across a surface to machine that part to a desired contour. Teeth of the broach increase in height progressively as they approach the finishing end of the tool, so that each tooth removes an equal amount of metal. The last few teeth in the broach are used to bring the work to the desired size and finish.

Broaching has developed from the keyway cutter in a hand press to large machines capable of machining several surfaces simultaneously. For example, the flat surfaces on an internal-combustion engine block are surface-broached in one operation; likewise, a splined internal hole in a gear blank can be sized by one pass of a broach. There is practically no limit to the shape or contour of a broached surface.

Broaching is a fast way to remove metal either externally or internally. It produces a good finish to close tolerances.

Processes similar to broaching include milling, planing, and reaming, although none of these produce the close tolerances and fine finish with the speed of operation of broaching.

Description of process   Figure 13-36 illustrates a typical broach. At the starting end are the roughing teeth, which remove the bulk of the metal. Then follow the finishing teeth, which remove the final increment of stock, bringing the work up to size and finish. When exceptionally fine finishes are required, the last few teeth are hemispherical burnishing teeth which remove no metal, but serve only to eliminate surface irregularities. In a single pass, then, the workpiece is roughed, sized, and burnished, and the finishing operation required with other methods of machining is eliminated. Because little metal is removed by the teeth at the finishing end, there is little wear on these teeth and the broached surfaces can be held to fine tolerances and fine finish. Unlike other machining methods, all of the teeth of the broach cut simultaneously. The rise per tooth in a broach is fixed, and the tool pressure cannot be varied. Therefore, the workpiece must either be strong enough to withstand that pressure or must be supported properly in a fixture.

An accurately located drilled hole is satisfactory as a preliminary procedure for an internal broaching operation. Holes may also be prepared for broaching by punching, boring, and coring. The positions of internal broached holes can be adjusted slightly when the holes are broached. The amount of shifting of position to obtain accuracy depends on the strength of the broach and the rigidity of the machine. For external broaching, no preliminary machine operations are required. Properly designed, broaching tools may be used at high rates of speed and are capable of production rates in excess of competitive machining methods.

Depending on the design of the machine, the stroke of the broach is either horizontal or vertical. The broach may be pulled or pushed. Only one pass is made; the broach or piece must be removed before the machine returns to normal operating position. Key slotters or broaches are arranged so that on the back stroke the pressure on the cutting teeth is released. Machines range in size from 2-ton pull, 12-in stroke to 75-ton pull, 76-in stroke. Both mechanically and hydraulically actuated broaching machines are on the market and either can be made fully automatic.

Tolerances and finish   The following table gives the tolerances possible through broaching and the recommended design tolerances for minimizing cost when a broaching operation is specified:

Summary	Holes and splines, in	Gear teeth, in	External surfaces, in
Tolerances possible	0.0002	0.0005	0.0002
Tolerances, low cost	0.002	0.002	0.002

When two or more parts are broached simultaneously, high dimensional accuracy (0.002 in) between the parts can be depended upon.

Surfaces   A fine finish is produced since burnishing is part of the operation. Usually no further surfacing operations are necessary. The tool marks evidenced in broached holes are axial rather than radial as in drilled or reamed holes. This is an advantage in close-fitting reciprocating parts, where radial lines may be objectionable because the high spots tend to wear rapidly.

Materials suited for broaching   Steels, cast irons, bronzes, brasses, aluminum, and a broad range of other materials are successfully broached with proper broach design and setup conditions. The best range for broaching of steels lies between 25 and 35 Rockwell C hardness, although steels of higher or lower hardness have been broached successfully. Soft and nonuniform materials are subject to tearing when broached. On surfaces of high hardness, the first tooth of the broach should cut beneath the scale or surface material, thus assuring longer broach life.

Economical quantity   Except when standard broaches may be employed, broaching is economical only for large-quantity production (over 2,500 parts). This is true because of the relative high cost of the broaching tool rather than because of the cost of setup. Broaching setups are simple except for fully automatic operations and can be made relatively quickly.

Broaching tools   As mentioned, broaching tools are expensive and are usually made specially for a given job. The broaching tool is made from a tough, wear-resistant alloy usually containing about 5 percent tungsten and 5 percent chromium.

Broaching tools must be handled carefully in order to prevent nicks in the teeth which would cause scratches in the work.

Design factors   In designing, the engineer should see to it that the amount of stock to be removed should always be less than $\frac{1}{4}$ in. Good design allows between $\frac{1}{32}$ and $\frac{1}{16}$ in of stock to be removed from the workpiece. If less than $\frac{1}{64}$ in is removed, a clean surface cannot be assured.

Since the broach must be able to make an unobstructed pass through or across the workpiece, it is not possible to broach blind holes.

The chip space between successive teeth on the broach must provide sufficient reservoir for the chip. This chip space will limit the axial length of the surface to be broached with a single broach.

Several surfaces can be broached simultaneously. When multiple surfaces on the same workpiece are broached, care must be exercised in designing the fixture to provide adequate strength to withstand the combined cutting-tooth pressure. When gears or splines are to be cut, mechanical, hydraulic, or pneumatic indexing equipment is frequently provided.

Summary   Broaching provides high repetitive accuracy (applicable to production of large numbers of parts of close tolerances and fine finish) and close dimensional relationship of several surfaces broached simultaneously. Broaching is 15 to 25 times faster than other competitive machining methods. The process can be used to accurately produce internal and external surfaces that are difficult to machine by other methods.

The principal disadvantage of broaching is the high cost of special broaching tools. This cost usually does not permit the process to be used when production requirements are low. Then, too, it cannot be employed economically for the removal of large amounts of stock (more than $\frac{1}{4}$ in). Lastly, the process has application only on unobstructed surfaces permitting the pass of the broach through the workpiece.

Broachings' principal applications are for the production of almost any desired external or internal contour. This includes flat, round, and irregular external surfaces, round and square holes, splines, keyways, rifling, and gear teeth.

Sawing

Sawing is a multipoint cutting operation which may be related to planing in the case of reciprocating blades and band saws, or it may be related to milling as found in circular sawing. In either case sawing is a parting operation, so the body of the saw is kept as thin as is practical to avoid wasting material in the cutoff operation. Yet it must be rigid enough to support the teeth during the cutting operation. Details of saw teeth and nomenclature are summarized in Fig. 13-37.

Circular saws   Sawing or slitting is performed in milling machines by narrow milling cutters. These cutters are known as circular saws and are also used in sawing machines. The cutters may run slowly and may be of large diameter, approximately 6 ft in some cases. Inserted teeth are used for sawing large sections of metal such as steel forgings and billets. Other types are run at high speeds to cut wood, plastics, and nonferrous metals. The principles of single-point tool cutting apply to saws. Ample room must be provided for chips and side clearance to reduce rubbing friction. The machines may be hand fed or power fed. The material may be fed automatically and cut off. The operator only loads the raw material and removes other cutoff pieces. Sawing provides a fairly smooth and flat surface with a slight burr and does not distort the part as in shearing or slitting. Friction sawing leaves a heavier burr, which can be chipped off easily. Abrasive sawing cuts any type of material with a clean, smooth cut and practically no burr.

Hack saws   Hack saws and hand saws cut the material in the same manner

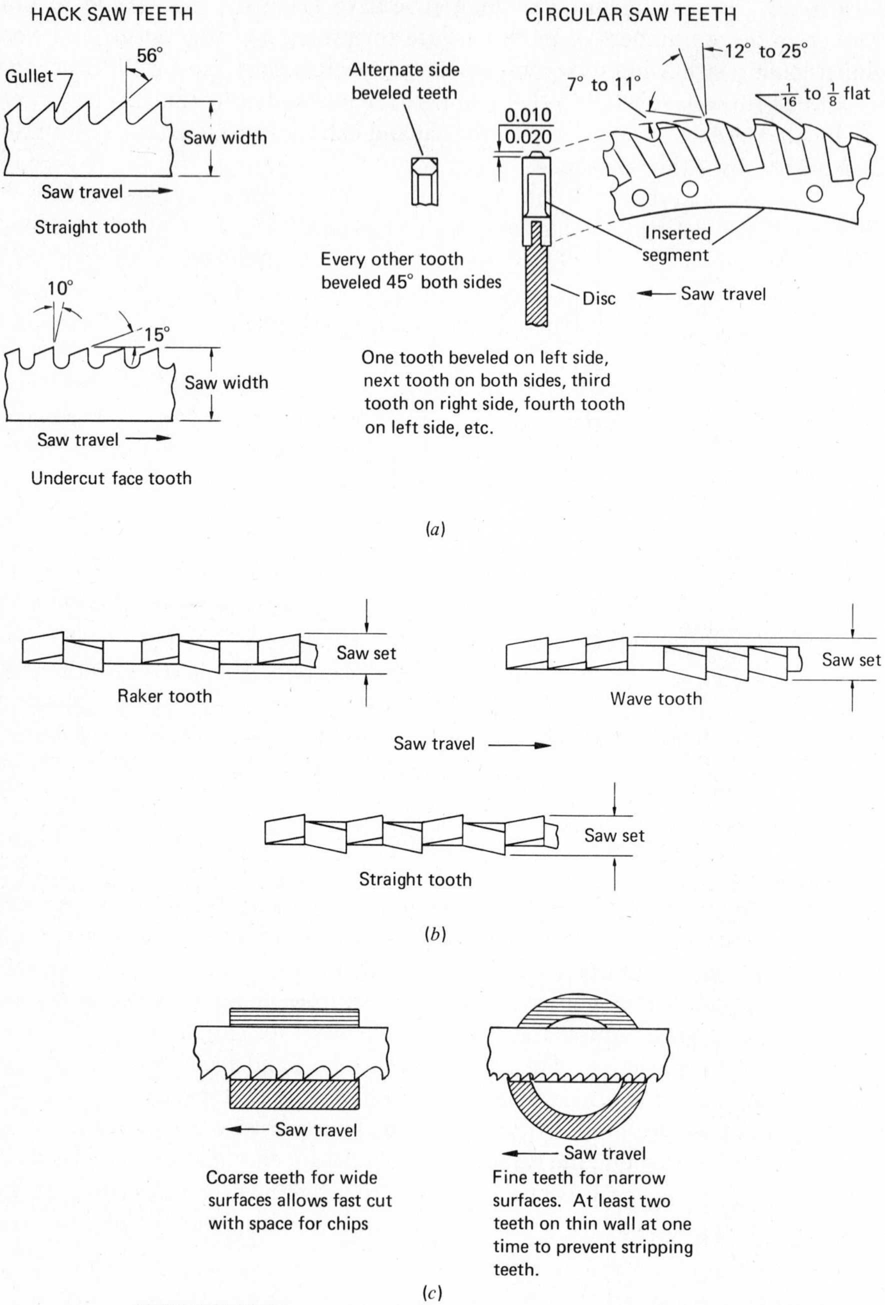

HACK SAW TEETH   CIRCULAR SAW TEETH

(a)

(b)

Coarse teeth for wide
surfaces allows fast cut
with space for chips

Fine teeth for narrow
surfaces. At least two
teeth on thin wall at one
time to prevent stripping
teeth.

(c)

FIGURE 13-37

Details of saw teeth. *(a)* Comparison of hacksaw teeth and circular saw teeth;
(b) types of set of saw teeth; *(c)* effect of tooth size and pitch.

as a circular saw, except that the teeth are in a straight line and act somewhat like a broach. Heavy pressures are required to obtain a cutting action in each tooth throughout the length of cut. Some hack saws are made to feed automatically, and more than one bar can be cut at the same time. Because more than one machine can be tended by an operator, it often costs less to saw than it does to shear, with the advantages of more accurate lengths and no distortion.

Contour sawing   Contour sawing has developed into an important process and has equipment especially designed to obtain maximum results. It is performed on jigsaws, continuous band saws, or long band saws wound on reels. The continuous band saws cut internal holes by separating the saw and inserting it through a pilot hole, and then rapidly welding the ends together on electrical-resistance welding equipment attached to the machine. The band saws are made in various thicknesses, widths, shapes, and forms of cutting edges. Carbide teeth and abrasive material are fastened to the edge of the band saw in some types. Glass, high-speed steel, armor plate, and all forms of metals and materials are successfully cut on contour band saws.

The engineer can use this process successfully in making model parts and low-quantity parts that are usually blanked, machined, or forged to shape. Dimensions can be held to $\pm \frac{1}{64}$ in. The part is not distorted and surfaces are smooth. Parts up to 6 in in thickness can be cut. Cut-off pieces can be used for other parts, for example, tool steel used in dies. By using multiple layers of sheets or plates, several parts can be cut out at one time, such as triggers, levers, cams, and cover plates. Speeds as high as 3,000 ft/min are used, depending on the type of material and its thickness. Filing, friction sawing, and diamond cutting are performed on these machines in a toolroom or model shop.

Friction sawing   Friction sawing is performed by high-speed circular saws and band saws. The saw has rather dull teeth, which strike the part at high speed. The heat of friction melts the material and the teeth remove the molten material. A spectacular shower of sparks, as well as noise, results. The saw is kept cool by water spray. The teeth gradually wear down and are easily sharpened by enlarging the space between them.

Friction sawing can be applied to any shape of part, angle, channel, I beam, bar, billet, gate, sprue, or casting. A fin forms on the exit edge, but can be chipped off easily. However, hack saws are more economical where they can be applied, because it is not necessary for the operator to tend the machine while sawing and the burr is not usually objectionable.

Summary   Sawing operations are required in many manufacturing processes. The cost of equipment is low; therefore, the cutoff operation should be located near the raw material and the user of the cut-off parts. The shop should not be depended upon to calculate developed lengths; drawings should specify exact lengths so that material can be cut to proper length by the storeroom.

Production Grinding

Traditionally grinding has been used successfully as a secondary machining operation for high-precision work or to finish hardened workpieces. In fact grinding is still the most economical way to machine external surfaces which are from R_c 45 to R_c 70 hardness. The burden of proof is on the competitive process. Grinding can produce a tolerance of ± 0.0001 in, but ± 0.0005 in is better.

Since World War II, the use of high-speed grinding wheels has been made possible by research and improvements in bonding and abrasive materials and in the grinding process itself. In some cases, grinding is economically competitive with primary machining methods. For example, finishing the flat surfaces of forgings or castings can be carried out with less total stock removal if parts are ground from the rough because the grinding wheel needs only to clean up the surface. There is no need to get under an oxidized or chilled skin by a deep initial cut. Abrasive belt finishing has also grown sufficiently to be recognized as a production finishing operation.

Grinding operations have many points in common. The machines must be particularly rigid because the wheels have a surface speed of at least 5000 ft/ min. The high speed makes dynamic balancing particularly important. Second, the wheel bearings must be particularly true, because a cylindrical workpiece has stock removed from both ends of a diameter, so any error in radial advancement is doubled when the work is measured. Third, wheel wear is continuous, and significant compensation must be provided either by having a large ratio of wheel diameter to work diameter, as in external grinding with automatic gaging, or there must be adaptive control to maintain a constant distance between the wheel surface and the centerline of the workpiece, as is the practice in internal grinding where the wheel must be smaller than the hole.

There are five major types of grinding operations: cylindrical (center or centerless), internal or hole, surface, tool and cutter, and abrasive belt grinding for precision and surface finishing operations.

Cylindrical grinding Cylindrical grinding can be done by grinding from center holes, or it can be centerless. In the former the machine is similar to a heavy-duty lathe except that both centers are fixed so that the workpiece can not run out at the headstock. To permit easy operator access to the work area the wheel is located behind the workpiece and is made large in relation to the workpiece diameter so that the wheel wear in one operation is small compared to the stock removed from the workpiece. A typical block diagram of a cylindrical grinding machine is given in Fig. 13-38. Typically a grinding wheel for such a machine is 2 to 3 ft in diameter and has a face width of 1 to 3 in. The feed rate is

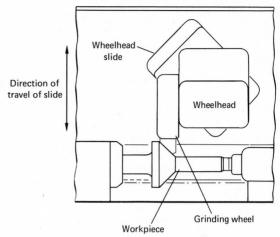

FIGURE 13-38
Block diagram of a typical cylindrical grinder.

one-fourth to three-fourths of the width of the wheel face, depending on the depth of cut, machine rigidity, and the power available at the machine. In plunge grinding a wheel at least equal in width to the diameter being produced is fed transversely until the proper diameter is reached.

Centerline grinding is a rapid, economical production operation for finishing shafts, pins, ball bearings, bearing races, and similar products to precise tolerances and good finish. When grinding on centers the work revolves on fixed centers; whereas in centerless grinding the diameter is determined by the periphery of the workpiece. In centerline grinding, time is required to locate and drill center holes, to clean them up after hardening, to clamp the dogs, and to move the piece into and out of the grinder, especially if the piece must be the same diameter from end to end. On the other hand, centerless grinders take more setup time but less operating time and little or no machine loading time.

In centerless grinding, the work is supported by three machine elements: the grinding wheel, the regulating wheel, and the work-rest blade which is beveled to push the workpiece toward the regulating wheel (Fig. 13-39). The rubber-bonded regulating wheel controls the rotational speed of the workpiece, and its rate of feed is determined by the angle of the horizontal axis of the regulating wheel with respect to that of the grinding wheel. Note that the faces of both wheels are still parallel.

There are two types of centerless grinding: through feed and infeed (Fig. 13-40). In through feed grinding the workpiece passes completely through the grinding zone and exits on the opposite side. Infeed grinding is similar to a plunge cut when grinding on centers. In this case the length of the section to be ground is limited by the width of the grinding wheel.

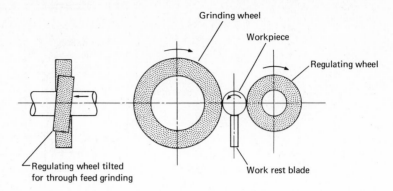

FIGURE 13-39
Schematic sketch showing the principle of centerless grinding.

The infeed method of centerless grinding differs from through feed in the way the workpiece is handled. The infeed method is used for grinding parts with several diameters or heads or shoulders which would prevent complete passage through the machine. The work rest and regulating wheel are withdrawn from the grinding wheel for loading or unloading. They then move toward the wheel to be finished to the final size.

In many cases special tooling makes centerless grinding a versatile process, but it is often less expensive to design and build new tools. Unless an economic analysis gives a strong indication that centerless grinding will be worthwhile, it is wise to stay with conventional grinding. In some cases it may be economical to centerless-grind as few as 25 or 100 parts, and in others as many as 5,000 pieces must be ground before centerless grinding can be justified.

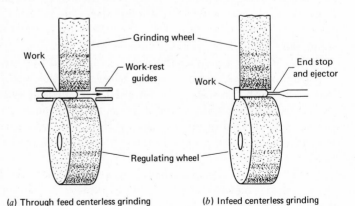

(a) Through feed centerless grinding   (b) Infeed centerless grinding

FIGURE 13-40

In centerless grinding the grinding wheel has a surface speed of 5,000 to 6,500 ft/min, whereas the regulating wheel operates at 50 to 250 ft/min. In through grinding the rate of feed depends on the speed of the regulating wheel and its angle of inclination; thus the feed in inches per minute is:

$$f = \pi dN \sin \delta$$

where d = diameter of regulating wheel, in

N = r/min of regulating wheel

δ = angle of inclination of the horizontal axis of the regulating wheel (which may vary from 0 to 8° but is usually 3 to 4°. For bent or warped pieces the angle should be 6 to 8°)

The feed force stems from the axis of the regulating wheel which tilts downward at the entrance to the grinding zone.

The advantages of centerless grinding are:

1 In through-feed grinding the cutting time approaches 100 percent of the operating time because the grinding action is almost continuous, with little loading and unloading time. Production rates of 150 or more pieces per hour can be readily achieved.

2 Since the workpiece is fully supported by the work-rest blade and regulating wheel, heavy cuts can be taken with minimum danger of distortion or overheating.

3 Centering errors do not exist; therefore the workpiece is rounded up with a minimum of excess stock allowance.

4 For every adjustment of 0.001 in of the work-rest blade or regulating wheel there is 0.001 in off the diameter.

5 Relatively unskilled personnel can be used as machine operators. The setup, specification, and design of the parts for the process requires the most time.

Internal grinding There are three types of internal grinding machines:

1 Machines in which the workpiece rotates slowly and the wheel spindle rotates and reciprocates the length of the hole

2 Machines in which the work rotates and reciprocates while the wheel spindle rotates only

3 Machines in which the wheel spindle rotates and has a planetary motion while the workpiece reciprocates

Hole grinding machines require high-speed spindles (up to 100,000 r/min for small holes) because grinding speeds should be in the order of 5,000 ft/min and the wheel must be smaller than the hole. Only light feed forces can be used because otherwise the cantilever-beam effect from the spindle overhang would

A. Grinding Wheel
B. Grinding Face
C. Wheel Spindle
D. Work Piece
E. Work Table

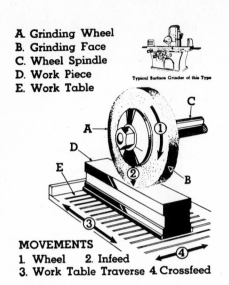

MOVEMENTS
1. Wheel   2. Infeed
3. Work Table Traverse 4. Crossfeed

A. Grinding Wheel
B. Grinding Face
C. Wheel Spindle
D. Work Pieces
E. Work Table

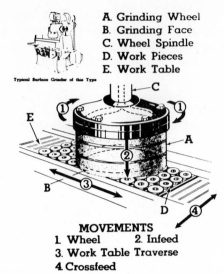

MOVEMENTS
1. Wheel   2. Infeed
3. Work Table Traverse
4. Crossfeed

A. Grinding Wheel
B. Grinding Face
C. Wheel Spindle
D. Work Pieces
E. Work Table

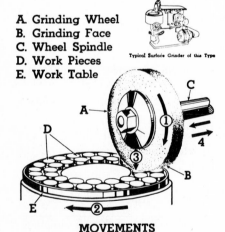

MOVEMENTS
1. Wheel   2. Work Table Rotation
3. Infeed   4. Crossfeed

A. Grinding Wheel
B. Grinding Face
C. Wheel Spindle
D. Work Pieces
E. Work Table

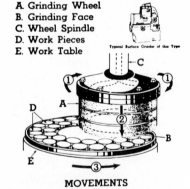

MOVEMENTS
1. Wheel   2. Infeed
3. Work Table Rotation

FIGURE 13-41

The principal kinds of surface grinders. Top, reciprocating table grinders; bottom, rotary table grinders; left, horizontal spindles; right, vertical spindles. (*Courtesy the Carborundum Co.*)

cause excessive taper. The wheel should not be permitted to leave the hole to minimize bell mouthing.

Wheel wear is rapid because of the small wheel size and the use of a soft grade to permit the wheel to break down under light cutting forces. Thus accurate hole sizing is difficult to obtain if tolerances are close. To meet this need production machines can be fitted with automatic gaging and feedback devices which assure accurate control of hole size. The grinding wheel keeps cutting until the hole is 0.001 in from the proper size and a special cam control is engaged to provide the fine feed which is continued until the hole is finished to the required tolerance.

Surface grinding   Surface grinding is primarily concerned with producing plane and frequently parallel surfaces on steel parts. For this reason magnetic chucks are standard equipment. As in milling the grinding wheel may have a vertical or horizontal axis and it may be considered to be equivalent to a face milling cutter when the axis is vertical and to a plain milling cutter when the axis is horizontal (Fig. 13-41). Both types of grinders can have either rotary or reciprocating tables. Surface grinders with rotary tables are particularly capable of grinding a number of small pieces simultaneously and at low cost, and they leave a characteristic crisscross pattern on the flat surfaces. The down feed on surface grinders can be automatic at 10^{-3} to 10^{-5} in per revolution of the work table.

Tool-and-cutter grinding   Tool-and-cutter grinders are not production machines and consequently will not be considered here. Any good text on machine tools describes such machines in detail. As a note in passing, the advent of throwaway carbide inserts for turning and milling operations has significantly reduced the work load in the tool-regrinding department, and the importance of these machines has been somewhat curtailed.

Abrasive-belt grinding   With the advent of the flexible plastic bond for applying abrasives to belts, the use of abrasive-belt grinding has increased because a coolant can be used and the bond is strong. The belt grinder is valuable in the model shop and toolroom, as well as in production. A platen in back of the belt applies pressure to the work and assures a flat or contoured surface as desired. Close dimensions can be held and a minimum of material (0.015 to 0.031 in) is allowed for machining. Coarse grit will remove material faster, but will not give as smooth a surface. The same general rules described at the beginning of the section on grinding apply to the selection of grain size, grade, and structure. The surface finish produced by belt grinding is better than machining. Simple fixtures, or merely holding the part against the belt, are all that are required for cleaning up surfaces, but water must be used to cool the workpiece.

Abrasive cutoff is accomplished by a thin disk-grinding wheel usually made with a rubber bond. The disk cuts principally on its edge, although slightly on the sides too. It gives a smooth surface and very little burr is produced. The process is as fast as sawing, and the part often requires no further machining operations when cut off. It is also accurate as to cutoff length. Water must be used as a coolant.

Any material that can be cut off by sawing, shearing, or flame cutting can also be cut off with an abrasive wheel properly selected for the purpose. This method is applicable to all metals, including hardened tool steels and stainless steel. Other materials that can be cut in this manner, many of which cannot be properly cut with a saw, are plastics in various forms and shapes, ceramic materials, carbon, hard rubber, slate, and casein.

In the foundry, cutoff wheels can be used to remove risers and gates from castings of all kinds, especially nonferrous ones. This can be done without damage to the casting, and the necessity for subsequent grinding may be eliminated. Furthermore, band and hack saws are slower and more easily dulled.

Cutting is performed dry on light, short cuts, and wet on long, heavy cuts. There is little discoloration due to heat when a coolant is used. In submerged cutting, both the work and the lower part of the wheel are completely submerged. At proper wheel speeds, submerged cutting is used to cut glass rods and tubing, plastics, copper tubing, heat-sensitive steels, and other materials that might be affected if the generated heat were not kept as low as possible.

The following are some common faults encountered when operating cutoff wheels; each one suggests its own correction.

1   Movement of work while wheel is in the cut as a result of improper clamping
2   Pinching of wheel caused by work vise high on one side
3   Wheel vibration caused by worn spindle bearings
4   Stalling of wheel in cut because of belt slippage or inadequate power
5   Excessive heating resulting from cuts too heavy for grade of wheel

Precision and surface-finishing operations   After secondary machining, either the precision or the surface finish may require further refinement. In this case fine abrasive machining is required. If both precision and finish are required, lapping or honing are indicated. If surface finish alone is desired, then fine abrasive blasting, buffing, tumbling, vibratory finishing, or polishing may be sufficient.

Lapping   Lapping is an abrading process which results in a wearing down of the ridges and high spots on a machined surface, leaving the valleys and a random array of fine scratches. Lapping is used to obtain fine dimensional accuracy, to correct minor imperfections in shape, to secure a fine surface finish, and to obtain a fit between two mating surfaces. In the lapping process a fine abrasive, carried in a light oil medium, is supplied to the work surface by a reciprocating type of motion which occurs in an ever-changing path. The resultant abrading action between the work surface and the softer lap such as brass, lead, cast iron, wood, or leather results in the removal of the high spots on the work surface. The cutting action occurs because the soft surface of the lap becomes charged with a layer of abrasive particles which provide a myriad of cutting edges while the lap surface maintains the proper contour.

The service life of lapped surfaces is usually substantially increased over that of the same parts when mated without lapping. Mating gears and worms are lapped to remove imperfections resulting from heat treatment or prior machining. Piston rings and gage blocks are customarily made parallel and to high precision by lapping.

The lapping process can be carried out economically by using the principles of mechanization and automation. If an eccentric spider is used to carry the parts which are to be made parallel in an ever-changing path between two rotating, parallel plates with a recirculating abrasive slurry, rapid and precise lapping will occur. Grooves in a checkerboard pattern in the lapping plate facilitate the delivery of the slurry and help flush away the surface particles. Wet lapping has been shown to be at least six times faster than dry lapping and there is no heat buildup.

Manual lapping is used to bring gage blocks to their final stage of dimensional accuracy and parallelism—a finish of 1 to 2 μin and a tolerance as small as $\pm 0.000,001$ in.

Honing   Honing may be defined as a production method for finishing internal or external surfaces of revolution after machining or grinding. Honing is usually accomplished by reciprocating several spring-loaded sticks of a fine abrasive material over the surface of the rotating workpiece. Honing provides a fine, finished texture in holes and on the exterior surface as well. Bored holes are frequently reamed or honed to obtain final dimensional accuracy and a characteristic figure-of-eight pattern which retains oil on the surface over an extended period of time. This is particularly fortunate in the case of internal-combustion engines, especially during the breaking in period. Typical honed cylinders are: internal-combustion engine cylinders, bearings, gun barrels, ring gages, piston pins, and shafts. Honing is a cutting operation and rarely exceeds a removal of more than 0.001 in of stock on a side. Boring and reaming should precede honing to assure surface shape and location. Although a 1-μin surface finish is possible, an 8- to 10-μin finish is more economical and probably just as good.

Honing machines can be either horizontal or vertical. The former are used for long holes such as cannon or rifle bores or in small models for ring gages. The latter type are more common, ranging from those capable of handling all eight bores of a V-8 engine simultaneously to large machines with an 8-ft stroke capable of handling bores up to 30 in in diameter.

A typical internal hone consists of a framework for supporting the abrasive stones which are mounted and internally spring-loaded so that when in the hole they can expand to make contact with the surface of the bore. A cutting fluid is always used to flush away the abraded particles, to improve the finish of the workpiece, and to keep the stones free cutting.

Superfinishing   Superfinishing is a proprietary name, but the process produces the ultimate in refinement of surfaces. Although it is similar to honing, its principles are basically different. A large area of abrasive is used, so that uneven projections and wavy surfaces are removed through a clean cut. Since the abrasive is moved in many directions, it is self-adjusting and becomes automatically a master shape that will correct the work surface. In this respect it is similar to lapping, but the surface is not charged with lapping material. The process does not remove major amounts of material (maximum of 0.0001 in) in the average production job. The abrasive may be stones, wheels, or belts supported by master platens. Superfinishing is in common use on automobile parts, and automatic machines have been developed to automate this process, which is effective in removing chatter marks and amorphous material until a true surface of the base metal remains. The abrasive is loosely bonded so that each reciprocating stone quickly wears to the contour of the part, but it is large enough so that a representative surface is encountered. Thus the stone bridges and equalizes a number of defects simultaneously and corrects the surface to an average profile. As the surface becomes smoother the unit pressure decreases until the stone rides on a fluid film and cuts very little. A 1-in wide ground surface may be refined to a 3-μin finish in less than 1 min.

Design for grinding   When a part must be ground, the designer should keep the following points in mind:

1   Avoid sharp corners at all shoulders to reduce both tool wear and stress concentrations.
2   Provide recesses if the wheel must grind square with a shoulder.
3   Avoid internal grinding of deep small holes, and be sure to provide an undercut at the base of any ground hole so the wheel can reverse.
4   Avoid narrow slots which require grinding on the sides.
5   All cold-worked metal should be stress-relieved before grinding.

Polishing and buffing   Polishing is the term applied to refining the surface through the cutting action of fine abrasive particles applied to the surface of resilient wheels made of cloth, felt, or wood. Polishing operations may be classified as roughing, dry fining, and finishing. The roughing and dry-fining operations are performed with dry wheels; 20 to 80 abrasives are used in roughing and 90 to 120 in dry fining. In finish polishing, oil, tallow, or beeswax is used with the fine abrasive (150), thus giving a fine finish.

Buffing is the smoothing and brightening of a surface by the rubbing action of fine abrasives applied periodically to a soft wheel. The wheels are spiral sewed buffs of wool, cotton, or fabric. Buffing permits mirrorlike finishes and may be thought of as refined polishing. The abrasive is applied in a lubricating binder commonly known as tripoli.

Barrel finishing   Barrel finishing is of interest because, by its use, many hand operations of burnishing and polishing can be eliminated. The use of tumbling instead of machining, grinding, or hand-finishing operations has resulted in a savings of up to 97 percent in direct labor.

The process differs from conventional tumbling in that, since the parts are always supported by a finishing medium, they are not normally permitted to drop during the rotation of the barrel. The abrading action that takes place is thus concentrated on exposed sharp edges, with material removal varying with the sharpness and relative exposure of these edges. On flat surfaces, while the material removal is practically negligible, there is an action that results in alteration of surface roughness. The degree of action in either case can be controlled accurately, and there is no tendency toward uncontrolled "nicking" of surfaces, which often results in tumbling without a proper abrasive media.

Barrel finishing can be used to improve the surface of a great variety of small parts made of ferrous, nonferrous, or plastic materials where the problem is to remove burrs or fins, apply a radius to edges, or polish the surface. Since the parts "slide" or "float" in the medium, fragile parts can be tumbled, as well as heavier sections. The process can be applied either wet or dry, depending on the application.

Some typical applications are:

1 Deburring and definning of metal and plastic parts
2 Abrading of desired radii on parts as a substitute for machining, grinding, or hand operations
3 Producing desired surface smoothness for either appearance or performance
4 Producing desired surfaces for subsequent finishing operations, such as plating or painting
5 Removal of scale, rust, and dirt

Almost any metal or plastic part of reasonable size which does not have too great a variation in section size may be tumbled advantageously. A large quantity of parts can be finished at one time, the actual number depending on the size of the part, the size of the barrel, and the surface smoothness desired. Economically and practically, this compares favorably with the individual finishing of pieces by other methods. The only labor involved in production setups is the loading and unloading of the barrel, and occasional screening of the abrasives.

Facilities and materials required   The barrel-finishing operation requires a multi-sided revolving barrel (preferably rubber-lined), motors and controls, handling facilities, water and drain facilities, and a supply of abrading media.

The selection of the proper abrasive for a given application is important.

Experimentation is usually necessary to determine the proper abrasive to use. Aluminum oxide chips, stone chips, steel balls, foundry stars and slugs, molded abrasives and iron slugs, plastic chips, sand, wood balls, and other items are used as abrasive media. In a vast majority of applications, wherever the rigidity of the parts permits, it is possible to use either aluminum oxide chips or granite chips. These materials are furnished in seven or eight sizes, ranging from $\frac{1}{16}$-in screen to $1\frac{1}{2} \times 2$ in.

The specific size of abrasive chip selected should be such as will not lodge in holes or recesses, or between protruding sections. Where action is required on holes or in cavities, the size of abrasive chips used must be such that they will pass through. Where this condition does not exist, the influence of the chip size on surface roughness is usually the deciding factor. Larger chips will cut faster, but will produce a coarser surface. The opposite is true of smaller chips.

On parts that have heavy outside burrs and small burrs around holes, it will be advantageous to use a combination of large and small chips. The large chips will remove the outside burrs faster, while the small chips will pass through the holes, removing the burrs around the hole edges.

Water, alkali compounds, soda ash, trisodium phosphate, burnishing soap, and alkali compounds are used as additives. Water as a coolant is usually added to the mixture of parts and abrasive. It helps prevent the abrasive from loading, provides some lubricating action, and keeps the parts clean. By adding burnishing soap (0.5 percent by weight of the abrasive) to the load, nearly all cutting action is stopped and is replaced by a polishing action. This makes it possible to abrade first and then polish parts without changing the contents of the barrel except for adding the burnishing soap.

Burnishing compounds are often used where high luster and smooth finish are desired. These burnishing soaps and compounds act as lubricants and cushioning agents by preventing abrasion and nicking. In some cases where unusual conditions cause loading of the chips, an abrasive alkali compound is often used to maintain the stones in a clean, sharp condition.

Barrel speed The vast majority of parts can be barrel-finished at a speed of 80 to 200 surface feet per minute (with a 32-in barrel, this varies from 10 to 25 r/min). This wide range indicates that each part should be considered separately when determining the proper speed. Generally speaking, to prevent drastic surface changes, the speed should be such that the parts do not drop when reaching the top of the turn. They should gently turn over and slide with relation to the medium.

Rotation time of barrel The material from which a part is made determines to a large extent the length of time of rotation. Soft materials, such as copper, brass, aluminum, and magnesium, can be abraded in a fraction of the time required for iron or steel parts, if the shapes are identical. A copper part of the same shape and dimensions as a steel part would be finished in one-half the time of the latter. Aluminum and magnesium require even less time than copper.

The following list gives data on results obtained from a particular operation.

1 Description of work to be done: remove all burrs from a 5-oz steel gear approximately 2 in in diameter by $\frac{1}{4}$ in thick, and produce uniform finish.

2 Former method: hand filing. Results: lacked uniformity. Time allowed —0.033 h; production rate per hour—30 pieces.

3 Barrel-finishing-method data:

a Operator allowed time per piece, 0.00083 h.

b Running time per piece (time not allowed), 0.0050 h.

c Total cycle time per piece, 0.00583 h.

d Production rate per hour (one two-compartment barrel), 171 pieces.

e Direct labor cost decrease, 97 percent.

f Production rate increase (one barrel), 470 percent.

g Barrel size, 32 in diameter by 48 in length. Lining, wood or rubber. Abrasive, alundum.

h Abrasive size, 8*T*. Quantity used, 100 lb for load.

i Pieces in barrel, 1,200. Speed of barrel, 15 r/min. Time in barrel, 7 h.

j Additives: water, level with mass; trisodium phosphate, 2 lb.

Hydrohoning   Hydrohoning is the smoothing of flat and irregular surfaces by the use of a stream of liquid filled with a concentrated mixture of fine abrasive. This gives a very smooth surface and removes burrs and tool marks. It is used on metal molds.

Sand and grit blasting and hydroblasting   These processes are not classed as machining processes, but are placed here because of their abrasive cutting action. The abrasive is carried by an air stream or water stream and strikes the surface at a high velocity. It removes scale, burrs, light fins, and rust. The surface is marked by the impression of the grit. Steel grit is used in most applications because its rate of recovery is higher. The abrasive is propelled by paddles, which throw the grit onto the work either by centrifugal action or by a blast of air which picks up the abrasive and carries it through pipe or hose on through a nozzle. Nozzles are guided by the operators so the abrasive can reach the proper place.

Blasting equipment ranges from the small bench sizes to rooms large enough to house the largest equipment. Castings and welded steel parts are economically cleaned by these processes before finishing with paint.

Shot peening   Shot peening is similar to shot blasting, except that the surface is peened by the impinging balls from the blast. This hard-works the surface of

Table 13-10 COMPARISON OF CHIP AND CHIPLESS MACHINING PROCESSES

Machining process	Power required			Material-removal rate, in³/h	Average power required, kWh/in³	Feed rate, in/min	Expected tolerance, in $\times 10^{-3}$	Expected surface finish, μin	Depth surface damage, mil
	Volts	Amp	Power, kWh						
Turning	220/440	30/15	6.6	360	0.018	3	±0.002	60–300	1
Electro-chemical (ECM)	12–15	10,000 (100–1,500 A/in²)	100–150	50–80	2	0.01–0.2	±0.002	5–200	0.2
Electric-discharge (EDM)	50	60	3	0.6–7 rough; 0.02–0.2 finish	0.4–20	0.005–0.03	±0.003–±0.001	10–300	5
Electron-beam (EBM)	150,000	0.001	0.15 av.	0.002–0.006	37.5	0.6	±0.001	20–100	10
Laser-beam (LBM)	4.500	0.001	0.0045 av.	0.0004	11,000	0.001	±0.001	20–50	5

the material and increases the residual compressive stresses at the surface and resistance to fatigue.

An experiment using shot peening on front-wheel suspension springs resulted in remarkable improvements. Without shot peening, a front-wheel spring made from well-ground steel bars averaged about 170,000 cycles of compression. With the introduction of shot peening, the average life of the springs was increased to 700,000 cycles.

Further research developed a special steel shot, so that the life has been increased to about 4,500,000 cycles. As a matter of fact, the springs wear out the test machine: they simply do not break.

Summary   Grinding has branched into so many fields, and has become so involved, that it seems at times to be an art that is difficult to master. The shop operator soon knows the intricacies of the machine he operates and some of its possibilities, but he seldom knows why the grinding wheel produces the desired results. The production-design engineer who knows why and how the desired finishes are obtained can help make correct decisions.

Chipless Material-Removal Processes

Material removal from ductile materials can be accomplished by using a tool which is harder than the workpiece. During World War II the widespread use of materials which were as hard or harder than cutting tools created a demand for new material-removal methods. Since then a number of processes have been developed which, although relatively slow and costly, can effectively remove excess material in a precise and repeatable fashion. There are two types of processes. The first type is based on electrical phenomena and is used primarily for hard materials; the second depends upon chemical dissolution. The first group consists of electric-discharge machining (EDM), electrochemical machining (ECM), electron-beam machining (EBM), laser-beam machining (LBM), and ultrasonic machining (USM). The second includes electrochemical machining and chemical milling. These processes are useful for machining hard materials because they depend on the physical and chemical properties of a material not its mechanical properties. The designation *chipless machining* has been chosen deliberately because the methods used are in direct contrast to chip-producing methods of machining.

A comparison of the various processes listed above with a typical production turning operation (Table 13-10) shows that there is little likelihood that even the best electrical or chemical machining processes are more than 1 percent as efficient as chip-forming machining processes. Therefore, they are only selected when conditions warrant the extra expense.

The tremendous rise in the use of tungsten carbide for cutting and forming tools has led to a search for methods for shaping it to exact sizes. Either electrical

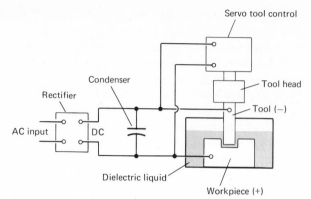

FIGURE 13-42
Electric-discharge machining.

or chemical means can be used, but hydrogen embrittlement and undercutting exclude chemical machining of tools when precise tolerances must be achieved. However, when electrical and chemical methods are combined, as in electrolytic grinding, good results are achieved.

Electric-discharge machining (EDM)   The EDM process has become the workhorse of the toolmaking industry for the precise machining of workpieces which can conduct electricity (Fig. 13-42). It can produce holes or cavities of complex cross section and to almost any depth in fully hardened steels or tungsten carbide with relative ease. The high-frequency discharge is developed between the negatively charged tool and the positive workpiece when they both are immersed in a dielectric fluid agent as it recirculates through a filter and cleans molten droplets and debris from the discharge zone.

At a potential of some 70 V a critical voltage gradient occurs so that the dielectric fluid is locally ionized for a few microseconds which permits a spark to pass from the work to the electrode with a consequent removal of a small droplet of the work surface, which then is left with a small pockmark. Within 10 to 20 μs the potential has dropped to 20 V and the capacitor is recharged. When the cycle is repeated at 20,000 to 300,000 Hz at a current density which may approach 10^6 A/in^2, significant metal-removal rates can be achieved. The gap between the tool and workpiece is critical so that it must be maintained by a servocontrol device that keeps a constant ratio between the average gap voltage and a suitable reference voltage.

The surface finish in EDM machining depends on the rate of metal removal. Good finishes require low-energy discharges that leave small craters, but the rate of removal is slow. Good practice indicates that roughing electrodes should be used first. In a typical forging die, finished with two electrodes, a tolerance of ±0.003 in was achieved; however, when seven electrodes were used, it dropped to ±0.001 in.

In perspective, EDM is only 1 percent as effective as grinding for removing hardened steel; but it is as good or better than grinding for use on carbide tools or space-age materials. It is definitely superior for shaping interior configuration in carbide dies. Frail pieces such as honeycomb can be cut without distortion. Again, placing numerous closely spaced holes in a hardened workpiece would be nearly impossible without EDM.

When the EDM process is used for machining dies, it would be advisable to polish or otherwise finish the surface, especially for alloy steels. The surface of the machined contour is carbon-enriched to a depth of 0.1 mil for finish and 5 mils for heavy cuts. There may also be residual-stress cracks from 1 to 30 mils deep, which may result in premature fatigue failure. The residual stresses should be relieved by annealing, and the cracked surface removed by lapping to restore the strength of the material at least in part.

Electrochemical machining   In the electrochemical machining (ECM) process, the material is removed by electrolytic deplating rather than by cutting action or solution. This process is being used successfully in drilling of holes from larger than 0.030 in up to several inches in diameter and in producing irregular slots and contours in solids. Sharp, square corners or sharp-corner, flat bottoms cannot be machined to accuracy.

A tool of conductive material such as lead, tin, zinc, etc., having the geometry of the inside diameter or surface contour desired in the work is fed to the workpiece. The tool retains its size and is not affected by the electrolyte. A precise gap is maintained at all times between the tool and the work (see Fig. 13-43). This gap is filled with an electrolyte and upon the flow of current, metal is electrochemically removed from the work. The maintenance of a precise gap and a uniform feed is the secret to accurate and rapid machining. If the machine

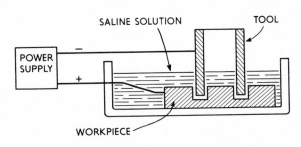

FIGURE 13-43
Electrochemical machining.

deflects, vibrates, or chatters as it moves into the cut, the electrode tool will ground and burn and nonuniform sizes will be obtained.

In accordance with Faraday's law, which states that the mass of any substance electrolyzed in a cell is directly proportional to the total quantity of electricity, the rate of metal removal will be proportional to the current flow. Thus, 96,540 C (A·sec) are required to remove one equivalent weight of metal, i.e., the atomic weight divided by the valence of the metal. For iron that would be 28 g ($\frac{56}{2}$), but only part of that energy is available at any one time because of heat and other losses. The amount of current flow also determines the quantity of electrolyte needed. It is common practice to feed the electrolyte through the tool.

As with Chem-milling and EDM, ECM is not affected in any way by the mechanical properties of the work. Irregular holes and contours can be produced in hard materials, such as Stellite turbine blades, free from burrs and with good surface finishes (less than 20 μin). The metal-removal rate is excellent. Holes can be machined at rates more than 0.5 in/min with accuracies of ±0.001 in.

Electrochemical grinding combines the anodic dissolution of a positive workpiece under a conductive rotary abrasive wheel with a recirculating conductive electrolytic. This process has made single-point carbide-insert tooling economically possible. The diamond abrasive in the wheel must remove only about 15 percent of the material, with a consequent saving of up to 50 percent in time and 80 percent for expensive diamond-grinding wheels.

Chem-milling In the Chem-milling (trade name) process, the material is removed by dissolution rather than cutting action. Here the work is masked with a paint in the areas where no metal removal is desired. The part is then immersed in an etching fluid and allowed to remain until the required depth has been obtained.

Three steps are involved in the process. First, the work is masked with a paint or vinyl-type plastic in the areas where no metal removal is desired. Often silk screening is used to apply the paint. The part is then immersed in an etching fluid, which may be acidic or basic, and it is allowed to remain until the required depth has been obtained. Lastly, the maskant is removed from the work.

This process was originally developed for alloys of magnesium and aluminum, but is now being extended to other materials. Dimensional control can be quite accurate. Production tolerances are ±0.005 in, although tolerances as close as ±0.003 in for aluminum and ±0.002 in for steel have been reported.

The principal advantages of this process include: low initial tooling costs; burr-free pieces; machining of extremely thin parts (it is possible to machine parts which are only 0.0005 in thick); and machining of hard and fragile workpieces and those of complex geometry at nominal cost.

In specifying this process, the production-design engineer should be aware

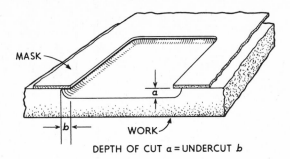

DEPTH OF CUT a = UNDERCUT b

FIGURE 13-44
Undercut in chemical machining.

of the fact that the template that outlines the position of the work that is to be machined must be smaller than the final size of cut. This is because material is removed under the maskant during the process. Since metal-removal rate is constant at all areas of the exposed work, the undercut will be equal to the depth of cut and this amount should be allowed for when producing the template (see Fig. 13-44).

Ultrasonic machining In ultrasonic machining, ultrasonic waves (those having a frequency exceeding the upper threshold of audibility of the human ear, approximately 20,000 Hz) are the means of transmitting energy. Here, a reciprocating tool of the desired shape drives abrasive grains suspended in a liquid which flows between the vibrating tool and the workpiece material. The abrasive granules remove small particles of the material until the desired shape of the workpiece is formed. See Fig. 13-45.

The abrasive particles contained in the liquid are particles of either aluminum oxide, boron carbide, or silicon carbide. As would be expected, coarser grits will increase the rate of metal removal, while finer grits are used when surface finish is the principal objective. The liquid containing the abrasive grains is recirculated until the cutting edges become dull and then fresh fluid-bearing material is introduced to the process. The abrasive grains are driven through movement of the tool which oscillates linearly approximately 20,000 times per second with a stroke of only a few thousandths of an inch.

This process is applicable for cutting hard, brittle materials. It can be used to machine holes of any shape and varying depth. The process is equally applicable to conductors or insulators. The process is not particularly suited to machining soft or tough materials. A further limitation is the taper that results when deep holes are drilled.

Tolerances as close as ± 0.0005 in may be readily attained. Surface finishes to 10 μin are achieved.

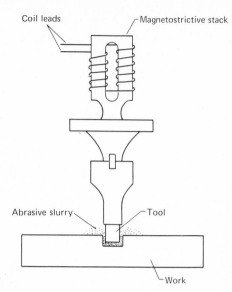

Coil leads

Magnetostrictive stack

Abrasive slurry

Tool

Work

FIGURE 13-45
Typical ultrasonic tooling setup.

Use of lasers The use of *l*ight *a*mplification by *s*timulated *e*mission of *r*adiation offers many possibilities in the machining of the "unmachinable." Already lasers have been used in the laboratory to burn a series of holes through $\frac{1}{2}$-in-thick asbestos. The procedure involves setting the workpiece in a fixture and setting a timer. The timer controls the amount of energy stored in the power supply, which in turn directly affects the power output of the laser gun. Upon the firing of the laser gun, the material upon which it was focused becomes vaporized in 0.002 s or less. The cycle is then repeated in order to machine the next piece or duplicate the previous cycle.

The principle of the laser is the emission of energy at a predetermined level of excitation. Since there does not exist an infinite number of excitation states for any given material, it is necessary for an electron to change its state from one level to a higher one by absorbing energy and thereby becoming excited. The laser beam is a very concentrated monochromatic beam of extremely high intensity.

The application of lasers offers distinct advantages over certain other processes. For example, contact between the laser and the workpiece is not required. The high-power densities available make it possible to vaporize any material, and the small spot permits the removal operation to be performed in microscopic regions. The limits on hole size today are between 0.001 and 0.010 in in diameter. Laser beams drill holes in diamond dies for wire drawing with ease.

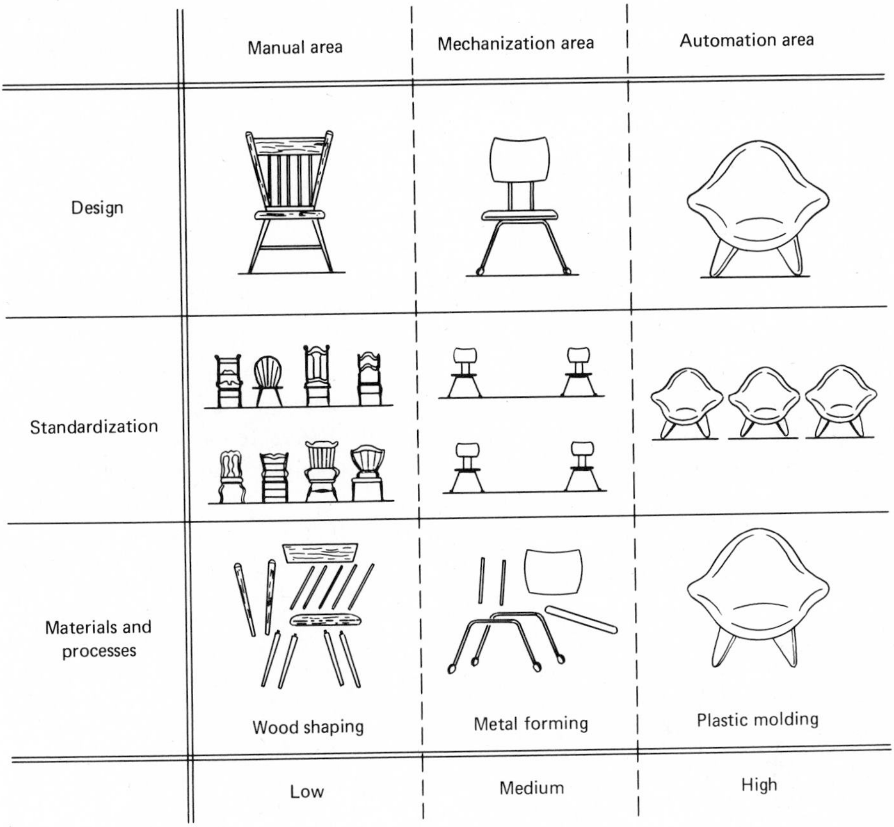

FIGURE 13-46
Volume and nature of product affect method of production.

AUTOMATION

Automation is not a new thing. It has been developing in industry for many years, although only in the last few years has there been a pronounced effort by both small and large industry to automate. The term "automation" was introduced by Mr. D. S. Harder of Ford Motor Company in 1947. It was so appropriate that it immediately became a part of the language. It may be defined as continuous automatic production with or without feedback (Fig. 13-46).

As more and more industries increase the degree of automation in their respective plants, there will be a continuing effort along the following lines:

1 Individual machine units will be connected by automatic handling mechanisms.

2 Provision will be made for stockpiling between machine units.

3 More and more safety controls will be integrated with the equipment. These will notify the operator that trouble has occurred.

4 Operators will cease to function as operators, but will assume the capacity of patrolmen. Instead of moving parts from one machine to another and manipulating machine controls, they will merely patrol the production line, watching for signs of impending trouble. Obviously, these men will be considerably more trained than production operators.

5 Maintenance and preventive maintenance procedures will be more important and will require upgraded personnel; however, normal maintenance, such as lubrication, will be done automatically.

6 Automatic inspection will be integrated with the automated process.

7 Equipment will automatically eject defective work and feed back the necessary adjustment information to the process to prevent subsequent rejects. Thus, the machines will have "toolometers" to tell the machine operator when to change tools.

It has been estimated that automation will eventually be applied to about 20 percent of all manufacturing operations in the metalworking industries. Perhaps the most fertile area for complete automation will be the fabrication of metal products such as toys, kitchenware, and household utensils.

The size of the plant has little bearing on the automation possibilities. A plant employing 300 persons may have just as much or more potential toward automation as a plant employing 3,000. The important criteria is to develop sufficient volume of a given design to justify the installation of an automated process. In general, it is probably less expensive to achieve automation by acquiring entirely new production equipment than to modify existing facilities.

Process types of industry have proved to be the easiest to automate. For example, in the typical steam electric plant, we have a high degree of automation. Here, the generator plant is started and shut down by automatic control; trouble is automatically sensed and corrective action taken.

Another familiar example of an automated process is the continuous-sheet rolling mill. Here, automatic controls govern all the steps from the time the billet enters the furnace and passes through the succeeding rolling mills until the final sheet is coiled at the end of the line. In order to make a conversion to a different size of sheet, new input data in the form of punched cards or tape is introduced in the controller. The roll speed and pressure is corrected automatically, the temperature is checked automatically and corrected if necessary, and the thickness of the sheet is gauged automatically to ensure that it is being produced to specifications.

Automatic transfer machines have been adopted by most high-produc-

FIGURE 13-47
Transfer machine to produce 1,000 connecting rods per hour, eight at a time through the operations shown in the diagram. An inspection operation at station 7 ensures that the holes have been drilled before reaming. (*Courtesy Greenlee Brothers and Co.*)

tion plants for machining automotive parts, engine blocks, compressor bodies, and similar items. Transfer machines provide straight-line flow and can be adapted to many types of layout. Usually the workpiece is securely clamped in a fixture which supports the part during its travel from station to station through the line. The parts can be reoriented at a station during the process. A typical automobile engine block requires from 80 to 100 stations depending on the complexity of the block and the number of inspection operations. Transfer machines eliminate many material handling expenses during operation, but the developing of proper fixtures and locating devices and maintaining proper alignment are major problems.

Transfer machines can cost as little as $100,000 for a simple machine, but may cost more than $2 million for a large one such as an automotive engine block. The major advantage of a transfer machine is greater output and reduced labor cost. Other advantages are better quality, less floor space, and reduced inspection cost.

A typical custom-made transfer line for automotive connecting rods and caps is shown in Fig. 13-47. Locators, guides, and strippers assure precise location and control of the connecting rod at all times during the machining and transfer cycles. Provisions are made for rapid checking of all locating nests and bushing plates.

NUMERICAL CONTROL

Today, a large percentage (approximately 22 percent) of industrial-type machine tools are equipped with numerical control (N/C) features. In a numerically controlled machine tool, manufacturing information portrayed on the designer's drawing is transferred to magnetic tape or punched tape, and, in this form, serves as the input or instructions to the machine. With these instructions the machine is able to automatically position the cutting tool, select the proper feed, select the proper speed, machine the part, move the machined part, and position the next part.

Of course, the information on the blueprint must be interpreted by a part programmer in order to prepare the input for the machine tool. In manual programming he assigns numbers to every machining operation on the design and then arranges these numbers (operations) in a logical sequence. Each operation must also be identified as to where it takes place with reference to an origin. A corner of the part or the corner of the work table is the customary point selected as an origin. The planning sheet, with the operation sequence and location identified, is then given to a typist who copies the information on a tape-punching typewriter (Fig. 13-48). The tape contains, in binary code, the instructions for producing the part (Fig. 13-49). Most control systems use the Electronic Industries Association standard 8-channel 1-in wide punched tape.

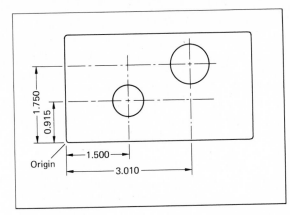

Planning sheet		
Operation Number	X	Y
1	1.500	0.915
2	3.010	1.750

FIGURE 13-48
Planning sheet for discrete controlled machining of two holes located in a plate.

The EIA standard binary coded decimal (BCD) system of punching, Standard R.S. 244, is the one used by 95 percent of the systems in the United States. Four of the eight channels are given values of 1, 2, 4, and 8 so that any digit from 0 through 9 can be identified on a line of the tape. For example, 5 is a combination of 4 and 1. The fifth channel is used for an odd parity; i.e., an odd number of holes must exist in every row of holes across the width of the tape. Holes 6 and 7 are not punched when numerical commands are used alone. They are needed for alpha characters, when alpha-numeric input is made to the system. The eighth channel is used at the end of each block of input.

The computer can also be used to assist in the programming of the N/C machine. In this case the part programmer writes out the computer instructions which may be in the form of English-like statements which specify the part geometry, cutting motion, tool geometry, feeds and speeds, etc. Then the computer generates the instructions for the N/C machine and prepares the N/C tape.

Computers can also assist the design engineer through the use of computer-aided design (CAD) programs. In this case the computer helps the engineer to develop the proper part geometry, which is then stored in the computer memory file for future input to be used in the generation of the N/C tape. The tape will give all the necessary instructions for producing the part, including positioning of the tool, turning of the coolant, establishing the correct feed and speed, retracting the tool, etc.

It will be noted that no jigs are required, since the table holding the part and the tool spindle will be automatically moved to the correct location with reference to the cutting tool; but generalized holding fixtures are often needed.

In the case of closed-loop systems, transducers are used to determine the location of the machine-tool table at any specific moment. The command issued

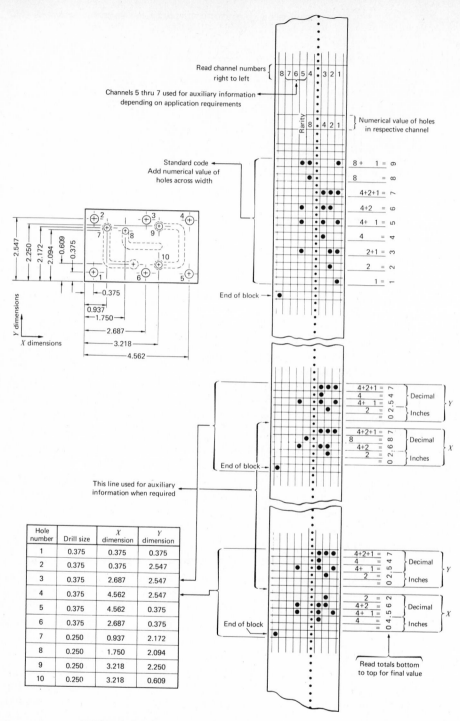

FIGURE 13-49
Punched tape used in numerical control.

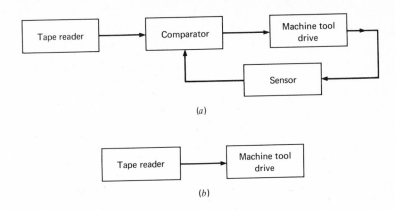

(a)

(b)

FIGURE 13-50
Comparison of (a) closed-loop and (b) open-loop control systems.

by the tape reader is compared with the feedback signal from the transducer telling the location of the table. If there is a difference between the present location of the table and where the command wishes the table to be, an "error" signal is released from the comparison unit which activates an electric motor which in turn rotates the lead screw to move the table (Fig. 13-50).

As the table moves toward the exact location requested by the "command," the error signal diminishes. When it reaches zero, the system is said to be *in position* and the table is at the desired location.

All machine-tool operations fall into one of two classes—point to point or continuous path. In point-to-point operations the table moves the work to the specific point at which the required operation is performed. The operation typically involves punching, drilling, tapping, chamfering, reaming, boring, etc. The table then positions the part for performing the next operation and so on until the part is machined.

In continuous-path operations the cutting tool performs work as it travels in a predetermined path. Typical continuous-path operations are grinding, milling, shaping, etc.

Both point-to-point and continuous-path N/C machines can be controlled on two or more axes. Two-axis machines typically control the X and Y axes, and the depth of cut is controlled by mechanical stops. A three-axis machine is more expensive, but depth of cut or feed of a drill can be controlled from tape.

Continuous-path control systems frequently are equipped with circle interpolators. Here the length of radius and center point is given as input and the interpolator controls the cutting tool in its circular path. The path along which the tool travels is really a large number of straight lines or parabolic spans. Each straight line (a pulse in the digital-type machine) is usually 0.0002 in. A sufficient number of these minute straight lines will take place to generate the desired arc.

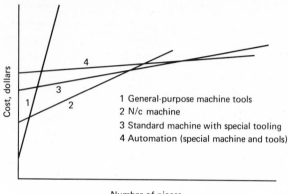

Number of pieces

FIGURE 13-51
General comparison of the machining cost of automation, N/C, and general-purpose machines with and without special tooling.

A major limitation to the more widespread use of N/C is its high initial cost, which is in excess of $12,000 to $70,000 for the control unit alone. A machining center would cost more than $0.25 million to $2 million dollars. Typically an N/C machine costs two to five times as much as its conventional counterpart, primarily because of the high cost of the sophisticated electronic control system. In addition, N/C machines require higher maintenance and support costs, which are reflected in high hourly rates. The latter are about twice as high as conventional machine hourly rates, i.e., $10 to $12 per hour versus $18 to $25 per hour on N/C machines.

In some cases, N/C is not the most suitable method (Fig. 13-51). For instance, for low-production items, general-production equipment is frequently more economical, whereas a part which has a large production volume can undoubtedly be produced at low cost if specialized machines and equipment are designed, built, and field-checked. Numerically controlled machine tools are most appropriate for lot sizes of 1 to 100 depending on part complexity and the specified tolerance.

SUMMARY

The production-design engineer should become familiar with the advantages and disadvantages of the five basic metal-cutting processes: drilling, turning, planing, milling, and grinding. Likewise, he should keep abreast of the special metal-removal processes such as Chem-milling, lasers, electrochemical, electric discharge, and ultrasonic machining so that he will be able to specify and design for the most economical technique.

With the development of ceramic cutting tools and the extended use of carbides, it is possible to remove metal at rates not considered at all feasible a few years ago. Today the machine-tool builder is finding it necessary to redesign so that his machine will be able to withstand the forces developed through the heavy feeds and fast speeds capable of the modern cutting tool.

The function of production engineering includes the work of tool engineering, which is today a study by itself. Machinability of metals and its relation to the form of the cutting tool is an area that engineers spend a lifetime studying and developing. The engineer that designs for production is vitally interested in all facets of the metal-removal processes in that they affect, to a large extent, the potential of his designs.

Material and machining is expensive; therefore parts should be preformed, or precast, closer to size in order to reduce the amount of material to be removed.

Finally, the production-design engineer must be cognizant of the limitations of the equipment in his plant and those that service his organization. He should avoid specifications that are impractical because of time and cost. However, he must realize that specifications that are impractical today may be realistic tomorrow. Many poor designs, from the standpoint of machining operations, can be avoided by giving consideration to the following Do Nots:

1 Do not specify tolerances closer than necessary.
2 Do not specify a thread to the bottom of a blind hole.
3 Do not specify a tapped hole unless the tap can cut the entire perimeter of the last thread.

The wide range of equipment necessary to accomplish the many methods of cutting and removing material will occupy the attention of the engineer during his entire career. Improved equipment will require changes in the design of his apparatus in order to take full advantage of the more efficient machinery. This procedure is necessary in order to meet competition satisfactorily.

The production-design engineer should follow the development of ceramic cutting tools so that he will be able to fully utilize the advantages offered. The trend will be to cut materials at faster speeds and heavier feeds. This will require more rigid machine tools. In order for a company to be competitive, it will have to understand and be able to make the most of the latest metal-removal techniques, tools, and equipment.

QUESTIONS

1 What are the characteristics of clean-cutting metals?
2 How is the property brittleness related to machinability?
3 What does the machinist mean when he speaks of a material as being "too ductile"?

4   What factors affect the cutting of all materials?

5   Make a sketch of a single-point tool and show the back rake angle, end relief angle, end cutting edge angle, side cutting edge angle, and built-up edge.

6   What steps can the production-design engineer take to decrease the amount of the built-up edge?

7   What would be the cutting speed in feet per minute of a 1-in-diameter drill running at 300 r/min?

8   Explain why large amounts of frictional heat are produced when machining very ductile metals.

9   What two pressure areas on cutting tools are subject to wear?

10   How does the rake angle affect the life of the cutting tool?

11   What is the 18–4–1 type of tool?

12   What is the future of the ceramic class of tools?

13   What does honing of the cutting and chip contact surfaces do to the tool life? Explain.

14   What are the major functions of the cutting fluid?

15   Upon what does the cooling ability of a cutting fluid depend?

16   What is the danger of multiple cuts on precision work?

17   The following measured data were taken from an orthogonal test cutting AISI 1015 steel using a tungsten carbide tool:

> Cutting speed—500 ft/min
> Feed—0.010 in/rev
> Chip width—0.100 in
> Chip thickness—0.022 in
> Cutting force—313 lb
> Feed force—140 lb
> Tool angle—10°

Compute:

a   Chip-thickness ratio
b   Resultant force
c   Shear angle
d   Length of shear plane
e   Friction angle
f   Friction force
g   Normal compressive force
h   Coefficient of friction
i   Shear force
j   Shear stress on shear plane
k   Shear velocity
l   Net horsepower
m   Specific cutting energy

PROBLEMS

1 Calculate the percent error made using the approximation r/min = $4CS/d$ in calculating a lathe setting compared with the exact expression r/min = $12\,CS/\pi d$. How does that error compare with the maximum error introduced when using a lathe with 16 spindle speeds varying from 31 to 2,400 r/min in a geometric progression.

2 Estimate the horsepower needed to turn an AISI 1045 steel shaft 6 in OD by 24 in long, with a triple carbide tool when given the following cutting parameters: speed = 300 ft/min; feed = 0.015 in/r; depth of cut = 0.125 in; the specific horsepower = 0.85 hp/(in³)(min) and the machine efficiency = 0.7.

3 Determine the size of the square shank for an SAE 1020 steel tool holder if it is to be used to hold the carbide tool bit in the operation above. The maximum allowable deflection at the tool tip is 0.002 in. Take the overhang of the tool holder beyond the toolpost as 3 in.

4 Compute the total time to face mill a cast iron block 5 in wide by 10 in long using an 8 in diameter inserted tooth cutter with 18 teeth. The cutting speed is 270 ft/min and the feed is 0.010 in per tooth. Allow $\frac{1}{16}$ in at each end for the overtravel. Note this is the minimum time to remove the stock.

5 Sketch the tools and turret layout complete with the work holding device and tool post turret to make the part shown in Fig. P13-5. Prepare detailed operation sheets specifying all tools, feeds, speeds, etc., and showing all overlapping operations. Lots of 250 pieces are required about every 2 months.

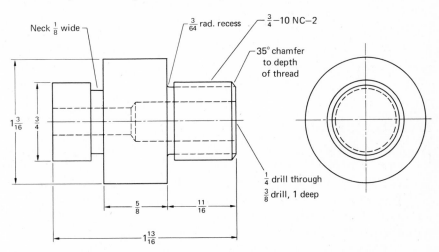

6 (*a*) Given a tangential cutting force of 525 lb, a cutting speed of 250 ft/min and a workpiece diameter of 4.5 in, find the horsepower required at the tool bit.

   (*b*) If the unit horsepower required for the above operation is 0.60 at a feed of 0.012 in/r, find the metal removal rate and the depth of cut.

7   Given the Taylor equation for a certain tool-work combination is $VT^{0.28} = 1,000$ at a feed of 0.020 in/r and a rake angle of 10°. The material has a dynamic shear strength of 1.5×10^5 lb/in. The chip thickness ratio varies approximately linearly from 0.4 at 200 ft/min to 0.6 at 800 ft/min and the data may be extrapolated as a linear function. How long should it take to machine a large steel forging 20 in in diameter and 10 ft long to 18 in in diameter using a heavy-duty lathe equipped with a 50-hp infinitely variable speed motor which is 70 percent efficient. Let the tool life = 16 min, tool changing time = 1 min.

8   A tool life of 16 min is obtained at a cutting speed of 500 ft/min for a given tool-work combination and depth of cut. In a second test a tool life of 4 min is found at 1,000 ft/min and the same depth of cut. What cutting speed should be specified to obtain a 30-min tool life for the same tool-work combination?

9   Determine the tool-life equations from the following data in which each value is the average of five experimental points. Use plotting and the method of least squares to determine the slopes and intercepts of the equations.

V, ft/min	Carbide K-6	Carbide K-3H	Oxide Al_2O_3
160	54.67	20.67	
200	25.56	8.60	
300	7.10	2.51	14.63
400	3.09	1.19	10.75
500	1.84	5.00	
600	1.03	0.41	6.57
800	.....	.....	4.93
1,000	.....	.....	3.25

SELECTED REFERENCES

AMBER, G. H., and P. S. AMBER: "Anatomy of Automation," Prentice-Hall, Englewood Cliffs, N.J., 1962.

AMERICAN SOCIETY FOR METALS: "Metals Handbook," 8th ed., "Machining," Metals Park, Ohio, 1967.

AMERICAN SOCIETY OF TOOL AND MANUFACTURING ENGINEERS: "Fundamentals of Tool Design," Dearborn, Mich., 1962.

———: "Nontraditional Machining Processes," Dearborn, Mich., 1967.

———: "Tool Engineers Handbook," 2d ed., McGraw-Hill, New York, 1959.

ARMAREGO, E. J. R., and R. H. BROWN: "The Machining of Metals," Prentice-Hall, Englewood Cliffs, N.J., 1969.

BHATTACHARYYA, A., and I. HAM: "Design of Cutting Tools," American Society of Tool and Manufacturing Engineers, Dearborn, Mich., 1969.

BLACK, P. H.: "Theory of Metal Cutting," McGraw-Hill, New York, 1961.

BOLZ, R. W.: "Manufacturing Processes and Their Influence on Design," Penton, Cleveland, 1949.

BOOTHROYD, G.: "Fundamentals of Metal Machining," Edward Arnold, London, 1965.

BOSTON, Q. W.: "Metal Processing," 2d ed., Wiley, New York, 1951.

CHILDS, J. J.: "Principles of Numerical Control," Industrial Press, New York, 1965.

COOK, N. H.: "Manufacturing Analysis," Addison-Wesley, Cambridge, Mass., 1966.

VIDOSIC, J. P.: "Metal Machining and Forming Technology," Ronald, New York, 1964.

WILSON, F. W. (ed.): "Numerical Control in Manufacturing," McGraw-Hill, New York, 1963.

WOODCOCH, F. L.: "Design of Metal Cutting Tools," McGraw-Hill, New York, 1948.

14

SECONDARY MANUFACTURING PROCESSES: FORMING

After a metal has been hot-worked to the desired shape, such as a structural shape or hot-rolled sheet, it is frequently descaled by pickling to make it ready for secondary forming processes which are carried out below the recrystallization temperature. These operations involve cold rolling into sheet or cold drawing into finished bars for machining. This chapter is concerned with the secondary forming processes which are common to sheet metal and plate. These cold-working operations include shearing, forming, squeezing, and drawing.

Cold-working operations enhance the strength and hardness of a material at the expense of ductility. The distorted crystal structure uniformly increases in strength in proportion to the degree of deformation, and its yield stress increases in proportion to the amount of reduction. A production engineer will take advantage of the increase in yield strength imparted by cold working when he designs a part to be made by cold-working operations. Examples of such techniques are seen in the design of curved trunk lids, hood designs, and formed patterns used in automobile bodies. The sheet-metal gage of the bodies is considerably reduced from that used in the 1920s.

COLD-WORKING FUNDAMENTALS

All operations that displace or move metal plastically require forces to move the metal, and the amount of displacement that can be sustained without failure is a function of the metal's ductility and the degree of hardening, strengthening, and embrittlement of the material as a result of the process.

Many metals exhibit an abrupt yielding (the stress in a material at which there occurs a marked increase in strain without an increase in stress[1]). In some materials the yield point may appear as little more than a "jog" in the stress-strain curve, while in others the yield-point behavior may extend over elongations of several percent.

Figure 14-1 illustrates a conventional stress-strain diagram. The plastic range is shown between the points A and B. It is within this area that the cold working of metals is effective. Figure 14-2, which was developed from the data

[1] Definition of yield point from ASTM Standards.

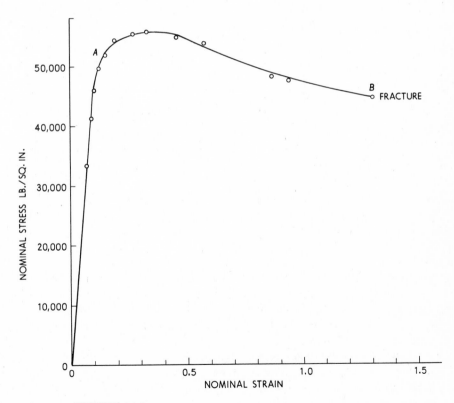

FIGURE 14-1
Conventional stress-strain diagram for cold-worked SAE 1020 semikilled steel.

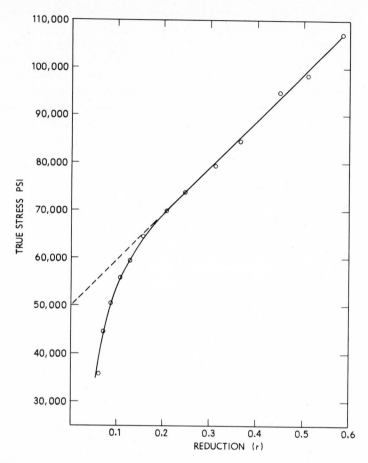

FIGURE 14-2
True stress versus reduction of area in plastic range of SAE 1020 semikilled steel.

used to plot Fig. 14-1, gives a useful straight-line approximation of the true stress versus the reduction of area in the plastic range of SAE 1020 semikilled steel. A study of this relationship provides information as to the amount of reduction of area possible with the material in question and the stress required to obtain such a reduction.

Figure 14-3 illustrates curves showing the commercial cold-working range for representative metals. The common origin of these graphs is at the theoretical limit of annealing at which the yield point is zero. The theoretical yield point of commercial, fully annealed metal is at the left end of the heavy line. The strain-hardened limit (at potential fracture) is at the right end of the heavy line.

The rate of strain hardening is indicated by the slope of the graph. Ductility of the annealed metal is shown by comparing the length of the heavy line with the total of the heavy line plus its right-hand extension.

In the process of cold working (e.g., deep drawing) a large-diameter work-piece is compressed to a smaller diameter. Therefore we need compression-test data rather than tensile-test data to predict the behavior of the work material. Steel, at least up to the yield point, has the same properties in both tension and compression, but in deep drawing the deformation takes place beyond the yield point in the plasticity range. Figure 14-4 gives a portion of the stress-strain diagram plotted on logarithmic coordinates.

If a metal is reduced by compression A percent and the load is removed, yield will not occur when a second load is applied until the stress reaches a new and higher stress σ_A lb/in². The difference between σ_A and σ_0 is known as the work hardening which occurs as the number of possible slip planes is reduced by

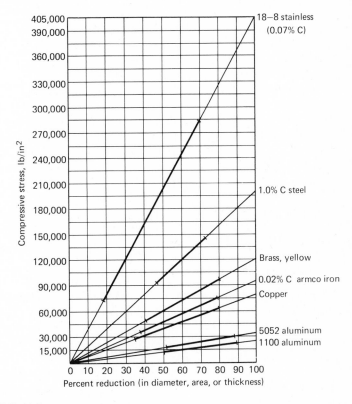

FIGURE 14-3
Tentative curves showing commercial cold-working range for representative metals.

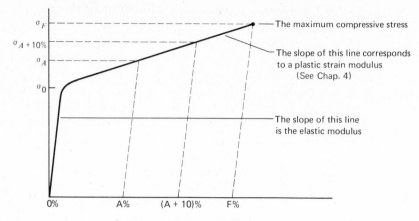

FIGURE 14-4
Compressive stress versus percent reduction curve for a given work-hardening alloy.

the plastic deformation process. The stress σ_0 is known as the flow stress of the annealed alloy, and σ_A is the flow stress of the deformed material.

At the beginning of the reloading operation we could use A as the starting point and have an additional 10 percent reduction based on A size (not percent). This then would be equivalent to a second draw operation of 10 percent when the first draw operation was A percent. The drawing force for the second draw would be determined by the yield stress at the beginning of the second draw operation which is the same as the stress σ_A of the stress at the end of the first draw.

The above data is replotted in Fig. 14-5.

MAXIMUM REDUCTION IN DRAWING

The effect of cold working may be estimated by an analysis of the compressive plastic deformation. The percent reduction may be defined as:

$$\text{Reduction} = \frac{\text{initial diameter} - \text{final diameter}}{\text{initial diameter}} \tag{1}$$

or

$$R = \frac{D - d}{D} \qquad R_T = \frac{d_0 - d_n}{d_0} \tag{2}$$

Figure 14-5 is a plot of the compressive yield strength of a material as a function

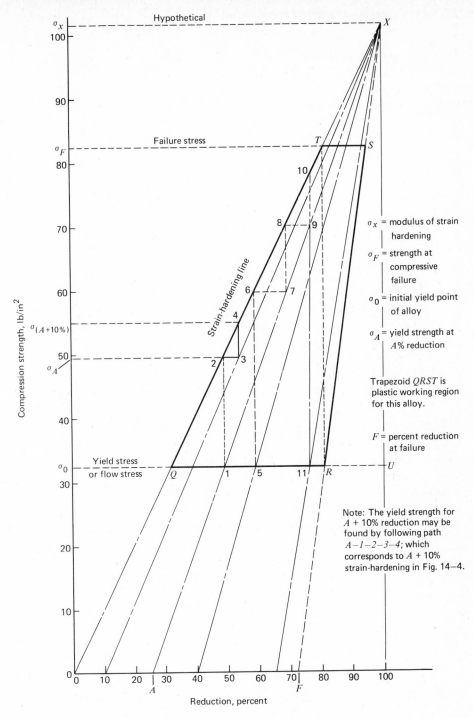

FIGURE 14-5
Typical work-hardinability graph.

of its percent reduction. Note that trapezoid $QRST$ encloses the area in which plastic deformation can occur without fracture of the material. Then from the diagram it can be seen that the theoretical yield strength at 100 percent reduction can be found as follows:

$$\frac{\sigma_F - \sigma_0}{F\% \text{ red.}} = \frac{\sigma_x - \sigma_0}{100\% \text{ red.}} \tag{3}$$

$$\therefore \sigma_x = \sigma_0 + \frac{100\% \text{ red.}}{F\% \text{ red.}} (\sigma_F - \sigma_0) \tag{4}$$

Strain hardening graphs can be constructed for the alloys shown in Fig. 14-3 as follows:

1   Plot the compressive yield stress as a function of the percent reduction from 0 to 100 percent as shown in Fig. 14-5.
2   Draw horizontal lines at σ_0 and σ_F, which are, respectively, the yield and failure strengths in compression.
3   Obtain σ_x by proportion using Eqs. (3) and (4) from triangles QRT and QUX. Note that any horizontal line within the plastic working region is divided similarly to the original X axis so that its length represents 0 to 100 percent reduction when it is extended to the 100 percent reduction line.
4   Construct the complete strain hardening graph $QRST$ as in Fig. 14-5.

The strain-hardenability graph is used as follows. If an annealed material is reduced A percent, to find its new yield point start at the σ_0 line and make a horizontal line to the right until it hits the construction line from σ_x to A percent reduction, then rise vertically to the $0 - \sigma_x$ line; that intersection will be σ_A, the new yield stress of the material. If a second draw of $A + 10$ percent is to be made, repeat the former procedure, as in Fig. 14-5. After the second draw the yield stress will have risen to $\sigma_A + 10$ percent as shown. To compute the value of σ_n, the unit stress at the end of any draw, use Eq. (4).

$$\sigma_n = \sigma_0 + R_T(\sigma_F - \sigma_0) \tag{5}$$

where

$$R_T = 1 - (1 - r_1)(1 - r_2) \cdots (1 - r_n) \tag{6}$$

and

$$r_i = \frac{d_{i-1} - d_i}{d_{i-1}}$$

PROPERTIES OF MATERIALS FOR SHEET-METAL WORKING

The operations described in this chapter are based on the assumption that the material has sufficient ductility so that it will not break or crack during the operation. If failures occur, intermediate annealing operations and additional steps in the process may be necessary. To avoid rupture, the material is usually bent across grain between 45 and 90° (Fig. 14-6). When it is necessary to bend with the grain, a larger radius of bend or a softer material is used. A smooth surface facilitates the bending of material to a small radius. The harder the material, the greater the inside radius required. Annealed stock usually can be bent satisfactorily with a radius equal to the thickness of material. Material with rough edges, such as sheared materials, may crack at the edge; this cracking will not occur if the edges are smoothed or rounded.

Work-hardening steels such as stainless steels limit the amount of drawing and bending that may be performed in one operation on conventional presses. High-energy forming and high-speed presses are permitting the forming of work-hardening materials over a larger range of sizes.

Three factors govern the choice of a material for press-working (stamping) operations: strength, wear resistance, and corrosion resistance. Typical materials for press working are listed in Table 14-1. Once the alloy has been selected to meet the anticipated service demands, the production engineer must specify the commercial stock dimensions and the proper initial hardness to meet the forming requirements. Finally, he must determine the correct slit or sheared width to obtain maximum material utilization in the blanking and forming operations. Supplier's handbooks and specification sheets can be consulted to determine the available sheet sizes, temper designations, and standard gages (thicknesses).

Low-carbon steel is by far the most commonly used material for press-working operations because of its good formability, economy, and weldability. For deep drawing, steels with 0.05 to 0.08 percent carbon and 0.25 to 0.50 percent manganese are suitable. In practice the type of drawing operation is specified and a suitable steel is supplied by the mill.

For large-volume production, automatic presses with coil stock and roll feeders are needed. This automated setup eliminates the press operator and can be used for materials up to 0.044 in thick on a routine basis. If roll straighteners are furnished, even thicker coil stock can be specified.

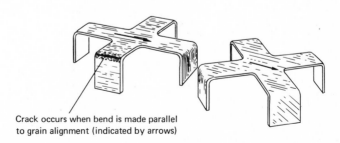

Crack occurs when bend is made parallel
to grain alignment (indicated by arrows)

FIGURE 14-6
Mutually perpendicular bends should be made 45° to the grain orientation.

Table 14-1 PROPERTIES OF SHEET METALS

	Theoretical yield points		Reduction in area (maximum annealed)	Shearing stress and percent penetration to fracture			
	Minimum* (commercially) annealed), lb/in²	Maximum (severely cold-worked), lb/in²	%	Annealed to soft temper, lb/in²	%	Cold-worked to temper noted, lb/in²	%
Aluminum*							
No. 25, commercially pure	(4-)† 8,000	H 21,000	80	8,000	60	H 13,000	30
No. 35, Mn alloy	11,500	H 25,000	80	11,000		H 16,000	
No. 175 for heat treatment	HT 30,000						
No. 525, Cr alloy	21,000	H 37,000	80	14,000		H 36,000	
Brass							
Yellow, for cold working	(10-) 40,000	95,000	75	32,000	50	? 52,000	20
Forging, beta							
Bronze, Tobin alloy	25,000	120,000	53	36,000	25	$\frac{1}{4}$H 42,000	17
Copper	(10-) 25,000	62,000	65	22,000	55		
Gold							
Iron							
Cast, gray	8,000 tension 30,000 compr.						
cast, high test	15,000 tension 40,000 compr.						
Lead	(1-) 3,000	4,000		3,500	50		
Monel metal	28,000	H 112,000	63				
Nickel	21,000	H 115,000	75	35,000	55		
Silver							
Steel							
0.03 C Armco	(20-) 35,000	70,000	76	34,000	60		
0.15 C	(32-) 55,000	110,000	60	48,000	38	61,000	25
0.50 C	(55-) 70,000		45	71,000	24	90,000	14
1.00 C	90,000	145,000	20–40	115,000	10	150,000	2
Electrical, high-silicon				65,000	30		
Stainless, low-carbon for dwg.							
Endure AA	50,000	120,000	60				
18-8	(35-) 50,000	165,000	71	57,000	39		
High-carbon for cutlery, etc.			32				
Tin	(2-) 4,000	4,500		5,000	40		
Zinc	(12-) 28,000		43	14,000	50	19,000	25

* Old and new designations of wrought aluminum alloys: 2S = 1100; 3S = 3003; 17S = 2017; 52S = 5052.

Tensile strength (nominal) annealed, lb/in²	Elongation in 2 in (average) annealed, %	Elasticity, Young's modulus, lb/in²	Modulus of strain hardening (tentative), lb/in²	Bulk modulus volume, lb/in²	Annealing temperature, approx. commercial, °F	Forging temperature, approx., °F	Thermal coefficient of expansion, (in/in) (°F)	Specific weight, lb/in³
13,000	35–45	10,300,000	25,000	10,200,000	650		0.000,010	0.0963
16,000	30–40	10,300,000	29,000		750			
					630–650	400		
29,000	25–30	10,300,000	38,000		650			
50,000	65	13,400,000	120,000	8,800,000	1100			0.303
						12–1400		0.301
60,000	46		250,000					0.315
33,000	68	14,500,000	81,000	17,400,000		16–1800	0.000,009	0.3195
		11,600,000		23,200,000	6–800		0.000,008	0.6949
23,000		12,000,000		13,900,000			0.000,005,5	0.260
45,000		20,000,000						
		2,470,000		1,100,000	Below room temp.		0.000,017	0.4106
70,000	30–50				11–1200			0.318
70,000	47	32,000,000		24,600,000	12–1300		0.000,007	0.3175
		10,900,000		14,500,000	6–800		0.000,010	0.3791
42–48,000	48	30,000,000	100,000		13–1400			0.2834
57,000	35	30,000,000	150,000	23,200,000	13–1400		0.000,009–16	
85,000	25	30,000,000			13–1400	18–2100		
120,000	10	30,000,000	200,000					
80,000	25	30,000,000			13–1550	19–2100		
95,000	57				13–1900	2100		0.285
260,000	11				1450–1600	20–2100	0.000,006–7	
		7,260,000		7,260,000	Below room temp.		0.000,015	0.2652
30–37,000	27	13,100,000		5,100,000	2–300		0.000,014–5	0.256

† Figures in parenthesis represent laboratory test minimum, which might prove misleading for estimating loads. All physical values are subject to some variations with testing methods and analysis of material.
SOURCE: E. V. Crane, "Plastic Working of Metals," 3d ed., Wiley, New York, 1941.

LUBRICATION

It is important that lubrication take place in press-working operations in order to reduce the friction between the work material and the working surfaces of the dies. A correctly lubricated setup reduces the press tonnage required, increases die life, and gives more uniform work which is free from abrasions and scratches.

When specifying a lubricant, the production engineer should select one that has sufficient surface tension and viscosity so that it will adhere well and will spread evenly over the die surfaces. The lubricant may be either oil- or water-soluble but should not have a corroding or abrasive effect on the dies. It should be able to maintain a protective film during the highest pressure to which the work and dies will be subjected. Lubricants must be removable in commercial washing equipment; if not, cleaning costs will outweigh savings made in the press-working operations.

In deep drawing, the selection of the proper lubricant is especially important in order to ensure uniform work and minimize rejects. Generally speaking, deep draws and draws entailing heavy gages of steel require a viscous lubricant, such as equal parts of white lead and animal or mineral oil, while shallow draws may be lubricated with light grease or mineral oil. If annealing is required after drawing, then in order to avoid burning of an oil residue into the metal surface, water-soluble lubricants are often preferred. These may include mixtures of water and chalk, soap, or zinc oxides.

Copper and its alloys are usually lubricated with water-soluble mixtures such as soap chips and hot water. The deeper or more severe the draw, the greater the concentration of soap required.

GENERAL PRINCIPLES OF SHEARING

Shearing operations, including blanking, notching, parting, piercing, nibbling, trimming, and perforating, involve the shear strength of a material, which is related to but is less than its tensile strength. When a punch penetrates a sheet-metal workpiece and passes into the die (Fig. 14-7), a section through the workpiece shows a smoothly cut area and an angular fractured area (Fig. 14-8). The slug has four features: a plastically deformed and rounded edge, the burnished cut area, a fractured area which corresponds to that of the sheet stock, and a burr at the periphery of the side toward the punch (Fig. 14-8).

Clearance

There is a clearance c between the punch and die at any radial point so that the difference between the diameter of the punch and die is $2c$. The magnitude of

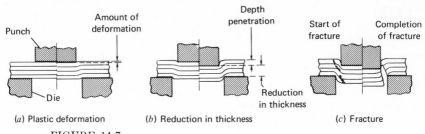

(a) Plastic deformation   (b) Reduction in thickness   (c) Fracture

FIGURE 14-7
Several steps in a sheet-metal punching operation. (*From Society of Manu-facturing Engineers, "Fundamentals of Tool Design," Prentice-Hall, Englewood Cliffs, N. J., 1962.*)

the clearance varies with the hardness, thickness, and properties of the alloy sheet. With too little clearance the initial fractures from the corners of the punch and die fail to meet (called secondary shear) and a greater shearing force is required. Too much clearance results in excessive plastic deformation. The ideal clearance minimizes plastic deformation and creates no secondary shear.

Since there is a difference in the diameter of the punch and die as a result of the clearance, the size of the slug is determined by the die and the hole is sized by the punch. Thicker and softer sheet metals require a clearance of 10 to 18 percent of t; harder metals need less, that is, 6 to 10 percent of t. Harder metals have less plastic deformation and a smaller burnished area than soft metals.

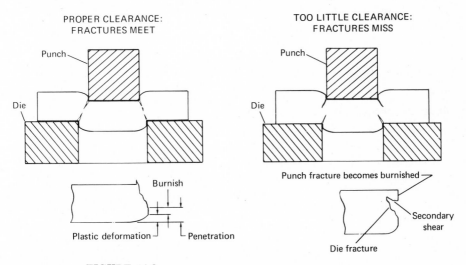

FIGURE 14-8
Cross sections of a typical punching operation showing the effect of clearance and the characteristics of the sheared slug.

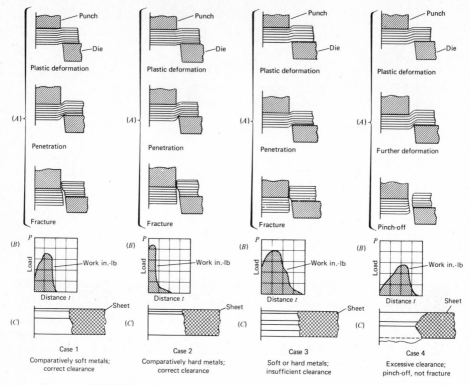

FIGURE 14-9
Effect of different clearances when punching hard and soft alloys. (*From Society of Manufacturing Engineers, "Fundamentals of Tool Design," Prentice-Hall, Englewood Cliffs, N. J., 1962.*)

For a given metal thickness, the angle of the fracture area decreases with the increasing hardness of the workpiece (Fig. 14-9).

Shearing Force

The shear strength of sheet metal is generally about 60 to 80 percent of its tensile strength. In plain carbon steels the shearing strength is directly proportional to the carbon content (Table 14-2).

The force P required to shear sheet metal is

$$P = \pi DSt \quad \text{(for round holes)}$$

$$P = \pi SLt \quad \text{(for other contours)}$$

where S = shear strength of material, lb/in²
D = die diameter, in
L = length of shear (i.e., circumference in a round punch and die set)
t = thickness of material, in

Table 14-2   TYPICAL SHEARING STRENGTH FOR STEEL

Carbon content	Annealed	% Penetration	Cold worked	% Penetration
1003	34,000	60		
1010	35,000	50	43,000	38
1015	39,000	45	51,000	33
1020	44,000	40	55,000	28
1030	52,000	33	67,000	22
1050	71,000	24	90,000	14
1% C	110,000	10	150,000	2
Fe 3% Si	65,000	30		
18–8 St. Steel	57,000	39		
SAE 4130	55,000			

Thus the force required to punch a 2-in-diameter hole in annealed SAE 1020 steel plate $\frac{1}{4}$ in thick is:

$$P = 2\pi(44,000)(\tfrac{1}{4})$$

$$= 69,000 \text{ lb}$$

Penetration

The penetration of the punch into the workpiece before fracture occurs is the sum of the plastic deformation and the burnished thickness (Fig. 14-8). It is usually expressed as a percentage of the original sheet-metal thickness.

Reduction in Press Load

In many cases the calculated punch load will exceed the capacity of the available press, especially if several punches are shearing simultaneously. Since the press loading is characterized by high force demand in a short time span, the cutting force may be reduced by spreading the energy required over a longer time span, i.e., a longer portion of the total ram stroke. A reduction in the energy demand may be accomplished in two ways:

1 Stagger the length of small punches by increments of the sheet thickness so that only a few are punching at any one time.
2 Add shear (Fig. 14-10) to the punch or die, thereby reducing the area in shear at any one time. The punch load may be cut in half. The change in punch face is about $1\frac{1}{2}$ times the stock thickness. If a punch is large a double shear is preferable to a single shear because the lateral forces are balanced. Apply the shear to either the punch or die in such a way that

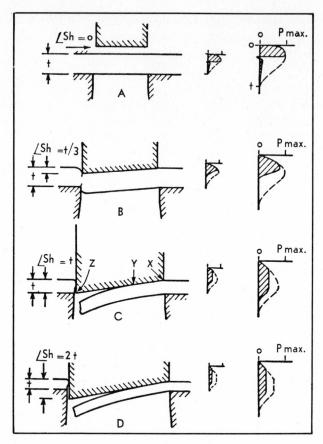

FIGURE 14-10
Effect of shear on the maximum force required for punching sheet metal.

any distortion will appear in the scrap from the punching operation; i.e., the shear will be on the punch. In piercing, the direction of the shear angles should be such that cutting is initiated at the outer extremities of the contour and move toward the center in order to avoid stretching the material before it is cut free.

The blanking force in practice is reduced by the penetration allowance for various alloys and states of cold work as given in Table 14-1. It can be reduced even more by grinding an angle called shear on the face of the punch or die. When shear is used, the magnitude of the maximum force is reduced up to 50 percent (Fig. 14-10), but the cutting operation requires a longer time so that the total energy expended is the same.

Shear on the die transfers distortion to the scrap and the blank is un-deformed, whereas shear on the punch results in distortion of the slug.

Shear when defined as kt is determined by the thickness of the sheet metal where k usually ranges from 1.5 to 2. If the thickness of the sheet metal is more than $\frac{3}{32}$ in, shear is usually applied in two directions to balance the forces on the workpiece and then should be equal to $2t$. Analysis shows that the magnitude of the punching force is the same for either single or double shear.

Consider the previous example in which a 2-in-diameter hole is to be punched in annealed SAE 1015 steel, $\frac{1}{4}$ in thick. Determine the punching force if penetration is considered. From Table 4-1, the shearing strength is 48,000 lb/in² and the penetration is listed as 38 percent. Thus the punching force is:

$$P = \pi DStp$$

$$= (3.14)(2)(48,000)(\tfrac{1}{4})(0.38)$$

$$= 28,700 \text{ lb}$$

If double shear is used in addition to the penetration allowance, the required punching force becomes:

$$P = \frac{\pi DStp}{k}$$

$$= \frac{(3.14)(2)(48,000)(\tfrac{1}{4})(0.38)}{2}$$

$$= 14,350 \text{ lb}$$

Thus the required press would be reduced from 75 to 15 tons by applying good design and engineering analysis.

Location of Blanks on the Stock for Economy

The die designer and production engineer must consider the location and spacing of the blanks on the stock. Three factors must be considered: (1) blank location for economy (best use of material), (2) proper location for bending, if required, and (3) location, so that the summation of the shearing forces is symmetrical about the centerline of the ram stroke.

Cardboard templates have proven to be helpful in laying out the strip stock to obtain the maximum economy, but blanks cannot be closer than $\frac{1}{32}$ in or the stock thickness whichever is greater. A number of cost-saving layouts are possible (Fig. 14-11). The waste skeleton should be stiff enough to stick together; otherwise waste disposal will be difficult.

The location of the center of pressure may be carried out mathematically as would be done for calculating the center of gravity, but calculations are usually long and tedious. It is adequate to locate the center of pressure within

$\frac{1}{2}$ in of the true center of a punch configuration. A simple procedure may be used to determine the center of pressure experimentally. Balance a wire loop which has been bent to the configuration of the blank or blanks which are made, in one stroke, across a pencil in both the x and y coordinate directions. The intersection of the two axes is approximately the center of pressure.

PUNCHING, BLANKING, AND STAMPING OPERATIONS

Production Shearing

Power shears In shearing material by means of shear blades, the edge of the movable blade is usually at a slight angle to the edge of the stationary blade.

The movable blade usually strikes the material at one edge and advances progressively to the opposite edge. Thus only a portion of the piece is being sheared at one time. In this way the blade does not carry the load of shearing the entire cross section at once. If this were done, the operation would be called "blanking" or "straight shearing." In all shearing operations the part under the movable blade is distorted, twisted, or bent. The portion lying on the stationary blade remains straight and may have a burr on the lower edge. When the tool designer says that shear is being placed on a die or punch, it means that the material is being cut progressively along the edge of the die and punch. This lightens the load on the punch and press but distorts the blank or part, depending upon where the shear is placed.

In sheet-metal work and in the making of formed stampings, it is often necessary to cut sheets into blanks having irregular contours or circular shapes. To avoid expensive blanking dies, blanks can be made by slitting, uni-shearing, nibbling, dinking, contour sawing, turning, routing, and milling.

Slitting Slitting is shearing by circular shears, 4 to 6 in in diameter, which draw the material through the shears as they rotate. The rollers overlap just enough to shear the material. Excessive overlap distorts the material. Shearing rolls are spaced any desired distance apart on rotating shafts and can shear a sheet or a strip into as many widths as desired. With continuous strip mills furnishing sheet and strip in long lengths, many parts (such as transformer punchings) can be made continuously from slit material. Slitting is usually performed on metal less than $\frac{1}{8}$ in thick. Circular blanks are sheared by circular shears mounted in a heavy frame. The plate is pivoted at the center and the circular shear blades are forced into the material. The edge is sheared or trimmed

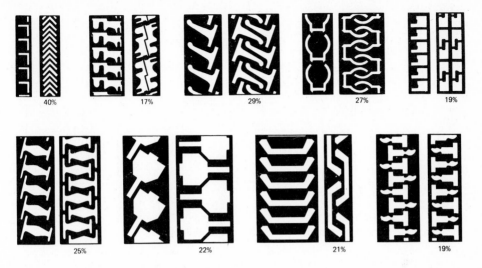

FIGURE 14-11
Several examples of nesting for material saving.

to form a circular blank. Depending on the capacity of the equipment, the thickness of material sheared may be as high as $\frac{3}{8}$ in. The material trimmed off is distorted.

Job-Shop Methods of Shearing

Uni-shearing   Uni-shearing is a trade name given to a machine that has a short shear mounted on a portable hand tool or in a frame having a deep throat. The Uni-shear can shear light metal to any shape by guiding the material through the shear or by guiding the portable shear through the material. The material touched by the movable blade is distorted. This is a handy machine for sheet-metal work of gages less than $\frac{1}{16}$ in thick.

Nibbling   Nibbling is separating material by continuously punching a slot in any direction desired. The nibbling punch and die (Fig. 14-12) are mounted in a press with a deep throat. A support for a circular blank center is usually provided. Any shape hole may be punched. Some nibblers require a starting hole when the hole to be cut is within the sheet. The nibbler consists of a step punch and die. The tip of the punch remains in the die and the rest of the punch cuts a slug. As the punch moves up and down, the tip, which remains in the die,

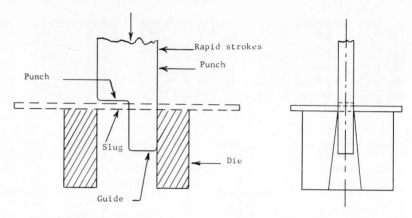

FIGURE 14-12
Typical step punch and die for use in nibbling.

guides the material. Thus a fairly smooth edge is produced and there is no distortion of either part.

Job-Shop Punching and Blanking

Dinking   Dinking or ruler dies are made with steel-cutting flexible-rule material. The die functions like a cookie cutter (Fig. 14-13). Thin aluminum, stainless steel up to 0.040 in thick, leather, celluloid, linoleum, and cardboard are blanked by this method. The steel rule used to blank metal is hardened after it is bent to shape and then clamped firmly by plywood pieces previously sawed with a jigsaw having the thickness of the cutting rule. The rule is supported its full width except for a 0.02-in extension. A layer of cork is glued on the inside of the cutter to strip the material from the blade. The assembled die is placed face up on the bed of the press. The material to be cut is placed on top, and the steel platen of the press comes down and forces the material onto the cutter. Proper spacers are placed between the bed of the press and the platen to prevent the platen from striking the cutters. Material is cut all the way through except for a slight amount which is sheared at the end of the stroke. Wherever these dies can be used, they have proved to be the least expensive method of blanking material where quantity is limited.

Quick-process or continental method   The production-design engineer can have tool-made samples furnished by the continental or quick process in his own toolroom, or he can purchase them from companies that specialize in making

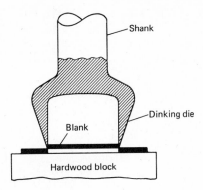

FIGURE 14-13
Dinking die and hardwood die block.

small quantities of tool-made parts. Their operators are trained in devising ingenious ways of producing blanks that can be punched and formed to the desired shape. Holes are pierced singly by setting gages at the proper distance and punching each piece separately. Standard notching dies for notching 90, 45, and 30° corners are available as are oblong hole punches and dies, so that a wide range of holes can be made in blanks. It is not difficult to make a single odd-shaped punch and die for a hole or blank. The blanking die is made by sawing the desired shape in a steel plate used as a die. The punch is made from the portion sawed out. A guide plate for the punch is sawed to the same shape. It is sometimes sawed together with the die plate. The die is relieved and brought to final shape by peening and by filing. The punch blank is peened out and filed to fit the die. Both may be hardened by applying a slight case. The guide plate is mounted above the die so that the material may be placed between them. The punch is placed in the guide plate on top of the material and the whole unit placed in a press with a flat bed and platen. The flat platen strikes the punch and drives it through the material. Both drop down into the die. The whole set is removed from the press, the blank and punch are removed, and the process is repeated (Figs. 14-14 and 14-15).

Punching and blanking structural steel and plate   There are two general classes of punching, blanking, and stamping work: structural and sheet metal. The first refers to work done on rough, hot-rolled bars, structural members, and thick plates. Extreme accuracy is usually not required or obtainable without excessive costs. Round or oblong holes can be punched in flanges of channels and I beams, the bevel portion resting on a bevel die, as shown in Fig. 14-16. Presses for these operations are called structural or plate punch presses. The dies and punches can be quickly changed. Some presses have more than one punch in the

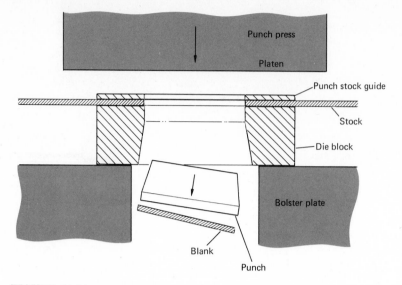

FIGURE 14-14
Simple punch-and-die tooling for the continental method of low-cost hole production in sheet metal.

press. The desired punch is actuated by moving a gag block between the ram and punch. In plate or girder punching the machines may have 50 or more punches which are set for actuation by an operator according to a template or drawing. The material is moved under and through the punch by a carriage which draws the plates or structural members the proper distance for each hole or series of holes. These gang punches make riveted structures economical to fabricate and place them in competition with welding.

Progressive Dies

When the production volume warrants the cost, a single progressive die should be designed and constructed to carry out all the operations required to complete a part automatically in one press. The aluminum cooling fins as used in the evaporator and condenser coils of an air-conditioning unit are a good example of the use of progressive dies. A cross section through a progressive die and the strip for making a notched blank (Fig. 14-17) shows the sequence of operations as a progressive die is used to carry out four cutting operations in sequence to make a notched blank at each stroke of the press.

The cost of a progressive-die setup is high because it must assume a share of the cost of a fully automatic press with cutoff, feed rolls, straightening rolls (if required), and a scrap reel if needed (Fig. 14-18). In addition, the die design

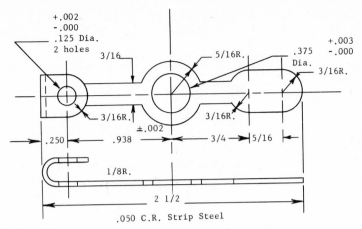

FIGURE 14-15
Typical part made by the continental process.

and construction cost is higher than the cost of a series of single-operation dies for making the same part. Provision must be made for obtaining the best sequence of operations at the design stage because once a die is built it is rather difficult to make changes.

On the other hand a number of savings are possible. An operator is required only when a coil of strip stock must be changed on the feed and scrap reels or when there is a malfunction such as when a punch breaks or the burrs become excessively large. The latter means that the dies need to be resharpened.

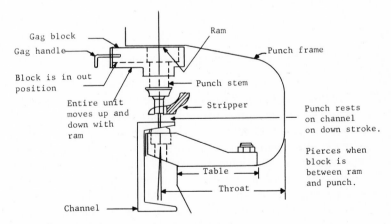

FIGURE 14-16
Heavy-duty press setup for punching holes in structural steel.

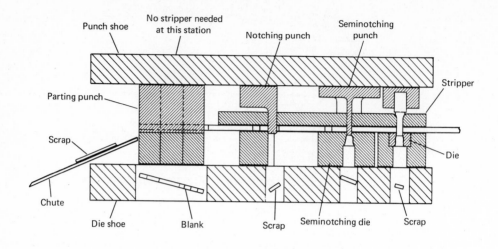

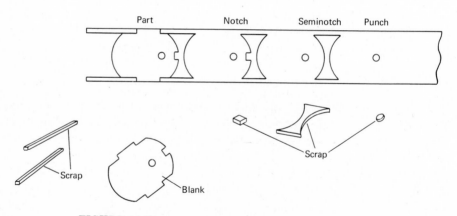

FIGURE 14-17

Typical four-station progressive die for producing a notched blank.

But the major saving in addition to direct labor is in the reduction in material handling and the amount of production control required. Since all four operations are combined the material is handled only once and in bulk as in the coiled raw material or in the tub of blanked parts to in-process stores. If made in four operations, three additional paperwork orders, three more setups, and three more material handlings would be required to make the same number of parts in the in-process storage area. There would be additional costs in inspection and supervision. Thus it is often prudent to consider the use of progressive dies for parts which are to be made in lots of even 10,000 provided there are anticipated to be several years of production before the part becomes obsolete.

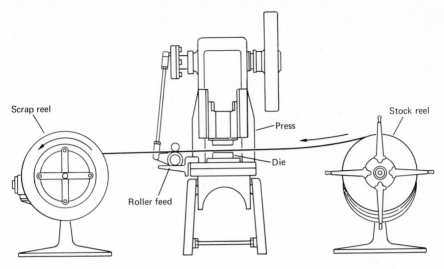

FIGURE 14-18
Roller-feed device for pulling strip off a stock reel and through a progressive die.

Dies for Punching and Blanking Operations

Considerable ingenuity and engineering development in tools and dies have made possible the high-quality and economical sheet-metal products of today.

The quality and the form of the part determine whether an expensive or simple tool will be required. This is under the control of the production-design engineer. The proper use of the following factors will aid the tool designer in producing an effective tool and quality part:

1 The punch determines the size of the hole and the die determines the the size of the blank.

2 Punches should always be as short as practical. Long production runs require punches long enough to grind from $\frac{1}{2}$ to 1 in.

3 All delicate projections or weak parts in a die should, if possible, be designed as inserts.

4 To minimize breaking of slender punches, taper the working end of the punch 0.001 to 0.002 in. This will facilitate stripping, and most breakage occurs during stripping.

5 A die will usually have a life of approximately two to three times the life of the punch or punches.

6 On a blanking die, the burr is toward the punch. Burrs from piercing are on the die side.

7 There is always clearance between the die and punch. Clearance in-

Table 14-3   COMPARISON OF DIE PRODUCTIVITY AND RELATIVE COST

Ranking	Type of die construction	Expected production (no. of pieces)	Relative cost (tool steel = 1)
1	Carbide construction	6×10^6	3.00
2	Tool steel (heavy duty, 2 in thick)	1×10^6	1.00
3	Tool steel (light duty, 1 in thick)	1×10^6	0.50
4	Flame hardened ($\frac{3}{8}$ in thick)	0.5×10^6	0.33
5	Template dies ($\frac{1}{16}$ in thick)	0.2×10^6	0.13
6	Rubber dies	0.1×10^6	0.06
7	Continental dies	1–1,000	0.01
8	Stock dies	1–1,000	0.01

creases with increase in thickness of material. The clearance should be subtracted from the die when designing the punch for blanking and should be added to the die opening when designing the punch for piercing. A good general clearance for blanking dies is 6 percent of the stock thickness on a side between punch and die.

8 The thickness of material should not exceed the diameter of the punch. Greater thicknesses can be punched by using good tool design but at a sacrifice to tool life.

9 Distortion of holes occurs if they are punched too close to the edge of the sheet, too close to each other, too close to a bend, or in a drawn portion. Holes should be punched after bending and drawing to ensure accurate size and location.

10 Material should be bent across the grain at least up to 45°, especially for sharp bends, so as to avoid cracking of the material. The "grain" is the direction the stock was drawn in passing through the rolls in manufacture.

11 The size of the sheet or strip material should be considered in designing the blank so that a minimum of loss in material is incurred.

12 The quantity of a part to be made determines the quality and complexity of the tools that can be built. A very active part—100,000 to 1,000,000 parts a year—may warrant elaborate progressive dies in which many operations, such as piercing, embossing, forming, blanking, and shearing, may be performed progressively in the die without the part being touched by the operator. The cost of picking up a part, placing it in the die, and removing it is much greater than the actual punching operation by the machine. However, as activity of the part decreases so does the economy of making progressive dies; operations are performed at a lower total cost by providing simpler tools or standard tools.

The cost of dies for punching and blanking can vary by as much as 300 to 1 (Table 14-3) when the simplest stock die setup is compared to a complex tungsten carbide die for blanking transformer laminations from 3 percent silicon iron.

FORMING PROCESSES

Forming processes are primarily bending or stretching operations that do not materially change the thickness of the metal. The original material remains about the same except that the thickness or diameter may be reduced by drawing or ironing. This lack of change in thickness is in contrast to forging operations which change the original shape of the plastic material by squeezing, pressing, or rolling.

Forming is based on two principles:

1 *Stretching and compressing the material beyond the elastic limit on the outside and inside of the bend.* When this condition of compression and tension exists, there is always some springback that frequently varies from piece to piece or from one lot of material to another, unless the shape of the part locks the stresses and does not permit movement after forming.
2 *Stretching the material beyond the elastic limit without compression, or compressing the material beyond the elastic limit without stretching.* When this is done, the springback is reduced to a negligible amount regardless of shape.

The first principle is used in pan- and press-brake forming, punch-press forming, and roll forming. The second principle is used in spinning, stretcher leveling, stretch forming, contour forming, and deep drawing. The processes described also may use a combination of these two principles.

Simple Bends

When a wire is bent by hand, it takes a natural bend which is governed by the shape of the thumb and finger and the forces applied to the wire. Usually when a bend is made its radius is governed principally by the tools making the bend.

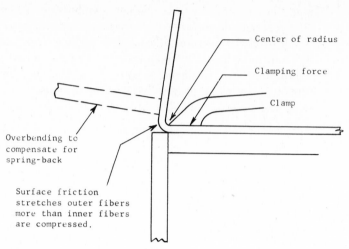

Center of radius

Clamping force

Clamp

Overbending to
compensate for
spring-back

Surface friction
stretches outer fibers
more than inner fibers
are compressed.

FIGURE 14-19
Pan-brake bender.

One leg may be clamped in stationary jaws and a form tool bends the material, as in tube bending, wire bending, pan-brake bending (Fig. 14-19), and dies with clamping pads.

Bending Allowance

The minimum bend radius varies with the alloy and its temper; most annealed sheet metals can be subjected to a bend which has a radius equal to the stock thickness without cracking. The more ductile metals can easily be bent back through a 180° bend, as in a hemming operation.

However, when sheet metal is bent, the total length including the bend is greater than the original stock. This change in length must be considered by the production engineer and die designer because the length of the sheared stock must be known in order to shear the stock.

In the flat position, the neutral axis of a piece of sheet metal coincides with its centerline. But in a bent position, the neutral axis has shifted to a position 0.33 to $0.40t$ from the inner radius (Fig. 14-20). A relationship for determining the length of the developed blank is given below in Fig. 14-20, where the length of the neutral axis is calculated from the circumference of the quadrant of a circle with a radius of $r + t/3$.

When developing the length of a complex part, first divide it into a series of straight sections, bends, and arcs. Trigonometry can be used to calculate unknown dimensions, but keep the legs of the triangle parallel to the dimension

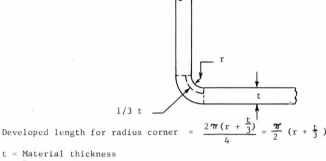

Developed length for radius corner $= \dfrac{2\pi(r + \frac{t}{3})}{4} = \dfrac{\pi}{2}(r + \frac{t}{3})$

t = Material thickness
r = Inside radius

FIGURE 14-20
Calculation of the length of blank for a part with a 90° bend.

lines, because the hypotenuse is then the bend angle and the length of the legs can easily be added to or subtracted from the blueprint dimensions.

The length of the bent metal can be calculated from the following empirical relationship:

$$B = \frac{A}{360} 2\pi(IR + kt)$$

where B = bend allowance, in (along the neutral axis)
   A = bend angle, °
   IR = inside radius of bend, in
   t = metal thickness, in
   $k = 0.33 \quad IR < 2t$
      $0.50 \quad IR > 2t$

The Glenn L. Martin Company has published a bend allowance chart (Table 14-4) which was developed primarily for the aluminum sheet used in the aircraft industry, but it should be applicable to other alloys. The notes at the bottom make the chart self-explanatory. The chart is based on the empirical relationship

$$B = (0.01743R + 0.0078T)90$$

Bending Force

The bending stress in a simply supported beam can be calculated from relationships developed in the field of strength of materials:

Table 14-4 TABLE FOR FINDING THE DEVELOPED LENGTH OF 90° BENDS*

Birmingham wire gage	Radius																				
	$\frac{1}{32}$	$\frac{3}{64}$	$\frac{1}{16}$	$\frac{5}{64}$	$\frac{3}{32}$	$\frac{7}{64}$	$\frac{1}{8}$	$\frac{9}{64}$	$\frac{5}{32}$	$\frac{11}{32}$	$\frac{3}{16}$	$\frac{13}{64}$	$\frac{7}{32}$	$\frac{15}{64}$	$\frac{1}{4}$	$\frac{9}{32}$	$\frac{5}{16}$	$\frac{11}{32}$	$\frac{3}{8}$	$\frac{7}{16}$	$\frac{1}{2}$
0.330																0.466	0.499	0.512	0.525	0.549	0.579
0.284															0.422	0.435	0.479	0.492	0.506	0.529	0.557
0.239														0.390	0.396	0.410	0.449	0.462	0.476	0.499	0.529
0.238													0.361	0.368	0.374	0.388	0.423	0.437	0.450	0.473	0.504
0.220												0.333	0.340	0.347	0.354	0.367	0.401	0.415	0.428	0.451	0.482
0.203											0.299	0.306	0.313	0.319	0.326	0.340	0.381	0.394	0.408	0.431	0.461
0.180										0.274	0.281	0.288	0.294	0.301	0.308	0.321	0.353	0.366	0.379	0.403	0.433
0.165									0.247	0.253	0.260	0.267	0.273	0.280	0.287	0.300	0.335	0.348	0.361	0.385	0.415
0.148								0.223	0.230	0.236	0.243	0.250	0.257	0.263	0.270	0.283	0.314	0.327	0.340	0.364	0.394
0.134							0.199	0.206	0.213	0.220	0.226	0.233	0.240	0.246	0.253	0.266	0.297	0.310	0.324	0.347	0.377
0.120						0.179	0.186	0.193	0.199	0.206	0.213	0.220	0.226	0.233	0.240	0.253	0.280	0.293	0.307	0.330	0.360
0.109					0.156	0.162	0.169	0.175	0.181	0.188	0.195	0.201	0.208	0.216	0.223	0.236	0.266	0.280	0.293	0.317	0.347
0.095				0.135	0.141	0.148	0.154	0.161	0.168	0.175	0.181	0.188	0.195	0.201	0.208	0.222	0.248	0.262	0.276	0.300	0.330
0.083			0.113	0.121	0.128	0.134	0.141	0.148	0.153	0.161	0.166	0.173	0.180	0.188	0.195	0.208	0.235	0.248	0.262	0.285	0.315
0.072		0.100	0.106	0.113	0.119	0.126	0.133	0.139	0.146	0.153	0.161	0.166	0.173	0.180	0.186	0.200	0.222	0.235	0.248	0.272	0.302
0.065	0.084	0.092	0.098	0.104	0.111	0.117	0.124	0.131	0.138	0.145	0.152	0.158	0.164	0.171	0.178	0.191	0.213	0.226	0.240	0.263	0.293
0.058	0.084	0.081	0.098	0.104	0.111	0.117	0.124	0.131	0.138	0.145	0.152	0.158	0.164	0.171	0.178	0.191	0.205	0.218	0.231	0.255	0.285
0.049	0.073	0.072	0.091	0.098	0.105	0.111	0.118	0.125	0.131	0.138	0.145	0.152	0.158	0.164	0.171	0.185	0.198	0.207	0.220	0.244	0.274
0.042	0.065	0.065	0.084	0.091	0.098	0.103	0.110	0.118	0.125	0.125	0.136	0.145	0.152	0.158	0.150	0.177	0.190	0.198	0.212	0.235	0.266
0.035	0.056	0.060	0.076	0.083	0.090	0.096	0.103	0.110	0.116	0.116	0.127	0.130	0.136	0.143	0.150	0.163	0.177	0.190	0.203	0.227	0.257
0.032	0.053	0.055	0.072	0.079	0.085	0.091	0.099	0.106	0.113	0.113	0.119	0.126	0.133	0.139	0.146	0.160	0.173	0.186	0.200	0.223	0.253
0.028	0.048	0.052	0.068	0.074	0.081	0.087	0.094	0.101	0.108	0.108	0.114	0.121	0.128	0.135	0.141	0.155	0.168	0.182	0.195	0.218	0.249
0.025	0.044	0.048	0.065	0.071	0.077	0.084	0.091	0.097	0.104	0.104	0.111	0.118	0.124	0.131	0.138	0.151	0.164	0.178	0.191	0.214	0.245
0.022	0.040	0.045	0.060	0.067	0.074	0.080	0.087	0.094	0.100	0.100	0.107	0.114	0.121	0.127	0.134	0.147	0.161	0.174	0.188	0.211	0.241
0.020	0.038	0.043	0.058	0.065	0.071	0.078	0.085	0.091	0.098	0.098	0.104	0.111	0.118	0.125	0.132	0.145	0.158	0.172	0.185	0.208	0.239
0.018	0.035	0.043	0.055	0.062	0.069	0.076	0.082	0.089	0.096	0.096	0.102	0.109	0.116	0.122	0.129	0.143	0.156	0.169	0.183	0.206	0.236

* Subtract the correct figure in the table from the sum of the length of legs.

$\frac{1}{2} + \frac{5}{8} = 1\frac{1}{8}$
$1.125 - 0.133 = 0.992$ length

$\frac{1}{2} + 1 + \frac{1}{2} = 2.000;\ 2 \times 0.106 = 0.212$
$2.000 - 0.212 = 1.788$ length

$\frac{1}{2} + \frac{3}{4} + 2 + \frac{3}{4} + \frac{1}{2} = 4\frac{1}{2};\ 4 \times 0.200 = 0.800$
$4.500 - 0.800 = 3.700$ length

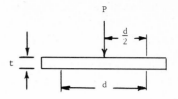

FIGURE 14-21
Bending force for sheet metal treated as a simply supported beam.

$$S = \frac{Mc}{I}$$

where S = stress, lb/in^2
   M = bending moment, in·lb
   I = moment of inertia at section involved, in^4
   c = distance from neutral axis to outermost fiber in,

For the workpiece shown in Fig. 14-21, the bending force P would be calculated as follows:

$$S = \frac{(Pd/2)\,(t/2)}{wt^3/12}$$

$$= \frac{3Pd}{wt^2}$$

$$P = \frac{0.33\ Swt^2}{d} \qquad \text{lb}$$

where S = nominal tensile strength, lb/in^2
   w = width of bend, in
   t = thickness of metal, in
   d = span of bend (approximately $8t$ for V dies), in

Although this is the stress to initiate bending, to make a good 90° V bend requires a certain amount of plastic deformation to eliminate undesirable spring-back. From experiment it has been found that an appropriate value of the total bending force at the bottom of the stroke should be four times greater than the calculated value:

$$P = \frac{1.33\ Swt^2}{d} \qquad \text{lb}$$

As the distance d decreases, the force P increases. The formula for the developed length of bend given in Fig. 14-20 is usually also applied to V dies.

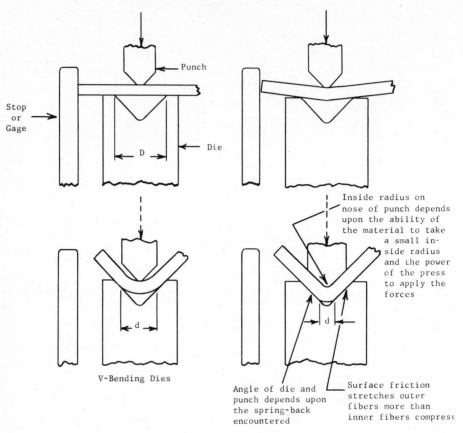

Punch

Stop
or
Gage

D

Die

V-Bending Dies

Inside radius on
nose of punch depends
upon the ability of
the material to take
a small in-
side radius
and the power
of the press
to apply the
forces

Angle of die and
punch depends upon
the spring-back
encountered

Surface friction
stretches outer
fibers more than
inner fibers compress

FIGURE 14-22
Details of making 90° bend in a V die in a press break.

V-Die Bending

V-die bending (Fig. 14-22) is performed easily in hydraulic or mechanical presses. The forces applied at the beginning of the bend are less than those required to complete the bend. V-die bending is suitable to the action of a mechanical crank press.

Punch-Press Bends

In a punch-press die the bend is usually made with a pressure pad holding the material while a punch forms it (Fig. 14-23).

The conditions that prevail in punch-press bending are similar to those in pan-brake and V-die bending, except that it is more difficult to obtain a right-

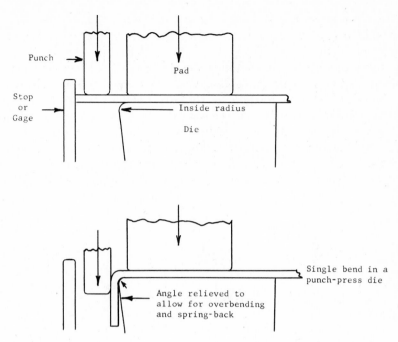

Punch →

Pad

Stop or Gage →

Inside radius

Die

Single bend in a punch-press die

Angle relieved to allow for overbending and spring-back

FIGURE 14-23
Bending a cantilevered workpiece in a punch press.

angle bend due to the ironing action on the material. Material thickness is important in punch-press bending as far as angle size goes. A few thousandths variation in material thickness will result in appreciable variation of the angle size in the piece being bent.

Although developed lengths may be computed rather closely (Fig. 14-20) it is wise to verify the developed lengths by experiment. For example, when a double bend is made on a punch press, the material may stretch between bends thus giving a smaller developed length than calculated (Fig. 14-24).

In pan- and press-brake bending of a wide sheet, the radius of bend may be larger in the center due to the greater deflection of the machine parts at the center (Fig. 14-25).

Limitations to Consider in Design for Bending

The production-design engineer should be aware of and understand the limitations of equipment and design so that interferences are avoided and the usual variations will be acceptable. The pan brake in Fig. 14-19 is subject to deflection in the center when bending wide sheets similar to the deflections of the press brake in Fig. 14-22. A craftsman can vary the radius of the bend by adjusting the machine and by using special tools. Although some pan-type brakes are made to form steel plates 150 in long and $\frac{3}{4}$ in thick, the most common sizes are smaller and are used for bending thin sheet metal (12 gage and less). The pan brake can bend the material nearly 180°. Clamps are then used to complete the bend to 180° and give a smooth edge or form a lock-joint as shown in Fig. 14-26.

On second and succeeding bends, the limitations indicated in Fig. 14-27a to d must be considered. Thus Fig. 14-27b illustrates the minimum length between bends, and Fig. 14-27c shows the relationship between adjacent lengths and the punch when a flange is "formed up" in the press. A layout of the bent

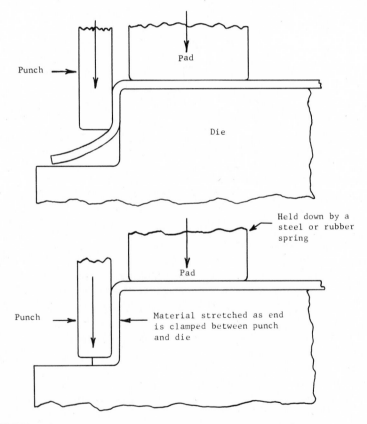

FIGURE 14-24
Case in which material stretches between bends, giving less developed length than expected.

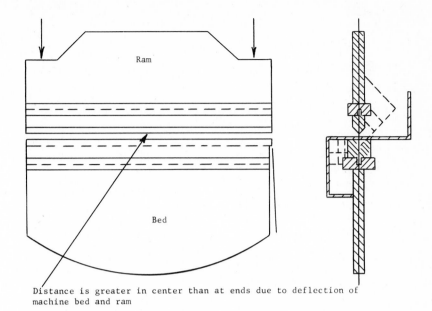

Distance is greater in center than at ends due to deflection of machine bed and ram

FIGURE 14-25
Typical press brake showing arrangement of ram and bed and possibility of deflection at center.

part with the forming tools or a paper representation of the bent part fitted into cardboard models of the dies helps to avoid gross errors in forming sheet metal. Dies are not expensive and can be modified into all forms and shapes (Fig. 14-28). Considerable time in the shop would be saved if a progressive picture of each bending operation were shown on the drawing when multiple bends are incorporated in the sheet-metal design. (See Fig. 14-25 and note that the gage stops either on the front or back may interfere and must be removed.)

Many ingenious tools and procedures are used in the trade, such as box-forming brakes (Fig. 14-29) and horn dies on press brakes (Fig. 14-30) which form closed sections such as boxes and channels and segmented punches that will permit forming the four edges of a sheet (Fig. 14-31).

FIGURE 14-26
Lock-type joints made on press-brake or pan-brake machines.

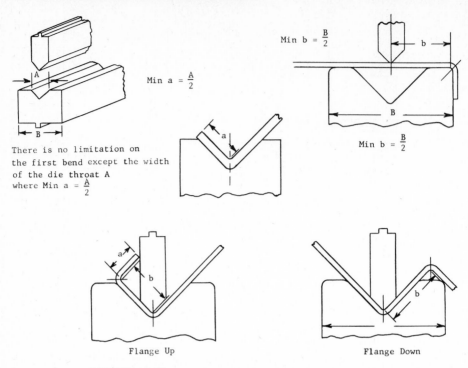

There is no limitation on
the first bend except the width
of the die throat A
where Min a = $\frac{A}{2}$

Min a = $\frac{A}{2}$

Min b = $\frac{B}{2}$

Min b = $\frac{B}{2}$

Flange Up

Flange Down

FIGURE 14-27
Limitations on the length between bends. *Top left,* no limitation on the first
bend; *top right,* width of die limits distance; *bottom left,* flange up length of
leg limits distance; *bottom right,* flange down width of die limits distance.

Tube Bending

A bend in a tube is difficult to make without buckling the walls on the inside of
the bend and causing the tube to collapse and become oval shaped. In order to
obtain a uniform cross section, the tube, prior to bending, may be filled with a
low-temperature ductile alloy such as lead, or it may be filled with sand and
the ends sealed. After the bend is made with regular bar-bending equipment,
the sand is removed or the tube is heated and the metal is poured out. Another
technique is to insert a coiled spring or a snake in the tube at the point of bend
in order to support the walls of the tube as it is bent. The snake is a series of
balls or lugs connected by flexible joints. Still another method is to insert a
mandrel into the tube which is held at the point of bend as the tube bender
rotates (Fig. 14-32). The end of the mandrel is hardened and shaped to the inside of the tube, which it supports and prevents from collapsing. The tube is
slightly stretched to prevent collapsing on the inside. Automatic machines are
able to make several bends in different directions and at different angles. Such

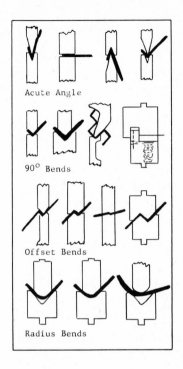

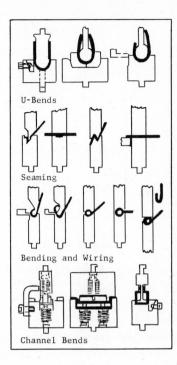

FIGURE 14-28

A group of representative press-brake forming die sets which illustrate some of the contours which can be produced.

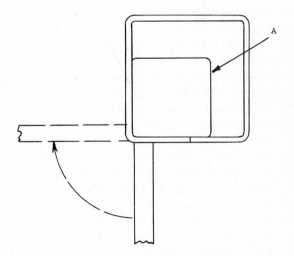

FIGURE 14-29

Box-forming brake. The outside support of bar A swings away so that the part can be slipped off. Otherwise the action is the same as a pan brake.

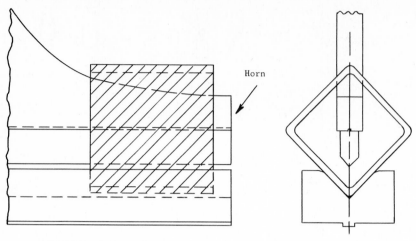

FIGURE 14-30
A horn die on the end of a press brake.

machines are used in the manufacture of furniture and automobile parts. The tools and accessories for bending and forming pipes, tubes, and structural shapes are shown in Fig. 14-33.

Flexible Die Forming

Rubber forming was used in only a few special applications until the aeronautical industry applied it extensively on aluminum sheet-metal parts. Rubber, when confined, is incompressible. This property permits the use of high pressure, advantageous in forming parts that would require a collapsible punch. For example, the threads on the inside of an electric light socket are formed by a rubber plunger which forces the metal to the internal form of the die. As the punch descends, the rubber expands; and, since it is confined, the pressure builds up. When the punch ascends, the rubber returns to the original shape and is easily withdrawn. The die then opens up and releases the socket. In this operation the punch is rubber and only a die is required.

The aeronautical industry developed a large hydraulic press with a sponge-rubber cushion on the plunger coming down from above. A typical size is 6 to 8 in deep, 5 ft wide, and 10 ft long. A platen is slid under this cushion from either side. One platen is prepared while the other is moved under the press and the parts are formed. The material is placed upon the die mounted on the platform and the rubber cushion enfolds the die parts. A part with multiple bends can be formed completely in one operation. A box with four sides and flanges, making eight bends in all, can be formed with one stroke. This process eliminates one-

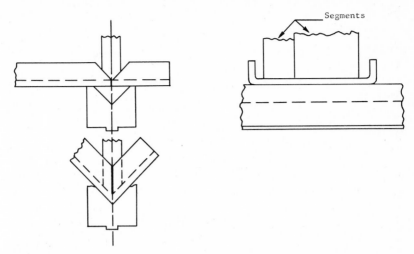

FIGURE 14-31
Segmented punches enabling bending sheets on one or two sides.

half the forming die. Either the punch or die can be replaced by rubber. (See Fig. 14-34 for a simple example.)

Another use of rubber is for blanking and piercing. The sheet metal is placed over a die and a rubber pad is pressed down on the sheet. The pressure is great enough to shear the material and form a blank. The presses consist of a

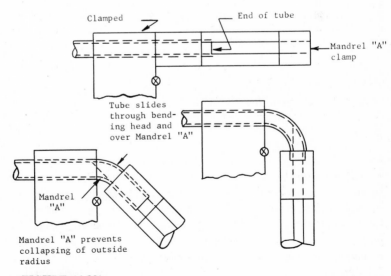

FIGURE 14-32
Tube bending by use of a mandrel.

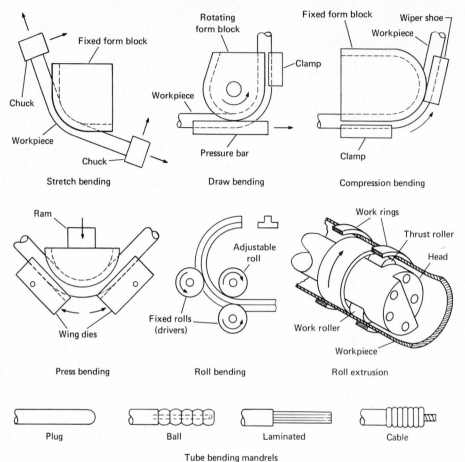

FIGURE 14-33

Operations and accessories for bending pipes, tubes, and structural shapes. (*Redrawn from Doyle et al., "Manufacturing Processes and Materials for Engineers," Prentice-Hall, Englewood Cliffs, N. J., 1969.*)

flat lower platen on which the thin blanking die is placed and an upper and moving platen on which the rubber pad is fastened. After the blank is made, the die and blank are moved out by sliding on the lower platen when the press is open. The blank is removed and the die and sheet are placed in position for the next operation. Holes can be pierced at the same time the sheet is blanked by placing punches in the required positions.

There are a number of flexible die-forming processes available, including the Guerin Process, Marforming, Hydroforming, and Hydrodynamic Drawing

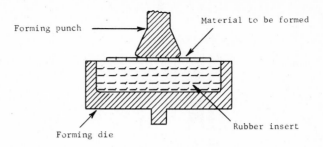

Beginning Stroke

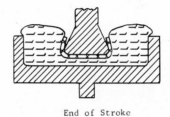

End of Stroke

Finished Part

FIGURE 14-34
Use of rubber-forming dies to produce a channel.

(Fig. 14-35). Note that only the first two processes use a rubber die, but all use the principle of having one die flexible. The flexible-die processes are slower than the steel-die processes when the latter are used on a suitable mechanical press and the production quantity exceeds several thousand pieces. At low production levels their short lead time and low tooling cost make them economically justifiable. However the original equipment cost for Marforming and Hydroforming is high, so that a steady work load is required to amortize the high equipment cost over a reasonable time period. Production rates vary from 50 to 250 parts per hour and draws as deep as 5 to 12 in can be achieved by Marforming or Hydroforming.

Roll Forming

Bends of large radii, such as circle bends for hoops, angle rings, or tanks, are made on a roll bender. The material is fed into the roll and bent continuously

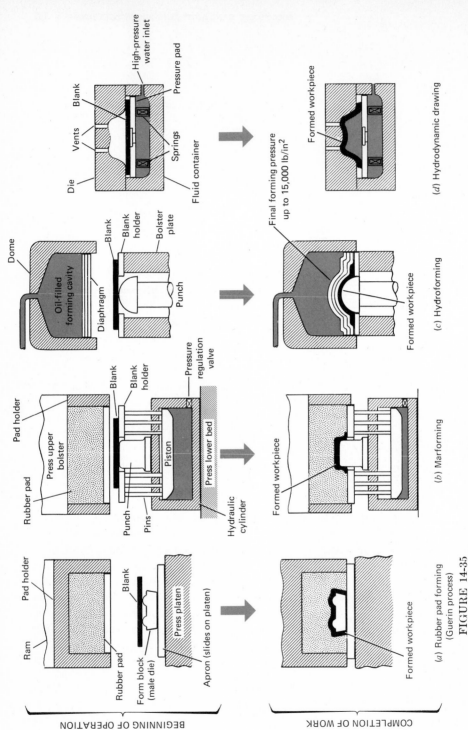

FIGURE 14-35

Diagram of common flexible die-forming processes. (*Redrawn from Doyle et al., "Manufacturing Processes and Materials for Engineers," Prentice-Hall, Englewood Cliffs, N. J., 1969.*)

(a) Rubber pad forming (Guerin process)

(b) Marforming

(c) Hydroforming

(d) Hydrodynamic drawing

BEGINNING OF OPERATION

COMPLETION OF WORK

Pad holder

Ram

Rubber pad

Blank

Form block (male die)

Press platen

Apron (slides on platen)

Formed workpiece

Pad holder

Press upper bolster

Rubber pad

Blank

Blank holder

Punch

Pins

Piston

Pressure regulation valve

Press lower bed

Hydraulic cylinder

Formed workpiece

Dome

Oil-filled forming cavity

Diaphragm

Blank

Blank holder

Bolster plate

Punch

Final forming pressure up to 15,000 lb/in^2

Formed workpiece

High-pressure water inlet

Blank

Vents

Die

Springs

Pressure pad

Fluid container

Formed workpiece

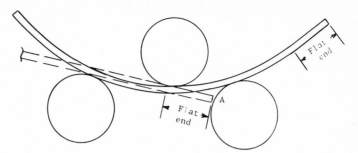

FIGURE 14-36
Use of pyramid rolls to bend plate.

into its final shape. One end is always flat when the inside roll is smaller than the desired radius. Here the work may have to be preformed to the radius desired (Fig. 14-36). Where there is a large number of rolled parts to be made, such as steel-plate motor frames, a roll having a diameter equal to the inside of the frame is used as the center roll. Thus, circular frames can be formed with no flat ends.

Roll-forming strip is the process in which a continuous strip passes through a series of roll stands in order to obtain the final form. The rolls on each roll stand are shaped and positioned to form the contours for each of the required steps (Fig. 14-37). The process is rapid, and the tool expense is moderate. Auxiliary operations such as punching, cutoff, and welding can be added. For example, tubing is roll-formed from strip and butt-welded by gas or electric current in the

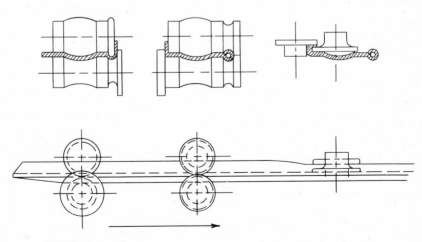

FIGURE 14-37
Continuous-roll forming of strip steel.

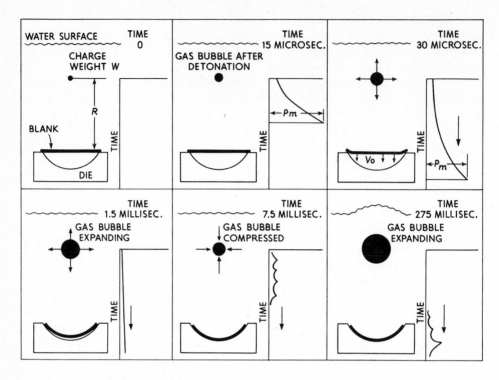

FIGURE 14-38
The development and transmission of explosive energy through water.

same machine. In roll forming, the bends allow for springback. Roll-formed parts compete with the extruded and molded parts used in building trim, metal window frames, doors, and metal furniture.

High-Velocity Forming

High-velocity forming, frequently called explosive forming, employs the principle of rapid release of energy in order to form or punch a metal part. It has been reported that pressures running as high as 2,000,000 lb/in^2 and speeds as high as 3,000 ft/s are used. The tooling required includes a female die of the correct geometry into which the flat material is forced by the impact of the explosion.

High-velocity forming has application in producing large parts where the cost of conventional equipment would be excessive, for forming materials that are difficult to produce by the usual techniques, and for forming small quantities where the cost of tooling would be high in proportion to the number of pieces to be produced.

High-velocity forming can be applied successfully to the production of such items as expanded tubes, ducts, drawing tubes, forming of sheet products, extruding and forging of steel, piercing of holes, and the compressing of metal and ceramic powders into shapes.

Both low and high explosives are being used in high-velocity forming. Low explosives include black and smokeless powder and other slow-burning, deflagrating explosives. High explosives include dynamite, TNT, RDX, and similar explosives that have a detonation speed of more than 5,500 ft/s.

As of this writing, high-velocity forming is not a mass-production process.

Figure 14-38 illustrates the manner of transfer of energy that takes place under high-velocity forming.

COMPRESSION PROCESSES

Cold Rolling

Cold rolling is a mill process, carried out below the recrystallization temperature, which is usually performed by the supplier of the sheet or strip material. Cold rolling entails the passing of hot-rolled metals such as bars, sheets, or strips of cleaned stock through a set of rolls. It is passed through the rolls many times, using light reductions each time, until the correct size is obtained. Cold rolling increases the tensile and yield strength and imparts a bright finish and close tolerances to the finished stock.

Cold Extrusion of Steel

When steel is cold-extruded, larger-tonnage presses are used and the blank must be homogenous and ductile. Also, the surface must be treated to reduce the friction and permit the flow of the material in the die (see Fig. 14-39). Cold-steel extrusions for making deep shell-shaped parts can be made with less heating and annealing than required in deep drawing. The Ford Motor Company is making steel-extruded axle spindles and other shapes. Here reductions of 30 percent in material, 65 percent in labor, 80 percent in floor space, and 40 percent in die cost have been obtained.

Since World War II, with information obtained from the Germans as a basis, and with additional research by Army Ordnance, knowledge of the cold-extrusion process has been considerably advanced.

Successful cold extrusion requires properly designed presses, dies, preparation of steel, and lubrication of the billet. Zinc phosphate coatings have met with much success as a suitable lubricant when reacting with other materials, such as soap, to form an integral continuous film. These may be applied either by immersion or spray.

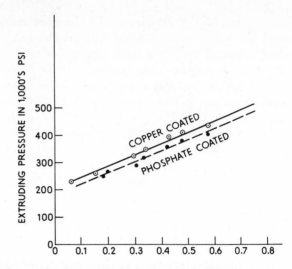

FIGURE 14-39
Relation between carbon content and extrusion pressure in the cold extrusion of steel.

The sequence of operations in treating the billet in preparation for extruding include: alkaline cleaning, rinse, pickling, rinse, phosphate coating, rinse, lubricate with reactive soap, dry.

Cold extrusion is particularly successful for:

1 A wide range of steels that are suitable for cold extrusion. A5120 and C1010 aluminum-killed in particular give high ultimate tensile strength to extruded products, with little loss of ductility.

2 The heat treatment of steel prior to extrusion, which has proved helpful. A normalizing procedure, followed by quenching in water and annealing, is suggested.

3 A zinc-phosphate coating of controlled formulation and application, which is the best surface preparation for extruding cold steel.

4 The processes of phosphatizing and the addition of a dilute emulsion type of compatible lubricant to increase the efficiency of the extrusion process.

Impact Extrusion

In impact extrusion, cold material with a low recrystallization temperature such as aluminum, copper, or lead is used to make the extruded parts (Fig. 14-40). The slug or disk is placed in the die, where it is struck by a plunger. The rapid travel of the plunger and resulting impact cause the material to heat up and become plastic. The material is thus quickly extruded through the space between the die and the plunger. The fact that the material passes through the opening between the plunger and the die reduces the abrasive wear on the tools and provides a slight clearance between the formed part and the plunger, so that the part can be slipped off easily. Thin-walled items such as toothpaste containers, radio shields, food containers, boxes for condensers, and cigarette lighter cases are made easily when they are symmetrical in shape. The process is very rapid; parts can be made as fast as material can be placed in the die and the cycle of the press completed. Large presses now make parts with thick walls for items such as paint spray-gun containers and grease-gun containers.

A second type of impact extrusion or cold extrusion is performed by forcing the material and plunger through the die. The material flows in the same direction as the plunger travels (Fig. 14-41). The proposed blank for making the part may have a hole with a solid bottom. The solid bottom remains untouched by the process. The plunger enters the hole in the blank, presses the walls of the blank between the plunger and the die, and forces the material from the bottom. This latter process is known as the Hooker process.

Coining

Coining is a cold-forging process where the material inserted into the die is at room temperature. In this process a blank is placed in a die and then forced under heavy pressure on a knuckle-joint or toggle-type press to conform to the

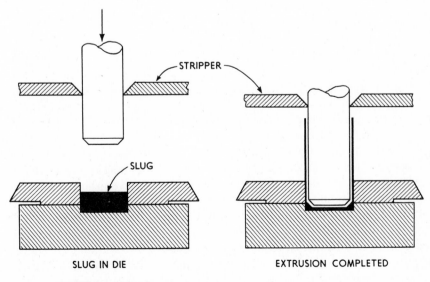

SLUG IN DIE   EXTRUSION COMPLETED

FIGURE 14-40
Backward impact extrusion.

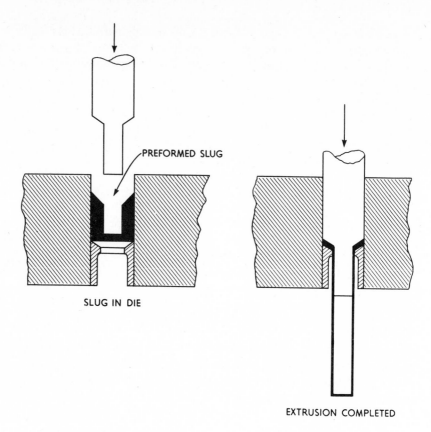

FIGURE 14-41
Forward (Hooker) impact extrusion.

design and shape of the die. The operation is frequently used for striking up jewelry and medals of all types where clear-cut designs are required in relief on the surface of the blank. Coining gets its name from the production of coins in the United States mint (see Fig. 14-42c).

Coining operations are performed with the same general type of press or hammer equipment as used in the production of die forgings. In coining operations, the blank is approximately the same size as the finished piece. Since large pressures are required in order to effect flow of cold materials, it is important that the amount of flow be kept as small as possible. Government tests record the following pressures required for different coins: silver dollar, 160 tons; half dollar, 98 tons; dime, 35 tons; penny, 40 tons.

Where accurate size is required, a subsequent shearing or sizing operation may be necessitated. For example, in the production of type bars used in book-

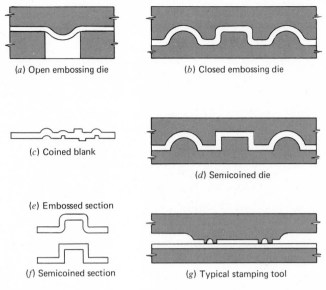

(a) Open embossing die   (b) Closed embossing die

(c) Coined blank

(d) Semicoined die

(e) Embossed section

(f) Semicoined section   (g) Typical stamping tool

FIGURE 14-42
Typical embossing dies.

keeping and adding machines, the following operations are required: (1) blank oversize, (2) coin configurations on type bar, and (3) trim flash.

Embossing Embossing is the placing of configurations on the surface of a part by means of dies which draw the material into position with very little effect on the thicknesses (Fig. 14-42a and b).

Swaging

Swaging is the shaping of material by a series of blows from a die made to the desired shape. It is usually used for making round shapes such as rod and wire. The outstanding application of swaging is the processing of tungsten into rods and wire by rotary-impact swaging equipment. Tungsten is a high-strength material and can be shaped by continuous blows. The swaging equipment consists of two or four dies distributed around the periphery of the material to be formed. Power-driven rollers mounted in a ring rotate around the outside of the die and cause the die to strike the material. The constant hammer blows gradually reduce the material in size. Then it is taken to another machine and reduced further until the desired size is obtained. Swaging equipment can quickly

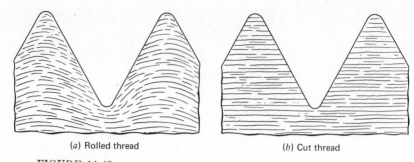

(a) Rolled thread   (b) Cut thread

FIGURE 14-43
Comparison of sections through a rolled (a) and a cut (b) thread. Note the
fiber flow lines in the former.

reduce or taper the diameter of wire in order to make possible its entrance into
drawing dies. Hollow parts, such as oxygen cylinders, can be reduced on a mandrel which holds the inside size. Cables can be attached to hollow shanks by the
swaging process; the cable is inserted into the shank and the shank is then
swaged. This firmly clamps the cable within the shank. Swaging operations are
performed on both hot and cold material.

Thread Rolling

In thread rolling, the screw thread is produced by the displacement of metal;
whereas in all other methods of producing threads, they are made by cutting.
To produce a screw thread by rolling, a cylindrical blank of a predetermined
diameter, approximately the pitch diameter of the screw being produced, is
placed between two hardened-steel dies. The dies, of course, have the same thread
pattern that is to be formed on the blank. The threading dies may be either flat
or circular. The circular dies rotate the part slightly more than one revolution
and then release; the flat dies rotate the part to the same degree by moving
parallel in opposite directions. As the blank rolls between the dies, the pressure
applied to the blank through the dies causes the metal to be displaced and to
follow the pattern of the dies. The resulting screw thread will have the same
form, pitch, and helix angle as the thread on the dies.

It can be seen that no cutting action takes place in this method, and consequently no material is removed from the original blank. The ridges on the
threads of the dies are fed into the blank approximately 50 percent of their
height. This action forms the roots of the thread being rolled, while at the same
time this displaced metal flows upward into the root of the die thread to form
the crest of the screw thread (see Fig. 14-43).

Due to the cold working of the material and the absence of rough surfaces caused by cutting, the rolled thread is stronger and tougher than cut threads. No difficulty is experienced in maintaining class 2 thread tolerances. Another advantage of the rolled thread is a savings in material. Since the blank that is used approximates the pitch diameter of the desired thread, a smaller-diameter stock is used than if the same thread were to be cut.

Although the life of thread rolls, when properly cared for, is quite long, the cost is relatively high. Care must be exercised to avoid getting a metallic chip between the roll and the blank, as this may result in chipping or cracking the thread rolls.

Thread-rolling machines today, utilizing the same principle of producing threads, are performing knurling, grooving, burnishing, and spinning operations. Pipe and tubing are formed to circular shapes; worm threads are formed on shafts in worm-gear units; grooves with long leads and grooves parallel to the length of the part can be made easily.

High-Pressure Extrusion of Metals

When a metal rod is drawn through a die from a container in which the rod is subjected to high hydrostatic pressures ranging from 100,000 to 600,000 lb/in², it is being cold-extruded under high pressure. That operation is called high-pressure forming. A true fluid flows when any shear force is applied to it; in contrast a Bingham solid does not flow under a low force, but flows viscously under a high shearing force. Metals with high ductility, i.e., face-centered cubic metals, behave like Bingham solids and are prime candidates for use in the high-pressure extrusion of wire.

The principal stresses in a solid are a set of three mutually perpendicular direct stresses which act on a plane of zero shear stress. The direction in which a principal stress acts is a principal axis; a plane normal to a principal axis is called a principal plane (Fig. 14-44). The average normal stress σ is the algebraic average of the principal stresses, $\sigma_{av} = (\sigma_1 + \sigma_2 + \sigma_3)/3$.

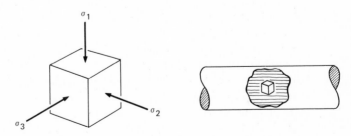

FIGURE 14-44
Principal stresses in a rod.

A hydrostatic state of stress is achieved when all three principal stresses are equal. Then the average normal stress is equal to the principal stress and there are no shear stresses. It is well known that high hydrostatic pressure increases the ductility, whereas hydrostatic tension reduces ductility. For example, TZM molybdenum alloy had a 63 to 65 percent reduction in area in the stress-relieved condition, but under 225,000 lb/in² its ductility improved to 97 percent reduction in area. Especially impressive is the case of beryllium, which metal has been successfully extruded under high pressure despite its usual rating as being nondeformable.

In normal wire-drawing practice using a small die angle, the wire is subjected to an environmental pressure of $(2C - T)/3$ rather than pure tensile stress. In hydrostatic extrusion the ductility of the rod increases with the environmental pressure, so that there is no necking failure even at large reduction ratios, i.e., at 2:1 or 5:1 change in diameter.

The Western Electric Company has developed two wire extrusion machines which use the principles of hydrostatic extrusion to achieve large reduction ratios when forming aluminum and copper rods into correctly sized wire in a single die. The secret of their success lies in the simultaneous solution of two major problems:

1 Designing a high-production device which is capable of maintaining consistently pressures up to 150,000 lb/in²

2 Designing a way to apply a film of lubricant between the die and the rod during the reduction process

The lubrication problem was solved by building a series of viscous drag-feed tube cells which are placed end to end along the rod and within the hydrostatic pressure chamber. By experiment it was found that 50 of the viscous drag-reversing cells were needed to increase the pressure from atmospheric to 150,000 lb/in². Lubricants and the pressure media included heated beeswax and silicone plus polyethylene mixtures. It was found that the pressure media had to have extremely high viscosity to be successful.

The conventional wire-drawing operation at Western Electric Company required from 9 to 12 carbide dies (Fig. 14-45).

Machine no. 1 Produced 14-gauge (0.065-in-diameter) wire using 9 to 12 carbide dies and 400 hp, starting with $\frac{5}{16}$-in-diameter annealed copper rod. End speed was 1,000 ft/min.

Machine no. 2 Produced 17-gauge (0.045-in-diameter) wire using 10 diamond dies and 150 hp, starting with $\frac{3}{8}$-in-diameter annealed aluminum rod. The end speed was 3,000 to 4,000 ft/min.

With the prototype models of the hydrostatic extrusion machine (Fig. 14-46), the Western Electric Company engineers have attained a 360:1 reduction of area in a single die and one machine. The resultant saving of 65 percent

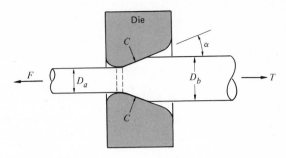

FIGURE 14-45
Conventional wire-drawing die.

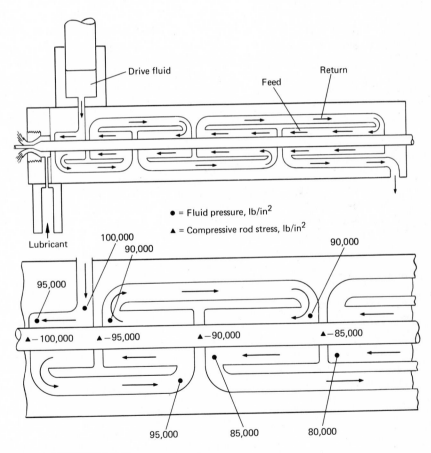

FIGURE 14-46
Schematic diagram of the hydrostatic extrusion of aluminum or copper rod
into wire at a reduction ratio of 360:1.

more than justifies the considerable expense of the engineering and toolmaking man-hours that were required to make the venture a practical success.

DRAWING PROCESSES

Cold Drawing

Cold drawing is the process of pulling material through a die in order to reduce it in size and to form the desired shape. Rounds, hexagons, squares, small angles, channels, and miscellaneous shapes, including tubing, are cold-drawn. The material must be clean and the surface must be free of scale and large defects before drawing. The resulting material is cold-worked and has a smooth finish. The size and tolerances of the finished part depend upon the ability of the dies to maintain their size. There is considerable abrasion between the material and die, and various lubricants are used to lengthen the life of the die. The dies are made of cemented carbide materials on high-activity sizes. Otherwise tool steel is used. The material is pulled through by means of a windlass or drawbench. Drawbench work usually straightens the material, thus eliminating straightening operations. Cold-drawn material is in demand because of its accuracy of size, good surface, and adaptability to many shapes and forms. Most bar stock used in screw machines is cold-drawn. In order to obtain the preliminary shape and size before drawing, the raw material may be rolled or extruded to shape. The drawing operations may be done in series. When this is done, the wire passes from one die to the other until it is necessary to anneal or until it is the correct size. Sometimes a back pull is applied to the bar or wire as it is passed through the die in order to stretch the wire as it is being drawn.

Deep Drawing

Deep drawing primarily uses the principle of stretching the material beyond the elastic limit without compression, but may have a combination of compressive and tensile stresses locked in the part. Due to the shape of the drawn part, these stresses do not cause springback. In deep-drawing shells, the problem is to obtain the final shape by the minimum number of draws and anneals. When too great a draw is made, the material may buckle, wrinkle, split, or tear. Buckling and wrinkling are sometimes the result of incorrect design of blank and too little hold-down pressure on the outer slide (which holds the blank). Then, too, incorrect lubrication and die design may result in faulty work.

The parameters for deep drawing using mechanical dies include the material, blank diameter, material thickness, and the flange overhang. The ratio of the flange overhang to the material thickness is called the buckling ratio.

When this ratio for sheet buckling is plotted against the ratio of overhang flange and blank radius, a theoretical curve may be obtained for a given material for predicting plastic and elastic buckling limits in the rim.[1] For circular sheet buckling (Fig. 14-47), we have the relationship:

$$S_b = K \frac{E}{(B/2t)^2}$$

where S_b = critical buckling stress

$$K = \frac{k}{12(1 - \mu^2)}$$

E = Young's modulus
μ = Poisson ratio
k = constant based on geometry, edge restraints, and type of loading

B, R, and t are defined in Fig. 14-47.
Values of k for various ratios of B and R are:

$\dfrac{B}{2R}$	k
1	2.4
2	3
3	3.2

The final piece may have portions which are thicker than the original material where sections were compressed. Other areas where the material was stretched will be thinner than the original material.

Drawing and hold-down forces   An empirical estimate of the drawing force F for a cupping operation can be computed from

$$F = \pi dt S \frac{B}{d} - C$$

where S = tensile strength of sheet metal, lb/in^2
d = average diameter of cup, in
B = blank diameter, in
C = constant to account for friction and other losses, with a value of 0.7 appropriate for average conditions

[1] Taken from W. W. Wood, R. E. Goforth, and R. A. Ford, "Theoretical Formability," *Product Engineering*, October, 1961.

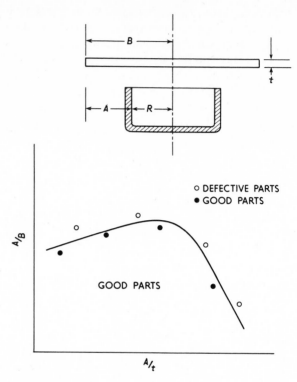

FIGURE 14-47
Curve for a given material whose upper boundary is characterized by plastic
buckling and boundary to right by elastic buckling.

The hold-down pressure on the rim of the blank should be from $\frac{1}{4}$ to $\frac{1}{3}$
of the drawing force. Too high hold-down forces will result in rupture at the
bottom radius or in the tube wall.

Drawing reduction and number of drawing operations required Before
the proper tooling can be designed for the drawing operations, the production
engineer must estimate the number of draws required. Because the metal flow in
one operation has a practical limit, on the first draw the area of the blank should
be limited to no more than $3\frac{1}{2}$ to 4 times the cross-sectional area of the punch.
Also it is usual practice to limit the reduction in the first draw to 40 percent, with
reductions of 20 to 25 percent or less on the next two draws.

The strain-hardening graph (Fig. 14-5) can be used to estimate the number
of reductions possible in a series of drawing operations. During a 40-25-25 series
of reductions, the mechanical properties of a cup will follow the dashed lines from

40 to point 5 to point 6 during the first draw. The second draw will start at the new yield point of about 60,000 lb/in^2 and proceed through points 7 and 8. Likewise the third draw will follow points 9 and 10 to end with a material with about 79,000 lb/in^2, but almost full hard. By moving vertically downward to point 11 on the σ_0 line, we find that the total reduction is about 65 percent.

Note that two successive draws at 40 percent reduction each, would have resulted in about the same total reduction, but in practice the second draw would probably have ruptured the bottom. An annealing operation between the two 40 percent reductions would have been a viable alternative. Thus we must

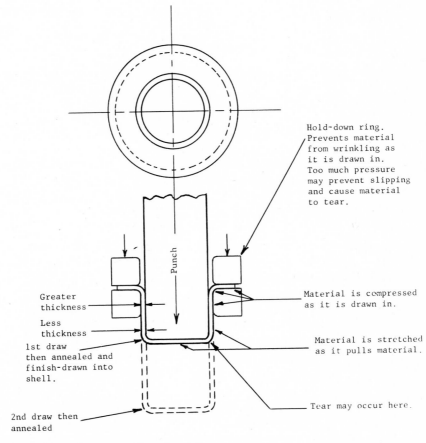

Hold-down ring. Prevents material from wrinkling as it is drawn in. Too much pressure may prevent slipping and cause material to tear.

Punch

Greater thickness

Less thickness

1st draw then annealed and finish-drawn into shell.

2nd draw then annealed

Material is compressed as it is drawn in.

Material is stretched as it pulls material.

Tear may occur here.

3rd draw may be to a smaller diameter

FIGURE 14-48
Typical cupping operation showing hold-down ring, vent hole, and drawing punch.

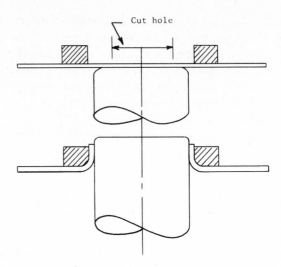

FIGURE 14-49
Holes can be flanged to give material for a threading operation or to stiffen the material near the hole.

make a judgment with regard to the tradeoff between adding an annealing operation and designing and building the tooling for a third drawing operation. This decision can only be resolved by estimating the probable activity of the part in the future and the cost and lead time of the tooling compared with the cost and delays involved when using the annealing operation.

Design details for drawing tools The magnitude of the corner radii for drawing punches and dies usually ranges from four to eight times the metal thickness. Radii should be as large as possible, but if they are too large, wrinkling will become a problem.

The normal clearance between the punch and die should be about $1.1t$, that is, the die opening would be the punch diameter plus $2.2t$. Ironing occurs when the cup wall thickness exceeds the clearance between the punch and die. This can happen by design if a specified wall thickness is required or it can occur at the top of the cup where the metal has been thickened by the drawing operation or wrinkling.

Air vents are needed to avoid collapse of the cup walls during the stripping operation. The drawing lubricant tends to seal the space between the punch and cup so that atmospheric pressure is applied to only the outside of the hollow cup during the stripping operation. Air vents should be about $\frac{3}{16}$ in diameter for punches up to 2 in diameter. Larger punches need several vents.

The production-design engineer can visualize the stresses encountered, and in many cases they can be calculated empirically with reasonable accuracy.

The forces depend on unknown friction between material and die, which can be estimated, plus the forces of compression (Fig. 14-48). Drawing operations performed on large parts are done in heavy duty presses. Successive draws may require annealing and cost is increased proportionally. However, small cups and cylinders that can be drawn cold in one operation cost approximately the same as most stamping operations. Often cups, or shells, need to be trimmed. This may be done in a lathe although it can also be done in a press. If the part has a flange, the edge can be accurately trimmed in a blanking die. Holes can be flanged by drawing (Fig. 14-49).

Determination of Blank Size

Graphical methods may be employed in order to compute the blank diameter of the cup (Fig. 14-50). Thus, for a cup of height H and diameter D, we need a blank diameter B which will equal $\sqrt{D^2 + 4DH}$ as found from Fig. 14-51. Thus

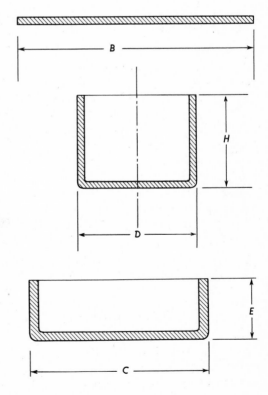

FIGURE 14-50
Standard designations for the dimensions referred to in Fig. 14-51.

	AREA OF BODY $A=$	DIA. OF BLANK $D=$		AREA OF BODY $'A'=$	DIA. OF BLANK $'D'=$
	$\dfrac{\pi d^2}{4} + \pi dh$	$\sqrt{d^2 + 4dh}$		$\dfrac{\pi}{4}(d_1^2+4h^2) + \pi f \dfrac{d_1+d_2}{2}$	$\sqrt{d_1^2+4h^2 + 2f(d_1+d_2)}$
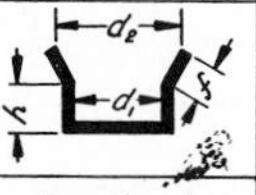	$\dfrac{\pi d_1^2}{4}+\pi d_1 h+\pi f \dfrac{d_1+d_2}{2}$	$\sqrt{d_1^2+4d_1h+2f(d_1+d_2)}$		$\dfrac{\pi}{4}(d^2+4h_2^2) + \pi d h_2$	$\sqrt{d^2+4(h_2^2+dh_2)}$
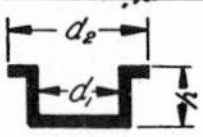	$\dfrac{\pi d_1^2}{4}+\pi d_1 h+\dfrac{\pi}{4}(d_2^2-d_1^2) = \sqrt{d_2^2+4d_1h}$	$\sqrt{d_1^2+4d_1h+d_2^2-d_1^2}=\sqrt{d_2^2+4d_1h}$		$\dfrac{\pi}{4}(d_1^2+4h_1^2)+\pi d_1 h_2 + \pi f \dfrac{d_1+d_2}{2}$	$\sqrt{d_1^2+4\left[h_1^2+d_1 h_2 +\tfrac{f}{2}(d_1+d_2)\right]}$
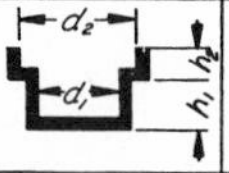	$\dfrac{\pi d_1^2}{4}+\pi d_1 h_1+\dfrac{\pi}{4}(d_2^2-d_1^2)+\pi d_2 h_2$	$\sqrt{d_2^2+4(d_1h_1+d_2h_2)}$		$\dfrac{\pi}{4}(d_1^2+4h_1^2)+\pi d_1 h_2+\dfrac{\pi}{4}(d_2^2-d_1^2)$	$\sqrt{d_2^2+4(h_1^2+d_1 h_2)}$
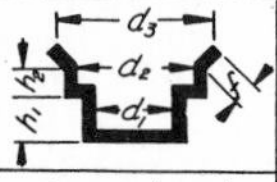	$\dfrac{\pi d_1^2}{4}+\pi d_1 h_1+\dfrac{\pi}{4}(d_2^2-d_1^2)+\pi d_2 h_2+\pi f \dfrac{d_2+d_3}{2}$	$\sqrt{d_3^2+4(d_1h_1+d_2h_2)+2f(d_2+d_3)}$		$\dfrac{\pi d_1^2}{4}+\pi f S \dfrac{d_1+d_2}{2}$	$\sqrt{d_1^2+2S(d_1+d_2)}$
	$\dfrac{\pi d_1^2}{4}+\pi d_1 h_1+\dfrac{\pi}{4}(d_2^2-d_1^2)+\pi d_2 h_2+\dfrac{\pi}{4}(d_3^2-d_2^2)$	$\sqrt{d_3^2+4(d_1h_1+d_2h_2)}$		$\dfrac{\pi d_1^2}{4}+\pi S \dfrac{d_1+d_2}{2}+\pi f \dfrac{d_2+d_3}{2}$	$\sqrt{d_1^2+2\left[S(d_1+d_2)+f(d_2+d_3)\right]}$
	$\dfrac{\pi d^2}{2}$	$\sqrt{2d^2}=1.414d$		$\dfrac{\pi d_1^2}{4}+\pi S \dfrac{d_1+d_2}{2}+\dfrac{\pi}{4}(d_3^2-d_2^2)$	$\sqrt{d_1^2+2S(d_1+d_2)+d_3^2-d_2^2}$

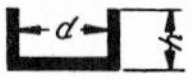

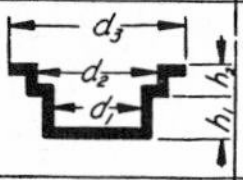

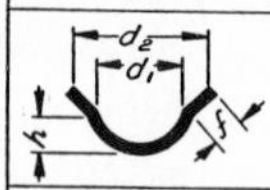

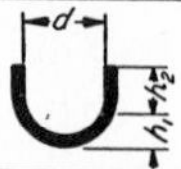

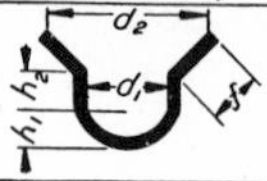

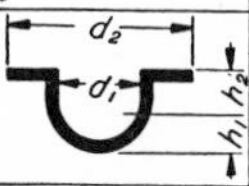

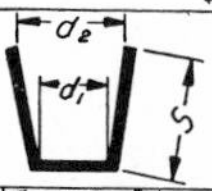

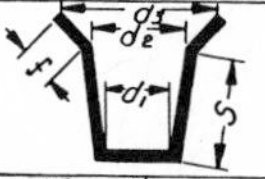

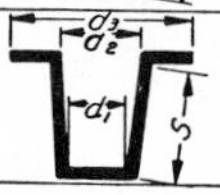

	AREA OF BODY $'A'=$	DIA. OF BLANK $'D'=$		AREA OF BODY $'A'=$	DIA. OF BLANK $'D'=$
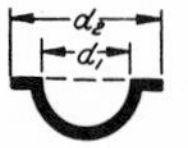	$\dfrac{\pi d_1^2}{2}+\dfrac{\pi}{4}(d_2^2-d_1^2)$	$\sqrt{d_1^2+d_2^2}$		$\dfrac{\pi d_1^2}{4}+\pi S \dfrac{d_1+d_2}{2}+\pi d_2 h$	$\sqrt{d_1^2+2[S(d_1+d_2)+2d_2 h]}$
	$\dfrac{\pi d_1^2}{2}+\pi f \dfrac{d_2+d_1}{2}$	$1.414\sqrt{d_1^2+f(d_2+d_1)}$		$\dfrac{\pi d_1^2}{4}+\dfrac{\pi^2 r}{2}(d_1+1.274r)$ OR, $\dfrac{\pi}{4}(d_2-2r)^2+\dfrac{\pi r}{2}(d_2-0.726r)$	$\sqrt{d_1^2+6.28rd_1+8r^2}$ OR $\sqrt{d_2^2+2.28rd_2-0.56r^2}$
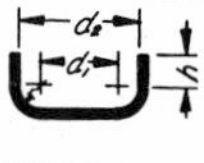	$\dfrac{\pi d^2}{2}+\pi d h$	$1.414\sqrt{d^2+2dh}$		$\dfrac{\pi d_1^2}{4}+\dfrac{\pi^2 r}{2}(d_1+1.274r)+\pi h d_2$ OR $\dfrac{\pi}{4}(d_2-2r)^2+\dfrac{\pi r}{2}(d_2-0.726r)+\pi h d_2$	$\sqrt{d_1^2+4(1.57r_1^2+2r^2+hd_2)}$ OR $\sqrt{d_2^2+4d_2(h+0.57r)-0.56r^2}$
	$\dfrac{\pi d_1^2}{2}+\pi d_1 h+\dfrac{\pi}{4}(d_2^2-d_1^2)$	$\sqrt{d_1^2+d_2^2+4d_1h}$		$\dfrac{\pi d_1^2}{4}+\dfrac{\pi^2 r}{2}(d_1+1.274r)+\pi f \dfrac{d_1+d_3}{2}$ OR $\dfrac{\pi}{4}(d_2-2r)^2+\dfrac{\pi r}{2}(d_2-0.726r)+\pi f \dfrac{d_2+d_3}{2}$	$\sqrt{d_1^2+6.28rd_1+8r^2+2f(d_1+d_3)}$ OR $\sqrt{d_2^2+2.28rd_2+2f(d_2+d_3)-0.56r^2}$
	$\dfrac{\pi d_1^2}{2}+\pi d_1 h+\pi f \dfrac{d_1+d_2}{2}$	$1.414\sqrt{d_1^2+2d_1h+f(d_1+d_2)}$		$\dfrac{\pi d_1^2}{4}+\dfrac{\pi^2 r}{2}(d_1+1.274r)+\dfrac{\pi}{4}(d_3^2-d_2^2)$ OR $\dfrac{\pi}{4}(d_2-2r)^2+\dfrac{\pi r}{2}(d_2-0.726r)+\dfrac{\pi}{4}(d_3^2-d_2^2)$	$\sqrt{d_1^2+6.28rd_1+8r^2+d_3^2-d_2^2}$ OR $\sqrt{d_3^2+2.28rd_2-0.56r^2}$
	$\dfrac{\pi}{4}(d^2+4h^2)$	$\sqrt{d^2+4h^2}$		$\dfrac{\pi d_1^2}{4}+\dfrac{\pi^2 r}{2}(d_1+1.274r)+\pi d_2 h+\dfrac{\pi d_3^2-d_2^2}{4}$ OR $\dfrac{\pi}{4}(d_2-2r)^2+\dfrac{\pi r}{2}(d_2-0.726r)+\pi d_2 h+\dfrac{\pi}{4}(d_3^2-d_2^2)$	$\sqrt{d_1^2+6.28rd_1+8r^2+4d_2h+d_3^2-d_2^2}$ OR $\sqrt{d_3^2+4d_2(0.57r+h)-0.56r^2}$
	$\dfrac{\pi}{4}(d_1^2+4h_1^2)+\dfrac{\pi}{4}(d_2^2-d_1^2)$	$\sqrt{d_2^2+4h_1^2}$		$\dfrac{\pi d_1^2}{4}+\dfrac{\pi^2 r}{2}(d_1+1.274r)+\pi d_2 h+\pi f \dfrac{d_3+d_4}{2}$ OR $\dfrac{\pi}{4}(d_2-2r)^2+\dfrac{\pi r}{2}(d_2-0.726r)+\pi d_2 h+\pi f \dfrac{d_3+d_4}{2}$	$\sqrt{d_1^2+6.28rd_1+8r^2+4d_2h+2f(d_3+d_4)}$ OR $\sqrt{d_3^2+4d_2(0.57r+h)+2f(d_3+d_4)}$

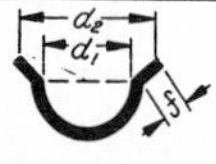

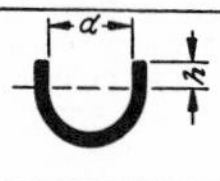

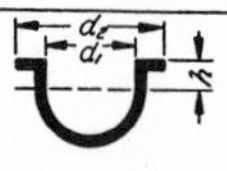

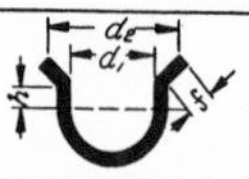

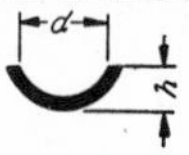

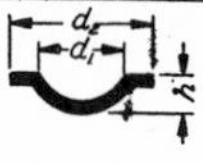

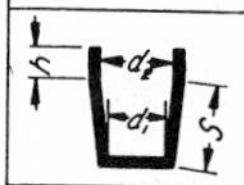

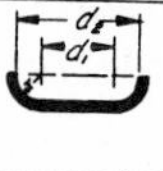

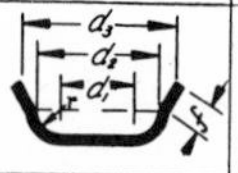

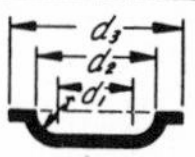

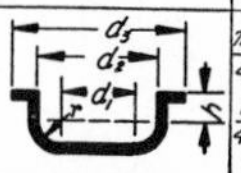

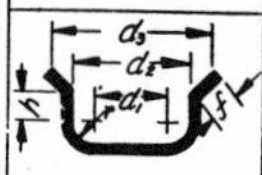

FIGURE 14-51

Formulas for areas and blank diameters of drawn shells.

the dimensions of the first draw are as follows: $C = 0.60B$ and $E = 0.267B$. The estimated percentage reduction of diameter C in the first redraw approximates the values listed in Table 14-5.

The geometrical relationships illustrated in Fig. 14-51 will assist the production-design engineer in estimating the material usage required for producing different designs of drawn shells.

Draw and trimming dies are expensive tools and the operations are costly. By the proper choice of material and intermediate drawn shapes, the number of drawing operations frequently can be reduced. Likewise, by careful designing, it is often possible to eliminate the necessity of blanking operations and final trimming.

Considerable interest and development activity is now evident in the hydraulic methods of drawing material. The material is formed by applying hydraulic pressure directly on the material either over a punch or within a die without its companion die or punch. The liquid is sealed in the cavity by clamping the material around the edge tightly enough to prevent the escape of the liquid. The fluid exerts a steady, uniform pressure on every part of the blank. This makes possible the formation of involved contours (as in embossing) and unusual draws (such as cone-shaped or tapered sections). For the ordinary mechanically drawn sections, the mechanical press and tools are less expensive.

The Hydroform press uses a thin flexible diaphragm sealed against high pressure and backed by hydraulic fluid. The material blank is held in position and the punch is moved up by hydraulic pressure into the die cavity, forcing the diaphragm pad and the blank to take the shape of the punch.

When compared with conventional tooling, Hydroforming tooling is relatively simple since only a punch and nest ring is required. The punch should have a good finish since the part being drawn will reproduce the punch. For short runs Kirksite, brass, and wood punches have proved successful.

Table 14-5   TYPICAL REDUCTION PERCENTAGES
FOR THE FIRST REDRAW OPERATION

Thickness of stock	Steel	Aluminum	Brass
$\frac{1}{16}$	20	25	30
$\frac{1}{8}$	15	20	24
$\frac{3}{16}$	12	15	18
$\frac{1}{4}$	10	12	15
$\frac{5}{16}$	8	10	12

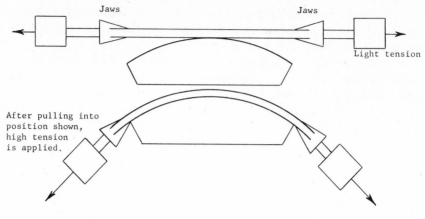

FIGURE 14-52
Schematic representation of a stretch bending operation.

Stretch Forming

The simplest use of the principle of stretching material beyond the elastic limit without compression is in straightening wire or bars by stretching. The process of stretcher-leveling sheets is slightly more involved, but uses the same concept. The sheet is clamped firmly on each end and stretched well beyond the elastic limit. This results in a flat or level sheet that is used for furniture and cabinet manufacture. In forming contours, the same principle is applied, with the addition of a form on which the sheet or part is stretched. The material takes the shape of the form without any springback. Several machines use this process to greater or lesser degrees. Some machines cannot form reverse bends or complete circles; others can to a limited extent. In all cases allowance must be made for extra material to be clamped by jaws when it is stretched. This portion cannot have a contour and may be removed or used as a part of the piece. For example, electric-stairway track angles that were made on forming rolls (Fig. 14-52) had to be straightened on a surface plate and had one tangential end. With contour forming, the exact radius was obtained without straightening, and it was possible to make both ends tangential. The time saved by stretch bending was from $\frac{1}{2}$ to $1\frac{1}{2}$ hours per piece, and better quality was obtained.

A stretch-bending problem It is desirable to purchase a stretch bender for forming irregular curves in odd-shaped members. A young engineer is asked to give an engineering explanation of why the parts do not spring back, but take the exact form of the die. An analysis of the stresses incurred during the process and the residual stresses is requested. This information will determine the extent of the application of the equipment and process to materials and parts.

The process consists of shaping a part over a form and pulling at the ends to stretch the metal. No springback occurs. The finished piece has the radius of the die and the part may be a strap or an angle with its flange in or out, or any odd shape.

Discussion of Fig. 14-52, Case 1   Consider a beam supported at each end and loaded in the center. If the load is released, springback will occur (Fig. 14-53a).

If tension is applied (Fig. 14-53b), the fiber stress will be affected as shown. If the tension applied is great enough, there are no compression stresses. If the load is released, springback will occur.

If further tension is applied (Fig. 14-53c), the elastic limit will be exceeded on the top side; and, when the load is released, some permanent set will occur, but there will be springback due to unequal forces on each side of the neutral axis.

If further tension is applied, as in stretch bending, the maximum strength of the fibers will be exceeded and the top fibers will not carry as much as those

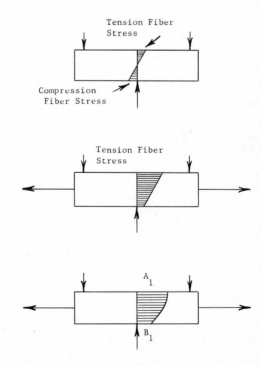

FIGURE 14-53
Schematic representation of the internal stress distribution within a typical work period. *Top*, springback can occur; *middle*, stress when tension is applied; *bottom*, sufficient stress is applied to cause permanent set and to eliminate springback.

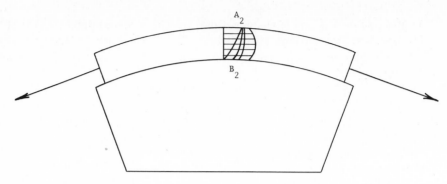

FIGURE 14-54
Typical stretch forming workpiece and die.

in the center. When the stress is fairly uniform across the section, the maximum stress will be in the center. When the load is released, there will be no springback, as the forces are balanced about the neutral axis (Fig. 14-54).

The foregoing explanation comes from analyzing what occurs within the part and is based on the stress-strain diagram information obtained in the testing laboratory.

Discussion of Fig. 14-52, Case 2   The stress-strain diagram (Fig. 14-55) can also be used to explain the reason for no springback after stretch bending.

1   As long as the stress in the material remains on line OD, the stresses will return to zero and the part returns to its original shape when the load is released.

2   When the stress in the outer or top fibers A_1 in Fig. 14-55 exceeds the elastic limit, the stress is indicated as progressing beyond line OD until it reaches point A_1. Stresses in the bottom fiber are shown reaching point B_1 on line OD. When the load is released, A_1 will travel down the dotted line to zero stress and indicate a strain S_{A_1} which is greater than strain S_{B_1} which is zero because B_1 was on the elastic portion of the stress-strain curve.

3   When the stress is increased, the stress in fiber A_1 progresses along the line over the maximum stress to point A_2. At the same time, fiber stresses in B_1 progress along to point B_2, both at about the same value of stress. When the load is released, the strains S_{A_2} equal strains S_{B_2}, and therefore there is no springback. It can also be seen that the material is cold-drawn and will have the characteristics of cold-drawn material.

The shape of the stress-strain curve of the material indicates whether it is suitable for stretch bending. Hot-rolled angles bend to shape easily. Thus unsatisfactory results can be avoided by giving thought to the shape of the stress-strain curves of the material in relation to the fundamental principles of stretch bending.

An analysis of the residual stresses remaining in the stretch-bent bar or sheet can be made to determine other limitations of the process.

Spinning

Spinning usually is performed on a lathe. The mold is fastened to the lathe head and the material fastened to the mold. As the mold and material are spun, a tool (resting on a steady rest) is forced against the material until the material contacts the mold. There is considerable friction between the tool and the material. The skilled operator applies his weight to the overhanging handle of the tool and presses with the proper force so that the tool will not dig into the material. The tools are hardened and round-nosed so they will slide easily and will not wear. Lubricants are used to reduce friction. The material is stretched as it is formed until it is snug against the mold. Brass, aluminum, and copper are the usual materials used for spinnings. Heated steel is spun by mechanical means in large diameters (up to 240 in) and in thicknesses as great as $\frac{1}{4}$ in on large vertical boring mills.

A variety of designs can be made. The material can be trimmed and the edge beaded. Parts with narrow openings, such as pitchers, vases, and lamp bases, are made on collapsible molds or by mechanical spinning.

Mechanical spinning operations in which rollers are used to form the material are utilized extensively in making tin cans, galvanized tanks, and large steel tanks. Spinnings are economical for experimental models and preliminary production runs before the final design is established. Later more expensive

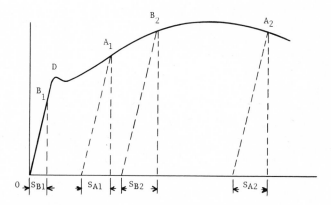

FIGURE 14-55
Typical stress-strain diagram for hot-rolled steel.

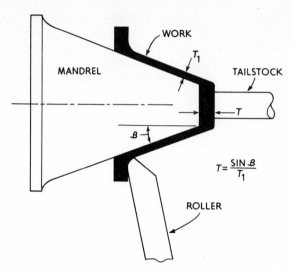

FIGURE 14-56
Roll flowing a conical part.

tools can be made for more economical operation costs. Spinnings are also frequently used in medium-volume production parts, because they are less expensive than drawn parts. Typical spinnings are lamp reflectors, mixing bowls, and cooking utensils.

Roll flowing is similar to spinning because the operation is usually performed on a lathe (Fig. 14-56). It also is known by the trade names Floturning, Shear Spinning, and Hydrospinning. This plastic working of metal process produces parts to a predetermined geometrical shape by displacing metal in advance of a roller that is fed along a mandrel machined to the desired inside diameter of the part. Usually, hydraulic pressure is used to force the roller against the part and for clamping the blank to the tailstock. In this process, the metal is displaced in the direction of the feed of the roller along the length of the mandrel.

From the production designer's viewpoint, the principal problems associated with this process are blank development and proper feed pressure. Blank development requires determination of wall thickness and diameter of the original blank so as to provide predetermined final size for the part allowing a specified amount of material for trimming.

Roller pressures are approximately 200,000 lb/in². This cold working will, of course, result in considerable work hardening. Tensile strengths are doubled frequently.

This process is being used where production requirements usually are not large, such as duct work or dairy equipment (cream separators, etc.).

Limitations of Press Techniques

The engineer, in order to reduce costs, may become overenthusiastic about using stampings and press techniques to make his parts. The successful production-design engineer must always consider the limitations of every process. For example, a punched hole does not make a good bearing. An inside diameter, to be used as a bearing, should be drilled, reamed, and, in some cases, bushed. It is difficult to have holes in alignment when a blank is formed after piercing the holes. Holes required to be in alignment should be punched, drilled, or line-reamed after forming. The punched hole is not always suitable for tapping. Sheet metal tolerances are usually $\pm\frac{1}{32}$ in, which is large compared to machining.

Die Designing

Die designing is a field in itself. The types of dies and presses run the full range of man's ingenuity and abilities. Problems of stripping and use of springs, rubber, air cushions, movable pads, automatic stops, feeders, and ejectors should be understood, but space is not available to cover these items.

Some practical hints for the die designer, as developed by the Royal Typewriter Company, are:

1 Always check die drawing with part prints to make certain that the die will produce the part correctly with reference to burr side, relation of grain direction to forms, shear forms, embossments, blanking pressure required, and strength of fragile projections.

2 Dimension the detail drawing with the information in the form required by the diemaker so that he has a minimum amount of translating to do.

3 As soon as the design of a die is decided upon, the material for the die should be ordered so that there are no unnecessary delays. Tryout material for the part should be ordered at the same time.

4 Always conform to customer or company specifications for type of die construction, kind of steel, die sets, and press data.

5 Whenever possible specify standard die parts from catalogs or technical literature of reputable manufacturers.

6 All die parts should be designed to be easily removable with minimum disassembly.

7 The safety and the relation of efficiency to fatigue of the press operator should be considered in the planning of every die design.

8 In designing a die it is essential to know the type of press in which it is to be used. In selecting the proper size and type of press for a specific job, the designer should consider the following factors:

a The size and type of die required

b The amount of stroke necessary

c The press pressure required for doing the work

d The distance above the bottom of the stroke where the pressure first occurs

e Any additional pressure required due to attachments, such as the blank holder, ironing wrinkles, or stretching the materials in drawn work

f The method of feeding, the direction of the feed, and the size of the sheet, blank, or workpiece

9 Remember that the die opening sizes the outside contour of the blank and that the perforators size the openings in the blank.

10 Split the punch pad if there are a great number of punches; it simplifies the lining-up process.

11 Pilots and punches must be correct in the lineup with back gage and blanking center of die.

12 Whenever possible removable pilots should be used to facilitate grinding of the punches.

13 Punches should always be as short as practical. Long production runs require punches long enough to grind from $\frac{1}{2}$ to 1 in.

14 Pilots must be long enough to locate the strip accurately before punches start cutting or forming.

15 On complicated forms always use a template to check the parts and the layout of punch and die.

16 The die thickness should be determined by the requirements: production run, degree of precision, and budget of both time and cost of building.

17 All delicate projections or weak parts in a die should be designed as inserts if possible.

18 If stock guides are indicated, their length should be made twice the width of the stock before the first operation.

19 The spring stripper should be of sufficient length and width to provide for additional springs if necessary. It should be thick enough to avoid damage against the stripper bolts, which sometimes loosen.

20 When punches are guided in a spring stripper, the stripper must be supported on guide posts for accurate alignment. The stripper should be bushed, with dowels in the die placed to engage the stripper before the punches enter the die. When punches are not guided in the stripper, ample clearance must be allowed around small punches to assure that they will not be thrown out of line. The stripper should be fitted close to one or two stronger, heavier punches (0.001- to 0.005-in clearance) and allow 0.010- to 0.015-in clearance on the smaller punches.

21 The solid or bridge stripper should always be well doweled to prevent it from shifting. This is of special importance when the stripper does heavy guiding.

22   The punch holder must be thick enough to allow proper travel on draw and form pads, strippers, spring pins, spring pilots, stripper bolts, and moving punches.

23   The shoulders of draw pads or shedders should never strike against the retaining sections. Make knockout pins the correct length.

24   Never form too close to the edge of any opening in the blank without proper support along the edge.

25   Use a standard commercial bolster plate whenever possible. These are usually torch-cut to any size or shape. The design of the bolster plate must be the most effective compromise possible between a strong, rigid support for the die and a means of locking the die assembly in the press.

26   Use a stationary stripper whenever possible. The stationary stripper screwed and doweled in the die is more positive and accurate for lining up punches going through bushing guides. It is also more economical to make than a spring stripper.

27   Spring strippers are used where visibility is an important factor and where flatness of parts is required.

28   Spring strippers are also needed when lugs, draws, or embossments necessitate the placing of the stripper at some distance from the die to allow space for the strip to feed through.

29   If possible, use the insert and split-die-section construction for small and difficult die openings.

30   Always determine if a more satisfactory design can be obtained by the use of an inverted die.

Practical hints for the tool and die maker are:

1   When dies for rectangular draws give excessive trouble in the corners, it is well to check the draw radii and fit between punch and die.

2   Tap holes in unhardened tool steel parts with tap 0.003 in oversize to allow for shrinkage and distortion.

3   Punches should be polished in their longitudinal direction (since stripping of the work takes place in this direction). In the case of slender perforating punches working in heavy stocks, annular marks from turning or polishing tend to weaken such punches.

4   The punch plate must be of adequate thickness to support its punch or punches. This is of particular importance when punches are not guided in the stripper.

5   Taper-ream all slug holes in the die shoe.

6   Normalize dies or sections after rough machining, particularly if for intricate shapes or when extreme accuracy is required; then finish-machine and harden.

7   A punch or shedder with a base only on one side, or with a split base, will usually warp in hardening; therefore, allow grinding stock to compensate for the warpage.

It must be remembered that the tool designer has only one chance to make his tool. He can modify and improve within the limits of his tool budget, but he cannot come back to his manager and say that he has a better way of doing the operation and request an appropriation for a new tool. He must be right the first time. This is in contrast to the designer who can make his models and test them and, if necessary, radically change them. The experienced tool designer usually obtains close to the ultimate results with the first set of tools. Of course, when major savings can be shown by a new method, money should be appropriated to make the saving.

High-Speed Operations

With the advent of cemented carbide and tough cast alloy tools, the speed of punching has been increased to such an extent that punching of material is not a shearing or tearing action, but a breaking action. Thick materials are being punched with cleaner holes and fewer burrs. Presses are now available that operate at 2,300 strokes per minute. Material is fed 5,000 to 10,000 in/min. Hexagonal nut blanks have been pierced, countersunk, blanked, and crowned at a rate of 1,150 nuts per minute. Ball-bearing and roller-bearing guide pins have been developed to withstand the high speeds of the press. There are presses that may operate rams in vertical and two horizontal side positions, enabling the forming of complex parts in progressive operations. The field of wire and strip forming is a specialty in itself.

Presses

Presses for sheet-metal work range from the massive machines used to make automobile body parts to the small bench types. They may be hydraulically operated or motor- or air-operated. A number of presses are compared in Table 14-6.

In addition to tonnage, the production-design engineer should consider the width through the back of the press, shut height, stroke, bolster size, and flywheel speed before selecting a given press to perform an operation. The shut height is important from the standpoint of placing work in the dies. The stroke is important when drawing operations are performed. Obviously deep-draw presses are those that are capable of having greater strokes than a press used for blanking and piercing. The flywheel speed determines the rate of production; the width through the back is important if the completed part or scrap is to be ejected through the back. The bolster size places a limitation on the die size and consequently the size of the work being processed.

Presses can be equipped with such auxiliary equipment as automatic feeding mechanisms, scrap outfits, and stock straighteners in one integrated automatic setup. Figure 14-18 illustrates an inclinable punch press equipped with a roll feed, scrap cutter, and a straightening machine for pulling the coiled stock from a brake reel.

Presses are usually classified by power-transmitting mechanisms, frame design, and purpose. Thus, under power transmitting mechanisms there are:

1 Crank, eccentric
2 Cam, cam and crank, cam and eccentric
3 Knuckle joint
4 Toggle
5 Rack and pinion
6 Hydraulic
7 Pneumatic

Frame design further classifies the style of press as:

1 C frame
2 Gap frame
3 Arbor
4 Horning

The purpose of the press will help identify special-purpose presses as:

1 Flat-edge trimming
2 Stretching
3 Quenching

Drawing Presses

The terms single, double, and triple action are often referred to in drawing operations. These terms refer to the number of separate movements in the tools. Thus, in a single-action press, work is done in one movement without any auxiliary movement to hold the blank. In drawing operations this can be done, without wrinkling, when the material is relatively thick in comparison to its blank diameter.

In a double-action setup, a separate movement of the press gives a blank-holding action on the periphery of the blank, while it is being drawn, so as to avoid the formation of wrinkles.

Where three successive and separate pressures are required, such as holding the blank, drawing, and redrawing, a triple-action press may be employed. The triple-action press is used for such work as automobile fenders, bathtubs, and refrigerators, In the triple-action setup, the metal must be able to withstand the second draw without annealing.

Table 14-6 STAMPING AND DRAWING PRESSES

Machine	Size	Operations	Features	Uses
Foot press	Small sizes only	Small work—single small holes, light stampings, embossing	Hand fed; may have adjustable bed or horn to accommodate large work	Jewelry, buttons, silverware, radio parts, etc.
Bench press	1,000 pounds to 12 tons	Embossing, stamping, etc., upon light material	May have roll feed, ratchet dials, magazines, etc.	Watch parts, novelties, jewelry, etc.
Inclinable press—open-back (OBI)	4 to 90 tons	Blanking, bending, stamping, forming, assembling	Inclined from vertical to 45° backward, may have drawing attachments	For light sheet metal
Open-back gap press (OBG)	1 to 225 tons	Punching, shearing, cutting-out, trimming, forming, etc.	Ram driven by cam or eccentric	Automobile parts, etc., for large or irregular work
End-wheel gap press	Small to 50 tons	Blanking, forming, notching, piercing, cutting	Flywheel at rear and crankshaft at right angles to bed	For work with long, narrow strips of metal
Deep-gap punch press	Medium sizes	Punching	Deep clearance in back frame	For wide sheets
Horning press	10 to 100 tons	Forming, stamping, blanking, wiring, punching, riveting	Horn bolted to frame; may have swinging table also	For hollow cylindrical work, as steel drums, etc.
Double-crank over-hanging press	Small	Blanking, cutting, piercing	Usually automatic or semi-automatic feed-overhanging frame gives large die space—flywheel—or gear-driven	For large, light sheet metal
Notching press	Small to medium	Cutting slots in edges of usually circular work	Usually short stroke machine, with work driven by rocker arm—450–650 strokes per min	Notching motor laminations

Type	Size	Operations	Construction	Application
Single- and double-action presses	Heavy	Shaping, blanking, forming, etc.	Frame consisting of a base, a crown, and two uprights tied together with steel tie rods	Heavy stock
Arch press	Light	Cutting, trimming, shaping, blanking	Offers large bed area	For light, large-area sheet metal
Double-crank straight-side press	To 2,000 tons	Punching, cutting, bending, blanking, shaping	Slide often counterbalanced by air cylinder—flywheel or geared	Wide variety of uses in many sizes
Four-point suspension press	100 to 1,500 tons	For large-scale deep drawing, etc.	Four corners of slide suspended, giving even pressure on work	Automobile body tops, etc.
Knuckle-joint press	25 to 250 tons	Coining, upsetting, swaging, embossing, extrusion	Knuckle-joint operating slide makes short stroke necessary	Coining money, light to medium thickness metals
Straight-side high-speed press	10 to 400 tons	Blanking, stamping, etc.	Heavy construction to eliminate vibration—high-speed roll feed, and scrap cutter, variable speed motor—about 400 rpm	For high production with comparatively light metal
Dieing machine	Small	Progressive die operation, etc.	Ram operated with a pulling stroke rather than a thrust, by vertical rods passing down through bed—about 350 strokes per min	Electrical appliance parts, small automobile parts, etc.
Oscillating-die press	Small	Blanking, cutting, etc.	Die plate moves horizontally back and forth; strip fed through continuously. About 1,000 strokes per min	Small parts of light-gage material

Table 14-6 STAMPING AND DRAWING PRESSES (*Continued*)

Machine	Size	Operations	Features	Uses
Multislide machine	Small	Combination of operations, as blanking, forming, bending, etc.	As many as 8 operations may be performed in succession by the different slides. Very accurate feeds. To 300 pieces per min	Operations upon light-gage metal
Double-action press	Small, medium, or large	Blanking, drawing, etc.	Stock is successively blanked and worked by an outer and an inner ram	Applications requiring two related operations
Pillar press	70 to 250 tons	Blanking, shaping, trimming, etc.	Subpress dies may be attached to the face of the ram for close-tolerance work	Light- to medium-gage work
Toggle press	Small	Usually double action, in which the outer slide holds the blank, and the punch, or drawing ram, forms the cup		Kitchenware
Hydraulic press	To 1,000 tons	High-speed pumping units can give this type of press operating speeds comparable to other large presses		Largest sizes of work
Horizontal draw press	Small to medium	Shell may be pushed through die for redrawing, or knockouts and stripper plates provided if shouldered		Cylindrical shells
Multiple-plunger eyelet machine	Small	Successive draws made by a row of plungers, the work transferred from station to station by finger conveyors		Light-gage shells
Rack-and-pinion deep-drawing press	10 to 30 tons	For long uniform redrawing operations with accurate length of stroke		Cylindrical shells

SOURCE: Kenneth Rose, Engineering editor, *Materials & Methods*, November 1943.

Turret-Type Punch Press

In the turret-type press, an inventory of different size and shape punches is maintained. The turret may contain 50 or more different punches. A circular plate above the material opening holds the punches and a similar plate below the material gap holds the dies. The plates may be rotated to select the correct punch and die.

Material is mounted on a movable table which can accurately locate the sheet in the x and y planes in preparation for punching.

Through the use of N/C operations, sheet metal may be processed automatically on this equipment.

SUMMARY

The range of application of cold metal-working processes has widened considerably in the last few years in view of added technology. Today, many parts that formerly were produced by other methods have become press-department products. The power press has always been a production machine and where quantities are high and price is an important criterion, the production-design engineer should always consider the feasibility of making the design using the cold-working operations explained in this chapter.

Today we find press work being used to form large parts, such as quarter panels for automobile bodies, as well as small brass eyelets. In the production of small items such as eyelets, as many as a quarter of a million pieces can be produced from one machine in a day. In order to be able to take full advantage of the possibilities of cold metal-working operations, the product-design function should be coordinated with the process-design function, since many of the answers concerning the possibilities of using to advantage a particular process are based on experience and empirical calculations.

QUESTIONS

1　Explain what occurs when a piece of hot-rolled steel passes through the yield point.
2　What is the "yield-point elongation"? Why is the yield-point elongation an important factor in the plastic working of metals?
3　How do the work-hardening characteristics of a metal affect its deep-drawing capabilities?
4　How much tonnage would be required to pierce and blank a 10-gage mild steel washer of 5 in OD and 3 in ID?
5　What are the desirable characteristics of cold-drawn bar stock?
6　What type lubricant would you recommend in the cold extrusion of steel?
7　Make a sketch illustrating the impact-extrusion process.

8 What is the Hooker process?

9 Describe the knuckle-joint press.

10 How does coining differ from embossing?

11 Describe the swaging process.

12 What advantages does thread rolling have over thread cutting?

13 Define pitch diameter.

14 Upon what two principles are forming processes based?

15 What would be the developed length of a bend in $\frac{3}{32}$-in aluminum alloy steel with a $\frac{3}{4}$-in inside radius?

16 Make a simple sketch showing a single bend being made in a punch-press die. Show the punch, pad, stop or gage, inside radius, die, and angle relieved to allow for overbending.

17 Illustrate by sketch how lock-type joints may be made on a pan brake.

18 Give several applications of rubber forming.

19 What is a box-forming brake?

20 How can a tube be bent without its walls buckling?

21 Indicate by a sketch a roll-forming operation. Show the position of the rolls.

22 Explain why no springback occurs during the stretch-forming process.

23 What would be the important criteria to determine if it would be more advantageous to spin or to deep-draw a part?

24 What are the principal causes of buckling or wrinkling in the deep-drawing process?

25 What is nibbling?

26 What determines the size of hole in a piercing die?

27 Why is it advisable to taper the working end of the punch? How much taper is practical?

28 How much clearance should there be between the punch and the die?

29 What is the "grain" in sheet stock? Why should material be bent across the grain?

30 What is the continental method of tool construction?

31 Outline some practical hints for the die designer.

32 If a material continually tears when being deep-drawn, outline the possible courses of action.

33 What lubricant would you recommend for deep drawing aluminum? Why?

PROBLEMS

1 A power shear cuts a length of 5 ft from a sheet of $\frac{1}{8}$-in steel, SAE 1020 (annealed). The shear blade has a slope of $\frac{1}{8}$ in/ft. Calculate the shearing load. Assume penetration is 40 percent.

2 Find the maximum press capacity in tons for cutting a blank of 10 in diameter from annealed aluminum of $\frac{1}{16}$ in in thickness. The shear strength of the metal is 8,000 lb/in.

3 How much energy will be required in the operation given in Question 2? Assume 60 percent penetration.

4 Given material of the washer is cold-worked brass; 20 percent penetration, ID = 6 in, OD = 9 in, six holes each ¾ in diameter, thickness of sheet 16 gage. Determine (see Fig. P14-4):

(*a*) Characteristics of the tooling detail as to the shears, etc. Disregard the detail of the accessories such as springs, etc.

(*b*) Calculate all pressures and work per stroke.

(*c*) Specify the size of equipment—using a safety factor of 2.

(*d*) Specify the stroke for one washer at a time.

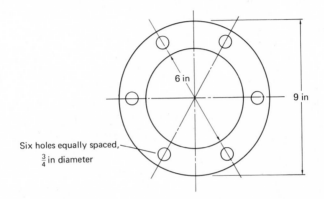

5 The sheet metal department of a plant has been requested to make the part shown in Fig. P14-5. The department has a punch with a ¾-in-thick blade and an included angle of 89° as shown, however the lower die must be designed to meet the require-

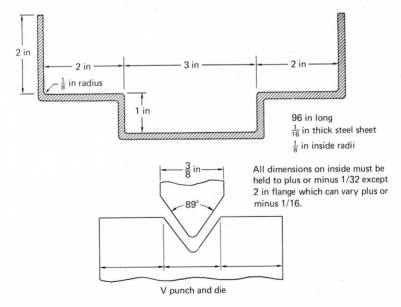

ments of this particular part. The chief industrial engineer has asked you to answer the following questions:

(a) Show the dimensions of the lower die.
(b) What is the developed width of the sheet?
(c) Determine the minimum number of forming operations required on a press brake. Sketch each one in sequence.
(d) Show the gaging points and position of the tools at each operation. Give appropriate dimensions for proper location of stops and the workpiece.
(e) Show one additional method to reduce costs by changing tools or using another process.
(f) What standard width of sheet should be used? (Nearest $\frac{1}{2}$ in)

SELECTED REFERENCES

AMERICAN SOCIETY FOR METALS: "Metals Handbook," 8th ed., "Forming," Metals Park, Ohio, 1969.

AVITZUR, B.: "Metal Forming: Processes and Analysis," McGraw-Hill, New York, 1968.

CRANE, E. V.: "Plastic Working of Metals and Nonmetallic Materials in Presses," 3d ed., Wiley, New York, 1944.

EARY, D. F., and E. A. REED: "Techniques of Pressworking Sheet Metal," Prentice-Hall, Englewood Cliffs, N.J., 1958.

E. W. BLISS COMPANY: "Computations for Metal Working in Presses," Canton, Ohio.

SACHS, GEORGE: "Principles and Methods of Sheet Metal Fabricating," Reinhold, New York, 1951.

SOCIETY OF MANUFACTURING ENGINEERS: "Die Design Handbook," McGraw-Hill, New York, 1955.

————: "Fundamentals of Tool Design," chaps. 3, 4, Prentice-Hall, Englewood Cliffs, N.J., 1962.

————: "Tool Engineers Handbook," 2d ed., McGraw-Hill, New York, 1959.

DECORATIVE AND PROTECTIVE COATINGS

In most merchandised products, the cost of finish is an appreciable percentage, amounting in some products to as high as 35 percent of total factory cost. Finishes include the coverage type, such as paint, lacquer, and varnish; the chemical-changing-of-the-surface type such as anodizing, sheradizing, or bonderizing; the plating type, which may be polished and buffed after the plating process; and the temporary or permanent type that covers materials in order to protect them against corrosion during storage or shipment.

The key to successful finishes is a clean surface to assure good adhesion. The control of temperature, humidity, and constituents of the finish is also important. There is considerable difference in cost of finishing materials and processes for applying the various finish types. There are competitive means of cleaning, such as water-base solvents versus chemical solvents; competitive means of drying, such as infrared heat versus steam, oil, or gas heat; and competitive means of application, such as dip, spray, flow, and centrifuge. Cementation, metal cladding, cathode sputtering, and vapor coating in a vacuum are specialized processes for applying finishes.

Finishes are used for the purposes of decoration, surface protection, corrosion resistance, and the providing of a hard surface. The colors should not fade.

The covering should be uniform and free from runs, checks, or peelings. It should be pliable so that it will expand or contract under weather or operating conditions. The surface should be hard so that it can be cleaned and will not permit imbedding of dust or grit or staining by oil. The time taken for drying often determines the choice of finish. The development of quick-drying finishes has eliminated the high cost of storing for several days automobile parts and bodies finished with slow-drying paints and varnishes.

CLEANING

Cleaning operations are performed both preparatory to finishing operations and after finishing operations. These rather common operations are worthy of careful analysis in that they frequently consume considerable time and cost. They are used to remove dirt, oil, oxides, scale, and other harmful ingredients that affect the operation of the equipment and the life of protective finishes. Cleaning is not a simple matter. It is often expensive and it requires considerable ingenuity and knowledge of cleaning processes to remove unwanted surface contamination effectively. Chemical processes, such as pickling, solvent, alkaline, electrolytic, and emulsion cleaning, are used to ensure clean parts and surfaces before the finish is applied. These processes have been highly developed through the use of many combinations of chemicals and types of equipment.

Acid Pickling

The most common method of removing unwanted oxides is by acid pickling. Either sulfuric or hydrochloric acid is sprayed on the part, or the part is dipped into a tank, agitated, and then washed and rinsed thoroughly. Acid solutions are difficult to maintain because of carryover into rinsing tanks or dilution from previous cleaning operations. Splash and vapors from the acid solution corrode equipment and tanks. The maintenance cost is high and working conditions are often disagreeable. Acid cleaning of steel parts creates hydrogen, which is absorbed by the steel and causes "hydrogen embrittlement." This is often the cause of excessive spring breakage. Other means of cleaning steel springs should be used. The hydrogen in the steel can be reduced by heating the parts after pickling.

Solvent Cleaning

Solvent cleaning is designed to remove oil and grease. Some processes involve soaking the part in a liquid solvent, and others use a spray, but most processes use a vapor. All such processes use a chlorinated solvent. The vapor type consists of a boiling liquid giving off vapors into which the cold pieces are dipped. The vapor condenses on the part and penetrates and dilutes the oil or grease, which runs off the part. The part is then removed, with a very light coat of oil remaining. This acts as a rust preventive for a brief period of time until the next operation. The vapors are condensed by cooling coils at the top of the tank. The solvent can be reused by removing the oil and dirt. In order to obtain chemically clean surfaces, other cleaning processes, such as electrolytic cleaning, are necessary.

Alkaline Cleaning

The most prevalent type of cleaning is with alkali. It is used when soluble compounds are not present. Its cost is low because it uses water and inexpensive cleaning compounds. Parts must be rinsed thoroughly after exposure to the alkaline solution to prevent residue on the metal and lack of adhesion of organic finishes. It should not be used on aluminum, zinc, tin, or brass.

Electrolytic Cleaning

In electrolytic cleaning, an alkaline cleaning solution is used with electric current passing through the bath, causing the emission of oxygen at the positive pole and hydrogen at the negative pole. This action breaks up the oil film holding dirt to the metal surface and results in chemically clean surfaces suitable for plating. The material from which the part is made and the cleaning action desired determine whether the part should be made the anode or the cathode.

Emulsion Cleaning

Emulsion cleaning combines the action of a solvent and a soap. It is the least expensive cleaning process and can be applied on most materials at room temperature by either spraying or dipping. The parts should be so constructed as to be easily rinsed and dried; therefore, the process is not suitable to parts having deep pockets that will trap the liquid.

Ultrasonic Cleaning

When ultrasonic vibrations of sufficient power level are transmitted in a liquid, cavitation takes place. This action in liquids has two effects: *bulk*, which refers to those effects within the liquid, and *surface*, which occurs between the liquid and the solid. Many organic compounds are broken down by cavitation; thus the dirt and grease clinging to solid articles placed in an ultrasonic cleaning tank are ripped apart and emulsified.

Frequencies of approximately 30,000 Hz are characteristic of ultrasonic cleaning. Typical fluids used are water to which has been added a detergent or solvents such as cyclohexane and trichloroethylene.

Ultrasonic cleaning is adaptable to complete assemblies since cavitation tends to occur in crevices and corners that under other cleaning methods would be difficult if not impossible to reach. Thus one of the main advantages of ultrasonic cleaning is reaching inaccessible areas.

A typical ultrasonic cleaning facility is composed of a generator, a transducer, and a cleaning tank. The generator, of course, produces the ac electric energy. The transducer converts the electric impulses into high-frequency sound waves and the tank holds the cleaning fluid into which the transducer transmits its sound energy.

FINISH–TYPES

The principal types of finishes applied to metal products are:

1 Organic finishes
2 Inorganic finishes
3 Metallic coatings
4 Conversion coatings

Types of Organic Finishes

If an opaque, organic coating is applied so that the metallic coating or base metal is not discernible, the finish is then classified as an organic finish.

Organic coatings may be applied directly to the parent metal, although they frequently are applied after a metal and/or phosphate coating has been given to the base metal so as to increase the resistance to corrosion of the part. Organic finishes usually include two coats: first, a priming coat, then the second coat. Thickness of organic coats will vary from 0.0002 to 0.002 in.

Oil paint Exterior surfaces are often finished with oil paint. It completely hides the surface to which it is applied. Paint requires a relatively long drying time, and painted articles cannot be stacked because they will stick together and spoil the finish when separated. The drying time, hardness, and elasticity of paint films depend principally on the drying oil or combination of drying oils used. Slow-drying oils are soybean and linseed; the faster-drying oils are tung and castor.

Enamels Where paints may be classified as organic finishes where a pigment is dispersed in a drying oil vehicle, enamels represent those organic finishes where the pigment is dispersed in either a varnish or a resin or a combination of both. Thus enamels may dry by either or both oxidation and polymerization. Both air-drying and baking-type enamels are available.

Enamels, because of their availability in all colors, ease of application, and ability to resist corrosive atmospheres and attack of most usually encountered chemical agents, are the most widely used organic coating in the metal-processing industry.

Baked enamels (baking time may vary from a few minutes to half an hour) usually provide a finish that is harder and more abrasion resistant than the typical air-drying enamels. The automotive and electrical-appliance industries are large users of baked enamels.

Varnishes Varnishes can be clear or may contain dye, but they do not hide a surface when applied to it. The adding of pigment to a varnish results in an enamel, and this type of finish does hide the surface to which it is applied. The oleoresinous-type varnishes and enamels are fast drying, produce glossy, hard films, and have various degrees of elasticity, toughness, and durability. On the whole, the chemical resistance of this group can be as good as that of the alkyds.

An alkyd coating is one of the most popular finishes for metal products. There are several types of alkyd coatings. The oil-modified alkyds are resistant to heat and solvents; however, the phenolic coatings are more resistant to water and alkalies. The alkyd-amine baking enamels are quite resistant to alkali, and give a hard-wearing, glossy surface. These enamels are used for refrigerators, stoves, and other indoor appliances. The alkyd-phenolic coatings have greater resistance to water and alkali; consequently, they are used on tank liners for such things as washing-machine bowls. Alkyd-silicone finishes have better heat resistance, but they are more costly. High-silicone coatings are resistant to temperatures up to 700°F. Alkyd-styrene enamels are fast drying and so are in demand to provide a good, one-coat, fast-drying finish.

The alkyd types of varnish are quick drying, especially under the application of heat, and have good adherence to smooth surfaces. Their exterior durability is very satisfactory, making them suitable for vehicle finishes.

Lacquers Lacquers are noted primarily for their property of drying in a few minutes. The main drawback of this group is the small coverage obtainable from one unit as compared to that obtained from an equal unit of paint, varnish, or enamel. At least two coats of lacquer are required to give the protection that one coat of varnish or enamel affords. A unit of lacquer will cover only 50 to 60 percent as much area as an equal unit of the other finishes. Their poor adherence to metal surfaces requires the use of a priming coat for best results. Poor exterior durability is another drawback of this group, although, when well pigmented, lacquers have better exterior durability than oleoresinous varnishes. They are still inferior to the alkyd group, but the extremely fast-drying property of this group sometimes outweighs its drawbacks. Clear lacquers are used for protection against indoor atmospheres, while pigmented lacquers are suitable for outdoor protections as well.

Vinyl lacquers have properties that make them useful for lining food and beverage containers. They are impermeable to water, chemical resistant, and free from odor, taste, and toxicity.

Shellac   Shellac dries quickly and does not penetrate wooden surfaces deeply. Quite often it is used as a sealing coat on wood, since it gives a durable film. Since shellac is soluble only in alcohol, lacquers and varnishes can be applied over it without causing running together of the two.

Stain   Stain is not a paint, but is nevertheless very important in finishing wood. Its primary purpose is to color, not to cover; hence, dyes and not pigments are used to supply the color.

Luminescent paints   Luminescent paints are coming into more prominence, and many unusual effects can be produced by them for both daylight and night applications. They are suitable for both indoor and outdoor exposure.

Metal-flake paints   Metal-flake paints are made by adding very small polished flakes of a metal such as brass, copper, or aluminum to a so-called "bronzing" liquid. The hiding and sealing properties of this type of paint are very good.

Pearl essence   Pearl-essence finishes originally used pigment obtained from fish scales, but are now made synthetically. They have little hiding value, but leave a beautiful pearl-like irridescent finish such as that found on jewelry.

Crystal finish   By dissolving chemicals in clear lacquer which crystallize upon drying, finishes known as crystal finishes are obtained. These form beautiful, regularly shaped crystal-type patterns. Special varnishes dried in atmospheres containing products of combustion form frosted surfaces in regular patterns which are also very decorative.

Wrinkle finish   Another of these special finishes is the wrinkle finish. Ordinarily wrinkling is not wanted, but, if controlled, it can produce a very effective finish. The surface to which this type of finish is to be applied need not be smooth, which is sometimes a definite advantage in hiding surface defects. Only one coat is required to produce a good covering coat. The film is usually applied by spraying and is then baked. Some types of wrinkle finishes available can be applied by dipping. These are air-dried. This type of finish requires special materials and techniques for good results.

Crackle finish Closely akin to the wrinkle finish is the crackle finish. However, a crackle finish usually employs two different colors of lacquer enamel. The surface is first sprayed with a color, such as yellow, and then sprayed by a crackle lacquer containing a high percentage of pigment, which in this case might well be brown. When this second coat dries, it shrinks and leaves cracks, the sizes of which are determined by the thickness of the crackle finish applied. Usually this process is followed by a coat of gloss or clear lacquer for protection.

Hammered finish A finish similar to the crackle finish is the hammered finish. Two coats of the finish are applied in the same manner as the crackle finish; but the second coat, which is called a spatter thinner, hits the surface in the form of droplets. This finish is customarily baked.

Flock finishing Flock finishing is used when it is desired to give the article the feel of fabric. The flock adhesive, which is either a clear varnish or lacquer, is applied and either the flock is sprayed on or the article is dropped into a pile of it. Flock is actually fibers of fabric about 0.02 in in length.

Organic Finishes—Method of Application

Organic finishes are applied by brushing, spraying, flow coating, dipping, tumbling, and the centrifugal process.

Brushing Brushing requires the greatest amount of labor and the least amount of material.

Spraying Spraying is rapid and gives a smooth coat, but is wasteful of material (Fig. 15-1). Booths are required to collect the overspray, to prevent contamination of the air, and to remove fire hazard. Some material can be recovered by water screens in the booths. Electrostatically charging the part or the sprayed paint or lacquer will cause the spray to collect on the part and in this way reduces waste. Heating the sprayed materials increases their fluidity, and so decreases the amount of solvent required. These methods have reduced the cost of material sprayed per part and have increased productivity per man-hour.

Flow coating Flow coating paint, enamel, or varnish over the surface has limited applications on large tanks and frames which permit the paint to run off evenly (Fig. 15-2). The finishing material sets and some solvent escapes as the paint drains from the collecting racks to the accumulating tank and pump.

Dipping Dipping can be done by hand or automatically on conveyors. The part must drain easily and must not accumulate paint and cause fatty edges and tears. Dipping is economical in the use of both labor and material.

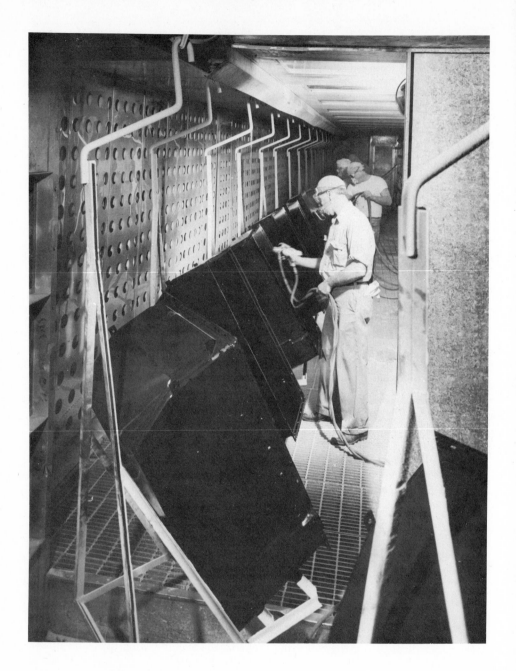

FIGURE 15-1
Spraying enamel on refrigerator shells.

FIGURE 15-2
Flow coating transformer frames.

Roller coating   Roller coating can be applied to sheets only.

Tumble finishing   Tumble finishing is used to finish many small parts (Fig. 15-3). The parts are put into the tumbler along with the finishing substance and tumbled until completely covered. They can then be removed for drying or can remain in the tumbler after draining.

Centrifugal finishing   Centrifugal finishing is also used for small parts. Such parts are placed in a strongly made wire basket, dipped, and then put in the centrifuge and whirled. This throws the excess finish material from the surface of the parts. The parts are finally removed and hung to dry. Parts treated by this method cannot have pockets that will accumulate the paint.

FIGURE 15-3
Modern-type tumble barrel suitable for tumble finishing.

Air drying and baking Some of the finishes are air-dried, but this process cannot be successfully accomplished under 70°F, and is usually done between 80 and 100°F. Some finishes are force-dried in ovens between 100 and 200°F; most synthetic primers and enamels are baked between 200 and 450°F. Baking is done only with nonporous articles and is to be preferred (except with lacquers) as it ordinarily gives a harder and more durable finish than air drying.

All enamels have a "smooth" look, such as that found on automobiles, by rubbing the finish with fine abrasives, applying wax, and polishing.

Preliminary treatment   The above finishes should be applied on a sound foundation. The surface for the finish should be pretreated as carefully as the finish is applied. It should be free of dirt, rust, scale, grease, oil, or flux residue, and should have a large surface area of a texture that will provide mechanical adhesion. The chromate-phosphate finishes, known in the trade as *parkerizing* or *bonderizing* finishes, are good foundations.

Summary   Organic coatings represent that family of coatings made up of a vehicle, that consists of either a drying oil or a resin, and a pigment. This family includes paints, enamels, varnishes, and shellacs, with vehicles of synthetic resins, rubber, linseed, and tung oils. As a family, these finishes are relatively inexpensive, provide the opportunity for a variety of pleasing colors, and give good resistance to corrosion. They do not allow the holding of close dimensional tolerance and they have only average resistance to abrasion. Their resistance to elevated temperature is poor when compared to other coating families. Table 15-1 provides details of the more important organic coatings.

Inorganic Coatings

Inorganic coatings are made up of refractory compounds. As a class, they are harder, more rigid, and have greater resistance to elevated temperatures than the organic coatings. The principal characteristics that lead the production-design engineer to specify an inorganic coating are eye-appealing finish, resistance to corrosion, a protection against elevated-temperature oxidation, and a surface that provides thermal insulation.

Porcelain enamels   Porcelain enamels are equal or superior to the organic finishes in beauty and permanence. Porcelain-enameled surfaces combine the strength and stability of steel with the beauty and utility of glass. They are easily cleaned, have color stability, and are durable. The cost has been reduced by using one coat for some applications, by using a less expensive steel, and by using conveyorized equipment for spraying, dipping, drying, and firing at 1500 to 1600°F. Porcelain enamels will resist temperatures up to 1000°F. Where appearance is not critical, a single coat of 0.003- to 0.004-in thickness has sufficient opacity for eye appeal and for protection.

Design considerations that will provide the best conditions for a durable porcelain finish are:

1   Avoid sharp edges; make radii as large as possible; avoid reentrant corners.

Table 15–1   ORGANIC COATINGS

	Alkyd			Acrylic	Cellulose		Epoxy		
	Alkyd	Alkydamine	Styrenated alkyd		Nitro-cellulose	Butyrate	Epoxy	Epoxy-urea	Epox
Cost, ¢/(ft²) (mil), dry	Inexpensive (1.50–3.0)	Inexpensive (1.75)	Inexpensive (1.75)	Moderate (2.75)	Moderate (2.50–5.0)	Moderate (2.75–7.0)	Moderate (5.0–7.0)	Inexpensive 2.00	Inexsive
Appearance (choice of color)	Unlimited	Unlimited	Unlimited	Unlimited	Unlimited	Unlimited	Somewhat limited	Somewhat limited	Some limit
Maintenance of dimensional tolerance	0.0005	0.0005	0.0005	0.0005	0.0005	0.0005	0.0005	0.0005	0.000
Typical thickness, mils	1.5	1.5	1.5	1.0	1.0	1.0	1.8	1.8	1.5
Resistance to atmosphere (salt spray)	Good	Good	Good	Excellent	Excellent	Excellent	Excellent	Very good	Exce
Resistance to elevated temperature (maximum service temp., °F)	200	250	200	180	180	180	400	400	300
Abrasion resistance	Fair	Good	Good	Fair	Fair	Fair	Good	Good	Good
(Tabor GS-10 wheel cycles)	3,500	5,000	5,000	2,500	2,500	2,500	5,000	5,000	5,000
Surface finish	Smooth or wrinkle	Smooth or wrinkle	Smooth or wrinkle	Smooth or wrinkle	Smooth or wrinkle	Smooth or wrinkle	Smooth or wrinkle	Smooth or wrinkle	Smoo wrink
Color retention	Good	Good	Good	Excellent	Very good	Excellent	Good	Good	Good
Gloss	Excellent	Excellent	Excellent	Excellent	Excellent	Excellent	Very good	Very good	Excel
Gloss retention	Excellent	Good	Good	Excellent	Very good	Excellent	Fair	Fair	Good
Exterior durability	Excellent	Excellent	Fair-good	Excellent	Good-excellent	Good-excellent	Good	Good	Good-excell
Resistance to alcohols	Fair	Good	Good	Poor	Good	Good	Excellent	Excellent	Fair
Resistance to gasoline	Good	Excellent	Fair-excellent	Good	Good	Good	Excellent	Excellent	Excel
Resistance to hydrocarbons	Good	Excellent	Excellent	Fair	Fair	Fair	Excellent	Excellent	Very
Resistance to esters, ketones	Poor	Poor	Poor	Poor	Poor	Poor	Fair	Very good	Fair
Resistance to chlorinated solvents	Poor	Poor	Poor	Poor	Poor	Poor	Excellent	Excellent	Fair
Resistance to salts	Very good	Excellent	Excellent	Very good	Good	Very good	Excellent	Excellent	Excell
Resistance to alkalies, concentrated	Poor-good	Good	Very good	Fair	Poor	Poor	Excellent	Excellent	Excel
Resistance to acids (mineral), concentrated	Poor	Poor	Poor	Poor	Poor-fair	Poor	Good	Fair	Poor
Resistance to acids (oxidizing), concentrated	Poor	Poor	Poor	Poor	Poor	Poor	Poor	Poor	Poor
Resistance to acids (organic as acetic, formic), concentrated	Poor	Poor	Poor	Poor	Poor	Poor	Poor	Poor	Poor
Resistance to acids (organic, as oleic, stearic), concentrated	Fair	Good	Fair	Fair	Fair	Fair	Excellent	Excellent	Excell
Resistance to acids (phosphoric)	Poor	Poor	Poor	Poor	Poor	Poor	Poor	Poor-good	Poor
Resistance to H₂O (salt and fresh)	Fair	Good	Good	Excellent	Fair-good	Excellent	Very good	Good	Very
Rockwell R	24	30	28	24	26–30	26	36	34	30
Flexibility	Excellent	Very good	Good	Excellent	Excellent	Excellent	Excellent	Very good	Good-excelle
Toxicity	None	Slight	Slight	None	None	None	Slight	Slight–none	Slight none
Impact resistance	Very good	Excellent	Good	Excellent	Excellent	Excellent	Good	Good	Excell
Dielectric properties	Good	Good	Good	Very good	Poor	Good	Very good	Very good	Very
Adhesion to ferrous metals	Excellent	Excellent	Excellent-fair	Very good	Excellent	Very good	Excellent	Excellent	Excell
Adhesion to nonferrous metals	Fair	Excellent	Excellent-fair	Very good	Good	Good	Excellent	Excellent	Very g
Adhesion to old paints	Very good	Good	Very good	Poor	Poor	Poor	Poor	Poor	Fair-very good
Ease of application	Excellent	Bake required	Excellent	Very good	Very good	Very good	Catalyst required	Bake required	Excell
Surface preparation	Primer	No primer	No primer	Primer	Primer	Primer	No primer	No primer	Primer
Bake-drying time	30 min (275°F)	20 min (320°F)	15 min (300°F)	....	....	....	....	30 min (350°F)	30 mir (320°F)
Air-drying time	2 h	....	10 min	5 min	5 min	5 min	45 min	....	1 h
Coverage, ft²/(gal) (mil)	450	450	400	350	200	200	450	500	450

| Fluorocarbon (air-dried) | Phenolic | Polyamide | Rubber | | | | Urethane | Vinyl | Linseed and tung oils |
			Chlorinated rubber	Neoprene	Hypalon	Silicone			
High 15.00	Inexpensive (1.75–4.0)	Moderate	Inexpensive (1.50–5.0)	Inexpensive	Moderate	Moderate (6.00)	Moderate–high (9.00)	Moderate	Inexpensive (0.50)
Unlimited	Limited	Limited	Limited	Limited	Unlimited	Unlimited	Unlimited	Unlimited	Unlimited
0.0005	0.0005	0.0006	0.0006	0.0007	0.0006	0.0006	0.0006	0.0005	
1.0	1.5	2–30	1.5	2–10	2	1.0	1–2	1.0	1.5–2.5
Excellent	Excellent	Fair	Excellent	Excellent	Excellent	Excellent	Excellent	Excellent	Good
220–550	350	300	200–250	200	250	550–1,200	300	150–180	
Poor	Good	Good	Good	Good		Fair	Good	Good	Fair
1,000	5,000	5,000	5,000	5,000	5,000	2,500	5,000	5,000	
Smooth or wrinkle	Smooth or wrinkle	Smooth	Smooth	Smooth	Smooth	Smooth or wrinkle	Smooth or wrinkle	Smooth or wrinkle	Smooth
Good	Poor	Very good	Good	Good	Excellent	Excellent	Good	Very good	
Excellent	Very good	Good	Fair	Poor	Poor	Excellent	Excellent	Good	
Fair	Fair	Fair	Fair	Fair	......	Excellent	Fair	Good	
Excellent	Excellent	Poor	Excellent	Excellent	Excellent	Excellent	Excellent	Excellent	
Fair–good	Excellent	Good	Fair–excellent	Excellent	......	Poor–fair	Fair–excellent	Fair–excellent	
Fair–excellent	Excellent	Good	Good	Fair	Poor	Fair–good	Fair–good	Excellent	
Poor–excellent	Excellent	Excellent	Poor	Poor	......	Very good	Fair–excellent	Good	
Poor	Fair–excellent	Good	Poor	Good	Poor	Poor	Fair	Poor	
Poor–excellent	Fair–good	Excellent	Poor	Poor	Poor	Poor	Poor–excellent	Poor	
Excellent	Excellent	Very good	Excellent	Excellent	Excellent	Good	Excellent	Excellent	
Excellent	Poor	Good	Fair–excellent	Excellent	Fair	Fair	Fair	Excellent	
Excellent	Excellent–poor	Poor	Poor–good	Poor	Excellent	Poor	Poor	Poor	Good
Excellent	Good–poor	Poor	Poor-fair	Poor	Poor	Poor	Poor	Poor	Good
Excellent	Poor	Poor	Poor	Poor	Poor	Poor	Poor	Poor	
Excellent	Excellent	Very good	Excellent–fair	Fair–poor	Fair	Poor–good	Fair–excellent	Excellent	
Excellent	Fair	......	Good	Very good	Excellent	Poor–fair	Fair–excellent	Excellent	
Excellent	Excellent	Fair	Very good	Excellent	Excellent	Excellent	Excellent	Excellent	
20	38	......	24	10	10	16	35–65	20	
Good	Good	Good	Very good	Excellent	Excellent	Fair	Excellent	Excellent	
Slight	None	......	Slight	None	......	Slight–none	Slight	None	
Excellent	Good	Very good	Good	Excellent	Excellent	Good–fair	Excellent	Excellent	Very good
Excellent	Excellent	Good	Excellent	Fair	Very good	Excellent	Excellent	Excellent	Fair
Very good	Excellent	Very good	Fair	Very good	Very good	Good–fair	Excellent	Good	Fair
Very good	Excellent	Very good	Very good	Very good	Very good	Fair–excellent	Excellent	Very good	
Poor	Good	......	Excellent–fair	......	Fair	Excellent	Fair–good	Good–fair	
Very good	Excellent	Good	Excellent	Very good	Very good	Excellent	Excellent	Very good–fair	
Primer	No primer	No primer	Primer	No primer	Primer	Primer	Primer	Primer	Primer
15 min (300°F)	30 min (350°F)	......	15 min (300°F)	15 min (300°F)	15 min (300°F)	1 h (400°F)	30 min (325°F)	15 min (300°F)	
5 min	10 min	......	45 min	15 min	15 min	45 min	45 min	15 min	
200	350	......	450	350	300	350	300–750	250	

2 Strengthen edges by flanging and stiffen large flat areas by embossing. This applies especially to porcelain-enameled parts. The firing temperatures cause expansion and contraction which require that the metal be distributed uniformly and not be restricted.

3 If there must be enclosed spaces which can trap and hold finish or cleaning materials, provide holes for adequate drainage. They will also prevent dragover of the various solutions used.

Ceramic coatings   Ceramic coatings are vitreous and metallic oxide coatings that are more refractory than the porcelain enamels. Typical ceramic coatings have a higher alumina content than porcelain enamels and, in addition to protecting metal surfaces from oxidation and corrosion, increase their strength and rigidity. This latter characteristic is especially important when the part is subjected to elevated temperatures.

Summary   Inorganic coatings represent that family of protective coatings that have a glasslike finish. They include the porcelain enamels and ceramic coatings composed of inorganic mineral materials which are fused to base metals. They may be readily applied to both ferrous and nonferrous surfaces. These coatings provide excellent resistance to corrosion and elevated temperatures. They also provide good appearance and resistance to abrasion. As a family, their cost is high and they afford limited ability to maintain close tolerances. Table 15-2 summarizes the properties of the most important inorganic coatings.

Metallic Coatings

Metallic coatings may be applied by electroplating, hot immersion (galvanizing), chemical deposition, or spraying of molten metal (metallizing). They are used to provide a decorative finish, protection against corrosion, and resistance to wear; they serve as a base for painting to provide a reflectant surface, and to provide a thermally or electrically conductive surface.

Electroplating   Electroplating is the process of passing current through the material to be deposited, the solution or bath, and the part on which the material is to be plated.

The base metal is made the cathode in an aqueous solution of a salt of the coating metal. Anodes of the coating material are used to complete the circuit and replenish the solution. Also, in order to increase its conductivity, other chemicals that will ionize highly are added (such as sulfuric acid to an acid copperplating bath). Plating solutions attack metals and containers, and plating equipment is expensive to maintain. The cost of plating is small compared to cost of surface preparation, cleaning, and handling of the parts. Polishing

Table 15-2 INORGANIC COATINGS

Property	Porcelain enamel	Ceramic coating		
		Enamel refractory oxide	Refractory oxide	Cermet
Cost	High	High	High	High
Appearance (choice of colors)	Wide range of colors	Dull	Dull	Dull
Maintenance of dimensional tolerance	0.0002	0.0002	0.0002	0.0002
Typical thickness, mils	3 to 5 coat; 3 coats typical	2–3	2–50	2–10
Resistance to atmosphere	Excellent	Good	Good	Good
Resistance to elevated temperature, °F	700–2000°F	. . . .	2500°	4000°
Abrasion resistance	Good	Good	Excellent	Excellent
Surface finish	Smooth, glassy appearance	Coarse–smooth	Coarse–smooth	Coarse–smooth
Base materials	Sheet iron or steel or cast iron with low carbon content; few impurities; aluminum; copper; gold	. . . .	. . . .	Refractory and high-temp. materials, columbium, molybdenum, tantalum, tungsten
Number of coats	1–3	1	1	1
Method of application	Dipping or spraying	Dipping or spraying	Spraying	Spraying or plating
Chemical resistance	Good	. . . .	Very good	
Impact resistance	Fair to excellent			
Torsion resistance	Fair to excellent			
Dielectric strength, V/mil	500–1,000		1,400	1,350
Hardness (Vickers)	4–7 mohs			

Table 15-3   HARDNESS OF ELECTRO-
DEPOSITED METALS

Metal	Brinell hardness
Cadmium	35–50
Chromium	700–1000
Copper	60–150
Gold	5
Iron	150–500
Lead	5
Nickel	150–500
Rhodium	400–800
Silver	50–150
Tin	5
Zinc	40–50

SOURCE: *Engineering Materials Manual*, Reinhold Publishing Company.

also adds considerably to the cost of plating; therefore, the advantages of improving appearance and corrosion resistance must be balanced against the increased cost.

The properties of the coating will vary with the composition of the plating solution, current density, agitation, solution pH, and solution temperature. Generally, the three properties usually sought when specifying an electroplate are hardness, resistance to corrosion, and appearance. Brinell hardness values of electrodeposited metals are shown in Table 15-3. Resistance-to-corrosion characteristics will vary with the thickness of the plate and, of course, the plating material. Appearance is usually thought of in terms of brightness of the finished coat. Bright plating can be realized by minimizing the buffing coat. The usual procedure is to buff the softer underlying metal (such as copper or nickel) and to follow with a hard bright coat (such as chromium) that requires little or no buffing.

The development of "periodic reverse current electroplating" has reduced the cost of polishing after plating. At the same time the plated deposit shows superior qualities of strength, elasticity, density, and freedom from flaws like porosity. It involves a novel plating cycle in which plating current is reversed briefly at short periodic intervals to deplate what may be unsound and inferior metal deposited in the previous plating period. Many microscopically thin increments of sound metal are built up to make a deposit more dense and of greater homogeneity than is possible with conventional, continuous-current methods. Work has been done with silver, copper, brass, zinc, cadmium, gold, nickel, and iron. Equipment for timing the cycles is available in mechanical and electronic types.

The protective value of electroplated surfaces is dependent on the thickness and porosity of the plate and its uniformity. The uniformity of thickness of plate on a given part is dependent upon the shape of the piece and the "throwing power" of the bath—the ability to deposit material in remote areas, such as thread roots and in deep holes.

The throwing power of cyanide and alkaline baths is good; thus parts including complex geometry should be plated in this type of bath rather than acid. The throwing power of the chromium bath is poor.

An electroplate may be applied directly to the base metal or it may be applied over another electroplate. Thus a nickel plate may be applied over a copper plate and chromium over a nickel plate. Lamination of plates allows an accumulation of specific desirable characteristics. For example, a base metal may be copper-plated and so acquire a greater polishability. The copper plate may be covered with nickel which is less porous, and thus has greater resistance to corrosion. A final lamination of chromium may then be applied for appearance.

Electroplating theory   Electroplating theory is based upon the work done by Michael Faraday in 1833. The laws developed by Faraday state that the quantity of material liberated at either the anode or cathode during electrolysis is proportional to the quantity of electricity that passes through the solution. Also the amount of a given element liberated by a given quantity of electricity is proportional to its weight. Thus in silver plating if a coulomb (C) deposits 1.118 mg of silver on the cathode, 5 C will deposit 5.590 mg.

By knowing the quantity of one metal that is liberated by a given amount of electricity we are able to determine the amount of other metals deposited. For example, if silver, having an atomic weight of 107.88, is deposited at the rate of 1.118 mg/C, cadmium, having an atomic weight of 112.41 and a valence of 2, will be deposited at a rate of

$$\frac{112.41/2}{107.88} \times 1.118 = 0.586 \text{ mg/C}$$

Of course, the actual amount of material deposited will never quite equal the theoretical value because of such factors as solution variations and current leakage. The amount actually deposited divided by the theoretical amount is known as the *cathode efficiency*. Average cathode-current efficiencies of common plating solutions are shown in Table 15-4.

Zinc plating   Zinc plating is one of the most common and widely used metallic coatings. It is the lowest-priced metal used for electrodeposited coatings. Zinc coatings provide galvanic resistance to steel, are resistant to ordinary at-

Table 15-4

Metal	Type of bath	Usual cathode efficiency
Cadmium	Cyanide	88–95
Chromium	Chromic acid-sulfate	12–16
Copper	Acid sulfate	97–100
Copper	Cyanide	30–60
Copper	Rochelle—cyanide	40–70
Lead	Fluoborate	100
Lead	Fluosilicate	100
Nickel	Acid sulfate	94–98
Silver	Cyanide	100
Zinc	Acid sulfate	99
Zinc	Cyanide	85–90

SOURCE: *Engineering Materials Manual*, Reinhold Publishing Company.

mospheric corrosion, and are inexpensively applied. Zinc tarnishes readily and its vapors are toxic; therefore, when welding zinc-plated parts, adequate ventilation should be provided. Representative applications of zinc-plated commodities are switch boxes, hardware, screw machine parts, and conduits. The appearance of electrodeposited zinc on smooth steel varies from a dull, light-gray surface to a mirror-bright, gray appearance.

Cadmium plating   Cadmium plates are more expensive than zinc, as cadmium is the highest priced of the metals commonly used in electrodepositing protective coatings on metals; however, cadmium plates provide greater protection to corrosion in saline atmospheres. Furthermore, cadmium plating is resistant to staining, as contrasted with zinc coatings. Cadmium is usually applied from 0.0002 to 0.0005 in thick for ordinary commercial parts.

Thickness of plate is readily controlled since both the covering power and throwing power of the cyanide bath, from which the cadmium deposit is obtained, is good. Cadmium-plated parts have a lower coefficient of friction than zinc; they are solderable, although they do make spot welding erratic. Since a cadmium-plated surface has a dull, silverlike appearance that is readily tarnished by many chemicals, it is not used as a decorative plate. Cadmium plating is specified on many aircraft and marine parts and is used on electronic instruments and miscellaneous hardware.

Tin plating   Tin plates are quite good but are more expensive. Consequently, tin plating of ferrous parts is seldom resorted to strictly for the prevention of atmospheric corrosion; however, since tin plates are especially resistant to tarnishing, and their oxides are neither toxic nor do they have an objectionable taste, they are used on "tin cans," kitchenware, and food containers. Also, since tin-plated parts solder quite easily, they are used on radio, television, and electronic equipment.

Copper plating Copper coatings upon steel have resistance to corrosion, but once the base metal is laid bare by a pinhole or scratch, strong voltaic action oxidizes the base metal at the break and in a fanning-out development beneath the coating. This results in a peeling away of the protective plate. For this reason, copper plates should be rather heavy (0.001 in). Since copper-plated parts polish well, this plate is frequently used as a base in preparation for subsequent coats of nickel or chromium. Copper plates have decorative value if protected from oxidation. This can be readily accomplished by coating the plated part with clear lacquer.

Brass plating Brass plates are frequently used as a base for bonding rubber and rubberlike materials to metal. Since brass tarnishes, it must be protected with lacquer when using it for decorative purposes. Brass plates are used on steel, zinc, aluminum, and can be applied on top of a copper plate.

Brass is deposited from a cyanide bath and, consequently, thickness of plate can be controlled on intricate sections.

Lead plating Lead plates provide good atmospheric protection and also give an excellent base to which paints can be applied. The plate holds well to the base metal and can be severely deformed without stripping off. Typical applications include cable sheathing, roofing, and paint cans.

Silver plating Silver plates are used principally for decorative purposes; they give a pleasing effect when polished to a bright luster. They also have high electrical conductivity, good bearing qualities, and resistance to oxidation at elevated temperatures, which may result in their selection for a certain process. The principal disadvantages of silver plate are its high cost and susceptibility to tarnishing. Typical silver-plate items are cutlery, jewelry, and musical and surgical instruments.

Gold plating Gold plates, similar to silver, find their greatest use in decorative parts. Gold is extremely expensive; consequently, the thickness of plate used for appearance is very thin. In addition to eye appeal, gold is resistant to most chemicals, is a good electrical conductor, and resists tarnishing.

Chromium plating Chromium plating is performed for two reasons: for its decorative and protective properties, and for its high wear-resistant properties.

Chromium plates have high reflectivity, giving an attractive, bright appearance.

Thick or "hard" chromium plates are applied to ferrous materials as well as aluminum and zinc where the predominant purpose is to provide a hard, wear-resistant surface. Typical applications are plug gages, taps, reamers, and dies.

Since the ratio of cost for equal thickness of plate of chromium to nickel is about 20:1, thin thicknesses of chromium are used when the object is for decorative purposes. If high resistance to corrosion is necessitated in addition to eye appeal, as in automobile hardware, then the chromium is applied over coats of nickel and/or copper. The sequence of operations performed on a typical automobile bumper might be:

1   Clean.
2   Nickel plate (should be at least 0.0008 in thick).
3   Wash.
4   Copper plate (should be at least 0.001 in thick).
5   Wash.
6   Nickel plate (should be at least 0.0008 in thick).
7   Wash.
8   Dry.
9   Polish and buff.
10   Chromium plate. (Chromium plate runs from 0.000,02 to 0.000,03 in for standard chrome plate. Duplex chrome deposits are about 0.000,06 in total, evenly divided between two layers.)

Since the chromium bath has poor throwing power, it is difficult to maintain uniformity of plating thickness on complex shapes. To remedy this, especially shaped anodes can be designed to surround or insert in the plated object so as to reduce the irregularity of "throwing" distance.

Plating of Plastics

Plated plastics—knobs, handles, buttons, and other parts made to look like metal parts—has been one of the fastest growing businesses in recent years. The biggest consumer of plated plastic components is the automobile industry. Also growing in importance are such nonauto products as TV, radio, and appliance knobs and trim. In the plating of plastic parts (acrylonitrile-butadiene-styrene and polypropylene are the plastics most used), it is necessary to first deglaze or etch the surface of the parts. Then onto this conditioned surface a very thin layer of precious metal (silver or gold) is applied. This forms the base for a thin layer of either nickel or copper which provides a conductive surface for the final layer which usually is chromium.

Metallizing   The metallizing process is accomplished by using a spray gun that feeds the wire through a nozzle surrounded by an oxyacetylene flame. A blast of air breaks the melted metal into globules and sprays it onto the surface. The wire is fed by a turbine driven by air pressure. The globules are oxidized on their surfaces and are molten on the inside. They strike the surface and flatten out. The oxidized surface opens and the molten material adheres to the oxidized surface of the flattened globules previously deposited. There are voids between the surfaces that serve to absorb oil. The surface acts as an oil reservoir similar to powdered metal materials and forms an excellent bearing material. The surface on which the metal is deposited should be rough like that obtained from a very rough turn on the lathe, and it must be clean and free of moisture or oil. This provides a mechanical lock for a deposited material. A metallized coating adheres to steels, stainless steels, Monel, nickel, iron manganese, and most aluminum alloys. It saves machining and can be used as a base for other less-expensive materials. The properties of porosity and ability to hold oil have led to spraying babbitt in bearing sleeves. This sprayed-babbitt material is superior to cast-babbitt bearings because of its oil-absorption ability.

A technique known as spray welding is used to hard-surface materials. The alloy is in powdered form held by a binder that is extruded into a wire form. The metallizing process vaporizes and consumes the binder so that the pure molten material is deposited without any trace of the binder. An even and uniform coating can be applied, as compared to the rough and thick coating of the arc-welding process of hard surfacing. This coating is then heated with a torch or in a furnace which fuses it to the base material and forms a hard, dense, homogeneous surface. The surface is free from porosity and the uniformity of the deposit reduces machine-finishing time.

Metallizing is extensively used for building up worn parts, for protection from corrosion, and for improving wearing surfaces. Cloth and paper are coated for use in electrical condensers. Thus metallizing is a production tool as well as a repair tool.

Solder sealing   Solder sealing of metal parts to glass or ceramic parts presupposes that the metal and nonmetal parts will have approximately the same coefficient of expansion during the wide variation of temperature necessary in the sealing process or in the operation of the apparatus. The procedure followed is to deposit chemically a metal on the glass or ceramic. The metallic compounds make the surface of the ceramic or glass chemically pure and deposit the metal at the same time. This simultaneous action enables the metal to adhere firmly to the glass or ceramic. A soft solder is added to this metal deposit. The metal container receiving the glass or ceramic is also prepared for soldering and the two are joined by the usual soldering operations.

Hot-dip coating The process of hot-dip coatings is quite common. Galvanizing is a hot-dip process by which the base material is immersed in a tank of molten zinc. Adhesion results from the tendency of the molten zinc to diffuse into the base metal.

Hot dipping is a rapid, inexpensive process which allows the coating of corrosion-resistant metals onto base metals at less cost than by electroplating; however, the process is limited to shapes that will not trap the molten metal upon extraction from the dip tank.

Tin, lead, and aluminum coatings are applied to the base metals in addition to zinc. Base metals are restricted to the materials with higher melting temperatures such as cast iron, steel, and copper.

Immersion coatings Immersion coatings are applied by dipping a base metal having a higher solution potential into an aqueous solution containing ions of the coating metal. No electric current is used. Deposition occurs while the base metal is in the solution. The coat is usually quite thin and can be controlled quite closely. Nickel, tin, zinc, gold, and silver are used as immersion-coating materials.

Nickel immersion coatings are used on steel parts that are constructed so that it would be difficult to maintain a uniform electrodeposited coating, such as valves, gears, threaded parts, and other items having deep recesses.

Tin immersion coatings provide a bright, decorative, and protective coating on such items as paper clips, pins, and needles.

Vapor-deposited coatings As the name implies, vapor-deposited coatings result from the condensation of a metal film on the base metal. Aluminum is the most widely applied vapor film. The aluminum is heated and vaporized in a vacuum. The aluminum vapor then condenses on the surfaces of the base metal. As would be expected, vapor-deposited films are very thin. Consequently, they have had little application where extreme resistance to corrosion is required. The principal use of this method of deposition has been for decorative purposes such as on trim for television and radio sets, for other household furniture, and on costume jewelry.

Summary Metallic coatings as a family refer to those coatings where a metal is deposited on a surface either by electroplating, hot dipping, immersion, metallizing, flame spraying, or electroless methods. Since most plating, dipping, and immersion tanks are limited in size metallic coatings are usually applied to smaller parts. An upper constraint is approximately 2,000 in in most plants.

This family of coatings provides for bright metallic surfaces with close tolerance control and good resistance to abrasion. The cost of metallic coatings as a family is somewhat higher than organic coatings and less than inorganic coatings. Metallic coatings as a family provide adequate resistance to corrosion and elevated temperatures. Table 15-5 provides detailed information as to design characteristics of the principal electroplated coatings.

Conversion Coatings

Conversion coatings are those produced when a film is deposited on the base material as a result of a chemical reaction. The most widely used conversion coatings are:

1 Phosphate coatings
2 Chromate coatings
3 Anodic coatings

Phosphate coatings Phosphate coatings are principally used as a base for the application of paint or enamel. The process itself does tend to rust-proof the base material, which is usually iron, steel, or zinc. A typical phosphate coating process consists of spray cleaning, phosphatizing, water-spray rinse, and chromic acid–spray rinse. The phosphate coating is usually 0.0001 to 0.003 in in thickness. The process is very rapid since the capacity of the equipment limits the rate of production rather than the cycle time. This process is widely used in the automotive and electrical-appliance industries to prepare automobiles, washing machines, refrigerators, and similar products to receive an organic finish.

Chromate coatings Chromate coatings are used on nonferrous materials including aluminum, magnesium, zinc-coated materials, and cadmium-coated materials. Chromate coatings are generally quite thin, being less than 0.00002 in thick, and are used chiefly for added resistance to corrosion and as a base for paint. The process can give either of two types of films: a yellow iridescent film or a clear film. Through the use of acidified organic dyes, the process can be used to impart a variety of popular colors to the treated part, including red, blue, lemon, violet, orange, green, and black.

Chromate films are less expensive to apply than anodized coatings in that the process is faster and less overhead is involved. They also give greater resistance to corrosion than anodic films; however, anodic films offer superior wear-resistant qualities.

A representative operation breakdown of a chromate treatment might be:

1 Degrease.
2 Rinse.
3 Clean in alkali.
4 Rinse.
5 Soak in hydrofluoric acid.

Table 15-5 METALLIC COATINGS (ELECTROPLATED)

Characteristic	Cadmium	Chromium	Copper	Gold	Iron	Lead	Nickel	Rhodium	Silver	Tin	Zinc
Cost	0.055–0.075/ft² moderate (3)	Moderate (4)	Inexpensive (2)	Very high (10)	Fairly expensive (6)	Fairly expensive (5)	Moderate (3)	Very high (10)	High (9)	Moderate (4)	0.0075–0.01/ft² inexpensive (1)
Appearance	Bright white	White, mirror-like, highly decorative	Bright or semi-bright pink, red	Bright yellow, highly decorative	Matte gray	Matte gray	White, either dull or highly decorative bright	Mirror-bright white, highly decorative	Bright white, highly decorative	Dull white, highly decorative	Matte gray to attractive bright
Maintenance of dimensional tolerance, decimal in	0.0003	0.0003	0.0003	0.0002	0.001	0.0005	0.0003	0.0002	0.0002	0.0003	0.0003
Typical thickness, mils	0.1–0.2 (indoor), 0.3–0.7 (outdoor)	0.01–0.05 (decorative), 0.05–2.0-up (hard)	0.3–0.5, (undercoat), 2.0–3.0 (functional topcoat)	0.002–0.01 (decorative); 0.01–2.0 (functional); 0.004–0.15 (electroforming)	2.0–10.0	0.5–8	0.1–1.5 (decorative), 5–20 (industrial)	0.001–1	0.1 (with undercoat) to 1.0	0.015–0.5	0.1–0.2 (indoor), 0.3–0.7 (outdoor)
Resistance to atmospheric corrosion	Very good (3)	Good (when over copper and nickel) (5)	Fair (6)	Excellent (1)	Poor (10)	Good	Very good (4)	Very good (4)	Good (5)	Good (5)	Very good (5)
Resistance to elevated temperature; melting point, °F	610	2939	1931	1944	2795	621	2651	3553	1760	448	786
Abrasion resistance	Fair (8)	Excellent (1)	Poor (9)	Poor (8)	Good (5)	Poor	Good (5)	Good (4)	Good (5)	Poor (8)	Poor (8)
Surface finish	Smooth (bright to dull)	Very smooth	Very smooth (bright to dull)	Very smooth (bright to dull)	Moderately smooth	Smooth	Very smooth (bright to dull)	Smooth	Very smooth	Smooth	Smooth (bright to dull)
Reflectivity at 6000 Å	Fair	Very good (66%)	Fair (44%)	Good (47%)	Poor (60%)	Poor	Good (60%)	Excellent (76%)	Excellent (90%)	Fair	Fair (55%)
Throwing power	(1)	(1)	(2)	(1)	(3)	(5)	(5)	(5)	(1)	(1)	(1)
Base metals	Steel, stainless steel, wrought iron, gray iron, copper and alloys	Ferrous, nonferrous metals	Most ferrous and nonferrous metals	Copper, brass, nickel, silver	Ferrous metals, copper and alloys	Ferrous metals, copper and alloys	Most ferrous, nonferrous metals	Most ferrous, nonferrous metals	Most ferrous, nonferrous metals	Most ferrous metals and nonferrous metals	Ferrous metals excepting east iron, copper and its alloys
Specific gravity	8.65	7.10	8.93	19.30	7.85	11.35	8.90	12.50	10.50	6.75	7.14
Resistivity, $\mu\Omega\cdot cm$	7.54 at 18°C	2.6 at 20°C	1.72 at 20°C	2.44 at 20°C	0.6 at 20°C	22 at 20°C	7.8 at 20°C	5.11 at 0°C	1.63 at 20°C	11.5 at 20°C	5.75 at 0°C
Thermal conductance, cal/((cm²)(°C)) (see)	0.217	0.65	0.023	0.707	0.190	0.083	0.210	0.210	0.974	0.157	0.268
Brinell hardness	2–22	400–1,000	40–130	(65–325 Knoop)	140–350	5	125–550	594–641	60–79	8–9	40–50
Adhesion	Good	Excellent	Excellent	Excellent	Good	Good	Very good	Good	Good	Good	Excellent
Type bath and cathode efficiency (%)	Cyanide 88 to 95	Chromic and sulfate 12 to 16	Acid sulfate cyanide 97 to 100, cyanide 30 to 60, rochell-cyanide 40 to 70	Cyanide 70 to 90	Acid chloride 90 to 98, acid sulfate 95 to 98	Fluoborate 98, fluo-silicate 98	Acid sulfate 94 to 98	Acid sulfate 10 to 18, acid phosphate 10 to 18	Cyanide 99	Acid sulfate 90 to 95, stannate 70 to 90	Acid sulfate 98, cyanide 85 to 90

Grams deposited per A·h	Solubility	Preparation for	Remarks
2.0968	Soluble in acids; ammonium nitrate	Grit blast or polish, clean chemically with organic solvent, alkali cleaner, or electrolytic alkali cleaner, prepare chemically by acid nickel or dip	Piece of cadmium is 8–10 times cost of zinc. A 0.0005-in coat will withstand 96 h of salt spray without showing iron rust or white salts
0.3233	Soluble in HCl, dilute H_2SO_4; insoluble HNO_3		Usually applied from 0.00001 to 0.00005 in over Cu and nickel. Composite thickness runs from 0.00041 min for noncorrosive applications to 0.0020 for outdoor service. Hard chromium deposits from 0.0001 to 0.010 in without undercoating
2.371 (ous) 1.186 (ic)	Soluble in HNO_3; hot H_2SO_4; slightly soluble in HCl and ammonium hydroxide		
7.356 (ous) 2.450 (ic)	Soluble in potassium cyanide aqua regia; hot selenic acid, insoluble most acids		
1.042			
8.865	Soluble in HNO_3 and hot concentrated H_2SO_4		Cathodic to iron and steel and must be used in thick deposits to prevent galvanic corrosion
1.095	Soluble in dilute HNO_3, slightly soluble in HCl and H_2SO_4, insoluble in ammonia		Cathodic to steel and therefore used over copper
1.280	Soluble in sulfuric-hydrochloric acid; hot concentrated H_2SO_4, slightly soluble in acids		
4.025	Soluble in HNO_3; hot H_2SO_4, potassium cyanide, insoluble in alkalies		Tarnishes in the vapors of sulfur compounds
2.214 (ous) 1.107 (ic)	Decomposes in HCl, H_2SO_4, dilute HNO_3, aqua regia, hot potassium hydroxide		Cathodic to iron
1.102	Soluble in acids, alkalies, acetic acid		

Note: Numbers in parentheses indicate relative ranking on a scale from 1 to 10: 1 represents most attractive and 10 the least.

6 Rinse.

7 Place in dichromate bath (45-min soak).

8 Rinse.

9 Dry.

Anodic coatings   Anodic coatings are those resulting when an oxide is applied to aluminum and magnesium and their alloys. Here the base metal is connected as an anode in an electrolytic immersion, and a film is deposited on the base metal that increases its resistance to atmospheric as well as galvanic corrosion and offers a good foundation for painting.

A typical operation breakdown of applying an anodic coating to aluminum is as follows:

1 Degrease.

2 Rinse.

3 Clean in alkali.

4 Rinse.

5 Apply anodic coating (3 to 10 percent solution of chromic acid).

6 Rinse.

7 Dry.

The process outlined above, using chromic acid, will give a coating of aluminum oxide and chrome salts varying in thickness from 0.00003 to 0.0002 in. The color will range from a light gray to a black. It provides excellent corrosion protection and has been used extensively by the aircraft industry.

Anodic coatings are applied also by sulfuric acid, oxalic acid, and boric acid bath processes.

It is possible to impart excellent colored coatings by immersing the parts in warm dye solutions, and then sealing the dye in the porous oxide coatings by dipping in dilute nickel acetate.

Summary   Conversion coatings represent those coatings used on ferrous or nonferrous metals where either a phosphate, chromate, or oxide salt represents the protective coating. As a family, these coatings are inexpensive and allow the maintenance of close tolerances. However, in view of the fact that they are very thin, they do not in themselves provide a great amount of resistance to corrosion or abrasion. Table 15-6 provides the principal properties of phosphate, chromate, and oxide conversion coatings.

SUMMARY

The production-design engineer has a vast number of finishes from which to choose. For a given application, one finish may be superior to another from the standpoint of adaptability, service, eye appeal, and cost.

Table 15-6 CONVERSION COATINGS

Property	Phosphate	Chromate	Oxide
Cost	Low (0.002–0.01/ft²)	Medium (0.005–0.015/ft²)	High (0.01–0.20/ft²)
Appearance (choice of colors)	Poor (preparatory to painting)	Wide choice: clear, golden iridescent, olive drab	Unlimited in anodizing process
Maintenance of dimensional tolerance	No appreciable change	No appreciable change	0.0001
Typical thickness, mils	0.01–0.02	0.01–0.02	0.04–0.5 (2.0–4.0 for wear, abrasion resistance)
Resistance to atmosphere	Fair	Good	Good
Resistance to elevated temperature, °F	Good	Good	Good
Abrasion resistance	Poor	Poor	Good
Surface finish	Thin amorphous film	Thin amorphous film	Thin crystalline structure
Base materials	Iron, steel, zinc, aluminum, cadmium, and tin	Zinc, copper, tin, magnesium, silver, aluminum	Nonferrous (aluminum and magnesium)
Number of coats	1	1	1
Method of application	Dipping, brushing, or spraying	Dipping or spraying	Dipping, electrolytic
Procedure	Pretreatment: phosphating, rinsing, drying	Degrease and etch-rinse chromate; rinse warm; rinse dry	Degrease; rinse; anodize; rinse; dry
Types	Zinc, iron, and magnesium	Chemical polishing; non-polishing colored coating applied	Sulfuric, chromic
Purpose	Paint adhesion and oil-retention, corrosion resistance	Paint adhesion, decoration, corrosion resistance	Decoration, corrosion resistance
Coefficient of friction	Medium	Medium	Low

The following characteristics represent the principal criteria for selection of the decorative and protective coating to be applied:

Cost Cost considerations enter into the selection process of all designs. The analyst is interested in the total cost resulting from a given coating. Thus, cost must include cost of surface preparation prior to receiving the coating, cost of coating the surface, and cost of any recommended treatment after coating.

Appearance (color) The characteristic appearance (color) has to do with the clarity of colors that the coating is capable of achieving. Not only is the range of colorability significant but also the durability of the color in the coating. Thus color clarity, range, and durability should be considered in this characteristic.

Appearance (bright-metallic) This characteristic relates to the ability of the coating to provide and retain a bright, reflective, metallic appearance.

Resistance to corrosion This characteristic is based on the ability of the coating to protect the covered surface from typical corrosive agents including salt spray, ozone, oxygen, and SO_2.

Ability to maintain close tolerance This characteristic refers to the tolerance capability, under normal conditions, of the coating process.

Abrasion resistance This characteristic is related to the ability of the coating to resist galling, wear, and scraping when a like or disimilar material is repeatedly rubbed against the coated surface.

Resistance to elevated temperature This characteristic represents the ability of a coating to permit the part to provide service as temperatures increase significantly above normal room temperature.

By identifying the pertinent characteristics desired of the decorative and protective coating, the well-informed production-design engineer, with the help of the guide sheets provided, should be able to select the most appropriate finish for his company's products.

QUESTIONS

1 What are the principal ways in which a metallic surface can be cleaned?
2 When would you advocate solvent cleaning? Alkaline cleaning? Electrolytic cleaning? Emulsion cleaning?
3 What four types of finish may be applied to metal products?
4 What advantages do oleoresinous varnishes have over oil paint?
5 What is the main drawback of lacquers?
6 What products have made use of vinyl lacquers?
7 In what type of products would you advocate the use of porcelain enamels?

8 How may metallic coatings be applied?

9 Why is cadmium plating very popular?

10 Why is a copper plate not particularly suitable to parts subject to atmospheric corrosion?

11 Where is metallizing used?

12 How are conversion coatings classified?

13 What is the future of anodic and chromate coatings?

14 With a cathode efficiency of 94 percent, what would be the rate of deposition of zinc?

15 What seven criteria are usually considered in the selection of a decorative and protective coating?

16 Explain the procedure used in the chrome plating of plastic components used on the dashboard of the modern automobile.

17 What type of plating would you recommend for a design having deep recesses and subject to atmospheric corrosion? Why?

PROBLEMS

1 A Unified National Coarse (UNC) thread is to be plated to a specific tolerance on the pitch diameter after plating. Determine the limits for the pitch diameter after machining, if the plating thickness between upper and lower limits are 0.0005 in and 0.0002 in, respectively. The final limits on the pitch diameter are: $\dfrac{0.2854 \text{ in}}{0.2830 \text{ in}}$. How much of the tolerance does the plating tolerance use?

SELECTED REFERENCES

AMERICAN SOCIETY FOR METALS: "Metals Handbook," 8th ed., vol. 2, "Heat Treating, Cleaning, and Finishing," Metals Park, Ohio, 1964.

BLOOM, W. and G. B. HOGABOOM: "Principles of Electroplating and Electroforming," McGraw-Hill, New York, 1930.

DUMOND, T. C.: "Engineering Materials Manual," Reinhold, New York, 1951.

DeKAY THOMPSON, MAURICE: "Theoretical and Applied Electrochemistry," Macmillan, New York, 1939.

16

JOINING PROCESSES

INTRODUCTION

Early man attached handles to his hunting clubs and cooking vessels, heads and feathers to his arrow shafts. His descendants forge welded Damascus swords, chain links, and devices needed by armorers and shipbuilders. Spinning, knitting, weaving, felting, and matting—of textiles and paper—are not new processes. But only in the twentieth century has man advanced in threaded fasteners, welding, rubber cements, synthetic resins, and other organic cements. The major industrial joining processes are presented in Table 16-1.

Today some $2 billion worth of adhesive bonded products per year are sold in the United States alone, rivaling or exceeding the dollar volume of welded products. Synthetic resins are used in plywood, rubber cements in tires and footwear, epoxy resins in metal assemblies, paints, and Fiberglas-reinforced pleasure craft and automobile bodies.

Table 16-1   PRODUCTION JOINING PROCESSES

Arc welding	Resistance welding
Shielded metal-arc Gas metal-arc (MIG) Submerged arc Gas tungsten-arc (TIG) Flux cored-arc Plasma-arc Arc-spot Stud	Spot Projection Seam Flash Upset
Gas welding	Solid state welding
Oxyacetylene	Friction Ultrasonic Diffusion bonding
Brazing	Other welding processes
Furnace brazing Gas brazing Induction brazing Vacuum brazing	Electron beam Electro slag Thermit Laser Explosive
Related processes	Soldering
Arc cutting Oxyacetylene cutting Hard-facing Plasma-arc cutting	Hand soldering Dip soldering Wave soldering
Adhesive bonding	Mechanical fastening
Synthetic resins Elastomeric adhesives Natural adhesives	Rivets Screws, nuts, and bolts Special fasteners Staking Lock seam, etc. Stitching

FIGURE 16-1
Typical single-V-butt welded joint showing edge preparation and filler metal.

WELDING: GENERAL CONSIDERATIONS

Welding is the joining of two polycrystalline workpieces—usually of metal—by bringing their fitted surfaces into such intimate contact that crystal-to-crystal bonding occurs. Industrial welding usually entails local heat from a burning gas or an electric arc, or heat from resistance; the fitted surfaces may melt together or a filler rod may melt between them to form a connecting bridge.

Oxides impede welding. A small disk of indium and another of silver will bond at room temperature when pinched between thumb and forefinger—but only if the surfaces are first abraded. To the same point is the phenomenon of welding that is *not* wanted. A weld deposit builds up on the edge of a cutting tool, causing chatter or poor finish on the workpiece. Bearings gall when overloaded or underlubricated. The parts of an instrument that rub together while unfolding from a satellite in space bond together despite the cold. One remedy is to pair a metal with a nonmetal.

Some welding involves metal in addition to the workpiece, as in brazing or soldering. More important is the joining of steel plates with a consumable electrode which penetrates the joint and deposits a weld bead as in Fig. 16-1. Welding equipment can be simple, such as that used in a farm shop or in automotive repair shop or complex, as in an establishment which offers production joining as a business. Such establishments have positioners, shears, annealing furnaces, and booths for sand blasting and painting.

The Effect of Heat on Welding

A designer must anticipate two problems inherent in the welding process (1) the effect of localized heating and cooling on the microstructure and properties of the base metal, and (2) the effect of the residual stresses which are locked in the weldment as a result of the uneven cooling of the weld deposit. In general, weldments have poorer fatigue and impact resistance than correctly designed castings or forgings. For example, critically stressed joints in hopper cars for railroads are riveted rather than welded to avoid fatigue failure in service.

The heat-affected zone is the region of the base metal, adjacent to the weld zone, where the temperature has caused the microstructure to change (Fig. 16-2). Note the difference in microstructure between the hot-worked metal near

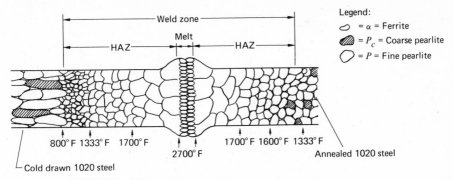

FIGURE 16-2

Microstructure of heat-affected zone of flash butt weld.

(*Redrawn from Joseph Datsko, "Material Properties and Manufacturing Processes," John Wiley & Sons, Inc., New York, 1966.*)

the weld joint and the cold-worked structure in more remote areas. The weld is a flash butt weld in which there was no weld metal added. In the case of fusion-welded steel, the weld deposit is usually stronger than the base metal because the grain growth in the heat-affected zone makes that metal somewhat weaker than the fine-grained dendritic structure which is typical of the weld deposit. Therefore a tensile test specimen taken from the weld area in low-carbon steel will break adjacent to the weld deposit rather than in the weld deposit itself.

In carbon and alloy steels the heat-affected zone is particularly important because of the metallurgical changes which occur in steels when subjected to heating to a high temperature and then to rapid cooling. In Fig. 16-2, the steel has been heated above 1200°F throughout the heat-affected zone. A large part of the metal within this zone has transformed to austenite because of the intense heat from the welding arc. Upon subsequent cooling, the properties of the metal within the heat-affected zone are determined by the cooling rates and consequent decomposition of the austenite in relation to the TTT curves and the iron–iron carbide phase diagram for the steel which was welded.

A profile curve of the maximum temperatures reached at various locations within the heat-affected zone of a weldment from 0.3 percent carbon steel may be correlated with the iron–iron carbide phase diagram (see Fig. 16-3). The structures found in the various regions of the heat-affected zone may be analyzed as follows:

Point 1 The steel in that zone has been heated to excess of 2400°F, so the austenite grains have grown to a large size.

Point 2 The alloy in this zone has been heated to 1800°F and is fully austenitized but there has been little grain growth.

Point 3 The steel in this region has been heated far enough above the austenite transformation temperature so that the austenite has achieved a homogeneous structure.

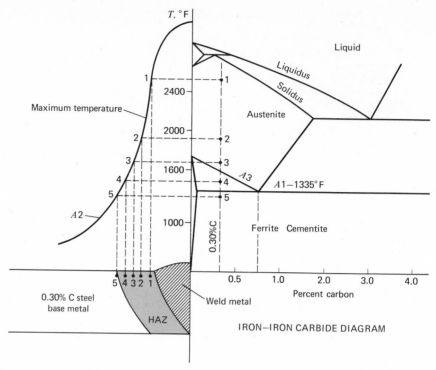

FIGURE 16-3
Relation between peak temperature in weld and iron-carbon phase diagram.

Point 4   The metal in this region has been heated to 1400°F. Only part of the alloy has changed to austenite, so the resultant mixture of structures has poor mechanical properties.

Point 5   No change has occurred at this point, which is in the base metal outside the heat-affected zone.

Residual Stresses

Residual stresses result from the restrained expansion and contraction that occur during the localized heating and cooling in the region of the weld deposit. The magnitude of such stresses depends upon the design of the weldment, the support and clamping of the components, their material, and the welding process used. The relationship between the thermal and shrinkage stresses is highly complex in both the stress direction and time phase because of the steep thermal gradients and the moving heat source in welding. Part of a weld bead is solidifying while a short distance away the molten pool is forming and the base metal is still gaining heat. The sequence can be analyzed by studying a simple bar heated in the center. In Fig. 16-4a the heated bar expands at either end. However, if

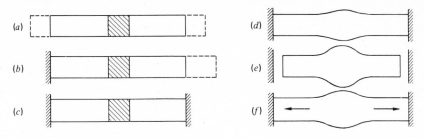

FIGURE 16-4
Comparison of effects on a heated bar with various constraints

one end is fixed (Fig. 16-4b), the other end moves twice as far relative to the original centerline. No lengthwise movement can occur if each end is restrained (Fig. 16-4c), so the center must bulge (Fig. 16-4d). However when the bar cools again, it is shorter than before (Fig. 16-4e). If, on the other hand, the bar is fully restrained, where it is heated a bulge occurs at the center and residual tensile stresses build up in the bar as it tries to contract, but the walls will not permit it to return to its shortened length (Fig. 16-4f).

In fusion welding the molten pool solidifies as a casting poured into a metal mold. It is restrained from contracting by an amount which varies with the welding process. The cooling rate has a great influence on the amount and nature of the residual stress. In general there are residual tensile stresses in the weld deposit which counterbalance compressive stresses in the base metal (Fig. 16-5). They are proportional in intensity to the weld size, and the maximum stress occurs in the direction of welding, with the transverse stress the next highest in intensity. The stress in the thickness direction is least, because there is the least hinderence to contraction in that direction.

The stress distribution around the weld deposit can be represented schemat-

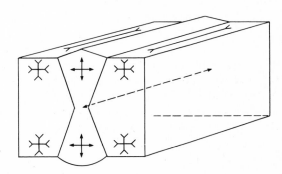

FIGURE 16-5
Residual stresses in the region of a welded joint.

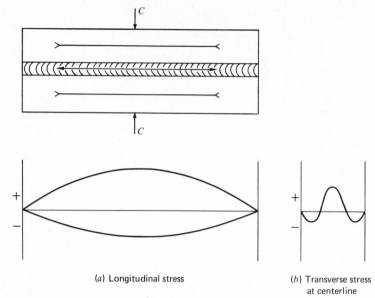

(a) Longitudinal stress

(b) Transverse stress at centerline

FIGURE 16-6
Typical stress distribution for a single pass weld.

ically as in Fig. 16-6. Note that the center of the weld is the point of maximum stress.

Distortion

Welding sequence is important in minimizing distortion and residual stresses. For example in ship construction welding is started at the keel at midship and then continues simultaneously fore and aft and upward. It is wise in any complex weldment to balance heat input and distortion by welding first on one side and then the other. In fact one of the greatest cost items in a custom-welded frame is the straightening operation. In production-welded structures such as railroad cars balanced welding sequences and proper fixtures reduce distortion to a minimum.

Welding fixtures can be used to reduce distortion especially on thin pieces. The thermal input to the workpiece can be lowered and the mechanical restraint permits movement only in one direction (Fig. 16-7). Proper design can also reduce residual stresses and their resultant distortion.

The transverse shrinkage for a butt weld may be estimated from the following empirical relationship:

$$S_t = 0.18 \frac{A_w}{t} + 0.05d$$

where S_t = transverse shrinkage, in
A_w = cross-sectional area of weld, in²
t = thickness of material, in
d = root opening, in

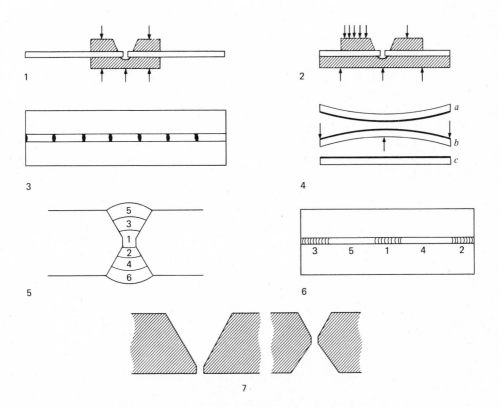

1 Restriction of heat input by means of massive or water-cooled chill blocks.
2 Controlled restraint to permit some movement during heating and cooling.
3 Tack welding to provide restraint during welding.
4 Prestraining to expected distortion level (*a*) distortion to be expected
   without prestrain, (*b*) welding in prestrained condition, (*c*) and release
   of prestrain to obtain undistorted product.
5 Bead sequence as shown, to allow opposing stresses to correct distortion
   obtained due to prior bead.
6 Skipping sequence of welding to prevent stress accumulation of continuous
   weld and allow opposing stresses to correct previous distortion.
7 Root gap to permit the weld metal to shrink transversely and reduce
   rotational distortion, the double V and sequence (no. 5) produce
   minimum distortion.

FIGURE 16-7
Methods for reducing the distortion which occurs in welding.

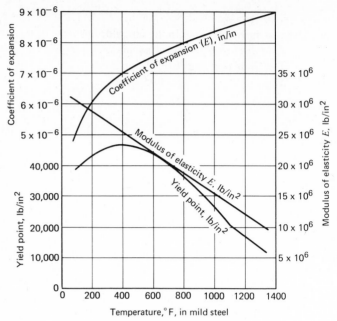

FIGURE 16-8
Change in the properties of mild steel when subjected to heat.

The lengthwise contraction for a V-butt weld (in/in) may be estimated as

$$S_1 = 0.025 \frac{A_w}{A_p}$$

where A_w = cross-sectional area of weld, in²
A_p = cross-sectional area of weldment, in²

Methods for reducing distortion are given in Fig. 16-7. The numbers in the figure are explained as follows:

1 Restriction of heat input by means of massive or water-cooled chill blocks
2 Controlled restraint to permit some movement during heating and cooling
3 Tack welding to provide restraint during welding
4 Prestraining to expected distortion level: (a) distortion to be expected without prestrain, (b) welding in prestrained condition, (c) release of prestrain to obtain undistorted product
5 Bead sequence as shown, to allow opposing stresses to correct distortion obtained due to prior bead

6   Skipping sequence of welding to prevent stress accumulation of continuous weld and allow opposing stresses to correct previous distortion

7   Root gap to permit the weld metal to shrink transversely and reduce rotational distortion; double V and sequence control

8   Angle of single V and number of passes

Heat changes the mechanical and physical properties of a metal, which in turn affects the heat flow and the uniformity of heat distribution. As the temperature increases, the yield strength, Young's modulus, and thermal conductivity decrease but the coefficient of thermal expansion increases (Fig. 16-8). If one can find suitable values for those properties of an alloy at elevated temperature, he should be able to estimate the severity of the distortion which results from welding operations. A high coefficient of thermal expansion leads to high shrinkage on cooling and the possibility of distortion, especially if the alloy has a low yield strength as would be the case for aluminum alloy, whereas high thermal conductivity would serve to counteract the effect of the high coefficient of thermal expansion. The higher the yield strength of an alloy in the weld area, the greater the residual stress; conversely, the lower the yield strength the less the severity of the distortion.

Longitudinal bending or cambering of a rolled section such as an I beam or angle results from contraction which occurred at some distance from the neutral axis. The amount of distortion is a function of the shrinkage moment and the resistance of the member to bending as given by its moment of inertia. The following formula can be used to estimate the amount of distortion in such cases:

$$\Delta_{\text{long}} = 0.005 \frac{A_w l^2 d}{I}$$

where A_w = total cross-sectional area within fusion lines of all welds, in^2

d = distance between center of gravity of weld to neutral axis of beam, in

l = full length of weld (should be length of structural member), in

I = moment of inertia of the structural member about its neutral axis, in^4

Some typical experimental values and calculated values for the same beams are given in Fig. 16-9.

Relieving Stress

The type of material, weld, and process (usually arc welding), and the service requirements of the part may require improvement of the metallurgical properties and the removal of residual stresses by stress relieving or annealing. The

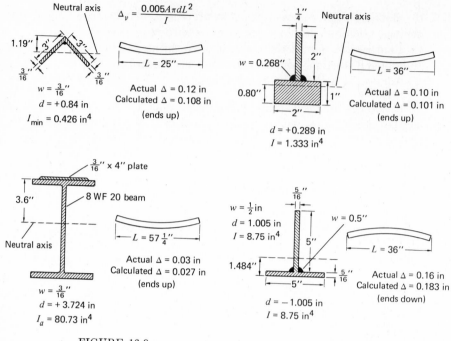

FIGURE 16-9
Longitudinal distortion in weldments. (*From Omar Blodgett, Welding Engineer, Feb. 1972.*)

kinds of weldments for which stress relief is advisable and the kinds that need no stress relief are listed below.

1 In general, statically loaded structures composed of light sections, nonrigid connections, and members containing only a small amount of welding need not be stress-relieved. Buildings, bed plates, housings, tanks, certain classes of pressure vessels, motor frames, etc., are typical structures of this class.

2 Work that requires accurate machining must not become distorted during machining or warp later as a result of the redistributing of the stresses. Jigs, fixtures, lathes, planer beds, etc., are structures of this class, and require stress relief.

3 Parts subjected to dynamic loads should, in general, be stress-relieved because experience has shown residual stresses to be more serious for loadings of this type than for static loadings. Rotors for electrical machinery, diesel engine crankcases, gears, etc., are in this class. There are some outstanding exceptions in this class of structures, such as bridges, freight cars, locomotive cabs, etc. The latter are generally composed of thin plates or long members containing small amounts of welding; consequently they can be considered as belonging to the first classification.

4 Structures composed of heavy or large parts and containing large amounts of welding should preferably be annealed, because they generally contain large residual stresses. The Boulder Dam cylinder gates for the intake towers are a typical example of this class of structure.

5 Materials that undergo metallurgical changes in the base or weld metal and produce hard zones and high residual stresses should be strain-relieved by heating. These materials include high-carbon and alloy steels.

From the previous classifications, it might appear at first sight that the major portion of all welded structures should be annealed. This, however, is not the case, because most of the structures fall within the first classification. The others actually comprise only a small percentage of all welded structures.

The process of stress relieving consists of slowly heating parts to a suitable temperature, allowing them to soak at this temperature for a definite period of time, and then slowly cooling them in the furnace. For steel the annealing temperatures generally used range from 1100 to 1200°F. (This temperature range has been established experimentally and is recommended by all welding codes.) These temperatures are not high enough to produce grain refinement, but they do have a tempering effect, which tends to increase the ductility of the weld and slightly lower the tensile strength. The tempering effect which results from strain annealing is often found beneficial in softening hard areas adjacent to the weld, thereby facilitating machining operations.

The theory of strain annealing to relieve residual stresses is dependent upon three things. First, within certain limits of strain, a ductile metal will not develop a stress greater than its yield point. Second, the yield point of metals decreases at high temperatures. Third, the phenomenon of creep, or plastic flow, takes place at high temperatures. The function of these factors during the annealing operation is best explained by an example.

Assume that a weldment containing residual stresses close to the yield point (30,000 lb/in²) of the base metal is to be annealed at 1100°F. When the temperature of the structure increases above 600°F, the yield point of the material decreases. The residual stresses also decrease (provided the amount of plastic strain is small) because the metal cannot develop a stress greater than the yield point. When the annealing temperature (1100°F) is reached, the remaining residual stress will be approximately 12,000 lb/in², which is the yield point of the base metal and the weld deposit at this temperature.

If the structure is cooled uniformly immediately after 1100°F is reached, the final residual stress will be 12,000 lb/in². If, however, instead of cooling the structure immediately after it has reached the annealing temperature, it is allowed to soak at this temperature, the phenomenon of creep occurs, and the metal flows plastically, thus further reducing the residual stresses. The longer the soaking time, within limits, the lower the final residual stresses will be. Note that subcritical stress relief does not wipe out residual stress completely but can reduce it to a minimum level of 5,000 lb/in².

Data obtained from welded test plates show that residual stresses, regardless of their initial values, can be reduced to a negligible amount, and that their final value decreases as the annealing temperature increases and the soaking time increases.

Annealing temperatures above 1200°F can be used, but they introduce excessive distortion and consequently are seldom used unless required to produce metallurgical changes in the weld or the heat-affected areas.

Weldability

Weldability denotes the relative ease of producing a weld which is free from defects such as cracks, hard spots, porosity, or nonmetallic inclusions. Weldability depends on one or more of five major factors.

Melting point   When welding low-melting-point alloys such as aluminum in thin sections, care must be taken to avoid melting too much base metal.

Thermal conductivity   Alloys with higher thermal conductivity, such as aluminum, are difficult to bring to the fusion point. Welds in certain alloys may cool quickly and crack. A high-intensity heat source is very important. For example, for a given size weld, aluminum alloy requires up to three times as much heat per unit volume as does steel.

Thermal expansion   Rapid cooling for alloys with high thermal coefficients of expansion results in large residual stresses and excessive distortion.

Surface condition   Surfaces coated with oils, oxides, dirt, or paint hinder fusion and result in excessive porosity.

Change in microstructure   Not only are steels above 0.4 percent C subject to grain growth in the HAZ (heat affected zone) but martensite is also formed wherever the temperature exceeds 1330°F for a sufficient time.

Steels   Steels vary widely in their weldability especially according to their carbon content. Plain low-carbon steels have excellent weldability, but for steels above 0.3 percent carbon, special operations such as preheating and postheating are needed. The amount of preheating and postheating increases with carbon content. Alloy steels and quenched and tempered steels also require special preheat interpass temperature control and/or postheat cycles.

Since steels are particularly susceptible to hydrogen embrittlement, care must be exercised to avoid introducing moisture into the weld zone. Low-hydrogen electrodes should be used; moreover, they should be stored in special warming ovens to prevent moisture absorption from the atmosphere.

Stainless steels   Stainless steel is a special case. The molten pool must be as free as possible from oxides if the stainless properties are to be retained in the weldment. The gas tungsten-arc process is best for high-quality work. Only low-carbon grades or grades with special carbide stabilizers should be specified if service in corrosive atmospheres is expected. Standard grades of stainless steel may fail by migration of carbides to the grain boundaries of the heat-affected zone where intergranular corrosion occurs. To minimize carbide precipitation:

1 Use a low-carbon less than 0.03 percent C base metal such as 304L stainless steel.

2 Use a columbium stabilized filler rod with stabilized base metal such as 316.

3 Heat above 1800°F and quench immediately after welding.

Generally the 400 series of martensitic stainless steels is more weldable than the 400 series of ferritic stainless steels because the latter have grain growth which leads to lower ductility, toughness, and corrosion resistance. Austenitic types of stainless steel (the 300 series) make up the bulk of stainless steel weldments.

Stainless steels can be spot-welded easily because of their low thermal conductivity, which is about 17 percent as large as that of steel. The low conductivity and absence of a high-resistance surface coating make the fabrication

Table 16-2   WELDABILITY OF ALUMINUM IN VARIOUS METHODS

Type of aluminum	Gas shielded-arc	Shielded metal-arc	Resistance	Gas
1060	A	A	B	A
1100	A	A	A	A
3003	A	A	A	A
3004	A	A	A	A
5005	A	A	A	A
5050	A	A	A	A
5052	A	A	A	A
2014	C	C	A	X
2017	C	C	A	X
2024	C	C	A	X
6061	A	A	A	A
6063	A	A	A	A
6070	A	B	A	C
6071	A	A	A	A
7070	A	X	A	X
7072	A	X	A	X
7075	C	X	A	X

A—Readily weldable
B—Weldable in most applications; may require special techniques
C—Limited weldability
X—Not recommended

of steel by resistance spot-welding possible at much lower current settings than needed for aluminum alloys.

Aluminum alloys   Many aluminum alloys are easily welded by the gas tungsten-arc and gas metal-arc processes, but their weldability varies considerably according to the alloying element and its amount. The low melting point, high thermal conductivity (about 10 times that of steel), high chemical activity, and high thermal expansion made aluminum alloys rather difficult to weld in the past. But today, the TIG process provides high unit energy and good protection from oxidation. See Table 16-2 for a comparison of the weldability of aluminum alloys.

Other alloys   Many other metals and alloys can be welded but most of them are of limited use in production. Zirconium, beryllium, and titanium can all be welded if the gas tungsten-arc process is used. For welding titanium the root side of the weld must also be protected. Nickel, and most magnesium- and copper-base alloys, can be welded successfully. Cast irons can be welded provided preheat and postheat treatments are used, but high-nickel electrodes are required, especially if the weldment is to be machined. Cast iron can also be braze-welded, but then service temperatures should not exceed 500°F. Most cast-iron welds are repair welds rather than welds for production joining.

Welding Versus Other Processes

Since 1920 welding has competed with riveting, bolting, and casting. Without it neither high-pressure boilers nor nuclear power plants would be economically possible. A welded joint permits no play; nor, if continuous, will it allow leakage of fluid. The demand for sheet-metal products, including items too thin in cross section to be cast, has increased markedly with the development of resistance welding. The cost of fixtures and of cleaning surfaces in preparation for finishing is generally less for welding than for casting, especially if the production quantity is low. Yet welding a given product can cost more in labor than casting it. Sometimes the expense of strain relief precludes a welded design.

In a number of cases, steel weldments can replace gray iron castings on an economic basis, but good engineering design practice would suggest that a valid decision could only be made if a casting redesign were carried out at the same time. A good example of weldment superiority would be motor generator bases for dc power supplies or air conditioning units. Such welded steel bases are lighter and less expensive for about the same rigidity. On the other hand, most machine tool bases are cast if they are complex and if there are more than a few to be built.

The strength of cast components is equivalent to that of weldments provided equivalent designs and alloys are compared. In tension, gray iron is weaker and less stiff than steel; but in compression, the opposite is true; therefore, when there are combined stresses as there are in most engineering applications, the strength advantage of steel is considerably lessened. Thus it is wise to analyze each design application carefully before deciding which production method is superior.

On balance, the designer of a product must consider both the properties of materials and the characteristics of available equipment. While welding is more costly than casting, bending, or cold forging in some cases, it is often the most useful—particularly if the material in question is easily welded and if suitable welding equipment has already been installed.

WELDING PROCESSES

Arc Welding—Consumable Electrodes

Shielded metal-arc welding Manual arc welding is widely used in the construction and fabrication of metal sheets, plates, and roll-formed products. The equipment includes a source of direct or alternating electric current, a ground, an electrode holder, and proper safety equipment. The latter consists of a helmet with dark eye protection, long sleeves, and a leather apron. The ultraviolet light from the welding arc can cause the equivalent of sunburn or snow blindness.

A conventional electrode forms a molten pool in the joint area. A gaseous shield and slag protect the weld deposit from oxidation and rapid loss of heat (Fig. 16-10a). Unskilled operators find the drag-type electrode, with large amounts of iron powder in the electrode coating, much easier to use (Fig. 16-10b). The iron powder increases the rate of deposition, but reduces the penetration and permits the core to burn away so that the coating can drag along the surface and the arc length stays constant. Thereby a good deposit can be made by an operator with relatively little skill.

In shielded-metal-arc welding the arc is started by momentarily striking the electrode against the base metal and quickly withdrawing to form an arc. The arc must not be too long, as this gives an opportunity for contamination by the atmosphere, and it is more difficult to control its application to the joint. The current and voltage must be under close control; they are governed by the quality of equipment and its inherent regulating characteristics.

When the arc forms between the base metal and the electrode, the immediate surface is melted, and, with the use of an electrode that cannot conduct the heat away rapidly, some of the metal is vaporized. These droplets and the vaporized metal flow along the stream of the arc path to the base metal where they condense, build up, and solidify. (Motion-picture studies of this action have been made and are available from leading welding equipment suppliers.) There-

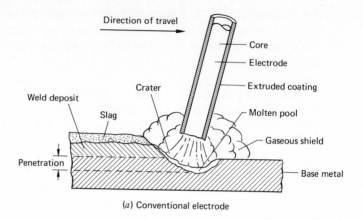

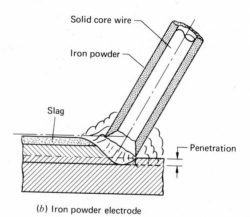

FIGURE 16-10
Shielded metal-arc welding with conventional and iron-powder electrodes.

fore, the arc process is primarily a localized casting process that is influenced by the action of the electrode, current, flux, and operator. In the liquid and gaseous state, it is essential that no harmful chemical action (such as oxidation and forming of nitrides) occur, that gas occlusions escape, that flux inclusions be avoided, and that the material cool without tearing or cracking.

Shielding   The molten metal is protected from contamination by four general methods:

   1   The electrode is coated with a flux which melts and forms a gaseous envelope around the arc and a liquid covering over the molten metal.

2 An inert gas envelope is blown around the electrode as it melts. The inert gas is argon, helium, or mixtures of argon and helium. This method has proven especially satisfactory in the welding of aluminum, magnesium, and copper alloys.

3 The end of the electrode is submerged in a powdered flux. This surrounds the arc and protects it and the molten metal.

4 Carbon dioxide is used as a shielding gas to weld low-carbon and alloy ferritic steels. Carbon dixoide is less expensive than inert gases and gives deeper penetration into the base metal. When carbon dioxide is used with filler metals of proper chemistry, it produces welds of high quality and soundness. A mixture of carbon dioxide and argon increases the stability of the arc over either a pure carbon dioxide or pure argon gas. Pure carbon dioxide causes splatter in and about the weld. Pure argon prevents oxidation of the metal with resulting surface tension and uneven surfaces. With a combination of carbon dioxide and argon, a slight oxidation takes place and the result is a smooth weld surface. Argon plus 25 percent carbon dioxide is used for welding steel. Especially for stainless steel argon plus 1 to 5 percent oxygen may be used. Shielding gases are used at rates of 10 to 35 ft³/h.

There are two basic types of coated stick electrodes: low-hydrogen types and cellulosic-coating types.

Low-hydrogen coating This is an all-mineral coating, containing no material that will form hydrogen. The coating is baked at high temperature during manufacture to remove moisture. Electrodes are sealed in containers by the manufacturer and stored under elevated temperatures by the user to prevent moisture absorption.

Cellulosic coating This is not a low-hydrogen coating. It can contain moisture. It essentially is a mixture of binders with cellulose and carbon as well as the necessary minerals for alloying.

Only coated electrodes with low hydrogen content that are kept in a drying oven can be used for welding low-alloy and high-tensile-strength steels. This type of electrode depends upon the transfer of the molten mineral slag coating for shielding. Therefore, it is important that an extremely close or short arc be used when welding with low-hydrogen electrodes.

For welding low-carbon mild steels, electrodes which contain cellulose in the coating can be used. The burning of the cellulose furnishes a gas shield which is primarily carbon dioxide. Therefore, maintenance of a short arc is not critical with this type of electrode.

Gas shielding In the gas-metal-arc welding process, various inert gas combinations can be used to protect the weld metal and molten pool. The process was originally called MIG welding for metal-inert-gas welding.

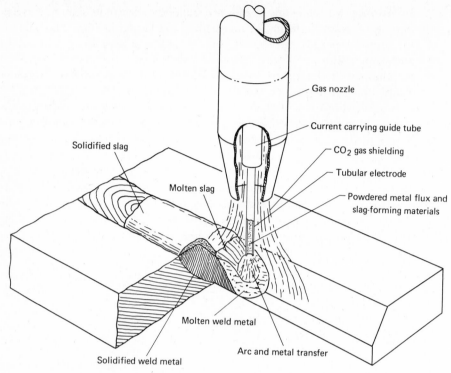

FIGURE 16-11
Schematic flux core electrode and gas shield.

But carbon dioxide and other gases are not inert; the process has changed but the name remains. Carbon dioxide is a slightly oxidizing gas and therefore electrodes should be selected with sufficient deoxidizer content to produce sound welds. Deoxidizers include silicon, aluminum, manganese, and others. The following combinations are used:

CO_2 + solid wires with sufficient deoxidizers in the wire chemistry
CO_2 + cored wires having sufficient deoxidizers as a flux in the core with a low-carbon steel shell (Fig. 16-11) and self-shielded fluxed cored wire welding (Fig. 16-12)
CO_2 + argon with solid wires having sufficient deoxidizers in wire chemistry

The CO_2 with solid wire is more economical because of no slag removal and CO_2 is cheaper than argon.

More effective welds can be obtained by proper selection of alloy electrodes, proper flux or gas protections, and automatic equipment to reduce variations caused by manual operation.

Flux cored electrodes   Many gas shielded–metal-arc applications are being replaced by semiautomatic welding facilities which use flux cored wire electrodes. The operation can be fully shielded by the flux in the core of the continuously fed electrode, or auxiliary carbon dioxide gas shielding can be applied through the electrode holder. The flux materials are sealed into the center of the electrode during its manufacture. Therefore it is always dry and ready for use. The use of an internal flux makes the continuous feeding of the electrode simple and ensures good electrical contact between the power source and the electrode.

Compared with an equivalent stick-electrode operation, the flux-cored wire facility gives deeper penetration, faster deposition rates, and continuous operation; and, because of the lightness of the equipment, the operator can make out-of-position welds with considerably less effort than he can with the conventional setup. In addition the process is easily learned and can use the same power source as is used for stick-electrode welding.

Power supplies for arc welding   The power supply to the arc may be dc or ac, and the polarity of the electrode may be positive or negative when dc is used. The type of electrode, the base material, and the position of the joint determine

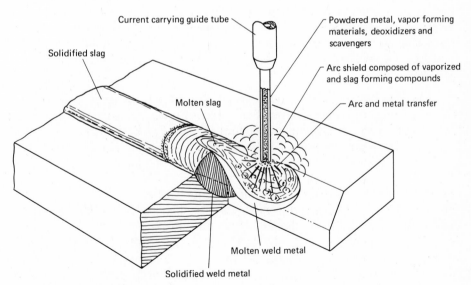

FIGURE 16-12
Schematic flux core electrode and no gas shield.

the polarity that should be used for the electrode. With electrode-negative polarity, more heat is concentrated at the tip of the electrode and more electrode is usually melted off per minute than with electrode-positive polarity. There is usually less tendency for burn-through, and better fill-in on poor fit-up when a negative electrode is used. With electrode positive and work negative, the base metal is hotter than the electrode.

Current control Too much current tends to burn and scatter the metal. Too little current does not fuse the base material to the welding material. Welding equipment should include voltage- and current-regulating controls to compensate for variations in the supply voltage caused by sudden overloads or reduction in loads on the power circuit to which the welding equipment is connected. An experienced operator can determine by observation whether he is getting proper penetration and fusion without burning the material. It is helpful experience for the designer to learn to weld, so that he can better visualize the problems of the welder and detect improper welds.

The operator is blind for the interval between the lowering of the helmet and the striking of the arc. After he strikes the arc, he can see the welding process through his dark glasses. Precautions should always be taken to shield the eyes and exposed portions of the skin from the arc flash to avoid a severe burn.

Arc blow The arc itself is a flexible gaseous conductor of current and is subject to deflection by outside magnetic forces. The passing of current creates magnetic lines of force which pass through the base material. When the base material is magnetic (steel, for instance), the phenomenon of arc blow[1] occurs, and it is often difficult to get an arc to penetrate into a corner or fillet. Alternating current has less arc blow than direct current. The following factors are related to arc blow:

1 If the direction of current changes, the arc will change its position.
2 Magnetic material around the arc will cause a change in direction of the arc when its position and volume changes.
3 Eddy currents of ac or pulsating dc welding affect the arc according to the position of hot-welded material and magnetic shunts near the arc.

Semiautomatic welding In semiautomatic welding flexible rubber hose, metal tubes, and wires carry the electrode wire, flux, gas, and electric current to a portable nozzle which can be manipulated by the operator about as easily as a stick-electrode holder (Fig. 16-13). Thus, the parts to be joined need not be positioned on the machine, but can be welded in any position provided that the proper small diameter wire and machine settings are used. In semiautomatic

---

[1] The term *arc blow* is used when the arc is unstable and cannot be directed properly by the operator. Sometimes it is impossible to deposit metal in a joint due to arc blow. Arc blow is not limited to magnetic base materials.

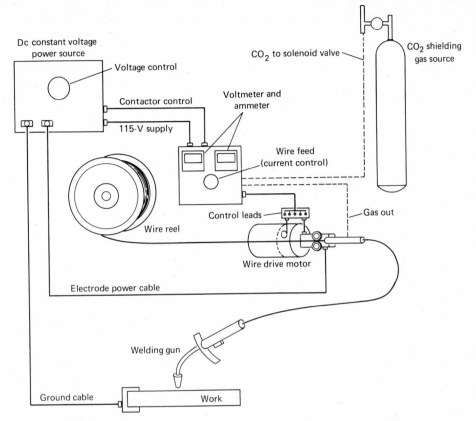

FIGURE 16-13
Schematic layout of typical flux-cored electrode welding facility.

welding the wire is fed through the hand-guided gun automatically and the arc length is always maintained.

The distinguishing feature of gas metal-arc welding is the mode of metal transfer, which can be either (1) globular, (2) spray, or (3) short circuit. The type which occurs depends on the diameter of the electrode wire, the arc voltage, the welding current, the type of shielding gas, and travel speed.

1   *Globular transfer* occurs at currents which are lower than those used for spray transfer. The molten drop on the end of the electrode wire grows to several times the wire diameter before it passes to the molten pool. Globular transfer is not a practical welding process because it results in too little penetration and poor arc stability.

2   *Spray transfer* occurs in the presence of argon gas which permits high current density and a large electrode diameter. The resulting deep penetra-

tion is good for heavy gage metals but it causes burn-through on thin gage alloys. The bead appearance is excellent with very little spatter.

3 *Short-circuiting transfer* occurs because as a molten droplet forms at the end of the wire, it bridges the arc when it contacts the molten pool and shorts it out from 20 to 200 times a second depending on the control setting. A number of gas mixtures can be used from 25 percent CO_2 balance argon to all CO_2. Short-circuit transfer results in a smaller, cooler molten pool which is particularly adaptable to welding out of position or for joining thin sheets.

Usually dc power sources with constant potential and slope control are used for MIG welding. Reversed polarity (wire positive) and current settings of about 100 to 400 A for thicknesses of 0.050 to $\frac{1}{2}$ in provide for good melting, deep penetration, and proper cleaning action during the spray-transfer process. In the short-circuit arc process, only shallow penetration can be achieved.

Automatic welding using consumable electrodes In automatic welding the machine provides the wire feed, the correct arc length, and the weld travel speed (Fig. 16-14). The bare-wire electrode is rarely used without a flux or gas envelope. Cored wire contains flux and alloys within a central core. The wire is used alone or with a gas shield. Some of the advantages of automatic welding over hand welding are:

1 A much higher rate of welding speed can be maintained.
2 Welding is continuous from the beginning to end of a seam, and intermittent craters are eliminated.
3 A more steady and uniform arc can be maintained.
4 Better fusion is possible because higher welding currents can be used.
5 No welding wire is lost in the stub ends. This means higher deposition efficiency and no time lost for electrode changes and slag removal.
6 A welding operator of limited experience can handle the welding machine and produce satisfactory results.
7 Slag removal is eliminated when solid wire is used or much reduced when using fluxed-cored wire.

Submerged arc welding Automatic welding in conjunction with proper tooling has three broad types of applications: circular welds, linear welds, and mass production of identical parts. Automatic welding has been made possible by the development of welding heads that strike the arc, feed the electrode, and maintain an arc of proper length and current. The metal electrode wire is coiled on reels and fed continuously. These heads are mounted on adjustable supports that may move along the welded joint, or the part to be welded may move under the head (Fig. 16-14). Various kinds of equipment are available for clamping the parts into position, for positioning the part, and for feeding the wire electrode.

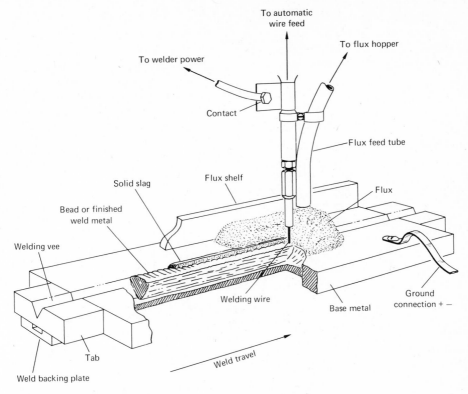

FIGURE 16-14
Submerged arc welding operation. (*Courtesy Hobart Bros., Troy, Ohio.*)

The submerged arc is an important process used with automatic welding equipment. The automatic submerged arc processes include:

1 Single ac or dc—straight or reverse polarity (this is the most common method)
2 Series arc—ac (used for cladding)
3 Three-phase, two-wire—ac, dc, or combination ac and dc
4 Multiwire—ac for high deposition rates

Submerged arc welding is used in connection with automatic and semi-automatic welding equipment. The submerged arc process uses a mineral powdered flux that surrounds the electrode and arc. The electrode can melt rapidly and fuse with the parent metal under a protective atmosphere. The flux is easily removed. No arc flash appears. The operator judges the proper location by observing the general direction of the wire and the welded material. High current densities (up to 40,000 A/in²) and high rates of metal deposition are possible with high quality, deep penetration, and high welding speed.

THE MECHANICS OF THE ELECTROSLAG WELDING EQUIPMENT

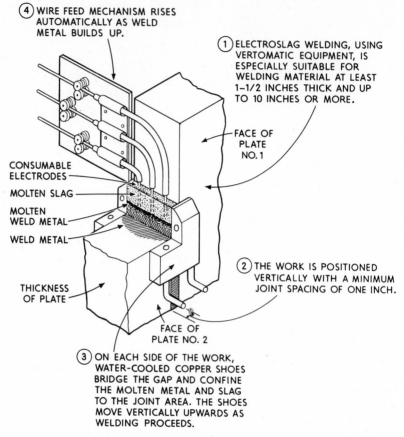

(4) WIRE FEED MECHANISM RISES AUTOMATICALLY AS WELD METAL BUILDS UP.

(1) ELECTROSLAG WELDING, USING VERTOMATIC EQUIPMENT, IS ESPECIALLY SUITABLE FOR WELDING MATERIAL AT LEAST 1–1/2 INCHES THICK AND UP TO 10 INCHES OR MORE.

FACE OF PLATE NO. 1

CONSUMABLE ELECTRODES

MOLTEN SLAG

MOLTEN WELD METAL

WELD METAL

(2) THE WORK IS POSITIONED VERTICALLY WITH A MINIMUM JOINT SPACING OF ONE INCH.

THICKNESS OF PLATE

FACE OF PLATE NO. 2

(3) ON EACH SIDE OF THE WORK, WATER-COOLED COPPER SHOES BRIDGE THE GAP AND CONFINE THE MOLTEN METAL AND SLAG TO THE JOINT AREA. THE SHOES MOVE VERTICALLY UPWARDS AS WELDING PROCEEDS.

FIGURE 16-15
Electroslag welder. (*Courtesy Arcos. Corporation.*)

Direct-current automatic welding seldom uses currents above 600 to 1,000 A, and is used usually for alloy and stainless steels. Alternating-current welding, which predominates, uses currents up to 2,000 A for standard equipment, and more for special equipment; ac welding is used usually for low-carbon steels.

A manually operated submerged arc welder has been developed that has the flexibility of hand operation and the advantages of automatic welding. The electrode wire ($\frac{1}{8}$ in) is fed through a flexible tube up to 55 in/min, and with 450 A of current applied at the nozzle. The powdered flux is fed around the electrode at the nozzle by compressed gas which carries the powder through a tube connected to the nozzle. The operator is much more comfortable because of the absence of smoke, spatter, and visible arc rays.

Electroslag welding   The electroslag welding process is unique in that it joins plates by casting the filler metal between the butted edges of the parent metal. The joint is made in one pass. The filler metal can be alloyed as desired through the introduction of alloy granules on top of the molten flux or through the cored consumable electrode. The process is explained in Fig. 16-15, and is applied to vertical joints. Copper bottom and side plates are needed to start the process. Special fixtures and procedures can produce circumferential welds to join thick-walled pipe of large diameter.

Arc Welding–Nonconsumable Electrodes

Gas tungsten-arc welding (TIG)   In this process, the arc usually passes between a tungsten electrode and the metal joint. High temperature (up to 10,000°F) is concentrated at the end of the arc, where a small pool of molten metal is formed. The arc passes from the electrode to the work and is shielded by an inert gas such as helium or argon or a mixture of the two. Formerly a carbon electrode was also used, but in the 1950s the (TIG) tungsten–inert-gas process proved to be economically superior. The TIG name is still used in the shop but the AWS designation is technically gas tungsten-arc welding (Fig. 16-16).

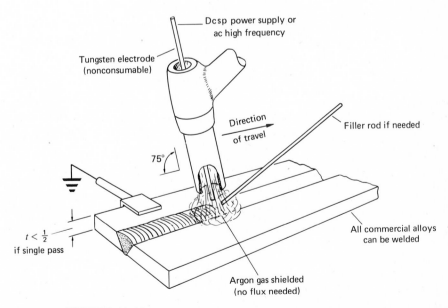

FIGURE 16–16
Gas tungsten-arc (TIG) welding operation. (*Courtesy of Linde Division, Union Carbide Corp., New York.*)

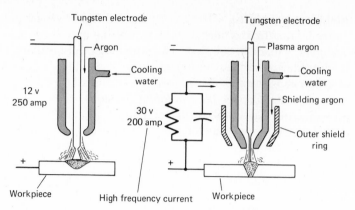

FIGURE 16-17
Schematic comparison of TIG and plasma arc. (*Courtesy of AWS.*)

TIG welding was originally developed for joining magnesium alloys, but it is now used for all alloys. It is particularly adapted to welding dissimilar metals and for hard-facing worn or damaged dies. It can also be adapted to welding light-gage sheet.

In general an ac power source is best for TIG welding nonferrous alloys except deoxidized copper. For ferrous alloys the dc power source with straight polarity (electrode negative) is better for gas tungsten-arc welding because it greatly reduces the volumetric loss from the tungsten electrode. For example, a TIG torch which has a rating of 250 amp when used with straight polarity (dcsp) must be derated to 15 to 25 amp when used with reverse polarity (dcrp-electrode positive).

Arc spot welds can also be made with a TIG torch fitted with special adaptors. In this case a $\frac{1}{8}$-in electrode is often used for a total cycle time of $\frac{1}{2}$ to 3 s depending on the alloy and its thickness.

Plasma-arc welding Plasma welding provides temperatures of 28,000°F or higher, as compared with 10,000°F for an electric arc. Therefore, the high-temperature flame can be used (1) to melt any metal or ceramic powder for surfacing metals or other materials, (2) to cut any material that will melt, and (3) to weld metals by heating the joint with or without filler metal.

A nonconsumable tungsten electrode within a water-cooled nozzle is enveloped by a gas (see Fig. 16-17). The gas is forced past an electric arc through a constricted opening at the end of the water-cooled nozzle. As the gas passes through the arc, it is dispersed. This releases more energy and raises the temperature of the gases, leaving the nozzle to 28,000°F or more. The arc may pass between the electrode and the workpiece, in which case it is called a transferred arc; or it may pass between the electrode and the nozzle, in which case it is called a nontransferred arc.

The types of arc welding are similar to gas tungsten-arc welding in that the heat is applied at the joint to obtain a weld and, when required, a filler rod can introduce material into the joint.

Electron-beam welding   In this process the heat for coalescence is derived from the impingement of a stream of high-velocity electrons on the area to be joined. The process requires a focused electron beam, control circuits, and a high vacuum or at least low pressure to avoid collisions between the electrons and molecules of air. Most electron-beam welders are powered by a triode gun and accessory circuits (Fig. 16-18) and are available up to 1 A at 60,000 V. The trend is toward machines with even greater power and larger vacuum chambers.

After the electron beam is aligned with the joint seam, its power, spot size, and time of application can be easily varied to control the width and depth of the weld. The great concentration of energy in the beam (up to 0.5 to 10 kW/mm²) permits the width of the heat-affected zone to be extremely narrow when compared with any other welding process (Fig. 16-19). Thus heat dissipation is minimized so that even pieces up to 5 in thick can be welded from one side. However, when specifying electron-beam welding, it is important that the

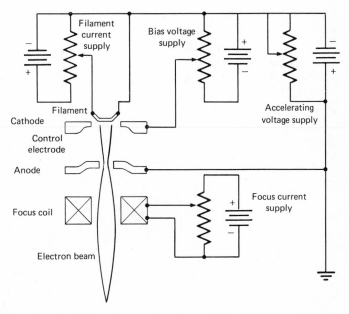

FIGURE 16-18
Triode electron-beam gun. (*Courtesy Sciaky Bros., Chicago, Ill.*)

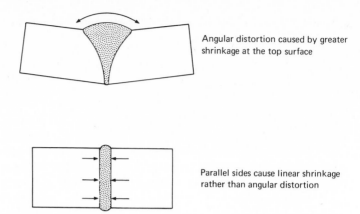

Angular distortion caused by greater shrinkage at the top surface

Parallel sides cause linear shrinkage rather than angular distortion

FIGURE 16-19
Effect of weld geometry on weldment distortion in electron-beam welding. (*Courtesy Sciaky Bros., Chicago, Ill.*)

mating parts fit together accurately to ensure ease of proper alignment of the mating parts. It is common practice to specify that faying surfaces be machined to close tolerances prior to assembly and joining by electron-beam welding.

The gun-transport system and the work-holding device must be designed and built with the accuracy of a machine tool. In fact optical aligning devices and closed-circuit television are integral parts of a large machine. Modern welders have specially designed electronic scanning devices which continuously locate the center of the weld within ± 0.0005 in on a routine basis. Feedback circuits correct for any misalignment and recenter the beam if there is any tendency to drift. The same device can be used to follow broken surfaces or circular or irregular contours.

The process usually requires a vacuum of about 10^{-4} Torr (1 Torr = $\frac{1}{760}$ standard atmosphere) for optimum quality. However, commercial welding can

FIGURE 16-20
Closeup view of the distributor cam piece parts and the welded assembly. (*Courtesy Sciaky Bros., Chicago, Ill.*)

be successful in vacuums of as little as 0.1 Torr. Such vacuums can be achieved in only a few seconds with roughing pumps alone.

Commercial applications   Electron-beam welding has found wide application in aerospace and defense applications, but it is also useful under low-vacuum conditions for high-volume production in the automotive and appliance industries. Electron-beam welding has been successfully applied to the production of automotive distributor cams (Fig. 16-20). In this case, an automated dual-sliding-seal, 12-station, dial-feed welder equipped with a soft vacuum produced 1,400 welded assemblies per hour (Fig. 16-21). No production time was needed for the pump-down operation because while one assembly was being welded in the inner chamber, the next fixture was loaded in the outer chamber and evacuation was begun there. Although the complete installation cost more than several hundred thousand dollars, it paid for itself in less than a year because not only was operation time saved but valuable space was made available when the former furnace brazing equipment was retired.

Another production application is a retainer ring for a roller bearing assembly for nuclear reactors welded at 200 pieces per hour with no damage to the rolls and fuel elements, and assembled without danger to the operator.

Typical joint designs   Joint design—material, preparation, and fixturing—are important factors in the economic application of electron-beam welding of

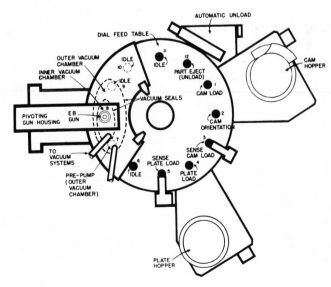

FIGURE 16-21
Schematic of high-speed dial-feed unit.

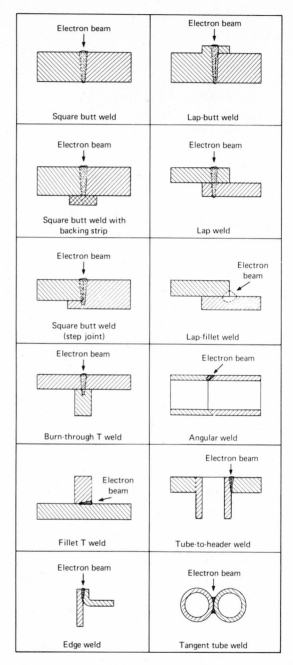

FIGURE 16-22

Typical joint configuration for electron-beam welding. (*Courtesy Sciaky Bros., Chicago, Ill.*)

Table 16-3   JOINT DESIGN

Do's	Don'ts
1 Linear joints should be machined square within 0.005 in.*	*1* Avoid poorly machined weld joints.
2 Gapping should be held to a maximum of 0.005 in.*	*2* Avoid partial penetration annular welds in areas of high restraint.
3 For annular welds, a 0.0005 to 0.0001 in press fit should be considered.	*3* Avoid loose fitting assemblies for annular welds in areas of close final tolerances.
4 For partial penetration annular welds, a relief groove in the root area should be considered to equalize shrinkage.	*4* Avoid partial penetration annular or girth welds in areas of high stress.
5 For girth welds a self-locating step should be used whenever possible in low stress areas.	*5* Avoid sharp corners when using a step joint.
6 Self-locating step joints can be used to eliminate underbead on the ID in low stress.	*6* Avoid welding parts having internal cavities which are not vented properly.
7 For linear or girth welds which employ a step joint, 0.005 in gap should be allowed in the lower portion of the step to accommodate lateral shrinkage.	*7* Avoid making square butt welds on sheet-metal parts with sheared edges.
8 Step joint corners should be chamfered.	*8* Avoid using backing bars or strips which are not metallurgically compatible with the base material.
9 For combination of annular and girth welds on the same assembly, the girth welds should be made first.	*9* Avoid lap weld joints in areas of high stress and fatigue.
10 Parts having internal cavities must be vented properly.	*10* Avoid gapping between sheets of material when using lap weld joint.
11 If backing bars or strips are used, they should be of the same alloy as the base material.	
12 Utilize lap-weld joints for use with sheet metal and thin-gage materials wherever possible.	
13 Consider seal pass techniques for annular welds to eliminate the possibility of gapping when other than interference fits are employed.	

* Larger values are feasible. Usable values of joint squareness and fit depend on the joint thickness, the material being welded, and the weld quality that is acceptable. In joints of more than 0.25 in thickness, gaps of 0.010 to 0.015 in may be acceptable in some instances; in 1-in-thick joints, gaps of 0.030 in can be tolerated if filler metal is wire-fed into the joint during its welding.

Table 16-4 MATERIAL WELDABILITY

Do's	Don'ts
1 High-quality wrought or forged materials should be considered whenever possible.	*1* Castings of poor density should be avoided.
2 Aluminum-killed steel should be considered as a first-choice material.	*2* Avoid rimmed steel unless maximum rim cannot be mixed into the weld zone.
3 Silicon-killed steel material should be considered as a second-choice material.	*3* Avoid case-hardened materials unless minimum amounts of case can be mixed into weld zone.
4 Rimmed steel can be used as long as maximum amounts of "rim" and minimum amount of "core" are mixed into the weld zone.	*4* Do not use parts that are case-hardened with any process using ammonia (this leaves a high nitrogen content in the surface-hardened area).
5 Minimum amounts of case should be mixed into the weld zone when welding case-hardened materials.	*5* Avoid cast materials unless slow welding speeds can be used.
6 Utilize slow welding speeds when welding cast materials.	*6* Avoid free-machining steels which contain high percentages of sulfur or phosphorus.
7 Utilize aluminum spray, shims, or wire when welding nonkilled steels.	*7* Avoid welding materials such as brass, bronze, zinc, magnesium, and powdered metals containing high percentages of nonmetallic binders.
8 Consider preheating and/or postheating when welding materials containing more than 0.35% carbon.	

quality assemblies. It also has a major influence on the production rate which can be achieved.

Typical joint designs are given in Fig. 16-22, and checklists of do's and don'ts have been borrowed from the Sciaky Bros., Chicago, and are given in Tables 16-3 to 16-6.

Resistance Welding

Resistance welding is the heating of material at the junction to be welded by local resistance to passage of electric current. Spot, projection, seam, flash, percussion, and butt welding are forms of resistance welding. The material is raised to a temperature which causes it to become plastic and, under pressure, it is fused or forged together. The principle is the same as that used in any black-smith-forged joint.

The amount of heat depends upon the amount of current and the length of time it is applied ($H = I^2RT$). The amount of current depends upon the voltage applied and the total resistance or impedance of the circuit; therefore, voltage must be consistent regardless of variations in the power required. Some welding equipment, especially that used for welding aluminum, places heavy

Table 16-5 MATERIAL PREPARATION

Do's	Don'ts
Minimum requirements	
1 Materials should be free of contaminants such as oil, grit, and heavy scale.	1 Avoid grinding of nonferrous materials.
2 For material having irregular surfaces, such as corrugations, pickling can be used.	2 Avoid wire-brushing of nonferrous materials.
3 Clean with acetone or equivalent solvent.	3 Avoid solvents of oil when machining nonferrous materials.
4 Demagnetize ferrous materials before welding.	4 Avoid use of solvents containing chlorides.
	5 Avoid use of solvents or oils when machining case materials.
Treatment of the joint area for optimum weld quality	
1 Remove all traces of scale and surface oxides.	
2 Materials should be machined to approximately 32 rms finish.	
3 Steel alloys can be machined or surface-ground.	
4 Nonferrous materials should be machined dry (no solvents or oils).	
5 Nonferrous materials should be hand scraped or chemically cleaned to remove surface oxides.	
6 Cast materials should be machined dry if possible.	

Table 16-6   WELDMENT TOOLING

Do's	Don'ts
1 Tooling should be made of nonmagnetic materials in areas of close proximity to the beam.	*1* Avoid use of magnetic materials in close proximity to the beam.
2 Use forged-wrought or densely cast materials.	*2* Avoid use of porous cast materials.
3 Blind holes or grooves should be vented.	*3* Avoid materials which contain lead or zinc plating.
4 Tooling fixture and clamps must have a good ground.	*4* Avoid use of nonmetallics, such as ceramics, rubber, and plastic which are porous.
5 Parts which are insulated for preheating purposes must have good ground return.	*5* Avoid severe restraint tooling when making annular welds.
6 Tooling should provide sufficient lateral pressure to minimize gapping during welding.	*6* Avoid unsupported edges when welding sheet metal or thin gage material.
7 When making girth welds on light-gage materials, spring-load for follow-up pressure.	*7* Avoid use of plastics and other electrically insulating or nonconductive materials in close proximity to the beam.
8 For girth welds in heavy thicknesses of material use rigid clamping.	
9 Maximum support and clamping close to the weld joint should be considered when welding sheet metal or light-gage material.	

demands on power lines and often requires special feeders and transformers to maintain suitable electrical capacity and voltage. The total resistance or impedance of the welding or welding equipment's circuit depends upon the following factors.

1 The impedance of the welding circuit varies as the position of the part within the welder changes. If the part is magnetic, the lines of force will pass through the material and reduce the current. Therefore, if a resistance weld is made when a small portion of the part is near the electric circuit, it will receive more current than it does when the part is moved so that a large portion is included in the electric circuit.

2 The resistance or impedance of the electrical equipment producing the current influences the amount of current. These parts can be designed with suitable electronic control, so that variations in current can be compensated for to a great extent—even variations in position of part, line voltage, and resistance of the joint.

3   Resistance of the joint is composed of: (a) resistance between electrodes or clamps and material; (b) resistance at joint of mating parts; (c) resistance of mating material.

Each one of the resistances at the surface of the material is affected by the size of electrode, uniform condition of the surface (cleanliness), and pressure applied by the electrodes or clamps.

Successful application of resistance welding to designs depends upon consistency of material composition, surface, pressure and current applied, and time of their application. Low-carbon steel is the most common material welded by resistance welding and is assumed in all data given in this chapter.

Resistance welding is usually performed with alternating current. The greatest advance in the use of various forms of resistance welding came in the early 1920s when engineers realized that consistency in each of the related factors would assure good welds. First, mechanical devices were developed to apply proper pressure and control the length of time the current was applied. Recently, with air and hydraulic systems for applying pressure at the correct time and in the right amount through electronic controls, resistance welding has been advanced. The length of time for applying current can be controlled from one-half cycle to as many cycles as desired. A shot of high current for a short time produces the best weld, since the heat is concentrated at the joint and does not have time to spread, cause distortion, or affect the material adjacent to the joint. Resistance welds in automobiles, airplanes, passenger cars, and all forms of sheet-metal equipment are accepted without question on the part of the public and in many cases have replaced arc welding and riveting. Equipment designed for resistance welding (such as electrical switch gear) has reduced scrap, weight, and labor, and has increased the use of standard parts.

Spot welding   Spot welding, the most common form of resistance welding, consists of joining two pieces of material by placing them between two electrodes (Fig. 16-23) and passing a current to heat the material sufficiently at the joint to cause plastic flow and the union of the two parts. The parts are held together while they cool sufficiently to regain mechanical strength. As outlined under the general principles of resistance welding, the greatest resistance should be between the two parts to be joined. The initial pressure should be great enough to obtain contact and then provide a forging action. The current must be controlled to give sufficient heat yet not melt or burn the material.

The diameter of the electrode end must remain the same and not mushroom and increase the area of contact. When this happens, the current density

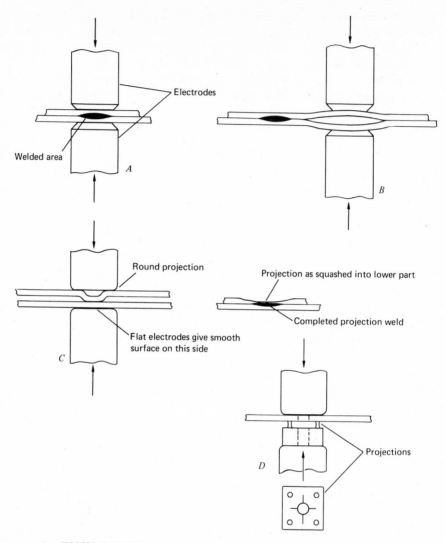

FIGURE 16-23
Resistance welding. *A*, spot welding; *B*, pressure is sometimes necessary to force parts together before welding; *C*, projection welding; *D*, projection welding of a special nut to a sheet.

is reduced. The area increases in proportion to the square of the diameter, and the heat generated is reduced in proportion to the square of the current (I^2R); therefore, the electrodes should be machined to size and should not be filed by an operator. The contact surface should meet the material surface evenly. The electrodes are usually water-cooled to prevent softening. Electrodes of special alloys are stronger, but have more electrical resistance. Copper is the best electrode material for general application.

Spot welds, unlike rivets, require no holes or riveting, heading, or squeezing operations. The spot can be placed easily because the operator can see the work as the weld is made. The time for assembling and positioning the parts in the spot welder is greater than the few seconds required for making a spot weld; therefore, it is common practice to make multiple spot welds. Strength equivalent to that of riveted joints can be obtained by an equivalent spot-welded structure. Since there is no movement of joints, squeaks are avoided. Shear and tensile strength are close to that of the material welded. The strength of spot welds should be determined by experiment, using material and equipment suppliers' data as a guide. Portable spot welders are used to join members in large structures. They permit the same flexibility that is obtained with riveting air hammers.

Sometimes the parts do not have proper contact at the joint of the weld and considerable pressure is required to force the two surfaces together and make a weld at the place desired (Fig. 16-23A). This condition may cause a poor weld. Multiple welds made with electrodes in series or parallel are difficult to make unless all factors are controlled. For this reason, projection welding has come into general use.

The weldability of metals for spot and projection welding is shown in Table 16-7.

Projection welding Projection welding is the use of a projection on one or both pieces to be welded, which forms a spot weld. More than one projection weld can be made at a time, as the area and location of the spot weld and contact pressures can be controlled. The current will be distributed uniformly between the multiple spots and good welds will result (Fig. 16-23C). For example, in Fig. 16-23D a special nut is shown projection-welded to a sheet. Spot- and projection-welded joints, like riveted joints, are not liquid- or gas-tight; therefore, seam welding was developed to make a continuous joint.

In all projection welding, it is necessary to have point or line contact in order to start a weld. When joining dissimilar sections, the projection should be placed on the heavier part. Heavy sections lend to joining by projection welding. The American Welding Society's "Recommended Practices for Resistance Welding" include recommended projections for sheet and plate sections up to 0.50 in. It should be remembered that as sections increase in thickness, the diameter and height of projections are increased to develop greater strength.

Only clean, scale-free surfaces should be used in projection welding. A dirty surface will cause much variation in the resistance between the parts being joined, with resulting variation in current flow and weld strength.

Line projection welds (see Fig. 16-24) are recommended over point welds when sections are subject to heavy static or dynamic loads.

Table 16-7 SPOT WELDING—WELDABILITY OF MATERIALS

Blank—Combination not tried
+ —Good weld
Ø—Good weld with limitations
X—Completely miscible but brittle weld
—Poor weld
0 —No weld

	Aluminum	Brass	Bronze	Copper	Cop. sil.	Iron-steel (low C†)	Iron-galv.* (hot dip)	Kovar	Lead	Magnesium	Molybdenum	Monel	Nickel	Nickel silver	Silver	Steel, cobalt	Steel, copper plated*	Steel, nickel plated*	Steel, stainless (austenitic)	Steel, tin plated*	Tin	Zinc
Aluminum	+																					
Brass	0	+																				
Bronze	0	+	+																			
Copper	0	Ø	Ø	Ø																		
Copper silicon	0	Ø	+	+	+																	
Iron, steel (low carbon)†	0	Ø	Ø	Ø	#	+																
Iron, galv. (hot dip)*	0	#	#	#	#	#	#															
Kovar						Ø		+														
Lead		#	0	#	Ø		0		+													
Magnesium	+									+												
Molybdenum	0	#	#			X																
Monel	0	Ø	Ø	#	#	+	#	Ø			X	+										
Nickel	0	Ø	Ø	Ø	Ø	+	#	Ø	0			+	+									
Nickel silver	0	Ø	Ø	Ø	Ø	+	#				X	+	+	+								
Silver		#	#	Ø	Ø	Ø	#	Ø				#	#	#	Ø							
Steel, cobalt	#	Ø	Ø	#	#	+	#	Ø			#	#	#	#	#	#						
Steel, copper plated*	0	Ø	Ø	Ø	Ø	+	#	Ø				+	+	+	Ø	+	+					
Steel, nickel plated*	0	Ø	Ø	Ø	Ø	+	#	Ø	#			+	+	+		#	Ø	+				
Steel, stainless (austenitic)	0	Ø	Ø	Ø	Ø	X	#	Ø				+	+	+	Ø	+	+	+	+			
Steel, tin plated*	#	Ø	Ø	Ø	Ø	Ø	#		Ø							#			Ø	Ø		
Tin	#	Ø	Ø	Ø	Ø	#	#	+	+			+	0	+		#			#	0	+	
Zinc	+	X	X	X	X	0	0		0			X	0	X			0	0	0	0	0	+

* In the course of spot welding coated materials the coating frequently dissolves in the other metals present or burns away.

† Limited to 0.15% carbon content for single thicknesses of less than $\frac{1}{8}$; limited to 0.25% carbon content for single thicknesses of $\frac{1}{8}$ to $\frac{1}{2}$.

This table should be taken only as a general guide, bearing in mind that some of the pairs marked "∅" call for special manipulation or equipment and that others so marked may produce unsatisfactory welds in certain varieties, as for instance, high-carbon stainless steel. Some of the pairs marked "0" can be welded by means of special techniques. Welding copper to copper becomes very difficult when heavy sections are involved.

In many cases, the success of a weld between these pairs depends upon the rating, timing control, and other characteristics of the available welding equipment.

It is probable that in time to come, some of the "0" signs will be changed to "+." It will be noted that most combinations with aluminum are marked "0." This means that, as a regular shop process, aluminum cannot be welded to these materials.

By special techniques, such as interposing thin pieces of high-resistance material, it is sometimes possible to accomplish some of these welds.

In case of doubt the designer should consult the specialist of the process or material company.

Limitations refer to equipment (size and type of control equipment), thickness of metal, and application of special technique. The welding section of the division involved should recommend use of limited applications before starting production.

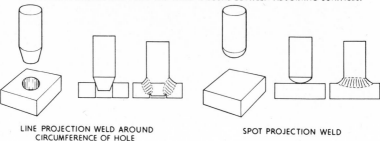

FIGURE 16-24
Line-type projection welds.

Seam welding A seam weld is a joint being continuously welded by the resistance-welding process. The electrodes are disks which are driven as the two pieces to be welded pass between them. Pressure and current are applied to the joint, as in spot welding. When seam or continuous welding was first developed, the current was continuous, but the heat was difficult to control. It was soon discovered that overlapping spot welds was more successful. By making a series of spot welds in rapid succession, the operator finds that slight variations in contact pressure, surface conditions, and electrode contact resistance result in a better weld. Representative seam welds are 12 spots per inch on stock 0.010 in thick at a speed of 100 in/min, and 5 spots per inch for $\frac{1}{8}$ in thickness at a speed of 25 in/min. Water is sprayed on the electrodes to cool them and the weld material. Sometimes only one roller is used and a bar is substituted for the lower roller. More than one seam weld can be made at a time on special machines. An example of this is welding parallel seams in refrigerator radiator shells. Most seam welding is limited to sheet metals 0.010 to 0.125 in thick (Fig. 16-25). Intermittent spots are made rapidly on seam-welding equipment—600 spots per minute, $\frac{1}{2}$ in apart.

Seam welding is applied to lap-joint seams of cylinders and cabinets and to circular seams for welding bottoms in ends of cylindrical tanks.

Flash-butt welding Flash welding is a resistance-welding process where joining is produced simultaneously over the entire area of abutting surfaces.

The flash-butt welding process produces a homogeneous weld between two sheets, wires, or bars without overlapping and without the addition of any materials (Fig. 16-26). Dissimilar materials may be flash- and butt-welded. Flash welding is limited only by the amount of current available to heat the surfaces and the pressure available for forging the parts together. Flash welding joins sheets together in fabricating the typical automobile body. Some companies

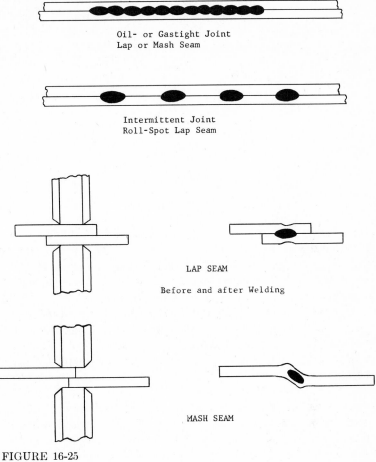

Oil- or Gastight Joint
Lap or Mash Seam

Intermittent Joint
Roll-Spot Lap Seam

LAP SEAM

Before and after Welding

MASH SEAM

FIGURE 16-25
Types of seam-welded joints.

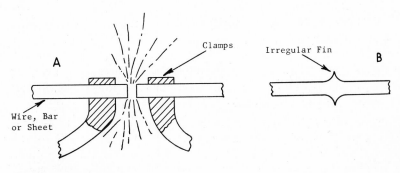

A

Clamps

Irregular Fin

B

Wire, Bar
or Sheet

FIGURE 16-26
Flash-welding operation.

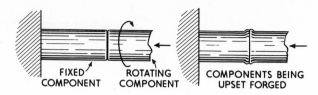

FIGURE 16-27
Friction welding.

find it economical to weld small sheets into a large sheet from which the car top is made.

Percussion or stored-energy welding   Percussion welding is similar to flash, stud, and spot welding in that a very high current is passed instantaneously through the surfaces to be joined, and the parts are joined immediately thereafter. The parts are moved toward one another rapidly. Just prior to contact the current is passed through the two conductors being joined. This usually is accomplished by the discharge of a condenser. Dissimilar metals are welded with very little penetration of the heat within the parts.

Stored-energy welding involves an electrical means of storing and releasing large amounts of energy. Spot or percussion welding utilizes a condenser or a transformer circuit, or a combination of the two. Thus less capacity is required in feeder lines and transformers. Stored-energy systems are used frequently for welding aluminum.

Stud welding   In the past it was difficult to flash-weld a stud to a flat plate because the stud would burn away and the plate would not heat sufficiently to permit a forged weld. A stud welder has been developed that is a combination arc welder and flash welder. The stud welder holds the stud as an electrode within an insulated tube. The end of the tube is placed over the location of the stud, and the stud is momentarily allowed to strike the plate. An arc is struck and passes between the stud and plate. The plate is heated by the arc, the stud is melted, and molten material is built up on the plate as in arc welding. After sufficient time, the stud welder cuts off the current, releases the stud, and drives the stud and built-up molten material together to make a forged weld similar to a flash weld. A strong weld results. This is a great saving over fastening studs to plates or flanges by drilling and tapping, welding threaded lugs to the plate, or arc welding studs to plates.

Friction Welding

Friction welding has been used successfully to join the ends of two components. One of the two components is rotated rapidly about an axis of symmetry and against the other component. Friction heat is developed at the interface. After the correct temperature is reached, rotation is stopped and the two components are upset forged together (see Fig. 16-27).

This process is applicable to steels, brass, aluminum, titanium, and practically all other metals, as well as ceramics and thermoplastics. Carbon steel can be joined to stainless steel or aluminum. It can be used to join tubes, solid bars, studs, etc., either to each other or to plates. Today, it can be considered as a substitute for both resistance and gas-pressure butt welding. This method eliminates the need for flux and special gaseous atmospheres.

The American method uses high speeds and low pressures (12,000 r/min and 2,000 lb/in²) as compared to the Russian method (1,200 r/min and 14,000 lb/in²). The higher rotational speeds ensure better cleansing of the abutting surfaces. The lower pressures cause less deformation, less chance for cracking, and less likelihood of premature distortion of the joined parts.

BRAZING, SOLDERING, AND COLD WELDING

Brazing

Brazing is the joining of metallic parts by a filler metal (such as copper and silver alloys) that has a melting point of 800°F or greater, which is lower than the melting point of the metals joined. The metals are usually at red heat and a flux is needed. The filler material adheres intimately to the surface of the metal and, with copper filler, forms an alloy with iron that is very near the strength of the iron itself. It should be recognized that this alloying action is at the surface only; therefore, it is necessary to keep the thickness of the filler metal to a minimum—a few thousandths of an inch thick is adequate. The strength of a brazed joint depends upon the shear area, the shear strength of the filler material, and the triaxiality of the applied stress. Therefore it is desirable that the filler metal cover the entire area without blowholes or gaps. The ability of the filler material to travel between the mating parts depends upon a very clean surface, proper fluxes, and the capillary attraction provided in the joint. The fluid metal cannot travel within a joint if the separation is excessive. The parts must be machined accurately, prepositioned, and held firmly in place without distortion while being brazed. Typical brazed joints are shown in Fig. 16-28.

The principal advantage of brazing lies in the fact that a strong joint can be made without undue distortion of parts. In continuous or batch furnaces, more than one joint can be brazed at the same time. The parts are heated and allowed to cool gradually, usually in a nonoxidizing atmosphere. Gas heat, oxyacetylene heat, and induction heating can be used to advantage in brazing operations. In all of these methods, even heating, clamping the parts to be brazed between jaws that are heated by the current passing through them enables the brazing of copper bars and leads on electrical equipment.

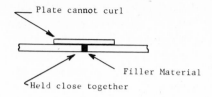

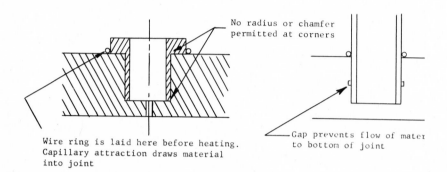

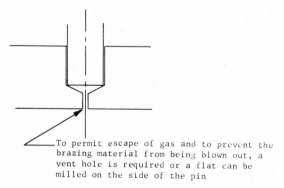

FIGURE 16-28
Typical brazed joints.

Induction brazing   Induction brazing refers to that process where the parent metals are brought to the fusion temperature of the filler metal by placing them in an electromagnetic field which occurs when an alternating current passes through an induction coil. The current flowing through the inductor sets up magnetic lines of force which pass through the surface of the parent materials and induce a flow of energy. Heat is developed in the parent materials due to hysteresis and eddy currents—and this heat in turn melts the filler metal.

It should be recognized, since induction brazing depends upon heat being applied by way of magnetic lines of force, that ferrous materials which are magnetic are well suited to the process. However, when ferrous materials lose their magnetic properties, such as when heated above the magnetic transformation point, they become increasingly difficult to braze. Copper and aluminum are unfavorably suited to induction welding in view of their magnetic properties.

In induction brazing, there are two considerations relative to the thermal aspects of the process which the production-design engineer should be cognizant of. These are the power necessary to heat the work to the required temperature and the generator rating necessary to get this power. The power in kilowatts re-

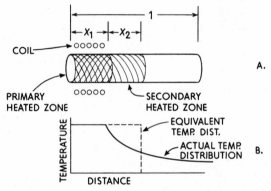

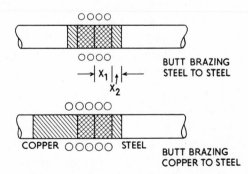

FIGURE 16-29
Brazing of similar and dissimilar alloys.

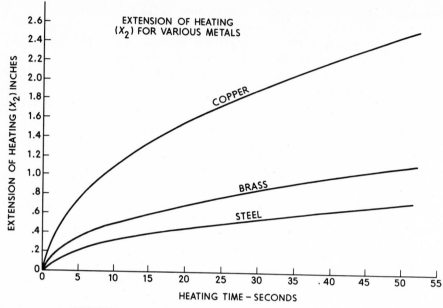

FIGURE 16-30
Extension of heating for various alloys.

quired to raise the temperature of a given amount of material to a given temperature may be calculated from the following relationship.

$$P = \frac{1.06\ WCT}{t}$$

where P = power, kW
W = weight of material, lb
C = specific heat of material, Btu/lb-°F
T = temperature rise, °F
t = time, s

After computing the thermal power needed, giving due consideration to the difficulties of bringing the power into the work, the production-design engineer will be able to determine the generator rating required. Figure 16-29 illustrates the temperature distribution within an induction-heated rod and Fig. 16-30 shows the extension of heating, x_2, for copper, brass, and steel.

The configuration of the work should be such that the inductor coil can be located to give close coupling to the section to be joined. Figure 16-31 illustrates some typical induction-coil configurations.

Braze Welding

Braze welding should be distinguished from brazing as a process used to repair a part by adding a nonferrous filler metal, such as naval brass. Such items as gears and machine frames made of steel or cast iron may be joined or built up by preparing the surface to be repaired and braze welding.

Nonferrous filler materials offer the advantage of accomplishing the work at lower heat input and consequent development of less thermal stress than if a ferrous filler material is used.

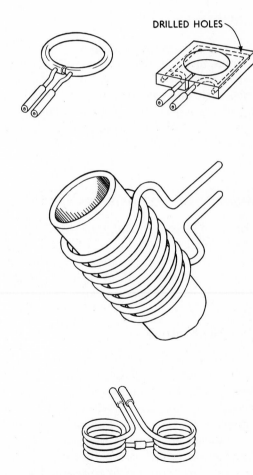

FIGURE 16-31
Typical coils for induction heating.

Soldering

Soldering is one of the most familiar joining processes. The filler material (solder) has low strength, less than 800°F melting temperature, and results in less alloying between the base metal and the filler. The strength of a soldered joint depends upon adhesion of the solder to the metals being joined and the type of mechanical joint. As in brazing, the thickness of filler material should be kept to a minimum. The same factors apply to solders as to brazing except that the soldering operation is performed at lower temperatures. Noncorrosive fluxes should be used wherever possible, since it is difficult to thoroughly clean the flux from the parts.

Cold Welding

Cold welding is the welding of metals by pressure at room temperature. Nonferrous metals, such as aluminum, cadmium, lead, copper, nickel, zinc, and silver, can be welded by this process. The surface must be clean of any contamination, including oxidation. The cleaning is best done by the use of rotating wire brushes before pressure is applied. Due to the high pressure the metal is squeezed together and the resulting joint is thinner than the sum of the two original thicknesses.

DESIGN OF WELDMENTS

About 70 percent of all fabricated steel is joined by arc welding—in production most often by MIG process, which stands for metal–inert-gas arc welding. But a design engineer is most interested in the major design factors which influence the economy and service life of typical weldments. Therefore a review of weld design fundamentals has been prepared.

Welding Symbols

Until recently there were no standards regarding how information should be conveyed from the designer to the weld shop. The American Welding Society has gained the cooperation of many of the leading industrial organizations who specialize in welding fabrication, and a set of conventional symbols has been developed and is now widely adopted. The AWS drawing nomenclature and standard symbols are given in the five-volume set of the American Welding Society "Handbook," which covers the full range of joints, weld sizes, and edge preparation for each major welding process. The basic welding symbol is an arrow connected to a reference line. All pertinent data for the desired weld are designated

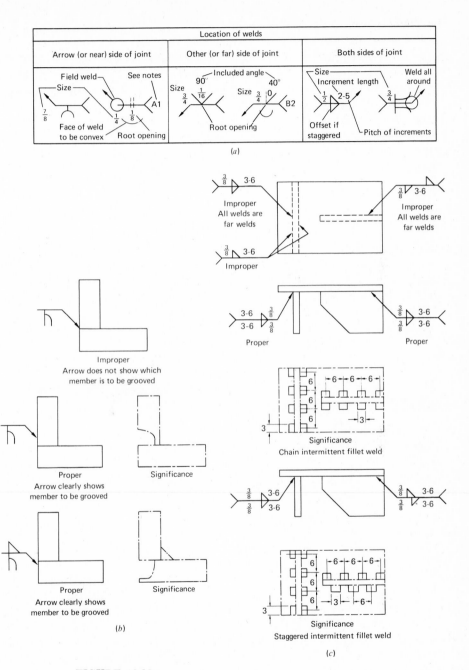

FIGURE 16-32
AWS standard joint symbols.

by symbols around the reference line. In the case of fillet, groove, flange, and flash welding, the arrow is drawn to touch the weld joint (Fig. 16-32). For plug, spot, seam, and projection welding the arrow should contact the centerline of the weld. The location of the weld can be represented by the location of the arrow with respect to the weld. If the weld is to be on the arrow side of the joint, the weld symbol is on the bottom of the reference line. If the joint is to be on the other side the weld symbol is above the reference line. If both sides are welded the symbol is located at both the top and bottom of the reference line. Further refinements and more detailed information can be obtained from the AWS "Handbook."

Joint Design

The American Welding Society has developed standard symbols for most types of welds, including four basic weld types (butt, fillet, flange, and plug), seven kinds of edge preparation, and the major resistance welds (Fig. 16-33). A weld bead is used for joining two pieces in the same plane end to end which results in a butt weld (Fig. 16-34).

Three types of edge preparation are commonly used in butt welding: (1) the *square edge*, the simplest and most economical, used for plates up to $\frac{1}{8}$ to $\frac{1}{4}$ in thick, depending on the process; (2) bevel or J joints for thicker plates where only one plate must be prepared; and (3) V or U joints, which require edge preparation on both plates (Fig. 16-35). Frequently it is advantageous to use double-edge preparation so that welding can be carried out on both sides. For joints over $\frac{1}{2}$ in

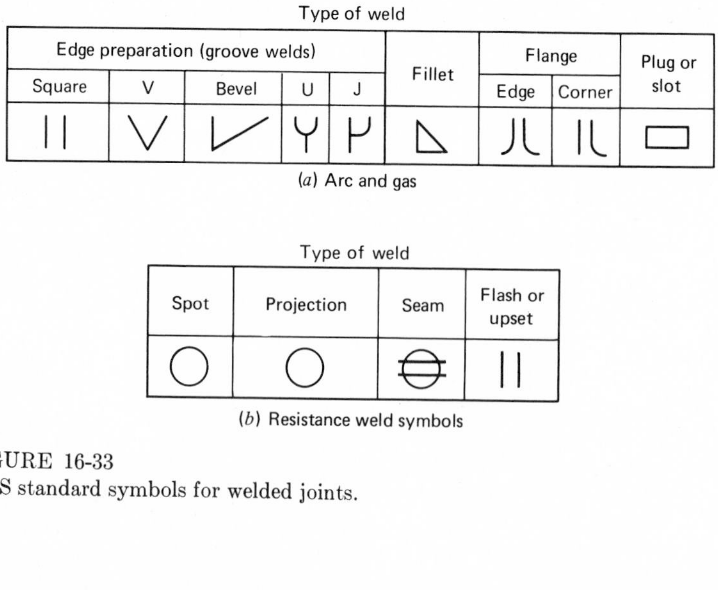

Type of weld

Edge preparation (groove welds)					Fillet	Flange		Plug or slot
Square	V	Bevel	U	J		Edge	Corner	
||	V	&#10697;	Y	P	&#9700;	JL	IL	▭

(a) Arc and gas

Type of weld

Spot	Projection	Seam	Flash or upset
○	○	⊖	||

(b) Resistance weld symbols

FIGURE 16-33
AWS standard symbols for welded joints.

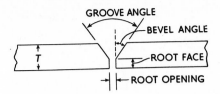

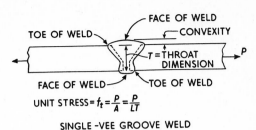

EDGE PREPARATION FOR A
SINGLE-VEE GROOVE WELD

SINGLE-VEE GROOVE WELD

UNIT STRESS = $f_t = \dfrac{P}{A} = \dfrac{P}{LT}$

FIGURE 16-34
AWS butt joint details.

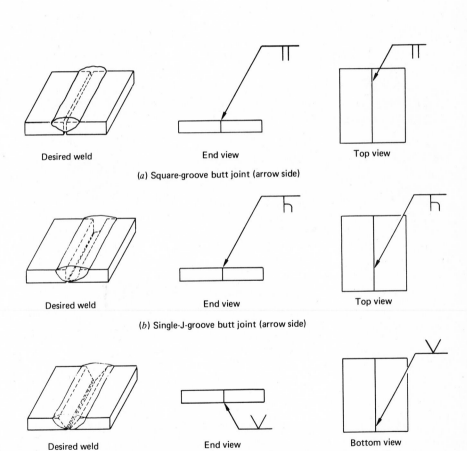

Desired weld   End view   Top view

(a) Square-groove butt joint (arrow side)

Desired weld   End view   Top view

(b) Single-J-groove butt joint (arrow side)

Desired weld   End view   Bottom view

(c) Single-V-groove butt joint (other side)

FIGURE 16-35
Typical joints for butt welds.

(Redrawn from Joseph Datsko, "Material Properties and Manufacturing Processes," John Wiley & Sons, Inc., New York, 1966.)

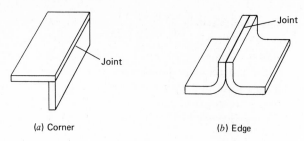

(a) Corner   (b) Edge

FIGURE 16-36
Weld bead as used for corner or edge welds.

thick designers should carefully investigate the economics of welding on one or both sides of the joint. Weld beads may also be used for corner and edge welds (Fig. 16-36).

In the T joint two members are joined by depositing two welds of triangular cross section, called fillet welds, on either side of the vertical member of the T (Fig. 16-37).

Edge preparation and typical joint details for a number of types of welds are given in Fig. 16-38.

The lap joint, as the name implies, refers to those joints where one member laps the other and a weld is made along the exposed ends of one or both members. Figure 16-39 illustrates a double fillet-welded lap joint. Fillet welds are stressed usually in shear at the throat area. The throat T is defined as the perpendicular to the hypotenuse of the largest isosceles right triangle which can be inscribed within the weld bead. For example, in Fig. 16-39 the weld stress would be

$$\sigma = \frac{P}{2LT} = \frac{P}{2(0.707)\ SL} = \frac{0.707P}{SL}$$

Corner joints are made usually with either a fillet or groove weld. Typical corner joints are illustrated in Fig. 16-40.

Weld Stress Calculations

The stresses in typical welded joints may be calculated by using the relationships given in Fig. 16-41. The direction of the applied load is shown by the arrow P. The size of a weld can be calculated if the load P is known or reliably estimated and the code requirement or factor of safety (usually 3, but up to 5 in some cases) is known.

Ever since welding has been used, it has had to meet the opposition of

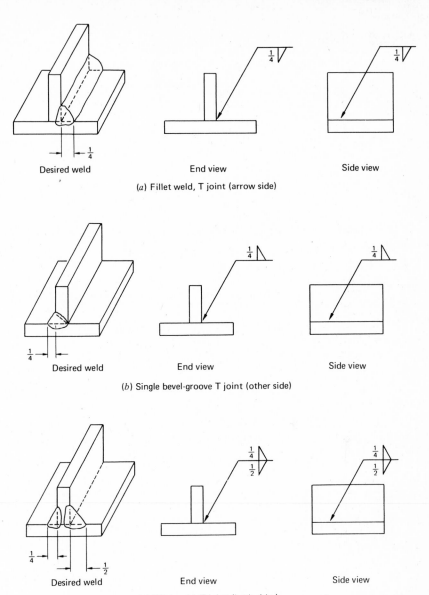

(a) Fillet weld, T joint (arrow side)

Desired weld   End view   Side view

(b) Single bevel-groove T joint (other side)

Desired weld   End view   Side view

(c) Fillet weld, T joint (both sides)

Desired weld   End view   Side view

FIGURE 16-37
Typical fillet welds.

(*Redrawn from Joseph Datsko, "Material Properties and Manufacturing Processes," John Wiley & Sons, Inc., New York, 1966.*)

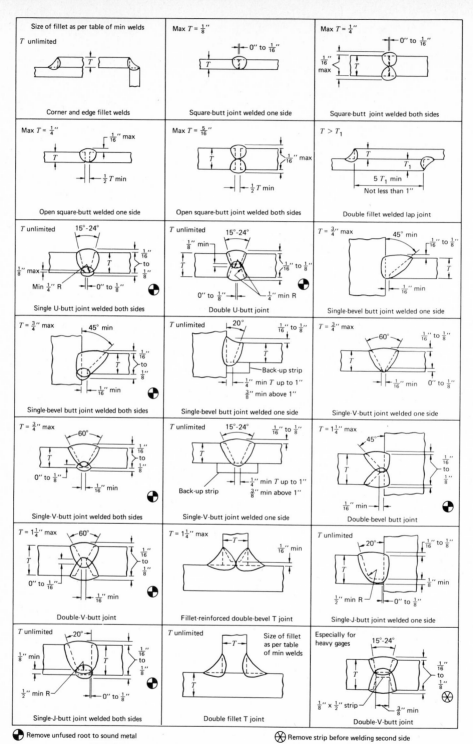

FIGURE 16-38
Edge preparation and typical joint details.

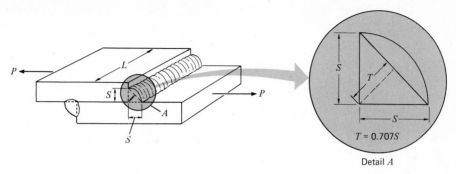

FIGURE 16-39
Double fillet weld.

conservative engineers who knew the reliability and uniformity of the material used in riveted and bolted joints; therefore, welded joints have always been calculated carefully and made under proper controls. Many experiments and extensive studies have been made to prove their reliability and agreement with calculated values. The mathematical derivation of formulas and the tests results made to prove their authenticity are shown in Fig. 16-42. An explanation of the figure is as follows.

The stress-concentration factor for welded joints on mild steel is

$$\text{Actual stress} = K \times \text{calculated stress}$$

For reinforced butt welds, $K = 1.2$; for toe of transverse fillet weld, $K = 1.5$; for end of parallel fillet weld, $K = 2.7$; for T butt with sharp corners, $K = 2.0$.

Stress-concentration factors are only important under conditions of dynamic loading and in cases where the joint is subjected to a static load with a

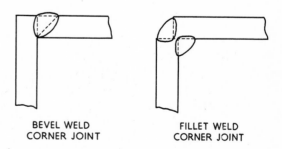

BEVEL WELD
CORNER JOINT

FILLET WELD
CORNER JOINT

FIGURE 16-40
Typical corner joints.

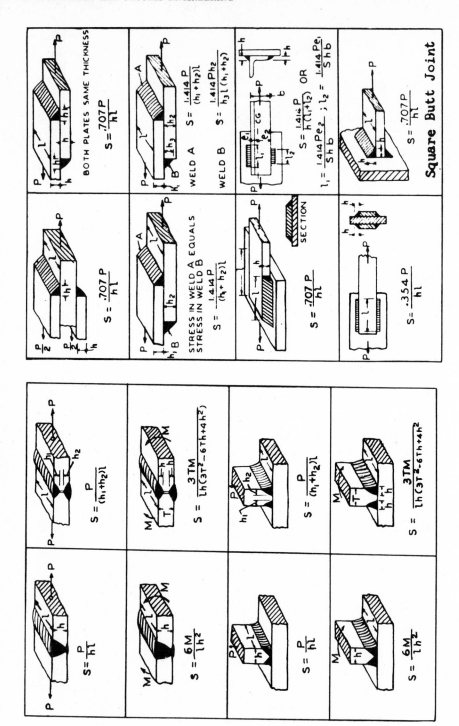

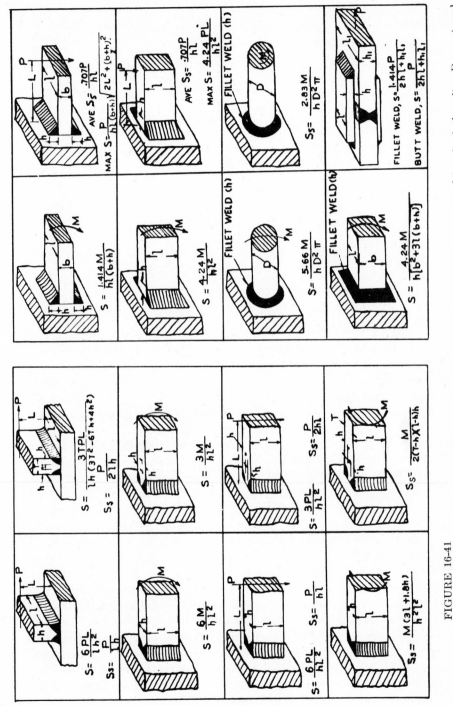

FIGURE 16-41

Stress formulas for typical welded joints. S = normal stress, lb/in²; I = moment of inertia, inch units; D = external load, lb; h = size of weld; Ss = unit stress, psi; M = bending moment, in·lb; L = linear distance; l = length of weld, in. (*Courtesy Charles Jennings Texas Testing Laboratory, Inc., Dallas, Texas.*)

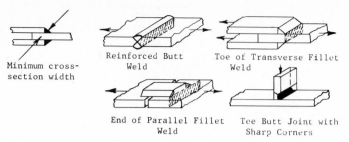

FIGURE 16-42
Working stresses for mild steel welds.

superimposed dynamic load. The following formula will be helpful:

$$\frac{K \ (S_{max} - S_{min})}{2 \ Swd} + \frac{(S_{max} + S_{min})}{2 \ Sws} = 1$$

where Sws = static design working stress
Swd = dynamic design working stress

A welded joint can be made equal or greater in strength, ductility uniformity, and reliability than the base material. In butt joints where no x-ray of the joint is specified, it is common practice to use 80 percent of the strength of the base material for the strength of the joint. If x-ray inspection is specified, then the strength of the base material can be used.

Arc-welded steel structures are three to six times as strong as ordinary gray cast iron, and the material is more uniform and reliable. They are 2 to $2\frac{1}{2}$ times as stiff as cast iron and 5 times as great in impact strength. Material can be added where it is most effective in meeting the functional requirements of the part. High-strength materials (alloy steels) can be incorporated where additional strength is desired. Steel-plate or sheet-metal walls are thinner and weigh less than cast walls. The cost of a pattern is eliminated; however, in repetitive parts or parts that are difficult to fabricate, welding fixtures and jigs may be necessary. Fixture costs should be compared with the pattern cost as well as comparative labor costs in making the choice between welding and casting (Fig. 16-43). Welding construction permits homogeneous base materials and the absence of sand inclusions; blowholes are not encountered in machining. The General Motors locomotive diesel-engine crankcases are welded in the production line. Cast construction was discarded because of the above factors. Hard surfacing materials can be added where wear due to abrasion may be encountered. Parts are stronger and consequently can be subjected to heavier machine cuts. Frequently, welded construction allows holding parts to closer tolerances, and less allowance for material to be removed by machining is necessary. Because of the smooth surface of steel-mill plates and bars, it is often not necessary to specify

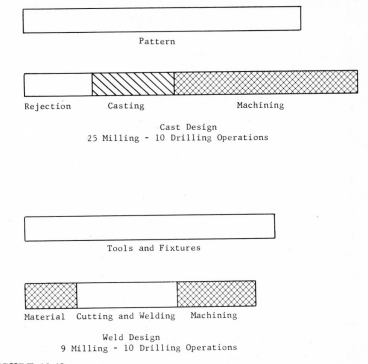

FIGURE 16-43
Cost comparisons of cast and welded designs for a part.

machined surfaces. Changes in design can be made without expensive pattern alterations. The production-design engineer should recognize that, as the structure becomes complicated and more forming and welding are required, the cost may exceed casting costs. Steel castings have about the same strength as welded structures of equivalent section thickness. Therefore, it is often profitable to cast such items as complex brackets and housings and combine them with structural steel members by welding them into the completed part. A part may be advantageously designed for welded construction when the quantity requirements are small, but, when the quantity increases, other processes and materials may be more economical. Thus quantity of manufacture is an important factor in determining whether welding will be economical.

Design of a Weld Size

Before the proper sizes and length of a weld bead can be specified for a given weldment, considerable time must be devoted to analyzing the forces on the welded joints with respect to their magnitude and direction. Low-cost welding

Table 16-8   MINIMUM WELD SIZES FOR
THICK PLATES (AWS)

Thickness of thicker plate joined t, in	Minimum leg size of fillet weld,* in
To $\frac{1}{2}$ incl.	$\frac{3}{16}$
$\frac{1}{2}-\frac{3}{4}$	$\frac{1}{4}$
$\frac{3}{4}-1\frac{1}{2}$	$\frac{5}{16}$
$1\frac{1}{2}-2\frac{1}{4}$	$\frac{3}{8}$
$2\frac{1}{4}-6$	$\frac{1}{2}$
Over 6	$\frac{5}{8}$

* Minimum leg size need not exceed thickness of the thinner plate.

can only be achieved by specifying and depositing the correct size weld bead. For applications which require maximum strength, groove welds must be used to assure 100 percent penetration. There is no need to calculate the strength of a groove weld, because it is more than equal to the strength of the plates being joined. Fillet weld sizes must be determined because they can be either too large or too small.

As a rule of thumb, the leg size of a fillet weld should be three-fourths of the plate thickness to develop the full strength of the plate. This rule of thumb is based on the assumption that:

1   The fillet weld is on both sides of the plate.
2   The fillet weld is the full length of the plate.
3   The thinner plate is the controlling factor if the plates are of unequal thickness.

If the weld is designed for rigidity only, then the weld size may be $\frac{3}{8}t$. For thick plates, i.e., those $\frac{3}{8}$ in and over, the fillet weld sizes recommended by AWS are given in Table 16-8.

Estimation of weld size by treating the weld as a line   In this method, as given by O. W. Blodgett of the Lincoln Electric Co.,[1] the weld is considered to be a line with no cross-sectional area, having a definite length and outline as shown in Fig. 16-44. Then use the information in Table 16-9 to find the properties of the weld when treated as a line. Insert the appropriate property of the

[1] Used by permission of the James F. Lincoln Arc Welding Foundation, Cleveland, Ohio.

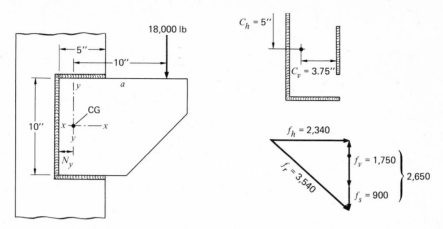

FIGURE 16-44
Determination of weld size by treating the weld as a line.

weld connection into standard design formulas from strength of materials to find the forces on the weld joint when subjected to the application of a particular load (Table 16-10). Thus the force on the welded connection is found in terms of pounds per lineal inch of weld.

For example, for tension or compression we have the standard design formula, which is the same formula as used for weld, treating weld as a line. That is,

$$\sigma = \frac{P}{A} \quad \text{lb/in}^2 \quad \text{or} \quad f = \frac{P}{A_w} \quad \text{lb/in}$$

where σ = normal stress in standard design formulas, lb/in^2
   f = load per inch of weld bead, lb/in
   P = tensile or compressive load, lb
   A = area of flange material held by welds in horizontal shear, in^2
   A_w = length of the weld measured along the contour of the weld, in

After determining the force, the weld size is determined by dividing the resultant force by the allowable strength of the appropriate weld (groove or fillet) as found in Table 16-11 for constant loads or in Table 16-12 for fatigue loads. Consider the following example, which illustrates the use of the procedure for determining the weld size required for the joint in Fig. 16-44, which is to be subjected to a load of 18,000 lb.

Example   *Step 1*   Find the properties of the weld, treating it as a line (use Table 16-9).

Table 16-9   PROPERTIES OF WELD TREATED AS LINE

Outline of welded joint b-width d-depth	Bending (about horizontal axis xx)	Twisting
	$S_w = \dfrac{d^2}{6}$ in²	$J_w = \dfrac{d^3}{12}$ in³
	$S_w = \dfrac{d^2}{3}$	$J_w = \dfrac{d(3b^2 + d^2)}{6}$
	$S_w = bd$	$J_w = \dfrac{b^3 + 3\,bd^2}{6}$
	$S_w = \dfrac{4bd + d^2}{6} = \dfrac{d^2(4b + d)}{6\,(2b + d)}$ top bottom	$J_w = \dfrac{(b + d)^4 - 6\,b^2 d^2}{12\,(b + d)}$
	$S_w = bd + \dfrac{d^2}{6}$	$J_w = \dfrac{(2b + d)^3}{12} - \dfrac{b^2(b + d)^2}{(2b + d)}$
	$S_w = \dfrac{2bd + d^2}{3} = \dfrac{d^2(2b + d)}{3\,(b + d)}$ top bottom	$J_w = \dfrac{(b + 2d)^3}{12} - \dfrac{d^2(b + d)^2}{(b + 2d)}$
	$S_w = bd + \dfrac{d^2}{3}$	$J_w = \dfrac{(b + d)^3}{6}$

$$S_w = \frac{2bd + d^2}{3} = \frac{d^2(2b + d)}{3(b + d)}$$

top   bottom

$$J_w = \frac{(b + 2d)^3}{12} - \frac{d^2(b + d)^2}{(b + 2d)}$$

$$S_w = \frac{4bd + d^2}{3} = \frac{4bd^2 + d^3}{6b + 3d}$$

top   bottom

$$J_w = \frac{d^3(4b + d)}{6(b + d)} + \frac{b^3}{6}$$

$$S_w = bd + \frac{d^2}{3}$$

$$J_w = \frac{b^3 + 3bd^2 + d^3}{6}$$

$$S_w = 2bd + \frac{d^2}{3}$$

$$J_w = \frac{2b^3 + 6bd^2 + d^3}{6}$$

$$S_w = \frac{\pi d^2}{4}$$

$$J_w = \frac{\pi d^3}{4}$$

$$S_w = \frac{\pi d^2}{2} + \pi D^2$$

Table 16-10   DETERMINING FORCE ON WELD

Type of loading		Standard design formula, stress, lb/in²	Treating the weld as a line, force, lb/in
		Primary welds (transmit entire load at this point)	
	Tension or compression	$\sigma = \dfrac{P}{A}$	$f = \dfrac{P}{A_w}$
	Vertical shear	$\sigma = \dfrac{V}{A}$	$f = \dfrac{V}{A_w}$
	Bending	$\sigma = \dfrac{M}{S}$	$f = \dfrac{M}{S_w}$
	Twisting	$\sigma = \dfrac{Tc}{J}$	$f = \dfrac{Tc}{J_w}$
		Secondary welds (hold section together—low stress)	
	Horizontal shear	$\tau = \dfrac{VAy}{It}$	$f = \dfrac{VAy}{In}$
	Torsional horizontal shear*	$\tau = \dfrac{T}{2A_t}$	$f = \dfrac{T}{2A}$

A = area contained within median line.
* Applies to closed tubular section only.

The following definitions are needed to estimate the weld size using the concept of defining the weld as a line:

n = number of welds—shear stress in standard design formulas
y = distance between the center of gravity of the flange material and the neutral axis of the whole section, in
I = moment of inertia of the whole section, in⁴

c = distance to the outermost fiber from center of gravity, in

t = thickness of plate, in

S = section modulus of the section, in³

J = polar moment of inertia of the section, in⁴

b = width of the connection, in

d = depth of the connection, in

V = vertical shear load, lb

M = bending moment, in·lb

T = twisting moment, in·lb

S_w = section modulus of the weld, in²

J_w = polar moment of inertia of the weld, in³

N_x = distance x axis to face, in

N_y = distance y axis to face, in

σ_y = yield strength, lb/in²

τ = horizontal shear stress, lb/in²

Table 16-11   ALLOWABLE STEADY LOADS (Lb/Linear in of Weld)

Fillet weld (for 1-in weld leg)	Groove weld (for 1-in weld thickness)	Partial-penetration* groove weld.† (for 1-in weld thickness)
	Parallel load	
E60 or SAW—1 weld 9,600 (AWS)	$\tau = 0.40\,\sigma_y$ of base metal (shear) (AWS)	E60 or SAW—1 weld 13,600 (AISC)
E70 or SAW—2 weld 11,200 (AWS)		E70 or SAW—2 weld 15,800 (AISC)
	Transverse load	
E60 or SAW—1 weld 11,200	$\tau = 0.60\,\sigma_y$ of base metal (tension) (AWS)	E60 or SAW—1 weld 13,600 (AISC)
E70 or SAW—2 weld 13,100		E70 or SAW—2 weld 15,800 (AISC)

* For tension transverse to axis of weld or shear-use table; for tension parallel to axis of weld or compression, weld same as plate.

† For bevel joint, deduct first ⅛ in for effective throat if done by manual electrode.

Table 16-12 ALLOWABLE FATIGUE STRESS FOR A7, A373, AND A36 STEELS AND THEIR WELDS

	2,000,000 cycles	600,000 cycles	100,000 cycles	But not to exceed
Base metal in tension connected by fillet welds, but not to exceed →	$\sigma = \dfrac{7,500}{1 - \frac{2}{3}K}$ lb/in² P_t	$\sigma = \dfrac{10,500}{1 - \frac{2}{3}K}$ lb/in² P_t	$\sigma = \dfrac{15,000}{1 - \frac{2}{3}K}$ lb/in² P_t	$\dfrac{2P}{3K}$ lb/in²
Base metal compression connected by fillet welds	$\sigma = \dfrac{7,500}{1 - \frac{2}{3}K}$ lb/in²	$\sigma = \dfrac{10,500}{1 - \frac{2}{3}K}$ lb/in²	$\sigma = \dfrac{15,000}{1 - \frac{2}{3}K}$ lb/in²	P_c lb/in² $\dfrac{P_c}{1 - K/2}$ lb/in²
Butt weld in tension	$\sigma = \dfrac{16,000}{1 - \frac{8}{10}K}$ lb/in²	$\sigma = \dfrac{17,000}{1 - \frac{7}{10}K}$ lb/in²	$\sigma = \dfrac{18,000}{1 - K/2}$ lb/in²	P_t lb/in²
Butt weld compression	$\sigma = \dfrac{18,000}{1 - K}$ lb/in²	$\sigma = \dfrac{18,000}{1 - 0.8K}$ lb/in²	$\sigma = \dfrac{18,000}{1 - K/2}$ lb/in²	P_c lb/in²
Butt weld in shear	$\tau = \dfrac{9,000}{1 - K/2}$ lb/in²	$\tau = \dfrac{10,000}{1 - K/2}$ lb/in²	$\tau = \dfrac{13,000}{1 - K/2}$ lb/in²	13,000 lb/in²
Fillet welds, ω = leg size	$f = \dfrac{51,000\,\omega}{1 - K/2}$ lb/in	$f = \dfrac{7,100\,\omega}{1 - K/2}$ lb/in	$f = \dfrac{8,800\,\omega}{1 - K/2}$ lb/in	8,800 ω lb/in

K = min/max
P_c = allowable unit compressive stress for member
P_t = allowable unit tensile stress for member
SOURCE Adapted from AWS "Bridge Specifications."

For the weld in this example the following calculations are needed:

$$N_y = \frac{b^2}{2b + d} = \frac{5^2}{2(5) + 10} = 1.25 \text{ in}$$

$$J_w = \frac{(2b + d)^3}{12} - \frac{(b)^2(b + d)^2}{(2b + d)}$$

$$= \frac{(2 \times 5 + 10)^3}{2(5 + 10)} - \frac{(5)^2(5 + 10)^2}{2(5) + 10}$$

$$= 385.9 \text{ in}^3$$

$$A_w = 5 + 10 + 5 = 20 \text{ in}$$

Step 2 Find the various forces on the weld, inserting the properties of the weld found above.

The forces are a maximum at point *a*. The twisting force per inch of weld is broken into horizontal and vertical components by proper values of *c*.

Twisting (horizontal component):

$$f_{t_h} = \frac{Tc_h}{J_w} = \frac{(18{,}000 \times 10)(5)}{385.9} = 2340 \text{ lb/in}$$

Twisting (vertical component):

$$f_{t_v} = \frac{Tc_v}{J_w} = \frac{18{,}000 \times 10(5.00 - N_y)}{385.9} = 1{,}750 \text{ lb/in}$$

Vertical shear:

$$f_{s_v} = \frac{P}{A_w} = \frac{18{,}000}{20} = 900 \text{ lb/in}$$

Step 3 Determine the resultant force on the weld.

$$f_r = \sqrt{f_{t_h}^2 + (f_{t_v} + f_{s_v})^2}$$

$$= \sqrt{(2{,}340)^2 + (1{,}750 + 900)^2} = 3{,}540 \text{ lb/in}$$

Step 4 Now find the required leg size of the fillet weld connecting the bracket, using Table 16-11.

$$\omega = \frac{\text{actual force}}{\text{allowable force}}$$

$$= \frac{3{,}540}{9{,}600}$$

$$= 0.368 \qquad (\text{approx. weld size} = \tfrac{3}{8} \text{ in})$$

Therefore make the leg $\tfrac{3}{8}$ in.

FASTENERS

"Fasteners" is a general term including such widely separated and varied materials as nails, screws, nuts and bolts, locknuts and washers, retaining rings, rivets, and adhesives, to mention but a few.

Fasteners are of many different types. Some fasteners, like nuts, bolts, and washers, have been in use for years. Other fasteners, such as Rivnuts and retaining rings, are fairly new in the field. A large number of different fasteners are used in most products. As an example, 80 fasteners are used in a telephone receiver and 136 different kinds of fasteners are needed throughout the country to erect, brace, and guy telephone poles.

Fasteners can be divided into two classifications: those that do not permanently join the pieces, and those that do. Included in the first category are such fasteners as nuts and bolts, lock washers, locknuts, and retaining rings. The second classification includes such fasteners as rivets, metal stitching, and adhesives.

Nuts, Bolts, and Screws

Nuts, bolts, and screws are undoubtedly the most common means of joining materials. Since they are so widely used, it is essential that these fasteners attain maximum effectiveness at the lowest possible cost. Bolts are, in reality, carefully engineered products with a practically infinite use over a wide range of services.

An ordinary nut loosens when the forces of vibration overcome those of friction, the latter being the basic principle of the nut and bolt. In a nut and lock washer combination, the lock washer supplies an independent locking feature preventing the nut from loosening. The lock washer is useful only when the bolt might loosen because of a relative change between the length of the bolt and the parts assembled by it. This change in the length of the bolt can be caused by a number of factors—creep in the bolt, loss of resilience, difference in thermal expansion between the bolt and the bolted members, or wear. In the above static cases, the expanding lock washer holds the nut under axial load and keeps the assembly tight. When relative changes are caused by vibration forces, the lock washer is not nearly as effective.

The slotted nut and cotter pin provides an assembly that locks. Since the diameter of the pin is smaller than the hole in the bolt and the slot in the nut, the nut can back off under vibration until these clearances are reduced and the cotter pin jams the assembly. Another method of locking is the use of a self-locking nut for which the top threads are manufactured at a reduced pitch diameter. During assembly, they clamp the threads of the bolt and produce greater frictional forces than an ordinary nut does. Another type of locknut that has received considerable attention is the nonmetallic-insert type. The elastic locking medium in this nut is independent of bolt loading, and seals the bolt threads against external moisture or leakage of internal pressure or liquids. The primary limitation in the use of this type nut is the inability of the insert to maintain its elasticity and locking characteristics at elevated temperatures.

When dissimilar materials, such as sheet metal and plastics or wood are to be joined, or when one side of the work is blind and the application of nuts to bolts becomes practically impossible, self-tapping and drive screws are available. They are used to fasten nearly everything from sheet metal and castings to plastics, fabrics, and leather.

Lead holes are necessary for the application of self-tapping or drive screws. Self-tapping screws have specially designed threads suitable to their applications. There are cutting threads and forming threads. For varied applications, they are available in all the standard head forms, either slotted or with a recess. Drive screws are usually self-tapping and are driven with a hammer. The holding power is greater than the nail and in many cases they may be backed out with a screw driver.

A special type of screw is the self-piercing screw. This is often used for attaching wood panels to light structural parts without having to drill lead holes. The screw is hammered through the wood to the metal and is then turned down with a screw driver.

For fastening metallic elements, machine screws, setscrews, and cap screws frequently are used. Machine screws are usually threaded the full length of the shaft, and are applied with a screwdriver. Cap screws have square, hexagonal, or knurled heads, and are threaded for only part of the shank. Setscrews are threaded for the length of the shaft and may be either headed or headless. The headless setscrew is usually fluted or provided with a slot or hexagonal socket for driving. Setscrews are used to secure machine tools or some machine element in a precise setting. Many types of points are provided on these screws for various applications.

Rivets

Rivets are permanent fasteners. They depend on deformation of their structure for their holding action. Rivets are usually stronger than the thread-type fastener and are more economical on a first-cost basis. Rivets are driven either hot or cold, depending upon the mechanical properties of the rivet material. Aluminum rivets, for instance, are cold-driven, since cold working improves the strength of aluminum. Most large rivets are hot-driven, however.

A hammer and bucking bar are used for heading rivets. The bar is held against the head of the inserted rivet, while the hammer heads the other end. Squeeze heading usually replaces hammer and bucking-bar methods. In this method, the rivet is inserted and brought between the jaws of a compression tool which does the setting by mechanical, hydraulic, or pneumatic pressure. Where production runs are larger, riveting machines are used, exerting pressure on the rivet to head it rather than heading it by hammering. Improperly formed heads fail quickly when placed under stress.

Riveting is the most common method of assembling aircraft. A medium bomber requires 160,000 rivets and a heavy bomber requires 400,000 rivets. Some of the forms of rivets are solid rivets with chamfered shanks, tubular or hollow rivets, semitubular rivets, swaged rivets, split rivets, and blind rivets. Solid rivets are used where great strength is required. Tubular rivets are used in the fastening of leather braces. Split rivets are used frequently in the making of suitcases. The materials used in making rivets are aluminum alloys, Monel metal, brass, and steel. Aluminum rivets make possible the maximum saving in weight, and are also quite resistant to corrosion. Anodic coating of aluminum improves the resistance of the rivets to corrosion and also provides a better surface for painting. Steel rivets are stronger than aluminum rivets and offer certain advantages in ease of driving from the standpoint of equipment required; however, their use is limited to those applications in which the structure can be protected adequately against corrosion by painting. Some of the types of rivet heads are button, mushroom, brazier, universal, flat, tinners, and oval. Brazier and mushroom heads are used for interior work. The flat head is used for streamlined pieces. Aluminum-alloy rivets can be distinguished by their physical appearance: 2017 aluminum has a small dimple and is the most used of the aluminum alloys, while 2024 has two small raised portions on either side of the head. It is poor practice to use a hard rivet in a soft alloy. For example, if a 2017 rivet should be driven into soft plate, such as 3003, the plate will be distorted and the resulting appearance will be poor. If the joint is not to be highly stressed, a soft rivet may sometimes be used in a hard alloy plate. In general, it is most advantageous and practical to use a rivet having similar properties to the material in which it is driven.

In determining the length, $1\frac{1}{2}$ times the diameter should extend below the piece being riveted. In addition to this, for a $\frac{1}{2}$- to 1-in rivet add $\frac{1}{16}$ in; for a 1- to $1\frac{1}{2}$-in rivet add $\frac{1}{8}$ in; and for a $1\frac{1}{2}$- to 2-in rivet add $\frac{3}{16}$ in. Rivet holes may be punched, drilled, or subpunched and reamed to size. The clearance between rivets and rivet holes should be 0.004 in on an average. Many times blind rivets are specified. For this purpose, cherry rivets and explosive rivets are used frequently. The cherry rivet is a hollow rivet that is fastened by pulling a pin partly through it.

The explosive rivet contains a small charge that is set off as the temperature is raised by an electrical contact. Also included in the category of rivets is the Rivnut. It resembles a hollow rivet, but has threads on its inner diameter. A screw is turned into the rivet and then pulled, while the head is held stationary. The sides give way, thus holding the Rivnut.

Metal Stitching

Metal stitching is used to join thin-section metals and nonmetals at high production rates. A big advantage is that this can be done without precleaning, drilling, punching, or hole alignment. In comparison with nut-and-bolt and riveting methods of assembly, metal stitching has increased production as much as 700 percent and effected material savings up to 50 percent. These stitches are formed in as little as $\frac{1}{5}$ s, and have high strength and durability. Wire cost is low, being less than $0.040 per hundred stitches in some instances. High material savings are possible, because flange widths need be only $\frac{1}{4}$ in for stitches, compared to $\frac{1}{2}$ in for rivets. To the cost-conscious production engineer, metal stitching is an important process worthy of consideration in many designs.

Adhesives

Adhesive bonding is becoming more and more important, particularly in manufacturing and construction. There are two types of adhesive action: (1) specific adhesion that results from interatomic or intermolecular action between the adhesive and the assembly, and (2) mechanical adhesion produced by the penetration of the adhesive into the assembly after which the adhesive hardens and is anchored. Since many adhesives are proprietary chemical products, their effective use depends upon consultation with the manufacturer.

In many applications, adhesive bonding is replacing rivets, bolts, nails, and other types of mechanical fasteners.

It is important for the production-design engineer to be familiar with the nomenclature of adhesives. The following definitions have appeared in the *Materials and Methods* magazine:

Tack   The characteristic that causes one surface, coated with an adhesive, to adhere to another on contact. It might be regarded as the essential stickiness of the adhesive.

Wet-strength   The bond strength that is realized immediately after adhesive-coated surfaces are joined, and before curing occurs.

Ultimate bond strength   The strength of the bond after the cure has been completed, or substantially completed.

Nonlocking properties   This is opposed to tack, and indicates the freedom of the adhesive-coated materials from sticking to unwanted material, as the hands of the operator.

Can stability   The length of time the adhesive can be held in storage without deterioration. Pot life is the length of time the adhesive remains usable after being put into serviceable condition.

Specific service conditions   Such items as color, nonstaining, thermal range, resistance to solvents, and cleanup.

Adhesives can be used to join practically all industrial materials. They can be obtained in solid, liquid, powder, or tape form. The elimination of rivets and other fasteners results in smoother surfaces and considerable savings in weight and space—features that are very important in airplane construction. In addition, adhesive joints prevent fluid leakage and bimetallic corrosion at joints as well as providing some degree of thermal and electrical insulation. Adhesive bonding is suited to mass-production techniques. There are many types of high-speed hand and mechanical applicators, including brushes, sprays, rollers, scrapers, extruding devices such as pressure guns, and tapes. Some machines automatically apply the correct bonding pressure and cause adhesion quickly with radio frequency heating.

Most adhesives have a limited temperature range of service. The surfaces to be bonded must be very clean and most glues require clamping and heat for proper setting because the bond is not instantaneous. Another disadvantage is that most adhesives are good for only one specific set of service conditions.

Elastomeric adhesives   Rubber, polyurethane, and silicones are the essential elements in adhesives used to bind such materials as felt to metals. There are two main types of rubber adhesives: rubber dissolved in an organic solvent and water dispersion of finely divided rubber. Organic solvent rubber adhesives are easy to handle and apply and give strong bonds, but they release inflammable vapors and attack certain materials. Rubber adhesives of the water-dispersion type possess immediate tack and are therefore good for such applications as cloth backing. They have the disadvantages of being subject to destruction by freezing and of being limited in use to porous materials, because the water must either be absorbed or removed.

Synthetic-resin adhesives   These are excellent for bonding plastics of the same composition. Thermoplastic-resin adhesives include cellulosics, vinyls, acrylics, and styrenes. They are usually of formaldehyde composition. These have a limited field of application, having been used most successfully for wood, rubber, thermosetting resins, and ceramics. Their cost is moderate and they produce a high-strength, waterproof bond which is resistant to attack by fungi. Their shelf life is limited, and since they have little initial tack, they must be clamped until the glue is set.

Design Suggestions

Poor engineering design causes most of the failures in adhesive application. Some important points to remember in designing for adhesive bonding are:

1   The assembly should be designed so that a sufficiently large bonded area is obtained thus preventing failures.

2   The bond should be stressed in shear rather than tension.

3   Joints should be tapered for greater strength.

4 The adhesive should be applied in uniform thin layers so as to avoid irregularities that may create points of stress concentration.

5 Combinations of materials should be selected so that stresses arising from differences in coefficients of thermal expansion are kept to a minimum.

Joining Plastics

Plastics are joined in one of four ways: (1) adhesive bonding, (2) mechanical fastening, (3) solvent cementing, and (4) thermal welding.

Solvent cementing is applicable only to certain thermoplastic resins (acrylics, polystyrenes, cellulosics, and some vinyls). The production-design engineer should observe the following pointers when specifying the joining of thermoplastic materials by solvent cementing:

1 The surfaces to be joined should be smooth, clean, and dry.

2 After application of a suitable solvent, constant pressure must be applied until joint has become dry and hard.

3 Additional processing on the joint parts should be delayed until joint has had ample time to set

If the above three pointers are followed and a suitable solvent is selected and applied, the joint properties will be similar to those of the parent material.

All thermoplastics may be joined by thermal methods. Heat is applied to the area to be joined either by burning gases, induction means, heated tool, or friction. Upon joining, the welded section is cooled thus hardening the materials and completing the joining process.

In gas welding, the procedure is similar to gas welding of metals. A welding rod of the same composition as the parent plastic is brought to the fusion temperature with the parent metal in order to provide filler material. Joint preparation in plastics is equally important as in metals. Butt welds, for example, should be shaped so as to have an included angle of 60° at the weld.

Heated tool joining is done without a filler material. Here heat and pressure are directly applied with a tool until fusion takes place. The heating tool may take the form of a simple soldering iron or hot plate or it may involve a resistance-heated form. In this category falls the heat-sealing of plastic film characteristic of many packaging applications.

In induction-heating joining methods, a metallic conductor is placed adjacent to the areas where the parent plastics are to be joined. Heat from the conductor brings the plastics to the fusion temperature.

SUMMARY

There is a large number of joining processes from which the production-design engineer may choose. Most products include several distinct varieties of joining.

For each set of conditions, a particular process is superior to competing techniques. Some general rules that the production-design engineer should consider before deciding on a specific process are:

1   Do not overjoin. For example, use intermittent instead of continuous welds; however, do not sacrifice quality for simplicity.

2   Consider relative costs of supply materials in addition to time involved when comparing joining methods, for example, cost of silver brazing alloys. versus cost of explosive rivets per assembly.

3   Consider overhead costs before deciding to standardize on a specific joining process.

4   Fabricated construction is not always the most economical. Perhaps a casting or forging would be a better solution.

5   Factors of safety at joined sections should be higher than at other portions, since joined areas can readily lead to points of stress concentration.

6   Keep appearance in mind when specifying a joining process. A projection-welded nut gives a much more pleasing appearance than one that has been arc-welded to a surface.

7   Consider adhesive bonding for sheet-metal assemblies of aluminum or stainless steel.

8   Contact cement is finding many uses such as the application of veneers to table tops or desks.

9   Specify joining methods that will be satisfactory for all conditions to which the end product will be subjected.

10   Keep maintenance in mind when specifying joining operations. Some parts may be interchangeable and may have to be replaced at times.

QUESTIONS

1   What problems are presented when aluminum alloys are welded?
2   Give the spot-welding symbol. The brazing symbol. The seam-welding symbol.
3   When should welded assemblies be stress-relieved?
4   What methods are used to protect the molten metal from contamination in arc-welding processes?
5   What gases are usually used in the gas shielded-arc-welding processes?
6   What is the danger of a long arc in electric welding?
7   What is the effect of too much current in electric welding?
8   What determines the polarity that should be used with electric welding?
9   Why are long seams more difficult to weld than short seams?
10   What effect does peening have on welding beads?
11   What nine points should be observed in the design of joints?
12   How is the stress in a square butt joint computed?

13 What is the relationship in strength of arc-welded steel structures to similar structures composed of gray cast iron?

14 What are some of the advantages of automatic welding? Semiautomatic welding?

15 What is meant by submerged arc welding?

16 List the arc-welding processes and give major features of each.

17 Why is stud welding usually thought of as an arc-welding process?

18 In resistance welding, what is the composition of the resistance of the joint?

19 When would you recommend projection welding?

20 Would you recommend the spot welding of Monel sheet to copper plate? Why?

21 What is flash welding?

22 Upon what does the strength of brazing depend?

23 What is cold welding?

24 What is the advantage of the explosive rivet?

25 When is metal stitching practical?

26 What are two types of adhesives? When would you recommend the use of each type?

PROBLEMS

1 A welded butt joint in an SAE 1020 steel plate is to carry a total force of 30,000 lb. The width of the plate is 4 in and the working stress is specified to be 13,000 lb/in². What standard plate thickness should be used? What process would you select for the welding operation?

2 A rectangular bar of machine steel which is $\frac{3}{8}$ in thick is to be welded to a machine frame by side fillet welds. The estimated tensile force which will be applied to the bar is 35,000 lb. Calculate the length of each $\frac{1}{4}$-in side fillet weld, if the allowable shear stress in the weld is 13,600 lb/in².

3 The tensile force P applied to a double fillet welded lap joint is 120,000 lb and the length of the joint is 10 in. The plate thickness is sufficient to cause failure to occur in the weld rather than in the plate. If the allowable shear stress on the nominal throat section of a fillet weld is 13,600 lb/in², find the required size of the fillet weld to the nearest $\frac{1}{16}$ in.

4 Calculate the amount of transverse shrinkage for the butt welded assembly shown in Fig. P16-4.

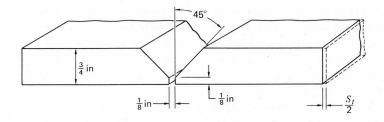

5   Two structural steel plates 2 in wide by $\frac{1}{2}$ in thick are joined by a single rivet in a lap joint. The rivet is 1 in diameter. If the joint is subjected to a tensile load of 20,000 lb, determine the factors of safety with respect to tension, shear, and bearing. The ultimate strengths for tension, shear, and bearing are respectively: 55,000 lb/in; 44,000 lb/in; and 95,000 lb/in. Specify a competitive welded joint design and welding process. Under what conditions would a riveted joint be used rather than a weldment?

6   A steel pipeline is being field-welded. A number of short pieces of random lengths have accumulated during the construction period. Now the field engineer must either use the pieces by welding them into longer lengths or scrap them. He has called on you to help him to determine the minimum length of pipe which can be economically salvaged based on the following data: Pipe diameter, 8 in; wall thickness, $\frac{1}{8}$ in; specific weight of steel, 0.287 lb/in³; welding rod cost, $0.04/in of weld; average welding time, 1.7 min/in deposit; cost of pipe, $0.38/lb; scrap value, $0.03/lb; weldor's pay, $4.38/h; overhead, 240 percent.

7   A firm is considering welding drive pulleys to automotive generator shafts. Specify the size weld needed between the pulley and the shaft. The shear strength of the weld is 38,000 lb/in² maximum, acceleration 0 to 9,000 rev/min in 2 s with 5.0 in·lb electrical torque load under the most severe conditions. Use a factor of safety of 2 and neglect friction.

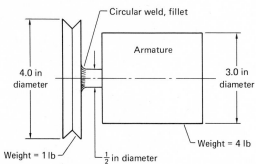

8   An elliptical flange with a major diameter of 100 in and a minor diameter of 80 in on its outer surface is to be cut from a plate $\frac{3}{8}$ in thick. The inner major and minor diameters are to be 80 and 60 in, respectively. Determine the scrap weight if the flange is flame cut as a unit from a rectangular plate. Calculate the saving if the flange is cut as four pieces and welded into a unit after flame-cutting. Cut the sections along axes at 45° to the major axis of the ellipse. Determine the best plate size graphically and use current prices for labor and material. A lot size of 100 pieces is needed.

SELECTED REFERENCES

AMERICAN SOCIETY FOR METALS: "Metals Handbook," 8th ed., vol. 6, "Welding and Brazing," Metals Park, Ohio, 1971.

AMERICAN WELDING SOCIETY: "Welding Handbook," 6th ed., Miami, Fla., 1973.

BLODGETT, O. M.: "Design of Weldments," James F. Lincoln Arc Welding Foundation, Cleveland, Ohio, 1963.

BRUCKNER, W. H.: "Metallurgy of Welding," Pitman, New York, 1954.

GIACHINO, J. W., W. WEEKS, and G. S. JOHNSON: "Welding Technology," American Technical Society, Chicago, 1968.

LANCASTER, J. F.: "The Metallurgy of Welding, Brazing and Soldering," Allen and Unwin, 1965.

LINCOLN ELECTRIC COMPANY: "Procedure Handbook of Arc Welding Design and Practice," Lincoln Electric Company, Cleveland, Ohio, 1957.

LINNERT, G. E.: "Welding Metallurgy," 3d ed., vols. 1 and 2, American Welding Society, Miami, Fla., 1965.

PATTON, W. J.: "The Science and Practice of Welding," Prentice-Hall, Englewood Cliffs, N.J., 1967.

ROSSI, B. E.: "Welding and Its Applications," McGraw-Hill, New York, 1954.

WEST, E. G.: "The Welding of Nonferrous Metals," Wiley, New York, 1952.

17

RELIABILITY AND QUALITY CONTROL

As the public escalates its demands for product performance and safety, the production-design engineer is faced with the task of incorporating the most effective product-assurance procedures with due regard for cost. The fact that foreign competition is not only equaling United States goods from the standpoint of quality but exceeding it in some instances is placing the burden of "proof" on the production-design engineer. Japan has been referred to as the "quality miracle of the century" and they may well lead the world market from the standpoint of product performance in the 1970s. Today, many consumers feel that the workmanship in Europe at least equals that in the United States.

In this chapter, we will briefly discuss reliability and quality control from the standpoint of the production-design engineer. By reliability, we mean the probability that a system (e.g., component, part, equipment, etc.) will operate a period of time without failure under specified usage conditions. For example, a part-failure rate of 0.0001 percent per 1,000 h of service is about as good a reliability performance as can be expected in electronic equipment today. The 2,000-part 1974 automobile has an average part-failure rate of 1.5 percent per 1,000 h of service.

"Quality control" may be defined in its broadest sense as all those activities within an organization that affect the quality of the products produced. Thus, quality control includes those processes or operations of testing, measuring, and comparing the manufactured part or apparatus with a standard, and then determining whether it should be accepted, rejected, adjusted, or reworked. Other quality-control activities include the initial specifications applying to allowable tolerances and the extent of inspections considered essential or feasible. Every successful product is a reflection of the effectiveness of quality control. Consequently, the production-design engineer is vitally interested in this activity. The high reliability of ball and roller bearings is taken for granted today because of many years of careful control of materials, processes, and workmanship.

Controls are exercised by federal, state, and local governments through codes and inspectors. The armed services and insurance companies place their own inspectors in their plants to control the product at its source.

In most organizations, the quality-control and reliability department manager reports to the manager of manufacturing; in other cases he reports directly to the plant manager. He usually has equal status with the manufacturing and engineering managers.

Members of the quality-control and reliability department contact suppliers of material in order to explain the requirements of purchase specifications and to assist in eliminating substandard material. They also contact customers who report failures or have difficulty with the equipment. They obtain firsthand information on the failure and endeavor to determine its cause. The quality-control and reliability department usually allocates charges to the department responsible for that failure.

As the representative of the customer, the inspector or tester must ask himself, "Would I be satisfied with this equipment if I were purchasing it?" As a representative of the engineer, he must understand the standards set for performance and appearance and should be able to exercise good judgment when variations from these standards occur. As a representative of the accounting department, he is responsible for recording the number of pieces produced on each operation and for determining whether the work is acceptable or whether the piece should be rejected or salvaged. This same responsibility applies to purchased material. The action of the quality-control department affects every operation from the first to the last.

All the activities of the quality-control and reliability department are promoted through others. If the operators, material handlers, production clerks, supervisors, and engineers are quality conscious, the work of the department is less difficult and can be directed toward more constructive and creative efforts. A newly designed line of equipment can be ruined if proper controls are not established and maintained in the shop. The production-design engineer relies on the quality-control and reliability man to assist in educating the shop personnel as to the importance of various features of the design.

The supervisors and engineers may be consulted, but the decision as to acceptance, rejection, adjustment, reoperation, or repair is the responsibility of the quality-control and reliability department. Its decisions should be such that delivery on time of the product is assured, product quality is maintained, and costs are minimized.

QUALITY-CONTROL PROCEDURES—GENERAL

The inspector for the quality-control and reliability department may inspect in any of the following ways.

1 Inspect each piece on each operation. This is called "100 percent inspection" and is quite expensive.

2 Inspect the first piece, and other pieces selected at random. This is called random inspection.

3 Check the final and complete piece, using random or 100 percent inspection.

4 Inspect the partial assembly or the final assembly only, using random or 100 percent inspection.

5 Inspect a part or complete apparatus taken from the production line to a laboratory or inspection station, where critical dimensions and functions are carefully checked. A refrigerator is carefully checked for color, placed in a tumbling machine to determine its ability to stand shipment, and given an accelerated-life test under extreme conditions. The door may be opened and closed until failure occurs. Critical parts are removed and carefully measured. Such control checks assure a product with practically no hidden defects.

One hundred percent inspection is practically never 100 percent. If there are 100 known defective in a batch of 5,000 pieces, the ordinary inspector will miss 5, 10, 20, or more of them. At the very best, 100 percent inspection is seldom more than 98 percent efficient. Monotony, fatigue, and ineptitude take their inevitable toll.

In one study made under particularly ideal conditions, it was found that batches containing not more than 2 percent defective parts could be screened so that about 0.1 percent of the parts shipped were defective, but as the work coming into inspection increased to 5, 8, and 10 percent or more defective, the substandard pieces slipping by inspectors increased to 2 percent and even 5 percent.

Quality and Wage Payment

Under incentive wage payment plans, the operator is usually paid for the good pieces he makes. In some organizations, the operator's pay is determined according to the inspector's records at the end of the day or at the completion of the order. Later, if the part proves defective due to improper workmanship, the operator usually cannot be penalized because more than one operator may have performed the operation. Therefore, the responsibility of the quality-control man in approving an operator's pay is very important.

Rejections and Inspection Orders

When a part is rejected, it can be reworked by the operator (usually at his expense), salvaged, repaired in another department, or used as a substandard or special part. Usually, the part or assembly is removed from its original production order and is made a part of a separate order. For convenience in describing this procedure, the term "IO order" will be used, meaning an inspection order.

The IO order can be used: (1) to deliver the material or parts to the salvage department where the assembly may be dismantled, the good parts salvaged, and the material sold for scrap; (2) to return to supplier; (3) to reoperate and return to the original order or deliver to its proper section.

The IO order must carry the necessary charges and time values for performing the additional work. Tags to identify the parts are provided and serve to notify supervisors, operators, and production control people that trouble has occurred. These tags can be removed only by an inspector. The production-design engineer is often involved in the IO order procedure, since newly developed items on first production runs are usually substandard and are often chargeable to the accounts for which the engineer is responsible. Copies of IO orders are distributed to: (1) the person receiving the charge, (2) the production control department, to notify them of trouble that may delay their orders, and (3) the accounting department. The original copy of the IO order travels with the work to notify the operators and supervisors of the action to be taken.

Inspection Equipment

In order to detect errors, measure performance, and check materials and parts with standards, inspection and testing facilities must be available. Tensile-strength test machines, hardness-testing machines, and other equipment is set up in laboratories that check materials received, made, or treated in the shop. Elaborate checking gages and fixtures are designed to inspect complicated parts and assemblies. Complete sets of gages and tools for measuring critical dimensions should be available to the operator, as well as the inspector. Where controls are extensive, it is necessary to have a gage section that checks, repairs, and adjusts gages by using master gages of the highest accuracy. Thus duplicate and triplicate sets of gages are sometimes necessary to maintain control.

Errors in measurement depend upon the accuracy of the measuring equipment, which is subject to errors. These errors are the result of one or more of the following:

1 Inherent errors in the measuring instrument
2 Errors in the "master gage" used to set the instrument
3 Errors resulting from temperature variation and different coefficients of linear expansion of instrument and part being gaged
4 Errors due to the "human element" of the inspector

The engineer should not specify a dimension, characteristic, or function of a part or apparatus that cannot be measured. Fortunately, due to the extensive development of inspection and testing apparatus, it is now possible to measure to a very high degree of accuracy. Instruments for measuring roughness of surfaces are available, and through the use of oscilloscopes, oscillographs, x-rays, and other types of sensitive measuring equipment, the quality of products can be controlled. Some of these instruments, such as air gages for measuring close dimensions, are rugged enough to use in production operations and thus the operator has a means of controlling the quality of the part he is making.

Inspection operations such as detection of surface defects and blowholes can be a part of the operator's responsibility. Checking gages can be built into jigs, fixtures, and equipment that can be used by the operator and inspector. This is especially valuable on low-activity and complicated parts.

Inspection and testing equipment may be stationary or portable and destructive or nondestructive of the material or part.

The ability to measure and control dimensions has progressed. A few years ago it was said, "We can work to 0.001 in and talk about holding to 0.0001 in; now it can be said, we can work to 0.0001 in and talk about holding to 0.00001-in tolerance." The basic equipment and standards for measurement are measuring blocks, known as Jo blocks, invented by Johansson. Interchangeable manufacture would not be possible without these carefully maintained standards.

Gage blocks Gage blocks are accurate in height, flatness, and parallelism and have a Rockwell hardness of over C65. The following set of gage blocks with accuracy tolerances indicated are available today.

	Grade	Tolerance, in
Laboratory set	AA	±0.000001
Inspection set	A	±0.000004
Working set	B	±0.000008
Working set	C	±0.000010

Dimension can be obtained by combining various blocks. These blocks are so accurate that they cling together because of the surface tension of the adsorbed water film and must be slid or pulled apart. They can be combined to check snap gages, height gages, micrometers, and verniers.

Optical flats Optical flats made of fused quartz are available in two grades: AA Grade, ±0.000001 in; A Grade, ±0.000002 in. One millionth of an inch can be measured with these flats by means of interference of light waves.

Air and electric comparator Air and electric comparator gages can be set by Jo blocks or other standards so that deviations from standard can be measured quickly by the operator. Large scales that can be read easily are provided. Range of variations measured is between 0.000001 and 0.001 in.

Optical comparators throw a profile of the part on a screen so that it can be compared to a master. The profile of the part is blown up many times so that deviations can be seen and easily measured.

Automatic Gaging and Controls

With millions of precision parts being made in automotive, space, and electrical apparatus, there is a great demand for automatic inspection devices. Automatic machines that sort and reject by dimension have been made; however, the ball-bearing industry still inspects every ball by eye for surface defects. The field of automatic gaging needs the production-design engineer to create new and economical devices.

There are many control instruments and continuous recording instruments for voltage, current, power, temperature, pressure, and chemical composition that are invaluable for process control.

Strain gages Strain gages give a picture of the stresses existing in parts under static or dynamic load. There are mechanical, magnetic, electrical-resistance, and electrical-capacity types of strain gages which record strains where the gage can be fastened to the surface of the part. This equipment is valuable in designing apparatus.

Photoelasticity Photoelasticity is the process of using special transparent material having the profile of the shape to be studied. Polarized light is passed through the part. When it is strained, colors indicate where stresses occur.

Stress paints Stress paints are another means of obtaining a picture of strains existing in parts under stress. Stress paints cling to the surface of the part and crack when stressed because they are brittle. The pattern of cracks indicates the direction and magnitude of the strain. Some quantitative values can be obtained by measuring the photograph of the part.

Hardness testers The Brinell, Rockwell, and Scleroscope type of hardness testers are universally used to control hardness and measure strength of materials. Hardness and strength tend to have a direct relation to each other in the same materials. Conversion tables are available. The Rockwell and Scleroscope hardness testers may be portable, and can measure hardness on castings, billets, and parts that would be difficult to transport to a machine. Hardness testers leave a mark on the surface, which limits their use.

Nondestructive Inspection

X-ray Powerful x-ray equipment is now available for medical, industrial, and military use. Parts and apparatus can be radiographed by exposing objects to x-ray and recording on a photographic film. Irregularities in the material, such as blowholes, slag, cracks, and nonuniform section, are revealed and then identified by the expert. Mines and ammunition may be disarmed by observing the device by x-ray and determining its design and method for disarming. X-rays can be focused so that any plane passing through the object can be observed with the other parts in the background. X-rays are used to control continuously the thickness of sheets rolled in steel, brass, or aluminum mills. The control is sensitive enough to measure variations of 0.001 in.

Supersonic sound equipment Supersonic sound equipment with electronic transmitters and receivers is available for detecting flaws in material. Fine cracks within material that cannot be detected by x-ray or magnetic means are picked up by the supersonic method.

Fluorescent penetrants A penetrating fluid that carries highly fluorescent dyes and is able to enter any minute crack shows up under ultraviolet light. This process and magnaflux are often used to inspect each part in a production line, such as a piston rod in an airplane, a gas engine housing, or any other vital part.

Test Equipment

Test equipment is usually designed to check, adjust, and determine the extent of the functional characteristics of the apparatus such as strength, capacity, and size. The equipment may include proving grounds where actual service conditions can be simulated. It may be individual equipment designed to check a specific characteristic of material or operation. For example, the automobile manufacturer has engineering laboratories equipped with testing equipment that can test and measure the performance of each part of a car or truck—rear-axle housings and differentials, transmissions, carburetors, generators, suspension springs, shock absorbers, and engines. These testing laboratories are sometimes supervised by the engineering department, which uses them to check current production assemblies and parts and to determine the characteristics of new equipment.

Such test equpiment, including electrical test panels for transformers, circuit breakers, motors, radio, and radar, is usually so complicated and specialized that the quality-control department designs and builds its own. Fortunately, standard test equipment and components are available, such as dial gages, meters, relays, tensile and torsional strength-testing machines, hardness testers, and electronic units. These can be built into special test equipment.

Test operations are expensive and the equipment is costly. To govern test practices, manuals are written for the guidance of testers and engineers, and are useful for instructing engineering graduates who serve their apprenticeship on the test floor.

Testing operations can be included in the production line and so enable the operator to adjust, reject, or accept the part or assembly before it is built into the product. The testing of samples of steel or alloys in a steel mill before pouring is an example of controls by testing. The production-design engineer can assist in determining these points of test in order to save disassembly and repair.

The errors which are discovered on test can be reduced by removing the possibility of error by the operator. For example, pegboards for wiring electrical panels can be equipped with guides and markings so that the wires can be bent to shape, laid into position, and tied together as a unit. They are then placed in the radio or on the back of a control panel and can be easily connected to the proper terminal. Each wire can be identified by a tag or color code or by position. The use of wiring pegboards reduces the skill required and the errors of connection, and gives a neat appearance.

The supervisor can determine his best operators by observing inspection and tests records. By studying tests records, the engineer can determine components that can be improved in design or manufacture. One manufacturer of small motors has the following facilities to maintain quality.

Electrical laboratory

1 Curve test equipment—to test all forms of motors and generators in order to determine their characteristics

2 Strobotac—to observe objects rotating at high speed as though they were standing still

3 Cathode-ray oscillograph—to observe electric energy in forms of curves that can be photographed and measured

4 Weather-duplicating equipment (sandstorm box, humidity cabinet, salt-spray cabinet)

5 Radio-frequency test equipment

6 Dynamometer

7 Motor-life test for aircraft—to duplicate high altitude, temperature, and humidity

8 Sound-level room

Physics laboratory

1 Vibration-testing equipment.

2 High-speed camera.

3 Equipment to test:

a Tensile, compressive, shear, and torsional strength

b Ductility

c Resistance to abrasion

d Hardness and depth of case-carburizing

e Depth of decarburization

f Response to heat treatment

4 Identometer—to identify steels by checking against known master specimens. Operation depends on the thermoelectric properties of metals that conduct electricity. An identity can be established in 1 or 2 min.

5 Taber abrader—to evaluate the resistance of a plated, painted, or plastic finish to abrasion. Samples are rotated on the spindle, while an abrasive wheel is allowed to roll on the surface in question. The number of cycles the surface will withstand before a breakdown is measured with an automatic counter.

6 Durometer (used principally on resilient materials such as rubber).

7 Magnaflux—to quickly determine whether subsurface imperfections exist in a magnetic metal. The equipment employs the wet process under which very fine iron particles suspended in kerosene are poured over a specimen held in a powerful magnetic field. Any break in the continuity of the metal causes the iron filings to assume a definite pattern at the surface.

Chemical laboratory

1 Portable gas analyzer

2 Electroanalyzer

3 Saybolt viscosimeter

4 Ph meter

5 Carbon combustion train

6 Spectrophotometer

7 Complete facilities for chemical analysis of inorganic compounds

8 Paint-spray booth and complete spraying equipment

9 Centrifuges for various purposes

10 Facilities for determining properties of greases, oils, and solvents

11 Salt-spray cabinet—to determine the resistance to corrosion of various finishes

12 Humidity cabinet—to study the effect of excess humidity on materials or parts

13 Cold box—to conduct tests at −70°F

Metallurgical laboratory

1 Marco etching
2 Bend testing
3 Fracture testing
4 Polishing
5 Profilometer
6 X-ray
7 Metallograph
8 Binocular microscope

Product-performance test laboratory Equipment for testing apparatus for service life and performance during operation.

Test equipment facilities Sufficient equipment is available to test and inspect mechanical and electrical apparatus on the production line. The manufacture of this equipment and its maintenance and calibration require machine tools and other equipment as well as master gages and electrical calibrating equipment.

QUALITY CONTROL—SURFACE FINISHES

Surface finishes have been discussed under the subjects of machining, honing, grinding, superfinishing, lapping, polishing, buffing, plating, and painting. In the control of surface finish, methods of measuring and comparing with samples are used. This discussion will be confined to the controlling of surfaces that can be

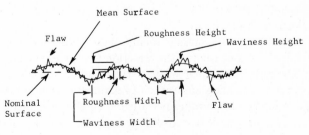

FIGURE 17-1
Enlarged surface profile with some surface properties indicated schematically.

measured and designed on drawings. Some of the instruments for measuring are the profilometer and the Brush surface analyzer. They have a tracer or stylus which moves over the surface. The up-and-down movement is recorded on an electric meter or on paper or film that moves at a uniform rate as the measurement is taken (Fig. 17-1).

In the past it has been difficult to duplicate, measure, and designate desired surfaces. The degree of surface roughness must be controlled and maintained because it affects:

1 The life of the product by controlling friction, abrasive wear, corrosion, and galling.

2 The function of the product by permitting the smooth surfaces of the parts to slide freely, fit properly, serve as bearings, reduce leakage, and rub against packings.

3 The appearance of plated and other decorative surfaces.

4 The safety of the part by preventing stress concentration, fatigue, and notch sensitivity. For these reasons, airplane engine manufacturers are especially conscious of the value of controlling surface finish.

5 The heat transfer, because smooth surfaces offer better heat conductivity.

Since it costs more to produce smooth surfaces, they should be specified only when desired. Progress has been made in advancing the methods of controlling surface finishes through the adoption of national standards.

Measurement or Evaluation

For compliance with specified ratings, surfaces may be evaluated by comparison with specified reference surfaces or observational standards, or by direct instrument measurements.

In many applications, these comparisons can be made by sight, feel, or instrument. In making comparisons, care should be exercised to avoid errors due to differences in material, contour, and type of operation represented by the reference surface and the work.

In using instruments for comparison, or for direct measurement, care should be exercised to insure that the specified quality or characteristic of the surface is measured.

Roughness measurements, unless otherwise specified, are taken across the lay of the surface or in the direction that gives the maximum value of the reading. The physical measurement of the roughness height value shall be the maximum *sustained* reading of a series of readings. It shall be the minimum *sustained* reading in case a minimum permissible value is specified also.

The physical measurement of the waviness height shall be the algebraic difference between the maximum and minimum readings of a dial gage over a distance not exceeding 1 in, if no other definite waviness width is specified. The waviness height can also be determined by means of a straight edge. The recommended values for roughness classification, in microinches, are:

1	16	125
2	32	250
4	63	500
8		1,000

The use of only one number or class to specify the height or width of irregularities shall indicate the maximum value. Any lesser degree or class on the actual surface of the part shall be satisfactory. When two numbers are used on the drawing or specification, they shall specify the maximum and minimum permissible values.

Surface Symbol

The symbol used to designate surface irregularities is the check mark and a horizontal extension. The point of the symbol may be on the line indicating the surface (on the witness line) on the drawing or on an arrow pointing to the surface. The long leg and extension shall preferably be to the right, as the drawing is read.

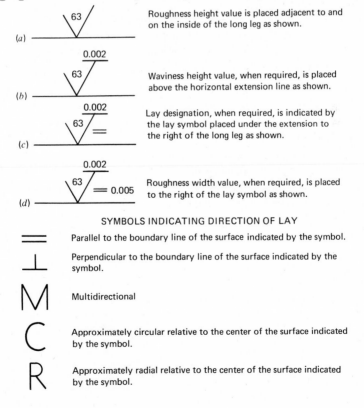

(a) Roughness height value is placed adjacent to and on the inside of the long leg as shown.

(b) Waviness height value, when required, is placed above the horizontal extension line as shown.

(c) Lay designation, when required, is indicated by the lay symbol placed under the extension to the right of the long leg as shown.

(d) Roughness width value, when required, is placed to the right of the lay symbol as shown.

SYMBOLS INDICATING DIRECTION OF LAY

= Parallel to the boundary line of the surface indicated by the symbol.

⊥ Perpendicular to the boundary line of the surface indicated by the symbol.

M Multidirectional

C Approximately circular relative to the center of the surface indicated by the symbol.

R Approximately radial relative to the center of the surface indicated by the symbol.

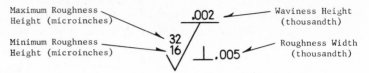

FIGURE 17-2
The surface-quality symbols conform to the "Aeronautical Standard" published by the Society of Automotive Engineers and they meet the ASA standard on "Surface Roughness, Waviness, and Lay," part I, B46.1.

Application

Figures 17-2 to 17-5 show the proper method of designating surface qualities on the drawings with examples for different ratings. The new symbol will replace the old finish mark.

The roughness of natural surfaces is shown in Fig. 17-6 and characteristic maximum surface roughness in common machine parts is illustrated in Fig. 17-7.

Statistical Control

Statistical quality control is an analytical tool that can be used to evaluate machines, materials, and processes by observing capabilities and trends in variation so that comparisons and predictions may be made to control the desired quality level. This tool makes possible:

1 Decreased inspection costs
2 Reduced rejects, scrap, and rework
3 More uniform product
4 Greater quality assurance
5 Anticipation of production trouble
6 More efficient use of materials
7 Rational setting of tolerances
8 Better purchaser-vender relationships
9 Improved and concise reports to top management on the quality picture

FIGURE 17-3
Designation of surface quality on external surface.

32

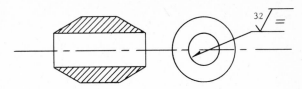

FIGURE 17-4
Designation of surface quality on internal surface.

1000
.0002
1000
63
63
63

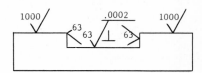

FIGURE 17-5
Designation of surface quality on adjoining surfaces.

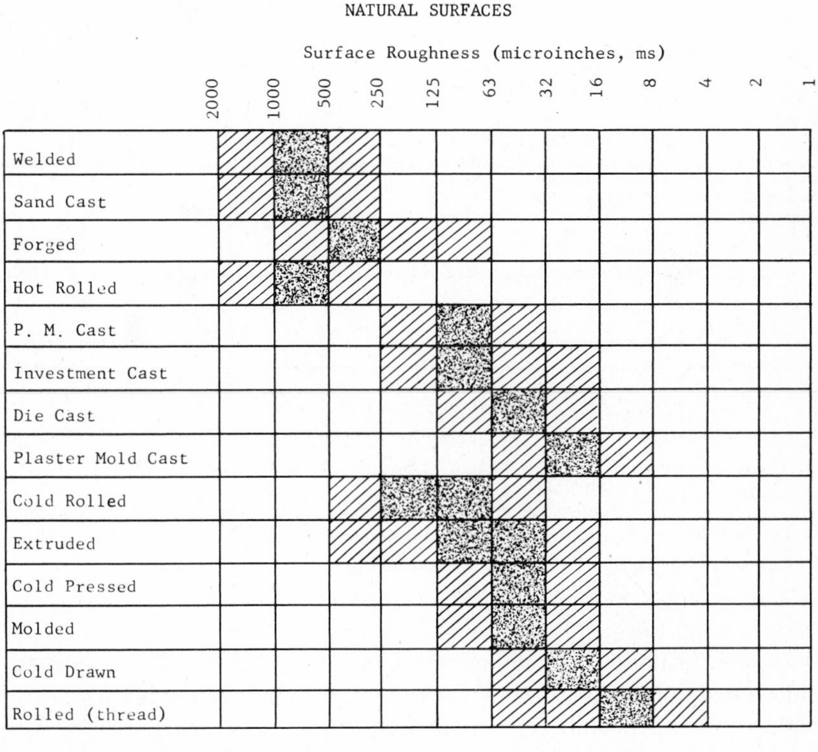

NATURAL SURFACES
Surface Roughness (microinches, ms)
2000
1000
500
250
125
63
32
16
8
4
2
1
Welded
Sand Cast
Forged
Hot Rolled
P. M. Cast
Investment Cast
Die Cast
Plaster Mold Cast
Cold Rolled
Extruded
Cold Pressed
Molded
Cold Drawn
Rolled (thread)

FIGURE 17-6
Chart showing overall and average range of natural surface-roughness characteristics with the nonmachining processes.

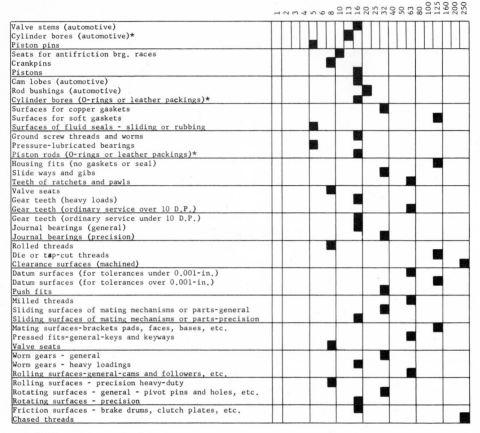

*With regular cross-hatch pattern.
  Smoothest possible finish.
  Waviness not considered.

FIGURE 17-7
Characteristic maximum surface roughness for common machine parts. Charted from various manufacturers' practices on such parts, these data will assist in judging practical finishes.

Mathematicians who had studied the laws of variation in nature became interested in the possibility of controlling quality by applying the laws of probability and dispersion. For example, they knew that in firing artillery weapons the shots fall in a pattern around the target in spite of the greatest care in obtaining uniform powder, projectile, gun, and accuracy of sighting. This pattern of shots was true to mathematical laws that had been worked out in other probability cases. That is, the mathematicians could predict the number of shots that would

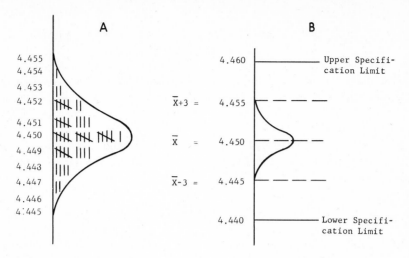

FIGURE 17-8
(A) A frequency curve of 50 measurements on screw machine part. (B) Relative position of frequency curve of the process relative to the specification limits shows good statistical control.

hit the target, the number that would fall 5 ft away, 10 ft away, and 20 ft away. In studying a screw machine, they found similar data. The machine capable of producing uniform parts would make a certain quantity that would fall on the dimensions required, so many would be 0.001 above or below, so many 0.002 above and below, and so on. As the tool wore, the number below size decreased and the number above size increased. By studying this pattern, the observer could make the amount of correction needed to bring the machine back to best performance (Figs. 17-8 and 17-9).

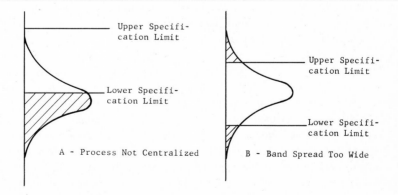

FIGURE 17-9
(A) Frequency curve of a process that is not centralized. (B) Frequency curve of a process that has too large a spread of variation.

In statistical quality control, observable features of a product or process are measured. A study of these measurements usually determines the ability of the operator or equipment to produce within the desired specifications. The percent of parts falling within limits can be determined by the sampling method and gives the same story as if the product were inspected or tested 100 percent. The size of sample is determined by the nature of the product and the standard of quality required.

The two statistics that the production-design engineer will use most frequently are the mean and the standard deviation. We define the mean, $\bar{x}$ of the set of numbers x_i, as

$$\bar{x} = \frac{1}{n} \sum_{i=1}^{n} x_i$$

The standard deviation is a most useful measure of the extent to which an individual item may deviate from the expected mean. To employ the standard deviation, we must also adopt confidence limits; that is the percentage of probability that the product will fall between the "upper and lower" limits. No result can be absolutely assured and nothing is perfect at least in production or anticipated sales. The standard deviation can be defined as the root-mean-square deviation in that it is the square root of the mean of the squares of the deviations. The standard deviation of a sample of data is usually symbolized by s. The value of s is given by the equation:

$$s = \sqrt{\frac{1}{n-1} \sum_{i=1}^{n} (x_i - \bar{x})^2}$$

The analysis of random errors of observation in gathering statistics is based on probability theory. The production-design engineer should be cognizant of the fact that the mean of a sample of data represents only an estimate of the mean of the parent distribution. Similarly, the standard deviation of the sample is an estimate of the standard deviation of the population. For most distributions, the precision of the estimate of both the mean and the standard deviation increases with the size of the sample of data taken.

Therefore the operator can select at random parts to check, plot the results as shown in Fig. 17-8A, and determine the performance of his machine. The tool may be wearing gradually. As it approaches the limit, it can be adjusted to compensate for wear. If it is chipped and producing defective parts, the sample immediately shows this and the entire lot will be rejected. Then 100 percent inspection of the lot will be necessary to salvage as many good parts as possible. When statistical control is established for an operation, the average size of part is plotted on one chart for each set of samples and the range of size is plotted on another chart for each set of samples. By observing the two charts, the operator can see the trends and know when the machine is exceeding limits. Figure 17-10 illustrates how such a chart revealed the possibility of reducing limits of variation

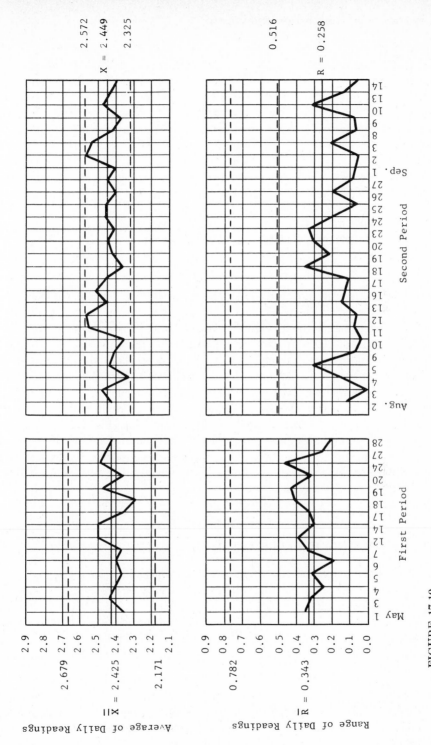

FIGURE 17-10
Chart reflecting average and range over a period of time.

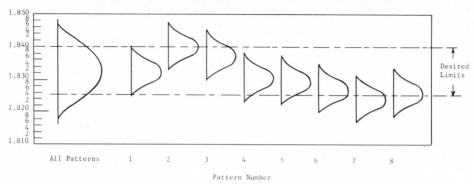

FIGURE 17-11
Frequency distribution of OD of hub diameter.

without increasing costs and resulted in a better product which assisted succeeding operations.

The patternmaker, mold maker, toolmaker, and designer have recently realized the value of making patterns, molds, and dies the same size for multiple patterns and cavities. For example, an electric-motor bracket hub varied to such an extent that a special machining chuck was required to hold the part. A study of deviations revealed that the patterns differed from each other. When they were made alike, the normal casting deviation permitted the use of a standard chuck and the reduction of material allowed for machining (Fig. 17-11).

The production-design engineer will benefit by studying the current articles and textbooks on statistical control. The subject has been reduced to "everyday" understandable terms that can be applied by quality-control men, shop supervisors, and engineers in any organization.

MANUFACTURING RELIABILITY

The concepts of reliability need to be incorporated in the functional design of the product and into all manufacturing operations in view of the long life requirements for many of today's products.

When we design for reliability, the product must withstand actual service conditions for a given period of time. We can prescribe the time period but we can only estimate the service conditions. Generally speaking, there are four types of failure while in service. These are:

1 *Infant mortality.* These are the early failures due to faulty material, or manufacturing errors such as sharp corners, scratches, heavy undercuts, etc., that cause points of stress concentration. Sound quality-control procedures should minimize infant mortality.

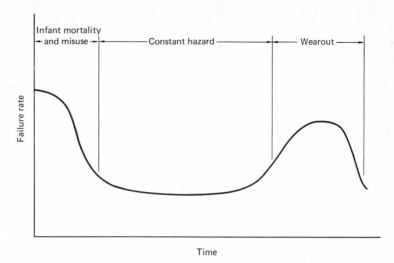

FIGURE 17-12
Mortality curve based on Robert Lusser's concepts.

2   Chance failures. Chance failures may be referred to as random or constant hazard failures. These are failures that take place during the product's life due to chance. A specific cause could be inadequate lubrication due to a clogged wick, leaking due to the development of excessive porosity, a burnout due to a surge of power, etc. Chance failures are often mistaken for wear-outs.

3   Abuse or misuse. These failures are due to using the design in installations beyond the intended purpose of the design. Known excessive loads are typical examples.

4   Wear-out These failures are due to aging, fatigue, corrosion, etc. Wear-out failures are usually progressive until the product capability is such that it is retired because of inefficiency. Wear-out usually can be postponed by proper maintenance and preventive maintenance.

Figure 17-12 illustrates a theoretical curve, that is seldom fulfilled in practice, of the relation in time and the magnitude of the failure rate of these four types of failure. Through a well-conceived reliability and quality control program a reliability performance similar to that shown in Fig. 17-13 can be achieved. Here, it can be seen that both the failure rate and the period of infant mortality has been reduced and the start of wear-out failures has been deferred.

Probability, the Tool of Reliability

As has already been explained, reliability is expressed as a probability, and this probability usually must be expressed as a function of time.

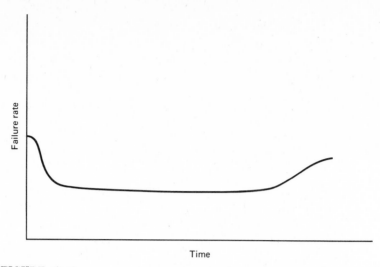

FIGURE 17-13
Mortality curve on product designed and built under a well-conceived quality-control and reliability program.

Typical industrial practice is to specify a safety factor which is the average design parameter (strength, for example) minus the maximum operating parameter (stress, for example) divided by the standard deviation of the design parameter (strength). Another typical method is to specify the mean time to failure. As has been shown (Fig. 17-12) wear-out failures of most designs take a normal probability distribution with a mean failure time equal to G and a standard deviation equal to σ. Figure 17-14 shows this distribution and the resulting reliability R which provides the expected percentage of designs operating after t hours of operation. Figure 17-15 illustrates the possible failure zone on a design that is subjected to an environment that is normally distributed. Figure 17-16 illustrates how the failure zone can be reduced by narrowing the tolerance of the design.

Where constant-hazard-rate failures take place, the failures when plotted as a percentage of the maximum ordinate will plot as an exponential curve. Here

$$y = \frac{1}{T} e^{-t/T}$$

where

$$y = \text{the number of failures at time } t$$

If a product represents the assembly of several components and each component has its own pattern of failure, then obviously the probability distribution

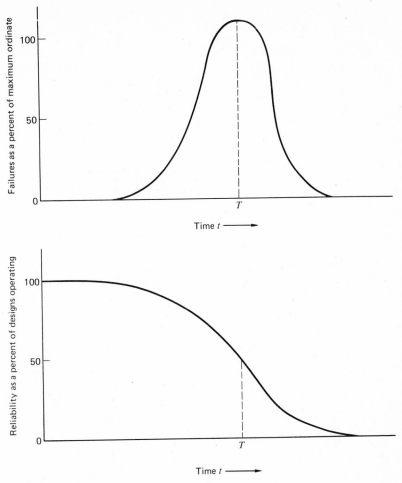

FIGURE 17-14
Typical wear-out failure with mean failure time t.

of failure will be different for each component. The reliability of the assembly will be equal to the product of the separate reliabilities of the components going into the assembly. For example, if an assembly is made up of four components with the following reliabilities: $R_1 = 0.95$; $R_2 = 0.98$; $R_3 = 0.97$; $R_4 = 0.90$, the reliability of the assembly would be:

$$R_a = R_1 \times R_2 \times R_3 \times R_4$$

$$= (0.95)(0.98)(0.97)(0.90)$$

$$= 0.813$$

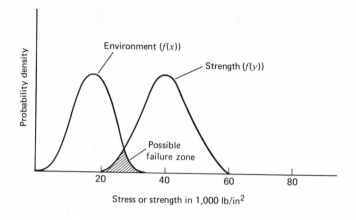

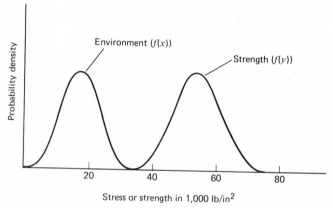

FIGURE 17-15
The possible failure zone can be removed by increasing the strength of the material.

Reliability Operations

In order to maintain a successful reliability program, a "reliability mindedness" attitude must be developed in design, manufacturing, purchasing, and management.

It will be necessary to review or develop standards for procured parts and material, manufactured parts, and assemblies. Test and inspection procedures for maintaining these standards will need to be developed and implemented.

Reliability data, including operational life, failures, consumption, etc., will need to be gathered both from a field and within the plant and be analyzed in a way so that significant feedback can be realized.

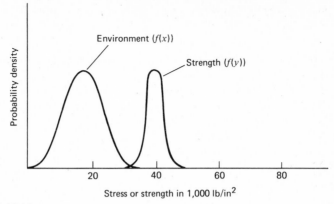

FIGURE 17-16
The possible failure zone can be reduced or removed by narrowing the tolerance of the material.

The work of the reliability personnel will continually involve the review, study, and analysis of equipment, specifications, control drawings, and manufacturing and inspection procedures, so as to assure reasonable product life. Product life can be short for cheap noncritical products; it must be long for items like costly machines (e.g., generators). But, infant mortality must be practically zero for aircraft or space vehicles. Reliability personnel will make improvement recommendations, and follow up such recommendations to verify that corrective action was taken resulting in improved product reliability.

SUMMARY

The cost of quality control and reliability is proportional to the level of reliability and accuracy desired in materials, processes, and functions of the apparatus. The accuracy desired is based upon the judgment of management, which depends upon the advice of the sales, engineering, and manufacturing departments. The level of reliability established usually emanates from the customer. The ultimate decision as to acceptance or rejection rests upon the quality-control and reliability department. In general, the company that is willing to maintain quality at considerable expense survives competition that has lower quality standards. Over the long pull, the high-quality, reliable product wins because customers are satisfied. Through good engineering, a good-quality product can be obtained at less than the former poor-quality product produced under uncontrolled conditions. By going to the source of trouble and eliminating the cause, by providing the shop operators and members of the quality-control department with adequate equipment, by improving maintenance of tools and machines, and inspecting in the critical places, quality control has become an asset to competitive industry.

The quality-control and reliability departments of industry are becoming effective in building quality into the product with minimum overall costs of production, scrap, reoperation, and service adjustments. Goodwill is maintained by verifying the quality and reliability before the product is shipped. Years of control have brought about uniform materials and parts which are produced in enormous quantities. These consistent materials make possible our mass-production lines producing reliable products today.

The steps involved in developing a sound quality-control and reliability program include:

1 Initiate, plan, and direct pre-prototype manufacturing and assembly work.

2 Determine and delineate manufacturing and assembly problem areas that affect product quality and product life.

3 Provide critical analysis of manufacturing operations to determine the relative merits of "make or buy" not only from a price standpoint, but also from a quality-control and reliability standpoint.

4 Identify the areas where process control should take place. Determine the "how" of this process control.

5 Determine the degrees of environmental control required for the various manufacturing operations and specify the specific conditions required in order to maintain the required quality and assure the desired reliability. The conditions specified may include:

a Humidity level
b Dust count and particle size
c Temperature
d Differential pressure
e Wearing apparel
f Use of cosmetics

6 Review and analyze critical storage and handling requirements for materials, parts, and components.

7 Specify functional and maintenance characteristics of tools, fixtures, and handling equipment to ensure precision and repeatability.

8 Review manufacturing plans and procedures and check all development hardware for compliance to precision.

9 During the manufacturing cycle, audit manufacturing operations and initiate corrective action where needed so as to help assure meeting long life requirements.

10 Evaluate vendor facilities, methods, and capabilities. Provide assistance where necessary.

QUESTIONS

1 Why is it that 100 percent inspection seldom ensures that a shipment of parts will be 100 percent to specifications?
2 In what way does incentive wage payment improve the quality of a product?
3 What is the purpose of Jo blocks?
4 For what type of work is the optical comparator used?
5 What advantage does the Brinell hardness tester have over the Rockwell tester?
6 What is the purpose of Magnaflux inspection?
7 What does the durometer measure?
8 What equipment would you recommend to test the ability of a part to resist corrosion?
9 For what reasons is it advisable to control the surface roughness of a part?
10 What is meant by the root-mean-square deviation?
11 What is the value of 1 μin?
12 Give the sequence in ascending order based on surface roughness of the following processes: cold rolled, die cast, plaster cast, hot rolled, and cold drawn.
13 What do we mean by statistical quality control?
14 Define standard deviation.
15 What is a normal distribution?
16 Why is it desirable to plot the "range" of successive samples?
17 Why do chance failures plot as an approximate straight line when failure rate is plotted against time?
18 Show graphically the effect of a strong maintenance and preventive maintenance program on the relation between failure rate and time.

PROBLEMS

1 A motor has two brushes, two ball bearings, and a small fan on the shaft. An analysis of electrical failures is known to be of the constant hazard type, with a mean time to failure of 20,000 h. The brushes have a peaked wear-out failure centering on 1,000 h and a standard deviation of 200 h. The bearings have a more spread-out failure centering on 1,800 h with a standard deviation of 600 h. Mechanical failures are of the constant hazard variety and have a mean time to failure of 10,000 h. What percentage of motors will still be running at the end of 500 h?
2 The weights of ceramic compacts were taken at random and recorded as follows:

Specimen weights, kg

1.24	1.31	1.29	1.34	1.26
1.32	1.41	1.37	1.28	1.31
1.40	1.34	1.29	1.32	1.30
1.31	1.26	1.28	1.34	1.31
1.25	1.29	1.34	1.31	1.32

Calculate the mean and standard deviation of the samples. Assume that the mean and standard deviation of the samples is representative of the population. Find the upper and lower control limits and plot an $\bar{x}$ control chart. Is the process in statistical control? Explain why.

3 Given a process in statistical control with a population mean of 5.000 in and a population standard deviation of 0.001 in, determine (a) the natural tolerance limits of the process and (b) if the specification limits are 5.001 ± 0.002, what percentage of the product is defective, assuming that the process output deviations are normally distributed.

4 Given a process with a population mean of 10.00 and a population standard deviation of 1, what can be said about the interval 10 ± 2 if the distribution is normal.

5 Determine the surface roughness values of the partial surface trace given in Fig.

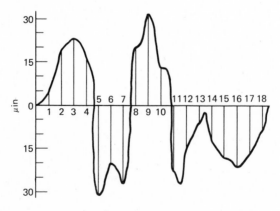

Partial trace of a surface

P17-5 in terms of the maximum peak to valley roughness, the arithmetic average AA and the root-mean-square rms.

Point	Deviations from the mean at Δx intervals, μin	Point	Deviations from the mean at Δx intervals, μin
1	4	10	13
2	19	11	23
3	23	12	15
4	16	13	6
5	31	14	12
6	20	15	18
7	27	16	21
8	20	17	17
9	31	18	9

6 On an interference fit the basic size of a hole is 3.5000 in. The interference between the shaft and the hole must be at least 0.0015 in. The tolerance on the shaft and the hole is 0.0009 in using the basic hole system and unilateral tolerances. Divide the tolerances on each item into three groups so that the small shafts mate with the small holes, the medium shafts mate with the medium holes, etc. Thereby there will be as nearly uniform as possible interference between the mating parts. This procedure is called selective assembly and is used when more precise metal fits are needed than can be obtained by conventional interchangeable manufacture.

7 A wrist pin $\frac{7}{16}$ in diameter is designed to have a medium force fit in the cast iron piston bosses of a small internal engine and a snug fit in the steel connecting rod.

(a) Using ASA standards, determine the limits for the pin and the hole in the connecting rod.

(b) Calculate the limits for the pin and the holes in the piston bosses.

(c) Devise an assembly method which would double the machining tolerance but still maintain the same average interference.

(d) Specify the sizes of the parts and how they will be prepared for assembly using the improved method of assembly.

(e) Using the same components, develop a functional design which will provide better wear characteristics than the present design and which will require only minor changes in tooling.

SELECTED REFERENCES

DUNCAN, ACHESON J.: "Quality Control and Industrial Statistics," Irwin, New York, 1959.

GRANT, EUGENE L.: "Statistical Quality Control," 3d ed., McGraw-Hill, New York, 1964.

LANDERS, RICHARD R.: "Reliability and Product Assurance," Prentice-Hall, Englewood Cliffs, N.J., 1963.

SHOOMAN, MARTIN L.: "Probabilistic Reliability: An Engineering Approach," McGraw-Hill, New York, 1968.

PLANNING THE OPTIMUM OPERATION SEQUENCE

Determining the best sequence of operations is an important step in the realization of a product that is designed for production. Both product cost and product quality are related to operation sequence. A different sequence of operations performed will result in different operational times, different transportation times to the work center, different tooling in view of different locating and clamping surfaces.

In this chapter we will explore some of the fundamentals of operation-sequence planning of the secondary operations.

CLASSIFYING THE SECONDARY OPERATIONS

Secondary operations are all those operations performed on the basic part to transform it to the finished specifications called for on the engineering drawing. The basic part is usually a forging, casting, extrusion, etc., which must be shaped and given a decorative or protective coating. Secondary operations involve such work as drilling, reaming, tapping, burring, boring, turning, facing, threading, milling, shaping, heat treating, enameling, plating, galvanizing, etc.

All secondary operations are classified in one of four categories: critical operations, placement operations, tie-in operations, and protection operations.

Critical manufacturing operations may be defined as those operations that are applied to areas of the part where dimensional and/or surface specifications are sufficiently exacting so as to require regular quality control and/or are used for locating the workpiece in relation to other areas and/or mating parts. A critical manufacturing area can be recognized in one of three ways:

1 The surface or area represents a location which other surfaces or areas are shown as having a relationship to.

2 The surface or area has close dimensional tolerances. Close tolerances are usually thought of as those being within 0.004 in. Tolerance can be applied not only to size but also to roundness, straightness, and concentricity.

3 The surface or area has specified conditions such as surface finish, flatness requirements, or squareness requirements.

Placement operations are those whose method of procedure and sequence are determined principally by the nature and occurrence of the critical operations. Placement operations are of two types: (1) those operations that take place so as to prepare for a critical operation; and (2) those operations that take place to correct the workpiece so as to return it to its required geometry or characteristic.

An example of a type 1 placement operation would be machining a flat on a piece so as to provide a stable locator for a subsequent critical operation. Another example would be the removal of burrs so a part could be placed in a fixture for a subsequent critical operation.

An example of a type 2 placement operation would be the flattening of a part that had curled slightly in view of a previous blanking operation. Another typical example of a placement operation would be the annealing of a part that had been work-hardened because of some previous operation or operations.

Tie-in operations are those productive operations whose sequence and method are determined by the geometry to be accomplished on the work as it comes out of a basic or critical operation. Tie-in operations always advance the shape of the basic component (casting, forging, extrusion, etc.) so as to bring it more closely to the geometry of the finished part called for on the engineering drawing.

Tie-in operations may be thought of as those secondary productive operations which are necessary to produce the part, but which are not thought of as being critical. Tie-in operations usually are performed to standard machine tolerances and do not identify a mastering surface from which other surfaces are located to close tolerances.

Protection operations are those operations, nonproductive in nature, that are performed to protect the product from the environment and handling during its progress through the plant and to the customer and also those operations that control the product's level of quality. Broadly speaking, all protection operations may be classified into one of three groups: (1) application of protective coatings; (2) inspection and test; and (3) packaging for shipment.

Protective coatings may be applied at several stages in the manufacture of a product. Frequently a rust-preventive coating is applied to castings, forgings, and similar raw material as soon as they are received so as to protect them against corrosion in the processor's plant. Similarly, semifinished products may be given a protective coating prior to being sent to temporary storage, and the finished product frequently is treated to protect it from the elements until it arrives safely at the customer's plant.

The reader should recognize that the application of protective and decorative coatings specified on the drawings are not operations falling under the classification of protection operations. The application of these decorative and protective coatings is usually a tie-in operation that is performed late in the process.

OPERATION-SEQUENCE RULES

Several general rules can be established that will assist the production-design engineer in planning the optimum operation sequence. The more important of these are:

1 Critical operations involving close dimensional tolerances or fine surface finishes are usually done late in the process sequence.

2 Critical operations involving the establishment of a locating surface are usually done early in the process.

3 Placement operations take place either immediately before a critical operation or immediately after a critical or tie-in operation.

4 Tie-in operations are governed by the logical process order and consequently can occur early and/or late in the process.

5 Protection operations are governed by the environment of the plant, the total cycle time, the other operations performed, and the final surface finish required. These operations can occur at different times during the operation sequence. Frequently protection operations are overlooked in the initial planning, and experience then requires the addition of this kind of operation.

6 Internal operations are usually performed in advance of external operations.

7 Roughwork involving heavy cuts and liberal tolerances should be performed early in the sequence.

8 The physical location of the facilities of the plant should be taken into consideration when planning the operation sequence.

9 The existing load of specific physical facilities should be taken into consideration when planning the operation sequence.

10 The nature of the tooling should influence the sequence of the operations performed.

11 The quantity to be produced and the delivery requirements will affect the operations and their sequence.

Critical operations involving close operations are done late   Since both tolerance and surface finish of surfaces can be adversely affected by subsequent clamping, material handling, and processing, those operations that provide fine finishes and/or close tolerances should be performed as late as possible in the operation sequence. A part held to 0.0001-in tolerance can subsequently be rejected if it is merely dropped or severely bumped. The impact can cause the part to lose its dimensional accuracy. This is especially true on long slender parts or parts having thin walls.

Critical operations involving locating surfaces should be done early   A locating surface should be established as soon as possible in the process sequence so that a point or area is established from which other parts of the design can be completed in accordance with drawing requirements. It is usually easier to maintain control from a large plane surface than from either a curved, irregular, or small surface. The production-design engineer should decide what surface is best qualified as a point of location and then finish that surface early in the sequence. The operation involved in the finishing of the locating surface would be classified as being critical by definition.

In the development of the engineering drawings of a product, it is desirable for the designer to specify critical dimensions with relation to locating points or surfaces rather than from some other baseline that might be convenient to the draftsman.

Placement operations take place before a critical operation or after a critical or tie-in operation   Placement-type operations might more correctly be referred to as correction operations because the operation is introduced to either prepare for a subsequent operation or to bring the material back to a specification that was altered because of a previous operation. Placement-type operations are frequently overlooked by the inexperienced production-design engineer. So as to avoid the mistake of not planning for the appropriate placement operations, the production-design engineer should ensure that the planned tooling will accept the work in the condition it is after the basic process or after whatever previous critical or tie-in operations have been performed. Likewise, he should be sure what geometrical variations may take place after a certain operation.

Locked-in stresses may relieve, causing considerable distortion in a part. The student should perform the simple experiment of cutting a circular disk out of a thin plate of steel or aluminum using a pair of "tin snips." After this disk is cut, note the lack of flatness and how the edges of the disk tend to curl. If a flat disk were required, a "placement" operation would need to take place after cutting the disk.

Another illustrative experiment that the student can use to demonstrate the need for placement operations is to take a small piece of annealed aluminum sheet metal and bend it back and forth several times. He will note that the metal becomes progressively harder, as evidenced by the greater difficulty in bending the stock with each successive fold. The part has become work-hardened due to the cold working that has taken place. In order to restore it to its original hardness a heat-treating (annealing) operation will need to be added as a placement operation.

Tie-in operations are governed by the logical process order   Logical process order is the basis for determining the sequence of most of the tie-in operations. For example, the reaming of a hole cannot take place until after the hole is drilled. Likewise, the burring of the periphery of a hole cannot be accomplished until the hole is produced. Final painting or enameling operations are performed late in the process, and, of course, rough machining operations are done early.

Protection operations are governed by several factors   As mentioned, protection-type operations include the application of protective coatings, during processing and temporary storage, that are not called for on the engineering drawing. When a protective or decorative coating is applied that has been specified on the drawing, we classify the operation as a tie-in unless close tolerances are specified. If close tolerances are called for, it will be classified as critical.

Protection-type operations are usually introduced when temporary storage periods are relatively long and consequently corrosion or damage may take place on the work prior to shipment to the customer. They are usually applied after a surface is finished so that the finished surface is protected.

Inspection and test operations are also referred to as protection operations. An inspection operation usually follows a critical operation involving close tolerances. Final inspection and test operations are performed late in the sequence.

Packaging for shipment is the third group of operations referred to as protection type. Packaging, of course, represents the final operation performed prior to shipment to the customer.

Internal operations are performed in advance of external operations   This is not a rule that need always be observed. In many instances it is not necessarily advantageous to perform external operations last. The principal reason for performing internal operations early is that internal surfaces are less likely to be damaged in material handling and subsequent processes so their surfaces can be completed more early in the sequence. Another reason is that internal surfaces frequently provide a better means of holding the work and thus help ensure concentricity between inside and outside diameters.

When internal work is performed, the logical sequence of operation is drilling, boring, recessing, reaming, and tapping. The logical sequence of external work is turning, facing, grooving, forming, and threading.

Rough work is performed early It is almost always desirable to perform rough work involving heavy cuts and liberal tolerances early in the sequence. Heavy cuts will reveal defects in the raw material (usually castings or forgings) much more readily than light cuts. Furthermore, heavy cuts involving coarse finishes are usually performed faster with less-expensive workmen than fine finishes. It is always advantageous to find out that work is being performed on defective material as soon as possible with the least investment in secondary processes.

Consider the physical location of production facilities A certain amount of flexibility exists in the assignment of most of the operations required to produce a part. This is especially true of the tie-in operations and some of the protection operations. In order to reduce distances traveled and material handling, the production-design engineer should take into consideration the location of physical facilities and the general layout of the plant. For example, if a surface grinder is physically located next to a drill press, tie-in type operations involving these two facilities might well be made in sequence so as to minimize backtracking.

Consider existing load of physical facilities in process planning Since it is common practice to assign machine rates (dollars per hour) to capital equipment in a plant, the production-design engineer will customarily assign the facility with the least machine rate that still has the capability to efficiently perform the operation at the quality level desired. For example, if he computes that a 22.4-ton force is needed to blank and pierce a part, he will probably assign this work to a press capable of about 25 tons of force. Even though a 100-ton press has the capability to do the job, he would not schedule the work to this larger facility in view of its higher rate.

However, in many instances a given facility has a large backlog of work and additional work scheduled to it could result in a delay that would not be satisfactory. Consequently, a different facility involving a different process may be the best way to plan the job under study. For example, broaching could be substituted for reaming; grinding for turning; milling for shaping; piercing for drilling; tumbling for burring; brazing for welding; thread grinding for tapping; etc.

Nature of tooling will influence sequence of operations performed   In planning the optimum operation sequence, consideration should be given to the tooling. For example, it usually is desirable to do as much work as possible once the work is located and held in a jig or fixture. Thus, if some holes of a certain size are to be drilled on one surface of a part and a hole of a much larger diameter is drilled on another surface, it is desirable to keep the part in the same drill jig during the drilling operations of both holes, and the one drilling operation should follow the other in order to minimize the tie-up of tooling. If the jig is not designed so as to accommodate the drilling of the two different-size holes, then it may not be advantageous to have the one drilling operation follow the other. Jig and fixture design is exceedingly important for production economy.

Quantity and delivery affect the planning   The production-design engineer will need to consider the number of parts ordered and their required delivery schedule in order to effect the best planning. Larger volume permits more mechanization, the combining and consequent reduction of numbers of operations while smaller volume usually results in more separate manual type operations. The larger the volume, the more costly the set-up permitted and the more advanced the tooling can be. Advanced tooling refers to the utilization of both multiple and combined cuts. Multiple cuts refer to the taking of two or more cuts from one tool post or station such as the hexagon turret station on the turret lathe. Combined cuts represent cuts taken at the same time from two or more different stations, as for example, the square turret and the hexagon turret.

DEVELOPING THE OPERATION SEQUENCE

When the production-design engineer is planning the operation sequence, it is a good idea for him to write down on a 3- by 5-in card, while he is studying the engineering drawing, each operation that must be performed. On this card, he should indicate the class of operation (critical, tie-in, placement, or protection), the best surface for location in order to perform the work, the equipment recommended, and any other pertinent information related to the operation. These cards can then be arranged in sequence so as to arrive at the most favorable order in which the work should be performed. Figure 18-1 illustrates a typical card that can prove helpful in operation planning.

A sequence of planned operations may appear as follows:

① ← Basic operation
② ← Tie-in operation

```
┌─────────────────────────────────────────────────────────────────┐
│  PART NO. ___J-1410_____  OPERATION NO. ___7_____  │
│  OPERATION ___Drill two 13/32" lug holes_____ │
│  _____  │
│                                                                   │
│  _____  │
│  OPERATION CLASS ___Tie-in_____   │
│  PERFORMED ON ___Cincinnati-Bickford 21" Drill Press_____   │
│  LOCATION OF EQUIPMENT ___Dept. 11_____   │
│  LOCATE FROM ___Housing face. Use three points of location of this surface, │
│    two points on ribbed face, and one point on back surface._____ │
│                                                                   │
│  _____  │
│  ESTIMATED SET—UP TIME ___15 min._____  EACH PC. TIME ___2 min. │
└─────────────────────────────────────────────────────────────────┘
```

FIGURE 18-1
Operation planning card.

③ ← Placement operation
④ ← Critical operation
⑤ ← Tie-in operation
⑥ ← Protective operation
⑦ ← Critical operation
⑧ ← Tie-in operation
⑨ ← Critical operation
⑩ ← Protective operation

LOCATING THE WORK IN THE TOOLING

At the same time the production-design engineer is planning the operations to be performed, he should be giving consideration to how the work should be located and held while the various operations are being performed. He will need to determine:

1 Those points or areas that are best suited for locating the workpiece while it is being processed
2 That portion of the workpiece that is suited to supporting or holding it while it is being processed
3 That portion or area that is best suited for clamping so that the workpiece is securely held during the processing

The design of jigs, fixtures, dies, gages, and the like is beyond the scope of this text. However, this design effort can be considered a function of the production-design engineer. So that the student will have an appreciation for this work, the use of locators, supports, and clamps will be briefly discussed.

Locators

Locators refer to those points on the work which contact the holding device so that the work has a known relationship with the cutting or forming tools. Three locators are needed to locate a plane, two are needed to determine a line, and one will determine a point.

A workpiece can move in either of two opposed directions along three perpendicular axes (x, y, and z). Also the work may rotate around each of these three axes, either clockwise or counterclockwise. If we consider each of these possible movements as a degree of freedom, it can be understood that 12 degrees of freedom can exist. These 12 degrees of freedom are illustrated in Fig. 18-2.

Work can be positively located by six points of contact in the tooling. These six points are known as the 3-2-1 points. Three points are in one plane—two

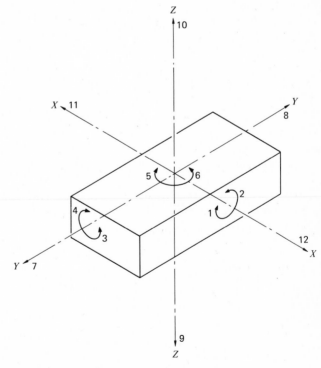

FIGURE 18-2
The 12 degrees of freedom applicable to a workpiece.

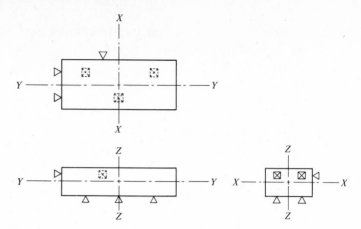

FIGURE 18-3
A workpiece and 3-2-1 system of locators.

points in a plane perpendicular to the first, and one point in a plane perpendicular to the second. This 3-2-1 system of support allows the work to be readily introduced and removed from the holding device, while providing positive location during the operation cycle. Figure 18-3 illustrates a block with 3-2-1 locators. The three locators on the bottom prevent movements 1, 2, 3, 4, and 9, as illustrated in Fig. 18-2. The two locators on the end prevent movements 5, 6, and 7. The single locator on the back side prevents movement 11. Directions 8, 10, and 12 have not been restricted, thus allowing easy loading and unloading of the block in its fixture.

In selecting a surface to be used as a subsequent locating surface, the production-design engineer should, if possible, select a surface that is flat and does not cross a parting line. The greatest dimensional variation in castings and forgings is across parting lines or flash lines. The surface selected should be machined to a good finish and should be able to be machined from the rough state in a comparatively short time.

The locators should be chosen at positions so that when the work is clamped in the holding device it will be stable and will not lift or rock away from the locators during the processing operation. Instability of the work in the fixture may be due to: insufficient locators; inadequate support of the workpiece; inadequate holding or clamping force; or positioning of the locators too close together.

It is always a good idea to place locators on one of the two surfaces that identify a close tolerance. If close dimensions are shown to centerlines, it is a good idea to place locators approximately equal distance apart on either side of the centerlines from which dimensioning takes place.

When a surface has a close tolerance on squareness, parallelism, or concentricity, then this surface should have either two or three locators identified with it. An effort should be made to avoid placing locators where they would come in contact with significant surface irregularities such as parting lines, flash lines, gates, burrs, etc.

SUPPORTING THE WORK

In addition to indicating points of location, it is usually desirable to specify points of support when ordering new tooling. It is important that the work is rigidly supported during the processing cycle, so that it does not deflect because of holding forces, tool forces, and its own static weight. The workpiece should not contact the supports until the tool or clamping forces are applied, since the only purpose of the support is to avoid or limit deflection and distortion.

CLAMPING THE WORK

In order to hold the work securely against all locators during the processing cycle, adequate clamping force is needed. Yet the clamping force should not be so great that it will mar the surface of the work or distort the part.

The holding force is applied directly opposite the locators. In many designs it is necessary to apply holding forces at more than one location in order to ensure that all locators are in contact with the work.

SUMMARY

It is not sufficient to identify the secondary operations that should take place to transform raw material into a completed functional part. Good planning must also identify the best operation sequence. Closely related to identification of operations and operation sequence is tooling. The engineer in charge of process planning should provide as much helpful information as possible to the tool designer. Valuable information that should be provided includes number and position of locators; number and position of supports; and number, size, and location of clamps. With this information the tool designer will be able to develop the jig, gage, fixture, dies, etc., that will produce the work in the manner desired.

QUESTIONS

1 In what ways can operation sequence affect the cost of a product? The quality of a product?

2 Explain what is meant by basic operations? By secondary operations?

3 What are the four categories of secondary operations?

4 Identify the following operations by their appropriate categories:

    (*a*) Milling a locating slot in a brass casting

    (*b*) Inspecting a reamed hole with a Go-NoGo plug gage

    (*c*) Drilling and tapping four corners of a steel plate

    (*d*) Zinc plating an alloy-steel-machined forging

    (*e*) Roll straightening bar stock before being machined on a six-spindle automatic screw machine

    (*f*) Grinding the flash from steel forgings after press trimming

5 Explain why critical operations involving close tolerances are done late in the process.

6 Why are internal operations frequently performed in advance of external operations?

7 When does the work come in contact with supports in jigs and fixtures?

8 What do we mean by the 3–2–1 system of location?

SELECTED REFERENCES

BUCKINGHAM, EARLE: "Production Engineering," Wiley, New York, 1942.

EARY, DONALD F., and GERALD E. JOHNSON: "Process Engineering for Manufacturing," Prentice-Hall, Englewood Cliffs, N.J., 1962.

The purpose of this chapter is to familiarize the prospective production-design engineer with United States patent laws so that he will be able to appreciate the desirability and understand the procedure for protecting a company's products.

Congress enacted the first patent law on April 10, 1790 (see Fig. 19-1), and the patent laws have remained substantially the same from 1870 to the present day. The United States Patent Office, an agency of the Department of Commerce, issues patents in accordance with the patent laws and established rules.

Here, on file in the Search Room, are approximately 4 million United States patents and 8 million foreign patents which are indexed and cross-indexed into over 360 major classes and more than 60,000 subclasses.

Prior to World War II, all United States industry spent about $300 million annually on research. Today, total monies spent on research in the United States are over $12 billion. The result of this momentous research effort is, of course, more patents. About 2,000 new applications are received in the Patent Office each

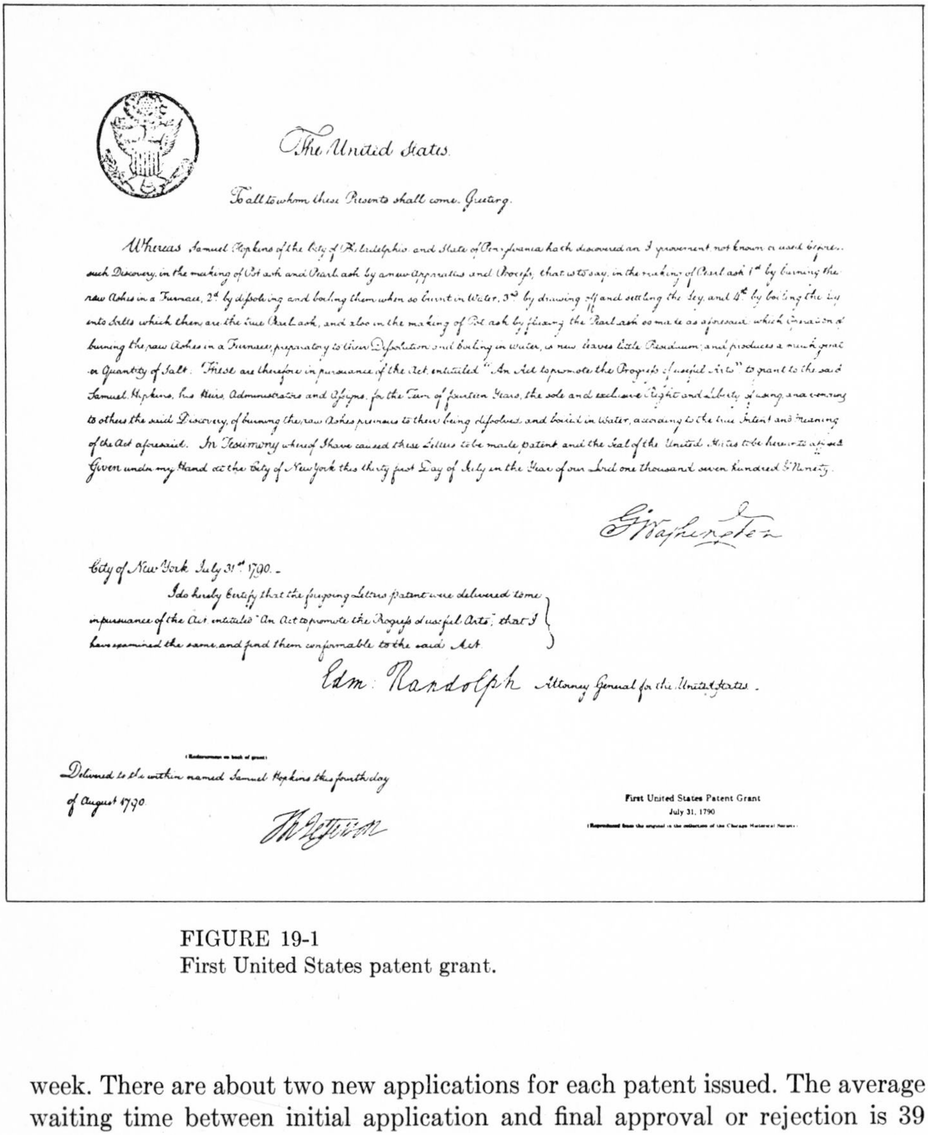

The United States.

To all to whom these Presents shall come. Greeting.

Whereas Samuel Hopkins of the City of Philadelphia and State of Pensylvania hath discovered an Improvement, not known or used before such Discovery, in the making of Pot ash and Pearl ash by a new Apparatus and Process; that is to say, in the making of Pearl ash 1st by burning the raw Ashes in a Furnace, 2d by dissolving and boiling them when so burnt in Water, 3d by drawing off and settling the Ley, and 4th by boiling the Ley into Salts which then are the true Pearl ash; and also in the making of Pot ash by fluxing the Pearl ash so made as aforesaid; which Operation of burning the raw Ashes in a Furnace, preparatory to their Dissolution and boiling in Water, is new, leaves little Residuum, and produces a much greater Quantity of Salt: These are therefore in pursuance of the Act, entituled "An Act to promote the Progress of useful Arts", to grant to the said Samuel Hopkins, his Heirs, Administrators and Assigns, for the Term of fourteen Years, the sole and exclusive Right and Liberty of using, and vending to others the said Discovery, of burning the raw Ashes previous to their being dissolved and boiled in Water, according to the true Intent and meaning of the Act aforesaid. In Testimony whereof I have caused these Letters to be made patent, and the Seal of the United States to be hereunto affixed. Given under my Hand at the City of New York this thirty first Day of July in the Year of our Lord one thousand seven hundred & Ninety.

G. Washington

City of New York July 31st 1790. –

I do hereby Certify that the foregoing Letters patent were delivered to me in pursuance of the Act, entituled "An Act to promote the Progress of useful Arts"; that I have examined the same, and find them conformable to the said Act.

Edm: Randolph Attorney General for the United States.

(Endorsement on back of grant)

Delivered to the within named Samuel Hopkins this fourth day of August 1790.

Th: Jefferson

First United States Patent Grant
July 31, 1790

FIGURE 19-1
First United States patent grant.

week. There are about two new applications for each patent issued. The average waiting time between initial application and final approval or rejection is 39 months.

The Patent Office supplies both a combination of incentives and safeguards to the inventor. An American patent grants an inventor exclusive rights to his invention in the United States for 17 years. It also assures him almost automatic patent approval in the 46 other member nations of the International Convention for the Protection of Industrial Property. (Russia is a nonmember.) There also are "patents" for unique designs not classed as inventions. These are *design patents*, with uniqueness, applying to the appearance of the product. Such patents have a life of $3\frac{1}{2}$ to 14 years.

The importance to the United States patent system was so evident to the founders of the United States that its principles were incorporated in the Constitution.

Article 1, Section 8 of the Constitution of the United States reads as follows: "The Congress shall have power . . . To Promote the Progress of Science and useful Arts, by securing for limited Times to Authors and inventors the exclusive Right to their respective Writings and Discoveries."

The fundamentals have been compressed into Sections 101 and 102 of the United States Code, Title 35—Patents:

. . . Whoever invents or discovers any new and useful process, machine, manufacture, or composition of matter, or any new and useful improvement thereof, may obtain a patent therefor, subject to the conditions and requirements of this title.

A person shall be entitled to a patent unless—

(*a*)   the invention was known or used by others in this country, or patented or described in a printed publication in this or a foreign country, before the invention thereof by the applicant for patent, or

(*b*)   the invention was patented or described in a printed publication in this or a foreign country or in public use or on sale in this country, more than one year prior to the date of the application for patent in the United States, or

(*c*)   he has abandoned the invention, or

(*d*)   the invention was first patented or caused to be patented by the applicant or his legal representatives or assigns in a foreign country prior to the date of the application for patent in this country on an application filed more than twelve months before the filing of the application in the United States, or

(*e*)   the invention was described in a patent granted on an application for patent by another filed in the United States before the invention thereof by the applicant for patent, or

(*f*)   he did not himself invent the subject matter sought to be patented, or

(*g*)   before the applicant's invention thereof the invention was made in this country by another who had not abandoned, suppressed, or concealed it. In determining priority of invention there shall be considered not only the respective dates of conception and reduction to practice of the invention, but also the reasonable diligence of one who was first to conceive and last to reduce to practice, from a time prior to conception by the other.

Notes—Section 4(*b*) of the Act of July 19, 1952 provides:

"Section 102(*d*) of Title 35, as enacted by section 1 hereof, shall not apply to existing patents and pending applications, but the law previously in effect, namely the first paragraph of R. S. 4887 (U.S. Code, title 35, sec. 32, first paragraph, 1946 ed.), shall apply to such patents and applications."

Section 4(*d*) of the Act of July 19, 1952 provides:

"The period of one year specified in section 102(*b*) of Title 35 as enacted by section 1 hereof shall not apply in the case of applications filed before August 5, 1940, and patents granted on such applications, and with respect to such applications and patents, said period is two years instead of one year."

CLASSES OF PATENTS

Exclusive right is granted by the laws of the United States under four classifications:

		Life from date of issuance	Fee
1	Patents	17 years. Patent term may be extended only by special act of Congress.	$65 plus $10 for each claim in independent form which is in excess of one plus $2 for each claim (dependent or independent) in excess of 10. Issuance fee $100 and $10 for each page of specifications as printed.
2	Design patents	3½, 7, 14 years.	$20 plus issuing fee of $10 for 3½ years; $20 for 7 years; $30 for 14 years.
3	Trademarks	20 years—may be renewed for 20 years.	$35
4	Copyrights	28 years—may be renewed for 28 years.	$2

The public as well as the holder of each of these exclusive rights is benefited. The public benefits by the increased probability of the invention or work being completed and being available for public consumption. If it were not for the protection given by law to these exclusive rights, many inventors would be unable to develop their ideas and would have to abandon their projects. The issuance of patents motivates others to make investigations resulting in further development of ideas. Without the patent system, inventors would keep their ideas secret, and advancement in the arts would be retarded. The patent system is one of the vital reasons for the technological advancement and high standard of living in the United States.

PATENTS DEFINED

A patent consists of four parts: (1) the grant from the government, (2) drawings illustrating at least one embodiment of the invention, (3) a specification describing the embodiment and explaining the drawings, and (4) a set of claims, each one of which asserts a particular field from which competitors of the patentee are to be excluded. The printed copies of the patents, on sale by the government at $0.50 each, include the drawings, the specification proper, and the claims.

Design patents are $0.20 each. Expedited service on patent copy orders including airmail delivery in the United States may be obtained on payment of an additional fee of $0.50 for each patent copy ordered.

Ownership Confers No Right to Make the Patented Device

It is important to realize that a patent is not a license to manufacture. In fact, many owners of patents cannot manufacture their own patented device without infringing on an earlier patent issued to someone else, because the product may embody other patented details.

A valid patent gives the inventor the right, for 17 years, to exclude all others from making, using, or selling anything within the scope of the patent claims; but it does not give to him the right to make, use, or sell his own product if it happens to infringe on the claims of some prior patent, even though his invention may be a decided improvement on the device illustrated in the prior patent.

A patent can be thought of as being a piece of real property such as a building or a plant site. Since it represents property it is entitled to, and receives, constitutional protection. Like other property, patents may be assigned or licenses may be granted under the patent. The patent can be leased or it can be sold. It is subject to the government's right of eminent domain.

A patent may be for a process, a machine, a composition of matter, new and useful improvements, a design, or plants. Examples are given in the booklet *Rules of Practice in the United States Patent Office*, which may be obtained from the Patent Office. The patent must be a concrete, workable expression of an idea that is new and useful. Old and new facts can be combined, but it must be more than an ordinary combination resulting from unusual skills and previous knowledge of the art.

Thomas Jefferson expressed the principle as follows:

As a member of the patent board for several years, while the law authorized a board to grant or refuse patents, I saw with what slow progress a system of general rules could be matured. Some, however, were established by that board. One of these was, that a machine of which we were possessed might be applied by every man to any use of which it is susceptible, and that this right ought not to be taken from him and given to a monopolistic, because the first perhaps had occasion so to apply it. Thus a screw for crushing plaster might be employed for crushing corn cobs. And a chain-pump for raising water might be used for raising wheat; this being merely a change of material should not give title to a patent. As the making a ploughshare of cast rather than of wrought iron; a comb of iron instead of horn or of ivory, or the connecting of buckets by a band of leather rather than of hemp or iron. A third was that a mere change of form should give no right to a patent, as a high-quartered shoe instead of a low one; a round hat instead of three-square; or a square bucket instead of a round one. But for this rule, all the changes of fashion in dress would have been under the tax of patentees.

AGGREGATION

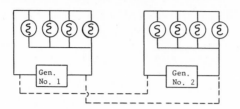

COMBINATION

is made by parallel connections
represented by dotted lines

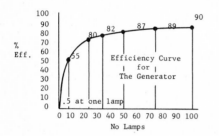

No. 1 Gen.		No. 2 Gen.		Total		
Lamps	Eff.	Lamps	Eff.	Lamps	Eff.	
50	87	1	5	51	65.8	
50	87	10	55	60	79.3	
100	90	1	5	101	77.0	Aggregate
100	90	10	55	110	85.2	
100	90	50	87	150	89.0	
100	90	100	90	200	90.0	
25	80	25	80	51	80.0	
30	82	30	82	60	82.0	
50	87	51	87	100	87.0	Combination
55	87.4	55	87.4	110	87.4	
75	89	75	89	150	89.0	
100	90	100	90	200	90.0	

FIGURE 19-2
Illustration of difference between aggregation and combination.

Combination Versus Aggregation

There is an important legal distinction between "combinations" (which are patentable) and "aggregations" (which are not). Suppose a given invention consists of putting together two or three already known elements. The mere combining of old machine parts, each operating in the old way and accomplishing the old result, is an aggregation, and hence unpatentable; whereas, if a new result is produced by the joint action of the elements, and if such result is not the mere adding together of the contributions of the separate elements, then a patentable combination exists. The test is as follows: Is the function of the whole equal to, or greater than, the sum of the functions of the parts? If equal to, then it is mere aggregation. If greater than, it is combination. In other words, a patentable combination is one which violates that fundamental axiom of geometry: The whole is equal to the sum of the parts.

To illustrate the distinction between "aggregation" and "patentable combination," suppose it is desired to supply electric current to 200 incandescent lamps, and there are at hand for this purpose two generators having a capacity for 100 lamps each. If 100 lamps are connected to one generator and the other 100 lamps to the other generator, the assemblage of the two circuits is a mere "aggregation"; each performs merely its own separate and expected function (Fig. 19-2).

On the other hand, if cross-connections for the two circuits are provided, so that lamps and generators are all in parallel, the assemblage becomes a patentable combination. The new function, over and above that of the two separate circuits, which raises the new assemblage to a patentable combination will become clear when the result of shutting off 50 lamps on one circuit is considered. With the two separate generators, one would then carry the current of 100 lamps; the other would carry current for but 50 lamps. Where the cross-connections are provided, as shown by the dotted lines in Fig. 19-2, the current for 150 lamps would divide equally between the two generators. A calculation will show that, under such circumstances, the heat losses in the two armatures would be much less than with the two separate generators. (See efficiency curve and tables in Fig. 19-2.) The new assemblage would thus perform the lamp-supplying functions of the first assemblage of two separate generators, and, in addition, function to reduce substantially electrical losses. It therefore fulfills the requirements of a patentable combination.

Another good example of aggregation is cited in the training manual for patent examiners. Here it is pointed out that an eraser on a pencil represents aggregation in that whenever the eraser is used the pencil is not; and when the pencil is used for writing, the eraser is not used. Thus, mere convenience of location of elements does not present patentability.

Novelty and Utility

The invention must be new and useful. Novelty exists if the invention has not been used by others in the United States, or patented, or described in any printed publication in this or any foreign country. Yet, prior knowledge or use abroad, unknown to the inventor, does not prevent the invention from being "new," even if such foreign use were known in this country.

An invention is "useful" if it works and if it is not frivolous or injurious to morale, health, or good order. Mere curiosities that do not have any intelligent purpose are not useful in a patentable sense.

An invention is inoperative if a model or sample built exactly as described in the patent will not work, or if further experiment and invention are required to make it work.

Mere ideas are not patentable   A mere idea is not patentable; there must also be the means for utilizing it practically. Newly discovered laws of nature are not patentable; invention consists not in discovering them, but rather in applying them to useful purposes. A good definition from a recent court decision is "Invention consists in the conception of a function, and the selection of means whereby the function can be operatively carried out."

For example, a time-study man may have an idea that he thinks will be patentable. He knows that, during a time study, a record is made on the study sheet of the total time elapsed as the operations occur. These times are read from a standard time-study stopwatch. The hands run continuously throughout the time study. After the study is finished, it is necessary to subtract the starting time from the finishing time of each element in order to get the elapsed time for each element. Therefore, there is a need for some means of reading directly the time that has elapsed for each operation. The time-study man conceives the idea of a time-study watch with three hands; one ready to start, one in motion, and one stopped. The stopped hand can be read as the actual time of the operation just completed. At the completion of the next element, the hand in motion is stopped and read, and the hand ready to start goes into motion. The hand that was stopped returns to start. Thus, there is no lost motion or time. Although this idea would be attractive and useful to other time-study men, the idea cannot be patented. It is necessary to design a watch with proper mechanism to accomplish the result. This mechanism can be patented provided other patents have not been issued having the same claims.

The actual exercise of inventive ingenuity is necessary Novelty and utility alone are not enough to make an idea patentable; more ingenuity must be displayed than could be expected of the average person skilled in that particular field when confronted with the same situation.

When a toolmaker makes a progressive die with the usual material guides, stops, and safety guards, he is applying knowledge of the toolmaking trade, and the die is not patentable. However, if the toolmaker uses ingenuity in making a feeding mechanism to feed parts into the die, permitting high-speed operation and guarding against parts doubling up and locating in the wrong position, he may have a claim for an invention. This device would increase production and avoid damage to the tool, press, or operator. It is not necessary for the inventor to understand why his device works, provided he understands and explains how it works.

The purpose of the patent laws is to reward those who make some substantial invention which adds to our knowledge and efficiency. It was never the object of the patent laws to grant a monopoly for every trifling device, every idea, which would naturally and spontaneously occur to any skilled mechanic or operator in the ordinary process of manufacture. Such an indiscriminate creation of exclusive privileges would tend to obstruct invention rather than to stimulate it. It would create a class of speculative schemers who would make it their business to watch for improvements and patent trivialities. This would enable them to lay a heavy tax upon the industries of the country without contributing anything to the real advancement of the sciences.

DESIGN PATENTS

In contrast to mechanical patents, which protect functional novelties, the design patent protects any new, original, and ornamental design for an article of manufacture. A design patent must be (1) new, (2) ornamental (attractive), (3) original, and (4) on a manufactured article.

A design patent applies to overall appearance and can even be applied to a line of products such as silverware. For example, design of women's dresses are protected by design patents. It should be noted that design patents protect only the appearance of an article and do not protect its structure or utilitarian features.

The attitude of the courts on design patents is expressed by the statement of the Supreme Court:

We hold therefore that if in the eye of the ordinary observer giving such attention as a purchaser usually gives, the two designs are substantially the same, if the resemblance is such as to deceive such an observer inducing him to purchase one instead of the other, the first one patented is infringed by the other.

Design patents are valuable and often can be obtained when mechanical patents are not possible. A governor for an elevator uses the conventional governor flyballs for its basic principle of operation. The arrangement of safety switches, overspeed devices, cable clamps, tripping devices, housing, cores, and drive for the flyballs can be designed into the governor in a distinctive manner. The position and relationships of the various parts can be a decided advantage. Such a design has a distinctive appearance and appeal to a customer and a design patent can be obtained.

The drawings submitted with the application for a design patent conform to the same rules as those that apply for mechanical patents. Only one claim is permitted in the application for a design patent.

THE PATENT DISCLOSURE

A patent disclosure is the information given to the attorney and patent office in order to obtain a patent. It is initiated by the inventor and is of vital importance to the engineer. The disclosure should include:

   1   Petition or request for a patent

2   Specifications:

a   Title of the invention
b   Brief summary of the invention, indicating its nature, substance, and object
c   Detailed description of the invention, indicating its nature, substance, and object
d   A claim or claims particularly pointing out and distinctly claiming the subject matter which the applicant regards as his invention

3   Drawings whenever the nature of the invention admits of drawings, which includes practically all inventions
4   Oath wherein applicant states or affirms that he believes himself to be the original or first inventor of the process, machine, manufacture, composition of matter, or improvement thereof, for which he solicits a patent

The responsibility of the engineer is to have all facts recorded and available for patent litigation. Changes, improvements, and information concerning existing patents and equipment should be transmitted to the patent attorney.

It is management's responsibility to determine whether the idea is worth spending the money to patent. The cost of obtaining the patent will be much less if all records are available. An inventor applying for a patent and personally paying for the expense will find that it will cost a minimum of $200. Since the returns from a patent are very uncertain, an engineer should investigate carefully the value of the project before proceeding with a patent application.

For full protection of an invention it is essential to:

1   Preserve the evidence of conception
2   Be diligent in reducing to practice, by drawings or working model
3   Apply for a patent at the earliest possible moment
4   Label the invented device with the patent number and date
5   Assert the patent against infringers with reasonable promptness

PATENT APPLICATION STEPS

Conception of Inventor

The production-design engineer is usually concerned principally with completing the project of design, process, or method and sometimes forgets to provide for patent protection. For some reason, such as modesty or lack of knowledge, the engineer does not like to have his notes witnessed or dated. He believes that others may think he is making a big fuss over nothing; however, much trouble could be avoided and lost business prevented if the engineer would take the proper precautions.

From the start of an idea—even in its embryonic stage—the thoughts should be placed on paper, dated, duly witnessed (full signature) by two or more people who understand the subject matter, and mailed to the patent council. Each sheet must be witnessed. The mailing of the information to the patent council avoids any accusation of collusion and misrepresentation of records. One or more friends in a department might act as witnesses for each other. This would limit the embarrassment of requesting signatures frequently. More than one witness is required because of the possibility that the witness will not be available when necessary. The inventor should not wait for the complete invention to develop but should disclose his idea the moment he has the first hazy outline. The most valuable part of any patent is found in the broad claims which come usually from the first general idea.

The law demands that diligence be exercised in reducing the idea to practice; therefore, models, tests, and experiments should be made. Records of tests should be observed by witnesses and a full description of test setup and results should be recorded, witnessed, and dated. Tests on equipment should be made according to standard practice, such as that required by Underwriter Regulations, to avoid questioning the test results. Photographs of witnesses observing tests, and closeups of equipment, dated and witnessed, are valuable evidence for patent litigations.

A patentable design cannot maintain priority if kept in cold storage and claimed to be *original* later—after another person has applied for such a patent. The valid date of conception must be justified by *diligent* development work, reducing the invention to practice, or by formal application for patent.

Evaluation by Management or Owner

The only way an invention can realize a profit is if it covers some feature which provides for a need or improvement for which the potential market is willing to pay. The production-design engineer should try to make sure that his invention will provide for this advantage. Management of his company will make an economic analysis before risking capital to pursue an invention.

Many patents, however, are applied for and *subsequently* granted even when the applicant does not guarantee that he will use it in his product or its manufacture. Such patents look to *possible* future use, to provide obstacles against competitors, or for possible assignment to another who may desire the patent right. A designer should not discard a patentable idea, in favor of what seems to be a preferable alternative, if the prospective patent is of interest to management.

Novelty Search to Determine Patentability

It is not possible to obtain a valid patent if the device has already been in commercial use in the United States or if it has already been publicly disclosed in a publication or a patent in a foreign country. To ensure that your device does represent novelty and patentability a search should be made. The least expensive way to find out that an invention does not qualify for a patent in view of prior art is through the patent search.

The novelty search should be made in the Search Room of the Patent Office in Washington, D.C. or in any one of 22 libraries located throughout the United States. Although the inventor can make the search himself, it is usually more prudent to employ an expert to make the search. The Patent Office has a roster of all registered practitioners who are available to prepare and prosecute patent applications for inventors. Anyone that is so listed is completely qualified to make the patent search.

Many engineering departments or legal departments of a firm maintain extensive files of patents, or indexes of patents applicable to their line of products. This provides for a preliminary search that may prevent unprofitable development work on a seemingly novel idea.

Study Patents Found in Search

In many instances the patent search will reveal several patents which are intended for the same purpose as the one on which the production-design engineer is working, but which are different in several ways. Thus patentability may exist. If your idea provides some important advantage, there probably is the opportunity to acquire a patent. You will have to decide whether to try to *get* a patent. The inventor should understand that no patent can cover old features. Thus you cannot obtain a patent which will prevent others from using inventions shown in prior patents.

If the search reveals that any former patent is exactly like your invention, then of course you will not be able to obtain a patent.

Development of the New Design

Once a decision has been made to follow up on an idea, research and experimental work will need to be carried out and the results reduced to practice.

Sketches and drawings should be in sufficient detail so that if the invention is made according to the specifications shown on the drawing it will work. It usually is a good idea to make a model of the device. This procedure not only will help uncover weak portions of the design but will ensure that the design will work as intended. The model itself can prove helpful in future discussions with the Patent Office relative to the novelty of the design.

Preparation of the Patent Application

Although every inventor may file his own patent application, it is advisable to obtain legal assistance for this task. It is possible to obtain help in the preparation and prosecution of the patent application from a registered patent practitioner who can be selected from a roster of names supplied by the Patent Office.

A patent application consists of these parts: formal papers, specifications, drawings, and claims. The formal papers include a petition, a power of attorney, and an oath or declaration.

The specification involves a complete and detailed description of the invention, ending with a set of claims. If the invention leads to being illustrated, a set of drawings should be included.

The claims should distinguish the invention from other patents that were identified in the search. The patent law requires that the patent specification must provide a description of the invention "which is sufficiently full and clear to teach a person skilled in the field of the invention to make and use it."

Patent Office Prosecution

When the patent application is received at the Patent Office it is examined by a Patent Office examiner who reads the application to ascertain if the invention has been properly described. He will also make a search and identify those patents that are related to the features covered in the claims of the submitted patent.

All correspondence is kept in what is known as a *file wrapper*. This is valuable information which, in case of litigation, is used to interpret the claims of the patent where the language is not adequate, and to prevent a patentee from applying to a claim a meaning inconsistent with its history. The file wrapper is destroyed after 20 years.

After the examiner reviews the application, he will advise the inventor as to which claims he is rejecting in view of earlier patents or applications, which claims he rejects because of "being obvious to a person having skill in the art," and which claims if any are allowable. In almost every instance, the examiner will find earlier patents close enough to the one under study to cause him to reject some of the claims. This response to the initial examination by the examiner is known as an *office action*.

The inventor is allowed 6 months to respond to Patent Office action unless a shorter period is prescribed by the examiner. The response is in a form of a letter and is known as an amendment. The amendment will point out the reasons that invention has occurred and that a patent should be granted.

A second office action takes place after the amendment is studied. This action may result in granting some of the claims and rejecting others, rejecting all the claims, or accepting all the claims.

An exchange of amendments and office actions continue until the examiner states that the rejection is final or until one or more of the claims on the application is allowed.

At any step until its final issue, the patent may be rejected for any of the following reasons:

1 Interferences may develop. Two inventors may make disclosures at the same time. It is then necessary to prove who had the idea first. The quality and accuracy of the original records now become valuable and an experienced attorney is required.

2 One inventor may have been more diligent in reducing his ideas to practice than the other. Here again records are valuable.

3 Prior publications may be discovered. A publication may have been found describing the patent 1 year or more before application.

4 Patents overlooked previously may be discovered to have prior claims.

5 The invention may have been sold publicly 2 years before patent was applied for.

An inventor may make an appeal to the Board of Appeals in the Patent Office if his patent application is rejected. This board, made up of the Commissioner of Patents, the Assistant Commissioners, and not more than 15 examiners-in-chief, will hear the case upon receiving a brief supporting the inventor's position, as well as an appeal fee. Approximately 5 percent of applications filed are appealed.

In the event the decision of the Board of Appeals supports the examiner, the inventor has a choice of appealing to the Court of Customs and Patent Appeals or of filing a civil action against the Commissioner of Patents in the United States District Court for the District of Columbia. In the past, applicants have sought further consideration of patentability through courts at the rate of 1 out of each 450 applications filed.

Sale of Rights

When working for a company, the engineer is faced with the problem of who has the rights to his patents. In general, one of the following is true.

1 If any work is done on company time in connection with a patent or knowledge used which has been obtained from the company, the patent automatically belongs to the company.

2 Usually a patent contract is signed by an engineer on entering the employment of a company. These contracts usually cover any contemplated fields of activity. For large corporations, this can cover about every phase of manufacturing. The employer has first rights to the patent, and if not interested will give a release, provided the privilege is granted to license under the patent at any time in the future.

A typical contract is shown below.

Agreement made by and between the _____ Corporation, a

_____ corporation, having a place of business at _____

_____, (hereinafter called the _____ Corporation) and . (employe) .

.... of ..... (address) .... (hereinafter called the Employe)

WITNESSETH:

In consideration of the covenants herein recited and the Employe's affiliation and activities with the Corporation, it is agreed as follows:

1. The Employe will perform for the Corporation such duties as may be designated by the Corporation from time to time, and will assign to the Corporation all inventions and improvements made by him while in its employ that shall be within the existing or contemplated scope of the Corporation's business, together with such patent or patents as may be obtained thereon, in this and all foreign countries, and upon request by the Corporation, will at any time during his employment with the Corporation and after its termination for any reason, execute all proper papers for use in applying for, obtaining, and maintaining such United States and foreign patents as the Corporation may desire, and will execute and deliver all proper assignments thereof, when so requested, but at the expense of the Corporation.

2. It is understood and agreed that the Employe will not, without written approval of the Corporation, publish or authorize any one else to publish, either during his term of employment or subsequent thereto, any technical information including secret processes and formulas acquired in the course of his employment under this agreement.

IN WITNESS WHEREOF, the Corporation has caused these presents to be executed and the Employe has hereunto set his hand and seal this .... day of ...., 19. .

_____(Name of corporation)_____

Witness ....   By .... (Seal)

Witness ....   By .... (Seal)

Employe

TRADEMARKS

A *trademark*, as defined in Section 45 of the Trademark Act of 1946, "includes any word, name, symbol, or device, or any combination thereof adopted and used by a manufacturer or merchant to identify his goods and distinguish them from those manufactured or sold by others." Its main purpose is to indicate the origin of the product, which will give the customer an assurance of a quality level. Well-known trademarks are valuable assets to the manufacturer. Consider the prestige enjoyed by such trademarks as Louisville Slugger, Craftsman, and Chesterfield. Advertising with a trademark tends to maintain a demand for the product. The owner of a trademark (it has a life of 20 years and may be renewed for another 20 years) can prevent the use of any similar mark which may likely be misinterpreted by consumers as being the mark of the party holding the trademark. Trademark rights are acquired only by use, and the use must continue if the rights are to be preserved during the renewal period.

Registration of Trademarks

The Patent Office points out in its general information on trademarks bulletin that a trademark cannot be registered if it:

1 Consists of or comprises immoral, deceptive, or scandalous matter or matter which may disparage or falsely suggest a connection with persons, living or dead, institutions, beliefs, or national symbols, or bring them into contempt or disrepute

2 Consists of or comprises the flag or coat of arms or other insignia of the United States, or of any State or municipality, or of any foreign nation, or any simulation thereof

3 Consists of or comprises a name, portrait, or signature identifying a particular living individual except by his written consent, or the name, signature, or portrait of a deceased President of the United States during the life of his widow, if any, except by the written consent of the widow

4 Consists of or comprises a mark which so resembles a mark registered in the Patent Office or a mark or trademark previously used in the United States by another and not abandoned, as to be likely when applied to the goods of another person, to cause confusion, or to cause mistake, or to deceive

In applying to register a trademark, the owner must include: a written application; a drawing of the mark; five specimens or facsimiles; the required filing fee ($35.00). There is no time limitation within which an application should be filed. It may be filed at any time after the mark has been used on the product and the product is sold or shipped in interstate, foreign, or territorial commerce.

It is advisable to make a search in the United States Patent Office before making application for this exclusive right. This will prevent infringement on an already existing mark and, of course, ensure that the proposed mark will identify, without confusion, the product for which it is proposed. The search is made in the Public Search Room of the Trademark Examining Operation of the Patent Office, located in the Longfellow Building, 1741 Rhode Island Avenue, N.W., Washington, D.C. It should be pointed out that the Patent Office will not make a search for the applicant.

The production-design engineer should also realize that he need not be represented by an attorney in the course of obtaining a trademark. He can file the application himself or an attorney may be appointed for this purpose.

When designing new trademarks, the inventor should try and use an adjective or descriptive form that will assist in the identification of the make of product. If this is not done, the trademark may become the name of the product and the exclusive right may be lost. For example, such products as "aspirin" and "cellophane" suffered this fate.

COPYRIGHTS

Copyrights are granted for a 28-year period, with a right of renewal for an additional term of 28 years. Material that is protected by this exclusive right must bear either the word "Copyright," its abbreviation "Copr.," or its symbol ©, the year the material was published, and the name of the copyright holder.

By having his material copyrighted, the author protects his work from being copied for the life of the copyright. The certificate of copyright is necessary in order to bring suit against anyone who has copied verbatim an author's work. A copy of the work (book, musical, dramatic, literary, etc.) is placed on file in the Library of Congress after a copyright has been obtained.

It should be understood that knowledge and ideas cannot be protected by copyright. It is the words or drawings that express or illustrate the knowledge or idea that is protected. Old material, expressions, etc. (e.g., *biblical quotations*) cannot be copyrighted except in combination with an author's original material.

The right to use copyrighted material can often be obtained from the owner, if appropriate credit is given.

SUMMARY

The patents in the Patent Office Search Room in Washington are one of the best sources of design information available to the production-design engineer. This vast wealth of technical information and suggestions is organized in a manner which allows the engineer to study those patents most closely related to his invention in a short period of time. In the event the engineer finds it difficult to make the trip to Washington, he can make a search or obtain technical information from patents at the following libraries: State University of New York, Albany, N.Y.; Georgia Tech Library, Atlanta, Georgia; Public Library, Boston; Buffalo and Erie County Public Library, Buffalo; Public Library, Chicago; Public Library, Cincinnati; Public Library, Cleveland; Ohio State University Library, Columbus, Ohio; Public Library, Detroit; Linda Hall Library, Kansas City, Montana; Public Library, Los Angeles; State Historical Society of Washington, Madison, Wisconsin; Public Library, Milwaukee; Public Library, Newark, N.J.; Public Library, New York, N.Y.; Franklin Institute, Philadelphia; Carnegie Library, Pittsburgh; Public Library, Providence, Rhode Island; Public Library, St. Louis; Oklahoma A & M College Library, Stillwater, Oklahoma; Public Library, Sunnyvale, California; Public Library, Toledo, Ohio.

As previously stated private concerns may maintain their own patent files on topics of interest. The monthly *Patent Gazette* (which lists brief descriptions of all new patents) can be reviewed to identify patents that should be obtained for the company's file.

Today, patents offer prestige, recognition, personal profit, and individual satisfaction for helping to provide better ways of living. The production-design engineer should always be on the alert to protect his ideas and his company's products through patent applications. The following can serve as a useful guide. It is *not* an invention to make:

1   A substitution of materials, such as plastic for metal

2   A substitution of an equivalent, such as a lever for a wheel

3   A change of size or degree, such as the diameter of a filament

4   A change of form or shape, such as the cross section of an orifice

5   A reversal of parts as by positioning a cutter before a grinding wheel instead of behind it

6   An omission or addition of parts, such as adding or eliminating a wheel on an auto

7   A change of conditions in a process, such as using higher temperature or pressure

8   A change of proportions in a composition, as a change in a cleaning bath

9   A device automatic, as by use of an electric motor

10   A device portable, as by mounting a machine on castors, or by making a television set small enough to carry

11   A similar use of a material, part, device, or composition

However, invention may be present:

1   If some new unusual and disproportionate or unobvious result is obtained or where a new property of the substituted material is used for the first time

2   If the change makes the difference between failure and success

3   In the means for making the device automatic

4   In the means for making the device portable

QUESTIONS

1   According to Sections 101 and 102 of the United States Patent Code, when shall a person not be entitled to a patent?

2   What is the maximum life of patents? Of design patents? Of trademarks? Of copyrights?

3 What is the object of the patent system?

4 What are the four parts of a patent?

5 Does ownership confer the right to manufacture the patented device? Why?

6 What is the difference between "combinations" and "aggregations"?

7 Outline the information required on a patent disclosure.

8 What are the steps that should be followed in applying for a patent?

9 What is the ratio of favorable disposition of patent applications to rejected applications?

10 Summarize the procedure followed in the Patent Office after the petition for a patent is received.

11 Of what value are trademarks? Give five trademarks with which you are familiar.

12 Can you successfully bring suit against someone who has copied some of your material that has not been copyrighted?

13 If two or more persons work together to make an invention, to whom will the patent be granted?

14 What was the nature of the first patent issued in the United States? Who was the recipient?

15 How does the public benefit from each of the four exclusive rights?

16 What do design patents protect?

17 Why is it advisable to obtain legal help in filing a patent application?

18 When does a product not fulfil the requirements for trademark registration?

SELECTED REFERENCES

PATENT OFFICE SOCIETY: *Journal of the Patent Office Society*, Box 685, Washington.

U.S. DEPARTMENT OF COMMERCE PATENT OFFICE: "General Information Concerning Patents," Washington, 1971.

————: *Official Journal of the United States Patent Office*, Washington.

APPENDIX

COMPARATIVE PRICES OF SELECTED ENGINEERING MATERIALS

Just as appliances and groceries may be purchased at retail or wholesale prices, commercial products and engineering materials can be purchased in different price ranges depending on the value added. The lowest prices are for *primary* materials, i.e., cast ingot or blast furnace pig. Today aluminum and steel can even be purchased in the molten state and trucked in specially insulated ladles to the point of use. *Mill products* are sold in minimum lot sizes of 30,000 to 40,000 lb and have a cross section of a standard shape such as a bar, sheet, I beam, or rail. They may also be further formed by cold working and with added cost. Typically mill products cost 60 to 100 percent more than primary metals. *Warehouse* stock sell at about 200 percent of the primary material cost and add *place utility* and availability in small lots (1 ton minimum in many cases). Finally the do-it-yourself worker can buy *mill* or *cast* products from a hardware wholesaler if he is lucky or knows someone who is in the hardware business. Otherwise he buys at retail at the local hardware stores in a lot size of one to ten but with a maximum amount of place utility. Such prices are 400 percent or more of the cost of the primary raw material.

For example, consider eight penny nails. With a primary cost at the blast furnace of \$0.038/lb, the mill cost of the cold-drawn wire would be \$0.010, but the retail cost of the cold finished nails would be $0.25 or more, or a markup of more than 600 percent. The wholesale cost of the same nails by the hundred weight in 1-ton lots would be $0.12 plus transportation from the wholesaler.

The following tables give comparative prices for selected primary metallic and polymeric materials, the mill prices for the same metals, and comparative prices for unalloyed metals, and metal powders. Prices of the same materials are given for 5 years of the past decade, so that the relative change in prices over time can be seen. In several instances prices of the same material in different shapes have been given in the same year to show the effect of shape and number of rolling operations on the cost of a material.

Table A-1   COMPARATIVE PRICES OF PRIMARY ALLOYS, $/LB

	Specific weight, lb/in^3	1963	1966	1968	1972	1973
Ferrous alloys						
Pig iron	0.250	0.029	0.030	0.035	0.041	0.043
Scrap						
No. 1 cupola iron	0.250	0.013	0.016	0.021	0.024	0.041
No. 1 heavy melt (steel)	0.083	0.018	0.022	0.011	0.017	0.037
Steel rails	0.283	0.022	0.025	0.021	0.024	0.049
Nonferrous alloys						
Aluminum alloys						
Ingot (99.5% pure)	. . . . .	. . . . .	. . . . .	. . . . .	. . . . .	0.263
Al-Cu (No. 43)	0.098	0.229	0.249	0.254	0.273	0.273
Al-Si (No. 356)	0.098	0.249	0.265	0.270	0.274	0.274
Copper alloys						
Electrolytic copper	0.320	0.31	0.036	0.423	0.526	0.603
Yellow brass (Nos. 1; 405)	0.310	0.275	0.039	0.345	0.445	0.550
Red brass (No. 115)	0.320	0.32	0.50	0.44	0.533	0.635
Tin bronze (No. 225)	0.320	0.413	0.685	0.593	0.705	0.745
Zinc alloys						
Pig	0.250	0.134	0.156	0.144	0.18	0.233
Die cast alloy (No. 3)	0.250	0.148	0.183	0.165	0.223	0.280
Die cast alloy (No. 5)	0.250	0.152	0.193	0.170	0.23	0.300
Magnesium						
Pig (AZ913—die-casting pig)	0.065	0.36	0.36	0.368	0.373	0.36
Pig (AZ63A—sand-casting pig)	0.065	0.41	0.41	0.430	0.42	0.45
Nickel						
Pig	. . . . .	. . . . .	0.80	. . . . .	1.33	
Titanium						
Commercially pure 8-in diameter billet	0.163	2.40	2.50	2.53	2.81	2.81
Ti-5Al—2.5 Sn billet	. . . . .	. . . . .	. . . . .	2.53	2.15–2.30	2.35
Ti-6Al—4 V billet	. . . . .	. . . . .	. . . . .	. . . . .	. . . .	2.50
Ti-6Al—6V 2Sn billet	. . . . .	. . . . .	. . . . .	. . . . .	. . . .	2.55

Table A-2 COMPARATIVE PRICES OF MILL PRODUCTS, $/LB

	1963	1966	1968	1972	1973
Steel—mill (0.283 lb/in³), 30,000-lb lots					
Carbon steel, H.R.: bar, strip, sheet	0.057	0.05	0.063	0.0865	0.083
Low alloy, H.R.: plate, structural shapes	0.055	0.06	0.062	0.0871	0.113
Carbon steel, C.R.: bars					
Galvanized steel sheet: hot-dipped	0.072	0.075	0.078	0.1202	0.120
Low alloy, C.F.: bars	0.0925	0.0938	0.0958	0.1385	0.1385
Tool steel, C.F.: bars H-13 die steel	0.550	0.570	0.575	0.796	0.847
Tool steel, T-1: 18-4-1	1.840	1.920	2.050	.....	2.82
Tool steel, M-2	1.345	1.280	1.360	1.880	1.970
Steel—warehouse (Cleveland—1-ton lots)					
H.R. carbon sheet: 10 gage × 36 to 120 in long (SAE 1020)	0.105	0.097	0.1055	0.1389	0.1389
C.R. carbon sheet: 20 gage × 36 to 120 in long (SAE 1020)	0.112	0.106	0.1290	0.1635	0.1635
H.R. carbon: round bars ¾ in OD (SAE 1020)	0.098	0.010	0.1119	0.1271	0.1440
C.F. carbon: bars 1 × 1 in (AISI C1018)	0.113	0.112	0.1143	0.1641	0.1705
H.R. carbon: plate (SAE 1020)	0.0989	0.118	0.1193	0.1495	0.1544
Structural shapes (low alloy) I beams 6 × 12¼ in	0.1035	0.1050	0.1130	0.1485	0.1536
Galvanized sheet: 10 gage × 120 in				0.1671	0.1671
H.R. 4140 2-in-diameter bar (annealed)	0.1621	0.1580	0.1709	0.2289	0.2336
H.R. 4615 2-in-diameter bar	0.1721	0.1640	0.1752	0.2446	0.2495
C.D. 4140 2-in-diameter bar (annealed)	0.2056	0.197	0.2105	0.2796	0.2856
C.D. 4615 2-in-diameter bar	0.2131	0.200	0.2132	0.2994	0.3052
Stainless steel plate (304) ½ in thick × 84 in wide	0.388	0.360	0.415	0.495	0.520
Stainless steel sheet (304) 11 gage × 48 in wide	0.480	0.360	0.440	0.493	0.525
Stainless steel bar (304) ½-in-diameter H.R.	0.468	0.515	0.543	0.633	0.694
Stainless steel bar (304) C.F.				0.754	0.804
Aluminum alloys—mill (0.098 lb/in³), 30,000 lb min.					
Extrusions (6061/6063)	0.453–0.618	0.365–0.51	0.371–0.495	0.460–0.495	0.460–0.521
Flat sheets (1100, 3003, 5005) 0.076 to 0.250 in thick	0.449–0.470	0.41	0.400–0.446	0.443–0.468	0.443–0.540
Plate (1100, 3003, 5005) 0.251 to 3.000 in thick	0.445–0.500	0.46	0.450	.....	0.468–0.494
Screw machine stock (2011-T3) ¼ to ⅜ in diameter	0.600	0.615	0.665	0.645	0.645

	C1	C2	C3	C4	C5
Screw machine stock (2011-T3) ½ to 1.5 in diameter	0.553	0.568	0.618	0.584	0.584
7075-T6 plate (¼ to ¾ in thick)					0.577
Aluminum alloys—warehouse (0.098 lb/in³), ton lots					
6063-T5 bar (extruded)	....	0.780			
1100-F bar (½ × 1 in)	....	0.900			
2024-T4 bar (½ × 1 in)	....	1.000			
2011-T3 bar (½ in diameter)	....	1.200			
Copper—mill (0.32 lb/in³)					
Drawn tubing	0.580	0.6523	0.765	0.974	1.09–1.17
Sheet (over 24 in wide)	0.565	0.6725	0.753	0.9603	1.05–1.09
Copper alloys—mill (0.305 to 0.320 lb/in³)					
Free cutting brass rod (round, hexagonal, square)	0.499	0.429	0.462	0.6499	0.7705
Yellow brass (360)	0.500	0.800	0.648	0.838	0.90–0.949
Red brass (230)	0.537	0.900	0.699	0.891	0.969–1.03
Silicon bronze (655)	0.617	0.620	0.790	....	1.14
Cupronickel (10% Ni)	0.660	0.660	0.826	....	1.23–1.29
Cupronickel (30% Ni)	....	0.810	1.03–1.19	....	1.45–1.50
Beryllium-copper (1.9% Be)	2.01	1.980	2.39	....	3.05
Zinc—mill (0.250 lb/in³)					
Plate	0.215	0.240	0.24		
Sheet	0.280	0.320	0.365		
Strip	0.225	0.250	0.223		
Die cast alloys 3 and 7 (40,000 lb minimum lot)	0.148	0.183	0.165	....	0.34
Die cast alloy 5 (40,000 lb minimum lot)	0.150	0.193	0.170	....	0.36
Magnesium alloys—mill (0.065 lb/in³), AZ 31B					
Tooling plate: 0.250 to 3.50 in	0.730	0.730	0.730	0.706	1.106
Sheet and plate:					
0.500 to 3.00 thick	0.948	0.948	0.948	0.702	0.702
0.190 in thick	1.008	1.008	1.008	0.725	0.725
0.125 in thick	1.095	1.095	1.045	0.818	0.818
0.080 in thick	1.192	1.192	1.192	1.083	1.083
0.032 in thick	1.716	1.716	1.716		
Nickel and its alloys—mill (0.321 lb/in³), 30,000-lb base					
Monel 400: rod, plate, sheet, strip	0.96–0.126	1.00	1.20–1.45	1.70–1.98	1.83–2.06

Table A-2 COMPARATIVE PRICES OF MILL PRODUCTS, \$/LB (*Continued*)

	1963	1966	1968	1972	1973
Inconel 600: (Ni–Cr–Fe): rod, plate, sheet	1.16–1.45	....	1.37–1.63	1.87–2.13	2.00–2.26
Nickel: rod, plate, sheet, strip	1.16–1.47	1.00	1.46–1.66	2.11–2.29	2.33–2.51
Waspalloy	....	5.15			
Titanium and its alloys—mill (0.163 lb/in³), 30,000-lb base					
Commercially pure: 8-in-diameter billet	3.20	2.50	2.53	2.81	2.31
Commercially pure: bar, rod, sheet, strip, wire	3.60–14.25	4.90–14.25	4.70–5.35	4.00–5.35	2.31–4.12
Ti–5 Al–2.5 Sn: billet	....	....	2.53	....	2.35
Ti–6 Al–4 V: billet	....	....	2.53	....	2.53
Ti–6 Al–6 V–2 Sn: billet	....	....	2.53	....	2.55
Ti–6 Al–4 V: bar—2 in diameter	....	....	3.63–4.13	....	3.20
Ti–6 Al–4 V: sheet—$\frac{1}{8}$ in thick	....	....	7.60–8.00	....	6.85
Ti–8 Al–1 Mo–1 V: strip, 0.040 in thick	....	....	9.10–9.50		

SOURCE: "Materials Selector," Reinhold, Stanford, Conn.; *Steel* now *Industrial Week*, Penton, Cleveland, Ohio.

Table A-3   COMPARATIVE COST OF CASTINGS, $/LB

	1965	1968	1973
Sand castings			
Gray iron	0.10–0.25	0.16–0.35	0.24–0.48
Ductile iron	0.12–0.28	0.20–0.42	0.34–0.60
Malleable iron	0.11–0.25	0.17–0.42	0.32–0.55
Steel	0.15–0.35	0.30–0.50	0.40–0.68
Aluminum	0.50–1.00	0.65–1.50	1.75–2.25
Copper base	0.50–1.00	0.85–2.40	1.50–3.00
Metal mold castings			
Aluminum (die cast)	0.50–0.80	0.60–1.00	0.80–1.20
Aluminum (permanent mold)	0.75–1.15	1.00–1.50	1.35–1.75
Zinc (die cast)	0.30–0.50	0.35–0.65	1.00–1.50

The lower cost figures apply to castings of simple shape bought in large quantities as in the automotive or appliance field. The higher figure applies to smaller quantities or where the shape is complex and coring is required.

The sources of the data are given so that if an update is required, those sources can be consulted and used in conjunction with the tables, to determine if there has been a major trend over time and if the relative prices of the various materials have changed with time.

It is hoped that a historic perspective of prices can give a young engineer a

Table A-4   COMPARATIVE PRICES OF METALLIC POWDERS, $/LB, IN TON LOTS

	1963	1966	1968	1972	1973
Iron					
Sponge (98.2% pure)	0.985	.....	0.128	0.099	0.109
Electrolytic (99.4% pure)	.....	0.120	0.278	0.365	0.410
Aluminum (atomized)	0.430	.....	....	0.418	0.418
Brass (80% Cu-20% Zn)	0.380	.....	0.567	0.630	0.735
Copper (spherical)	.....	.....	0.660	0.730	0.835
Stainless steel (304)	0.885	.....	0.870	0.870	0.870
Bronze (90% Cu-10% Sn)	0.572	.....	0.795	0.886	1.011
Nickel silver (64% Cu, 18% Ni, 18% Zn)	0.562	.....	0.854	0.917	1.046
Titanium (sponge)	1.32	1.32	1.32	1.32	1.42
Cobalt	1.820	1.500	....	....	3.200
Molybdenum	3.900	.....	3.280	3.280	3.280
Tungsten (99.9% pure, hydrogen reduced)	2.400	2.500	4.500	5.430	5.430
Zirconium (sponge)					5.500
Beryllium (200)	.....	.....	....	48.50	48.50
Rhenium (99.99% pure)	.....	.....	575.00	1400.00	900.00

Table A-5 COMPARATIVE PRICES OF UNALLOYED METALS, $/LB

Metal	Form	Specific weight, lb/in³	1963	1966	1968	1972	1973
Commercial metals							
Iron	Pig	0.283	0.029	0.030	0.035	0.041	0.043
Lead	Pig	0.41	0.096	0.150	0.195	0.158	0.165
Zinc	Pig	0.258	0.125	0.140	0.190	0.245	0.258
Aluminum	Pig	0.098	0.240	0.245	0.260	0.254	0.263
Manganese	....	0.270	0.325	0.325	0.360	0.423	0.526
Antimony	Pig	0.239	0.33	0.44	0.44	0.57	0.526
Copper	....	0.324	0.310	0.36	0.746	1.00	1.07
Magnesium	Pig	0.065	0.36	0.36	0.37	0.378	0.405
Calcium	....	....	4.55	0.95	0.96	1.25	1.25
Nickel	Pig (10 lb)	0.321	0.867	0.770	1.56	1.32	1.53
Tin	Pig	0.264	1.16	1.25	1.40	1.74	2.12
Chromium	....	0.263	1.31	1.31	0.97	1.15	1.56
Cobalt	Shot	0.322	1.50	1.70	1.85	2.30	3.20
Cadmium	....	0.313	2.35	2.40	2.65	2.60	3.75
Bismuth	....	0.354	2.25	4.00	4.00	3.50	5.05
Titanium	8-in-diameter billet	0.163	1.55	1.32	2.53	2.81	2.81
Mercury	Flask, 76 lb	0.489	2.75	4.00	4.50	5.89	5.89
Tungsten	Powder	0.697	2.40	2.50	4.50	5.43	5.43
Vanadium	....	0.220	3.45	3.45	3.65	2.19	2.18
Zirconium	Sponge	0.234	5.00	5.00	5.00	5.50	5.50
Molybdenum	Billet	0.369	7.50	8.00	30.	30.	30.
Other metals							
Sodium	....	0.035	....	....	....	....	0.118
Tellurium	....	0.225	6.00	6.00	6.00	6.00	6.00
Thallium	99.7%	0.428	....	....	....	....	7.50
Uranium	....	0.689	....	....	....	....	8.00
Selenium	....	0.174	....	....	....	....	10.00

Columbium		0.310	. . . . . .	. . . . . .	. . . . . .	. . . . . .	10.00
Boron		0.085	. . . . . .	. . . . . .	. . . . . .	. . . . . .	18.00
Silicon	99.9%	0.084	. . . . . .	. . . . . .	. . . . . .	. . . . . .	27.00
Indium		0.264	32.80	48	48	. . . . . .	48
Beryllium		0.067	62.00	62.00	62.00	69.00	61.00
Hafnium		0.473	. . . . . .	. . . . . .	. . . . . .	. . . . . .	100.00
Cesium		0.688	. . . . . .	. . . . . .	. . . . . .	. . . . . .	120.00
Germanium		0.192	135.00	125.00	125.00	125.00	125.00
Rhubidium		0.055	. . . . . .	. . . . . .	. . . . . .	. . . . . .	300.00
Gallium		0.213	. . . . . .	. . . . . .	. . . . . .	. . . . . .	750.00
Precious metals							
Silver	Pig	0.379	22	18	31	26	42
Gold	Pig	0.698	510	510	567	962	1,436
Rhenium		0.756	. . . . . .	. . . . . .	. . . . . .	1,400	900
Palladium		0.434	490	580	660	540	1,108
Platinum		0.775	1,550	1,750	1,746	1,785	2,224
Iridium		0.813	1,018	2,473	3,191	2,223	3,245
Rhodium		0.447	2,450	3,500	3,568	2,877	3,245

SOURCE: *Iron Age* and *American Metal Market*.
Note: Conversion from Troy ounce to pound avoirdupois is 14.58 (Troy ounce). This factor was used for precious metals and indium.

Table A-6 COMPARATIVE PRICES OF POLYMERS, $/LB

Polymer	Density, lb/in³	1963	1966	1968	1972	1973
Polyethylene, low density	0.032	0.235	0.190	0.13	0.14	0.125
Polystyrene, general purpose	0.038	0.175	0.145	0.145	0.13	0.145
Polystyrene, medium high impact	0.037	0.285	0.25	0.215	0.205	0.170
Polyethylene, high density	0.034	0.280	0.250	0.175	0.13	0.145
Urea, filled	0.054	0.260	0.260	0.26		
Phenolic filled	0.05	0.270	0.270	0.31		
Polypropylene	0.035	0.480	0.240	0.235	0.21	0.42
Polyvinyl chloride (PVC)	0.051	0.24	0.280	0.260	0.26	0.55
ABS	0.038	0.47	0.430	0.360	0.37	0.43
Polyester	0.048	0.32	0.260	0.44	0.68	0.67
Alkyd	0.685	0.52	0.510	0.550		
Melamine	0.052	0.435	0.435	0.460		
Ionomer	0.034	.....	0.50	0.47	0.47	0.47
Acrylic	0.043	0.51	0.460	0.455	0.455	0.505
Cellulose acetate butyrate	0.042	0.51	0.620	0.620	0.62	0.46
Acetal	0.051	0.52	0.650	0.650	0.65	0.625
Nylon 66	0.041	0.98	0.87	0.80	0.73	0.73
Polycarbonate	0.043	1.20	1.01	0.80	0.75	1.00
Polyurethane	0.048	1.40	1.40	1.30		
Polysulphone	0.045	0.93	1.00	1.00	1.00	1.21
Epoxy, filled	0.065	0.62	0.530	1.30		
Silicone	0.058	2.98	2.92	3.45		

SOURCE: *Materials Selector*, Reinhold Publishing Company, Stanford, Conn.

better appreciation of the economic aspect of material selection. Once the functional, load-carrying, and style decisions have been made, the final material will be chosen on the basis of price and intangible factors. The figures given in the charts, plus a reasonable extrapolation to account for inflation, can be used for solving classroom problems. However, in practice, current costs must be obtained from vendors at the time process engineering decisions are made.

ABS materials, 234
Absorption of light, 24
Acetal, 235
Acid pickling, 644
Acrylic resins, 230
Actual cost, 42
Adhesives, 743
Age hardening, 124–125
Aggregation, 796–797
Air comparator, 755
Air drying, 652
Alkaline cleaning, 645
Alloy steel, 153–156
   aluminum, 189
   copper, 208
   low-melting, 217–220

Alloys:
   nickel, 210–213
   steels, 153–156
   tin-base, 217–218
   zinc-base, 217
Alpha iron, 138
Aluminum:
   alloys, 189
   castings, 188–189
   corrosion resistance of, 198
   designation system, 186–189
   electrical properties, 199
   emissitivity, 196
   fabrication of, 197–198
   foil, 193
   forming, 197

heat reflectance, 196
heat treatment of, 189
light reflectance, 196
machining of, 197
plate, 193
properties, 193–199
sheet, 193
toxicity, 194
weight, 196
welding, 686
Analysis of cutting forces, 481–483, 486–489
Aniline-formaldehyde resins, 231
Annealing, 106, 127
Anodic coatings, 668
Antioch process, 325
Appearance, 325
  balance, 25
  color, 25–27
  interest, 24
  laws of, 23
  surface treatment, 25–27
  unity, 23–24
  value of, 21–23
Arc blow, 692
Arc welding, 687–704
Atomic structure, effect of, on machining, 480
Austenite, microstructure of, 111–124
Austenite stainless steel, 159
Austenitizing, 103–104
Automatic gaging and controls, 755
Automatic screw machines, 517–522
Automatic welding, 694
Automation, 553–556

Baking, 652–653
Balance, 25
Barrel finishing, 543–545
Basic machine tools:
  drilling machines, 508
  lathes, 507–508
    milling machines, 508–512
    planers, 508
Bending recommendation, 164, 592–602
Bernoulli's theorem, 360
Beryllium bronze, 210
Blank size, 623–625
Blanking, 582–591
Block diagram of basic machine tools, 506
Blow molding, 457
Bolts, 740–741
Borosilicate glass, 277
Bosses in plastic design, 241
Brass, 208–209
Brass plating, 661
Braze welding, 719
Brazing, 715–719
Break-even chart, 40
Brinell test, 81–83
Broaching, 528–531
Bronze, 209–210
Brushing of organic materials, 649
Buffing, 542
Built-up edge, 483
Butyl rubber, 267

Cadmium plating, 660
Carbides, 476–477
Carbon, 282–284
Carbon steel, 153
Carburizing, 117
Case hardening, 116–123
Cast iron:
  castability, 147
  centrifugal, 337–342
  compression strength, 145
  corrosion resistance, 148
  ductile iron, 138–141, 149
  ductility, 146
  endurance limit, 146
  gray iron, 137–138

hardness, 146–147
machinability, 147
malleable iron, 141–142
mechanical properties of, 142–147
modulus of elasticity, 145
shrinkage rules, 149
tensile properties, 142–144
vibration damping, 148
wear resistance, 148
weldability, 148–149
yield strength, 145
Casting:
  aluminum, 188
  Antioch process, 325
  centrifugal, 337–342
  centrifuged, 340
  ceramic mold, 325–327
  continuous, 342–344
  cores, 353–358
  costs, 389–393
  design, 345–351, 394–409
  die, 334–337, 381–387
  fluidity, 313–314
  full-mold process, 321–324
  fundamentals, 298–314
  gating systems, 359–364
  gray iron, 137–138
  hot tears, 311–313
  inserts, 358–359
  investment, 327–329, 370–379
  materials, 298–304
  nonferrous tools, 476–477
  permanent mold, 331–334, 374–381
  plaster mold, 324, 368–370
  plastics, 454–455
  precision processes, 324–329
  process of, 304–306
  risers, 310–311
  shell mold, 317–319
  shrinkage defects, 311
  slush, 342
  steel, 400–402

  vacuum, 344–345
  weight, 304
  whitewares, 272
Catalyzed-resin bonding, 319–320
Cellulose acetate butyrate, 231
Cellulose nitrate, 231
Cellulose plastics, 230–231
Cemented carbides, 477
Centrifugal casting, 337–342
Centrifugal finishing, 651
Centrifuged, 340
Ceramic coatings, 656
Ceramics:
  casting of, 272
  design considerations, 275–276
  enamels, 272
  extruding of, 272
  glass, 276–279
  jiggering, 272
  molds, 326–327
  oxides, 477–478
  pressing of, 272–273
  properties, 273–274
  refractory types, 279–283
  shell casting, 329–331
  structural clay products, 272
  whitewares, 272–276
Chemical stability of plastics, 230
Chemically bonded sands, 316–317
Chem-milling, 530
Chip-thickness ratio, 483–485
Chipless material-removal processes, 547–552
Chromate coatings, 665–668
Chromium plating, 661–662
Classifying operations, 779–781
Cleaning, 644–646
Clearance in press work, 576–578
Climb milling, 525–527
Coated electrodes, 689
Coating of plastics, 252–253, 662
Cobalt, 214

Coining, 611–613
Cold-box cores, 319–320
Cold-chamber die casting, 335–337
Cold extrusion, 609
Cold rolling, 609–610
Cold welding, 720
Cold work, effect of, 171–173, 479–480
Cold-working fundamentals, 567–570
Color, 25
Color wheel, 27
Combination, 796–797
Comparator, air and electric, 755
Comparison of materials, 93–95
Compression molding, 445–448
Continental method, 584–585
Continuity, law of, 362
Continuous casting, 342–344
Contract negotiation, 48
Conventional milling, 525–527
Conversion coatings, 665–670
Cooling rates, 125
Copper:
　alloys, 208
　plating, 661
　properties, 203–207
　uses, 203–208
Copyrights, 807
Core making, 355
Cores in casting, 353–358
Corrosion resistance:
　cast iron, 148
　steel, 164–165
Cost:
　actual, 42
　casting, 389–393
　comparison of materials, 93–95
　control, 47–48
　development, 38
　elements of, 38–41
　estimation, 41
　influence of activity, 48–50
　labor, 43–44

　material, 42–43, 92–95
　product, 39
　recorded, 41–42
　reduction, 46–48
　standard, 41–42
　system, 45–46
Crackle finish, 649
Creative thinking, 5–17
Creep, 87–88
Criteria for product success, 18–35
Critical operations, 782
Crystal finish, 648
Curiosity, 7–8
Current control in welding, 692
Cutting tool:
　broaching, 528–531
　cemented carbides, 477
　diamonds, 478–479
　drills, 466–468
　energy considerations, 490–491
　fluids, 499–502
　forces, 481–483, 486–489
　gear, 527–528
　geometry, 465
　grinding, 468–470, 534–540
　honing, 541
　lapping, 540–541
　life, 492–502
　materials, 474–476
　milling cutters, 468–470
　power requirements, 491–492
　sawing, 531–533
　superfinishing, 541–542
　velocity relations, 489–490
　wear, 492–498
Cyaniding, 117

Decorative coatings, 643–670
Dendrites, 307–310
Design:
　functional, 20–21

incorporating quality, 27–30
patents, 799
Designing for production:
adhesives, 744–745
broaching, 530
carbon, 284
casting, 345–351, 394–397
ceramics, 275–276
forging, 435–441
function of, 1–2
grinding, 542
permanent mold casting, 374–381
plaster-mold casting, 368–370
press work, 631–634
welding, 722–724
Diamonds, 478–479
Die casting, 334–337, 381–387
Die design:
casting, 381–387
forging, 435–441
plastic components, 239–253
powdered metals, 286–289
press work, 631–634
whitewares, 275–276
Dinking, 584
Dipping, 649–650
Distortion in welding, 678–681
Draft, plastics: casting design, 352
plastic design, 240–241
Drawing:
blank size, 623–625
cold, 618
deep, 618–625
design details, 622–623
effect of, 171–173, 479–480
fundamentals, 567–570
hold-down force, 619–620
maximum reduction, 570–572
presses, 635
properties of material for, 572–575
reduction in size, 620–622
Drilling, 466–468, 508

Drying, 652–653
Ductile iron, 138–141, 149
Durometer test, 81–83

Economic analysis, 37
Economy of material in press work, 581–583
Ejector pins in molds, 244–245
Elastomeric adhesives, 744
Electric comparator, 755
Electric-discharge machining (EDM), 548–549
Electrical laboratory for inspection, 757–758
Electrical properties of aluminum, 199
Electrical resistivity of materials, 70–71
Electrochemical machining (ECM), 549–550
Electrodes, 689
Electrolytic cleaning, 645
Electron-beam welding, 699–704
Electroplating, 656–663
Electroslag welding, 697
Elements of cost, 38–41
Embossing, 613
Emulsion cleaning, 645
Enamels, 646–647
Endurance limit, 86–87
Energy considerations, 490–491
Engine lathe, 507–508
Engineering stress-strain relationships, 75–77
Epoxies, 239
Equilibrium diagram, iron-carbide, 135–137
Estimating cost, 41
Ethyl cellulose, 231
Extruding:
cold, 609
high-pressure, 615–618
hot, 432

impact, 610–612
metals, 432–433
plastics, 453–454
whitewares, 272
Extrusion molding, 453–454

Fabrication of plastics, 228–229
Fasteners, 740–747
Fatigue properties, 86–87
Feeds in metal cutting, 462–464
Ferritic iron, 138
Ferritic stainless steel, 159–160
Ferrous metals:
  availability, 133
  cast irons, 137–149
  ductile iron, 138–141
  formability, 174
  heat treatment of, 104–123
  iron-carbon equilibrium diagram,
    135–137
  malleable iron, 141–142
  steels, 152–180
Fillets in plastic design, 241
Finishes:
  conversion, 665–670
  inorganic, 653–656
  metallic, 656–665
  organic, 646–653
  surface, 759–762
Finishing:
  barrel, 543–545
  grit blasting, 545
  hydroblasting, 545
  hydrohoning, 545
  sand blasting, 545
  shot peening, 545, 547
Flame hardening, 121–123
Flash-butt welding, 712–714
Flaskless molding, 320–321
Flexible die forming, 602–605
Flock finishing, 649
Flow coating, 649

Fluidity in casting, 313–314
Fluids, cutting, 499–502
Fluorescent penetrants, 756
Fluorocarbons, 234–235
Flux cored electrodes, 691
Force relations, cutting, 486–489
Force shearing, 578–579
Forgeability, 424–425
Forging:
  advantages of, 437
  design factors, 435–441
  die, 429–430
  effect of temperature, 421–423
  failures, 423–424
  forgeability, 424–425
  hammer, 427–429
  high-velocity, 434–435
  hot extrusion, 432
  hydraulic-actuated press, 429
  materials, 420–421
  mechanical actuated press, 429
  press, 427–429
  roll die, 430–431
  specifications, 437
  upsetting, 431–432
Formed glassware, 278
Forming:
  bending, 592–602
  flexible die, 602–605
  high-velocity, 608–609
  principles, 591
  roll, 605–608
  stretch, 626–629
  tube bending, 600–602
Friction sawing, 533
Friction welding, 714–715
Full-mold process, 321–324
Functional design, 20–21

Gage blocks, 754–755
Galling, 167–168
Gas shielding, 689–690

Gas tungsten-arc welding, 697–698
Gating systems for sand molds, 359–364
Gear cutting, 527–528
Geometry, cutting tool, 465
Glass:
  borosilicate, 277
  design considerations, 279
  forms, 277–278
  laminated, 278
  lead, 277
  properties, 278–279
  silica, 276
  window, 277–278
Gold plating, 661
Golden rectangle, 22
Grain size, 101, 114–115
Gray iron, 137–138
Grinding:
  abrasive belt, 539–540
  cutter, 539
  cylindrical, 534–537
  design for, 542
  internal, 537–539
  process, 470–474
  production, 534
  surface, 539–540
  tool, 539
Grit blasting, 545

Hack sawing, 531–533
Hamilton Watch Company, 22
Hammer forging, 427–429
Hammered finish, 649
Hardening:
  age, 124–125
  carburizing, 117
  case, 116–123
  cyaniding, 117
  flame, 121–123
  induction, 120–121
  precipitation, 123–124
  steels, 173–174
  strain, 99–101
  surface, 116–123
  through, 115–116
Hardness, 80–85, 107–110
Hardness testers, 756
Heat, effect of, in welding, 674–676
Heat capacity of materials, 68–69
Heat treatment:
  aluminum, 123–128, 189
  annealing, 106–127
  austenitizing, 103–104
  carburizing, 117
  cyaniding, 117
  equipment, 127–128
  ferrous materials, 104–123
  flame hardening, 121–123
  grain size, 101–114
  hardening, 115–123
  homogenizing, 127
  induction hardening, 120–121
  microstructure of heat treated steel, 110–114
  nitriding, 117
  normalizing, 106–107
  procedure, 128
  spheroidizing, 107
  stress relieving, 104–105
  Woodvine diagram, 119–120
High-pressure extrusion, 615–618
High-speed steels, 476
High-velocity forging, 434–435
High-velocity forming, 608–609
Holes, molded, 247–249
Homogenizing, 127
Honing, 541
Hooke's law, 75–76
Hot-box cores, 319–320
Hot-chamber die casting, 335
Hot-dip coating, 664
Hot extrusion, 432
Hot tears, 311–312

Hot working, 416–443
Human dimensions, 31–33
Humidity, effects on plastics, 227–228
Hydraulic-actuated press forging, 429
Hydroblasting, 545
Hydrohoning, 545

Ideas for new products, 3–4, 8, 797–798
Immersion coatings, 664
Impact extrusion, 610–612
Impact properties, 85–86
Indentation tests, 81–82
Induction brazing, 716–718
Induction hardening, 120–121
Infusion of carbides into high-speed
    steels, 476
Ingenuity in invention, 798–799
Injection molding, 450–453
Inorganic finishes, 653–656
Inserts:
    in casting, 358–359
    in plastic molding, 249–251
Inspection:
    equipment, 753–759
    orders, 753
    procedures, 752
    wage payment, 752–753
Insulating materials, 228
Interest, 24
Investment casting, 327–329, 370–374
Iron-carbon equilibrium diagram 135–
    137
Isothermal transformation, 111–114

Jiggering, 272
Job cost system, 45–46
Joining processes, 672–746
Joint design in welding, 722–724
Jominy hardenability test, 109

Knoop test, 81–83

Lacquers, 647–648
Laminated glass, 278
Laminated-type phenolics, 236–238
Lapping, 540–541
Lasers, 552
Latex, 253–254
Lathe, engine, 507–508
Lead-base alloys, 218
Lead glass, 277
Lead plating, 661
Letters in molds, 246
Light, effect on plastics, 228
Loading of plastics, 229
Locating work in the tooling, 786–787
Locators, 787–789
Low-melting alloys, 217–220
Lubrication in press working, 576
Luminescent paints, 648

Machinability, 479
Machine tools:
    drilling, 466–468, 508
    grinding, 470–474
    lathe, 507–508
    milling, 468–470, 508–512
    planers, 508
Machining:
    allowance on castings, 352–353
    allowance on forgings, 437
    automatic screw, 517–522
    drilling, 466–468
    feeds, 462–464
    grinding, 468–470
    magnesium, 202
    milling, 468–470
    multiple-spindle, 517–522
    principles of, 461–464
    speeds, 462–464
    surface finish, 512–514
    turning, 514–517
Magnesium:
    alloys, 201

characteristics, 200–201
forms, 201–202
joining, 202–203
machining, 202
uses, 199–200
Magnetic properties of materials, 73–74
Maintenance, ease of, 32–34
Malleable iron, 141–142
Man-machine considerations, 30–32
Martensitic stainless steel, 161–162
Material cost, 42–43
Materials:
    aluminum, 193–199
    brass, 208–209
    bronze, 208–209
    casting, 298–304
    comparison of, 93–95
    comparison of properties, 88–92
    costs, 92–95
    creep properties, 87–88
    cutting tool, 475–479
    electrical resistivity, 70–73
    endurance limit, 86–87
    fatigue properties, 86–87
    hardness, 80–85
    heat capacity, 68–69
    high-speed steels, 476
    impact properties, 85–86
    magnetic properties, 73–74
    mechanical properties of, 74–88
    microstructure of steel, 110–114
    physical properties of, 61–63
    quantum mechanics concepts of, 63–68
    semiconductors, 71–73
    stress-strain relationships, 75–80
    tensile properties, 75–80
    thermal conductivity, 69–70
Mechanical actuated press forging, 429
Mechanical properties of materials:
    aluminum, 193–199

brass, 208–209
cast iron, 142–147
ceramics, 273–274
materials, 74–78
plastics, 226–230
steels, 162–165
Melamine formaldehyde, 238
Metal-flake paints, 648
Metal stitching, 743
Metalizing, 663
Metallic finishes, 656–665
Microstructure, effect of, on machining, 480
Microstructure of heat treated steel, 110–114
MIG welding, 689–690
Milling, 468–470, 508–512
Modifying agents for plastics, 235
Moh test, 81–83
Molded-in inserts, 249–250
Molds:
    die casting, 381–387
    permanent, 374–381
    plastic, 239–253
    sand, 315–317
Molybdenum, 213–214
Monel, 212–213
Multiple-spindle machining, 517–522
Multipoint cutting tools, 468–470

Natural rubber, 253–264
Natural stress-strain relationships, 77–80
NBR, 265–266
Neoprene, 266
Nesting of blanks in press work, 581–583
Nibbling, 583
Nickel, 210–212
    alloys, 210–213
Nitriding, 117
Nondestructive inspection, 756

Normal distribution, 766
Normalizing, 106–107
Novelty, 797
Numerical control, 556–562
Nuts, 740–741
Nylon, 231–232

Oblong holes, 246–249
Oil paint, 646
Operation sequence, 781–786
Operations:
   critical, 782
   placement, 782–783
   protection, 783
   tie-in, 783
Opportunities for the production-
   design engineer, 11–15
Optical flats, 755
Organic finishes, 646–653
Oxides, ceramic, 477–478
Oxygen, effect on plastics, 229

Parting lines in molds, 244–245
Patent disclosure, 799–800
Patents, 791–808
Patterns, 345–352
Pearl essence, 648
Pearlite, microstructure of, 110–114
Penetration in shearing, 579
Percussion welding, 714
Permanent mold casting, 331–334,
   374–381
Phase diagram, 124
Phenolic resins, 236
Phosphate coatings, 665
Photoelasticity, 755
Pickling, 644
Pitting, 168
Placement operations, 782–783
Planers, 508

Planning the operation sequence, 779–
   789
Plasma arc welding, 698–699
Plaster-mold casting, 324, 368–370
Plastics:
   ABS materials, 234
   acetal, 235
   acrylic resins, 230
   aniline-formaldehyde resins, 231
   baseline for tolerances, 246–247
   blow molding, 457
   bosses, 241
   casting, 454
   cellulose acetate butyrate, 231
   cellulose nitrate, 231
   chemical stability, 230
   coating of, 252–253, 662
   cold molding, 455
   compression molding, 445–448
   design suggestions, 239–253
   draft, 240–241
   ejection pins, 244–245
   epoxies, 239
   ethyl cellulose, 231
   extrusion, 453–454
   fabrication, 228–229
   fillers, 235
   fillets, 241
   fluorocarbons, 234–235
   heat insulation capability, 228
   holes, 247–249
   humidity, effects of, 227–228
   injection molding, 450–453
   inserts, 249–251
   joining, 745
   letters, 246
   light, effects of, 228
   loading, effects of, 229
   lugs, 244
   machining, 457
   melamine formaldehyde, 238
   modifying agents, 235

molding, cold, 455
nylon, 231–232
oxygen, effects of, 229
phenolic resins, 236
plating, 662
polyamide resins, 231–232
polyesters, 238–239
polyethylene, 233–234
polymerization, 225
polypropylene, 234
polystyrenes, 233
projections, 244
properties, 226–230
recesses, 247–249
resistivity, 228
ribs, 241
shellac compounds, 233
silicone resins, 239
slots, 247–249
temperature, effects of, 227
thermoforming, 455
thermosetting resins, 235–239
threads, 245–246
tolerance specifications, 251
transfer molding, 448–450
urea formaldehyde, 238
vinyl resins, 233
vinylidene chloride resins, 233
wall thickness, 240
weight, 228
Plate glass, 278
Polishing, 542
Polyamide resins, 231–232
Polyesters, 238–239
Polyethylene, 233–234
Polymerization, 225
Polypropylene, 234
Polystyrene, 233
Polysulfide, 267
Porcelain enamels, 653–656
Powder metallurgy:
advantages of, 285
compacting, 289–290
designing for, 286–289
forging, 292
limitations of, 286
sintering, 290–291
Power requirements:
in metal cutting, 491–492
in welding, 691–692
Precipitation hardening, 123–124
Press forging, 427–429
Press work:
blanking, 582–591
clearance, 576–578
coining, 611–613
cold drawing, 618
compression processes, 609–618
deep drawing, 618–625
dinking, 584
drawing, 618–626
effect of, 171–173, 479–480
embossing, 613
forming, 591–609
fundamentals, 567–570
location of blanks, 581–582
lubrication, 576
nibbling, 583
penetration, 579
progressive dies, 586–589
punching, 582–591
reduction in load, 579–581
shearing, 576–591
stamping, 582–591
swaging, 613–614
whitewares, 272–273
Pressed casting, 342
Pressed-in inserts, 251
Pressed whitewares, 272–273
Presses, 634–639
Product design engineer:
opportunities, 11–15
scope, 5
Progressive dies, 586–589

Projection welding, 709–712
Protection coatings, 643–670
Protection operations, 781–786
Punching, 582–591
Purchased parts, 50–51

Quality:
    incorporating in the design, 27–30
    relationship between cost, 28
Quality control, 751–777
Quantum mechanics, 63–68
Quenching, 128
Quick process, press work, 584–585

Ratio gating, 364
Raw materials, 811–817
Reactions, solid state, 101–103
Reaming, 466
Recorded cost, 41–42
Recovery in metals, 100–101
Recrystallization, 101
Rectangle, golden, 22
Reflection of light, 24
Refractory ceramics:
    alumina-silica, 279
    basic, 279
    design of, 279–281
    silica, 279
Reliability, 751–777
Residual stresses, 676–678
Resilience, 77
Resistance welding:
    flash-butt, 712–714
    percussion, 714
    projection, 709–712
    seam, 712
    spot, 707–709
    stud, 714
Rhythm in design, 24
Ribs in plastic design, 241
Risers in casting, 310–311, 364–368

Riveting, 741–742
Rockwell test, 81–83
Roll die forging, 430–431
Roll flowing, 651
Roll forming, 605–608
Roller coating, 651
Rolling, cold, 609–610
Roughness, 761
Rubber:
    compression designs, 256
    coral, 254
    design fundamentals, 256–259
    natural, 253–256
    shear designs, 256
    synthetic, 264–268
Rubber forming, 602–604
Rupture, 77

Sand blasting, 545
Sand casting, 315–316
Sawing, 531–533
SBR, 264–265
Scleroscope test, 81–83
Scoring, 167–168
Screws, 740–741
Seam welding, 712
Secondary operations, 779–781
Semiautomatic welding, 692–694
Semiconductors, 71–73
Shear strain in metal cutting, 489
Shearing, 576–591
Shell-mold casting, 317–319
Shellac compounds, 233, 648
Shielding in arc welding, 688–691
Shot peening, 545, 547
Shrinkage allowance, 352–353
Shrinkage defects, 311
Silica, 276
Silicone resins, 239
Silicone rubber, 267–268
Silver plating, 661
Slitting, 531

Slots in molding, 247–248
Slush casting, 342
Solder, 218
Solder sealing, 663
Soldering, 720
Solidification of alloys, 306–308
Solidification shrinkage, 365
Solvent cleaning, 644–645
Speeds, machining, 462–464
Spheroidizing, 107
Spinning, 629–630
Spot welding, 707–709
Spraying, 649
Stain, 648
Stainless steel, 157
  welding of, 684–686
Standard cost, 41–42
Statistical quality control, 762–768
Steels:
  alloy, 153–156
  bending recommendations, 164
  carbon, 153
  corrosion resistance of, 164–165
  endurance limit, 164–167
  flow, 168–169
  formability, 174
  galling, 167–168
  hardenability of, 173–174
  machinability, 164, 175–178
  mechanical properties of, 162–165
  modulus of elasticity, 164
  pitting, 168
  properties, as purchased, 169–171
  scoring, 167–168
  selection factors, 165–180
  stainless, 157
  tool, 156–157
  torsion of, 164
  welding of, 175, 684
Stick electrodes, 689
Stitching, 743
Strain gages, 755

Strain hardening, 99–101
Stress analysis:
  in metal cutting, 488–489
  in welding, 724–731
Stress paints, 755
Stress relieving, 104–106, 681–684
Stress rupture properties, 87–88
Stress-strain relationships, 75–80
Stretch forming, 626–629
Stud welding, 714
Submerged arc welding, 694–696
Superfinishing, 541–542
Supersonic sound equipment, 756
Surface finishes, 759–762
Surface hardening, 116–123
Surface treatment, 25–27
Swaging, 613–614
Symbols:
  for surface finish, 761
  in welding, 720–722
Synthetic resin adhesives, 744
Synthetic rubber:
  butyl rubber, 267
  NBR, 265–266
  neoprene, 266
  polysulfide, 267
  SBR, 264–265
  silicone, 267–268
System of cost control, 45–46
Systematic procedure for product development, 1–9

Temperature, effects on plastics, 227
Tensile properties of materials:
  cast iron, 143–144
  materials, 75–80
Thermal conductivity of materials, 69–70
Thermoforming, 455
Thermoplastics, 230–235
Thread rolling, 614–615
Threads in plastics, 245–246

Through hardening, 115–116
Tie-in operations, 783
Time to think, 9
Tin-base alloys, 217–218
Titanium, 214–216
Tolerance, plastic design, 251
Tool signature, 465–466
Tool steel, 156–157
Toolwear, 492–502
Torsion, 164
Toxicity of aluminum, 194
Trademarks, 805–807
Transfer molding, 448–450
Transformations, solid state, 103–105
Tube bending, 600–602
Tumble finishing, 651
Turning, 514–517
Turret-type punch press, 639
Twist drill, 466–468

Ultrasonic cleaning, 645–646
Ultrasonic machining, 551–552
Uni-shearing, 583
Unity in design, 23–24
Up-milling, 525–527
Upsetting, 431–432
Urea formaldehyde, 238
Utility of design, 797

Vacuum casting, 344–345
Value of appearance, 21–23
Vapor-deposited coatings, 664
Varnishes, 647
Velocity within gating system, 360–361
Velocity relations in metal cutting, 489–490
Venting of molds, 353–358
Vibration damping, 148
Vickers test, 83
Vinyl resins, 233
Vinylidene chloride resins, 233

Vulcanized rubber:
    abrasion resistance, 257
    atmosphere effects, 261–262
    bondability, 258–259
    cut-growth resistance, 257
    elasticity, 256
    extrudability, 258
    fatigue resistance, 256–257
    friction properties, 257–258
    moldability, 258
    resilience, 256
    solvent effects, 263
    temperature effects, 262–263
    tensile strength, 256

Wage payment and quality control, 752–753
Wall thickness of plastics, 240
Wear of cutting tools, 492–502
Weight:
    of castings, 304
    of plastics, 228
Welding:
    aluminum alloys, 686
    arc, 687–704
    arc blow, 692
    automatic, 694
    braze, 719
    cold, 720
    current control, 692
    distortion, 678–681
    electron-beam, 699–704
    electroslag, 697
    friction, 714–715
    gas shielding, 689–690
    gas tungsten arc, 697–698
    heat, effect of, 674, 676
    joint design, 722–724
    plasma-arc, 698–699
    power supply, 691–692
    relieving stress, 681–684
    residual stresses, 676–678

resistance, 704–714
semiautomatic, 692
shielded metal arc, 688–691
stainless steels, 684–686
steels, 684
stress calculation, 724–731
submerged arc, 694–696
symbols, 720–722
weldability, 684–687
Weldments, design of, 720–739
Whitewares:
casting, 272
design considerations, 273
extruding, 272
jiggering, 272
pressing, 272–273
production methods, 272–273
properties, 273–274
types, 274–275

Window glass, 277–278
Woodvine diagram, 119–120
Work:
clamping, 789
locating in tooling, 786–787
supporting, 789
Wrinkle finish, 648

X ray:
equipment, 756
inspection of castings, 387–389, 409

Yield point, 76–77
Young's modulus, 76

Zinc-base alloys, 217
Zinc plating, 659–660